To
My Daughters and Their Husbands

Molly and Phillip
Janis and Masuo
Linda and Harry
Debra and Mark

MATHEMATICAL FUNCTIONS and their APPROXIMATIONS

YUDELL L. LUKE
University of Missouri
Kansas City, Missouri

Academic Press Inc.
NEW YORK SAN FRANCISCO LONDON 1975
A Subsidiary of Harcourt Brace Jovanovich, Publishers

ACADEMIC PRESS, INC.
111 Fifth Avenue, New York, New York 10003

United Kingdom Edition published by
ACADEMIC PRESS, INC. (LONDON) LTD.
24/28 Oval Road, London NW1

Library of Congress Cataloging in Publication Data
Main entry under title:

Mathematical functions and their approximations.

An updated version of part of Handbook of mathematical functions with formulas, graphs, and mathematical tables, edited by M. Abramowitz and I. A. Stegun.
Bibliography: p.
Includes index.
1. Mathematics–Tables, etc. I. Luke, Yudell I. II. Abramowitz, Milton, 1915-1958. Handbook of mathematical functions, with formulas, graphs, and mathematical tables.
QA55.M418 515'.021'2 75-22358
ISBN 0-12-459950-8

AMS (MOS) 1970 Subject Classifications: 26-A86, 33-02, 41-02, 42A16, 42A56, 44A20, 65D20

PRINTED IN THE UNITED STATES OF AMERICA

CONTENTS

IV. Incomplete Gamma Functions

X. Lommel Functions, Struve Functions, and Associated Bessel Functions

XI. Orthogonal Polynomials

XII. Computation by Use of Recurrence Formulas

XIII. Some Aspects of Rational and Polynomial Approximations

XIV. Miscellaneous Topics

PREFACE

On September 15-16, 1954, I was privileged to attend a Conference on Mathematical Tables held at Cambridge, Massachusetts under the auspices of the National Science Foundation and the Massachusetts Institute of Technology. The purpose of the meeting was to determine the need for mathematical tables in view of the availability of high speed computing machinery. It was the consensus of the participants that the need for tables would continue to exist. Furthermore, since mathematical functions including the so-called special functions of mathematical physics arise in numerous contexts in the natural and social sciences, the Conference recognized a pressing need for a modernized version of the classical "Tables of Functions with Formulae and Curves" by E. Jahnke and F. Emde. A revision of this volume by F. Lösch appreared in 1960.

An outgrowth of the above Conference was the production of the "Handbook of Mathematical Functions with Formulas, Graphs and Mathematical Tables" (edited by M. Abramowitz and I. Stegun), National Bureau of Standards Applied Mathematics Series 55, U.S. Government Printing Office, Washington, D.C., 1964. For short, we refer to this tome as AMS 55. This huge volume of 1043 pages which is divided into 29 chapters composed by 28 authors contains a massive amount of data designed to meet the needs of applied workers in all fields. A vast portion of the special functions and much more are covered. The volume is immensely popular. To date there have been nine printings and about 200,000 copies have been sold (including sales of a softback Dover edition).

The cutoff date for much of the material in AMS 55 is about 1960. In the past 15 years much valuable new information on the special functions has appeared. In some quarters, it has been suggested that a new AMS 55 should be produced. This is not presently feasible. The task would be gigantic and would consume much time. Most certainly the economics of the situation forbids such a program. A feasible approach is a handbook in the spirit of AMS 55, which, in the main, supplements the data given there.

The present volume can be conceived as an updated supplement to that portion of AMS 55 dealing with mathematical functions. I have made no attempt to reproduce many of the tabular and descriptive properties of the mathematical functions given in AMS 55. Sufficient basic descriptive material is provided, however, to make the present volume complete in itself.

The nature of mathematical tables in the classical sense are primarily for the occasional computer. To impress tables in the memory of a computer and then program for table look up and interpolation is not economical. A computer requires efficient algorithms and schemes for the evaluation of functions on demand. My philosophy of approximation is global in character. Numerical values of functions are but a facet of the overall problem. We desire approximations to compute functions and their zeros, to simplify mathematical expressions such as integrals and transforms, and to facilitate directly the mathematical solution of a wide variety of functional equations such as differential equations, integral equations, etc. So the main thrusts of my volume are on the development of analytical expansions and approximations of functions for universal use.

We now turn to some of the principal features of the volume. Machinery for the expansion of the generalized hypergeometric function, call it ${}_pF_q$ for short, and other functions in infinite series of Jacobi and Chebyshev polynomials of the first kind is provided. Numerical coefficients for Chebyshev espansions of the more common functions are tabulated.

Polynomial and rational approximations for the ${}_pF_q$ and a certain class of G-functions which generalizes the ${}_pF_q$ are set forth. In some cases, the rational approximations are of the Padé type. The coefficients in the early polynomials of these rational approximations for many common functions are given.

A striking virtue of the Chebyshev expansions and polynomial and rational approximations cited is that they have better convergence properties than their Taylor series counterparts. Better yet, they converge in domains where their series counterparts diverge but are asymptotic to some function is an appropriate region. Also truncated Chebyshev expansions can be handled simply as an ordinary polynomial without the necessity of first converting the weighted sum of Chebyshev polynomials into an ordinary polynomial. A virtue of the polynomial and rational approximations, especially the latter, is that the polynomials in the approximations satisfy simple recursion formulas, which can be used in the forward direction to generate values of the polynomials. In a number of cases, we can prove that such a computational scheme is stable. In those instances where proofs are not available, extensive numerical tests serve to confirm stability. The coefficients in the Chebyshev expansions for many functions also satisfy simple recurrence formulas. In these situations the coefficients can be readily determined by use of the backward recursion process.

Padé approximations for many of the elementary functions and the incomplete gamma functions are known in closed form. The main diagonal and first subdiagonal approximations give rise to two-sided inequalities for these functions. They can be used as building blocks to obtain similar inequalities for numerous other functions. These and many other inequalities are treated.

Unlike AMS 55, we do not give tables of mathematical functions in the classical sense. To trace the existence of mathematical tables, there is the valuable second edition of "An Index of Mathematical Tables" by A. Fletcher, J.C.P. Miller, L. Rosenhead and L.J. Comrie, Addison-Wesley, Reading,

Massachusetts, 1962. Call this FMRC for short. It appears that the cutoff date for much of this material is also about 1960. In the present volume, we briefly describe and give references to numerical tables of pertinent functions that have appeared since about 1960 and which are not recorded in FMRC. The journal *Mathematics of Computation* gives detailed reviews of such material as it appears.

We also briefly describe and give references to the development of analytic approximations and expansions by other authors. Though our interest in approximations is primarily on those of an analytic character as opposed to numerical curve fits such as "best" Chebyshev polynomial and rational approximations, description of and references to the latter type approximations are also provided.

There are a number of topics that are only briefly treated herein in view of time and space. In later works, we hope to deal in extenso with subjects like computation by use of recurrence formulas, Padé approximations, approximations of functional equations, and elliptic functions and integrals.

To assist the reader in the use of this volume, detailed table of contents, notation, and subject indices are provided. Since the tome is a handbook, there are virtually no proofs, but extensive references are supplied. The bibliography is composed of more than 600 items, most of which (about 85%) have appeared during the past 15 years.

YUDELL L. LUKE
Kansas City, Missouri
April, 1975

CHAPTER I THE GAMMA FUNCTION AND RELATED FUNCTIONS

1.1. Definitions and Elementary Properties

$$\Gamma(z) = p^z \int_0^\infty e^{-pt} t^{z-1} dt, \; R(p)>0, \; R(z)>0;$$

$$R(p) = 0 \;\; if \;\; 0 < R(z) < 1. \tag{1}$$

$$\Gamma(z+n) = \left[\prod_{k=0}^{n-1}(z+k)\right]\Gamma(z). \tag{2}$$

$$\Gamma(z)\Gamma(1-z) = \pi \csc \pi z. \tag{3}$$

$$\Gamma(mz) = (2\pi)^{\frac{1}{2}(1-m)} m^{mz-\frac{1}{2}} \prod_{r=0}^{m-1}\Gamma(z + r/m). \tag{4}$$

For the logarithmic derivative of the gamma function, we have

$$\Psi(z) = (d/dz)\ln\Gamma(z) = \Gamma'(z)/\Gamma(z) \; or \; \ln\Gamma(z) = \int_1^z \Psi(t)dt. \tag{5}$$

$$\Psi(mz) = m^{-1} \sum_{k=0}^{m-1} \Psi(z + k/m) + \ln m. \tag{6}$$

$$\Psi(z+n) = \sum_{k=0}^{n-1} (z+k)^{-1} + \Psi(z), \tag{7}$$

$$\Psi(1) = -\gamma, \; \gamma = 0.57721\;56649\;01532\;86061. \tag{8}$$

$$\Psi(z) - \Psi(1-z) = -\pi \cot \pi z. \tag{9}$$

1.2. Power Series and Other Series Expansions

$$\ln \Gamma(z+1) = \sum_{k=1}^{\infty} (-)^k S_k z^k/k, \; |z| < 1,$$

$$S_1 = -\Psi(1) = \gamma, \;\; S_k = \sum_{r=1}^{\infty} r^{-k}, \; k > 1. \tag{1}$$

$$1/\Gamma(z+1) = \sum_{k=0}^{\infty} a_k z^k, \;\; a_0 = 1, \; |z| < \infty,$$

$$ra_r = \sum_{k=1}^{r} (-)^{k+1} S_k a_{r-k}, \;\; r > 0. \tag{2}$$

TABLE 1.1

COEFFICIENTS FOR THE TAYLOR SERIES EXPANSIONS OF $[\Gamma(z+1)]^{-1}$ AND $\Gamma(z+3)$

$$\left[\Gamma(z+1)\right]^{-1} = \sum_{n=0}^{\infty} a_n z^n, \quad |z| < \infty \qquad\qquad \Gamma(z+3) = \sum_{n=0}^{\infty} b_n z^n, \quad |z| < 3$$

n	a_n	n	b_n
0	1.00000 00000 00000 00000	0	2.00000 00000 00000 00000
1	0.57721 56649 01532 86061	1	1.84556 86701 96934 27879
2	-0.65587 80715 20253 88108	2	1.24646 49959 51346 52897
3	-0.04200 26350 34095 23553	3	0.57499 41689 20612 22755
4	0.16653 86113 82291 48950	4	0.23007 49407 54114 06302
5	-0.04219 77345 55544 33675	5	0.07371 50466 16023 86878
6	-0.00962 19715 27876 97356	6	0.02204 11093 67516 96733
7	0.00721 89432 46663 09954	7	0.00544 87540 75820 30942
8	-0.00116 51675 91859 06511	8	0.00135 52208 60239 43520
9	-0.00021 52416 74114 95097	9	0.00026 47856 63045 49638
10	0.00012 80502 82388 11619	10	0.00006 12030 62819 20073
11	-0.00002 01348 54780 78824	11	0.00000 85055 79174 88135
12	-0.00000 12504 93482 14267	12	0.00000 24061 77240 13144
13	0.00000 11330 27231 98170	13	0.00000 00880 23909 90648
14	-0.00000 02056 33841 69776	14	0.00000 01142 22764 53422
15	0.00000 00061 16095 10448	15	-0.00000 00163 14752 10083
16	0.00000 00050 02007 64447	16	0.00000 00086 23497 38998
17	-0.00000 00011 81274 57049	17	-0.00000 00024 41104 02524
18	0.00000 00001 04342 67117	18	0.00000 00008 72915 06387
19	0.00000 00000 07782 26344	19	-0.00000 00002 83902 95131
20	-0.00000 00000 03696 80562	20	0.00000 00000 95608 89805
21	0.00000 00000 00510 03703	21	-0.00000 00000 31782 70013
22	-0.00000 00000 00020 58326	22	0.00000 00000 10610 93470
23	-0.00000 00000 00005 34812	23	-0.00000 00000 03536 84281
24	0.00000 00000 00001 22678	24	0.00000 00000 01179 38189
25	-0.00000 00000 00000 11813	25	-0.00000 00000 00393 18677
26	0.00000 00000 00000 00119	26	0.00000 00000 00131 08214
27	0.00000 00000 00000 00141	27	-0.00000 00000 00043 69853
28	-0.00000 00000 00000 00023	28	0.00000 00000 00014 56734
29	0.00000 00000 00000 00002	29	-0.00000 00000 00004 85607
		30	0.00000 00000 00001 61876
		31	-0.00000 00000 00000 53961
		32	0.00000 00000 00000 17987
		33	-0.00000 00000 00000 05996
		34	0.00000 00000 00000 01999
		35	-0.00000 00000 00000 00666
		36	0.00000 00000 00000 00222
		37	-0.00000 00000 00000 00074
		38	0.00000 00000 00000 00025
		39	-0.00000 00000 00000 00008
		40	0.00000 00000 00000 00003
		41	-0.00000 00000 00000 00001

TABLE 1.2

CHEBYSHEV COEFFICIENTS FOR $\Gamma(1+x)$ AND ITS RECIPROCAL

$$\Gamma(1+x) = \sum_{n=0}^{\infty} a_n T_n^*(x) \qquad \left[\Gamma(1+x)\right]^{-1} = \sum_{n=0}^{\infty} b_n T_n^*(x)$$

$0 \le x \le 1$

n	a_n	n	b_n
0	0.94178 55977 95494 66571	0	1.06377 30078 05261 97553
1	0.00441 53813 24841 00676	1	-0.00498 55872 86840 03595
2	0.05685 04368 15993 63379	2	-0.06419 25436 10915 82288
3	-0.00421 98353 96418 56050	3	0.00506 57986 40286 08725
4	0.00132 68081 81212 46022	4	0.00041 66091 38709 68886
5	-0.00018 93024 52979 88804	5	-0.00008 04814 12497 84711
6	0.00003 60692 53274 41245	6	0.00000 29600 11775 18802
7	-0.00000 60567 61904 46086	7	0.00000 02689 75996 44060
8	0.00000 10558 29546 30228	8	-0.00000 00333 96463 06868
9	-0.00000 01811 96736 55424	9	0.00000 00010 89653 86454
10	0.00000 00311 77249 64715	10	0.00000 00000 51385 01863
11	-0.00000 00053 54219 63902	11	-0.00000 00000 06600 74100
12	0.00000 00009 19327 55199	12	0.00000 00000 00247 69163
13	-0.00000 00001 57794 12803	13	0.00000 00000 00002 20039
14	0.00000 00000 27079 80623	14	-0.00000 00000 00000 67072
15	-0.00000 00000 04646 81865	15	0.00000 00000 00000 03132
16	0.00000 00000 00797 33502	16	-0.00000 00000 00000 00039
17	-0.00000 00000 00136 80782	17	-0.00000 00000 00000 00003
18	0.00000 00000 00023 47319		
19	-0.00000 00000 00004 02743		
20	0.00000 00000 00000 69101		
21	-0.00000 00000 00000 11856		
22	0.00000 00000 00000 02034		
23	-0.00000 00000 00000 00349		
24	0.00000 00000 00000 00060		
25	-0.00000 00000 00000 00010		
26	0.00000 00000 00000 00002		

TABLE 1.3

CHEBYSHEV COEFFICIENTS FOR $\Gamma(x+3)$ AND $\ln \Gamma(x+3)$

$$\Gamma(x+3) = \sum_{n=0}^{\infty} a_n T_n^*(x) \qquad 0 \le x \le 1 \qquad \ln \Gamma(x+3) = \ln 2 + \sum_{n=0}^{\infty} b_n T_n^*(x)$$

n	a_n	n	b_n
0	3.65738 77250 83382 43850	0	0.52854 30369 82234 59887
1	1.95754 34566 61268 26928	1	0.54987 64461 21414 11418
2	0.33829 71138 26160 38916	2	0.02073 98006 16136 65136
3	0.04208 95127 65575 49199	3	−0.00056 91677 04215 43842
4	0.00428 76504 82129 08770	4	0.00002 32458 72104 00169
5	0.00036 52121 69294 61767	5	−0.00000 11306 07585 70393
6	0.00002 74006 42226 42200	6	0.00000 00606 56530 98948
7	0.00000 18124 02333 65124	7	−0.00000 00034 62843 57770
8	0.00000 01096 57758 65997	8	0.00000 00002 06249 98806
9	0.00000 00059 87184 04552	9	−0.00000 00000 12663 51116
10	0.00000 00003 07690 80535	10	0.00000 00000 00795 31007
11	0.00000 00000 14317 93030	11	−0.00000 00000 00050 82077
12	0.00000 00000 00651 08773	12	0.00000 00000 00003 29187
13	0.00000 00000 00025 95850	13	−0.00000 00000 00000 21556
14	0.00000 00000 00001 10789	14	0.00000 00000 00000 01424
15	0.00000 00000 00000 03547	15	−0.00000 00000 00000 00095
16	0.00000 00000 00000 00169	16	0.00000 00000 00000 00006
17	0.00000 00000 00000 00003		

TABLE 1.4

CHEBYSHEV COEFFICIENTS FOR $\ln \Gamma(x)$

$$\ln \Gamma(x) = (x - \tfrac{1}{2}) \ln x - x + \tfrac{1}{2} \ln 2\pi + S(x)$$

$$S(x) = (12x)^{-1} \sum_{n=0}^{\infty} (-)^n c_n T_{2n}(1/x), \quad x \ge 1$$

n	c_n	n	c_n
0	0.98575 15540 05098	14	0.00000 00000 17553
1	0.01357 51199 40355	15	0.00000 00000 06710
2	0.00060 97577 84871	16	0.00000 00000 02651
3	0.00005 47947 47404	17	0.00000 00000 01079
4	0.00000 72256 71298	18	0.00000 00000 00451
5	0.00000 12227 85636	19	0.00000 00000 00193
6	0.00000 02472 06061	20	0.00000 00000 00085
7	0.00000 00571 76475	21	0.00000 00000 00040
8	0.00000 00147 06251	22	0.00000 00000 00017
9	0.00000 00041 24183	23	0.00000 00000 00008
10	0.00000 00012 43043	24	0.00000 00000 00004
11	0.00000 00003 98333	25	0.00000 00000 00002
12	0.00000 00001 34579	26	0.00000 00000 00001
13	0.00000 00000 47620		

These coefficients are due to Németh (1967).

TABLE 1.5

CHEBYSHEV COEFFICIENTS FOR $\Psi(x+3)$
AND THE FIRST SIX DERIVATIVES OF $\Psi(x+3)$*

$$\psi^{(m)}(x+3) = \sum_{n=0}^{\infty} c_n^{(m)} T_n^*(x), \quad m = 0,1,\ldots,6, \quad \psi^{(0)}(x+3) = \Psi(x+3)$$
$$0 \le x \le 1$$

n	$c_n^{(0)}$
0	1.09632 65312 32801 58001
1	0.16629 18012 71471 81741
2	-0.00685 27220 20053 29673
3	0.00037 33963 42378 60654
4	-0.00002 27095 69468 03564
5	0.00000 14623 88714 57953
6	-0.00000 00974 17753 95695
7	0.00000 00066 31970 83199
8	-0.00000 00004 58133 78140
9	0.00000 00000 31971 21417
10	-0.00000 00000 02247 37976
11	0.00000 00000 00158 81140
12	-0.00000 00000 00011 26604
13	0.00000 00000 00000 80152
14	-0.00000 00000 00000 05715
15	0.00000 00000 00000 00408
16	-0.00000 00000 00000 00029
17	0.00000 00000 00000 00002

n	$c_n^{(1)}$
0	0.33483 86979 10949 38576
1	-0.05518 74820 48730 09463
2	0.00451 01907 36011 50186
3	-0.00036 57058 88303 72083
4	0.00002 94346 27468 22336
5	-0.00000 23527 76815 15061
6	0.00000 01868 53176 63281
7	-0.00000 00147 50720 18379
8	0.00000 00011 57993 33714
9	-0.00000 00000 90439 17904
10	0.00000 00000 07029 62700
11	-0.00000 00000 00543 98873
12	0.00000 00000 00041 92525
13	-0.00000 00000 00003 21903
14	0.00000 00000 00000 24630
15	-0.00000 00000 00000 01878
16	0.00000 00000 00000 00143
17	-0.00000 00000 00000 00011
18	0.00000 00000 00000 00001

n	$c_n^{(2)}$
0	-0.11259 29353 45473 83037
1	0.03655 70017 42820 94137
2	-0.00443 59424 96027 28223
3	0.00047 54758 54728 92648
4	-0.00004 74718 36382 63232
5	0.00000 45218 15237 35268
6	-0.00000 04163 00079 62011
7	0.00000 00373 38998 16535
8	-0.00000 00032 79914 47410
9	0.00000 00002 83211 37682
10	-0.00000 00000 24104 02848
11	0.00000 00000 02026 29690
12	-0.00000 00000 00168 52418
13	0.00000 00000 00013 88481
14	-0.00000 00000 00001 13451
15	0.00000 00000 00000 09201
16	-0.00000 00000 00000 00741
17	0.00000 00000 00000 00059
18	-0.00000 00000 00000 00005

n	$c_n^{(3)}$
0	0.07601 26046 55110 38390
1	-0.03625 71864 81828 73882
2	0.00579 72023 38937 00231
3	-0.00076 96465 13610 48095
4	0.00009 14920 82189 88449
5	-0.00001 00971 31488 36381
6	0.00000 10557 77442 83084
7	-0.00000 01059 29577 48119
8	0.00000 00102 85494 20092
9	-0.00000 00009 72314 31000
10	0.00000 00000 89884 63545
11	-0.00000 00000 08153 17069
12	0.00000 00000 00727 57169
13	-0.00000 00000 00064 01023
14	0.00000 00000 00005 56152
15	-0.00000 00000 00000 47789
16	0.00000 00000 00000 04066
17	-0.00000 00000 00000 00343
18	0.00000 00000 00000 00029
19	-0.00000 00000 00000 00002

*For details on the computation of these coefficients, see Luke (1969).

TABLE 1.5 *(Concluded)*

n	$c_n^{(4)}$	n	$c_n^{(5)}$
0	-0.07723 47240 56994 79348	0	0.10493 30344 59278 63245
1	0.04786 71634 51599 46688	1	-0.07887 79016 52793 55735
2	-0.00944 07021 86674 63167	2	0.01839 74151 12159 39737
3	0.00148 95447 40103 44841	3	-0.00335 22841 59396 50400
4	-0.00020 49440 23348 86027	4	0.00052 28782 30918 01646
5	0.00002 56714 25065 29656	5	-0.00007 31797 85814 73963
6	-0.00000 30013 93581 58404	6	0.00000 94497 29612 08531
7	0.00000 03327 66437 35634	7	-0.00000 11463 39856 72257
8	-0.00000 00353 65412 11070	8	0.00000 01322 69366 10768
9	0.00000 00036 30622 92684	9	-0.00000 00146 46669 18029
10	-0.00000 00003 62096 95057	10	0.00000 00015 66940 74162
11	0.00000 00000 35237 50903	11	-0.00000 00001 62791 15730
12	-0.00000 00000 03357 44020	12	0.00000 00000 16490 34452
13	0.00000 00000 00314 06802	13	-0.00000 00000 01634 02764
14	-0.00000 00000 00028 90815	14	0.00000 00000 00158 80745
15	0.00000 00000 00002 62300	15	-0.00000 00000 00015 17110
16	-0.00000 00000 00000 23498	16	0.00000 00000 00001 42724
17	0.00000 00000 00000 02081	17	-0.00000 00000 00000 13243
18	-0.00000 00000 00000 00182	18	0.00000 00000 00000 01214
19	0.00000 00000 00000 00016	19	-0.00000 00000 00000 00110
20	-0.00000 00000 00000 00001	20	0.00000 00000 00000 00010
		21	-0.00000 00000 00000 00001

n	$c_n^{(6)}$ *
0	-0.17861 76221 42502 75322
1	0.15577 64622 00520 57932
2	-0.04172 36376 73831 27704
3	0.00859 71413 03245 40038
4	-0.00149 62277 61073 22905
5	0.00023 10896 08557 13710
6	-0.00003 26320 44778 43649
7	0.00000 42960 97867 08959
8	-0.00000 05345 28790 20447
9	0.00000 00634 78151 64391
10	-0.00000 00072 48699 71418
11	0.00000 00008 00521 97925
12	-0.00000 00000 85888 79302
13	0.00000 00000 08985 44247
14	-0.00000 00000 00919 35574
15	0.00000 00000 00092 22505
16	-0.00000 00000 00009 08955
17	0.00000 00000 00000 88173
18	-0.00000 00000 00000 08431
19	0.00000 00000 00000 00796
20	-0.00000 00000 00000 00074
21	0.00000 00000 00000 00007
22	-0.00000 00000 00000 00001

* These coefficients are accurate only to 19D.

$$\Gamma(z + 1) = \sum_{k=0}^{\infty} b_k z^k, \quad b_0 = 1, \quad |z| < 1,$$

$$r b_r = -\sum_{k=1}^{r} (-)^{k+1} s_k b_{r-k}, \quad r > 0, \tag{3}$$

$$\sum_{k=0}^{r} a_k b_{r-k} = 0, \quad r > 0. \tag{4}$$

Wrench (1968) has tabulated the coefficients in

$$1/\Gamma(z) = \sum_{n=1}^{\infty} f_n z^n \quad \text{and} \quad 1/\Gamma(z) = z(1 + z)\sum_{n=0}^{\infty} g_n z^n \tag{5}$$

to 31 D. See also Morris (1973). Notice that

$$a_n = f_{n+1},$$

$$f_n = g_{n-2} + g_{n-1}, \quad g_{-1} = 0, \quad n > 0. \tag{6}$$

$$\psi(z) + \gamma = -\frac{1}{z} + \sum_{k=1}^{\infty} \frac{z}{k(z + k)} = (z - 1)\sum_{k=0}^{\infty} \left[(k + 1)(z + k)^{-1}\right]. \tag{7}$$

See also 1.4(1). Further representations follow from (1)-(6) and the definition 1.1(5).

Rational approximations for

$$\frac{\Gamma(z + 1)}{A(z)}, \quad A(z) = (2\pi)^{\frac{1}{2}}(z + c + \tfrac{1}{2})^{z+1}\exp\left[-(z + c + \tfrac{1}{2})\right], \tag{8}$$

originally due to Lanczos (1964), have been examined in detail by Luke (1969). Approximations for $\Gamma(z + 1)$ based on the Padé approximants for two forms of the incomplete gamma function have also been treated by Luke (1970b). In this connection, see 4.5(10).

1.3. Asymptotic Expansions

$$\ln \Gamma(z) = (z - \tfrac{1}{2}) \ln z - z + \tfrac{1}{2} \ln 2\pi$$
$$+ \sum_{k=1}^{n} B_{2k}\left[(2k - 1)(2k)z^{2k-1}\right]^{-1} + R_n(z),$$
$$|\arg z| \le \pi - \varepsilon, \quad \varepsilon > 0, \quad R_n(z) = O(z^{-2n-3}), \tag{1}$$

where B_{2k} are the Bernoulli numbers, see 14.2. If z is real and positive, $R_n(z)$ is less in magnitude than the first term neglected and has the same sign as that term. For convenience in the further discussion of the remainder, let

$$S_n(z) = B_{2n+2}\left[(2n+1)(2n+2)z^{2n+1}\right]^{-1},$$

$$\arg z = \zeta. \tag{2}$$

Theorem 1.

$$R_n(z) = \theta S_n(z),$$

$$|\theta| \le 1 \quad \textit{if} \quad -\pi/4 < \zeta < \pi/4,$$

$$|\theta| \le \csc 2\zeta \quad \textit{if} \quad \pi/4 \le \zeta < \pi/2. \tag{3}$$

In fact, if $0 \le \zeta < \pi/2$, then $0 \le \arg \theta \le 2\zeta$ and θ lies on or above the real axis. Again, if $-\pi/2 < \zeta \le 0$, then $2\zeta < \arg \theta \le 0$ and θ lies on or below the real axis.

Theorem 1 is very poor for values of z near the imaginary axis. The next result corrects this deficiency.

Theorem 2. *Suppose that* $-\pi < \zeta < \pi$. *We can determine a number* ρ *such that*

$$-\pi/2 < \rho < \pi/2, \quad -\pi/2 < \rho - \zeta < \pi/2. \tag{4}$$

Then

$$|R_n(z)| = \sigma(\cos\rho)^{-2n-2}|S_n(z)|,$$

$$\sigma = 1 \quad \textit{for} \quad |\rho - \zeta| \le \pi/4,$$

$$\sigma = |\csc 2(\rho - \zeta)| \quad \textit{for} \quad \pi/4 \le |\rho - \zeta| < \pi/2. \tag{5}$$

We give two illustrations of (5).

Example 1. *Let* $\pi/4 \le |\zeta| < \pi/2$.

(a) If $\zeta \ge 0$, *choose* $\rho = \zeta - \pi/4$, *whence* $\sigma = 1$ *and* $\cos\rho = \sin(\zeta + \pi/4)$.

(b) If $\zeta < 0$, *choose* $\rho = \zeta + \pi/4$, *whence* $\sigma = 1$ *and* $\cos\rho = \sin(-\zeta + \pi/4)$.

For such ζ,

$$|R_n(z)| \leq \left[\sin\ (|\zeta| + \pi/4)\right]^{-2n-2} |S_n(z)|$$
$$\leq\ 2^{n+1} |S_n(z)|. \qquad (6)$$

Clearly (5) is sharper than (3) for z near the imaginary axis.

Example 2. *Let* $\pi/4 \leq |\zeta| < \pi$.

(a) *If* $\zeta > 0$, *choose* $\rho = 2\zeta/3 - \pi/6$, *whence* $\sigma = \csc\ (2\zeta/3 + \pi/3)$, $\sigma\cos\ \rho = 1$.

(b) *If* $\zeta < 0$, *choose* $\rho = 2\zeta/3 + \pi/6$, *whence* $\sigma = \csc(-2\zeta/3 + \pi/3)$, $\sigma\cos\ \rho = 1$.

It follows that

$$|R_n(z)| < \sin\ (2|\zeta|/3 + \pi/3)^{-2n-3}\ |S_n(z)|. \qquad (7)$$

Theorem 3.

$$|R_n(z)| \leq 2|B_{2n}/(2n - 1)|\left(I(z)\right)^{1-2n}\ \textit{for}\ R(z) < 0,\ I(z) \neq 0, \qquad (8)$$

$$|R_n(z)| \leq |B_{2n}/(2n - 1)|\,|z|^{1-2n}\quad \textit{for}\quad R(z) \geq 0. \qquad (9)$$

Theorems 1 and 2 and the examples following Theorem 2 are due to van der Corput (1955). Theorem 3 is due to Spira (1971).

Equation (1) is equivalent to

$$\Gamma(z) = e^{-z} z^{z-\frac{1}{2}} (2\pi)^{\frac{1}{2}} \left[\sum_{k=0}^{n} c_k z^{-k} + O(z^{-n-1})\right],$$

$$|\arg z| \leq \pi - \varepsilon,\ \varepsilon > 0. \qquad (10)$$

Exact rational values for c_k, $k = 1(1)20$ have been given by Wrench (1968). He also gives these coefficients to 50D, and these coefficients are recorded in Table 1.5 on the following page. For $k = 21(1)30$, Spira (1971) gives exact rational values for c_k and also their 45D equivalents.

TABLE 1.5

50D *values of the first twenty coefficients in the asymptotic series for* $\Gamma(z)$

c_1		$0.08\dot{3}$									
c_2		0.00347	$\dot{2}$								
c_3	−	0.00268	$13\dot{2}71$	60493	$\dot{8}$						
c_4	−	0.00022	94720	93621	39917	69547	32510	28806	$58\dot{4}$		
c_5		0.00078	40392	21720	06662	74740	34881	44228	88496	96257	10366
c_6		0.00006	97281	37583	65857	77429	39882	85757	83308	29359	63594
c_7	−	0.00059	21664	37353	69388	28648	36225	60440	11873	91585	19680
c_8	−	0.00005	17179	09082	60592	19337	05784	30020	58822	81785	34534
c_9		0.00083	94987	20672	08727	99933	57516	76498	34451	98182	11159
c_{10}		0.00007	20489	54160	20010	55908	57193	02250	15052	06345	17380
c_{11}	−	0.00191	44384	98565	47752	65008	98858	32852	25448	76893	57895
c_{12}	−	0.00016	25162	62783	91581	68986	35123	98027	09981	05872	59193
c_{13}		0.00640	33628	33808	06979	48236	38090	26579	58304	01894	09396
c_{14}		0.00054	01647	67892	60451	51804	67508	57024	17355	47254	41598
c_{15}	−	0.02952	78809	45699	12050	54406	51054	69382	44465	65482	82544
c_{16}	−	0.00248	17436	00264	99773	09156	58368	74346	43239	75168	04723
c_{17}		0.17954	01170	61234	85610	76994	07722	22633	05309	12823	38692
c_{18}		0.01505	61130	40026	42441	23842	21877	13112	72602	59815	45541
c_{19}	−	1.39180	10932	65337	48139	91477	63542	27314	93580	45617	72646
c_{20}	−	0.11654	62765	99463	20085	07340	36907	14796	96789	37334	38371

Note. The first four entries are shown as repeating decimals, the repetends being indicated by superior dots. This table is from John W. Wrench, Jr., "Concerning two series for the gamma function," reprinted with permission of the publisher, The American Mathematical Society, from MATHEMATICS OF COMPUTATION, Copyright © 1968, Volume 22, No. 103, page 620.

A generalization of (1) is

$$\ln \Gamma(z+a) = (z+a-\tfrac{1}{2})\ln z - z + \tfrac{1}{2}\ln 2\pi + \sum_{k=1}^{n} \frac{(-)^{k+1}B_{k+1}(a)}{k(k+1)} z^{-k} + O(z^{-n-1}),$$

$$|\arg z| \le \pi - \varepsilon, \quad \varepsilon > 0. \tag{11}$$

If $a = \tfrac{1}{2}$, $B_k(a)$ vanishes if k is odd and

$$\ln \Gamma(z+\tfrac{1}{2}) = z(\ln z - 1) + \tfrac{1}{2}\ln 2\pi + \sum_{k=1}^{n} \frac{B_{2k}(\tfrac{1}{2})}{2k(2k-1)} z^{1-2k} + O(z^{-2n-1}),$$

$$|\arg z| \le \pi - \varepsilon, \quad \varepsilon > 0. \tag{12}$$

$$\frac{\Gamma(z+a)}{\Gamma(z+b)} = z^{a-b}\sum_{k=0}^{N-1}\frac{(-)^k B_k^{(a-b+1)}(a)(b-a)_k}{k!}z^{-k} + z^{a-b}O(z^{-N}),$$

$$|\arg(z+a)| \le \pi - \varepsilon, \quad \varepsilon > 0, \tag{13}$$

where a and b are bounded complex numbers and $B_k^{(a-b+1)}(a)$ is the generalized Bernoulli polynomial. If $b - a = 1 - N$, then (13) is exact. That is, we can ignore the $O(z^{-N})$ term and the restriction on arg z. Further, if $b - a$ is a positive integer m, and $z > \max\{|a|, |b - 1|\}$, then with $N \to \infty$ the power series is convergent and correctly sums to $1/(z + a)_m$. Thus in this case also, the $O(z^{-N})$ term and the restriction on arg z can be omitted. A convergent factorial expansion for $\Gamma(z + a)/\Gamma(z + b)$ of the same general character as (13) has been given by Nörlund (1961, eq. (43)).

Fields (1966) has shown that

$$\frac{\Gamma(z+a)}{\Gamma(z+b)} = (z+a-\rho)^{a-b}\sum_{k=0}^{N-1}\frac{B_{2k}^{(2\rho)}(\rho)(b-a)_{2k}(z+a-\rho)^{-2k}}{(2k)!}$$
$$+ (z+a-\rho)^{a-b}O((z+a-\rho)^{-2N}), \tag{14}$$
$$2\rho = 1 + a - b, \quad |\arg(z+a)| \le \pi - \varepsilon, \quad \varepsilon > 0.$$

Note that this series is essentially an even one.

The formulas (9) and (10) are valid even though $\Gamma(z + a)/\Gamma(z + b)$ has poles at $z = -a - n$, n a positive integer or zero, if $|z|$ is sufficiently large, $|\arg(z+a)| < \pi$. But the approximation will break down when used for moderate values of z if z is near one of these poles. To obviate this difficulty, use the reflection formula for gamma functions to write

$$\frac{\Gamma(z+a)}{\Gamma(z+b)} = \frac{\sin\pi(z+b)}{\sin\pi(z+a)}\,\frac{\Gamma(1-z-b)}{\Gamma(1-z-a)},$$

and then use the asymptotic expansion for the ratio of the gamma functions on the right of the latter equation for $|\arg(-z)| \le \pi - \varepsilon$, $\varepsilon > 0$.

With $z = n$, $a = -x$, and $b = 1$, we get from (13) and (14) useful expressions for the binomial coefficient. Thus (13) gives

$$\binom{x}{n} \sim \frac{(-)^n n^{-(x+1)}}{\Gamma(-x)} \sum_{k=0}^{\infty} \frac{(x+1)_k B_k^{(-x)}}{k!\,n^k},$$

$$\binom{x}{n} \sim \frac{(-)^n n^{-(x+1)}}{\Gamma(-x)} \left[1 + \frac{(x)_2}{2n} + \frac{(x)_3(3x+1)}{24n^2} + \frac{(x)_4(x)_2}{48n^3} + \frac{(x)_5}{5760n^4}(15x^3 + 30x^2 + 5x - 2) + \cdots\right], \qquad (15)$$

and from (14),

$$\binom{x}{n} \sim \frac{(-)^n (n - x/2)^{-(x+1)}}{\Gamma(-x)} \sum_{k=0}^{\infty} \frac{B_{2k}^{(2\rho)}(\rho)(x+1)_{2k}}{(2k)!\,(n - x/2)^{2k}}, \quad \rho = -\frac{x}{2},$$

$$\binom{x}{n} \sim \frac{(-)^n (n - x/2)^{-(x+1)}}{\Gamma(-x)} \left[1 + \frac{(x)_3}{24(n - x/2)^2} + \frac{(x)_5(5x-2)}{5760(n-x/2)^4} + \frac{(x)_7(35x^2 - 42x + 16)}{2903040(n - x/2)^6} + \cdots\right]. \qquad (16)$$

It is clear that (14) is more powerful than (13). In illustration, put $x = -\frac{1}{2}$ in (15) and (16). We get the respective equations

$$\pi = \frac{(n!)^4 2^{4n}}{n\,[(2n)!]^2} \left[1 - \frac{1}{8n} + \frac{1}{128n^2} + \frac{5}{1024n^3} + O(n^{-4})\right]^2, \quad n \to +\infty, \qquad (17)$$

$$\pi = \frac{4(n!)^4 2^{4n}}{y\,[(2n)!]^2} \left[1 - \frac{1}{4y^2} + \frac{21}{32y^4} - \frac{671}{128y^6} + \frac{1\,80323}{2048y^8} - \frac{208\,98423}{8192y^{10}} + \frac{74263\,62705}{65536y^{12}} - \frac{187\,44094\,65055}{2\,62144y^{14}} + O(y^{-16})\right]^2,$$

$$y = 4n + 1, \quad n \to +\infty. \qquad (18)$$

If $n = 20$, use of the first three terms in (17) gives π with an error of $0.38 \cdot 10^{-5}$. If $n = 10$, use of the first three terms in (18) gives π with an error of $-0.69 \cdot 10^{-8}$, while use of all the terms in (18) gives π with an error of $0.56 \cdot 10^{-16}$.

1.4. Rational Approximations for $\Psi(z)$

$\Psi(z)$ can be represented in hypergeometric form as

$$\Psi(z) + \gamma = \frac{2(z-1)}{z}\,{}_3F_2\left(\begin{matrix}1,\ 1,\ 2-z\\ 2,\ 1+z\end{matrix}\,\middle|\,-1\right) \tag{1}$$

and so rational approximations follow from the developments in 5.12. Complete details are given in Luke (1971) where rational approximations for $\Psi(z+\frac{1}{2}) - \Psi(z)$ are also treated. We have

$$\Psi(z) + \gamma = \frac{2(z-1)A_n(z)}{z\,B_n(z)} + S_n(z), \tag{2}$$

$$A_n(z) = (3-z)_n \sum_{k=0}^{n} \frac{(-n)_k(n+1)_k(2-z)_k}{\left[(2)_k\right]^2(3-z)_k}\,{}_4F_3\left(\begin{matrix}-n+k, n+1+k, z+1+k, 1\\ 1+k, 2+k, 3-z+k\end{matrix}\,\middle|\,-1\right), \tag{3}$$

$$B_n(z) = (3-z)_n\,{}_3F_2\left(\begin{matrix}-n, n+1, 1+z\\ 2, 3-z\end{matrix}\,\middle|\,-1\right), \tag{4}$$

$$\begin{aligned} &A_0(z) = 1,\ A_1(z) = (z+6)/2,\ A_2(z) = (z^2+82z+96)/6,\\ &A_3(z) = (z^3+387z^2+2906z+1920)/12,\\ &A_4(z) = (3z^4+3643z^3+86068z^2+2\ 93508z+1\ 49760)/60,\end{aligned} \tag{5}$$

$$\begin{aligned} &B_0(z) = 1,\ B_1(z) = 4,\ B_2(z) = 4(2z+7),\\ &B_3(z) = 14z^2+204z+310,\ B_4(z) = 22z^3+864z^2+4958z+4956.\end{aligned} \tag{6}$$

Both $A_n(z)$ and $B_n(z)$ satisfy the same four-term recurrence formula

$$\begin{aligned}(2n-3)(n+1)B_n(z) &= (2n-1)\left[3(n-1)z+7n^2-9n-6\right]B_{n-1}(z)\\ &\quad -(2n-3)(z-n-1)\left[3(n-1)z-7n^2+19n-4\right]B_{n-2}(z)\\ &\quad +(2n-1)(n-3)(z-n-1)(z+n-4)B_{n-3}(z),\quad n \geq 4.\end{aligned} \tag{7}$$

When $z = 2$, $A_n(z) = B_n(z)$ and this provides a neat numerical check on the above and all subsequent polynomials. Further, if z is a positive integer m, $m \geq 2$,

$$ {}_3F_2\left(\begin{matrix}1, 1, 2-m \\ 2, 1+m\end{matrix}\middle| -1\right) = \frac{m}{2(m-1)}\sum_{r=1}^{m-1} r^{-1} = \frac{A_n(z)}{B_n(z)}, \quad n \geq m-2. \tag{8} $$

For the remainder, we have

$$ S_n(z) = \frac{O(1)}{f_n(z)}, \quad R(z) > 0, $$

$$ f_n(z) = \frac{\Gamma(3-z)}{(2\pi)^{\frac{1}{2}}\Gamma(1+z)}\left(\frac{N^2}{2}\right)^{z-7/4}\cosh\{N\xi+O(N^{-1})\}, \tag{9} $$

$$ \cosh \xi = 3, \quad e^{-\xi} = 3-2^{3/2} = 0.17157\ldots, \quad N^2 = n(n+1), $$

$$ \lim_{n\to\infty} S_n(z) = 0, \quad z \text{ fixed}, \quad R(z) > 0. $$

According to (9) convergence is improved if we take $R(z) > 7/4$. Notice that the rational approximations converge much more rapidly than the ${}_3F_2$ series expansion. In illustration, Table A which follows exhibits the rational approximations for the ${}_3F_2$ when $z = \frac{1}{2}$ and $z = 1$, respectively, for which the exact values are ln 2 and $\pi^2/12$, respectively. About 100,000 terms of the ${}_3F_2$ series for $z = \frac{1}{2}$ and $z = 1$ are required to produce ln 2 and $\pi^2/12$, respectively, with an accuracy of about 5 decimals and 10 decimals, respectively. The efficiency of the rational approximations is manifest. Following this is Table B which gives the evaluation of $\gamma + \Psi(z)$ for $z = 2.5 + 4i$. The imaginary part of $\Psi(z)$ for $z = 1 + Di$ is $(\pi D \coth \pi D - 1)/2D$ and this value is also recorded. In the tables, the notations $A(N,Z)$ and $B(N,Z)$ are used for $A_n(z)$ and $B_n(z)$, respectively. The machine evaluated $A_n(z)$ and $B_n(z)$ for $n = 0,1,2$ and 3 by means of (5) and (6), respectively, and got all subsequent values of these polynomials by use of the recursion formula (7). In the original calculations, Luke (1971), the machine calculated

$$ \delta_n(z) = |C(z)|\,|A_n(z)/B_n(z) - A_{n-1}(z)/B_{n-1}(z)|, \tag{10} $$

for $n = 1,2,\ldots,N$ where $C(z) = 1$ or $2(z-1)/z$ as appropriate, and N is such that $\delta_n(z) \leq 0.1\cdot 10^{-26}$. Only a portion of the original tables are given below.

TABLE A

EVALUATION OF $A_n(z)/B_n(z)$ FOR $z = \frac{1}{2}$ AND $z = 1$

Z = 5.0000000000000000000000000000000D − 01 0. Z = 1.0000000000000000000000000000000D + 00 0.

N	$A(N,Z)/B(N,Z)$	N	$A(N,Z)/B(N,Z)$
1	8.1250000000000000000000000000000D − 01	1	8.7500000000000000000000000000000D − 01
2	7.1484375000000000000000000000000D − 01	2	8.2870370370370370370370370370364D − 01
3	6.9592358604091456077015643802D − 01	3	8.2291666666666666666666666666666D − 01
4	6.9353054058468071206924710110D − 01	4	8.2250308641975308641975308648D − 01
5	6.9320316214931050879695672850D − 01	5	8.2247038917089678510998307962D − 01
6	6.9315565472978875374472655149D − 01	6	8.2246738636392172225110051413D − 01
7	6.9314849409236780351403667432D − 01	7	8.2246707391182694157354458494D − 01
8	6.9314738753724831290421294092D − 01	8	8.2246703835935517041566925571D − 01
9	6.9314721356667069137522060602D − 01	9	8.2246703405274466475587162974D − 01
10	6.9314718587150800717752853328D − 01	10	8.2246703350690207957580740730D − 01
11	6.9314718142079355217622304567D − 01	11	8.2246703343530682887274455002D − 01
12	6.9314718070026171946091597912D − 01	12	8.2246703342565934175426132486D − 01
13	6.9314718058292386228898045798D − 01	13	8.2246703342433059920392371467D − 01
14	6.9314718056372305936505710800D − 01	14	8.2246703342414424307534160725D − 01
15	6.9314718056056844195730624602D − 01	15	8.2246703342411770340454662436D − 01
16	6.9314718056004838592164015455D − 01	16	8.2246703342411387391170360411D − 01
17	6.9314718055996240196697686433D − 01	17	8.2246703342411331502837508557D − 01
18	6.9314718055994814987529553842D − 01	18	8.2246703342411323264869706829D − 01
19	6.9314718055994578234427851252D − 01	19	8.2246703342411322039879729626D − 01
20	6.9314718055994538829095250833D − 01	20	8.2246703342411321856294526147D − 01
21	6.9314718055994532259147097164D − 01	21	8.2246703342411321828588049469D − 01
22	6.9314718055994531162070640862D − 01	22	8.2246703342411321824380179178D − 01
23	6.9314718055994530978622951405D − 01	23	8.2246703342411321823737457919D − 01
24	6.9314718055994530947909395183D − 01	24	8.2246703342411321823638775615D − 01
25	6.9314718055994530942761371596D − 01	25	8.2246703342411321823623551998D − 01
26	6.9314718055994530941897597577D − 01	26	8.2246703342411321823621193208D − 01
27	6.9314718055994530941752529779D − 01	27	8.2246703342411321823620826263D − 01
28	6.9314718055994530941728144942D − 01	28	8.2246703342411321823620768962D − 01
29	6.9314718055994530941724042736D − 01	29	8.2246703342411321823620759986D − 01
30	6.9314718055994530941723352114D − 01	30	8.2246703342411321823620758575D − 01
31	6.9314718055994530941723235769D − 01	31	8.2246703342411321823620758350D − 01
32	6.9314718055994530941723216157D − 01	32	8.2246703342411321823620758315D − 01
33	6.9314718055994530941723212848D − 01		
34	6.9314718055994530941723212295D − 01		
35	6.9314718055994530941723212201D − 01		

This table and the one which follows are taken from
Luke (1971) "Rational Approximations for the Logarithmic
Derivative of the Gamma Function", *Applicable Analysis*,
1(1971), 65-73 (Gordon and Breach), with permission.

TABLE B

EVALUATION OF $\gamma + \Psi(z)$ FOR $z = 2.5 + 4i$

N	(2(Z-1)/Z)A(N,Z)/B(N,Z) (REAL)	(2(Z-1)/Z)A(N,Z)/B(N,Z) (IMAGINARY)
1	1.7064606741573033707865168539D+00	1.2696629213483146067415730337D+00
2	2.1000117041198501872659176027D+00	1.1087312734082397003745318351D+00
3	2.0712647591877834958918510965D+00	1.1049820733740498482429652941D+00
4	2.0739539567762721314823575619D+00	1.1056510099400000218945525061D+00
5	2.0738453559739868901275605649D+00	1.1054467111269958902843036618D+00
6	2.0738356152686577282107330596D+00	1.1054642344298933636818766424D+00
7	2.0738373874313878583713603884D+00	1.1054645618883228432227514933D+00
8	2.0738373928150860872828865867D+00	1.1054643973372486187883685677D+00
9	2.0738373773522742891263168979D+00	1.1054643968942664032287279395D+00
10	2.0738373771290208030649348766D+00	1.1054643983692476936171733204D+00
11	2.0738373772659847543482983821D+00	1.1054643984238233497986679847D+00
12	2.0738373772756103102744550999D+00	1.1054643984124270031338871359D+00
13	2.0738373772749177377048057366D+00	1.1054643984110480314919296883D+00
14	2.0738373772747547577207206951D+00	1.1054643984110462457963794309D+00
15	2.0738373772747455535927027496D+00	1.1054643984110611961140198015D+00
16	2.0738373772747463395021554933D+00	1.1054643984110629547887578608D+00
17	2.0738373772747465488655487447D+00	1.1054643984110630039265729954D+00
18	2.0738373772747465684058102116D+00	1.1054643984110629893066780347D+00
19	2.0738373772747465686821652637D+00	1.1054643984110629865079410469D+00
20	2.0738373772747465684722110170D+00	1.1054643984110629862548584140D+00
21	2.0738373772747465684335272232D+00	1.1054643984110629862500724409D+00
22	2.0738373772747465684296622706D+00	1.1054643984110629862527123005D+00
23	2.0738373772747465684295053457D+00	1.1054643984110629862532591437D+00
24	2.0738373772747465684295315995D+00	1.1054643984110629862533239837D+00
25	2.0738373772747465684295390229D+00	1.1054643984110629862533285120D+00
26	2.0738373772747465684295401127D+00	1.1054643984110629862533284393D+00
27	2.0738373772747465684295402220D+00	1.1054643984110629862533283545D+00
28	2.0738373772747465684295402278D+00	1.1054643984110629862533283378D+00
29	2.0738373772747465684295402273D+00	1.1054643984110629862533283356D+00
30	2.0738373772747465684295402272D+00	1.1054643984110629862533283354D+00

$$\Psi(z) = 1.49662\ 17123\ 73213\ 70782\ 30281\ 371 + 1.10546\ 43984\ 11062\ 98625\ 33283\ 354i.$$

1.5. Inequalities

Luke (1972b) has proved that

$$L(z) = \frac{2A(z)(5z^2+20z+19)}{(z+2)(z+7)} < \Gamma(z+1) < R(z) = \frac{2zA(z)(3z+4)}{(z+1)(z+13)}$$

$$+ 2A(z)\frac{(z^4+36z^3+266z^2+690z+579)}{(11z+71)(z+2)(z+3)} \; ,$$

$$A(z) = (z+1)^z e^{-(z+1)}, \quad 0 \leq z \leq 1. \tag{1}$$

Based on calculations for $z = 0(0.05)1.0$, $\Gamma(z+1) - L(z)$ goes monotonically from $1.47 \cdot 10^{-3}$ at $z = 0$ to $7.54 \cdot 10^{-3}$ at $z = 1$; $R(z) - \Gamma(z+1)$ equals $0.103 \cdot 10^{-4}$ and $1.61 \cdot 10^{-4}$ at $z = 0$ and $z = 1$, respectively, and has a relative maximum near $z = 0.25$, and for $z = 0.25$ the value of $R(z) - \Gamma(z+1)$ is $6.4 \cdot 10^{-4}$. Thus the inequality is quite sharp.

By integration of (11) ahead,

$$4z - 15 \ln(1+z/5) < \gamma z + \ln \Gamma(z+2) < 5z - \frac{18}{11} \ln (1+z/2)$$

$$- \frac{455}{11} \ln (1+z/13), \quad -1/2 \leq z \leq 1, \tag{2}$$

with equality if $z = 0$.

If $0 \leq v \leq 1$ and $0 \leq u \leq 1$, general two-sided inequalities for $\Gamma(v+1)/\Gamma(u+1)$ follow from (1). Gautschi (1959a) has shown that if n is a positive integer

$$n^{1-s} \leq \frac{\Gamma(n+1)}{\Gamma(n+s)} \leq (n+1)^{1-s}, \quad 0 \leq s \leq 1. \tag{3}$$

Keckic and Vasic (1971) prove that

$$\frac{x^{x-1}e^y}{y^{y-1}e^x} \leq \frac{\Gamma(x)}{\Gamma(y)} \leq \frac{x^{x-\frac{1}{2}}e^y}{y^{y-\frac{1}{2}}e^x}, \quad x \geq y \geq 1, \tag{4}$$

and with $x = n + 1$ and $y = n + s$, make a comparison of (3) and (4). For $s = 1$, the inequalities coincide. For $s = \frac{1}{2}$, $n = 1$, the left hand inequality of (4) is weaker than the corresponding inequality of (3). For $s = 3/4$, $n = 1$, the left hand inequality of (4) is sharper than the corresponding inequality of (3). Thus for x and y as above, the left hand inequalities of (3) and (4) cannot be effectively compared. The above named authors report that D. V. Slavic has made a comparison of (3) and (4) on a computer and that for a large number of s and n values, the right hand side of (4) is sharper than the right hand side of (3). Luke (1972b) has also proved (5) and (6) below.

$$\frac{(2z+1)}{3\left[\frac{z(4z+3)}{7}\right]^{\frac{1}{2}}} < \frac{(\pi/z)^{\frac{1}{2}}\Gamma(z+1)}{2\Gamma(z+\frac{1}{2})} < \left(\frac{4z+1}{5z}\right)^{\frac{1}{2}}, \quad z > 1, \tag{5}$$

with equality if $z = 1$. If $0 < z < 1$, (5) is valid with inequality signs reversed. If we put $s = \frac{1}{2}$ in (3) and compare it with (5) for z a positive integer, then the left hand inequality of (5) is superior to the left hand inequality of (3). The same is true of the right hand inequalities if $1 \leq z \leq 40$.

$$\frac{z+1}{2z^2+z+1} < \frac{2^{-2z}\pi^{\frac{1}{2}}\Gamma(z+1)}{\Gamma(z+\frac{1}{2})} < \frac{(1-z)(2z^2+5z+1)}{(z+1)^2(2z+1)} + \frac{8z^2(z+2)^2}{(z+1)^2(2z+1)(2z^2+5z+5)},$$
$$0 \leq z \leq 1. \tag{6}$$

For references and comments on other inequalities for gamma functions, their ratios and related matters, see Mitrinovic (1970), Keckic and Vasic (1971), Keckic and Stankovic (1972) and Luke (1972b).

Amos (1973) has proved that

$$\left[\frac{\Gamma(z+1)}{\Gamma(z+\frac{1}{2})}\right]^2 < z\left(1 + \frac{1}{4z} + \frac{1}{32z^2} - \frac{1}{128z^3} + \frac{6}{5z^4}\right), \quad z \geq 2, \tag{7}$$

an expression which is asymptotically correct in all terms except the last.

Some numerical examples follow. From (2), with $z = -\frac{1}{2}$, $\frac{1}{2}$ and 1, we get the respective inequalities

$$-0.41959 < \left(\ln(\pi/4) - \gamma\right)/2 = -0.40939 < -0.40693,$$

$$0.57035 < \left(\gamma + \ln(9\pi/16)\right)/2 = 0.57328 < 0.57378,$$

$$1.26518 < \gamma + \ln 2 = 1.27036 < 1.27114. \tag{8}$$

For $z = 1/2$, $3/2$, and $5/2$, we get from (5) and the remark following (5) the respective inequalities,

$$4/5 < \pi^2/12 = 0.82247 < \frac{112}{135} = 0.82963,$$

$$\frac{1792}{2187} = 0.81939 < \pi^2/12 < \frac{112}{135},$$

$$\frac{7168}{8775} = 0.81687 < \pi^2/12 < \frac{2816}{3375} = 0.83437. \tag{9}$$

If $z = 0$ or 1, (6) is exact. For $z = \tfrac{1}{2}$, (6) yields

$$3/4 < \pi/4 = 0.78540 < 57/72 = 0.79167. \tag{10}$$

The inequalities (11)-(13) below are also due to Luke (1972b).

$$\frac{4(z^2-1)}{z(z+4)} < \Psi(z) + \gamma < \frac{2(z-1)}{z} - \frac{3(z-1)(4-z^2)}{z(z+1)(z+12)},$$
$$1 < z < 2, \tag{11}$$

with equality if $z = 1$ or $z = 2$. If we reverse the inequality signs, then (11) is valid for $\tfrac{1}{2} \leq z < 1$.

$$\left(\frac{z+1}{2-z}\right) \ln \left(\frac{3}{z+1}\right) < \frac{z}{2(z-1)}\left[\gamma + \Psi(z)\right] < 1 - \frac{(4-z^2)}{(3-z)(z+1)}$$
$$+ \frac{(4-z^2)(z+2)}{(3-z)^2(z+1)} \ln\left(\frac{5}{z+2}\right), \quad \tfrac{1}{2} < z < 2, \tag{12}$$

with equality if $z = \tfrac{1}{2}$ or $z = 2$. Finally, for $z > 0$,

$$\frac{2z+1}{4z+1} < z\left[\Psi(z+\tfrac{1}{2}) - \Psi(z)\right] < \frac{4z^2+6z+3}{(2z+1)(4z+3)}, \tag{13}$$

with equality if $z = 0$ or $z = \infty$. See also Shafer (1973).

The following numerical examples manifest the sharpness of the inequalities (11)-(13). From (11) (see also the comment after this equation), with $z = 1/2$ and $z = 3/2$, we have the respective inequalities,

$$2/3 < \ln 2 = 0.69315 < 0.7,$$

$$20/33 = 0.60606 < 2 - 2\ln 2 = 0.61371 < 83/135 = 0.61482. \quad (14)$$

Divide (11) throughout by $z-1$ and let $z \to 1$. Then

$$4/5 < \pi^2/12 = 0.82247 < 43/52 = 0.82692. \quad (15)$$

Using (12) with $z = 1$ and $z = 3/2$, we get the respective inequalities

$$0.81093 < \pi^2/12 = 0.82247 < 0.82468,$$

$$0.91161 < 3(1 - \ln 2) = 0.92056 < 0.92171. \quad (16)$$

If $z = 1$, (13) yields

$$3/5 < 2 - 2\ln 2 = 0.61371 < 13/21 = 0.61905. \quad (17)$$

1.6. Bibliographic and Numerical Data

1.6.1. General References

For references on the gamma function and related functions, see Whittaker and Watson (1927), Erdélyi *et al.* (1953), Abramowitz and Stegun (1964), Artin (1964), Nielsen (1965), Campbell (1966) and Luke (1969) and references given in these sources. An historical profile of the gamma function is the subject of an article by Davis (1959). See also references given in 5.1.

1.6.2. DESCRIPTION OF AND REFERENCES TO TABLES

Gamma Function and Related Functions

Slavic (1971): $\Gamma(x)$, $1/\Gamma(x)$, $x = 1(0.01)2$, 30D.
Also $(-)^n\Psi(n+1)/\Gamma(n+1)$, $n = 0(1)30$, 30D.

Khamis (1965): $\Gamma(n)$, $n = 1(0.025)2$, 10D.

Galant and Byrd (1968): $\Gamma(p/q)$, $\ln \Gamma(p/q)$, $p = 1, 2, \ldots, q-1$, $2p \neq q$, $q = 3, 4, 5, 8, 10$, 60D.

Kireeva and Karpov (1959): $\Gamma\left(\frac{1-p}{2}\right)/2^{1+p/2}\pi^{\frac{1}{2}}$, $\Gamma\left(-\frac{p}{2}\right)/2^{1+p/2}(2\pi)^{\frac{1}{2}}$, $p = 0.05(0.05)0.95$, $1.05(0.05)1.95$, 9D.

Karpov and Cistova (1964): $\pi^{\frac{1}{2}}/2^{p/2}\Gamma(-p)$, $p = 0.05(0.05)1.95$, 10D.

Ascari (1968): See entry on p. 147.

Cody and Hillstrom (1970): Zero of $\arg \Gamma(1+iy)$, 22D.

Ratio of Gamma Functions

Smirnov (1961): $(\pi p)^{-\frac{1}{2}}\Gamma\left(\frac{p+1}{2}\right)\Big/\Gamma\left(\frac{p}{2}\right)$ and its common logarithm, $p = 1(1)24$, 6D.

Fettis and Caslin (1970): $\dfrac{\Gamma(m+n+\frac{1}{2})\Gamma(n-m+5/2)}{\Gamma(n+\frac{1}{2})\Gamma(n-\frac{1}{2})}$,
$m = 0(1)10$, $n = 0(1)450$, 16S.

Pearson (1968): See entry on p.279.

Osborn and Madey (1968): See entry on p.279.

Psi Function and Related Functions

Ditkin (1965): $\Psi(x+iy)$, $x = 1(0.01)2$, $y = 0(0.01)4$, 7S.
$\psi^{(n)}(x+iy)$, $n = 1(1)10$, $x = 1(0.1)2$, $y = 0(0.1)4$, 7S.

Kölbig (1971): $A_{np} = \int_0^{\frac{1}{2}\pi} (\ln \cos x)^n (\ln \sin x)^p dx$,
$n,p = 0(1)5$, 16S.

Cody, Strecok and Thacher (1973): Positive zero of $\psi(x)$, 36D.

Robinson (1974): $\psi'(x)$, $x = n+a$, $n = 0(1)50$, $a = 0, 1/2, 1/3, 2/3, 1/4, 3/4, 1/5, 2/5, 3/5, 4/5$, 58D.

Miscellaneous

Knuth and Buckholtz (1967): Exact values of Euler numbers for $n \leq 404$, tangent numbers for $n \leq 418$ and corresponding numbers C_{2n}, $n \leq 418$ from which $B_{2n} = C_{2n} - \sum 1/p$ where the sum is taken over all primes p such that $(p-1)|2n$. These primes are listed with each C_{2n}.

Boas and Wrench (1971): Let n_A be the smallest integer n such that $\sum_{k=1}^{n} k^{-1} > A$ $(A \geq 3)$. Let $e^{A-\gamma} = m + \delta$, m is an integer, $0 < \delta < 1$, γ = Euler's constant. Then $n_A = m$ if $\delta < \frac{1}{2} - (10n)^{-1}$, $n_A = m + 1$ if $\delta > \frac{1}{2} + n^{-1}$. The table gives $e^{A-\gamma}$ for $A = 1(1)20$, 100, 16S. Also $S_n = \sum_{k=1}^{n} k^{-1}$ is given to 10D for various n.

1.6.3. Description of and References to Other Approximations and Expansions

Clenshaw (1962): $\Gamma(x+1)$ and its reciprocal, $CT^*(x)$, $a = 1$, 20D.

Clenshaw, Miller and Woodger (1963): $1/\Gamma(x+1)$, $CT^*(x)$, $a = 1$, 15D.

Wimp (1961): $\ln \Gamma(x+b)$, $\Psi(x+b)$, $b = 2,3,4,5$, $CT(x)$. $a = 1$, 8D.

Fields and Wimp (1963): $\zeta(1+x)$, $\Gamma(x+3)$, $\{\Gamma(x)\}^{-1}$, $CT^*(x)$, $a = 1$, 10D. There the last five digits in the coefficient for h_3 should read 33909 instead of 23167.

Cody and Hillstrom (1967): Best Chebyshev coefficients in rational approximations for a function related to $\ln \Gamma(x)$. The ranges covered are $0.5 \leq x \leq 1.5$, $1.5 \leq x \leq 4.0$, and $4.0 \leq x \leq 12.0$. Accuracy up to about 20S is possible.

Hart *et al.* (1969): Best polynomial and rational approximations for $\Gamma(x)$ and $\ln \Gamma(x)$, various ranges with accuracy up to about 10^{-23}.

Werner and Collinge (1961): $\Gamma(x+2)$, $BCP(x)$, $a = 1$. $N = 7, 8, 10, 13, 15, 17, 20$. Accuracy goes from $0.25 \cdot 10^{-7}$ for $N = 7$ to $0.10 \cdot 10^{-18}$ for $N = 20$.

Luke (1969): Coefficients g_k in $\Gamma(z+1) = A(z)\left[\sum_{k=0}^{n-1} g_k H_k(z) + S_n(z)\right]$, $A(z) = (2\pi)^{\frac{1}{2}}(z+11/2)^{z+\frac{1}{2}}e^{-(z+11/2)}$, $R(z) > -11/2$, $H_k(z) = \left((z+1)_k(z+1)_{-k}\right)^{-1}$, $k = 0(1)15$, 15D. Some data relating to $S_n(z)$ are also given.

Cody and Hillstrom (1970): Nearly best Chebyshev rational approximations are presented for $\arg \Gamma(1+iy)$. Maximal relative errors range down to between $4.24 \cdot 10^{-19}$ and $1.09 \cdot 10^{-20}$.

Moody (1967): Best Chebyshev coefficients for a function simply related to $\Psi(x+1)$. Three sets of coefficients are given so that accuracy varies from about 6 to 8D.

Cody, Strecok and Thacher (1973): Rational Chebyshev approximations for $\Psi(x)$ are presented for $0.5 \leq x \leq 3.0$ and $x \geq 3.0$. Maximum relative errors range down to the order of 10^{-20}.

Khovanskii (1963): Continued fraction representations for $\Gamma(1+x)$, $\ln \Gamma(1+x)$ and $\psi(1+x)$.

Shenton and Bowman (1971): Continued fraction representations for $\ln \Gamma(x)$ and some of its lower derivatives. A few partial quotients are derived for the general case.

Luke (1969, v.2, p.306): $x\zeta(1+x)$, $CT^*(x)$, $a = 1$; $-\ln \xi(x)$, where $\xi(x) = \frac{1}{2}s(s-1)\pi^{-\frac{1}{2}s}\Gamma(\frac{1}{2}s)\zeta(s)$, $CT(x)$, $a = 1$, 20D; and a third function related to $\zeta(x)$, $CT^*(x)$, $a = 1$, 15D. Coefficients for the first function are given to 20D, but they are only correct to 16D provided also that the 16th digit in a_0 is increased by unity. In this connection, see Piessens and Branders (1972) who give $x\zeta(1+x)$ as above to 23D, and $\zeta(x+k)$, $k = 2(1)5, 8$, $CT^*(x)$, $a = 1$, 23D.

Cody, Hillstrom and Thacher (1971): Best Chebyshev rational approximations for $\zeta(x)$ over segments of the real axis to cover the range $0.5 \leq x \leq 55$. Accuracy from 8 to 22S.

CHAPTER II THE BINOMIAL FUNCTION

2.1. Power Series

$$(1 + z)^a = {}_1F_0(-a; -z) = {}_2F_1(-a, b; b; -z), \qquad |z| < 1. \tag{1}$$

$$(1 + z)^{2a} + (1 - z)^{2a} = 2\,{}_2F_1(-a, \tfrac{1}{2} - a; \tfrac{1}{2}; z^2), \qquad |z| < 1. \tag{2}$$

$$(1 + z)^{2a} - (1 - z)^{2a} = 4az\,{}_2F_1(\tfrac{1}{2} - a, 1 - a; \tfrac{3}{2}; z^2), \qquad |z| < 1. \tag{3}$$

$$[\tfrac{1}{2} + \tfrac{1}{2}(1 - z)^{1/2}]^{1-2a} = {}_2F_1(a, a - \tfrac{1}{2}; 2a; z), \qquad |z| < 1. \tag{4}$$

$$(1 + z)(1 - z)^{-2a-1} = {}_2F_1(a + 1, 2a; a; z), \qquad |z| < 1. \tag{5}$$

2.2. Expansions in Series of Jacobi and Chebyshev Polynomials

$$(a + x)^{-\mu} = \sum_{n=0}^{\infty} C_n R_n^{(\alpha,\beta)}(x),$$

$$C_n = \frac{(\mu)_n(-)^n}{a^{n+\mu}(n + \lambda)_n}\,{}_2F_1\left(\begin{matrix}\beta + 1 + n, \mu + n\\ \lambda + 1 + 2n\end{matrix}\,\middle|\, -\frac{1}{a}\right),$$

$$a \neq -1, \qquad |\arg(1 + 1/a)| < \pi, \qquad 0 \leqslant x \leqslant 1. \tag{1}$$

$$(a + x)^{-\alpha-3/2} = \sum_{n=0}^{\infty} C_n R_n^{(\alpha,\alpha)}(x),$$

$$C_n = \frac{4[(a + 1)^{1/2} - a^{1/2}]q^{\alpha+n}(-)^n(n + \alpha + \tfrac{1}{2})\Gamma(n + 2\alpha + 1)}{(a^2 + a)^{1/2}(\tfrac{3}{2})_\alpha \Gamma(n + \alpha + 1)},$$

$$q = 2a + 1 - 2(a^2 + a)^{1/2}, \tag{2}$$

$$(a + x)^{-1} = (a^2 + a)^{-1/2} \sum_{n=0}^{\infty} \epsilon_n(-)^n q^n T_n^*(x). \tag{3}$$

The conditions for validity of (2), (3) are the same as in (1).

$$x^s = \Gamma(\beta+1+s)\sum_{n=0}^{\infty}\frac{(-)^n(2n+\lambda)\Gamma(n+\lambda)(-s)_n}{\Gamma(n+\beta+1)\Gamma(n+\lambda+1+s)}R_n^{(\alpha,\beta)}(x),$$

$$\alpha > -1, \quad \beta > -1, \quad -R(s) < \min(0, \tfrac{1}{2}(\beta-\alpha)), \quad 0 \leqslant x \leqslant 1. \tag{4}$$

$$(1-k^2\sin^2\theta)^{-\omega} = \sum_{n=0}^{\infty}\epsilon_n h_n \cos 2n\theta,$$

$$h_n = \frac{(-)^n(q+1)^{2\omega}q^n(\omega)_n}{n!}\,{}_2F_1\left(\begin{matrix}n+\omega,\ \omega\\ n+1\end{matrix}\,\middle|\, q^2\right),$$

$$q = \frac{2-k^2-2(1-k^2)^{1/2}}{k^2}, \qquad k^2 = \frac{4q}{(q+1)^2}, \tag{5}$$

$$\sum_{n=0}^{\infty}\epsilon_n h_n = 1, \qquad \sum_{n=0}^{\infty}(-)^n\epsilon_n h_n = (1-k^2)^{-\omega} = [(1-q)/(1+q)]^{-2\omega}, \tag{6}$$

$$h_{n-1} = -\frac{\{[n(1+q^2)/q]h_n + (n+1-\omega)h_{n+1}\}}{n+\omega-1}. \tag{7}$$

Here and throughout the entire volume

$$\varepsilon_0 = 1, \ \varepsilon_n = 2 \text{ for } n > 0. \tag{8}$$

For $\omega = -\frac{1}{2}$ and $\omega = \frac{1}{2}$, 20D values of $h_n = C_n(k^2)$, $k^2 = 0.1(0.1)0.9$, are given in Tables 50 and 51 respectively in Volume 2 of Luke (1969).

TABLE 2.1

CHEBYSHEV COEFFICIENTS FOR $(1+x)^{-1}$, $(1+x)^{-\frac{1}{2}}$, AND $(1+x^2)^{-\frac{1}{2}}$

$$(1+x)^{-1} = \sum_{n=0}^{\infty} a_n T_n^*(x) \qquad (1+x)^{-\frac{1}{2}} = \sum_{n=0}^{\infty} b_n T_n^*(x)$$

$$(1+x^2)^{-\frac{1}{2}} = \sum_{n=0}^{\infty} b_n T_{2n}(x)$$

$$0 \leq x \leq 1$$

n	a_n	n	b_n
0	0.70710 67811 86547 52440	0	0.83462 68416 74073 18628
1	-0.24264 06871 19285 14641	1	-0.14373 41563 44519 99644
2	0.04163 05603 42615 82963	2	0.01851 87309 28697 86157
3	-0.00714 26749 36409 83137	3	-0.00264 94146 51037 73765
4	0.00122 54892 75843 15857	4	0.00039 78961 13409 89251
5	-0.00021 02607 18649 12008	5	-0.00006 14567 65156 74189
6	0.00003 60750 36051 56190	6	0.00000 96673 53519 58914
7	-0.00000 61894 97660 25132	7	-0.00000 15403 87437 40443
8	0.00000 10619 49909 94605	8	0.00000 02477 96599 15423
9	-0.00000 01822 01799 42497	9	-0.00000 00401 56585 74937
10	0.00000 00312 60886 60376	10	0.00000 00065 45740 80578
11	-0.00000 00053 63520 19759	11	-0.00000 00010 72084 16458
12	0.00000 00009 20234 58178	12	0.00000 00001 76284 90447
13	-0.00000 00001 57887 29310	13	-0.00000 00000 29083 61832
14	0.00000 00000 27089 17684	14	0.00000 00000 04811 92026
15	-0.00000 00000 04647 76796	15	-0.00000 00000 00798 10029
16	0.00000 00000 00797 43091	16	0.00000 00000 00132 65691
17	-0.00000 00000 00136 81751	17	-0.00000 00000 00022 09145
18	0.00000 00000 00023 47417	18	0.00000 00000 00003 68509
19	-0.00000 00000 00004 02753	19	-0.00000 00000 00000 61564
20	0.00000 00000 00000 69102	20	0.00000 00000 00000 10299
21	-0.00000 00000 00000 11856	21	-0.00000 00000 00000 01725
22	0.00000 00000 00000 02034	22	0.00000 00000 00000 00289
23	-0.00000 00000 00000 00349	23	-0.00000 00000 00000 00049
24	0.00000 00000 00000 00060	24	0.00000 00000 00000 00008
25	-0.00000 00000 00000 00010	25	-0.00000 00000 00000 00001
26	0.00000 00000 00000 00002		

2.3. Expansions in Series of Bessel Functions

$$\left(\frac{z}{2}\right)^{a} = \Gamma(a+1) \sum_{n=0}^{\infty} \frac{(z/2)^n}{n!} J_{n+a}(z). \tag{1}$$

$$\left(\frac{z}{2}\right)^{a} = \sum_{n=0}^{\infty} \frac{(-)^n (2n+a)\Gamma(n+a)}{n!} I_{2n+a}(z), \tag{2}$$

a not a negative integer.

$$1 = J_0(z) + 2\sum_{n=1}^{\infty} J_{2n}(z) = I_0(z) + 2\sum_{n=1}^{\infty} (-)^n I_{2n}(z). \qquad (3)$$

2.4. Padé Approximations

2.4.1. $(1 + 1/z)^{-c}$

Let

$$E(z) = {}_1F_0(c;-1/z) = (1+1/z)^{-c} \qquad (1)$$

and write

$$E(z) = \left[A_n(z,c)/B_n(z,c)\right] + R_n(z,c). \qquad (2)$$

Then

$$B_n(z,c) = {}_2F_1(-n,\ n+1-a;\ c+1-a;\ -z), \qquad (3)$$

$$A_n(z,c) = \left(\frac{nz}{1-c}\right)^a \frac{(1-c)_n}{(c+1-a)_n}\,{}_2F_1(\ a-n,\ n+1;\ 1+a-c;\ -z), \qquad (4)$$

$$B_n(-1,c) = \frac{(-)^n (1-c)_n}{(c+1-a)_n}, \quad A_n(-1,c) = (-)^n (n/c)^a. \qquad (5)$$

Both $B_n(z,c)$ and $A_n(z,c)$ satisfy the recurrence formula

$$\frac{(n+1-a)(n+c+1-a)}{(2n+2-a)(2n+1-a)}B_{n+1}(z,c) = \left\{z + \frac{2n^2+2n(1-a)-a(c+1-a)}{(2n-a)(2n+2-a)}\right\}B_n(z,c)$$

$$- \frac{n(n-c)}{(2n-a)(2n+1-a)}B_{n-1}(z,c). \qquad (6)$$

$$R_n(z,c) = \frac{(-)^{1-a} z^c (\sin \pi c) 2^{2c+1} \exp\left[-(2n+c+1-a)w\right]\left[1+O(n^{-1})\right]}{(1+e^{-w})^{2c}\{1+\exp\left[\pm i\pi(c+\tfrac{1}{2}-a)\right]\exp\left[-(2n+1-a)w\right]\}}, \qquad (7)$$

where e^{-w} and the sign conventions are as in 6.10(14-17). See also Table 2.4. Clearly for z fixed,

$$\lim_{n\to\infty} R_n(z,c) = 0, \ z \neq 0, \ z \neq -1, \ |\arg (1+1/z)| < \pi. \qquad (8)$$

Further results for the case $c = \pm\tfrac{1}{2}$ are given in 2.4.2.

In 2.4.1(2), put $z/(z+1) = x$. Then with $c = x$ and $c = 1/x$, we can obtain rational approximations for x^x and $x^{1/x}$, respectively. For details, see Luke (1969).

2.4.2. The Square Root

Consider the elementary identities

$$\tanh\theta = \left\{\frac{\cosh(2n+1)\theta}{\cosh\theta}\right\}\Big/\left\{\frac{\sinh(2n+1)\theta}{\sinh\theta}\right\} - \frac{e^{-(2n+1)\theta}\tanh\theta}{\sinh(2n+1)\theta}, \tag{1}$$

$$\coth\theta = \left\{\frac{\sinh(2n+1)\theta}{\sinh\theta}\right\}\Big/\left\{\frac{\cosh(2n+1)\theta}{\cosh\theta}\right\} + \frac{e^{-(2n+1)\theta}\coth\theta}{\cosh(2n+1)\theta}, \tag{2}$$

$$\tanh\theta = \left\{\frac{\sinh\theta\sinh 2n\theta}{\cosh\theta}\right\}\Big/\cosh 2n\theta + \frac{e^{-2n\theta}\tanh\theta}{\cosh 2n\theta}, \tag{3}$$

$$\coth\theta = \{\cosh 2n\theta\}\Big/\left\{\frac{\sinh\theta\sinh 2n\theta}{\cosh\theta}\right\} - \frac{e^{-2n\theta}\coth\theta}{\sinh 2n\theta}. \tag{4}$$

Now

$$\frac{\cosh(2n+1)\theta}{\cosh\theta}, \qquad \frac{\sinh(2n+1)\theta}{\sinh\theta}, \qquad \cosh 2n\theta, \qquad \frac{\sinh\theta\sinh 2n\theta}{\cosh\theta}$$

are polynomials in $\sinh^2\theta$ of degree n. So if

$$z = \sinh^2\theta, \qquad e^\theta = (1+z)^{1/2} \pm z^{1/2}, \tag{5}$$

where the sign is chosen so that $|e^\theta| > 1$ which is possible for all z except $-1 \leqslant z \leqslant 0$, then

$$\tanh\theta = (1 + 1/z)^{-1/2}, \tag{6}$$

and clearly (1), (3) and (2), (4) give rational approximations for $(1+1/z)^{-1/2}$ and $(1+1/z)^{1/2}$, respectively. If $R_n(z)$ represents any of the remainder terms in (1)–(4), then for z fixed,

$$\lim_{n\to\infty} R_n(z) = 0, \qquad z \neq 0, \quad z \neq -1, \quad |\arg(1+1/z)| < \pi. \tag{7}$$

Next we relate the above to the developments in 2.4.1. We write

$$\left(1+\frac{1}{z}\right)^{m-3/2} = \frac{A_n(z, \frac{3}{2}-m)}{B_n(z, \frac{3}{2}-m)} + R_n(z, \tfrac{3}{2}-m), \qquad m = 1 \text{ or } 2, \tag{8}$$

$$\left(1+\frac{1}{z}\right)^{1/2} = \frac{(z+1)A_n(z, \frac{1}{2})}{zB_n(z, \frac{1}{2})} + \frac{(z+1)}{z}R_n(z, \tfrac{1}{2}), \tag{9}$$

$$B_n(z, \tfrac{3}{2}-m) = {}_2F_1(-n, n+1-a; \tfrac{5}{2}-a-m; -z), \tag{10}$$

$$A_n(z, \tfrac{3}{2}-m) = \left(\frac{2nz}{2m-1}\right)^a \frac{(m-\frac{1}{2})_n}{(\frac{5}{2}-a-m)_n}\, {}_2F_1\left(\begin{matrix} a-n, n+1 \\ m+a-\frac{1}{2} \end{matrix}\,\middle|\, -z\right). \tag{11}$$

Observe that

$$A_n(z, \tfrac{3}{2} - m) = (2n+1)^{2m-3} B_n(z, m - \tfrac{3}{2}) \qquad \text{if} \quad a = 0, \tag{12}$$

$$B_n(z, \tfrac{1}{2}) = \frac{(-)^n T_{2n+1}(iz^{1/2})}{(2n+1)\, iz^{1/2}} = (2n+1)^{-1} U_{2n}([1+z]^{1/2}) \qquad \text{if} \quad a = 0, \tag{13}$$

$$A_n(z, \tfrac{1}{2}) = \frac{T_{2n+1}([1+z]^{1/2})}{(2n+1)[1+z]^{1/2}} = \frac{(-)^n U_{2n}(iz^{1/2})}{(2n+1)} \qquad \text{if} \quad a = 0, \tag{14}$$

$$B_n(z, \tfrac{1}{2}) = (-)^n T_n^*(-z) = (-)^n T_{2n}(iz^{1/2}) \qquad \text{if} \quad a = 1, \tag{15}$$

$$A_n(z, \tfrac{1}{2}) = 2z(-)^{n+1} U_{n-1}^*(-z) = (-)^n iz^{1/2} U_{2n-1}(iz^{1/2}) \qquad \text{if} \quad a = 1, \quad n > 0. \tag{16}$$

It is easy to show that (1) and (8) are the same if $m = 1$ and $a = 0$; that (2) and (8) are the same if $m = 2$ and $a = 0$; and that (3) and (8) are the same if $m = 1$ and $a = 1$. In these situations the remainder terms in (1)-(3) agree exactly with the results which follow from 2.4.1(7) if there the order term is suppressed. Also the approximations in (4) occupy the $(n+1, n)$ positions in the Padé matrix table.

Since the zeros of the Chebyshev polynomials are known, we can easily express the rational approximations as a sum of partial fractions. Let

$$z = p^2, \qquad \theta_k = k\pi/(2n+1), \qquad \varphi_k = (2k-1)\pi/4n. \tag{17}$$

Then

$$(1+p^2)^{-1/2} = (2n+1)^{-1}\left[\frac{1}{p} + 2\sum_{k=1}^{n} \frac{p}{p^2 + \sin^2\theta_k}\right] + \frac{R_n(p^2, \tfrac{1}{2})}{p}, \qquad a = 0, \tag{18}$$

$$\frac{(1+p^2)^{1/2}}{p} = 1 + 2(2n+1)^{-1} \sum_{k=1}^{n} \frac{\sin^2\theta_k}{p^2 + \cos^2\theta_k} + R_n(p^2, -\tfrac{1}{2}), \qquad a = 0, \tag{19}$$

$$(1+p^2)^{-1/2} = n^{-1} \sum_{k=1}^{n} \frac{p}{p^2 + \cos^2\varphi_k} + R_n(p^2, \tfrac{1}{2}), \qquad a = 1. \tag{20}$$

A certain Newton-Raphson process for finding the square root generates rational approximations of the Padé type developed here. For details, see Luke (1969).

2.4.3. Padé Coefficients

In Tables 2.2 and 2.3 of this section we give coefficients in the polynomials for the Padé approximations for the square and cube roots. The notation is that of 2.4.1 with $a = 0$. Further polynomials can be generated by use of 2.4.1(6). To estimate the error, use 2.4.1(7) and Table 2.4 below. Note that for $c = \pm\frac{1}{2}$, 2.4.1(7) is exact when $O(n^{-1})$ is omitted.

TABLE 2.2

Padé Coefficients for the Square Root

$$\left(1+\frac{1}{z}\right)^{1/2} = \frac{(2n+1)\mathcal{B}_n(z,\frac{1}{2})}{\mathcal{B}_n(z,-\frac{1}{2})} + R_n(z,-\tfrac{1}{2}), \quad \left(1+\frac{1}{z}\right)^{-1/2} = \frac{\mathcal{B}_n(z,-\frac{1}{2})}{(2n+1)\mathcal{B}_n(z,\frac{1}{2})} + R_n(z,\tfrac{1}{2}),$$

$$z \neq 0, \qquad z \neq -1, \qquad |\arg(1+1/z)| < \pi.$$

$$(2n+1)\mathcal{B}_n(z,\tfrac{1}{2}) = \sum_{k=0}^{n} a_k z^k \qquad\qquad \mathcal{B}_n(z,-\tfrac{1}{2}) = \sum_{k=0}^{n} b_k z^k$$

n	$a_0, a_1, \ldots, a_n$
0	1
1	3, 4
2	5, 20, 16
3	7, 56, 112, 64
4	9, 120, 432, 576, 256
5	11, 220, 1232, 2816, 2816, 1024
6	13, 364, 2912, 9984, 16640, 13312, 4096

n	$b_0, b_1, \ldots, b_n$
0	1
1	1, 4
2	1, 12, 16
3	1, 24, 80, 64
4	1, 40, 240, 448, 256
5	1, 60, 560, 1792, 2304, 1024
6	1, 84, 1120, 5376, 11520, 11264, 4096

TABLE 2.3

PADÉ COEFFICIENTS FOR THE CUBE ROOT

$$\left(1+\frac{1}{z}\right)^{1/3} = \frac{(\frac{4}{3})_n B_n(z, \frac{1}{3})}{(\frac{2}{3})_n B_n(z, -\frac{1}{3})} + R_n(z, -\tfrac{1}{3}), \qquad \left(1+\frac{1}{z}\right)^{-1/3} = \frac{(\frac{2}{3})_n B_n(z, -\frac{1}{3})}{(\frac{4}{3})_n B_n(z, \frac{1}{3})} + R_n(z, \tfrac{1}{3}),$$

$$z \neq 0, \qquad z \neq -1, \qquad |\arg(1 + 1/z)| < \pi,$$

$$B_n(z, \tfrac{1}{3}) = a_0^{-1} \sum_{k=0}^{n} a_k z^k, \qquad B_n(z, -\tfrac{1}{3}) = b_0^{-1} \sum_{k=0}^{n} b_k z^k.$$

n	$a_0, a_1, \ldots, a_n$
0	1
1	2, 3
2	14, 63, 54
3	7, 63, 135, 81
4	91, 1365, 5265, 7371, 3402
5	52, 1170, 7020, 16848, 17496, 6561
6	988, 31122, 2 66760, 9 60336, 16 62120, 13 71249, 4 33026

n	$b_0, b_1, \ldots, b_n$
0	1
1	1, 3
2	5, 45, 54
3	2, 36, 108, 81
4	22, 660, 3564, 6237, 3402
5	11, 495, 4158, 12474, 15309, 6561
6	187, 11781, 1 41372, 6 36174, 13 01265, 12 26907, 4 33026

2.4.4. THE FUNCTION e^{-w}

In a number of Padé approximations developed in this volume, estimation of the error is facilitated by having a numerical table of absolute values for a function e^{-w} defined as follows.

$$e^{-w} = 2z + 1 \mp 2(z^2+z)^{\frac{1}{2}} \tag{1}$$

where the sign is chosen so that $|e^{-w}| < 1$. This is possible for all z, $|\arg(1+1/z)| < \pi$. If $-1 \leq z \leq 0$, $|e^{-w}| = 1$. We have the expansions

$$|e^{-w}| = 1 + 2r\cos^2\theta/2 - r^{1/2}(r+2)\cos\theta/2 + O(r^{5/2}),\ z = re^{i\theta}, \tag{2}$$

$$e^{-w} = 1/4z - 1/8z^2 + 5/64z^3 + O(z^{-4}),\ |z| > 1. \tag{3}$$

The following approximations for e^{-w}, designated as $f(z)$, are based on second order Padé approximations for the square root. The domain of applicability for each formula is given. As a practical guide, formulas (4), (5), (6) and (7) work best when $|z| \geq 1$, $|z| \leq 1$, $|v| \leq 1$, and $|u| \leq 1$ respectively. This should be sufficient for practical purposes since further tests of reliability are easily determined by comparison with the absolute numerical values given in Table 2.4 below.

$$f(z) = \frac{4z+1}{16z^2+12z+1},\quad |\arg(1+1/z)| < \pi. \tag{4}$$

$$f(z) = 1 + 2z - 2z^{\frac{1}{2}}\frac{(16+20z+5z^2)}{16+12z+z^2},$$

$$|\arg(1+z)| < \pi,\quad -\pi < \arg z \leq \pi. \tag{5}$$

$$f(z) = -1 - 2v + 2v^{\frac{1}{2}}\frac{(16+20v+5v^2)}{16+12v+v^2},$$

$$z = -1 - v,\quad |\arg v| \leq \pi/2. \tag{6}$$

$$f(z) = -1 + 2u - 2iu^{\frac{1}{2}}\frac{(16-20u+5u^2)}{16-12u+u^2},$$

$$z = -1 + u,\quad |\arg u| \leq \pi/2. \tag{7}$$

TABLE 2.4

VALUES OF $|e^{-w}| = |2z+1\mp 2(z^2+z)^{\frac{1}{2}}|$

(The sign is chosen so that $|e^{-w}| < 1$. This is possible for all $z = re^{i\theta}$, $r \neq 0$, except $\theta = \pi$ and $0 < r \leq 1$.)

r/θ	0°	45°	90°	135°	150°
0.0	1.000(0)	1.000(0)	1.000(0)	1.000(0)	1.000(0)
0.1	0.537(0)	0.560(0)	0.635(0)	0.777(0)	0.842(0)
0.2	0.420(0)	0.443(0)	0.521(0)	0.689(0)	0.775(0)
0.3	0.351(0)	0.372(0)	0.446(0)	0.621(0)	0.718(0)
0.4	0.303(0)	0.322(0)	0.390(0)	0.562(0)	0.666(0)
0.5	0.268(0)	0.285(0)	0.346(0)	0.509(0)	0.616(0)
0.6	0.240(0)	0.255(0)	0.310(0)	0.461(0)	0.566(0)
0.7	0.218(0)	0.232(0)	0.281(0)	0.417(0)	0.516(0)
0.8	0.200(0)	0.212(0)	0.256(0)	0.378(0)	0.467(0)
0.9	0.185(0)	0.195(0)	0.235(0)	0.342(0)	0.420(0)
1.0	0.172(0)	0.181(0)	0.217(0)	0.311(0)	0.376(0)
2.0	0.101(0)	0.105(0)	0.120(0)	0.147(0)	0.158(0)
3.0	0.718(-1)	0.742(-1)	0.817(-1)	0.934(-1)	0.972(-1)
4.0	0.557(-1)	0.573(-1)	0.618(-1)	0.682(-1)	0.700(-1)
5.0	0.455(-1)	0.466(-1)	0.496(-1)	0.536(-1)	0.547(-1)
6.0	0.385(-1)	0.393(-1)	0.415(-1)	0.442(-1)	0.449(-1)
7.0	0.334(-1)	0.340(-1)	0.356(-1)	0.376(-1)	0.381(-1)
8.0	0.294(-1)	0.299(-1)	0.312(-1)	0.327(-1)	0.330(-1)
9.0	0.263(-1)	0.267(-1)	0.277(-1)	0.289(-1)	0.292(-1)
10.0	0.238(-1)	0.241(-1)	0.250(-1)	0.259(-1)	0.261(-1)

r/θ	165°	170°	175°	180°
0.0	1.000(0)	1.000(0)	1.000(0)	1.000(0)
0.1	0.917(0)	0.944(0)	0.971(0)	1.000(0)
0.2	0.878(0)	0.917(0)	0.957(0)	1.000(0)
0.3	0.844(0)	0.892(0)	0.945(0)	1.000(0)
0.4	0.810(0)	0.868(0)	0.931(0)	1.000(0)
0.5	0.774(0)	0.841(0)	0.917(0)	1.000(0)
0.6	0.734(0)	0.810(0)	0.899(0)	1.000(0)
0.7	0.687(0)	0.772(0)	0.876(0)	1.000(0)
0.8	0.632(0)	0.722(0)	0.843(0)	1.000(0)
0.9	0.565(0)	0.651(0)	0.784(0)	1.000(0)
1.0	0.493(0)	0.559(0)	0.661(0)	1.000(0)
2.0	0.168(0)	0.170(0)	0.171(0)	0.172(0)
3.0	0.100(0)	0.101(0)	0.101(0)	0.101(0)
4.0	0.713(-1)	0.716(-1)	0.717(-1)	0.718(-1)
5.0	0.555(-1)	0.556(-1)	0.557(-1)	0.557(-1)
6.0	0.454(-1)	0.455(-1)	0.455(-1)	0.455(-1)
7.0	0.384(-1)	0.385(-1)	0.385(-1)	0.385(-1)
8.0	0.333(-1)	0.333(-1)	0.334(-1)	0.334(-1)
9.0	0.294(-1)	0.294(-1)	0.294(-1)	0.294(-1)
10.0	0.263(-1)	0.263(-1)	0.263(-1)	0.263(-1)

2.5. Inequalities

In the notation of 2.4.1, let $B_n(z,c)$ be denoted by $B_n(z,c,a)$ and likewise for $A_n(z,c)$ and $R_n(z,c)$. Then

$$(1 + z)^{-c} = \frac{A_n(z^{-1},c,a)}{B_n(z^{-1},c,a)} + R_n(z^{-1},c,a). \tag{1}$$

If

$$z > 0, \quad c + 1 - a > 0, \quad n + 1 - a > 0,$$

then

$$\operatorname{sign} R_n(z^{-1},c,a) = \operatorname{sign} \left\{\frac{(-c)_{1-a}(1-c)_n}{(n-a)!}\right\}. \tag{2}$$

Thus if

$$z > 0, \quad 0 < c < 1, \quad r \text{ is an integer or zero,}$$

then

$$(1+z)^r \frac{A_n(z^{-1},c,1)}{B_n(z^{-1},c,1)} < (1+z)^{r-c} < (1+z)^r \frac{A_m(z^{-1},c,0)}{B_m(z^{-1},c,0)}, \tag{3}$$

$$m,\ n > 0,$$

$$(1+z)^r \frac{B_m(z^{-1},c,0)}{A_m(z^{-1},c,0)} < (1+z)^{r+c} < (1+z)^r \frac{B_n(z^{-1},c,1)}{A_n(z^{-1},c,1)}, \tag{4}$$

$$m,\ n > 0.$$

Further inequalities can be deduced from the following data.

If the hypotheses of (2) hold, and if with u a positive integer, $u < c < u+1$, and if $n+1-c > 0$, then

$$\operatorname{sign} R_n(z^{-1},c,a) = (-)^{u+1-a}.$$

Under these same conditions but $n+1-c < 0$, then

$$\operatorname{sign} R_n(z^{-1},c,a) = (-)^{n+1-a}.$$

In illustration,

$$(1+cz)^{-1} < (1+z)^{-c} < \frac{1-c}{1+c} + \frac{2c}{1+c}\left[1+\frac{(1+c)z}{2}\right]^{-1},$$

$$z > 0, \quad 0 < c < 1, \tag{5}$$

with equality if $c = 0$ or $c = 1$. Also

$$\frac{2}{z+2} < (1+z)^{-\frac{1}{2}} < \frac{z+4}{3z+4}, \quad z > 0, \tag{6}$$

$$\frac{4(z+2)}{z^2+8z+8} < (1+z)^{-\frac{1}{2}} < \frac{z^2+12z+16}{5z^2+20z+16}, \quad z > 0, \tag{7}$$

$$\frac{3}{z+3} < (1+z)^{-1/3} < \frac{z+3}{2z+3}, \quad z > 0, \tag{8}$$

$$\frac{3(5z+9)}{2z^2+24z+27} < (1+z)^{-1/3} < \frac{5z^2+45z+54}{14z^2+63z+54}, \quad z > 0. \tag{9}$$

In (5)-(9), we have equality if $z = 0$.

CHAPTER III ELEMENTARY FUNCTIONS

3.1. Logarithmic Functions

3.1.1. Power Series

$$\ln(1+z) = z\,{}_2F_1\left(\begin{matrix}1,1\\2\end{matrix}\middle| -z\right) = z\sum_{k=0}^{\infty}\frac{(-)^k z^k}{k+1},$$

$$|z| \le 1, \quad z \ne -1. \tag{1}$$

$$\ln(1+z) = \frac{z}{1+z}\,{}_2F_1\left(\begin{matrix}1,1\\2\end{matrix}\middle| \frac{z}{z+1}\right), \quad R(z) > -\tfrac{1}{2}. \tag{2}$$

$$\ln(z+a) = \ln a + 2x\,{}_2F_1\left(\begin{matrix}1/2,1\\3/2\end{matrix}\middle| x^2\right) = \ln a + 2x\sum_{k=0}^{\infty}\frac{x^{2k}}{2k+1},$$

$$x = \frac{z}{z+2a}, \quad a > 0, \quad R(z) \ge -a \ne z. \tag{3}$$

$$\ln(z+a) = \ln a + \frac{z}{\left(a(z+a)\right)^{1/2}}\,{}_2F_1\left(\begin{matrix}1/2,1/2\\3/2\end{matrix}\middle| -\frac{z^2}{4a(z+a)}\right)$$

$$= \ln a + \frac{z}{\left(a(z+a)\right)^{1/2}}\sum_{k=0}^{\infty}\frac{(\tfrac{1}{2})_k(-)^k}{k!\,(2k+1)}\left\{\frac{z^2}{4a(z+a)}\right\}^k,$$

$$a > 0, \quad \left|\frac{z^2}{4a(z+a)}\right| \le 1, \quad z + 2a \ne 0. \tag{4}$$

$$\ln(z+a) = \ln a + \frac{z(z+2a)}{2a(z+a)}\,{}_2F_1\left(\begin{matrix}1,1\\3/2\end{matrix}\middle| -\frac{z^2}{4a(z+a)}\right),$$

$$a > 0, \quad \left|\frac{z^2}{4a(z+a)}\right| \le 1, \quad z + 2a \ne 0. \tag{5}$$

$$\ln (z+a) = \ln a + \frac{2\left[(z+a)^{\frac{1}{2}}-a^{\frac{1}{2}}\right]}{\left[a(z+a)\right]^{\frac{1}{4}}}\,{}_2F_1\left(\left.\begin{matrix}1/2,1/2\\3/2\end{matrix}\right|-\frac{\{(z+a)^{\frac{1}{2}}-a^{\frac{1}{2}}\}^2a^{\frac{1}{2}}}{4(z+a)^{\frac{1}{2}}}\right),$$

$$a > 0, \quad |\{(z+a)^{\frac{1}{2}}-a^{\frac{1}{2}}\}^2a^{\frac{1}{2}}| \le 4|(z+a)^{\frac{1}{2}}|, \quad |\arg (z+a)| < \pi. \tag{6}$$

$$\ln\left(\frac{a+z}{a-z}\right) = \frac{2z}{a}\,{}_2F_1\left(\left.\begin{matrix}1/2,1\\3/2\end{matrix}\right|\frac{z^2}{a^2}\right), \quad a > 0, \quad |z| \le a, \quad z \ne \pm a. \tag{7}$$

$$\ln\left(\frac{a+z}{a-z}\right) = \frac{2z}{v} - \frac{4(a-v)^2}{3vz}\,{}_2F_1\left(\left.\begin{matrix}1/2,1\\5/2\end{matrix}\right|\left\{\frac{a-v}{z}\right\}^2\right),$$

$$v = (a^2-z^2)^{\frac{1}{2}}, \quad |\arg (a^2+z^2)| < \pi, \quad z^2 \ne a^2. \tag{8}$$

3.1.2. Expansion in Series of Chebyshev Polynomials

$$\ln(x+a) = \ln a - 2\ln(1-q) - 2\sum_{n=1}^{\infty} \frac{(-)^n q^n}{n} T_n^*(x),$$

$$q = 2a + 1 - 2(a^2+a)^{\frac{1}{2}},$$

$$a \neq -1, \quad |\arg(1+1/a)| < \pi, \quad 0 \leq x \leq 1. \tag{1}$$

TABLE 3.1

Chebyshev Coefficients for $\ln(1+x)$ and $x^{-1}\ln(1+x)$

$0 \le x \le 1$

$\ln(1+x) = \sum_{n=0}^{\infty} a_n T_n^*(x)$		$\ln(1+x) = x\sum_{n=0}^{\infty} b_n T_n^*(x)$	
n	a_n	n	b_n
0	0.37645 28129 19195 43163	0	0.82842 71247 46190 09760
1	0.34314 57505 07619 80479	1	-0.15104 29978 15598 46868
2	-0.02943 72515 22859 41438	2	0.01781 47481 69295 96133
3	0.00336 70892 55564 38925	3	-0.00233 55046 14431 11150
4	-0.00043 32758 88610 04446	4	0.00032 46180 81823 81869
5	0.00005 94707 11989 57983	5	-0.00004 68351 03656 70370
6	-0.00000 85029 67541 20286	6	0.00000 69349 73447 90804
7	0.00000 12504 67362 20057	7	-0.00000 10467 13403 92385
8	-0.00000 01877 27995 65082	8	0.00000 01603 22808 74191
9	0.00000 00286 30250 64840	9	-0.00000 00248 44196 16327
10	-0.00000 00044 20956 98068	10	0.00000 00038 86586 17822
11	0.00000 00006 89560 27323	11	-0.00000 00006 12804 11591
12	-0.00000 00001 08450 68551	12	0.00000 00000 97263 14651
13	0.00000 00000 17175 87317	13	-0.00000 00000 15524 91914
14	-0.00000 00000 02736 42009	14	0.00000 00000 02490 18446
15	0.00000 00000 00438 19577	15	-0.00000 00000 00401 13015
16	-0.00000 00000 00070 48360	16	0.00000 00000 00064 85891
17	0.00000 00000 00011 38172	17	-0.00000 00000 00010 52206
18	-0.00000 00000 00001 84431	18	0.00000 00000 00001 71208
19	0.00000 00000 00000 29978	19	-0.00000 00000 00000 27933
20	-0.00000 00000 00000 04886	20	0.00000 00000 00000 04568
21	0.00000 00000 00000 00798	21	-0.00000 00000 00000 00749
22	-0.00000 00000 00000 00131	22	0.00000 00000 00000 00123
23	0.00000 00000 00000 00021	23	-0.00000 00000 00000 00020
24	-0.00000 00000 00000 00004	24	0.00000 00000 00000 00003
25	0.00000 00000 00000 00001	25	-0.00000 00000 00000 00001

If x lies outside the range $0 \leqslant x \leqslant 1$, put $1 + x = 2^m(1 + y)$ where $0 \leqslant y \leqslant 1$ and m is a positive or negative integer. Then $\ln(1 + x) = \ln(1 + y) + m \ln 2$.

For other logarithmic functions, see Table 3.11.

3.1.3. Padé Approximations

TABLE 3.2

Padé Coefficients for $z \ln(1 + 1/z)$

These approximations follow from 6.10(1-5) with $b = c = 1$. We write

$$z \ln (1 + 1/z) = A_n(z)/B_n(z) + R_n(z),$$

$$z \neq 0, \quad z \neq -1, \quad |\arg (1 + 1/z)| < \pi.$$

The coefficients in the polynomials are recorded below for $n = 0(1)6$, $\alpha = 0$ and $\alpha = 1$. Further polynomials can be generated from 6.10(13). For approximate evaluation of the error, use 6.10(17) and Table 2.4.

$$A_n(z) = r_n^{-1} \sum_{k=0}^{n} a_k z^k, \quad a = 0$$

n	$r_n, a_0, a_1, \ldots, a_n$
0	1, 1
1	4, 1, 6
2	9, 1, 21, 30
3	48, 3, 140, 510, 420
4	150, 6, 505, 3360, 6510, 3780
5	180, 5, 672, 7035, 23520, 30870, 13860
6	490, 10, 1981, 29904, 1 50780, 3 31100, 3 28020, 1 20120

$$B_n(z) = b_0^{-1} \sum_{k=0}^{n} b_k z^k, \quad a = 0$$

n	$b_0, b_1, \ldots, b_n$
0	1
1	2, 3
2	3, 12, 10
3	4, 30, 60, 35
4	5, 60, 210, 280, 126
5	6, 105, 560, 1260, 1260, 462
6	7, 168, 1260, 4200, 6930, 5544, 1716

TABLE 3.2 *(Concluded)*

$$A_n(z) = s_n^{-1} \sum_{k=1}^{n} c_k z^k, \quad a = 1$$

n	$s_n, c_1, c_2, \ldots, c_n$
0	
1	1, 2
2	1, 3, 6
3	3, 11, 60, 60
4	6, 25, 260, 630, 420
5	30, 137, 2310, 9870, 15120, 7560
6	10, 49, 1218, 7980, 20720, 23100, 9240

$$B_n(z) = \sum_{k=0}^{n} d_k z^k, \quad a = 1$$

n	$d_0, d_1, \ldots, d_n$
0	1
1	1, 2
2	1, 6, 6
3	1, 12, 30, 20
4	1, 20, 90, 140, 70
5	1, 30, 210, 560, 630, 252
6	1, 42, 420, 1680, 3150, 2772, 924

It is of interest to illustrate the effectiveness of these Padé approximations and the striking realism of the error estimates. We take $z = 1$ so that

$$\ln 2 = A_n(1)/B_n(1) + R_n(1) = 0.69314\ 71806,$$

where the remainder is approximated by $2\pi(-)^{1-a} e^{-(2n+2-a)w}$.

	$a = 0$		
n	A_n/B_n	$-R_n$	-True Error
0	1.0	0.185	0.307
1	0.7	0.544(-2)	0.685(-2)
2	0.69333 33333	0.160(-3)	0.186(-3)
3	0.69315 24548	0.472(-5)	0.527(-5)
4	0.69314 73324	0.139(-6)	0.151(-6)
5	0.69314 71850	0.409(-8)	0.440(-8)

n	$a = 1$: A_n/B_n	R_n	True Error
0	----	----	----
1	0.66666 66667	0.317(-1)	0.265(-1)
2	0.69230 76923	0.934(-3)	0.839(-3)
3	0.69312 16933	0.275(-4)	0.255(-4)
4	0.69314 64174	0.809(-6)	0.763(-6)
5	0.69314 71579	0.238(-7)	0.222(-7)

Notice that the asymptotic estimates for the error are remarkably accurate even for small n. Also the approximations for $a = 1$ and $a = 0$ give lower and upper bounds for ln2 respectively. See 3.1.4(1).

3.1.4. INEQUALITIES

In 3.1.3 for $A_n(z)$ write $A_n(z,a)$ and likewise for $B_n(z)$. Then

$$\frac{A_n(z,1)}{B_n(z,1)} < z \ln (1 + 1/z) < \frac{A_n(z,0)}{B_n(z,0)}, \quad z > 0, \tag{1}$$

with equality as $z \to \infty$. In illustration,

$$\frac{2z}{2z+1} < z \ln (1 + 1/z) < \frac{6z+1}{6z+4}, \quad z > 0, \tag{2}$$

$$\frac{6z^2+3z}{6z^2+6z+1} < z \ln (1 + 1/z) < \frac{30z^2+21z+1}{30z^2+36z+9}, \quad z > 0. \tag{3}$$

Thus if $z = 1$,

$$2/3 < \ln 2 = 0.69315 < 0.7, \tag{4}$$

$$9/13 = 0.69231 < \ln 2 < 52/75 = 0.69333. \tag{5}$$

Let

$$F(z) = \frac{z\left[\ln (z+a) - \ln a\right]}{2(z+2a)y}, \quad y = \frac{z^2}{4a(z+a)}, \quad y > 0. \tag{6}$$

Then

$$\frac{3}{3+2y} < F(z) < \frac{15+2y}{15+12y}, \tag{7}$$

$$\frac{105+50y}{105+120y+24y^2} < F(z) < \frac{315+210y+8y^2}{315+420y+120y^2}. \tag{8}$$

For an example, let $z = 2$, $a = 2^{\frac{1}{2}} - 1$. So $y = 1$ and $F(z) = 2^{-\frac{1}{2}}\ln(1+2^{\frac{1}{2}}) = 0.62323$. Thus

$$3/5 < F(z) < 17/27 = 0.62963, \tag{9}$$

$$155/249 = 0.62249 < F(z) < 533/855 = 0.62339. \tag{10}$$

For more detail on these inequalities, see 6.11.

3.2. Exponential Function

3.2.1. Series Expansions

$$e^z = {}_0F_0(\ ;\ z) = \sum_{k=0}^{\infty} \frac{z^k}{k!}, \quad |z| < \infty. \tag{1}$$

$$e^{-z} = (1+z)^{-1}\,{}_1F_1(z;\ 2+z;\ -z), \quad R(z) > -1. \tag{2}$$

Actually the latter expansion is valid for all z, but for z a negative integer, limiting processes must be used in portions of the infinite series. For $R(z) > -1$, (2) converges better than (1).

3.2.2. Expansions in Series of Jacobi and Chebyshev Polynomials and Bessel Functions

$$e^{ax} = e^{-a} \sum_{n=0}^{\infty} \frac{(2a)^n}{(n+\lambda)_n}\, {}_1F_1\left(\begin{matrix}\beta+1+n \\ \lambda+1+2n\end{matrix}\,\middle|\, 2a\right) P_n^{(\alpha,\beta)}(x), \quad -1 \le x \le 1. \tag{1}$$

$$e^{ax} = \sum_{n=0}^{\infty} \frac{a^n}{(n+\lambda)_n} \; {}_1F_1\left(\begin{matrix}\beta+1+n \\ \lambda+1+2n\end{matrix}\middle| a\right) R_n^{(\alpha,\beta)}(x), \quad 0 \leq x \leq 1. \tag{2}$$

$$e^{ax} = \frac{2\Gamma(\frac{1}{2})}{(2a)^{\alpha+\frac{1}{2}}} \sum_{n=0}^{\infty} \frac{(n+\alpha+\frac{1}{2})\Gamma(n+2\alpha+1)}{\Gamma(n+\alpha+1)} I_{n+\alpha+\frac{1}{2}}(a) P_n^{(\alpha,\alpha)}(x), \quad -1 \leqslant x \leqslant 1. \tag{3}$$

$$e^{ax} = \sum_{n=0}^{\infty} \epsilon_n I_n(a) T_n(x), \qquad -1 \leqslant x \leqslant 1. \tag{4}$$

$$e^{ax} = e^{a/2} \sum_{n=0}^{\infty} \epsilon_n I_n(\tfrac{1}{2}a) T_n^*(x), \qquad 0 \leqslant x \leqslant 1. \tag{5}$$

$$a^x = a^{1/2} \sum_{n=0}^{\infty} \epsilon_n I_n(\tfrac{1}{2}\ln a) T_n^*(x), \qquad 0 \leqslant x \leqslant 1. \tag{6}$$

$$a^{-x} = a^{-1/2} \sum_{n=0}^{\infty} (-)^n \epsilon_n I_n(\tfrac{1}{2}\ln a) T_n^*(x), \qquad 0 \leqslant x \leqslant 1. \tag{7}$$

$$e^{-ax^2} = e^{-a/2} \sum_{n=0}^{\infty} \epsilon_n (-)^n I_n(\tfrac{1}{2}a) T_{2n}(x), \qquad -1 \leqslant x \leqslant 1. \tag{8}$$

$$e^{-z}(2z)^a = \frac{1}{(\frac{1}{2})_a} \sum_{n=0}^{\infty} \frac{(-)^n (2n+2a)\Gamma(n+2a)}{n!} I_{n+a}(z),$$

$2a$ not a negative integer. (9)

$$e^{-z} = I_0(z) + 2\sum_{n=1}^{\infty} (-)^n I_n(z). \tag{10}$$

$$1 - e^{-2x} = \sum_{n=0}^{\infty} M_n(\lambda) T_n^*(\lambda/x), \quad x \geq \lambda. \tag{11}$$

For complete details on (11), see Luke (1971-1972).

TABLE 3.3

CHEBYSHEV COEFFICIENTS FOR THE EXPONENTIAL FUNCTION

$$e^x = \sum_{n=0}^{\infty} C_n(a)\, T_n^*(x/a), \qquad e^{-x} = e^{-a} \sum_{n=0}^{\infty} (-)^n\, C_n(a)\, T_n^*(x/a), \qquad 0 \leqslant x \leqslant a.$$

To compute the exponential function outside the above range, employ the multiplication formula $e^{x+y} = e^x e^y$. To evaluate e^x, x real, in binary form, put $x \log_2 e = m - y$ where $0 \leq y \leq 1$ and m is a positive or negative integer. Then $e^x = 2^m 2^{-y}$ and use Table 3.4. For the evaluation of $\cosh x$ and $\sinh x$, we have

$$\cosh x = \sum_{n=0}^{\infty} D_n(a)\, T_n^*(x/a), \qquad \sinh x = \sum_{n=0}^{\infty} E_n(a)\, T_n^*(x/a), \qquad 0 \leqslant x \leqslant a,$$

$$D_n(a) = \tfrac{1}{2}[1 + (-)^n e^{-a}]\, C_n(a), \qquad E_n(a) = \tfrac{1}{2}[1 - (-)^n e^{-a}]\, C_n(a).$$

Chebyshev coefficients for $(2/\pi)^{\frac{1}{2}} \cosh x$ and $(2/\pi)^{\frac{1}{2}} \sinh x$ valid for $0 \leq x \leq 8$ are given in Table 9.19 of 9.7.

n	$C_n(a)$, a = 1	n	$C_n(a)$, a = 4
0	1.75338 76543 77090 39572	0	16.84398 36812 58988 06741
1	0.85039 16537 80810 96654	1	23.50660 99038 83644 88885
2	0.10520 86936 30936 92530	2	10.18135 74586 34331 24596
3	0.00872 21047 33315 56411	3	3.14389 49866 14982 39694
4	0.00054 34368 31150 15596	4	0.74967 24987 89384 05514
5	0.00002 71154 34913 06869	5	0.14520 49914 57446 17637
6	0.00000 11281 32888 78208	6	0.02364 75415 02153 17329
7	0.00000 00402 45582 29871	7	0.00331 97424 44527 13665
8	0.00000 00012 56584 41828	8	0.00040 93443 90463 21676
9	0.00000 00000 34880 91362	9	0.00004 49873 20821 40254
10	0.00000 00000 00871 52789	10	0.00000 44585 03070 59393
11	0.00000 00000 00019 79808	11	0.00000 04022 90115 46324
12	0.00000 00000 00000 41229	12	0.00000 00333 11800 49829
13	0.00000 00000 00000 00793	13	0.00000 00025 48509 48372
14	0.00000 00000 00000 00014	14	0.00000 00001 81177 20993
		15	0.00000 00000 12028 54472
		16	0.00000 00000 00749 03917
		17	0.00000 00000 00043 91802
		18	0.00000 00000 00002 43280
		19	0.00000 00000 00000 12771
		20	0.00000 00000 00000 00637
		21	0.00000 00000 00000 00030
		22	0.00000 00000 00000 00001

TABLE 3.3 *(Concluded)*

n	$C_n(a)$, a = 8	n	$C_n(a)$, a = 9
0	617.06403 04056 25796 11338	0	1573.60494 22133 66755 41881
1	1065.69748 54103 97316 30776	1	2770.58737 04720 49063 06815
2	701.27931 81060 52934 07289	2	1915.83771 97724 89482 80734
3	364.41816 73043 44382 23487	3	1067.62050 84520 58411 68385
4	154.65206 71495 36360 72058	4	492.34370 85030 78267 22887
5	55.11403 30052 71660 79371	5	192.34280 44465 85936 61030
6	16.86698 46363 57208 73630	6	64.91525 41773 31741 42822
7	4.51307 90962 00034 58481	7	19.23545 99737 01292 80172
8	1.07120 77996 57087 68945	8	5.07160 09258 16608 26732
9	0.22824 78975 71683 82701	9	1.20310 11263 53352 29567
10	0.04409 22605 84510 46791	10	0.25919 64204 03199 08464
11	0.00778 65946 49131 48746	11	0.05111 70356 72467 47507
12	0.00126 59900 14287 28686	12	0.00929 09126 71135 87319
13	0.00019 06545 63407 76628	13	0.00156 55014 26409 48470
14	0.00002 67353 52136 80606	14	0.00024 57933 18547 73935
15	0.00000 35070 98450 12386	15	0.00003 61207 77667 99542
16	0.00000 04321 13760 87713	16	0.00000 49881 34094 43654
17	0.00000 00501 88363 10681	17	0.00000 06496 01885 33560
18	0.00000 00055 12674 46923	18	0.00000 00800 30960 78976
19	0.00000 00005 74292 88370	19	0.00000 00093 54199 01752
20	0.00000 00000 56892 07408	20	0.00000 00010 39946 86403
21	0.00000 00000 05372 14292	21	0.00000 00001 10226 89282
22	0.00000 00000 00484 57339	22	0.00000 00000 11162 53107
23	0.00000 00000 00041 83565	23	0.00000 00000 01082 14458
24	0.00000 00000 00003 46336	24	0.00000 00000 00100 60868
25	0.00000 00000 00000 27538	25	0.00000 00000 00008 98530
26	0.00000 00000 00000 02106	26	0.00000 00000 00000 77204
27	0.00000 00000 00000 00155	27	0.00000 00000 00000 06391
28	0.00000 00000 00000 00011	28	0.00000 00000 00000 00510
29	0.00000 00000 00000 00001	29	0.00000 00000 00000 00039
		30	0.00000 00000 00000 00003

TABLE 3.4

CHEBYSHEV COEFFICIENTS FOR 2^x AND 10^x

$$2^x = 2\sum_{n=0}^{\infty}(-)^n a_n T_n^*(x) \qquad 10^x = \sum_{n=0}^{\infty} b_n T_n^*(x)$$

$$2^{-x} = \sum_{n=0}^{\infty} a_n T_n^*(x) \qquad 10^{-x} = 10^{-1}\sum_{n=0}^{\infty}(-)^n b_n T_n^*(x)$$

$0 \leq x \leq 1$

n	a_n	n	b_n
0	0.72849 99375 06481 47962	0	4.30022 91548 49946 95129
1	-0.24876 24339 05220 94259	1	4.27816 40812 99252 58760
2	0.02144 65559 94839 81983	2	1.16852 60969 60052 96801
3	-0.00123 57140 81997 92534	3	0.21828 85943 41690 87392
4	0.00005 34530 58179 04256	4	0.03090 77131 39888 00947
5	-0.00000 18506 90713 86222	5	0.00351 98061 03259 00425
6	0.00000 00534 11876 87701	6	0.00033 50657 79592 55033
7	-0.00000 00013 21516 38145	7	0.00002 73928 43619 26469
8	0.00000 00000 28613 23785	8	0.00000 19620 76422 96037
9	-0.00000 00000 00550 73811	9	0.00000 01250 36785 21154
10	0.00000 00000 00009 54096	10	0.00000 00071 76132 27776
11	-0.00000 00000 00000 15027	11	0.00000 00003 74599 22597
12	0.00000 00000 00000 00217	12	0.00000 00000 17931 70005
13	-0.00000 00000 00000 00003	13	0.00000 00000 00792 58354
		14	0.00000 00000 00032 53785
		15	0.00000 00000 00001 24697
		16	0.00000 00000 00000 04481
		17	0.00000 00000 00000 00152
		18	0.00000 00000 00000 00005

To compute for x outside the above range, put $x = N + y$ where N is an integer and $0 \leqslant y \leqslant 1$. Then $p^x = p^N p^y$, $p = 2$ or $p = 10$.

3.2.3. PADÉ APPROXIMATIONS

The $(n-a,n)$ approximations, $a = 0$ or $a = 1$, of the Padé matrix table for e^{-z} are given by

$$e^{-z} = \frac{A_n(z,a)}{B_n(z,a)} + S_n(z,a), \tag{1}$$

$$B_n(z,a) = (n+1-a)_n \; {}_1F_1^n(-n; -2n+a; z)$$

$$= z^n \; {}_2F_0(-n, n+1-a; -1/z), \tag{2}$$

$$A_n(z,a) = (n+1-a)_n \; {}_1F_1^{n-a}(\,a-n;\ -2n+a;\ -z)$$
$$= n^a(-z)^{n-a} {}_2F_0(\,a-n,\ n+1;\ 1/z). \tag{3}$$

$$B_n(z,0) = A_n(-z,0). \tag{4}$$

Both $B_n(z,a)$ and $A_n(z,a)$ satisfy the same recurrence formula

$$B_{n+1}(z,a) = \left\{1 - \frac{az}{(2n-a)(2n+2-a)}\right\}\frac{(2n+1-a)(2n+2-a)}{(n+1-a)} B_n(z,a)$$
$$+ \frac{n(2n+2-a)z^2}{(n+1-a)(2n-a)} B_{n-1}(z,a). \tag{5}$$

$$S_n(z,a) = \tag{6}$$
$$\frac{(-)^{n+1-a}\pi n!\,(n-a)!\,z^{2n+1-a}e^{-z+z(z-4a)/4(2n+1-a)}}{2^{4n-2a}(2n+1-a)\left(\Gamma(n+\frac{1-a}{2})\Gamma(n+\frac{2-a}{2})\right)^2}\left[1+O(n^{-3})\right],$$

$$S_n(z,a) = \frac{(-)^{n+1-a}\pi z^{2n+1-a}e^{-z}}{2^{4n+1-2a}n!\,(n-a)!}\left(1+O(n^{-1})\right). \tag{7}$$

Hence for z fixed,

$$\lim_{n\to\infty} S_n(z,a) = 0. \tag{8}$$

To facilitate a priori estimation of the error,

$$S_n(z,a) = \left|\frac{e^{-z}(4n/z)^{a+\frac{1}{2}}}{\pi^{\frac{1}{2}}}\right| R_n(z)\left(1 + O(n^{-3})\right) \tag{9}$$

and values of $R_n(z)$ are given in Table 4.6 of 4.8.3.

When $a = 0$, it is convenient to let

$$G_n(z) = B_n(z,0) = M_n(z^2) + zN_n(z^2). \tag{10}$$

Thus with $S_n(z) = S_n(z,0)$,

$$e^{-z} = \left(G_n(-z)/G_n(z)\right) + S_n(z). \tag{11}$$

In hypergeometric form

$$M_{2n+b}(z^2) = 2^b(n+\tfrac{1}{2}+b)_n(n+1)_{n+b}2^{2n}{}_2F_3^n\left(\begin{matrix}-n, \tfrac{1}{2}-b-n\\ \tfrac{1}{2}, -b-2n, \tfrac{1}{2}-b-2n\end{matrix}\middle| z^2/4\right),$$

$$b = 0 \quad \text{or} \quad b = 1, \tag{12}$$

$$N_{2n+b}(z^2) = (n+\tfrac{1}{2}+b)_n(n+b)_{n+b}2^{2n}{}_2F_3^{n-1}\left(\begin{matrix}1-n, \tfrac{1}{2}-n\\ 3/2, 1-b-2n, \tfrac{1}{2}-b-2n\end{matrix}\middle| z^2/4\right),$$

$$b = 0 \quad \text{or} \quad b = 1. \tag{13}$$

Computation of these rational approximations to e^{-z} are considerably simplified by evaluating the polynomials $M_n(z^2)$ and $N_n(z^2)$. In this connection, the polynomials $G_n(z)$, $M_n(z^2)$ and $N_n(z^2)$ satisfy the same recursion formula

$$G_{n+1}(z) = 2(2n+1)G_n(z) + z^2G_{n-1}(z), \tag{14}$$

which is stable when used in the forward direction.

TABLE 3.5

PADÉ COEFFICIENTS FOR THE EXPONENTIAL FUNCTION

(For notation, see 3.2.3.)

$$G_n(z) = \sum_{k=0}^{n} a_k z^{n-k}$$

n	$a_0, a_1, \ldots, a_n$
0	1
1	1, 2
2	1, 6, 12
3	1, 12, 60, 120
4	1, 20, 180, 840, 1680
5	1, 30, 420, 3360, 15120, 30240
6	1, 42, 840, 10080, 75600, 3 32640, 6 65280

For extended coefficients of the above to $n = 7(1)10$, see Luke (1969).

$$A_n(z,1) = \sum_{k=0}^{n-1} a_k z^{n-1-k}, \quad A_0(z,1) = 0$$

n	$a_0, a_1, \cdots, a_{n-1}$
1	1
2	-2, 6
3	3, -24, 60
4	-4, 60, -360, 840
5	5, -120, 1260, -6720, 15120
6	-6, 210, -3360, 30240, -1 51200, 3 32640

TABLE 3.5 *(Concluded)*

$$B_n(z,1) = \sum_{k=0}^{n} b_k z^{n-k}$$

n	$b_0, b_1, \cdots, b_n$
0	1
1	1, 1
2	1, 4, 6
3	1, 9, 36, 60
4	1, 16, 120, 480, 840
5	1, 25, 300, 2100, 8400, 15120
6	1, 36, 630, 6720, 45360, 1 81440, 3 32640

The following data illustrates the effectiveness of the Padé approximates for the exponential function and the striking realism of the error estimates. Let $z = 2$. So

$$e^{-2} = A_n(2,a)/B_n(2,a) + S_n(2,a) = 0.13533\ 52832,$$

where the remainder is approximated by

$$\frac{\pi(-)^{n+1-a}(2n)^a}{2^{2n}(n!)^2}\, e^{-2+(1-2a)/(2n+1-a)}.$$

	$a = 0$		
n	A_n/B_n	S_n	True Error
0	1.0	-1.156	-0.865
1	0	0.148	0.135
2	0.14285 71429	-0.811(-2)	-0.752(-2)
3	0.13513 51351	0.213(-3)	0.200(-3)
4	0.13533 83459	-0.322(-5)	-0.306(-5)
5	0.13533 52530	0.316(-7)	0.302(-7)

n	$a = 1$: A_n/B_n	S_n	True Error
0	---	---	---
1	0.33333 33333	-0.129	-0.198
2	0.11111 11111	0.207(-1)	0.242(-1)
3	0.13636 36364	-0.937(-3)	-0.103(-2)
4	0.13531 35314	0.203(-4)	0.218(-4)
5	0.13533 55580	-0.261(-6)	-0.275(-6)

If $n = 5$ and $a = 0$, the Padé approximate gives an accuracy of very nearly ten decimals at least for all z in the unit circle. Notice that the Padé approximates when used appropriately give upper and lower bounds. See 3.2.4(1,2).

3.2.4. Inequalities

$$\frac{A_{2n+1}(z,0)}{B_{2n+1}(z,0)} < e^{-z} < \frac{A_{2n+1}(z,1)}{B_{2n+1}(z,1)}, \quad z > 0. \tag{1}$$

$$\frac{A_{2n}(z,1)}{B_{2n}(z,1)} < e^{-z} < \frac{A_{2n}(z,0)}{B_{2n}(z,0)}, \quad n > 0,\ z > 0. \tag{2}$$

In particular,

$$\frac{2-z}{2+z} < e^{-z} < \frac{1}{1+z}, \quad z > 0. \tag{3}$$

$$\frac{6-2z}{6+4z+z^2} < e^{-z} < \frac{12-6z+z^2}{12+6z+z^2}, \quad z > 0. \tag{4}$$

If $z = 1$,

$$1/3 < e^{-1} = 0.36788 < 1/2, \tag{5}$$

$$4/11 = 0.36364 < e^{-1} < 7/9 = 0.36842. \tag{6}$$

$$\frac{4+2z-z^2}{(1+z)(4+2z+z^2)} < e^{-z} < \frac{z^3+8z^2+13z+6}{(z+1)^2(z+2)(z^2+2z+3)}, \quad z > 0. \tag{7}$$

3.3. Circular and Hyperbolic Functions

3.3.1. Power Series

$$\cos z = \tfrac{1}{2}(e^{iz}+e^{-iz}) = \cosh iz = \sum_{k=0}^{\infty} \frac{(-)^k z^{2k}}{(2k)!}, \tag{1}$$

$$\sin z = \frac{1}{2i}(e^{iz}-e^{-iz}) = -i \sinh iz = \sum_{k=0}^{\infty} \frac{(-)^k z^{2k+1}}{(2k+1)!}, \tag{2}$$

$$e^{iz} = \cos z + i \sin z = \sum_{k=0}^{\infty} \frac{i^k z^k}{k!}, \tag{3}$$

$$\cosh z = \tfrac{1}{2}(e^{z}+e^{-z}) = \sum_{k=0}^{\infty} \frac{z^{2k}}{(2k)!}, \tag{4}$$

$$\sinh z = \tfrac{1}{2}(e^{z}-e^{-z}) = \sum_{k=0}^{\infty} \frac{z^{2k+1}}{(2k+1)!}. \tag{5}$$

The above series converge for all z.

3.3.2. Expansions in Series of Jacobi and Chebyshev Polynomials and Bessel Functions

Expansions for $\cos ax$ and $\sin ax$ in series of Jacobi polynomials follow from 3.2.2(1,2) with a replaced by ia.

$$\cos ax = \frac{2\Gamma(\frac{1}{2})}{(2a)^{\alpha+\frac{1}{2}}} \sum_{n=0}^{\infty} \frac{(-)^n(2n+\alpha+\frac{1}{2})\Gamma(2n+2\alpha+1)}{\Gamma(2n+\alpha+1)}$$

$$\times J_{2n+\alpha+\frac{1}{2}}(a) P_{2n}^{(\alpha,\alpha)}(x), \qquad -1 \leqslant x \leqslant 1. \tag{1}$$

$$\sin ax = \frac{2\Gamma(\frac{1}{2})}{(2a)^{\alpha+\frac{1}{2}}} \sum_{n=0}^{\infty} \frac{(-)^n(2n+\alpha+\frac{3}{2})\Gamma(2n+2\alpha+2)}{\Gamma(2n+\alpha+2)}$$
$$\times J_{2n+\alpha+\frac{3}{2}}(a)\, P_{2n+1}^{(\alpha,\alpha)}(x), \qquad -1 \leqslant x \leqslant 1. \tag{2}$$

$$\cos ax = \sum_{n=0}^{\infty} \epsilon_n(-)^n J_{2n}(a) T_{2n}(x), \qquad -1 \leqslant x \leqslant 1. \tag{3}$$

$$\sin ax = 2\sum_{n=0}^{\infty} (-)^n J_{2n+1}(a) T_{2n+1}(x), \qquad -1 \leqslant x \leqslant 1. \tag{4}$$

Since

$$\tan\frac{\pi x}{4} = \frac{8x}{\pi}\sum_{n=0}^{\infty}\frac{1}{q^2-x^2}, \qquad q = 2(2n+1), \tag{5}$$

$$\frac{1}{q^2-x^2} = \frac{1}{q(q^2-1)^{1/2}}\sum_{k=0}^{\infty}\epsilon_k[q+(q^2-1)^{1/2}]^{-2k}T_{2k}(x), \tag{6}$$

we have

$$\frac{1}{x}\tan\frac{\pi x}{4} = \sum_{k=0}^{\infty} a_k T_{2k}(x), \qquad -1 \leqslant x \leqslant 1,$$

$$a_k = \frac{8\epsilon_k}{\pi}\sum_{n=0}^{\infty}\frac{[q+(q^2-1)^{1/2}]^{-2k}}{q(q^2-1)^{1/2}}. \tag{7}$$

The expansion for a_k converges quite slowly for small and moderate k. The situation can be remedied by using the Euler-Maclaurin summation formula and known values of $\tan \pi x/4$. For details see Luke (1969). A Chebyshev expansion for $\cot \pi x/2$ can similarly be deduced from

$$\frac{\pi x}{2}\cot\frac{\pi x}{2} = 1 + 2\sum_{n=1}^{\infty}\left(1 - \frac{4n^2}{4n^2-x^2}\right). \tag{8}$$

Chebyshev expansions for $\cosh x$ and $\sinh x$ follow from the comments in Table 3.3.

A Chebyshev expansion for

$$\tanh z = 1 - \frac{2x}{1+x}, \qquad x = e^{-z} > 0, \tag{9}$$

follows from Table 2.1.

TABLE 3.6

CHEBYSHEV COEFFICIENTS FOR $\cos x$ AND $\sin x$

$$\cos x = \sum_{n=0}^{\infty} C_n(a)\, T_{2n}(x/a), \qquad \sin x = \sum_{n=0}^{\infty} D_n(a)\, T_{2n+1}(x/a),$$

$$\sin x = (x/a) \sum_{n=0}^{\infty} E_n(a)\, T_{2n}(x/a), \qquad -a \leqslant x \leqslant a.$$

To evaluate the circular functions outside the above range, use the addition formulas:

$$\cos(\alpha \pm \beta) = \cos\alpha \cos\beta \mp \sin\alpha \sin\beta, \qquad \sin(\alpha \pm \beta) = \sin\alpha\cos\beta \pm \cos\alpha \sin\beta.$$

In particular, for the cases $a = \pi/2$ and π, we can proceed as follows. Write the above expansions in the form

$$\cos ax = \sum_{n=0}^{\infty} C_n(a)\, T_{2n}(x),$$

$$\sin ax = \sum_{n=0}^{\infty} D_n(a)\, T_{2n+1}(x) = x \sum_{n=0}^{\infty} E_n(a)\, T_{2n}(x), \qquad -1 \leqslant x \leqslant 1.$$

Let N be an integer and put

$$x = 2N + r + y, \quad -1 \leqslant y \leqslant 1, \qquad a = \pi/s, \quad s = 1, 2.$$

We then have the following table.

s	r	a	$\cos ax$	$\sin ax$
1	0	π	$\cos \pi y$	$\sin \pi y$
1	1/2	π	$-\sin \pi y$	$\cos \pi y$
2	0	$\pi/2$	$(-)^N \cos \pi y/2$	$(-)^N \sin \pi y/2$
2	1	$\pi/2$	$(-)^{N+1} \sin \pi y/2$	$(-)^N \cos \pi y/2$

Chebyshev coefficients for $(2/\pi)^{\frac{1}{2}}\cos x$ and $(2/\pi)^{\frac{1}{2}}x^{-1}\sin x$ valid for $-8 \leq x \leq 8$ are given in Table 9.13 of 9.7.

n	$C_n(a)$, $a = \pi/4$	n	$C_n(a)$, $a = 1$
0	0.85163 19137 04808 01270	0	0.76519 76865 57966 55145
1	-0.14643 66443 90836 86332	1	-0.22980 69698 63800 96094
2	0.00192 14493 11814 64680	2	0.00495 32779 28219 91009
3	-0.00000 99649 68489 82930	3	-0.00004 18766 76004 77854.
4	0.00000 00275 76595 60719	4	0.00000 01884 46883 45209
5	-0.00000 00000 47399 49808	5	-0.00000 00005 26123 02474
6	0.00000 00000 00055 49549	6	0.00000 00000 00999 94364
7	-0.00000 00000 00000 04710	7	-0.00000 00000 00001 37708
8	0.00000 00000 00000 00003	8	0.00000 00000 00000 00144

TABLE 3.6 *(Continued)*

n	$D_n(a)$, $a = \pi/4$
0	0.72637 56766 93734 66359
1	-0.01942 00290 53201 50631
2	0.00015 16929 22851 07399
3	-0.00000 05605 80468 41200
4	0.00000 00012 05324 16785
5	-0.00000 00000 01694 13931
6	0.00000 00000 00001 67781
7	-0.00000 00000 00000 00123

n	$D_n(a)$, $a = 1$
0	0.88010 11714 89867 03192
1	-0.03912 67079 65336 81184
2	0.00049 95154 60422 46886
3	-0.00000 30046 51634 87362
4	0.00000 00104 98500 35982
5	-0.00000 00000 23960 13493
6	0.00000 00000 00038 51234
7	-0.00000 00000 00000 04595
8	0.00000 00000 00000 00004

n	$E_n(a)$, $a = \pi/4$
0	0.74594 79604 57275 64210
1	-0.03914 45675 27081 95702
2	0.00030 45094 20678 94441
3	-0.00000 11235 74976 79642
4	0.00000 00024 14039 97241
5	-0.00000 00000 03391 63671
6	0.00000 00000 00003 35809
7	-0.00000 00000 00000 00247

n	$E_n(a)$, $a = 1$
0	0.91973 04100 89760 23931
1	-0.07925 84771 99786 41479
2	0.00100 50612 69112 79111
3	-0.00000 60303 48267 85339
4	0.00000 00210 44998 10616
5	-0.00000 00000 47997 38651
6	0.00000 00000 00077 11666
7	-0.00000 00000 00000 09199
8	0.00000 00000 00000 00008

n	$C_n(a)$, $a = \pi/2$
0	0.47200 12157 68234 76745
1	-0.49940 32582 70407 08740
2	0.02799 20796 17547 61751
3	-0.00059 66951 96548 84650
4	0.00000 67043 94869 91684
5	-0.00000 00465 32295 89732
6	0.00000 00002 19345 76590
7	-0.00000 00000 00748 16487
8	0.00000 00000 00001 93230
9	-0.00000 00000 00000 00391
10	0.00000 00000 00000 00001

n	$C_n(a)$, $a = \pi$
0	-0.30424 21776 44093 86420
1	-0.97086 78652 63018 21941
2	0.30284 91552 62699 42151
3	-0.02909 19339 65011 12115
4	0.00139 22439 91176 23186
5	-0.00004 01899 44510 75494
6	0.00000 07782 76701 18153
7	-0.00000 00108 26530 34186
8	0.00000 00001 13510 91779
9	-0.00000 00000 00929 52966
10	0.00000 00000 00006 11136
11	-0.00000 00000 00000 03298
12	0.00000 00000 00000 00015

n	$D_n(a)$, $a = \pi/2$
0	1.13364 81778 11747 87542
1	-0.13807 17765 87192 10354
2	0.00449 07142 46554 91791
3	-0.00006 77012 75842 15249
4	0.00000 05891 29533 02893
5	-0.00000 00033 38059 40892
6	0.00000 00000 13297 02838
7	-0.00000 00000 00039 27500
8	0.00000 00000 00000 08945
9	-0.00000 00000 00000 00016

n	$D_n(a)$, $a = \pi$
0	0.56923 06863 59505 51469
1	-0.66691 66724 05979 07078
2	0.10428 23687 34236 94948
3	-0.00684 06335 36991 57901
4	0.00025 00068 84950 38623
5	-0.00000 58502 48308 63914
6	0.00000 00953 47727 50299
7	-0.00000 00011 45638 44171
8	0.00000 00000 10574 27262
9	-0.00000 00000 00077 35271
10	0.00000 00000 00000 45960
11	-0.00000 00000 00000 00226
12	0.00000 00000 00000 00001

TABLE 3.6 *(Concluded)*

n	$E_n(a)$, $a = \pi/2$
0	1.27627 89624 02265 88021
1	-0.28526 15691 81036 00957
2	0.00911 80160 06651 80250
3	-0.00013 65875 13541 96667
4	0.00000 11849 61857 66169
5	-0.00000 00067 02791 60383
6	0.00000 00000 26672 78599
7	-0.00000 00000 00078 72922
8	0.00000 00000 00000 17923
9	-0.00000 00000 00000 00032

n	$E_n(a)$, $a = \pi$
0	1.34752 63146 73990 17123
1	-1.55659 12566 28969 31308
2	0.22275 79118 17011 17152
3	-0.01419 31743 48537 27256
4	0.00051 19072 74554 11454
5	-0.00001 18935 04653 34208
6	0.00000 01930 08036 06380
7	-0.00000 00023 12581 05781
8	0.00000 00000 21304 17439
9	-0.00000 00000 00155 62915
10	0.00000 00000 00000 92373
11	-0.00000 00000 00000 00454
12	0.00000 00000 00000 00002

n	$C_n(a)$, $a = 8$
0	0.17165 08071 37553 90609
1	0.22598 34408 48150 49999
2	-0.21071 48697 50777 87408
3	-0.67515 18002 27186 15493
4	0.44690 99727 02205 90857
5	-0.12153 40535 48502 31263
6	0.01924 76436 24363 26081
7	-0.00203 85123 27064 67258
8	0.00015 60127 90934 61655
9	-0.00000 90761 87888 00372
10	0.00000 04161 16592 79434
11	-0.00000 00154 49541 91187
12	0.00000 00004 74549 70621
13	-0.00000 00000 12269 03972
14	0.00000 00000 00270 83252
15	-0.00000 00000 00005 16620
16	0.00000 00000 00000 08604
17	-0.00000 00000 00000 00126
18	0.00000 00000 00000 00002

n	$D_n(a)$, $a = 8$
0	0.46927 26937 07829 24876
1	0.58226 44141 31904 49876
2	0.37154 95443 81126 62468
3	-0.64117 81559 59652 60771
4	0.25264 17894 44759 20942
5	-0.05119 33444 26496 57216
6	0.00654 95864 46593 21028
7	-0.00058 52066 98133 14377
8	0.00003 88444 65605 32242
9	-0.00000 19983 79890 69433
10	0.00000 00822 03073 27737
11	-0.00000 00027 69407 23793
12	0.00000 00000 77890 99935
13	-0.00000 00000 01857 75885
14	0.00000 00000 00038 06879
15	-0.00000 00000 00000 67771
16	0.00000 00000 00000 01058
17	-0.00000 00000 00000 00015

n	$E_n(a)$, $a = 8$
0	1.21074 68348 30450 16550
1	-1.48294 82822 45241 83347
2	2.64747 71105 09050 83099
3	-1.90437 80217 46797 58163
4	0.62202 17098 27492 36621
5	-0.11673 81309 37973 94736
6	0.01435 14420 84980 80303
7	-0.00125 22691 91794 38248
8	0.00008 18557 95528 09495
9	-0.00000 41668 64317 45012
10	0.00000 01701 04536 06146
11	-0.00000 00056 98389 50671
12	0.00000 00001 59575 03085
13	-0.00000 00000 03793 03215
14	0.00000 00000 00077 51445
15	-0.00000 00000 00001 37687
16	0.00000 00000 00000 02146
17	-0.00000 00000 00000 00030

TABLE 3.7

CHEBYSHEV COEFFICIENTS FOR $\tan \pi x/4$ AND $\cot \pi x/2$

$$\tan \pi x/4 = x \sum_{n=0}^{\infty} a_n T_{2n}(x) \qquad -1 \leq x \leq 1 \qquad x \cot \pi x/2 = \sum_{n=0}^{\infty} b_n T_{2n}(x)$$

n	a_n	n	b_n
0	0.88507 37113 31335 57028	0	0.33443 41422 00192 96061
1	0.10675 39285 70380 06708	1	-0.31720 38386 49646 28360
2	0.00758 61015 77835 02721	2	-0.01604 55338 22381 91347
3	0.00054 41703 81657 91150	3	-0.00110 04059 17927 72992
4	0.00003 90663 69546 88324	4	-0.00007 83173 07329 73714
5	0.00000 28048 16136 39434	5	-0.00000 56125 49263 33059
6	0.00000 02013 76576 94228	6	-0.00000 04027 99819 08717
7	0.00000 00144 58186 58815	7	-0.00000 00289 17123 36430
8	0.00000 00010 38051 08512	8	-0.00000 00020 76114 25478
9	0.00000 00000 74528 71466	9	-0.00000 00001 49057 62416
10	0.00000 00000 05350 72096	10	-0.00000 00000 10701 84507
11	0.00000 00000 00384 17884	11	-0.00000 00000 00768 35773
12	0.00000 00000 00027 58280	12	-0.00000 00000 00055 16560
13	0.00000 00000 00001 98036	13	-0.00000 00000 00003 96071
14	0.00000 00000 00000 14218	14	-0.00000 00000 00000 28437
15	0.00000 00000 00000 01021	15	-0.00000 00000 00000 02042
16	0.00000 00000 00000 00073	16	-0.00000 00000 00000 00147
17	0.00000 00000 00000 00005	17	-0.00000 00000 00000 00011
		18	-0.00000 00000 00000 00001

If $|x| > 1$, put $x = 2m + y$, m an integer, $-1 \leqslant y \leqslant 1$. Then

$$\tan \pi x/4 = \tan \pi y/4 \quad \text{if} \quad m \text{ is even}, \qquad \tan \pi x/4 = -\cot \pi y/4 \quad \text{if} \quad m \text{ is odd},$$
$$\cot \pi x/2 = \cot \pi y/2.$$

3.3.3. RATIONAL AND PADÉ APPROXIMATIONS

Rational approximations for the circular functions given below are based on the main diagonal Padé approximations for e^{-z}. See 3.2.3(11). We have

$$e^{-iz} = \left(U_n(z^2)/W_n(z^2)\right) - iz\left(X_n(z^2)/W_n(z^2)\right) + S_n(iz), \qquad (1)$$

$$\cos z = \frac{U_n(z^2)}{W_n(z^2)} + R\left(S_n(iz)\right), \qquad (2)$$

$$\sin z = \frac{zX_n(z^2)}{W_n(z^2)} - I\left(S_n(iz)\right), \qquad (3)$$

$$z^{-1}\tan z \sim \frac{X_n(z^2)}{U_n(z^2)}, \tag{4}$$

$$U_n(z^2) = [M_n(-z^2)]^2 - z^2[N_n(-z^2)]^2, \qquad X_n(z^2) = 2M_n(-z^2)\,N_n(-z^2),$$
$$W_n(z^2) = [M_n(-z^2)]^2 + z^2[N_n(-z^2)]^2. \tag{5}$$

The polynomials $U_n(z^2)$, $X_n(z^2)$, and $W_n(z^2)$ satisfy the same recursion formula

$$(2n-1)\,W_{n+1}(z^2) = [4(4n^2-1)-z^2][(2n+1)\,W_n(z^2) - z^2(2n-1)\,W_{n-1}(z^2)]$$
$$+ z^6(2n+1)\,W_{n-2}(z^2), \tag{6}$$

which is stable when used in the forward direction.

From 3.2.3 (8,9),

$$\lim_{n\to\infty} S_n(z) = 0, \quad z \text{ fixed}, \tag{7}$$

$$S_n(z) = 2|e^{-z}(n/\pi z)^{\frac{1}{2}}|R_n(z)\left(1+O(n^{-3})\right), \tag{8}$$

and values of $R_n(z)$ are provided in Table 4.6 of 4.8.3. Further coefficients as in Table 3.8 for $n = 7(1)10$ are available in Luke (1969).

TABLE 3.8

RATIONAL APPROXIMATIONS FOR THE CIRCULAR FUNCTIONS

$$U_n(z^2) = \sum_{k=0}^{n} b_k z^{2n-2k}$$

n	$b_0, b_1, \ldots, b_n$
0	1
1	-1, 4
2	1, -60, 144
3	-1, 264, -6480, 14400
4	1, -760, 69360, -13 10400, 28 22400
5	-1, 1740, -4 08240, 258 04800, -4318 27200, 9144 57600
6	1, -3444, 17 03520, -2578 86720, 1 35390 52800, -21 12397 05600, 44 25974 78400

TABLE 3.8 *(Concluded)*

$$X_n(z^2) = \sum_{k=0}^{n-1} c_k z^{2n-2k-2}$$

n	$c_0, c_1, \ldots, c_{n-1}$
0	0
1	4
2	-12, 144
3	24, -1680, 14400
4	-40, 8880, -3 69600, 28 22400
5	60, -31920, 37 90080, -1270 08000, 9144 57600
6	-84, 90720, -239 50080, 21388 14720, -6 37072 12800, 44 25974 78400

$$W_n(z^2) = \sum_{k=0}^{n} d_k z^{2n-2k}$$

n	$d_0, d_1, \ldots, d_n$
0	1
1	1, 4
2	1, 12, 144
3	1, 24, 720, 14400
4	1, 40, 2160, 1 00800, 28 22400
5	1, 60, 5040, 4 03200, 254 01600, 9144 57600
6	1, 84, 10080, 12 09600, 1270 08000, 1 00590 33600, 44 25974 78400

The above rational approximations are not of the Padé type. We next present the $(n-a,n)$ Padé approximations for $z^{-1}\tanh z$.

$$z^{-1}\tanh z = \frac{C_n(z,a)}{D_n(z,a)} + E_n(z), \tag{9}$$

$$b_n C_n(z,a) = 2N_{2n+1-a}(4z^2), \tag{10}$$

$$b_n D_n(z,a) = M_{2n+1-a}(4z^2), \tag{11}$$

$$b_n = 2^{2n+1-a}(n-a+3/2)_n(n+1)_{n+1-a}, \tag{12}$$

where $M_n(z^2)$ and $N_n(z^2)$ are defined by 3.2.3(10-13). See also Table 3.5.

$$E_n(z) = \frac{(-)^{1-a}(z/2)^{2n-a}I_{2n-a+3/2}(z)}{2(\tfrac{1}{2})_{2n+1-a}D_n(z,a)I_{-\frac{1}{2}}(z)}, \tag{13}$$

$$E_n(z) = \frac{(-)^{1-a}\pi(z/2)^{4n+2-2a}}{2\{(2n+1-a)!\}^2\cosh^2 z}\{1 + O(n^{-1})\}. \tag{14}$$

To illustrate, let $z = i$, $n = 2$ and $a = 0$. Then $\tan 1 \sim 1.55740\ 741$ and the true error is 0.317(-6). The error deduced from (14) with $O(n^{-1})$ neglected is 0.365(-6). The above $n = 3$, $a = 0$ approximation gives at least 10D accuracy for all $|z| \leq 1$.

3.3.4. Inequalities

With the notation of 3.3.3(9), we have

$$\frac{C_n(z,1)}{D_n(z,1)} < z^{-1}\tanh z < \frac{C_n(z,0)}{D_n(z,0)}, \quad z > 0. \tag{1}$$

In particular,

$$\frac{3}{3+z^2} < z^{-1}\tanh z < \frac{15+z^2}{15+6z^2}, \tag{2}$$

$$\frac{105+10z^2}{105+45z^2+z^4} < z^{-1}\tanh z < \frac{945+105z^2+z^4}{945+420z^2+15z^4}, \quad z > 0. \tag{3}$$

In illustration,

$$3/4 < \tanh 1 = 0.76159\ 41560 < 16/21 = 0.76190\ 47619, \tag{4}$$

$$115/151 = 0.76158\ 94040 < \tanh 1 < 1051/1380 = 0.76159\ 42029. \tag{5}$$

$$\frac{C_n(iz,a)}{D_n(iz,a)} < z^{-1}\tan z, \quad a = 0 \quad \text{or} \quad a = 1,$$

$$0 < z < \pi/2. \tag{6}$$

3.4. Inverse Circular and Hyperbolic Functions

3.4.1. Power Series

$$\text{arc sin } z = z(1-z^2)^{\frac{1}{2}}{}_2F_1(1,1;3/2;\ z^2), \quad |z| < 1, \tag{1}$$

$$\text{arc sin } z = z\,{}_2F_1(1/2,1/2;3/2;\ z^2), \quad |z| < 1, \tag{2}$$

$$\text{arc sin } z = \frac{z}{(1-z^2)^{\frac{1}{2}}}\ {}_2F_1(1/2,1;3/2;\ \frac{z^2}{z^2-1}), \quad R(z^2) < \tfrac{1}{2}, \tag{3}$$

$$\text{arc sin } z = \frac{\pi}{2} - (1-z^2)^{\frac{1}{2}}{}_2F_1(1/2,1/2;3/2;\ 1-z^2), \quad |z| < 1. \tag{4}$$

$$\text{arc cos } z = \frac{\pi}{2} - \text{arc sin } z. \tag{5}$$

$$\text{arc tan } z = z\,{}_2F_1(1/2,1;3/2;\ -z^2), \quad |z| \le 1,\ z^2 \ne -1, \tag{6}$$

$$\text{arc tan } z = \frac{z}{1+z^2}\ {}_2F_1(1,1;3/2;\ \frac{z^2}{1+z^2}),$$

$$z = x + iy, \quad y^2 - x^2 < \tfrac{1}{2}, \tag{7}$$

$$\text{arc tan } z = \frac{z}{w} + \frac{2(w-1)^2}{3wz}\ {}_2F_1\left(\frac{1}{2},1;\frac{5}{2};\ -\left\{\frac{w-1}{z}\right\}^2\right),$$

$$w = (1+z^2)^{\frac{1}{2}}, \quad |\arg\ (1+z^2)| < \pi, \quad z^2 \ne -1. \tag{8}$$

$$\text{arc tan } z = \frac{\pi}{2} - \text{arc tan } \frac{1}{z}. \tag{9}$$

$$\text{arc sinh } z = -i \text{ arc sin } iz, \tag{10}$$

$$\text{arc sinh } z = z(1+z^2)^{\frac{1}{2}} {}_2F_1(1,1;3/2;\ -z^2), \quad |z| < 1, \tag{11}$$

$$\text{arc sinh } z = \frac{z}{(1+z^2)^{\frac{1}{2}}}\ {}_2F_1(1/2,1;3/2;\ \frac{z^2}{z^2+1}),$$

$$z = x + iy,\ y^2 - x^2 < \tfrac{1}{2},\ |\arg\ (1+z^2)| < \pi. \tag{12}$$

$$\text{arc sinh } z = \ln\ \left(z+(z^2+1)^{\frac{1}{2}}\right),$$

$$|\arg\ (1+z^2)| < \pi, \quad -\pi < \arg\ \left(z+(z^2+1)^{\frac{1}{2}}\right) \le \pi. \tag{13}$$

$$\text{arc sinh } z = \ln\ 2z - \tfrac{1}{2}\sum_{k=1}^{\infty} \frac{(-)^k(\frac{1}{2})_k}{k!kz^{2k}}$$

$$= \ln\ 2z + \frac{1}{4z^2}\ {}_3F_2(1,1,3/2;2,2;\ -\frac{1}{z^2}),$$

$$-\pi < \arg z \le \pi,\ |z^2| > 1. \tag{14}$$

$$\text{arc cosh } z = \text{arc cos } iz, \tag{15}$$

$$\text{arc cosh } z = \ln\ \left(z+(z^2-1)^{\frac{1}{2}}\right),$$

$$|\arg\ (z^2-1)| < \pi, \quad -\pi < \arg\ \left(z+(z^2-1)^{\frac{1}{2}}\right) < \pi. \tag{16}$$

$$\text{arc tanh } z = -i \text{ arc tan } iz, \tag{17}$$

$$\text{arc tanh } z = \tfrac{1}{2} \ln\ \left((1+z)/(1-z)\right), \quad |\arg\ (1+z^2)| < \pi. \tag{18}$$

3.4.2. EXPANSIONS IN SERIES OF CHEBYSHEV POLYNOMIALS

Chebyshev coefficients for the various functions in 3.4.1 can be developed by use of the data in 6.9. Alternatively, in some instances, they can be built up by employment of results on integrals of Chebyshev polynomials given in 11.4 and 11.5. In other situations, the techniques in 11.6 are useful. In this connection, note that

$$\text{arc tan } z = \int_0^z \frac{dt}{1+t^2}, \quad \text{arc sin } z = \int_0^z \frac{dt}{(1-t^2)^{\frac{1}{2}}}, \tag{1}$$

and

$${}_3F_2(1,1,3/2;2,2;\ -z) = z^{-1}\int_0^z t^{-1}\left(1-(1+t)^{-\frac{1}{2}}\right)dt. \tag{2}$$

In illustration, by integration of 2.2(3) with a and x replaced by a^2 and x^2 respectively, we find

$$\text{arc tan}(x/a) = 2\sum_{n=0}^{\infty} \frac{(-)^n[(a^2+1)^{1/2}-a]^{2n+1}}{(2n+1)} T_{2n+1}(x),$$

$$a^2 \neq -1, \qquad |\arg(1+1/a^2)| < \pi, \qquad -1 \leqslant x \leqslant 1. \tag{3}$$

Also

$$\text{arc tan } x = \pi/4 + 2\sum_{n=0}^{\infty} \frac{(-)^n(2^{1/2}-1)^{2n+1}}{2n+1} T_{2n+1}\left(\frac{x-1}{x+1}\right),$$

$$0 \leqslant x \leqslant \infty. \tag{4}$$

Numerical values of Chebyshev coefficients for the principal functions of this section are given in the following tables. We also present coefficients for the dilogarithm and inverse tangent integrals.

TABLE 3.9

CHEBYSHEV COEFFICIENTS FOR arc tan x AND x^{-1} arc tan x

$$\arc \tan x = \sum_{n=0}^{\infty} a_n T_{2n+1}(x) \qquad \arc \tan x = x \sum_{n=0}^{\infty} b_n T_{2n}(x)$$

$-1 \leq x \leq 1$

n	a_n	n	b_n
0	0.82842 71247 46190 09760	0	0.88137 35870 19543 02523
1	-0.04737 85412 43650 16267	1	-0.10589 29245 46705 85526
2	0.00487 73235 27902 56610	2	0.01113 58420 59405 52991
3	-0.00059 77260 15160 92785	3	-0.00138 11950 03600 39771
4	0.00007 97638 88582 90437	4	0.00018 57429 73278 54201
5	-0.00001 11970 79759 12191	5	-0.00002 62151 96112 73327
6	0.00000 16255 58988 91589	6	0.00000 38210 36594 48945
7	-0.00000 02417 14919 00658	7	-0.00000 05699 18616 65767
8	0.00000 00365 92697 33444	8	0.00000 00864 88778 64451
9	-0.00000 00056 17439 10602	9	-0.00000 00133 03383 97562
10	0.00000 00008 72009 68575	10	0.00000 00020 68505 76359
11	-0.00000 00001 36603 36477	11	-0.00000 00003 24486 39209
12	0.00000 00000 21562 43750	12	0.00000 00000 51279 66256
13	-0.00000 00000 03425 49018	13	-0.00000 00000 08154 78756
14	0.00000 00000 00547 18870	14	0.00000 00000 01303 80719
15	-0.00000 00000 00087 82579	15	-0.00000 00000 00209 42978
16	0.00000 00000 00014 15528	16	0.00000 00000 00033 77821
17	-0.00000 00000 00002 28988	17	-0.00000 00000 00005 46765
18	0.00000 00000 00000 37164	18	0.00000 00000 00000 88789
19	-0.00000 00000 00000 06049	19	-0.00000 00000 00000 14460
20	0.00000 00000 00000 00987	20	0.00000 00000 00000 02361
21	-0.00000 00000 00000 00162	21	-0.00000 00000 00000 00386
22	0.00000 00000 00000 00026	22	0.00000 00000 00000 00063
23	-0.00000 00000 00000 00004	23	-0.00000 00000 00000 00010
24	0.00000 00000 00000 00001	24	0.00000 00000 00000 00002

For x outside the range $-1 \leq x \leq 1$, we have the formulas

$$\arc \tan x = \pi/4 + \sum_{n=0}^{\infty} a_n T_{2n+1}\big((x-1)/(x+1)\big), \quad -\infty \leq x \leq \infty,$$

$$\arc \tan x = \pi/2 - \arc \tan x^{-1}, \quad \arc \tan x = \arc \sin x(1+x^2)^{-\frac{1}{2}}.$$

In connection with the latter formula, see Table 3.10.

TABLE 3.10

CHEBYSHEV COEFFICIENTS FOR arc sin x AND x^{-1} arc sin x

$$\text{arc sin } x = \sum_{n=0}^{\infty} a_n T_{2n+1}(2^{\frac{1}{2}}x) \qquad \text{arc sin } x = (2^{\frac{1}{2}}x) \sum_{n=0}^{\infty} b_n T_{2n}(2^{\frac{1}{2}}x)$$

$$-d \le x \le d \; , \; d = 2^{-\frac{1}{2}}$$

n	a_n	n	b_n
0	0.76275 97635 01813 18806	0	0.74333 32466 43551 73448
1	0.02086 92375 69303 68915	1	0.03885 30337 16522 90716
2	0.00158 69316 27766 01239	2	0.00288 54414 22084 47113
3	0.00016 08227 52687 70322	3	0.00028 84218 33447 55366
4	0.00001 86910 74902 95281	4	0.00003 32236 71927 85279
5	0.00000 23540 64165 32512	5	0.00000 41584 77878 05283
6	0.00000 03125 75618 54557	6	0.00000 05496 50452 59742
7	0.00000 00430 86361 27501	7	0.00000 00755 00784 49372
8	0.00000 00061 07059 21779	8	0.00000 00106 71938 05630
9	0.00000 00008 84495 91957	9	0.00000 00015 42180 37928
10	0.00000 00001 30325 15812	10	0.00000 00002 26811 45985
11	0.00000 00000 19473 89696	11	0.00000 00000 33838 85639
12	0.00000 00000 02944 02572	12	0.00000 00000 05108 93752
13	0.00000 00000 00449 47589	13	0.00000 00000 00779 11392
14	0.00000 00000 00069 20379	14	0.00000 00000 00119 83786
15	0.00000 00000 00010 73296	15	0.00000 00000 00018 56973
16	0.00000 00000 00001 67523	16	0.00000 00000 00002 89619
17	0.00000 00000 00000 26295	17	0.00000 00000 00000 45428
18	0.00000 00000 00000 04148	18	0.00000 00000 00000 07162
19	0.00000 00000 00000 00657	19	0.00000 00000 00000 01134
20	0.00000 00000 00000 00105	20	0.00000 00000 00000 00180
21	0.00000 00000 00000 00017	21	0.00000 00000 00000 00029
22	0.00000 00000 00000 00003	22	0.00000 00000 00000 00005
		23	0.00000 00000 00000 00001

For other ranges and for arc cos x, we have

$$\text{arc sin } x = \text{arc cos}(1-x^2)^{\frac{1}{2}} = \pi/2 - \text{arc sin}(1-x^2)^{\frac{1}{2}}, \quad d \le x \le 1,$$

$$\text{arc cos } x = \pi/2 - \text{arc sin } x, \qquad -d \le x \le d,$$

$$\text{arc cos } x = \text{arc sin}(1-x^2)^{\frac{1}{2}}, \qquad d \le x \le 1.$$

TABLE 3.11

CHEBYSHEV COEFFICIENTS FOR THE INVERSE HYPERBOLIC FUNCTIONS

$\operatorname{arc\,sinh} x = \sum_{n=0}^{\infty} a_n T_{2n+1}(x)$, $-1 \le x \le 1$		$\operatorname{arc\,sinh} x = x \sum_{n=0}^{\infty} b_n T_{2n}(x)$, $-1 \le x \le 1$		$\operatorname{arc\,sinh} x = \ln 2x + \sum_{n=0}^{\infty} c_n T_{2n}(1/x)$, $x \ge 1$	
n	a_n	n	b_n	n	c_n
0	0.90649 39198 46333 18450	0	0.93589 98004 41309 06828	0	0.10037 21321 67212 85013
1	-0.02704 21478 78869 64300	1	-0.05881 17611 89951 76757	1	0.09350 60801 53666 81550
2	0.00211 68145 57973 55992	2	0.00472 74654 32212 48156	2	-0.00618 98182 58528 94325
3	-0.00021 76650 54603 40215	3	-0.00049 38363 16265 36172	3	0.00059 85212 42396 69563
4	0.00002 55196 04364 81302	4	0.00005 85062 07058 55741	4	-0.00006 79770 86241 56795
5	-0.00000 32329 14485 28777	5	-0.00000 74669 98328 93137	5	0.00000 84463 81136 59092
6	0.00000 04310 66959 88437	6	0.00000 10011 69358 35582	6	-0.00000 11116 41057 46485
7	-0.00000 00596 06134 55196	7	-0.00000 01390 35438 58708	7	0.00000 01522 82266 61319
8	0.00000 00084 69211 32069	8	0.00000 00198 23169 48317	8	-0.00000 00214 85481 00162
9	-0.00000 00012 29008 59356	9	-0.00000 00028 84746 84178	9	0.00000 00031 00806 17346
10	0.00000 00001 81376 78501	10	0.00000 00004 26729 65467	10	-0.00000 00004 55609 22836
11	-0.00000 00000 27138 45802	11	-0.00000 00000 63976 08465	11	0.00000 00000 67925 43622
12	0.00000 00000 04107 37046	12	0.00000 00000 09699 16861	12	-0.00000 00000 10249 60533
13	-0.00000 00000 00627 69516	13	-0.00000 00000 01484 42770	13	0.00000 00000 01562 38481
14	0.00000 00000 00096 72449	14	0.00000 00000 00229 03738	14	-0.00000 00000 00240 23094
15	-0.00000 00000 00015 01221	15	-0.00000 00000 00035 58840	15	0.00000 00000 00037 21486
16	0.00000 00000 00002 34467	16	0.00000 00000 00005 56397	16	-0.00000 00000 00005 80277
17	-0.00000 00000 00000 36824	17	-0.00000 00000 00000 87463	17	0.00000 00000 00000 91001
18	0.00000 00000 00000 05812	18	0.00000 00000 00000 13815	18	-0.00000 00000 00000 14344
19	-0.00000 00000 00000 00921	19	-0.00000 00000 00000 02192	19	0.00000 00000 00000 02271
20	0.00000 00000 00000 00147	20	0.00000 00000 00000 00349	20	-0.00000 00000 00000 00361
21	-0.00000 00000 00000 00023	21	-0.00000 00000 00000 00056	21	0.00000 00000 00000 00058
22	0.00000 00000 00000 00004	22	0.00000 00000 00000 00009	22	-0.00000 00000 00000 00009
23	-0.00000 00000 00000 00001	23	-0.00000 00000 00000 00001	23	0.00000 00000 00000 00001

Alternatively, to evaluate inverse hyperbolic functions, we can use Table 3.1 and the equations

$$\operatorname{arc\,sinh} x = \ln[x + (x^2 + 1)^{1/2}], \qquad \operatorname{arc\,cosh} x = \operatorname{arc\,sinh}(x^2 - 1)^{1/2} = \ln[x + (x^2 - 1)^{1/2}],$$

$$\operatorname{arc\,tanh} x = \operatorname{arc\,sinh} x(1 - x^2)^{-1/2} = \tfrac{1}{2}\ln[(1 + x)/(1 - x)].$$

TABLE 3.12

CHEBYSHEV COEFFICIENTS FOR THE DILOGARITHM AND INVERSE TANGENT INTEGRALS

$L(x) = \int_0^x t^{-1}\ln(1+t)dt = \sum_{n=0}^{\infty} a_n T_n^*(x)$ $\qquad$ $T(x) = \int_0^x t^{-1}\text{arc tan } t dt = \sum_{n=0}^{\infty} b_n T_{2n+1}(x)$

$0 \le x \le 1$		$-1 \le x \le 1$	
n	a_n	n	b_n
0	0.42996 69356 08136 97204	0	0.93432 00492 92895 95286
1	0.40975 98753 30771 05847	1	−0.01950 47944 34351 89753
2	−0.01858 84366 50145 91965	2	0.00125 17037 06300 59276
3	0.00145 75108 40622 67855	3	−0.00011 19241 41205 63855
4	−0.00014 30418 44423 40049	4	0.00001 17754 53855 07085
5	0.00001 58841 55418 79553	5	−0.00000 13652 83304 87376
6	−0.00000 19078 49593 86583	6	0.00000 01688 82892 73643
7	0.00000 02419 51808 54165	7	−0.00000 00218 80246 51007
8	−0.00000 00319 33412 74252	8	0.00000 00029 35063 60647
9	0.00000 00043 45450 62677	9	−0.00000 00004 04523 41419
10	−0.00000 00006 05784 80118	10	0.00000 00000 56976 00371
11	0.00000 00000 86120 97799	11	−0.00000 00000 08168 82727
12	−0.00000 00000 12443 31660	12	0.00000 00000 01188 68900
13	0.00000 00000 01822 55696	13	−0.00000 00000 00175 15916
14	−0.00000 00000 00270 06766	14	0.00000 00000 00026 09029
15	0.00000 00000 00040 42209	15	−0.00000 00000 00003 92271
16	−0.00000 00000 00006 10325	16	0.00000 00000 00000 59463
17	0.00000 00000 00000 92863	17	−0.00000 00000 00000 09079
18	−0.00000 00000 00000 14226	18	0.00000 00000 00000 01395
19	0.00000 00000 00000 02193	19	−0.00000 00000 00000 00216
20	−0.00000 00000 00000 00340	20	0.00000 00000 00000 00034
21	0.00000 00000 00000 00053	21	−0.00000 00000 00000 00005
22	−0.00000 00000 00000 00008	22	0.00000 00000 00000 00001
23	0.00000 00000 00000 00001		

For x outside the above range, use the following formulas together with the above expansions. For a reference to these and many other properties, see Lewin (1958).

$$L(x) + L(x^{-1}) = \pi^2/6 + \tfrac{1}{2}(\ln x)^2; \qquad T(x) = -T(-x).$$

$$L(-x) + L[x/(1-x)] = \tfrac{1}{2}[\ln(1-x)]^2, \quad x < 1; \qquad T(x) - T(x^{-1}) = (\pi/2)\ln x, \quad x > 0.$$

$$L(-x) + L(-x^{-1}) = \tfrac{1}{2}(\ln x)^2 - \pi^2/3 + i\pi \ln x, \quad x > 1.$$

Two other formulas useful for check purposes are

$$L(-x) + L(x-1) = (\ln x)(\ln(1-x)) - \pi^2/6, \qquad 0 < x < 1,$$

$$L(-x) + L[x/(1-x)] = \tfrac{1}{2}[\ln(x-1)]^2 - \pi^2/2 + 2i\pi \ln x - i\pi \ln(x-1), \qquad x > 1.$$

3.4.3. Padé Approximations

TABLE 3.13

Padé Coefficients for z^{-1} arc tan z

These approximations follow from 6.10(19-22) with $b = c = \frac{1}{2}$ and z replaced by z^2. We write

$$H(z^2) = z^{-1} \text{ arc tan } z$$

$$= A_n(z^2)/B_n(z^2) + V_n(z^2), \quad |\arg(1+z^2)| < \pi. \tag{1}$$

We also have

$$H(-z^2) = (2z)^{-1}\ln\left((1+z)/(1-z)\right). \tag{2}$$

The coefficients in the polynomials $A_n(z^2)$ and $B_n(z^2)$ are recorded below for $n = 0(1)6$, $a = 0$ and $a = 1$. Further polynomials can be generated from 6.10(26). For approximate evaluation of the error, use 6.10(30) and Table 2.4 with z replaced by $1/z^2$.

$$A_n(z^2) = a_0^{-1} \sum_{k=0}^{n} a_k z^{2k}, \quad a = 0$$

n	$a_0, a_1, \ldots, a_n$
0	1
1	15, 4
2	945, 735, 64
3	15015, 19250, 5943, 256
4	38 28825, 68 31825, 37 38735, 6 38055, 16384
5	3055 40235, 6983 77680, 5524 73922, 1758 55680, 192 25635, 3 27680
6	5 85618 78375, 16 31924 34405, 16 92692 90190, 8 01912 17106, 1 69794 77515, 12960 36105, 157 28640

TABLE 3.13 *(Concluded)*

$$B_n(z^2) = b_0^{-1} \sum_{k=0}^{n} b_k z^{2k}, \quad a = 0$$

n	$b_0, b_1, \ldots, b_n$
0	1
1	5, 3
2	63, 70, 15
3	429, 693, 315, 35
4	12155, 25740, 18018, 4620, 315
5	88179, 2 30945, 2 18790, 90090, 15015, 693
6	13 00075, 40 56234, 48 49845, 27 71340, 7 65765, 90090, 3003

$$A_n(z^2) = c_0^{-1} \sum_{k=0}^{n-1} c_k z^{2k}, \quad A_0(z) = 0, \quad a = 1$$

n	$c_0, c_1, \ldots, c_{n-1}$
1	1
2	21, 11
3	165, 170, 33
4	2 25225, 3 45345, 1 47455, 15159
5	13 22685, 26 91780, 6 00054, 4 37580, 27985
6	34236 17505, 86830 87875, 48296 46822, 15810 50394, 4991 68285, 218 18175

$$B_n(z^2) = d_0^{-1} \sum_{k=0}^{n} d_k z^{2k}, \quad a = 1$$

n	$d_0, d_1, \ldots, d_n$
0	1
1	3, 1
2	35, 30, 3
3	231, 315, 105, 5
4	6435, 12012, 6930, 1260, 35
5	46189, 1 09395, 90090, 30030, 3465, 63
6	6 76039, 19 26015, 20 78505, 10 21020, 2 25225, 18018, 231

TABLE 3.14

PADÉ COEFFICIENTS FOR $z^{-1}(1+z^2)^{-\frac{1}{2}}\text{arc sinh } z$

These approximations follow from 6.10(19-22) with $b = 1$, $c = \frac{1}{2}$ and z replaced by z^2. We write

$$H(z^2) = \frac{\text{arc sinh } z}{(1+z^2)^{\frac{1}{2}}}$$

$$= A_n(z^2)/B_n(z^2) + V_n(z^2), \quad |\arg(1+z^2)| < \pi, \tag{3}$$

$$H(z^2) = \frac{\ln\left[z+(1+z^2)^{\frac{1}{2}}\right]}{z(1+z^2)^{\frac{1}{2}}}, \quad H(-z^2) = \frac{\text{arc sin } z}{z(1-z^2)^{\frac{1}{2}}}. \tag{4}$$

The coefficients in the polynomials $A_n(z^2)$ and $B_n(z^2)$ are recorded below for $n = 0(1)6$, $\alpha = 0$ and $\alpha = 1$. Further polynomials can be generated from 6.10(26). For approximate evaluation of the error, use 6.10(30) and Table 2.4 with z replaced by $1/z^2$.

$$A_n(z^2) = a_0^{-1} \sum_{k=0}^{n} a_k z^{2k}, \quad \alpha = 0$$

n	$a_0, a_1, \ldots, a_n$
0	1
1	15, 2
2	315, 210, 8
3	15015, 17710, 4648, 80
4	7 65765, 7 23690, 6 48648, 93456, 896
5	1018 46745, 2230 92870, 1666 30464, 485 76528, 44 57024, 26880
6	83659 82625, 2 25323 79870, 2 23868 84520, 1 00034 28864, 19357 76128, 1233 70624, 5 06880

TABLE 3.14 *(Concluded)*

$$B_n(z^2) = b_0^{-1} \sum_{k=0}^{n} b_k z^{2k}$$

n	$b_0, b_1, \ldots, b_n$
0	1
1	5, 4
2	21, 28, 8
3	429, 792, 432, 64
4	2431, 5720, 4576, 1408, 128
5	29393, 83980, 88400, 41600, 8320, 512
6	1 85725, 6 24036, 8 13960, 5 16800, 1 63200, 23040, 1024

$$A_n(z^2) = c_0^{-1} \sum_{k=0}^{n-1} c_k z^{2k}, \quad A_0(z^2) = 0, \quad a = 1$$

n	$c_0, c_1, \ldots, c_{n-1}$
1	1
2	21, 10
3	165, 160, 28
4	2 25225, 3 30330, 1 32440, 12176
5	13 22685, 25 98960, 16 60932, 3 78144, 21472
6	1801 90395, 4439 47350, 3915 69024, 1477 80864, 219 64096, 8 46976

$$B_n(z^2) = d_0^{-1} \sum_{k=0}^{n} d_k z^{2k}, \quad a = 1$$

n	$d_0, d_1, \ldots, d_n$
0	1
1	3, 2
2	35, 40, 8
3	231, 378, 168, 16
4	6435, 13728, 9504, 2304, 128
5	46189, 1 21550, 1 14400, 45760, 7040, 256
6	6 76039, 21 16296, 25 19400, 14 14400, 3 74400, 39936, 1024

3.4.4. INEQUALITIES

In 3.4.3(1), let $A_n(z^2)$ now be designated as $A_n(z^2,a)$ and similarly for $B_n(z^2)$. Then

$$\frac{A_n(z^2,1)}{B_n(z^2,1)} < z^{-1} \text{ arc tan } z < \frac{A_n(z^2,0)}{B_n(z^2,0)},$$

$$n > 0, \ z > 0. \tag{1}$$

Thus for $z > 0$,

$$\frac{3}{3+z^2} < z^{-1} \text{ arc tan } z < \frac{15+4z^2}{15+9z^2}, \tag{2}$$

$$\frac{105+55z^2}{105+90z^2+9z^4} < z^{-1} \text{ arc tan } z < \frac{945+735z^2+64z^4}{945+1050z^2+225z^4}, \tag{3}$$

In particular, if $z = 1$,

$$3/4 < \pi/4 = 0.78540 < 19/24 = 0.79167, \tag{4}$$

$$40/51 = 0.78431 < \pi/4 < 436/555 = 0.78559. \tag{5}$$

Let

$$F(z) = \frac{3u(\text{arc tan } z - z/u)}{2v^2 z},$$

$$u = (1+z^2)^{\frac{1}{2}}, \quad v = (u-1)/z, \quad z > 0. \tag{6}$$

Then

$$\frac{5}{5+v^2} < F(z) < \frac{35+8v^2}{35+15v^2}, \tag{7}$$

$$\frac{315+147v^2}{315+210v^2+15v^4} < F(z) < \frac{3465+2457v^2+192v^4}{3465+3150v^2+525v^4}, \quad z > 0. \tag{8}$$

In illustration, if $z = 3^{\frac{1}{2}}$, we deduce that

$$29(3)^{\frac{1}{2}}/16 = 3.13934 < \pi < 653(3)^{\frac{1}{2}}/360 = 3.14175, \tag{9}$$

$$526(3)^{\frac{1}{2}}/29 = 3.141582 < \pi < 18664(3)^{\frac{1}{2}}/10290 = 3.1415934. \tag{10}$$

If $z \to \infty$, then

$$28/9 = 3.11111 < \pi < 236/75 = 3.14667, \tag{11}$$

$$424/135 = 3.1407 < \pi < 5608/1785 = 3.1417. \tag{12}$$

Next in 3.4.3(3), let $A_n(z^2)$ now be noted as $A_n(z^2,a)$, etc. Then

$$\frac{A_n(z^2,1)}{B_n(z^2,1)} < \frac{\text{arc sinh } z}{z(1+z^2)^{\frac{1}{2}}} < \frac{A_n(z^2,0)}{B_n(z^2,0)},$$

$$n > 0, \quad z > 0. \tag{13}$$

Thus for $z > 0$,

$$\frac{3}{3+2z^2} < \frac{\text{arc sinh } z}{(1+z^2)^{\frac{1}{2}}} = \frac{\ln\{z+(z^2+1)^{\frac{1}{2}}\}}{(1+z^2)^{\frac{1}{2}}} < \frac{15+2z^2}{15+12z^2}, \tag{14}$$

$$\frac{105+50z^2}{105+120z^2+24z^4} < \frac{\text{arc sinh } z}{z(1+z^2)^{\frac{1}{2}}} < \frac{315+210z^2+8z^4}{315+420z^2+120z^4}. \tag{15}$$

In particular, if $z = 1$,

$$3/5 < \frac{\ln(1+2^{\frac{1}{2}})}{2^{\frac{1}{2}}} = 0.62323 < 17/27 = 0.62963, \tag{16}$$

$$155/249 = 0.62249 < \frac{\ln(1+2^{\frac{1}{2}})}{2^{\frac{1}{2}}} < 533/855 = 0.62339. \tag{17}$$

Using 3.4.1(13,14) and 5.14(3), we find

$$\frac{8z^2}{8z^2+3} < 4z^2 \ln\left\{\frac{z+(z^2+1)^{\frac{1}{2}}}{2z}\right\} < \frac{72z^2+13}{72z^2+40}, \quad z > 0. \tag{18}$$

For $z = 1$ and 2 we have

$$8/11 = 0.72727 < 4 \ln\left(\frac{1+2^{\frac{1}{2}}}{2}\right) = 0.75291 < 85/112 = 0.75893. \tag{19}$$

$$32/35 = 0.91429 < 16 \ln\left(\frac{2+5^{\frac{1}{2}}}{4}\right) = 0.91746 < 301/328 = 0.91768. \tag{20}$$

3.5. Bibliographic and Numerical Data

3.5.1. Description of and References to Tables

Shanks and Wrench (1969): e computed to 100,000D.

Kölbig, Mignaco and Remiddi (1970). Let $S^*_{n,p}(x) = p!p^n x^{-p}\int_0^1 t^{-1}(\ln t)^{n-1}(\ln(1-tx))^p dt$, n and t are positive integers, $CT\left(\frac{4x+1}{3}\right)$, $-1 \leq x \leq \frac{1}{2}$, $p = 1(1)4$, $n = 1(1)(5-p)$, 15D. Zeros for $S^*_{n,p}(x)$ as a function of x are given for p and n as above to 14S. Let $S_{n,p}(x) = x^p S^*_{n,p}(x)/p!p^n = s_{n,p}(x) + \int_1^x t^{-1}s_{n-1,p}(t)dt$. Numerical data relating to $s_{n,p}(x)$ are presented.

Dimsdale and Inselberg (1972): $\overline{\Gamma}(t) = -\int_0^1 x^x(\ln x)^{t-1}dx$, $\Gamma(t)$, $\overline{\Gamma}(t)/\Gamma(t)$, $t = 0.5(0.5)10$, 8S.

3.5.2. Description of and References to Other Approximations and Expansions

Clenshaw (1962): This gives $CT(x)$ and $CT^*(x)$ as appropriate for $\sin ax$, $\cos ax$, $a = \pi/2$, π, $\tan \pi x/4$, $x \cot \pi x/2$, $\arcsin x$, $\arctan x$, e^x, 2^{-x}, $\ln(1+x)$, $\operatorname{arc\,sinh} x$, 20D.

Clenshaw, Miller and Woodger (1963): $CT(x)$ and $CT^*(x)$ as appropriate for $\sin \pi x/2$, $\tan \pi x/4$, $\arcsin x$, $\arctan x$, 2^{-x}, $\ln(1+x)$, 15D. Also e^x, 9D.

Lyusternik, Chervonenkis and Yanpol'skii (1963): Analytical description of elementary functions including square root, exponential function, logarithm function, trigonometric functions and their inverses, hyperbolic functions and their inverses. $CT(x)$ mostly to 11D for many of these functions, and coefficients for expansions in series of Legendre polynomials and for $BCP(x)$ are given. Continued fraction representations for many of the functions included.

Khovanskii (1963): Continued fraction representations for the elementary functions.

Hart *et al.* (1969): Best Chebyshev polynomial and rational approximations for the elementary functions, various ranges. Relative accuracy up to about 10^{-24}. In some cases, the approximations are best in the absolute error norm.

Spellucci (1971): Best Chebyshev rational approximations for $\ln x$, e^x, $\sin x$, $\cos x$ and $\tan x$ are presented both as a Thiele type fraction and as a Jacobi fraction. Accuracy is to at least 20D.

Spielberg (1961): The idea of this paper is to approximate $f(x)$ by the ratio of two polynomials each expressed as a finite sum of the Chebyshev polynomials $T_n(x)$. This is readily accomplished once $CT(x)$ is known. Coefficients are developed for a number of the elementary functions.

Simauti (1964): $\sin \pi x/4$, $x^{-1}\sin \pi x/4$, $\cos \pi x/4$, $BCP(x)$, $\alpha = 1$, 15D. With $x = w + u$ and

$$\tan x = \frac{\tan w + \tan u}{1 - \tan w \tan u},$$

similar expansions are developed for $\tan u$ for each $w = (2k-1)/32, k = 1(1)4$. With $x = w + u$, $v = u(1 + wx)^{-1}$,

$$\text{arc tan } x = \text{arc tan } w + \text{arc tan } v,$$

similar expansions are developed for arc tan v for each $w = (2k-1)/8$, $k = 1(1)4$. See this same reference for similar expansions relating to e^x, 10^x, and $\ln(1+x)$.

Vionnet (1960): $\ln(1+0.16x)$, $CT(x)$, $\alpha = 1$, 24D. $\text{arc sin } 2^{-\frac{1}{2}}x$, $CT(x)$, $\alpha = 1$, 23D. Truncated expansions of these functions in powers of x are also given.

Németh (1965c): $x^{-1}\ln(1+x)$, $CT^*(x)$, $a = 1$, 30D.

Spielberg (1962): This paper gives coefficients for polynomial and continued fraction approximations for logarithmic functions.

Carlson (1972): An algorithm for computing logarithms and arc tangents. Only 3 or 4 square roots needed for 10D accuracy.

Kelisky and Rivlin (1968): τ-method is used to get rational approximations for $\ln z$. Most results are a special case of 5.12.

Bulirsch (1969): arc sinh x, $CT(x)$, $a = 1$, 15D.

Cody, Meinardus and Varga (1969): Best Chebyshev rational approximations to e^{-x} in $[0,+\infty)$, $n = 1(1)14$. Accuracy ranges from $0.67\cdot 10^{-1}$ to $0.18\cdot 10^{-13}$.

Bellman, Kashef and Vasudevan (1972): Differential approximation to e^{-t^2} in the form $\sum_{i=1}^{N} b_i e^{-a_i t}$, $0 \le t \le 1$. Coefficients are given to 4D for $N = 3, 5$. Maximum magnitude of error for $N = 3$ and 5 is about $0.13\cdot 10^{-3}$ and $0.2\cdot 10^{-6}$ respectively.

Pierre (1964): Rational approximations for $\exp(-z^{1/2})$.

Booth (1955): $\sin \pi x/2$, $\cos \pi x/2$, $CT(x)$, $a = 1$, 11D. This reference also gives coefficients for expansion of these functions in series of Legendre polynomials.

Hornecker (1958): $x^{-1}\sin \pi x/4$, $\cos \pi x/4$, $(1+x)^{-1}$, $|x|$, $BCP(x)$, $a = 1$. Various N with accuracy up to 9D.

Vionnett (1959): $\sin \pi x/4$, $x^{-1}\sin \pi x/4$, $\cos \pi x/4$, e^x, $CT(x)$, $a = 1$, 24D. Truncated expansions of these functions in powers of x are also given.

Perlin and Garret (1960): arc tan x, $CT(x)$, $a = \tan \pi/24$, 22D.

Luke (1969, v.2, pp. 273-281): Padé approximations for $L(t) = t(e^t-1)^{-1}$ and use of same to get approximations for the Debye functions $(m/z^m)\int_0^z t^{m-1}L(t)\,dt$, m a positive integer.

CHAPTER IV INCOMPLETE GAMMA FUNCTIONS

4.1. Definitions and Series Expansions

$$\gamma(v,z) = \int_0^z t^{v-1}e^{-t}dt, \quad R(v) > 0, \tag{1}$$

$$\Gamma(v,z) = \Gamma(v) - \gamma(v,z), \tag{2}$$

$$\Gamma(v,z) = \int_z^{\infty e^{i\delta}} t^{v-1}e^{-t}dt, \quad \delta \text{ real},$$

$$|\delta| < \pi/2,\ R(v) > 0 \quad \text{or} \quad |\delta| = \pi/2,\ 0 < R(v) < 1. \tag{3}$$

Here the path of integration lies in the branch of the cut plane, $|\arg z| < \pi$, and is the ray $r\exp(i\delta)$, $r \to \infty$, except for an initial finite path. If $z \neq 0$, the integral in (2) exists without the restriction on v. If $v \to 0$ in (2), we get the exponential integral, and in this event, exclude the origin in the path of integration. We also have the representations

$$\gamma(v,z) = (2/\pi)^{\frac{1}{2}}\int_0^z t^{v-\frac{1}{2}}K_{\frac{1}{2}}(t)dt, \quad R(v) > 0, \tag{4}$$

$$\Gamma(v,z) = z^v e^{-z}\int_0^\infty e^{-zt}(1+t)^{v-1}dt, \quad R(z) > 0, \tag{5}$$

$$\Gamma(v,z) = \frac{e^{-z}}{\Gamma(1-v)}\int_0^\infty e^{-zt}t^{-v}(1+t)^{-1}dt, \quad R(z) > 0,\ R(v) < 1. \tag{6}$$

$$\gamma(v,z) = v^{-1}z^v{}_1F_1(\ v;\ v+1;\ -z), \tag{7}$$

$$\gamma(v,z) = v^{-1}z^v e^{-z}{}_1F_1(\ 1;\ v+1;\ z), \tag{8}$$

$$\Gamma(v,z) = e^{-z}U(1-v,\ 1-v;\ z) = z^{v}e^{-z}U(1,\ 1+v;\ z), \tag{9}$$

$$\Gamma(v,z) \sim z^{v-1}e^{-z}{}_2F_0(1,\ 1-v;\ -1/z), \tag{10}$$

$$|z| \to \infty, \quad |\arg z| \leq 3\pi/2 - \varepsilon, \quad \varepsilon > 0.$$

$$\Gamma(1-u,z) = \frac{z^{1-u}e^{-z}}{z+u}\left[\sum_{k=0}^{m-1} \frac{f_k(z)}{(z+u)^{2k}} + R_m(z,u)\right],$$

$$f_{k+1}(z) = (u-2kz)f_k(z) + z(z+u)f_k'(z), \quad f_0 = 1, \tag{11}$$

uniformly for all u and z, $|z+u|$ large. Further, for positive u and z,

$$a_m \leq u^m R_m(z,u) \leq b_m\left(1 + \frac{1}{z+u-1}\right), \tag{12}$$

where the a_m's and b_m's are lower and upper bounds, respectively, of $u^m f_m(x)/(x+u)^{2m}$ when $x \geq 0$. The first eight polynomials $f_m(x)$ and the corresponding values of a_m and b_m have been given by Gautschi (1959b).

Chebyshev expansions follow from the definitions in 4.1 and the material in 5.10.2. In a similar fashion, one can also derive expansions in series of Bessel functions. See also 5.11. Chebyshev expansions can also be deduced from known Chebyshev expansions for the exponential function. For details, see Luke (1969).

4.2. Differential Equations and Difference Equations

Let

$$g(v,z) = ve^{z}z^{-v}\gamma(v,z) = {}_1F_1(1;\ v+1;\ z), \tag{1}$$

$$G(v,z) = e^{z}z^{1-v}\Gamma(v,z). \tag{2}$$

Then

$$(zD + v - z)g(v,z) = v, \tag{3}$$

$$(zD + v - z - 1)G(v,z) = -z, \quad D = d/dz, \tag{4}$$

$$z^n g(n+v,z) = -(v+1)_n \, {}_1F_1^{n-1}(1;\ v+1;\ z) + (v+1)_n g(v,z), \tag{5}$$

$$z^n G(n+v,z) = (z/v)(v)_n \, {}_1F_1^{n-1}(1;\ v+1;\ z) + (v)_n G(v,z),$$
$$n = 0,1,2,\cdots. \tag{6}$$

In particular,

$$\bigl(z/(v+1)\bigr)g(v+1,z) = g(v,z) - 1, \tag{7}$$

$$G(v+1,z) = (v/z)G(v,z) + 1. \tag{8}$$

Both $\gamma(v,z)$ and $\Gamma(v,z)$ as well as $\Gamma(v)$ satisfy

$$h(v+2,z) - (v+z+1)h(v+1,z) + vz\,h(v,z) = 0. \tag{9}$$

4.3. Padé Approximations

4.3.1. ${}_1F_1(1;\ v+1;\ -z)$

It is convenient to treat

$$H(v,z) = {}_1F_1(1;\ v+1;\ -z)$$
$$= vz^{-v}e^{-z-iv\pi}\gamma(v,ze^{i\pi})$$
$$= vz^{-v}e^{-z}\int_0^z e^t t^{v-1}dt, \quad R(v) > 0. \tag{1}$$

The approximations for (1) in the $(n-a,n)$ positions of the Padé table are confluent forms of those for the ${}_2F_1$ given by 6.10(19-28). We have

$$H(\nu, z) = [A_n(\nu, z)/B_n(\nu, z)] + V_n(\nu, z), \tag{2}$$

$$B_n(\nu, z) = {}_1F_1(-n; -2n + a - \nu; z)$$

$$= \frac{z^n}{(n + \nu + 1 - a)_n} \, {}_2F_0\left(-n, n + \nu + 1 - a; -\frac{1}{z}\right), \tag{3}$$

$$A_n(\nu, z) = \left[\frac{n(n + \nu)}{z}\right]^a \frac{z^n}{(n + \nu + 1 - a)_n} \sum_{k=0}^{n-a} \frac{(a - n)_k \, (n + \nu + 1)_k}{(\nu + 1)_k \, (1 + a)_k}$$

$$\times \, {}_3F_1\left(\begin{matrix} -n + a + k, n + \nu + 1 + k, 1 \\ 1 + a + k \end{matrix} \middle| -\frac{1}{z}\right), \tag{4}$$

$$V_n(\nu, z) = P_n(\nu, z)/B_n(\nu, z), \tag{5}$$

$$P_n(\nu, z) = \frac{(-)^{n+1-a}\nu^a z^{-\nu} e^{-z}}{(\nu + 1 - a)_n \, (n + \nu + 1 - a)_n} \int_0^z (z - t)^n \, t^{n+\nu-a} e^t \, dt,$$

$$R(\nu) > a - 1,$$

$$= \frac{(-)^{1-a} z^{n+1} (\nu/z)^a \, e^{-z}}{(n + \nu + 1 - a)_n} \int_0^1 t^{\nu-a} e^{zt}$$

$$\times \, {}_2F_1\left(\begin{matrix} -n, n + \nu + 1 - a \\ \nu + 1 - a \end{matrix} \middle| t\right) dt, \qquad R(\nu) > a - 1,$$

$$= \frac{(-)^{n+1-a} n! \, z^{2n+1-a} e^{-z}}{(\nu + 1)^{1-a} \, (\nu + 2 - a)_{2n} \, (n + \nu + 1 - a)_n}$$

$$\times \, {}_1F_1(n + \nu + 1 - a; 2n + \nu + 2 - a; z). \tag{6}$$

From (4), with $A_n(\nu, z) = A_n(\nu, z)$ if $a = 0$, and $A_n(\nu, z) = O_n(\nu, z)$ if $a = 1$, we have

$$A_0(\nu, z) = 1, \qquad A_1(\nu, z) = 1 - [z/(\nu + 1)(\nu + 2)],$$

$$A_2(\nu, z) = 1 + \frac{(\nu - 2) \, z}{(\nu + 1)(\nu + 4)} + \frac{2z^2}{(\nu + 1)(\nu + 2)(\nu + 3)(\nu + 4)}, \tag{7}$$

$$O_0(\nu, z) = 0, \qquad O_1(\nu, z) = 1, \qquad O_2(\nu, z) = 1 + \frac{(\nu - 1) \, z}{(\nu + 1)(\nu + 3)}. \tag{8}$$

From (3),

$$B_0(\nu, z) = 1, \qquad B_1(\nu, z) = 1 + [z/(\nu + 2 - a)],$$

$$B_2(\nu, z) = 1 + \frac{2z}{(\nu + 4 - a)} + \frac{z^2}{(\nu + 3 - a)(\nu + 4 - a)}. \tag{9}$$

Both $A_n(\nu, z)$ and $B_n(\nu, z)$ satisfy the recurrence formula

$$B_{n+1}(\nu, z) = \left\{1 + \frac{(\nu - a)\, z}{(2n + \nu - a)(2n + \nu + 2 - a)}\right\} B_n(\nu, z)$$
$$+ \frac{n(n + \nu - a)\, z^2}{(2n + \nu - 1 - a)(2n + \nu - a)^2\,(2n + \nu + 1 - a)} B_{n-1}(\nu, z), \tag{10}$$

which is stable when used in the forward direction.

$$V_n(\nu, z) = \frac{\left[\begin{array}{l}(-)^{n+1-a}\pi\Gamma(\nu + 1)\, n!\, \Gamma(n + \nu + 1 - a)\, z^{2n+1-a} \\ \qquad \times \exp[-z + z(z + 4\nu - 4a)/4(2n + \nu + 1 - a)]\end{array}\right]}{2^{4n+2\nu-2a}(2n + \nu + 1 - a)\{\Gamma[n + (\nu + 1 - a)/2]\, \Gamma[n + (\nu + 2 - a)/2]\}^2} \times [1 + O(n^{-3})] \tag{11}$$

$$= \frac{\left[\begin{array}{l}(-)^{n+1-a}\pi\Gamma(\nu+1)\, z^{2n+1-a} n^{a-\nu} \\ \qquad \times \exp\left(-z+z(z+4\nu-4a)/4(2n+\nu+1-a)\right)\end{array}\right]}{2^{4n+2\nu+1-2a}(n!)^2} \times \left[1 + O(n^{-1})\right]. \tag{12}$$

It follows that for z and ν fixed,

$$\lim_{n\to\infty} V_n(\nu, z) = 0. \tag{13}$$

For the applications in the general case when ν is rational and $\nu \neq 0$, it is preferable to deal with polynomials in the approximation for $H(\nu, z)$, all of whose coefficients are integers. To this end, let

$$\begin{aligned} C_n(\nu, z) &= q^{2n}(\nu + 1 - a)_{2n}\, A_n(\nu, z), \\ D_n(\nu, z) &= q^{2n}(\nu + 1 - a)_{2n}\, B_n(\nu, z), \qquad \nu = p/q, \quad q \neq 0, \end{aligned} \tag{14}$$

where p and q are co-prime integers. It follows that both

$C_n(\nu, z)$ and $D_n(\nu, z)$ satisfy the same recurrence formula

$$D_{n+1}(\nu, z) = [(2n+\nu-a)(2n+\nu+2-a) + (\nu-a)\,z] \times \frac{q^2(2n+\nu+1-a)}{(2n+\nu-a)} D_n(\nu, z) + \frac{q^4 n(n+\nu-a)(2n+\nu+2-a)}{(2n+\nu-a)} D_{n-1}(\nu, z). \tag{15}$$

Padé coefficients for the configurations $\nu = \frac{1}{2}$, $a = 0$, and $a = 1$ are presented in Table 4.6 of 4.8.3. There we also give values of

$$R_n(z) = |\, 2z^{1/2} e^z \{V_n(\tfrac{1}{2}, z)\}_{a=0} \,|, \tag{16}$$

for a large range of complex z and n values. Here, $\{V_n(\frac{1}{2}, z)\}_{a=0}$ is given by (11) with $\nu = \frac{1}{2}$, $a = 0$, and $O(n^{-3})$ neglected. This is sufficient to estimate the error for general ν and $a = 0$ or $a = 1$, since from (11),

$$|\, V_n(\nu, z)| = \left| \frac{\Gamma(\nu+1)\, n^{a+\frac{1}{2}-\nu} e^{-z}}{\pi^{1/2} 2^{2\nu-2a-1} z^{a+\frac{1}{2}}} \right| R_n(z)[1 + O(n^{-1})]. \tag{17}$$

Note that $H(0,z) = e^{-z}$ and approximations for this are delineated in 3.2.3.

4.3.2. $z^{1-\nu} e^z \Gamma(\nu, z)$

The approximations for

$$z^{1-\nu} e^z \Gamma(\nu, z) = z^{1-\nu} e^z \int_z^\infty t^{\nu-1} e^{-t} dt \tag{1}$$

which lie in the $(n-a, n)$ positions of the Padé table are given by

$$z^{1-\nu} e^z \Gamma(\nu, z) = [E_n(\nu, z)/F_n(\nu, z)] + T_n(\nu, z), \tag{2}$$

$$F_n(\nu, z) = {}_1F_1(-n; 2-a-\nu; -z) = [n!/(2-a-\nu)_n]\, L_n^{(1-a-\nu)}(-z), \tag{3}$$

$$E_n(\nu, z) = \left[\frac{nz}{1-\nu}\right]^a \sum_{k=0}^{n-a} \frac{(a-n)_k\,(1-\nu)_k}{(2-\nu)_k\,(1+a)_k}\, {}_2F_2\left(\begin{matrix} -n+a+k,\ 1 \\ 2-\nu+k,\ 1+a+k \end{matrix} \middle| -z\right), \tag{4}$$

$$T_n(\nu, z) = [S_n(\nu, z)/F_n(\nu, z)], \tag{5}$$

$$S_n(\nu, z) = (\nu - 1)^{1-a} z^{1-\nu} e^z \int_z^\infty (t - z)^n \, t^{\nu+a-2-n} e^{-t} \, dt, \qquad z \neq 0, \ |\arg z| < \pi, \tag{6}$$

$$S_n(\nu, z) = (\nu - 1)^{1-a} z^a \int_0^\infty t^n (1 + t)^{\nu+a-2-n} e^{-zt} \, dt, \quad z \neq 0, \ |\arg z| < \pi, \tag{7}$$

$$S_n(\nu, z) = (\nu - 1)^{1-a} \, z^a n! \, U(n + 1; \nu + a; z). \tag{8}$$

Let $E_n(\nu, z) = E_n(\nu, z)$ if $a = 0$, and $E_n(\nu, z) = M_n(\nu, z)$ if $a = 1$. Thus,

$$\begin{aligned} &E_0(\nu, z) = 1, \qquad E_1(\nu, z) = (2 - \nu)^{-1} (1 + z), \\ &E_2(\nu, z) = [(2 - \nu)(3 - \nu)]^{-1} [2 + (5 - \nu) z + z^2], \end{aligned} \tag{9}$$

$$\begin{aligned} &M_0(\nu, z) = 0, \qquad M_1(\nu, z) = (1 - \nu)^{-1} z, \\ &M_2(\nu, z) = [(1 - \nu)(2 - \nu)]^{-1} [(3 - \nu) z + z^2]. \end{aligned} \tag{10}$$

Also,

$$\begin{aligned} &F_0(\nu, z) = 1, \qquad F_1(\nu, z) = 1 + \frac{z}{2 - a - \nu}, \\ &F_2(\nu, z) = 1 + \frac{2z}{(2 - a - \nu)} + \frac{z^2}{(2 - a - \nu)(3 - a - \nu)}. \end{aligned} \tag{11}$$

Both $E_n(\nu, z)$ and $F_n(\nu, z)$ satisfy the same recurrence formula

$$(n + 2 - a - \nu) \, F_{n+1}(\nu, z) = (z + 2n + 2 - a - \nu) \, F_n(\nu, z) - nF_{n-1}(\nu, z). \tag{12}$$

For the applications when ν is rational, it is preferable to tabulate polynomials in the Padé approximation, all of whose coefficients are integers. To accomplish this, let $\nu = p/q$, $q \neq 0$ where p and q are co-prime integers, and set

$$G_n(\nu, z) = q^n (2 - a - \nu)_n \, E_n(\nu, z), \quad H_n(\nu, z) = q^n (2 - a - \nu)_n \, F_n(\nu, z). \tag{13}$$

It follows that both $G_n(\nu, z)$ and $H_n(\nu, z)$ satisfy the same recurrence formula

$$H_{n+1}(\nu, z) = (z + 2n + 2 - a - \nu) q H_n(\nu, z) - n(n + 1 - a - \nu) q^2 H_{n-1}(\nu, z). \tag{14}$$

The error is given by

$$T_n(\nu, z) = \frac{(\nu - 1)^{1-a}\,(2\pi)\,z^{1-\nu}\exp[z - 4(kz)^{1/2}]}{\Gamma(2 - a - \nu)}\,[1 + O(k^{-1/2})]\,,$$

$$k = n + 1 - (a+\nu)/2,$$

$$z \neq 0,\ z \text{ and } \nu \text{ bounded},\ |\arg z| \leqslant \pi - \epsilon,\ |\arg(kz)^{1/2}| \leqslant \pi/2 - \epsilon,\ \epsilon > 0. \tag{15}$$

We therefore have convergence everywhere in the complex plane except on the negative real axis and except, possibly, in the neighborhood of the zeros of $F_n(\nu, z)$. If $\alpha > 1$, the zeros of $L_n^{(\alpha)}(x)$ are real and positive, and hence the restriction that z not be on the negative real axis is expected. A detailed description of the zeros of $L_n^{(\alpha)}(x)$ for real α is given by Szegö (1959). If α is complex, so are the zeros of $L_n^{(\alpha)}(x)$, and obviously, domains enclosing these points must be removed.

The formula (15) is quite effective in practice provided we fix z and take n sufficiently large. Now the rational approximation for $\Gamma(\nu, z)$ is based on its asymptotic series expansion. For a given number of terms, the error in the asymptotic expansion decreases as $|z|$ increases. Thus for a given n, we should expect the error in the Padé approximation to have a like behavior. This is true, but such is not reflected by (15). We now turn to asymptotic developments of the error, valid uniformly in z.

If z is real, say $z = x$, there are four cases to consider. They are $0 < x/4k < \infty$, $-1 < x/4k < 0$, $x/4k \sim -1$, and $x/4k < -1$. More generally, for z and k complex, we study the regions $|\arg z/k| < \pi$ and $|\arg z/k| = \pi$, where the latter is divided into three subregions. $x/4k \sim -1$ is sometimes called the transition region.

4.3.3. The Error $T_n(\nu, z)$ for $|\arg z/k| < \pi$

It is convenient to first give appropriate asymptotic forms for the general confluent hypergeometric functions ${}_1F_1(d; c; z)$ and $U(1-d, 2-c; z)$.

Let

$$k = c/2 - d, \tag{1}$$

$$z/4k = \sinh^2 \alpha. \tag{2}$$

Because z and k may be complex, in the determination of $\sinh \alpha$ and α, we choose the branch such that $\sinh\alpha$ is real and positive if $z/4k$ is real and positive. To be explicit, let

$$\begin{aligned} \sinh^2\alpha &= z/4k = \rho e^{i\theta}, \qquad \rho > 0, \qquad |\,\theta\,| < \pi, \\ \alpha &= \beta + i\delta, \qquad \beta \text{ and } \delta \text{ real}, \qquad \beta > 0, \qquad |\,\delta\,| < \pi/2, \end{aligned} \tag{3}$$

so that

$$\cosh\beta = (\tfrac{1}{2}[1 + \rho + \{(1+\rho)^2 - 4\rho\sin^2\tfrac{1}{2}\theta\}^{1/2}])^{1/2}, \tag{4}$$

$$\sin\delta\cosh\beta = \rho^{1/2}\sin\theta/2, \qquad \cos\delta\sinh\beta = \rho^{1/2}\cos\theta/2. \tag{5}$$

Then,

$$\begin{aligned} {}_1F_1(d;\, c;\, ze^{\mp i\pi}) \sim{}& \frac{e^{-\frac{1}{2}z} z^{-m-\frac{1}{2}} \Gamma(2m+1)(\tanh\alpha)^{1/2} \exp[m(4m^2-1)/24k^2]}{(2\pi)^{1/2}\, k^m} \\ &\times [\exp\{k(2\alpha + \sinh 2\alpha)\}\{A(\alpha, -k) + O(k^{-3})\} \\ &+ \exp\{\pm(2m + \tfrac{1}{2})\, i\pi\} \exp\{-k(2\alpha + \sinh 2\alpha)\} \\ &\times \{A(\alpha, k) + O(k^{-3})\}], \end{aligned} \tag{6}$$

where the upper sign is chosen if $0 \leqslant \arg z < \pi$ and the lower sign is chosen if $-\pi < \arg z \leqslant 0$. Here,

$$m = (c-1)/2, \qquad k = c/2 - d, \tag{7}$$

$$A(\alpha, k) = 1 + P_1(\alpha)\, k^{-1} + P_2(\alpha)\, k^{-2}, \tag{8}$$

and

$$\begin{aligned} P_1(\alpha) &= (96\coth^3\alpha)^{-1}\{3(16m^2-1)\coth^4\alpha - 6\coth^2\alpha + 5\}, \\ P_2(\alpha) &= (18432\coth^6\alpha)^{-1}\{9(16m^2-9)(16m^2-1)\coth^8\alpha \\ &\quad + 36(48m^2-7)\coth^6\alpha + 6(121-112m^2)\coth^4\alpha - 924\coth^2\alpha + 385\}. \end{aligned} \tag{9}$$

Actually, for $k \to \infty$, $|\arg k\,| < \pi/2$, we could omit the term involving $\exp(2m + \frac{1}{2})\, i\pi$ as it is a decay of exponential order. In this event, on the left-hand side of (6) write $ze^{\mp i\pi} = -z$. The resulting expression is then valid for $|\arg z\,| < \pi$. The form (6) is convenient since we can readily derive from it a representation of ${}_1F_1(d;\, c;\, z)$ for $0 < z/4k < 1$ {see 4.3.4(2)}. Further

$$\begin{aligned} U(1-d,\, 2-c;\, z) \sim{}& e^{z/2} z^{m-\frac{1}{2}} (\tanh\alpha)^{\frac{1}{2}} \\ &\times \exp\{k - k\ln k - k(2\alpha + \sinh 2\alpha) - (12m^2-1)/24k\}\{A(\alpha, k) + O(k^{-3})\}, \end{aligned} \tag{10}$$

where α is defined by (3) and $P_j(\alpha)$, $j = 1, 2$, is given by (9). For the construction of $T_n(\nu, z)$, we have

$$m = \tfrac{1}{2}(1 - a - \nu), \qquad k = n + 1 - \tfrac{1}{2}(a + \nu) = n + m + \tfrac{1}{2}. \tag{11}$$

Then under the conditions (3), we have

$$T_n(\nu, z) \sim \frac{(-)^{1-a}\, 2\pi e^z z^{1-\nu} \exp[-2k(2\alpha + \sinh 2\alpha)][A(\alpha, k) + O(k^{-3})]}{\left[\begin{array}{l}\Gamma(1-\nu)[A(\alpha, -k) + O(k^{-3}) + \exp\{\pm i\pi(2m + \frac{1}{2})\} \\ \qquad \times \exp\{-2k(2\alpha + \sinh 2\alpha)\}\{A(\alpha, k) + O(k^{-3})\}]\end{array}\right]}, \tag{12}$$

uniformly in z where the upper (lower) sign governs if $0 \leq \arg z < \pi$ $(-\pi < \arg z \leq 0)$.

If ν is fixed and $n \to \infty$, the conditions (3) and $|\arg kz| < \pi$ are the same, and

$$\lim_{n\to\infty} T_n(\nu, z) = 0, \qquad z \neq 0, \qquad |\arg z| < \pi \tag{13}$$

uniformly in z. It should be noted that under these circumstances the term involving $\exp(2m + \frac{1}{2})\, i\pi$ in (12) could be omitted. However we have used (12) as it stands in the development of numerics.

If z is small with respect to n, $z \sim 4k\alpha^2$ and from (12) we easily recover 4.3.2(15).

If n is large but fixed, and $|z|$ is large so that $|z/4k| \gg 1$, then

$$T_n(\nu, z) \sim \frac{(-)^{1-a}\, n^{1-\nu-a}(n!)^2}{\Gamma(1-\nu)\, z^{2n+1-a}}, \qquad |\arg k| < \frac{\pi}{2}, \tag{14}$$

which shows that for n fixed but sufficiently large, the Padé approximation has the same behavior as the asymptotic expansion for $\Gamma(\nu, z)$. Indeed, if $z = x$ with $x > 0$, then the error committed using $2n + 1$ terms of the asymptotic expansion of $z^{1-\nu}e^z\Gamma(\nu, z)$ [see 4.1(10)] is negative provided $\nu < 1$ and does not exceed $x^{-2n-1}(1-\nu)_{2n+1}$ in magnitude. For n sufficiently large, the ratio of the right-hand side of (14), $a = 0$ and sign omitted, to this last number is approximately $2^{-2n-1+\nu}(\pi/n)^{1/2}$. Under these conditions, it follows that for virtually the same number of operations, the Padé approximation is superior to the asymptotic expansion.

To facilitate approximate evaluation of the error, in Tables 4.3 and 4.7, we give values of $|e^{-z}z^{\nu-1}T_n(\nu, z)|$ for $a = 0$, $\nu = 0$, and $\nu = \frac{1}{2}$, respectively, where $T_n(\nu, z)$ is given by the right-hand side of (12) with $a = 0$ and $O(k^{-3})$ omitted. Actually, it is sufficient to have tables for $a = 0$ and $\nu = 0$ only. To see this, let $T_n(\nu, z)$, α, k, and m be defined as above with $a = 0$, and let $R_n(\nu, z)$, α_1, k_1, and m_1 be the corresponding values with $a = 1$. Then

$$R_n(\nu, z) \sim -T_n(\nu, z) \exp\left[2\alpha + \frac{\tanh\alpha}{4k} + \frac{(\tanh\alpha)(2\cosh^2\alpha + 1)}{48k^2\cosh^2\alpha} + O(k^{-3})\right]$$
$$\times \left[1 - \frac{(4m-1)\coth\alpha}{4k} + \frac{(4m-1)^2\coth^2\alpha}{32k^2}\right.$$
$$+ \frac{1}{k^2}\left\{P_1(\alpha) - \frac{(4m-1)\coth\alpha}{8} - \frac{1}{64\cosh^2\alpha}\right.$$
$$\left.\left.\times [(4m-1)(4m-3)\coth\alpha + 2\tanh\alpha - 5\tanh^3\alpha]\right\} + O(k^{-3})\right], \tag{15}$$

where the term involving $\exp[\pm i\pi(2m + \frac{1}{2})]$ in (12) and the like term arising in the expression for $R_n(\nu, z)$ have been omitted. In a similar fashion, we can show that with $a = 0$,

$$\Gamma(1-\nu)\, z^\nu T_n(\nu, z) \sim T_n(0, z) \exp\left[2\alpha_0\nu + \frac{\nu^2\tanh\alpha_0}{4k_0} + O(k_0^{-2})\right]$$
$$\times \left[1 - \frac{\nu(2-\nu)\cosh\alpha_0}{2(k_0 z)^{1/2}} + O(k_0^{-3/2})\right], \tag{16}$$

where α_0 and k_0 are the values of α and k, respectively, when $\nu = 0$ and $a = 0$.

We now present some numerics to illustrate the formulas (6) and (12). In Table A, we compare the exact values (rounded to 5d) of ${}_1F_1(-n; \frac{3}{2}; -z)$ for $n = 5$ and various $z = xe^{i\varphi}$, $x = 4$, with the approximate values given by (6). All data are for $a = 0$. In the notation of 4.3.2(3), the above ${}_1F_1$ is $F_n(\frac{1}{2}, z)$. We should remark that (6) is valid if $|\arg z| = \pi$, provided $|z/4k| < 1$. In this connection, see the remarks following (9).

In Table B we tabulate $\Gamma(\nu, z)$ for $\nu = 0, \frac{1}{2}$, its rational approximation, 4.3.2(2), the exact error, and the approximate error according to (12). The data are developed for $z = 2e^{i\varphi}$ and $n = 4$; also, $a = 0$. If $z = 2$, the ratio of $z^{-1/2}e^{-z}T_4(\frac{1}{2}, z)$ to $z^{-1}e^{-z}T_4(0, z)$ is 0.685, while the corresponding number deduced from (12) is 0.677.

TABLE A

$z = 4e^{i\varphi}$	$F_5(\frac{1}{2}, z)$	
φ	Eq. (6)	Exact
0	130.58725	130.58615
$\pi/4$	$-47.95306 + 84.34664i$	$-47.95274 + 84.34563i$
$\pi/2$	$-19.99510 - 32.27692i$	$-19.99471 - 32.27629i$
$3\pi/4$	$6.60906 - 0.98754i$	$6.60883 - 0.98770i$
π	0.09123	0.09110

TABLE B

$z = 2e^{i\varphi}$, φ	$\int_z^\infty t^{-1}e^{-t}\,dt$	$z^{-1}e^{-z}E_4(0, z)/F_4(0, z)$
0	$4.89005 \cdot 10^{-2}$	$4.89190 \cdot 10^{-2}$
$\pi/4$	$-3.95846 \cdot 10^{-2} - 8.22921i \cdot 10^{-2}$	$-3.95652 \cdot 10^{-2} - 8.22414i \cdot 10^{-2}$
$\pi/2$	$-4.22981 \cdot 10^{-1} + 3.46167i \cdot 10^{-2}$	$-4.23980 \cdot 10^{-1} + 3.43682i \cdot 10^{-2}$
$3\pi/4$	$-2.16947 + 0.31777i$	$-2.12827 + 0.36462i$

$z = 2e^{i\varphi}$	$z^{-1}e^{-z}T_4(0, z)$	
φ	Exact	Approximate
0	$-0.185 \cdot 10^{-4}$	$-0.184 \cdot 10^{-4}$
$\pi/4$	$-0.194 \cdot 10^{-4} - 0.507i \cdot 10^{-4}$	$-0.194 \cdot 10^{-4} - 0.507i \cdot 10^{-4}$
$\pi/2$	$0.999 \cdot 10^{-3} + 0.248i \cdot 10^{-3}$	$0.996 \cdot 10^{-3} + 0.249i \cdot 10^{-3}$
$3\pi/4$	$-0.412 \cdot 10^{-1} - 0.468i \cdot 10^{-1}$	$-0.412 \cdot 10^{-1} - 0.473i \cdot 10^{-1}$

$z = 2e^{i\varphi}$, φ	$\int_z^\infty t^{-1/2}e^{-t}\,dt$	$z^{-1/2}e^{-z}E_4(\frac{1}{2}, z)/F_4(\frac{1}{2}, z)$
0	$8.06471 \cdot 10^{-2}$	$8.06597 \cdot 10^{-2}$
$\pi/4$	$-2.03962 \cdot 10^{-2} - 0.146768i$	$-2.03888 \cdot 10^{-2} - 0.146732i$
$\pi/2$	$-0.560363 - 0.337570i$	$-0.560936 - 0.337918i$
$3\pi/4$	$-2.14095 - 2.19684i$	$-2.12898 - 2.16019i$

$z = 2e^{i\varphi}$	$z^{-1/2}e^{-z}T_4(\frac{1}{2}, z)$	
φ	Exact	Approximate
0	$-0.126 \cdot 10^{-4}$	$-0.126 \cdot 10^{-4}$
$\pi/4$	$-0.074 \cdot 10^{-4} - 0.36i \cdot 10^{-4}$	$-0.074 \cdot 10^{-4} - 0.36i \cdot 10^{-4}$
$\pi/2$	$0.573 \cdot 10^{-3} + 0.348i \cdot 10^{-3}$	$0.572 \cdot 10^{-3} + 0.348i \cdot 10^{-3}$
$3\pi/4$	$-0.120 \cdot 10^{-1} - 0.366i \cdot 10^{-1}$	$-0.119 \cdot 10^{-1} - 0.369i \cdot 10^{-1}$

4.3.4. The Negative Real Axis and the Zeros of $F_n(\nu, z)$

In this and the next section, we give uniform asymptotic formulas for the error and other data in regions which include the negative real axis. It is convenient to present formulas for $T_n(\nu, ze^{\pm i\pi})$ so that $z > 0$ now describes the negative real axis in the sense of our previous discussion. The results lead to useful approximations for the zeros of $F_n(\nu, z)$. Obviously, the rational approximations cannot be used in the vicinity of these zeros. Indeed, we have shown convergence requires, to use our previous notation, that $|\arg z/k| < \pi$. However, if $|\arg z/k| = \pi$, $|z/k| \geqslant \frac{3}{4}$ approximately, and n is fixed but large, we prove in 4.3.5 that $\lim_{z\to\infty} z^{2n} T_n(\nu, -z) = 0$. That is, the rational approximations for $\Gamma(\nu, -x)$, $x > 0$, behave like the asymptotic series for $\Gamma(\nu, -x)$.

We first write an expansion for ${}_1F_1(d; c; z)$ for $0 < z/4k < 1$. The result follows from 4.3.3(2-9) if there we replace α by $\mp i(\alpha - \pi/2)$ and z by $ze^{\pm i\pi}$. Then with

$$z/4k = \cos^2\alpha, \tag{1}$$

$$\begin{gathered}{}_1F_1(d; c; z) \sim e^{z/2} z^{-m-\frac{1}{2}}[(2/\pi)\cot\alpha]^{1/2} k^{-m}\Gamma(2m+1)\exp[m(4m^2-1)/24k^2] \\ \times [\{1 - S_2(\alpha)k^{-2} + O(k^{-4})\}\sin\omega + \{S_1(\alpha)k^{-1} + O(k^{-3})\}\cos\omega], \\ 0 < z/4k < 1,\end{gathered} \tag{2}$$

$$\omega = k(\sin 2\alpha - 2\alpha + \pi) - m\pi + \pi/4, \tag{3}$$

$$S_1(\alpha) = (96\tan^3\alpha)^{-1}\{3(16m^2-1)\tan^4\alpha + 6\tan^2\alpha + 5\}, \tag{4}$$

$$\begin{aligned}S_2(\alpha) = {} & (18432\tan^6\alpha)^{-1}\{9(16m^2-9)(16m^2-1)\tan^8\alpha - 36(48m^2-7)\tan^6\alpha \\ & + 6(121 - 112m^2)\tan^4\alpha + 924\tan^2\alpha + 385\}.\end{aligned}$$

This result is actually valid for all z, $|\arg z| < \pi$, with the proviso that we omit the origin and the line $1 \leqslant z/4k < \infty$. Here it is understood that α is real and positive if $z/4k$ is real and positive. Thus,

$$\cos^2\alpha = z/4k = \rho e^{i\theta}, \qquad 0 \leqslant \rho < 1, \quad |\theta| < \pi,$$

$$\alpha = \beta - i\delta, \quad \beta \text{ and } \delta \text{ real}, \quad \beta > 0, \tag{5}$$

$$\cos\beta = (\tfrac{1}{2}[1 + \rho - \{(1+\rho)^2 - 4\rho\cos^2\tfrac{1}{2}\theta\}^{1/2}])^{1/2}, \tag{6}$$

$$\cos\beta\cosh\delta = \rho^{1/2}\cos\tfrac{1}{2}\theta, \qquad \sin\beta\sinh\delta = \rho^{1/2}\sin\tfrac{1}{2}\theta. \tag{7}$$

If $0 < z/4k < 1$, the approximate zeros of ${}_1F_1(d; c; z)$ may be found from (2) by solving the transcendental equation

$$(1 - S_2(\alpha)\, k^{-2}) \sin \omega + S_1(\alpha)\, k^{-1} \cos \omega = 0 \tag{8}$$

for α, whence z is recovered by using (1). If k is sufficiently large, $\sin \omega = 0$ and ω is a multiple of π. An approximate closed form solution of (8) is not readily apparent. Luke (1969) has shown that if x_r is the r-th positive zero of ${}_1F_1(d; c; z)$ and $j_{2m,r}$ is the r-th positive zero of $J_{2m}(x)$, then with m and k as in 4.3.3(7),

$$x_r = \frac{j_{2m,r}^2}{4k}\left\{1 + \frac{2(4m^2 - 1) + j_{2m,r}^2}{48k^2}\right\} + O(k^{-4}),$$

$$j_{2m,r} = \pi(r + m - \tfrac{1}{4}) - [(16m^2 - 1)/8\pi(r + m - \tfrac{1}{4})] + O(r^{-2}), \tag{9}$$

and for n sufficiently large, the largest root of $F_n(\nu, -x)$ does not exceed $3k$.

To illustrate the above results, first suppose that $n = 5$ and $\nu = \frac{1}{2}$. In Table C we give the exact zeros x_r of $F_5(\frac{1}{2}, -x)$ and the approximate zeros as determined by (9). Also tabulated are the values of $\omega = \omega_r$ as given by (3) for the exact zeros x_r. All data are for $a = 0$. Note that ω_r is approximately $r\pi$ as expected. Also, the error of the approximate x_r increases as r increases. Even so, the values are sufficiently accurate for numerical work. More accurate approximate values for the two largest roots can be readily found as follows. Assune that the first $(r - 2)$ roots are known with sufficient accuracy. Since the exact sum and product of the roots are known a priori, approximate values

TABLE C

	x_r		
r	Exact	Eq. (9)	ω_r
1	0.43140	0.43138	3.14007
2	1.75975	1.75753	6.27986
3	4.10446	4.07453	9.41884
4	7.74671	7.54250	12.5556
5	13.45768	12.38561	15.6818

of the sum and product of the two largest roots are readily obtained, and approximate values of the two largest roots follow by solution of a quadratic. To illustrate, the sum and product of the first three x_r values in the third column of Table C are 6.26344 and 3.08913, respectively. The exact sum and product of the roots are 55/2 and 10395/32, respectively, whence the approximate sum and product of the two largest roots are 21.23656 and 105.15696, respectively. Solution of a quadratic leads to the improved approximate roots 7.86311 and 13.37345.

For the sake of completeness, we give the asymptotic representation for $T_n(\nu, ze^{\pm i\pi})$ when $0 < z/4k < 1$. The conditions of validity are the same as for (2) (see also the remarks following (4)). We have

$$T_n(\nu, ze^{\pm i\pi})$$
$$\sim -\frac{\pi e^{-z}z^{1-\nu}}{\Gamma(1-\nu)[\{1-S_2(\alpha)k^{-2}+O(k^{-4})\}\sin\omega+\{S_1(\alpha)k^{-1}+O(k^{-3})\}\cos\omega]}$$
$$\times[\{1-S_2(\alpha)k^{-2}+O(k^{-4})\}\cos\omega-\{S_1(\alpha)k^{-1}+O(k^{-3})\}\sin\omega$$
$$\mp i\{1-S_2(\alpha)k^{-2}+O(k^{-4})\}\sin\omega\mp i\{S_1(\alpha)k^{-1}+O(k^{-3})\}\cos\omega].$$

4.3.5. The Error $T_n(\nu, ze^{\pm i\pi})$ in the Transition Region and in the Region $|z/4k| > 1$

Next we consider the representation for $T_n(\nu, ze^{\pm i\pi})$ when $z/4k \sim 1$. Let

$$z - 4k = 2\epsilon = O[(z/6)^{1/3}]. \tag{1}$$

Then

$$T_n(\nu, ze^{\pm i\pi}) \sim \frac{3^{1/2}\pi e^{-z}z^{1-\nu}(f+g)}{\Gamma(1-\nu)(f-g)}\left\{1 \pm \frac{i(f-g)}{3^{1/2}}\right\}, \qquad f = 1 + \frac{\epsilon(\epsilon^2+\frac{1}{5})}{3z},$$

$$g = \frac{\Gamma(\frac{5}{6})}{\pi^{1/2}(z/3)^{1/3}}\left\{\epsilon + \frac{(\epsilon^4+\epsilon^2+12m^2-33/35)}{6z} + O\left[\frac{\epsilon^6}{(z/3)^2}\right]\right\},$$
$$|\arg z| \leqslant \frac{\pi}{2}. \tag{2}$$

If $z = x$ is real and positive, we take

$$2T_n(\nu, -x) = T_n(\nu, xe^{i\pi}) + T_n(xe^{-i\pi}).$$

Thus,

$$T_n(\nu, -x) \sim \frac{3^{1/2}\pi e^{-x}x^{1-\nu}(f+g)}{\Gamma(1-\nu)(f-g)}. \tag{3}$$

Finally, we give the asymptotic formula for $T_n(\nu, ze^{\pm i\pi})$ when $|z/4k| > 1$. Write

$$z/4k = \cosh^2 \alpha. \tag{4}$$

Then

$$T_n(\nu, ze^{\pm i\pi}) \sim \frac{2\pi e^{-z}z^{1-\nu}\exp[2k(\sinh 2\alpha - 2\alpha)]}{\Gamma(1-\nu)}$$
$$\times \frac{[1+Q_1(\alpha)k^{-1}+Q_2(\alpha)k^{-2}+O(k^{-3})]}{[1-Q_1(\alpha)k^{-1}+Q_2(\alpha)k^{-2}+O(k^{-3})]},$$
$$k\to\infty, \qquad |z/4k| > 1, \qquad |\arg z/k| \leqslant \pi/2. \tag{5}$$

Here, $Q_j(\alpha) = P_j(\alpha + i\pi/2)$, $j = 1, 2$. That is, $Q_j(\alpha)$ is of the same form as $P_j(\alpha)$ [see 4.3.3(9)] if in the latter, $\coth \alpha$ is replaced by $\tanh \alpha$. When z and k are complex, for the determination of α, we choose the branch such that α is real and positive when z and k are real and positive. Specifically,

$$z/4k = \cosh^2\alpha = \eta e^{i\tau}, \qquad \eta > 0, \quad \alpha = \mu + i\xi, \quad \mu \text{ and } \xi \text{ real}, \quad \mu > 0, \tag{6}$$

$$\cosh \mu = (\tfrac{1}{2}[\eta + 1 + \{(\eta-1)^2 + 4\eta \sin^2 \tfrac{1}{2}\tau\}^{1/2}])^{1/2}, \tag{7}$$

$$\cos\xi \cosh\mu = \eta^{1/2}\cos\tau/2, \qquad \sin\xi\sinh\mu = \eta^{1/2}\sin\tfrac{1}{2}\tau. \tag{8}$$

Now suppose $|z/4k| \gg 1$. Then

$$T_n(\nu, ze^{\pm i\pi}) \sim [n^{1-\nu-a}(n!)^2]/[\Gamma(1-\nu)\, z^{2n+1-a}], \qquad |\arg k| < \pi/2, \tag{9}$$

and so,

$$\lim_{z\to\infty} z^{2n}T_n(\nu, -z) = 0. \tag{10}$$

Thus, if $z = x$, $x > 0$, and we fix the order of the rational approximate, then the accuracy of the approximation increases as x increases, and thus the rational approximations and asymptotic expansion for $\Gamma(\nu, -x)$ [see 4.1(10)] exhibit the same behavior. However, no efficient estimate is available to assess the error when using the truncated asymptotic expansion. If we truncate the asymptotic expansion for $x^{1-\nu}e^{-x}\Gamma(\nu, -x)$ after $2n+1$ terms, then the error certainly exceeds $x^{-2n-1}(1-\nu)_{2n+1}$. The ratio of the right-hand side of the second expression in (9), for $a = 0$, to this last quantity is $2^{-2n-1+\nu}(\pi/n)^{1/2}$ for n sufficiently large, and the superiority of the Padé approximations is manifest.

TABLE D[a]

		$T_2(0, -x)$			$T_4(0, -x)$	
x	$E_2(0, -x)/F_2(0, -x)$	True	Approx.	$E_4(0, -x)/F_4(0, -x)$	True	Approx.
9	1.15151 515(0)	0.124(− 2)	0.116(− 2)[b]			
10	1.13043 478(0)	0.104(− 2)	0.104(− 2)[b]			
12	1.10256 410(0)	0.410(− 3)	0.410(− 3)[b]			
14	1.08474 576(0)	0.146(− 3)	0.145(− 3)[b]			
15	1.07801 418(0)	0.889(− 4)	0.947(− 4)[c]	1.07810 3837(0)	− 0.729(− 6)	− 0.445(− 6)[b]
16	1.07228 916(0)	0.558(− 4)	0.571(− 4)[c]	1.07234 3632(0)	0.134(− 5)	0.131(− 5)[b]
18	1.06306 306(0)	0.239(− 4)	0.242(− 4)[c]	1.06308 6548(0)	0.730(− 6)	0.732(− 6)[b]
20	1.05594 406(0)	0.118(− 4)	0.117(− 4)[c]	1.05595 5678(0)	0.227(− 6)	0.227(− 6)[b]
25				1.04366 1924(0)	0.12(− 7)	0.12(− 7)[b, c]

[a] All data are for $a = 0$. The numbers in parentheses following base numbers indicate the powers of 10 by which base numbers must be multiplied.

[b] Approximate formula (2).

[c] Approximate formula (5).

The statement analogous to 4.3.3(15) is

$$
\begin{aligned}
&R_n(\nu, ze^{\pm i\pi}) \\
&\sim T_n(\nu, ze^{\pm i\pi}) \exp\left[2\alpha + \frac{\coth\alpha}{4k} + \frac{(\coth\alpha)(2\sinh^2\alpha - 1)}{48k^2\sinh^2\alpha} + O(k^{-3})\right] \\
&\times \left[1 - \frac{(4m-1)\tanh\alpha}{4k} + \frac{(4m-1)^2\tanh^2\alpha}{32k^2} + \frac{1}{k^2}\left\{Q_1(\alpha) - \frac{(4m-1)\tanh\alpha}{8}\right.\right. \\
&\left.\left. + \frac{1}{64\sinh^2\alpha}[(4m-1)(4m-3)\tanh\alpha + 2\coth\alpha - 5\coth^3\alpha]\right\} + O(k^{-3})\right].
\end{aligned} \tag{11}
$$

We conclude this section with some numerical examples to illustrate the application of (2) and (5). Table D gives for various values of x and $n = 2, 4$ the rational approximate to $xe^{-x}\,\mathrm{Ei}(x)$, $T_n(0, -x)$ true, and $T_n(0, -x)$ approximate.

For a final example, Padé approximations to $e^{-x^2}\,\mathrm{Erfi}(x)$ are tabulated in Table E for several values of x. Also given are the true errors and the approximate errors as deduced from (2), (5) and (11) as appropriate.

TABLE E[a]

x	$(2x)^{-1}E_2(\frac{1}{2}, -x^2)/F_2(\frac{1}{2}, -x^2)$	Error	Approx. error
	$a = 0$		
3	0.17808 219(0)	0.740(− 4)	0.741(− 4)[b]
4	0.12934 631(0)	0.169(− 5)	0.158(− 5)[c]
5	0.10213 40744(0)	0.794(− 7)	0.793(− 7)[c]
	$a = 1$		
3	0.17808 219(0)	0.189(− 3)	0.189(− 3)[b]
4	0.12934 132(0)	0.668(− 5)	0.607(− 5)[d]
5	0.10213 34544(0)	0.620(− 6)	0.619(− 6)[d]

[a] Numbers in parentheses following base numbers indicate the powers of 10 by which base numbers must be multiplied.

[b] Deduced from formula (2).

[c] Deduced from formula (5).

[d] Deduced from formula (11).

4.4. Inequalities

4.4.1. $H(\nu,z)$

In 4.3.1(2) let $A_n(\nu,z,a)$ stand for $A_n(\nu,z)$, etc. Then

$$\frac{A_n(\nu,z,1)}{B_n(\nu,z,1)} >(<)\ H(\nu,z)\ <(>)\ \frac{A_m(\nu,z,0)}{B_m(\nu,z,0)} ,$$

$$z > 0, \quad \nu > -1, \quad m > n > 0, \tag{1}$$

where >(<) sign pertains if both m and n are odd (even). Further, if $z > 0$, $\nu < 0$, but ν is not a negative integer, $m + \nu + 1 - a > 0$, $n + \nu + 1 - a > 0$, and $r < -\nu < r+1$ where r is a positive integer or zero, then (1) holds where >(<) pertains if both $r + n$ and $r + m$ are odd (even). We have equality if $z = 0$.

If $\nu \geq 0$ and $z = -x$, $x > 0$, then from 4.3.1(6) $P_n(\nu, -x)$ is positive (negative) if n is even (odd). Thus additional inequalities can be written once the sign of $B_n(\nu, -x)$ is determined. It is known that if ν and x are fixed and restricted as above, $B_n(\nu, -x)$ is positive provided n is sufficiently large.

If ν is negative but not a negative integer, further inequalities follow from (1) and the relation

$$H(\nu,z) = \sum_{k=0}^{s-1} \frac{(-)^k z^k}{(\nu+1)_k} + \frac{(-)^s z^s}{(\nu+1)_s} H(\nu+s,z) . \tag{2}$$

For example, for $n = 1$ and 2, respectively, we have

$$\frac{(\nu+1)(\nu+2)-z}{(\nu+1)[(\nu+2)+z]} < \nu z^{-\nu} e^{-z} \int_0^z t^{\nu-1} e^t\, dt < \frac{\nu+1}{\nu+1+z}, \quad z > 0,\ \nu \geqslant 0, \tag{3}$$

$$\frac{(\nu+2)[(\nu+1)(\nu+3)+(\nu-1)z]}{(\nu+1)[(\nu+2)(\nu+3)+2(\nu+2)z+z^2]} < \nu z^{-\nu} e^{-z} \int_0^z t^{\nu-1} e^t\, dt$$

$$< \frac{(\nu+1)_4 + (\nu-2)(\nu+2)_2 z + 2z^2}{(\nu+1)_2 [(\nu+3)_2 + 2(\nu+3)z + z^2]}, \qquad z > 0,\ \nu \geqslant 0. \tag{4}$$

When $\nu = \frac{1}{2}$, from (3) and (4) respectively,

$$\frac{15 - 4z^2}{15 + 6z^2} < z^{-1}e^{-z^2} \int_0^z e^{t^2}\,dt < \frac{3}{3 + 2z^2}, \qquad z > 0, \quad (5)$$

$$\frac{105 - 10z^2}{105 + 60z^2 + 12z^4} < z^{-1}e^{-z^2} \int_0^z e^{t^2}\,dt < \frac{945 - 210z^2 + 32z^4}{945 + 420z^2 + 60z^4}, \quad z > 0. \quad (6)$$

In illustration, for $z = 1$, we get

$$0.52381 < 0.53808 < 0.6, \quad (7)$$

$$0.53672 < 0.53808 < 0.53825. \quad (8)$$

4.4.2. $\Gamma(\nu, z)$

In the notation of 4.3.2(2) let $E_n(\nu,z,a)$ stand for $E_n(\nu,z)$, etc. Then

$$\frac{E_n(\nu,z,1)}{F_n(\nu,z,1)} < z^{1-\nu}e^z \int_z^\infty t^{\nu-1}e^{-t}dt < \frac{E_n(\nu,z,0)}{F_n(\nu,z,0)}, \quad (1)$$

and equality obtains if $z \to \infty$. In particular, for $n = 1$ and 2, respectively, we have

$$\frac{z}{z + 1 - \nu} < z^{1-\nu}e^z \int_z^\infty t^{\nu-1}e^{-t}\,dt < \frac{z + 1}{z + 2 - \nu}, \quad (2)$$

$$\frac{z(z + 3 - \nu)}{z^2 + 2(2 - \nu)z + (1 - \nu)(2 - \nu)} < z^{1-\nu}e^z \int_z^\infty t^{\nu-1}e^{-t}\,dt$$

$$< \frac{z^2 + (5 - \nu)z + 2}{z^2 + 2(3 - \nu)z + (2 - \nu)(3 - \nu)}, \quad (3)$$

where z and ν are as in (1). Thus for $\nu = 0$, we have

$$\frac{z}{z + 1} < ze^z \int_z^\infty t^{-1}e^{-t}\,dt < \frac{z + 1}{z + 2}, \qquad 0 < z < \infty, \quad (4)$$

$$\frac{z(z + 3)}{z^2 + 4z + 2} < ze^z \int_z^\infty t^{-1}e^{-t}\,dt < \frac{z^2 + 5z + 2}{z^2 + 6z + 6}, \qquad 0 < z < \infty. \quad (5)$$

If $z = 2$, (5) gives

$$0.71429 < 0.72266 < 0.72727. \tag{6}$$

Gautschi (1959a) proved that

$$\tfrac{1}{2}\left((z^p+2)^{1/p} - z\right) < \exp(z^p)\int_z^\infty \exp(-t^p)dt \le c_p\left((z^p+c_p^{-1})^{1/p} - z\right),$$

$$c_p = \left\{\Gamma(1 + 1/p)\right\}^{p/(p-1)}, \quad 0 \le z < \infty. \tag{7}$$

4.5. Notes on the Computation of the Incomplete Gamma Function

The following discussion is designed to facilitate effective use of the tables provided herein to evaluate the incomplete gamma function throughout the complex z and ν planes, each suitably cut.

For the common special cases of the incomplete gamma functions, the given Chebyshev expansions afford the best method of computation. Since $|T_n^*(x)| \le 1$ for $0 \le x \le 1$, it is easy to estimate the error when these series are truncated. One merely sums the absolute values of the coefficients truncated. Further, the Chebyshev series are easily summed by the nesting procedure described in 11.8. For the values of ν not covered in the tables of this chapter, one can develop Chebyshev coefficients by appropriate use of recurrence relations in the backward direction.

For general ν and z, the rational and Padé approximations are superb for moderate values of ν. In fact, from 4.3.1, the Padé approximations for $H(\nu,z)$ mimic the behavior of this function for $|\nu|$ large, ν not a negative integer. Thus large values of $|\nu|$ as above are easily tolerated in 4.3.1. However, it is clear from 4.3.1(12) that for a given n, the error decreases as $R(\nu) \to -\infty$, ν not a negative integer.

It can be deduced from Gautschi (1967) that $\gamma(\nu,z)$ is a minimal solution of 4.2(9) whence the latter can be used in the backward direction. In this instance, we need two starting values which are readily obtained from the representations given in 4.1, 4.3, etc. as appropriate. Instead of using 4.2(9) because of 4.2(1), it would seem that 4.2(7) could be safely used in the backward direction to evaluate $\gamma(\nu,z)$. A simple examination of 4.2(7) shows that round off errors will certainly be damped provided $|z/(\nu+1)| < 1$. If $|z/(\nu+1)| > 1$, then certainly round off errors will increase, but the relative error in the desired solution will have little if any effect provided $g(\nu,z)$ is increasing. In 4.2(7), replace ν by $\nu+m-1$ where ν is fixed and m is an integer. Consider

$$\alpha(s,t) = \rho_t/\rho_s, \quad \rho_m = \left|\frac{{}_1F_1(1;\ \nu+1;\ z)\Gamma(m+\nu+1)}{{}_1F_1(1;\ m+\nu+1;\ z)z^m\Gamma(\nu+1)}\right|. \tag{1}$$

Then Gautschi (1972) shows that in going from stage s to stage t, the error is amplified if $\rho_t > \rho_s$ and damped if $\rho_t < \rho_s$. The recurrence relation 3.2(7) is said to be stable for the desired solution ${}_1F_1(1;\ \nu+1;\ z)$ if there is a positive constant C such that for $s < t$, $\sup \alpha(s,t) \leq C < \infty$. Otherwise, it is unstable. Ideally one should apply the recursion formula 4.2(7) in the direction of decreasing ρ_m. A complete analysis of (1) is wanting as it requires uniform asymptotic estimates of ${}_1F_1(1;\ \nu+1;\ z)$ for both ν and z large.

Next consider $\Gamma(\nu,z)$. For effective use of the Padé approximants, we want $|k|$ large as is seen from 4.3.2(15). Thus increasing $R(\nu)$ demands increasing n to maintain the same level of accuracy. For $|\nu|$ large, $R(\nu) < 0$, the Padé approximants are very useful in appropriate sectors of the complex plane, see 4.3.2. Clearly $\Gamma(\nu,z)$ is not the minimal solution of 4.2(9) and this should be used for $\Gamma(\nu,z)$ only in the forward direction if at all. In view of 4.2(2), we can possibly use 4.2(8) to evaluate $G(\nu,z)$. In 4.2(8), replace ν by $\nu+m-1$ where ν is fixed and m is an integer. In this event, we have a situation analogous to (1) with

$$\rho_m = \left|\frac{\Gamma(\nu,z)\Gamma(m+\nu)}{\Gamma(m+\nu,z)\Gamma(\nu)}\right|. \tag{2}$$

Again, ideally one should apply 4.2(8) in the direction of decreasing ρ_m. A complete analysis of (2) is not available owing primarily to lack of uniform asymptotic estimates of $\Gamma(\nu,z)$ for large ν and z.

Some valuable insights can be gained by examining a special case previously considered by Gautschi (1972). Suppose we desire to compute the exponential integrals

$$E_n(x) = \int_1^\infty t^{-n}e^{-xt}dt, \quad n = 1,2,\cdots, \quad x > 0, \tag{3}$$

$$E_n(x) = x^{n-1}\Gamma(1-n,x), \tag{4}$$

by use of the recursion formula

$$E_{n+1}(x) = \left(e^{-x} - xE_n(x)\right)/n. \tag{5}$$

Then ideally this recursion formula should be applied in the direction in which

$$\sigma_n = \frac{x^n\, E_1(x)}{n!\; E_{n+1}(x)} \tag{6}$$

decreases. If $\sigma_1 \leq 1$, then $x \leq 0.61006\cdots$, and it can be shown that σ_n decreases on the set of positive integers from 1 to 0 so that (5) is stable in the forward direction. If $x > 0.61001\cdots$, then σ_n first increases, and then after reaching a maximum at a point $n_0 = [x]$, tends monotonically decreasing to zero. Gautschi (1972), presents some graphs of σ_n for $x = 0.1, 0.5, 2, 5, 10, 15, 20$. If an error amplification of σ_{n_0} can be tolerated, then (5) can be used in the forward direction for all $n = 1, 2, 3,\cdots$. If not, one should generate E_n by backward recursion for $0 \leq n \leq n_0$ and by forward recursion for $n > n_0$. Here the initial value E_{n_0} must be evaluated by some other method.

For another illustration, consider evaluation of

$$\Gamma(m+1,z) = m!e^{-z}e_m(z) = z^m e^{-z} G(m+1,z),$$

$$e_m(z) = \sum_{k=0}^{m} z^k/k! \tag{7}$$

by use of 4.2(8) with $v = m$. In this instance, from (2) we get $\rho_m = \left(|e_m(z)|\right)^{-1}$ whence the recurrence formula 4.2(8) can be safely used in the forward direction if $R(z) \geq 0$. Gautschi (1972) presents graphs of ρ_m for $z < 0$ which show application of 4.2(8) in the forward direction is useful on the interval $1 \leq m \leq [-z]$, but only conditionally useful on the interval $[-z] \leq m \leq \infty$.

If values of $\Gamma(\nu)$ are available, it is natural to ask whether one should compute $\gamma(\nu,z)$ or $\Gamma(\nu,z)$ with the aid of related Padé approximants, see 4.3, in view of 4.1(2). For small to moderately large values of z, z fixed, $|\arg z| < \pi$, it is usually best to compute $\gamma(\nu,z)$ with the aid of the Padé approximants for $H(\nu,z)$ developed in 4.3.1 rather than by use of the Padé approxiamtions in 4.3.2. Indeed, the approximations for $H(\nu,z)$ hold for unrestricted z and ν, ν not a negative integer, while those for $\Gamma(\nu,z)$ work best for ν moderate, $|z|$ large and suitably restricted as in 4.3.2(15). From 4.3.1(11), $V_n(\nu,z)$ decreases rather rapidly as n increases because

$$\frac{V_{n+1}(\nu,z)}{V_n(\nu,z)} = -\frac{z^2(n+1)(n+\nu+1-a)}{(2n+\nu+1-a)(2n+\nu+2-a)^2(2n+\nu+3-a)} \times \exp\left[-\frac{z(z-4\nu+4a)}{2(2n+\nu+1-a)(2n+\nu+3-a)}\right]\left(1+O(n^{-3})\right), \tag{8}$$

while under the same conditions $T_n(\nu,z)$, see 4.3.2(15), decreases quite slowly because

$$\frac{T_{n+1}(\nu,z)}{T_n(\nu,z)} = \exp\left(-2(z/k)^{\frac{1}{2}}\right)\left(1+O(k^{-\frac{1}{2}})\right). \tag{9}$$

Indeed we have used the Padé approximations for both forms of the incomplete gamma function to evaluate $\Gamma(\nu)$ since

$$\Gamma(\nu) = \gamma(\nu,z) + \Gamma(\nu,z), \tag{10}$$

see Luke (1970b). Here z is free, $|\arg z| < \pi$, and we found $z = 8$ a very convenient value. The latter reference contains some tables (not given here) to further facilitate estimation of the error in the Padé approximations.

To complete our discussion, we need to consider the situation when ν is moderate and for convenience say ν is real, $|z|$ is large and z lies in a sector including $\arg z = \pi$, say $|\arg ze^{-i\pi}| < \theta$, $\theta = \pi/4$. Here we propose to use Taylor series expansions to compute $\Gamma(\nu,z+h)$ based on $\Gamma(\nu,z)$ and its derivatives. To evaluate $\Gamma(\nu,z)$, compute $\gamma(\nu,z)$ as previously discussed whence $\Gamma(\nu,z) = \Gamma(\nu) - \gamma(\nu,z)$, $\nu \neq 0$. For $\nu = 0$, use 4.6.1(1,8) or 4.6.1(2,9) and the approximations given in 4.6 as appropriate. Alternatively, when appropriate, we can evaluate $\Gamma(\nu,ze^{-i\pi})$ by use of a Padé approximant for n fixed and $|z|$ large, $|\arg ze^{-i\pi}| < \theta$.

The following discussion on Taylor series is valid for all ν, except as noted. The case $\nu = \frac{1}{2}$, in which event $\Gamma(\frac{1}{2},z)$ is related to the error function, is especially singular in view of 4.8.1(27) and further analysis for $\nu = \frac{1}{2}$ is deferred to the end of this section.

For the Taylor series expansion, we have

$$\Gamma(\nu,z+h) = \Gamma(\nu,z) + \sum_{r=0}^{N-1} \frac{h^{r+1}}{(r+1)!} D^{r+1}\Gamma(\nu,z) + R_N(z), \quad D = d/dz, \tag{11}$$

$$D^{r+1}\Gamma(\nu,z) = (-)^{r+1} z^{\nu-1-r} (1-\nu)_r e^{-z} {}_1F_1(-r;\ \nu-r;\ z) \tag{12}$$

$$= (-)^{r+1} z^{\nu-1} e^{-z} {}_2F_0(-r,\ 1-\nu;\ -1/z) \tag{13}$$

$$= (-)^{r+1} z^{\nu-1-r} (1-\nu)_r {}_1F_1(\nu;\ \nu-r;\ -z). \tag{14}$$

Here (13) is valid for unrestricted ν, and (12) and (14) are valid for all ν except $\nu = m + 1$ and $r \geq m + 1$ and except for $\nu = 1 - m$, $m > 1$, where m is a positive integer or zero. The case $\nu = m + 1$ is easily disposed of since $\Gamma(m+1,z)$ is given by (7). We also have the relation

$$e^{-z} e_m(z) = 1 - \frac{z^{m+1} e^{-z}}{(m+1)!} {}_1F_1(1;\ m+1;\ z), \tag{15}$$

and the latter ${}_1F_1$ can be readily evaluated by its series expansion or by the Padé approximations of 4.3.1. For the other two exceptions just noted, we have

$$D^{r+1}\Gamma(1-m,z) = \frac{(-)^{r+1} e^{-z} z^{-m-r} (m+r-1)!}{(m-1)!} {}_1F_1^r(-r;\ 1-m-r;\ z), \quad m > 1, \tag{16}$$

$$D^{r+1}\Gamma(m+1,z) = \frac{(-)^{r+1-m} r!\, e^{-z}}{(r-m)!} {}_1F_1(-m;\ r+1-m;\ z), \quad r \geq m, \tag{17}$$

$$= \frac{(-)^{r+1-m} r!}{(r-m)!} {}_1F_1(r+1;\ r+1-m;\ -z), \quad r \geq m. \tag{18}$$

It is easy to show that the Taylor series converges for all h when $\nu = m + 1$ and for $|h/z| < 1$ when ν is otherwise.

From (14) and 7.3.2(2), we get that $D^r\Gamma(\nu,z)$ satisfies the recurrence relation

$$zq_{r+2} + (r + 1 + z - \nu)q_{r+1} + rq_r = 0. \tag{19}$$

It can be shown that $D^r\Gamma(\nu,z)$ is a minimal solution of (19) if $\nu = m + 1$, $r \geq m$, and is not a minimal solution of (19) provided ν is not a positive integer. We omit further consideration of the case $\nu = m + 1$. Then if ν is not a positive integer, (19) cannot be used in the backward direction, but can safely be used in the forward direction in view of a theorem by Wimp (1972). The point is that the relative error involved in calculating $D^r\Gamma(\nu,z)$ by use of (19) in the forward direction will be small if the initial errors in $\Gamma(\nu,z)$ and $D\Gamma(\nu,z)$ are small and if the relative errors in $D^s\Gamma(\nu,z)$ are small, $s < r$.

The idea of using Taylor series to compute $Ei(\pm x)$ has been examined by Didry and Guy (1962). A related notion for the computation of Fresnel integrals has been studied by Fleckner (1968).

As remarked earlier, the error function is a singular case, for if

$$y(z) = e^{z^2}\operatorname{erfc}(z), \tag{20}$$

then by Taylor series and 4.8.1(26),

$$y(z+h) = \sum_{k=0}^{\infty} (-)^k (2h)^k i^k\operatorname{erfc}(z), \tag{21}$$

and this converges for all h in view of 4.8.1(29). Also $i^n\operatorname{erfc}(z)$ is the minimal solution of 4.8.1(27), and the latter together with the normalization relation $i^{-1}\operatorname{erfc}(z) = 2e^{-z^2}/\pi^{\frac{1}{2}}$ can be used in the backward direction to compute the necessary values of $i^k\operatorname{erfc}(z)$. A more detailed analysis of this entire schema has been given by Gautschi (1970a).

Finally, the trapezoidal rule approximations for the error function outlined in 4.8.4 provide a universal scheme for the evaluation of this function throughout the complex plane.

4.6. Exponential Integrals

4.6.1. Relation to Incomplete Gamma Function and Other Properties

$$E_1(z) = -Ei(-z) = \int_z^{\infty e^{i\delta}} t^{-1}e^{-t}dt = \Gamma(0,z) = e^{-z}U(1;\ 1;\ z). \quad (1)$$

For the path of integration and other remarks, see 4.1(3) and the comments following it.

$$Ei(x) = -\mathrm{P.V.}\int_{-x}^{\infty} t^{-1}e^{-t}dt$$

$$= \mathrm{P.V.}\int_{-\infty}^{x} t^{-1}e^{t}dt = -\Gamma(0,-x) = e^{x}U(1;\ 1;\ -x), \quad x > 0. \quad (2)$$

$$\mathrm{Ei}(z) = \mathrm{Ei}(-ze^{\pm i\pi}) \mp i\pi. \quad (3)$$

$$\mathrm{Ei}(\pm 2z) = e^{\pm z}(\pi z/2)^{1/2}\left[\frac{\partial I_\nu(z)}{\partial\nu}\bigg|_{\nu=-1/2} \mp \frac{\partial I_\nu(z)}{\partial\nu}\bigg|_{\nu=1/2}\right]. \quad (4)$$

$$\int_0^z t^{-1}(1-e^{-t})\,dt = \lim_{a\to 0}\{a^{-1}z^a - \gamma(a,z)\} \quad (5)$$

$$\int_0^z t^{-1}(1-e^{-t})\,dt = e^{-z}\sum_{n=1}^{\infty}\frac{[\psi(n+1)-\psi(1)]\,z^n}{n!}. \quad (6)$$

$$\int_0^z t^{-1}(1-e^{-t})\,dt = z\,{}_2F_2\left(\begin{matrix}1, 1\\ 2, 2\end{matrix}\,\middle|\,-z\right). \quad (7)$$

$$E_1(z) + (\gamma + \ln z) = \int_0^z t^{-1}(1-e^{-t})\,dt. \quad (8)$$

$$\mathrm{Ei}(z) - (\gamma + \ln z) = \int_0^z t^{-1}(e^{t}-1)\,dt. \quad (9)$$

$$\Gamma(-n,z) = \frac{(-)^n}{n!}\left[\Psi(n+1) - \ln z\right] - z^{-n}\sum_{\substack{k=0\\ k\neq n}}^{\infty}\frac{(-)^k z^k}{k!\,(k-n)}. \quad (10)$$

4.6.2. Expansions in Series of Chebyshev Polynomials

Expansions for $\int_0^z t^{-1}(1 - e^{\pm t})dt$ follow from 4.6.1(7) and 5.10.2(10). They can also be found from results given by 11.6.2(9-12) and known Chebyshev expansions for the exponential function. Expansions for $E_1(z)$ and $Ei(x)$ follow from 4.6.1(1,2) and the discussion surrounding 5.10.1(11).

TABLE 4.1

Chebyshev Coefficients for the Exponential Integral and Related Functions

$$\int_0^x t^{-1}(1-e^{-t})dt = \sum_{n=0}^{\infty} a_n T_n^*(x/8) \qquad \int_0^x t^{-1}(e^{t}-1)dt = \sum_{n=0}^{\infty} b_n T_n^*(x/8)$$

$0 \le x \le 8$

n	a_n	n	b_n
0	1.67391 43547 42720 57853	0	100.45508 96546 02520 21261
1	1.22849 44785 47155 95018	1	166.43057 54008 54275 91901
2	-0.31786 98982 98373 61369	2	98.99379 59517 45053 15793
3	0.09274 60391 97292 19112	3	46.29118 62994 06148 50738
4	-0.02602 73631 69001 19930	4	17.72313 53141 47393 99804
5	0.00674 81530 94642 28057	5	5.72909 85775 35024 78695
6	-0.00159 89706 89342 25285	6	1.60025 92468 73221 35378
7	0.00034 59445 38800 95403	7	0.39319 28655 59692 28983
8	-0.00006 85365 26820 11094	8	0.08619 01547 03137 60832
9	0.00001 24859 32391 73701	9	0.01704 87404 95589 22176
10	-0.00000 21013 44638 71704	10	0.00307 16972 52687 03632
11	0.00000 03281 84479 81081	11	0.00050 80630 37137 14702
12	-0.00000 00477 68786 45154	12	0.00007 76592 68861 29861
13	0.00000 00065 05816 88192	13	0.00001 10326 32657 22594
14	-0.00000 00008 32101 93692	14	0.00000 14639 27623 35879
15	0.00000 00001 00281 01125	15	0.00000 01822 17797 31061
16	-0.00000 00000 11422 29304	16	0.00000 00213 57358 09922
17	0.00000 00000 01233 07945	17	0.00000 00023 65156 95193
18	-0.00000 00000 00126 48523	18	0.00000 00002 48223 35860
19	0.00000 00000 00012 35706	19	0.00000 00000 24755 84899
20	-0.00000 00000 00001 15226	20	0.00000 00000 02351 98394
21	0.00000 00000 00000 10276	21	0.00000 00000 00213 34535
22	-0.00000 00000 00000 00878	22	0.00000 00000 00018 51442
23	0.00000 00000 00000 00072	23	0.00000 00000 00001 54002
24	-0.00000 00000 00000 00006	24	0.00000 00000 00000 12299
		25	0.00000 00000 00000 00945
		26	0.00000 00000 00000 00070
		27	0.00000 00000 00000 00005

TABLE 4.1 *(Concluded)*

$$\int_x^{\infty} t^{-1}e^{-t}dt = x^{-1}e^{-x}\sum_{n=0}^{\infty} c_n T_n^*(5/x)$$

$$x \geq 5$$

n	c_n
0	0.92078 51444 53893 91645
1	-0.07343 41178 31621 28775
2	0.00520 98119 67272 32977
3	-0.00050 21407 19895 99012
4	0.00005 92079 69379 26337
5	-0.00000 80856 45130 97880
6	0.00000 12372 03856 47926
7	-0.00000 02075 01768 60703
8	0.00000 00375 60054 74022
9	-0.00000 00072 54109 76004
10	0.00000 00014 81816 01182
11	-0.00000 00003 17956 82863
12	0.00000 00000 71269 35828
13	-0.00000 00000 16612 31319
14	0.00000 00000 04011 55069
15	-0.00000 00000 01000 38792
16	0.00000 00000 00256 93341
17	-0.00000 00000 00067 80374
18	0.00000 00000 00018 34785
19	-0.00000 00000 00005 08209
20	0.00000 00000 00001 43861
21	-0.00000 00000 00000 41560
22	0.00000 00000 00000 12238
23	-0.00000 00000 00000 03669
24	0.00000 00000 00000 01119
25	-0.00000 00000 00000 00347
26	0.00000 00000 00000 00109
27	-0.00000 00000 00000 00035
28	0.00000 00000 00000 00011
29	-0.00000 00000 00000 00004
30	0.00000 00000 00000 00001

$$\text{P.V.}\int_{-\infty}^{x} t^{-1}e^{t}dt = x^{-1}e^{x}\sum_{n=0}^{\infty} d_n T_n^*(8/x)$$

$$x \geq 8$$

n	d_n
0	1.08158 51832 53608
1	0.08967 51661 74404
2	0.00932 57117 93980
3	0.00134 14749 99410
4	0.00004 44875 08106
5	-0.00009 44120 60877
6	-0.00003 55365 08652
7	0.00000 13764 01560
8	0.00000 47035 53094
9	0.00000 04780 19500
10	-0.00000 06457 19093
11	-0.00000 01031 03516
12	0.00000 01066 10083
13	0.00000 00138 53456
14	-0.00000 00204 26935
15	-0.00000 00003 99461
16	0.00000 00040 73797
17	-0.00000 00006 14592
18	-0.00000 00007 33329
19	0.00000 00002 98349
20	0.00000 00000 88133
21	-0.00000 00000 92197
22	0.00000 00000 06840
23	0.00000 00000 19820
24	-0.00000 00000 08394
25	-0.00000 00000 01913
26	0.00000 00000 02849
27	-0.00000 00000 00634
28	-0.00000 00000 00485
29	0.00000 00000 00377
30	-0.00000 00000 00037
31	-0.00000 00000 00087
32	0.00000 00000 00051
33	-0.00000 00000 00001
34	-0.00000 00000 00014
35	0.00000 00000 00008
36	0.00000 00000 00000
37	-0.00000 00000 00002
38	0.00000 00000 00001

$$\int_x^{\infty} t^{-1}e^{-t}dt = \int_0^x t^{-1}(1 - e^{-t})dt = (\gamma + \ln x)$$

$$\text{P.V.}\int_{-\infty}^{x} t^{-1}e^{t}dt = \int_0^x t^{-1}(e^{t} - 1)dt + (\gamma + \ln x)$$

4.6.3. Rational and Padé Approximations

We first consider

$$z\,E(z) = \int_0^z t^{-1}(1 - e^{-t})dt, \tag{1}$$

which is of hypergeometric type, see 4.6.1(7). Consider 5.12(1-11) with z and γ replaced by their negatives, $z = \gamma$, and

$$\begin{gathered} p = q = 2, \quad \alpha_1 = \alpha_2 = 1, \quad \rho_1 = \rho_2 = 2, \quad a = 0, \\ f = g = 1, \quad c_1 = 2, \quad d_1 = 1, \quad \alpha = \beta = 0. \end{gathered} \tag{2}$$

Then

$$E(z) = V_n(z)/W_n(z) + R_n(z), \tag{3}$$

$$V_n(z) = z^n \sum_{k=0}^{n} \frac{(-n)_k(n+1)_k}{(2)_k k!} \, {}_4F_2\left(\begin{matrix}-n+k, \; n+1+k, \; 2+k, \; 1 \\ 1+k, \; 1+k\end{matrix}\,\middle|\, -1/z\right),$$

$$W_n(z) = z^n {}_3F_1(-n, \; n+1, \; 2; \; 1; \; -1/z), \tag{4}$$

$$R_n(z) = \frac{(z/4)^n}{n!} (\pi/n)^{\frac{1}{2}} e^{-nz/2(n+1)} O(n^2), \tag{5}$$

$$\lim_{n\to\infty} R_n(z) = 0, \tag{6}$$

uniformly on all compact subsets of the z-plane, $z \neq 0$. Both $V_n(z)$ and $W_n(z)$ satisfy the following four term recursion formula.

$$\begin{aligned} (n+3)(2n+3)W_{n+3}(z) = {} & (2n+5)\{(n+1)z + 2(n+4)(2n+3)\}W_{n+2}(z) \\ & + (2n+3)z\{(n+3)z - 2n(2n+5)\}W_{n+1}(z) \\ & - (n+1)(2n+5)z^3 W_n(z). \end{aligned} \tag{7}$$

The polynomials $V_n(z)$ and $W_n(z)$ are tabulated below for $n = 0(1)6$. Further coefficients for $n = 7(1)10$ have been given by Luke (1969). Values of $R_n(z)$ are also listed, and we note that the estimate (5) is very conservative. In overall accuracy, these approximations are somewhat inferior to Padé approximations for the form (1) given by Luke (1969). However, closed form recurrence relations for the Padé approximations are not known and from this point of view the approximations (3) are advantageous.

TABLE 4.2

COEFFICIENTS FOR RATIONAL APPROXIMATIONS TO

$$z^{-1}\int_0^z t^{-1}(1 - e^{-t})dt$$

AND THE ERROR IN THESE APPROXIMATIONS

$$E(z) = z^{-1}\int_0^z t^{-1}(1 - e^{-t})dt = V_n(z)/W_n(z) + R_n(z)$$

$$V_n(z) = \sum_{k=0}^{n} a_k z^{n-k}$$

n	$a_0, a_1, \ldots, a_n$
0	1
1	0, 4
2	0, 3, 36
3	0, 17/3, 60, 480
4	0, 25/6, 500/3, 1260, 8400
5	0, 197/30, 210, 4620, 30240, 1 81440
6	0, 49/10, 2058/5, 8190, 1 37760, 8 31600, 46 56960

TABLE 4.2 *(Continued)*

$$W_n(z) = \sum_{k=0}^{n} b_k z^{n-k}$$

n	$b_0, b_1, \ldots, b_n$
0	1
1	1, 4
2	1, 12, 36
3	1, 24, 180, 480
4	1, 40, 540, 3360, 8400
5	1, 60, 1260, 13440, 75600, 1 81440
6	1, 84, 2520, 40320, 3 78000, 19 95840, 46 56960

n/θ	$\lvert R_n(z)\rvert$, $z = re^{i\theta}$ 0	$\pi/2$	π	$\lvert R_n(z)/E(z)\rvert$ π
	r = 1			
3	0.593(-5)	0.177(-4)	0.487(-4)	0.369(-4)
4	0.188(-6)	0.287(-6)	0.754(-6)	0.571(-6)
5	0.376(-8)	0.632(-8)	0.125(-7)	0.947(-8)
6	0.760(-10)	0.125(-9)	0.225(-9)	0.170(-9)
7	0.144(-11)	0.236(-11)	0.407(-11)	0.308(-11)
8	0.259(-13)	0.420(-13)	0.712(-13)	0.539(-13)
9	0.439(-15)	0.709(-15)	0.119(-14)	0.902(-15)
10	0.658(-17)	0.114(-16)	0.188(-16)	0.142(-16)
	r = 2			
3	0.272(-4)	0.291(-3)	0.219(-2)	0.119(-2)
4	0.407(-5)	0.543(-5)	0.703(-4)	0.382(-4)
5	0.140(-6)	0.379(-6)	0.188(-5)	0.102(-5)
6	0.602(-8)	0.154(-7)	0.559(-7)	0.304(-7)
7	0.229(-9)	0.579(-9)	0.185(-8)	0.101(-8)
8	0.825(-11)	0.208(-10)	0.626(-10)	0.340(-10)
9	0.280(-12)	0.708(-12)	0.206(-11)	0.112(-11)
10	0.898(-14)	0.227(-13)	0.646(-13)	0.351(-13)
	r = 3			
3	0.134(-4)	0.148(-2)	0.317(-1)	0.115(-1)
4	0.226(-4)	0.155(-4)	0.180(-2)	0.655(-3)
5	0.918(-6)	0.449(-5)	0.674(-4)	0.245(-4)
6	0.665(-7)	0.253(-6)	0.233(-5)	0.847(-6)
7	0.370(-8)	0.137(-7)	0.936(-7)	0.340(-7)
8	0.201(-9)	0.759(-9)	0.431(-8)	0.157(-8)
9	0.102(-10)	0.391(-10)	0.205(-9)	0.745(-10)
10	0.492(-12)	0.189(-11)	0.952(-11)	0.346(-11)

TABLE 4.2 *(Continued)*

n/θ	$\|R_n(z)\|$, $z = re^{i\theta}$: 0	π/2	π	$\|R_n(z)/E(z)\|$: π
		r = 4		
3	0.870(-4)	0.444(-2)	0.284(0)	0.643(-1)
4	0.716(-4)	0.198(-3)	0.276(-1)	0.624(-2)
5	0.290(-5)	0.306(-4)	0.148(-2)	0.335(-3)
6	0.337(-6)	0.184(-5)	0.603(-4)	0.136(-4)
7	0.236(-7)	0.118(-6)	0.240(-5)	0.543(-6)
8	0.173(-8)	0.939(-8)	0.116(-6)	0.262(-7)
9	0.117(-9)	0.655(-9)	0.657(-8)	0.149(-8)
10	0.747(-11)	0.425(-10)	0.392(-9)	0.887(-10)
		r = 5		
3	0.286(-3)	0.975(-2)	0.175(1)	0.230(0)
4	0.166(-3)	0.102(-2)	0.326(0)	0.429(-1)
5	0.598(-5)	0.145(-3)	0.242(-1)	0.318(-2)
6	0.113(-5)	0.883(-5)	0.129(-2)	0.170(-3)
7	0.905(-7)	0.526(-6)	0.551(-4)	0.725(-5)
8	0.853(-8)	0.648(-7)	0.237(-5)	0.312(-6)
9	0.710(-9)	0.571(-8)	0.127(-6)	0.167(-7)
10	0.566(-10)	0.463(-9)	0.831(-8)	0.109(-8)
		r = 6		
3	0.566(-3)	0.172(-1)	0.719(1)	0.517(0)
4	0.315(-3)	0.317(-2)	0.374(1)	0.269(0)
5	0.916(-5)	0.504(-3)	0.314(0)	0.226(-1)
6	0.295(-5)	0.290(-4)	0.229(-1)	0.165(-2)
7	0.251(-6)	0.142(-5)	0.120(-2)	0.863(-4)
8	0.297(-7)	0.327(-6)	0.528(-4)	0.380(-5)
9	0.290(-8)	0.330(-7)	0.235(-5)	0.169(-6)
10	0.267(-9)	0.315(-8)	0.133(-6)	0.956(-8)
		r = 7		
3	0.895(-3)	0.259(-1)	0.206(2)	0.762(0)
4	0.524(-3)	0.724(-2)	0.274(3)	0.101(2)
5	0.107(-4)	0.136(-2)	0.318(1)	0.118(0)
6	0.649(-5)	0.936(-4)	0.343(0)	0.127(-1)
7	0.558(-6)	0.627(-5)	0.229(-1)	0.848(-3)
8	0.820(-7)	0.142(-5)	0.120(-2)	0.444(-4)
9	0.904(-8)	0.142(-6)	0.535(-4)	0.198(-5)
10	0.101(-8)	0.151(-7)	0.241(-5)	0.893(-7)
		r = 8		
3	0.125(-2)	0.345(-1)	0.490(2)	0.896(0)
4	0.787(-3)	0.133(-1)	0.773(2)	0.141(1)
5	0.860(-5)	0.298(-2)	0.216(2)	0.395(0)
6	0.126(-4)	0.288(-3)	0.465(1)	0.850(0)
7	0.105(-5)	0.341(-4)	0.367(0)	0.671(-2)
8	0.192(-6)	0.555(-5)	0.242(-1)	0.442(-2)
9	0.231(-7)	0.474(-6)	0.126(-2)	0.230(-4)
10	0.296(-8)	0.546(-7)	0.569(-4)	0.104(-5)

TABLE 4.2 *(Concluded)*

n/θ	$\lvert R_n(z)\rvert$, $z = re^{i\theta}$: 0	$\pi/2$	π	$\lvert R_n(z)/E(z)\rvert$: π
		r = 9		
3	0.160(-2)	0.413(-1)	0.110(3)	0.957(0)
4	0.110(-2)	0.209(-1)	0.126(3)	0.110(1)
5	0.117(-5)	0.553(-2)	0.848(2)	0.737(0)
6	0.221(-4)	0.778(-3)	0.825(2)	0.717(0)
7	0.172(-5)	0.130(-3)	0.493(1)	0.429(-1)
8	0.399(-6)	0.189(-4)	0.418(0)	0.363(-2)
9	0.509(-7)	0.124(-5)	0.266(-1)	0.231(-3)
10	0.742(-8)	0.154(-6)	0.139(-2)	0.121(-4)
		r = 10		
3	0.193(-2)	0.454(-1)	0.243(3)	0.976(0)
4	0.145(-2)	0.288(-1)	0.256(3)	0.103(1)
5	0.129(-4)	0.895(-2)	0.228(3)	0.916(0)
6	0.360(-4)	0.178(-2)	0.508(3)	0.204(0)
7	0.255(-5)	0.380(-3)	0.506(2)	0.203(0)
8	0.754(-6)	0.558(-4)	0.638(1)	0.256(-1)
9	0.995(-7)	0.283(-5)	0.484(0)	0.194(-2)
10	0.164(-7)	0.426(-6)	0.256(-1)	0.103(-3)

TABLE 4.3

COEFFICIENTS FOR PADÉ APPROXIMATIONS TO

$$\int_z^\infty e^{-t}t^{-1}dt$$

AND THE ERRORS IN THESE APPROXIMATIONS

For explanation of the data, see the discussion in 4.3.2-4.3.5 with $\nu = 0$. We write

$$\int_z^\infty e^{-t}t^{-1}\,dt = e^{-z}z^{-1}\left[\frac{G_n(0,z)}{H_n(0,z)} + T_n(0,z)\right], \qquad z \neq 0, \quad |\arg z| < \pi,$$

$$S_n(z) = |\,e^{-z}z^{-1}[T_n(0,z)]_{a=0}\,|.$$

The coefficients in the polynomials $G_n(0,z)$ and $H_n(0,z)$ are recorded below for $n = 0(1)6$. Further coefficients for $n = 7(1)10$ have been given by Luke (1969). Additional polynomials can be generated from 4.3.2(14). We also record $S_n(z)$ for a wide range of n and z values. Error coefficients follow from 4.3.2(15) or 4.3.3(12) as appropriate.

TABLE 4.3 *(Continued)*

$$G_n(0,z) = \sum_{k=0}^{n} a_k z^{n-k}, \quad a = 0$$

n	$a_0, a_1, \ldots, a_n$
0	1
1	1, 1
2	1, 5, 2
3	1, 11, 26, 6
4	1, 19, 102, 154, 24
5	1, 29, 272, 954, 1044, 120
6	1, 41, 590, 3648, 9432, 8028, 720

$$H_n(0,z) = \sum_{k=0}^{n} b_k z^{n-k}, \quad a = 0$$

n	$b_0, b_1, \ldots, b_n$
0	1
1	1, 2
2	1, 6, 6
3	1, 12, 36, 24
4	1, 20, 120, 240, 120
5	1, 30, 300, 1200, 1800, 720
6	1, 42, 630, 4200, 12600, 15120, 5040

$$G_n(0,z) = \sum_{k=0}^{n} c_k z^{n-k}, \quad a = 1$$

n	$c_0, c_1, \ldots, c_n$
0	0
1	1, 0
2	1, 3, 0
3	1, 8, 11, 0
4	1, 15, 58, 50, 0
5	1, 24, 177, 444, 274, 0
6	1, 35, 416, 2016, 3708, 1764, 0

TABLE 4.3 *(Continued)*

$$H_n(0,z) = \sum_{k=0}^{n} d_k z^{n-k}, \quad a = 1$$

n	$d_0, d_1, \ldots, d_n$
0	1
1	1, 1
2	1, 4, 2
3	1, 9, 18, 6
4	1, 16, 72, 96, 24
5	1, 25, 200, 600, 600, 120
6	1, 36, 450, 2400, 5400, 4320, 720

$\mathcal{S}_n(z)$, $z = re^{i\theta}$

r = 1

n/θ	0°	45°	60°	75°	90°
4	0.901(-3)	0.184(-2)	0.314(-2)	0.610(-2)	0.133(-1)
6	0.172(-3)	0.396(-3)	0.742(-3)	0.162(-2)	0.408(-2)
8	0.414(-4)	0.106(-3)	0.215(-3)	0.519(-3)	0.147(-2)
12	0.363(-5)	0.111(-4)	0.259(-4)	0.745(-4)	0.260(-3)
16	0.455(-6)	0.162(-5)	0.426(-5)	0.142(-4)	0.594(-4)
20	0.720(-7)	0.295(-6)	0.860(-6)	0.329(-5)	0.160(-4)

n/θ	105°	120°	135°	150°	165°
4	0.322(-1)	0.850(-1)	0.241(0)	0.729(0)	2.274(0)
6	0.116(-1)	0.361(-1)	0.121(0)	0.417(0)	1.324(0)
8	0.479(-2)	0.175(-1)	0.695(-1)	0.291(0)	1.199(0)
12	0.107(-2)	0.512(-2)	0.274(-1)	0.164(0)	1.164(0)
16	0.301(-3)	0.180(-2)	0.123(-1)	0.923(-1)	0.694(0)
20	0.973(-4)	0.710(-3)	0.602(-2)	0.568(-1)	0.556(0)

r = 2

n/θ	0°	45°	60°	75°	90°
4	0.184(-4)	0.543(-4)	0.121(-3)	0.326(-3)	0.103(-2)
6	0.184(-5)	0.635(-5)	0.161(-4)	0.506(-4)	0.193(-3)
8	0.250(-6)	0.995(-6)	0.280(-5)	0.102(-4)	0.458(-4)
12	0.823(-8)	0.419(-7)	0.143(-6)	0.657(-6)	0.395(-5)
16	0.444(-9)	0.280(-8)	0.112(-7)	0.635(-7)	0.490(-6)
20	0.332(-10)	0.253(-9)	0.117(-8)	0.800(-8)	0.770(-7)

n/θ	105°	120°	135°	150°	165°
4	0.370(-2)	0.147(-1)	0.627(-1)	0.280(0)	1.274(0)
6	0.868(-3)	0.444(-2)	0.252(-1)	0.156(0)	1.098(0)
8	0.250(-3)	0.159(-2)	0.114(-1)	0.882(-1)	0.668(0)
12	0.300(-4)	0.277(-3)	0.297(-2)	0.360(-1)	0.502(0)
16	0.495(-5)	0.627(-4)	0.952(-3)	0.165(-1)	0.302(0)
20	0.100(-5)	0.168(-4)	0.347(-3)	0.837(-2)	0.227(0)

TABLE 4.3 *(Continued)*

r = 4

n/θ	0°	45°	60°	75°	90°
4	0.651(-7)	0.357(-6)	0.124(-5)	0.567(-5)	0.321(-4)
6	0.264(-8)	0.177(-7)	0.724(-7)	0.406(-6)	0.295(-5)
8	0.163(-9)	0.131(-8)	0.619(-8)	0.418(-7)	0.381(-6)
12	0.136(-11	0.153(-10	0.927(-10)	0.865(-9)	0.117(-7)
16	0.226(-13)	0.337(-12)	0.256(-11)	0.317(-10)	0.604(-9)
20	0.587(-15)	0.114(-13)	0.106(-12)	0.169(-11)	0.438(-10)

n/θ	105°	120°	135°	150°	165°
4	0.212(-3)	0.156(-2)	0.121(-1)	0.956(-1)	0.694(0)
6	0.264(-4)	0.274(-3)	0.315(-2)	0.388(-1)	0.529(0)
8	0.443(-5)	0.623(-4)	0.100(-2)	0.176(-1)	0.313(0)
12	0.216(-6)	0.510(-5)	0.145(-3)	0.476(-2)	0.169(0)
16	0.166(-7)	0.611(-6)	0.285(-4)	0.157(-2)	0.980(-1)
20	0.171(-8)	0.940(-7)	0.675(-5)	0.592(-3)	0.585(-1)

r = 6

n/θ	0°	45°	60°	75°	90°
4	0.719(-9)	0.702(-8)	0.368(-7)	0.271(-6)	0.257(-5)
6	0.150(-10)	0.183(-9)	0.115(-8)	0.106(-7)	0.134(-6)
8	0.511(-12)	0.769(-11)	0.568(-10)	0.652(-9)	0.107(-7)
12	0.153(-14)	0.336(-13)	0.333(-12)	0.560(-11)	0.147(-9)
16	0.103(-16)	0.319(-15)	0.412(-14)	0.971(-13)	0.383(-11)
20	0.121(-18)	0.508(-17)	0.834(-16)	0.267(-14)	0.152(-12)

n/θ	105°	120°	135°	150°	165°
4	0.286(-4)	0.345(-3)	0.422(-2)	0.505(-1)	0.623(0)
6	0.213(-5)	0.387(-4)	0.762(-3)	0.154(-1)	0.308(0)
8	0.232(-6)	0.609(-5)	0.181(-3)	0.575(-2)	0.188(0)
12	0.551(-8)	0.271(-6)	0.163(-4)	0.111(-2)	0.822(-1)
16	0.232(-9)	0.196(-7)	0.215(-5)	0.280(-3)	0.411(-1)
20	0.141(-10)	0.194(-8)	0.362(-6)	0.835(-4)	0.221(-1)

r = 8

n/θ	0°	45°	60°	75°	90°
4	0.139(-10)	0.239(-9)	0.187(-8)	0.218(-7)	0.338(-6)
6	0.168(-12)	0.365(-11)	0.345(-10)	0.516(-9)	0.109(-7)
8	0.353(-14)	0.961(-13)	0.108(-11)	0.204(-10)	0.576(-9)
12	0.451(-17)	0.185(-15)	0.289(-14)	0.830(-13)	0.394(-11)
16	0.145(-19)	0.870(-18)	0.182(-16)	0.762(-15)	0.573(-13)
20	0.875(-22)	0.739(-20)	0.202(-18)	0.119(-16)	0.136(-14)

n/θ	105°	120°	135°	150°	165°
4	0.612(-5)	0.116(-3)	0.208(-2)	0.332(-1)	0.442(0)
6	0.290(-6)	0.866(-5)	0.267(-3)	0.801(-2)	0.227(0)
8	0.217(-7)	0.978(-6)	0.484(-4)	0.247(-2)	0.127(0)
12	0.276(-9)	0.255(-7)	0.284(-5)	0.354(-3)	0.469(-1)
16	0.691(-11)	0.119(-8)	0.266(-6)	0.702(-4)	0.206(-1)
20	0.268(-12)	0.805(-10)	0.332(-7)	0.170(-4)	0.100(-1)

TABLE 4.3 *(Concluded)*

r = 12

n/θ	0°	45°	60°	75°	90°
4	0.132(-13)	0.697(-12)	0.120(-10)	0.348(-9)	0.141(-7)
6	0.663(-16)	0.447(-14)	0.935(-13)	0.352(-11)	0.198(-9)
8	0.635(-18)	0.543(-16)	0.138(-14)	0.665(-13)	0.516(-11)
12	0.202(-21)	0.273(-19)	0.990(-18)	0.769(-16)	0.108(-13)
16	0.193(-24)	0.398(-22)	0.202(-20)	0.241(-18)	0.579(-16)
20	0.388(-27)	0.119(-24)	0.819(-23)	0.146(-20)	0.574(-18)

n/θ	105°	120°	135°	150°	165°
4	0.657(-6)	0.289(-4)	0.102(-2)	0.256(-1)	0.466(0)
6	0.139(-7)	0.102(-5)	0.675(-4)	0.367(-2)	0.165(0)
8	0.537(-9)	0.633(-7)	0.738(-5)	0.780(-3)	0.737(-1)
12	0.230(-11)	0.635(-9)	0.198(-6)	0.639(-4)	0.202(-1)
16	0.235(-13)	0.137(-10)	0.999(-8)	0.822(-5)	0.712(-2)
20	0.417(-15)	0.476(-12)	0.739(-9)	0.139(-5)	0.289(-2)

r = 16

n/θ	0°	45°	60°	75°	90°
4	0.254(-16)	0.416(-14)	0.158(-12)	0.114(-10)	0.122(-8)
6	0.638(-19)	0.132(-16)	0.609(-15)	0.569(-13)	0.842(-11)
8	0.324(-21)	0.858(-19)	0.479(-17)	0.580(-15)	0.119(-12)
12	0.334(-25)	0.142(-22)	0.116(-20)	0.232(-18)	0.896(-16)
16	0.118(-28)	0.786(-26)	0.919(-24)	0.293(-21)	0.204(-18)
20	0.955(-32)	0.979(-29)	0.161(-26)	0.796(-24)	0.957(-21)

n/θ	105°	120°	135°	150°	165°
4	0.145(-6)	0.147(-4)	0.984(-3)	0.355(-1)	0.560(0)
6	0.152(-8)	0.258(-6)	0.334(-4)	0.285(-2)	0.156(0)
8	0.324(-10)	0.918(-8)	0.223(-5)	0.410(-3)	0.560(-1)
12	0.531(-13)	0.385(-10)	0.287(-7)	0.194(-4)	0.115(-1)
16	0.247(-15)	0.416(-12)	0.819(-9)	0.166(-5)	0.329(-2)
20	0.222(-17)	0.803(-14)	0.377(-10)	0.201(-6)	0.112(-2)

r = 20

n/θ	0°	45°	60°	75°	90°
4	0.754(-19)	0.383(-16)	0.323(-14)	0.586(-12)	0.165(-9)
6	0.106(-21)	0.675(-19)	0.685(-17)	0.159(-14)	0.616(-12)
8	0.314(-24)	0.256(-21)	0.315(-19)	0.946(-17)	0.509(-14)
12	0.124(-28)	0.163(-25)	0.295(-23)	0.148(-20)	0.154(-17)
16	0.185(-32)	0.390(-29)	0.103(-26)	0.844(-24)	0.163(-20)
20	0.689(-36)	0.228(-32)	0.857(-30)	0.113(-26)	0.389(-23)

n/θ	105°	120°	135°	150°	165°
4	0.510(-7)	0.121(-4)	0.159(-2)	0.857(-1)	1.691(0)
6	0.283(-9)	0.111(-6)	0.274(-4)	0.340(-2)	0.191(0)
8	0.355(-11)	0.233(-8)	0.111(-5)	0.319(-3)	0.532(-1)
12	0.241(-14)	0.430(-11)	0.690(-8)	0.860(-5)	0.807(-2)
16	0.546(-17)	0.243(-13)	0.115(-9)	0.494(-6)	0.185(-2)
20	0.265(-19)	0.273(-15)	0.339(-11)	0.434(-7)	0.534(-3)

4.7. Cosine and Sine Integrals

4.7.1. Relation to Exponential Integral and Other Properties

$$\mathrm{Ci}(\alpha, z) + i\,\mathrm{Si}(\alpha, z) = \int_0^z t^{-\alpha} e^{it}\,dt = e^{i\pi(1-\alpha)/2}\gamma(1-\alpha, ze^{-i\pi/2}),$$
$$R(\alpha) < 1. \tag{1}$$

$$\int_0^z t^{-\alpha}\cos t\,dt = \frac{z^{1-\alpha}}{(1-\alpha)}\,{}_1F_2\left(\begin{matrix}\frac{1}{2}(1-\alpha)\\ \frac{1}{2}(3-\alpha), \frac{1}{2}\end{matrix}\,\middle|\, \frac{-z^2}{4}\right), \qquad R(\alpha) < 1. \tag{2}$$

$$\int_0^z t^{-\alpha}\sin t\,dt = \frac{z^{2-\alpha}}{(2-\alpha)}\,{}_1F_2\left(\begin{matrix}\frac{1}{2}(2-\alpha)\\ \frac{1}{2}(4-\alpha), \frac{3}{2}\end{matrix}\,\middle|\, \frac{-z^2}{4}\right), \qquad R(\alpha) < 2. \tag{3}$$

$$\mathrm{Ci}(z) + i(\pi/2 - \mathrm{Si}(z)) = -E_1(ze^{i\pi/2}) = \int_\infty^z t^{-1}e^{-it}\,dt. \tag{4}$$

$$\begin{aligned}\mathrm{Ci}(z) - (\gamma + \ln z) &= -\int_0^z t^{-1}(1-\cos t)\,dt \\ &= -(z^2/4)\,{}_2F_3\left(\begin{matrix}1, 1\\ 2, 2, 3/2\end{matrix}\,\middle|\, -z^2/4\right).\end{aligned} \tag{5}$$

$$\mathrm{si}(z) + \pi/2 = \mathrm{Si}(z) = \int_0^z t^{-1}\sin t\,dt = z\,{}_1F_2\left(\begin{matrix}1/2\\ 3/2, 3/2\end{matrix}\,\middle|\, -z^2/4\right). \tag{6}$$

$$\begin{aligned}\mathrm{Ci}(z) + i\,\mathrm{si}(z) \sim {}& -z^{-1}e^{iz}\{z^{-1}\,{}_3F_0(3/2, 1, 1; -4/z^2) \\ & + i\,{}_3F_0(1/2, 1, 1; -4/z^2)\}, \\ & |z| \to \infty, \qquad |\arg z| \leqslant \pi - \epsilon, \qquad \epsilon > 0.\end{aligned} \tag{7}$$

4.7.2. Expansions in Series of Chebyshev Polynomials

TABLE 4.4

CHEBYSHEV COEFFICIENTS FOR COSINE AND SINE INTEGRALS AND RELATED FUNCTIONS

$$\int_0^x t^{-1}(1-\cos t)dt = \sum_{n=0}^{\infty} a_n T_{2n}(x/8) \qquad \int_0^x t^{-1}\sin t\,dt = \sum_{n=0}^{\infty} b_n T_{2n+1}(x/8)$$

$-8 \le x \le 8$

n	a_n	n	b_n
0	1.94054 91464 83554 93374	0	1.95222 09759 53071 08224
1	0.94134 09132 86521 34390	1	-0.68840 42321 25715 44408
2	-0.57984 50342 92992 76547	2	0.45518 55132 25584 84126
3	0.30915 72011 15927 13017	3	-0.18045 71236 83877 85342
4	-0.09161 01792 20771 33969	4	0.04104 22133 75859 23964
5	0.01644 37407 51546 24963	5	-0.00595 86169 55588 85229
6	-0.00197 13091 95216 41024	6	0.00060 01427 41414 43021
7	0.00016 92538 85083 49925	7	-0.00004 44708 32910 74925
8	-0.00001 09393 29573 10627	8	0.00000 25300 78230 75133
9	0.00000 05522 38574 83778	9	-0.00000 01141 30759 30294
10	-0.00000 00223 99493 31410	10	0.00000 00041 85783 94210
11	0.00000 00007 46533 25345	11	-0.00000 00001 27347 05516
12	-0.00000 00000 20818 33157	12	0.00000 00000 03267 36126
13	0.00000 00000 00493 12353	13	-0.00000 00000 00071 67679
14	-0.00000 00000 00010 04784	14	0.00000 00000 00001 36020
15	0.00000 00000 00000 17803	15	-0.00000 00000 00000 02255
16	-0.00000 00000 00000 00277	16	0.00000 00000 00000 00033
17	0.00000 00000 00000 00004		

TABLE 4.4 *(Continued)*

$$\int_x^{\infty} t^{-1}e^{-it}dt = -ix^{-1}e^{-ix}\sum_{n=0}^{\infty} c_n T_n^*(5/x)\ ,\quad x \geq 5$$

$$c_n = R(c_n)+iI(c_n)$$

n	$R(c_n)$	n	$I(c_n)$
0	0.97615 52711 28712 28562	0	0.08968 45854 91642 30208
1	-0.03046 56658 03069 59120	1	0.08508 92472 92294 52754
2	-0.00578 07368 31483 85631	2	-0.00507 18267 77756 90802
3	0.00083 86432 56650 89313	3	-0.00033 42234 15981 73821
4	-0.00002 15746 20728 12156	4	0.00012 85606 50086 06518
5	-0.00001 56456 41351 02321	5	-0.00001 52025 51359 72619
6	0.00000 40400 10138 43204	6	-0.00000 05958 96122 75216
7	-0.00000 04349 85305 97434	7	0.00000 07134 72533 53084
8	-0.00000 00534 30218 60611	8	-0.00000 01760 03581 15661
9	0.00000 00385 02885 51259	9	0.00000 00192 57654 44417
10	-0.00000 00100 73535 82172	10	0.00000 00033 63591 94377
11	0.00000 00012 80496 19406	11	-0.00000 00024 25468 70827
12	0.00000 00001 86917 28895	12	0.00000 00007 13431 29834
13	-0.00000 00001 70673 48371	13	-0.00000 00001 14604 07035
14	0.00000 00000 58800 44115	14	-0.00000 00000 06784 17843
15	-0.00000 00000 12157 23809	15	0.00000 00000 12656 12487
16	0.00000 00000 00474 81418	16	-0.00000 00000 05323 09477
17	0.00000 00000 00905 90381	17	0.00000 00000 01400 46450
18	-0.00000 00000 00500 96832	18	-0.00000 00000 00180 45804
19	0.00000 00000 00166 16291	19	-0.00000 00000 00050 26164
20	-0.00000 00000 00034 84536	20	0.00000 00000 00046 00566
21	0.00000 00000 00000 57400	21	-0.00000 00000 00019 53107
22	0.00000 00000 00003 68837	22	0.00000 00000 00005 62862
23	-0.00000 00000 00002 17822	23	-0.00000 00000 00000 89561
24	0.00000 00000 00000 81978	24	-0.00000 00000 00000 16803
25	-0.00000 00000 00000 21304	25	0.00000 00000 00000 21351
26	0.00000 00000 00000 02270	26	-0.00000 00000 00000 10783
27	0.00000 00000 00000 01445	27	0.00000 00000 00000 03813
28	-0.00000 00000 00000 01220	28	-0.00000 00000 00000 00919
29	0.00000 00000 00000 00577	29	0.00000 00000 00000 00052
30	-0.00000 00000 00000 00199	30	0.00000 00000 00000 00099
31	0.00000 00000 00000 00046	31	-0.00000 00000 00000 00073
32	-0.00000 00000 00000 00001	32	0.00000 00000 00000 00034
33	-0.00000 00000 00000 00006	33	-0.00000 00000 00000 00012
34	0.00000 00000 00000 00005	34	0.00000 00000 00000 00003
35	-0.00000 00000 00000 00002		
36	0.00000 00000 00000 00001		

$$\int_x^{\infty} t^{-1}\cos t\,dt = \int_0^x t^{-1}(1-\cos t)dt-(\gamma+\ln x)$$

$$\int_x^{\infty} t^{-1}\sin t\,dt = \pi/2 - \int_0^x t^{-1}\sin t\,dt$$

TABLE 4.4 *(Concluded)*

$$\int_x^{\infty} t^{-1} e^{-it} dt = x^{-1} e^{-ix}\left[x^{-1}P(x) - iQ(x)\right]$$

$$P(x) = \sum_{n=0}^{\infty} e_n T_{2n}(8/x) \qquad x \geq 8 \qquad Q(x) = \sum_{n=0}^{\infty} f_n T_{2n}(8/x)$$

n	e_n
0	0.96074 78397 52035 96305
1	-0.03711 38962 12398 05610
2	0.00194 14398 88991 90367
3	-0.00017 16598 84251 47079
4	0.00002 11263 77532 31466
5	-0.00000 32716 32567 11532
6	0.00000 06006 92116 14777
7	-0.00000 01258 67944 03387
8	0.00000 00293 25634 57996
9	-0.00000 00074 56959 20628
10	0.00000 00020 41054 78359
11	-0.00000 00005 95022 30388
12	0.00000 00001 83229 67411
13	-0.00000 00000 59205 06078
14	0.00000 00000 19965 16518
15	-0.00000 00000 06995 11401
16	0.00000 00000 02536 85774
17	-0.00000 00000 00949 28512
18	0.00000 00000 00365 52312
19	-0.00000 00000 00144 48739
20	0.00000 00000 00058 51312
21	-0.00000 00000 00024 23290
22	0.00000 00000 00010 24704
23	-0.00000 00000 00004 41796
24	0.00000 00000 00001 93968
25	-0.00000 00000 00000 86623
26	0.00000 00000 00000 39309
27	-0.00000 00000 00000 18110
28	0.00000 00000 00000 08463
29	-0.00000 00000 00000 04009
30	0.00000 00000 00000 01923
31	-0.00000 00000 00000 00934
32	0.00000 00000 00000 00459
33	-0.00000 00000 00000 00228
34	0.00000 00000 00000 00114
35	-0.00000 00000 00000 00058
36	0.00000 00000 00000 00030
37	-0.00000 00000 00000 00015
38	0.00000 00000 00000 00008
39	-0.00000 00000 00000 00004
40	0.00000 00000 00000 00002
41	-0.00000 00000 00000 00001
42	0.00000 00000 00000 00001

n	f_n
0	0.98604 06569 62382 59766
1	-0.01347 17382 08295 21313
2	0.00045 32928 41165 22654
3	-0.00003 06728 86516 55165
4	0.00000 31319 91976 01087
5	-0.00000 04211 01964 96310
6	0.00000 00690 72448 30282
7	-0.00000 00131 83212 90423
8	0.00000 00028 36974 32997
9	-0.00000 00006 73292 34255
10	0.00000 00001 73396 86940
11	-0.00000 00000 47869 38904
12	0.00000 00000 14032 34652
13	-0.00000 00000 04334 95713
14	0.00000 00000 01402 72653
15	-0.00000 00000 00473 06173
16	0.00000 00000 00165 57878
17	-0.00000 00000 00059 93808
18	0.00000 00000 00022 37213
19	-0.00000 00000 00008 58807
20	0.00000 00000 00003 38292
21	-0.00000 00000 00001 36473
22	0.00000 00000 00000 56287
23	-0.00000 00000 00000 23698
24	0.00000 00000 00000 10171
25	-0.00000 00000 00000 04445
26	0.00000 00000 00000 01975
27	-0.00000 00000 00000 00892
28	0.00000 00000 00000 00409
29	-0.00000 00000 00000 00190
30	0.00000 00000 00000 00090
31	-0.00000 00000 00000 00043
32	0.00000 00000 00000 00021
33	-0.00000 00000 00000 00010
34	0.00000 00000 00000 00005
35	-0.00000 00000 00000 00002
36	0.00000 00000 00000 00001
37	-0.00000 00000 00000 00001

$$\int_x^{\infty} t^{-1}\cos t\,dt = \int_0^x t^{-1}(1-\cos t)dt - (\gamma + \ln x) \qquad \int_x^{\infty} t^{-1}\sin t\,dt = \pi/2 - \int_0^x t^{-1}\sin t\,dt$$

4.8. Error Functions

4.8.1. Relation to Incomplete Gamma Function and Other Properties

$$\mathrm{Erf}(z) = \int_0^z e^{-t^2}\,dt = \tfrac{1}{2}\gamma(\tfrac{1}{2}, z^2) = z\,{}_1F_1(\tfrac{1}{2}; \tfrac{3}{2}; -z^2) = ze^{-z^2}\,{}_1F_1(1; \tfrac{3}{2}; z^2). \quad (1)$$

$$\mathrm{Erfc}(z) = \int_z^\infty e^{-t^2}\,dt = \tfrac{1}{2}\pi^{1/2} - \mathrm{Erf}(z) = \tfrac{1}{2}\Gamma(\tfrac{1}{2}, z^2) = \tfrac{1}{2}e^{-z^2}U(\tfrac{1}{2}; \tfrac{1}{2}; z^2). \quad (2)$$

The notation $\mathrm{erf}(z) = (2/\pi^{1/2})\,\mathrm{Erf}(z)$ is often used.

$$\mathrm{Erfi}(z) = -i\,\mathrm{Erf}(iz) = \int_0^z e^{t^2}\,dt. \quad (3)$$

$$\mathrm{Erfc}(z) \sim (e^{-z^2}/2z)\,{}_2F_0(1, 1/2; -1/z^2)$$
$$|z| \to \infty, \qquad |\arg z| \leqslant 3\pi/4 - \epsilon, \qquad \epsilon > 0. \quad (4)$$

$$\mathrm{Erfi}(z) \sim (e^{z^2}/2z)\,{}_2F_0(1, 1/2; 1/z^2) - (i\pi^{1/2}\epsilon/2),$$
$$|z| \to \infty, \qquad -(3+2\epsilon)\pi/4 < \arg z < (3-2\epsilon)\pi/4, \qquad \epsilon = \pm 1. \quad (5)$$

Padé approximations for the error functions follow from the material in 4.3 and pertinent coefficients are given in 4.8.3. Inequalities are given in 4.4.2. It is convenient to record the connecting relations here.

$$\mathrm{Erf}(z) = ze^{-z^2}\left[\frac{C_n(\tfrac{1}{2}, -z^2)}{D_n(\tfrac{1}{2}, -z^2)} + V_n(\tfrac{1}{2}, z^2e^{-i\pi})\right], \quad (6)$$

$$\mathrm{Erfi}(z) = ze^{z^2}\left[\frac{C_n(\tfrac{1}{2}, z^2)}{D_n(\tfrac{1}{2}, z^2)} + V_n(\tfrac{1}{2}, z^2)\right], \quad (7)$$

$$\mathrm{Erfc}(z) = \frac{e^{-z^2}}{2z}\left[\frac{G_n(\tfrac{1}{2}, z^2)}{H_n(\tfrac{1}{2}, z^2)} + T_n(\tfrac{1}{2}, z^2)\right],$$

$$z \neq 0, \quad |\arg z| < \pi/2. \quad (8)$$

$$\mathrm{Erfi}(z) = -\tfrac{1}{2}i\pi^{\frac{1}{2}}\varepsilon + \frac{e^{z^2}}{2z}\left[\frac{G_n(\tfrac{1}{2}, -z^2)}{H_n(\tfrac{1}{2}, -z^2)} + T_n(\tfrac{1}{2}, z^2e^{i\pi})\right],$$

$$-(3+2\varepsilon)\pi/4 < \arg z < (3-2\varepsilon)\pi/4, \quad \varepsilon = \pm 1. \quad (9)$$

In some applications, it is desirable to deal with the function

$$w(z) = e^{-z^2}\left(1 + \frac{2i}{\pi^{\frac{1}{2}}}\int_0^z e^{t^2}\,dt\right) = e^{-z^2}\operatorname{erfc}(-iz), \tag{10}$$

$$w(z) = e^{-z^2} + 2i\pi^{-\frac{1}{2}}\operatorname{Erfi}(z). \tag{11}$$

$$w(z) = \frac{i}{\pi}\int_{-\infty}^{\infty}\frac{e^{-t^2}}{z-t}\,dt + 2ae^{-z^2},$$

$$a = 0 \text{ if } I(z) > 0; \quad a = 1 \text{ if } I(z) < 0. \tag{12}$$

With

$$z = x + iy, \quad y > 0, \quad w(z) = u(x,y) + iv(x,y), \tag{13}$$

$$u(x,y) = \frac{y}{\pi}\int_{-\infty}^{\infty}\frac{e^{-t^2}}{(x-t)^2 + y^2}\,dt, \quad v(x,y) = \frac{1}{\pi}\int_{-\infty}^{\infty}\frac{(x-t)e^{-t^2}}{(x-t)^2 + y^2}\,dt. \tag{14}$$

$$u(x,y) = \frac{2ye^{-x^2}}{\pi}\int_0^{\infty}\frac{e^{-t^2}\cosh 2xt}{y^2 + t^2}\,dt. \tag{15}$$

$$v(x,y) = \frac{2e^{-x^2}}{\pi}\int_0^{\infty}\frac{te^{-t^2}\sinh 2xt}{y^2 + t^2}\,dt. \tag{16}$$

Another form of interest is

$$W(p,t) = \frac{q}{(4\pi t)^{\frac{1}{2}}}\int_{-\infty}^{\infty}\frac{e^{-u^2}}{u^2+q^2}\,du = \left(\frac{\pi}{4t}\right)^{\frac{1}{2}}e^{q^2}\operatorname{erfc}(q), \quad q = (1-ip)/2t^{\frac{1}{2}}. \tag{17}$$

$$W(p,t) = U(p,t) + iV(p,t), \tag{18}$$

where

$$U(p,t) = (4\pi t)^{-\frac{1}{2}} \int_{-\infty}^{\infty} \frac{e^{-(p-y)^2/4t}}{1+y^2}\, dy, \tag{19}$$

$$V(p,t) = (4\pi t)^{-\frac{1}{2}} \int_{-\infty}^{\infty} \frac{ye^{-(p-y)^2/4t}}{1+y^2}\, dy, \tag{20}$$

are called Voigt functions. With

$$x + iy = iq = \frac{p+i}{2t^{\frac{1}{2}}},$$

$$W(p,t) = U(p,t) + iV(p,t) = (\pi/4t)^{\frac{1}{2}}\big(u(x,y) + iv(x,y)\big). \tag{21}$$

From the work of Thompson (1965) and Wood, Kenan and Glasser (1966),

$$\left(\frac{\pi}{4t}\right)^{\frac{1}{2}} e^{q^2} \operatorname{erfc}(q) \sim \sum_{n=0}^{\infty} \frac{(\frac{1}{2})_n (4t)^n}{(p^2+1)^{n+\frac{1}{2}}} \left[(-)^n T_{2n+1}(u) + iT_{2n+1}(up)\right],$$

$$u = (p^2+1)^{\frac{1}{2}}, \quad q = (1-ip)/2t^{\frac{1}{2}}, \quad p \text{ and } t \text{ are real.} \quad |q|\to\infty. \tag{22}$$

Repeated Integrals and Derivatives

$$i^{-1}\operatorname{erfc}(z) = \frac{2e^{-z^2}}{\pi^{\frac{1}{2}}}, \quad i^0\operatorname{erfc}(z) = \operatorname{erfc}(z) = 2\pi^{-\frac{1}{2}}\operatorname{Erfc}(z)$$

$$i^n\operatorname{erfc}(z) = \int_z^{\infty} i^{n-1}\operatorname{erfc}(t)\,dt. \tag{23}$$

$$i^n \operatorname{erfc}(z) = \frac{e^{-z^2}}{2^n\Gamma(n/2+1)}\, {}_1F_1\left(\left.\begin{matrix}(n+1)/2\\ 1/2\end{matrix}\right| z^2\right) - \frac{ze^{-z^2}}{2^{n-1}\Gamma[(n+1)/2]}\, {}_1F_1\left(\left.\begin{matrix}n/2+1\\ 3/2\end{matrix}\right| z^2\right). \tag{24}$$

$$i^n\operatorname{erfc}(z) = \frac{e^{-z^2}}{2^n\pi^{\frac{1}{2}}}\, U(\tfrac{1}{2}n+\tfrac{1}{2};\ \tfrac{1}{2};\ z^2). \tag{25}$$

$$\frac{d^n}{dz^n}\left(e^{z^2}\operatorname{erfc}(z)\right) = (-)^n n!\,2^n i^n\operatorname{erfc}(z). \tag{26}$$

$$i^n\operatorname{erfc}(z) + (z/n)i^{n-1}\operatorname{erfc}(z) - (2n)^{-1}i^{n-2}\operatorname{erfc}(z) = 0. \tag{27}$$

$$i^n \operatorname{erfc}(z) \sim \frac{2e^{-z^2}}{\pi^{1/2}(2z)^{n+1}}\,{}_2F_0((n+1)/2, n/2+1; -1/z^2),$$
$$|z| \to \infty, \qquad |\arg z| \leqslant 3\pi/4 - \epsilon, \qquad \epsilon > 0. \tag{28}$$

$$i^n \operatorname{erfc}(z) = \frac{e^{-z^2/2}e^{-z(2n)^{1/2}}}{2^n \Gamma(n/2+1)}[1 + O(n^{-1/2})], \qquad z \text{ bounded}, \quad n \to \infty. \tag{29}$$

4.8.2. Expansions in Series of Chebyshev Polynomials and Bessel Functions

$$\operatorname{Erf}(ax) = ae^{-a^2/2}\sum_{n=0}^{\infty}\frac{(-)^n}{2n+1}\left[I_n\left(\frac{a^2}{2}\right) + I_{n+1}\left(\frac{a^2}{2}\right)\right]T_{2n+1}(x),$$
$$-1 \leqslant x \leqslant 1. \tag{1}$$

$$\operatorname{Erf}(ax) = x\sum_{n=0}^{\infty} B_n T_{2n}(x), \qquad -1 \leqslant x \leqslant 1,$$

$$B_n = \epsilon_n a e^{-a^2/2}(-)^n\left[\frac{I_n(a^2/2)}{2n+1} + 4\sum_{k=0}^{\infty}\frac{(n+k+1)}{(2n+2k+1)(2n+2k+3)}I_{n+k+1}\left(\frac{a^2}{2}\right)\right]. \tag{2}$$

$$e^{a^2x^2}\operatorname{Erf}(ax) = \pi^{1/2}e^{a^2/2}\sum_{n=0}^{\infty} I_{n+\frac{1}{2}}(a^2/2)T_{2n+1}(x), \qquad -1 \leqslant x \leqslant 1. \tag{3}$$

$$e^{a^2x^2}\operatorname{Erf}(ax) = x\sum_{n=0}^{\infty} A_n T_{2n}(x), \qquad -1 \leqslant x \leqslant 1,$$

$$A_n = \pi^{1/2}e^{a^2/2}\epsilon_n\sum_{k=0}^{\infty}(-)^k I_{n+k+\frac{1}{2}}(a^2/2). \tag{4}$$

Expansions for $\operatorname{Erfi}(z)$ follow by an obvious change of variable.

Further expansions follow from the definition 4.8.1(2) and 5.10.1(11).

TABLE 4.5

Chebyshev Coefficients for the Error and Complementary Error Functions

$$\int_0^x e^{-t^2}dt = \sum_{n=0}^{\infty} a_n T_{2n+1}(x/3) \qquad \int_0^x e^{t^2}dt = \sum_{n=0}^{\infty} b_n T_{2n+1}(x/3)$$

$-3 \le x \le 3$

n	a_n	n	b_n
0	1.09547 12997 77623 19604	0	564.93377 09320 26671 65422
1	-0.28917 54011 26989 01480	1	427.37482 53497 79790 13041
2	0.11045 63986 33795 06164	2	254.46516 33961 29321 33705
3	-0.04125 31882 27856 54783	3	123.27359 99890 67173 81178
4	0.01408 28380 70651 63996	4	50.00015 06760 82055 10310
5	-0.00432 92954 47431 43677	5	17.37648 41276 25572 07028
6	0.00119 82719 01592 28759	6	5.27074 54850 34282 53383
7	-0.00029 99729 62353 24930	7	1.41638 59047 88468 45344
8	0.00006 83258 60378 87479	8	0.34133 82175 99699 05632
9	-0.00001 42469 88454 86775	9	0.07451 87925 75012 09561
10	0.00000 27354 08772 83989	10	0.01486 28131 95052 25783
11	-0.00000 04861 91287 19754	11	0.00272 77906 30521 62621
12	0.00000 00803 87276 21172	12	0.00046 35246 74683 58331
13	-0.00000 00124 18418 31213	13	0.00007 33171 17103 43030
14	0.00000 00017 99532 58879	14	0.00001 08451 31424 81434
15	-0.00000 00002 45479 48775	15	0.00000 15064 18237 43027
16	0.00000 00000 31625 08603	16	0.00000 01972 06009 50459
17	-0.00000 00000 03859 02200	17	0.00000 00244 10182 48054
18	0.00000 00000 00447 20291	18	0.00000 00028 65274 12590
19	-0.00000 00000 00049 33613	19	0.00000 00003 19778 92898
20	0.00000 00000 00005 19303	20	0.00000 00000 34014 14529
21	-0.00000 00000 00000 52258	21	0.00000 00000 03455 73355
22	0.00000 00000 00000 05037	22	0.00000 00000 00336 01288
23	-0.00000 00000 00000 00466	23	0.00000 00000 00031 32561
24	0.00000 00000 00000 00041	24	0.00000 00000 00002 80480
25	-0.00000 00000 00000 00004	25	0.00000 00000 00000 24157
		26	0.00000 00000 00000 02004
		27	0.00000 00000 00000 00160
		28	0.00000 00000 00000 00012
		29	0.00000 00000 00000 00001

TABLE 4.5 *(Concluded)*

$$\int_x^\infty e^{-t^2}dt = \frac{e^{-x^2}}{2x}\sum_{n=0}^{\infty} c_n T_{2n}(3/x) \qquad \int_0^x e^{t^2}dt = \frac{e^{x^2}}{2x}\sum_{n=0}^{\infty} d_n T_{2n}(3/x)$$

$x \geq 3$

n	c_n	n	d_n
0	0.97508 34237 08555 92854	0	1.03262 24550 64934
1	-0.02404 93938 50414 60496	1	0.03455 14313 66641
2	0.00082 04522 40880 43199	2	0.00215 99451 18650
3	-0.00004 34293 08130 34276	3	0.00026 45367 70758
4	0.00000 30184 47034 03493	4	0.00003 40993 28995
5	-0.00000 02544 73319 25082	5	-0.00000 20792 21959
6	0.00000 00248 58353 02051	6	-0.00000 35534 10422
7	-0.00000 00027 31720 13238	7	-0.00000 09174 72782
8	0.00000 00003 30847 22797	8	0.00000 01451 00081
9	-0.00000 00000 43505 49080	9	0.00000 01365 12657
10	0.00000 00000 06141 21457	10	0.00000 00018 58501
11	-0.00000 00000 00922 36928	11	-0.00000 00185 45233
12	0.00000 00000 00146 35665	12	-0.00000 00014 17393
13	-0.00000 00000 00024 39278	13	0.00000 00028 37303
14	0.00000 00000 00004 24976	14	0.00000 00001 92580
15	-0.00000 00000 00000 77084	15	-0.00000 00004 92284
16	0.00000 00000 00000 14507	16	0.00000 00000 04740
17	-0.00000 00000 00000 02824	17	0.00000 00000 90213
18	0.00000 00000 00000 00567	18	-0.00000 00000 13272
19	-0.00000 00000 00000 00117	19	-0.00000 00000 15602
20	0.00000 00000 00000 00025	20	0.00000 00000 05643
21	-0.00000 00000 00000 00005	21	0.00000 00000 02055
22	0.00000 00000 00000 00001	22	-0.00000 00000 01693
		23	-0.00000 00000 00015
		24	0.00000 00000 00384
		25	-0.00000 00000 00111
		26	-0.00000 00000 00055
		27	0.00000 00000 00045
		28	-0.00000 00000 00003
		29	-0.00000 00000 00010
		30	0.00000 00000 00005
		31	0.00000 00000 00001
		32	-0.00000 00000 00002

$$\int_0^\infty e^{-t^2}dt = \tfrac{1}{2}\pi^{\frac{1}{2}}$$

4.8.3. Padé Approximations

TABLE 4.6

Coefficients for Padé Approximations to $\int_0^z e^t t^{-1/2}\,dt$ and the Errors in These Approximations

These approximations follow from material in 4.3.1 with $\nu = \frac{1}{2}$. See also 4.8.1(6-7). We write

$$\int_0^z e^t t^{-\frac{1}{2}}dt = 2e^z z^{\frac{1}{2}}\left[\frac{C_n(\frac{1}{2},z)}{D_n(\frac{1}{2},z)} + V_n(\tfrac{1}{2},z)\right],$$

TABLE 4.6 *(Continued)*

$$|2e^{z}z^{\frac{1}{2}}V_n(\tfrac{1}{2},z)| = |n/4z|^{a}R_n(z)\left[1 + O(n^{-3})\right].$$

The coefficients in the polynomials $C_n(\frac{1}{2},z)$ and $D_n(\frac{1}{2},z)$ are recorded below for $n = 0(1)6$, $a = 0$ and $a = 1$. Further coefficients for $n = 7(1)10$ are available in Luke (1969). Additional polynomials can be generated from 4.3.1(15). We also record $R_n(z)$ for a wide range of n and z values.

$$C_n(\tfrac{1}{2},z) = \sum_{k=0}^{n} a_k z^{n-k}, \quad a = 0 .$$

n	$a_0, a_1, \ldots, a_n$
0	1
1	-4, 15
2	32, -210, 945
3	-128, 1932, -9240, 45045
4	2048, -43560, 5 40540, -22 52250, 114 86475
5	-8192, 2 81424, -38 91888, 453 33288, -1745 94420, 9166 20705
6	65536, -28 85792, 699 73904, -7914 94704, 89625 13560, -3 27946 51890, 17 56856 35125

$$D_n(\tfrac{1}{2},z) = \sum_{k=0}^{n} b_k z^{n-k}, \quad a = 0$$

n	$b_0, b_1, \ldots, b_n$
0	1
1	6, 15
2	60, 420, 945
3	280, 3780, 20790, 45045
4	5040, 1 10880, 10 81080, 54 05400, 114 86475
5	22176, 7 20720, 108 10800, 918 91800, 4364 86050, 9166 20705
6	1 92192, 86 48640, 1837 83600, 23279 25600, 1 83324 14100, 8 43291 04860, 17 56856 35125

TABLE 4.6 *(Continued)*

$$C_n(\tfrac{1}{2},z) = \sum_{k=0}^{n} c_k z^{n-k}\ ,\quad a = 1$$

n	$c_0, c_1, \ldots, c_n$
0	0
1	0, 3
2	0, -10, 105
3	0, 84, -420, 3465
4	0, -744, 23100, -90090, 6 75675
5	0, 5104, -82368, 17 29728, -61 26120, 436 48605
6	0, -25376, 15 53552, -171 53136, 3026 30328, -10184 67450, 70274 25405

$$D_n(\tfrac{1}{2},z) = \sum_{k=0}^{n} d_k z^{n-k}\ ,\quad a = 1$$

n	$d_0, d_1, \ldots, d_n$
0	1
1	2, 3
2	12, 60, 105
3	40, 420, 1890, 3465
4	560, 10080, 83160, 3 60360, 6 75675
5	2016, 55440, 7 20720, 54 05400, 229 72950, 436 48605
6	14784, 5 76576, 108 10800, 1225 22400, 8729 72100, 36664 82820, 70274 25405

TABLE 4.6 *(Continued)*

$R_n(z), \quad z = re^{i\theta}$

n/θ	0	π/2	π
	r = 1		
2	0.896(-3)	0.747(-3)	0.747(-3)
3	0.521(-5)	0.456(-5)	0.456(-5)
4	0.178(-7)	0.160(-7)	0.160(-7)
5	0.401(-10)	0.368(-10)	0.368(-10)
6	0.639(-13)	0.593(-13)	0.593(-13)
7	0.757(-16)	0.710(-16)	0.710(-16)
8	0.693(-19)	0.655(-19)	0.655(-19)
9	0.505(-22)	0.480(-22)	0.480(-22)
10	0.300(-25)	0.286(-25)	0.286(-25)
	r = 2		
2	0.509(-1)	0.295(-1)	0.354(-1)
3	0.111(-2)	0.746(-3)	0.853(-3)
4	0.147(-4)	0.107(-4)	0.119(-4)
5	0.130(-6)	0.998(-7)	0.109(-6)
6	0.812(-9)	0.650(-9)	0.700(-9)
7	0.380(-11)	0.313(-11)	0.334(-11)
8	0.138(-13)	0.116(-13)	0.123(-13)
9	0.400(-16)	0.343(-16)	0.361(-16)
10	0.943(-19)	0.820(-19)	0.860(-19)
	r = 3		
2	0.651(0)	0.219(0)	0.377(0)
3	0.294(-1)	0.132(-1)	0.197(-1)
4	0.833(-3)	0.443(-3)	0.608(-3)
5	0.160(-4)	0.949(-5)	0.123(-4)
6	0.220(-6)	0.141(-6)	0.176(-6)
7	0.228(-8)	0.155(-8)	0.188(-8)
8	0.184(-10)	0.131(-10)	0.155(-10)
9	0.119(-12)	0.872(-13)	0.102(-12)
10	0.625(-15)	0.473(-15)	0.544(-15)
	r = 4		
2	0.477(1)	0.774(0)	0.230(1)
3	0.344(0)	0.905(-1)	0.202(0)
4	0.162(-1)	0.567(-2)	0.107(-1)
5	0.532(-3)	0.223(-3)	0.375(-3)
6	0.127(-4)	0.603(-5)	0.941(-5)
7	0.228(-6)	0.120(-6)	0.176(-6)
8	0.322(-8)	0.182(-8)	0.256(-8)
9	0.364(-10)	0.218(-10)	0.296(-10)
10	0.337(-12)	0.217(-12)	0.280(-12)
	r = 5		
2	0.268(2)	0.175(1)	0.108(2)
3	0.264(1)	0.358(0)	0.136(1)
4	0.181(0)	0.373(-1)	0.108(0)
5	0.879(-2)	0.238(-2)	0.569(-2)
6	0.316(-3)	0.104(-3)	0.218(-3)
7	0.866(-5)	0.329(-5)	0.627(-5)
8	0.187(-6)	0.793(-7)	0.140(-6)
9	0.325(-8)	0.151(-8)	0.251(-8)
10	0.464(-10)	0.231(-10)	0.368(-10)

TABLE 4.6 *(Concluded)*

$R_n(z), \quad z = re^{i\theta}$

n/θ	0	π/2	π
		r = 6	
2	0.132(3)	0.290(1)	0.443(2)
3	0.160(2)	0.973(0)	0.719(1)
4	0.144(1)	0.158(0)	0.765(0)
5	0.949(-1)	0.153(-1)	0.563(-1)
6	0.470(-2)	0.993(-3)	0.302(-2)
7	0.180(-3)	0.465(-4)	0.122(-3)
8	0.547(-5)	0.165(-5)	0.388(-5)
9	0.134(-6)	0.458(-7)	0.987(-7)
10	0.272(-8)	0.102(-8)	0.206(-8)
		r = 7	
2	0.609(3)	0.375(1)	0.171(3)
3	0.838(2)	0.200(1)	0.330(2)
4	0.923(1)	0.484(0)	0.442(1)
5	0.774(0)	0.678(-1)	0.421(0)
6	0.498(-1)	0.625(-2)	0.296(-1)
7	0.250(-2)	0.411(-3)	0.159(-2)
8	0.101(-3)	0.203(-4)	0.674(-4)
9	0.329(-5)	0.783(-6)	0.230(-5)
10	0.891(-7)	0.242(-7)	0.643(-7)
		r = 8	
2	0.275(4)	0.395(1)	0.642(3)
3	0.402(3)	0.331(1)	0.138(3)
4	0.513(2)	0.116(1)	0.221(2)
5	0.520(1)	0.227(0)	0.259(1)
6	0.414(0)	0.287(-1)	0.229(0)
7	0.261(-1)	0.256(-2)	0.156(-1)
8	0.133(-2)	0.170(-3)	0.840(-3)
9	0.553(-4)	0.873(-5)	0.367(-4)
10	0.192(-5)	0.359(-6)	0.132(-5)
		r = 9	
2	0.125(5)	0.349(1)	0.243(4)
3	0.183(4)	0.454(1)	0.552(3)
4	0.259(3)	0.227(1)	0.100(3)
5	0.305(2)	0.609(0)	0.139(2)
6	0.288(1)	0.103(0)	0.148(1)
7	0.220(0)	0.121(-1)	0.123(0)
8	0.138(-1)	0.104(-2)	0.817(-2)
9	0.701(-3)	0.698(-4)	0.442(-3)
10	0.301(-4)	0.371(-5)	0.198(-4)
		r = 10	
2	0.578(5)	0.262(1)	0.938(4)
3	0.813(4)	0.531(1)	0.214(4)
4	0.123(4)	0.375(1)	0.428(3)
5	0.161(3)	0.135(1)	0.677(2)
6	0.176(2)	0.300(0)	0.841(1)
7	0.158(1)	0.455(-1)	0.829(0)
8	0.117(0)	0.503(-2)	0.659(-1)
9	0.716(-2)	0.427(-3)	0.429(-2)
10	0.370(-3)	0.287(-4)	0.232(-3)

TABLE 4.7

COEFFICIENTS FOR PADÉ APPROXIMATIONS TO $\int_z^\infty e^{-t}t^{-1/2}\,dt$ AND THE ERRORS IN THESE APPROXIMATIONS

These approximations follow from material in 4.3.2-4.3.5 with $\nu = \frac{1}{2}$. See also 4.8.1(8,9). We write

$$\int_z^\infty e^{-t}t^{-\frac{1}{2}}dt = e^{-z}z^{-\frac{1}{2}}\left[\frac{G_n(\frac{1}{2},z)}{H_n(\frac{1}{2},z)} + T_n(\frac{1}{2},z)\right], \quad z \neq 0, \quad |\arg z| < \pi,$$

$$V_n(z) = \left| e^{-z}z^{-\frac{1}{2}}\left[T_n(\tfrac{1}{2},z)\right]_{a=0} \right| .$$

The coefficients in the polynomials $G_n(\frac{1}{2},z)$ and $H_n(\frac{1}{2},z)$ are recorded below for $n = 0(1)6$, $a = 0$ and $a = 1$. Further coefficients for $n = 7(1)10$ have been given by Luke (1969). Additional polynomials can be generated from 4.3.2(14). We also record $V_n(z)$ for a wide range of n and z values. Error coefficients for $a = 1$ readily follow from 4.3.2(15) or 4.3.3(12) as appropriate.

$$G_n(\tfrac{1}{2},z) = \sum_{k=0}^{n} a_k z^{n-k}, \quad a = 0 .$$

n	$a_0, a_1, \ldots, a_n$
0	1
1	2, 2
2	4, 18, 8
3	8, 80, 174, 48
4	16, 280, 1380, 1950, 384
5	32, 864, 7504, 24360, 25290, 3840
6	64, 2464, 33120, 1 90512, 4 59060, 3 74850, 46080

TABLE 4.7 *(Continued)*

$$H_n(\tfrac{1}{2},z) = \sum_{k=0}^{n} b_k z^{n-k}, \quad a = 0$$

n	$b_0, b_1, \ldots, b_n$
0	1
1	2, 3
2	4, 20, 15
3	8, 84, 210, 105
4	16, 288, 1512, 2520, 945
5	32, 880, 7920, 27720, 34650, 10395
6	64, 2496, 34320, 2 05920, 5 40540, 5 40540, 1 35135

$$G_n(\tfrac{1}{2},z) = \sum_{k=0}^{n} c_k z^{n-k}, \quad a = 1$$

n	$c_0, c_1, \ldots, c_n$
0	0
1	2, 0
2	4, 10, 0
3	8, 56, 66, 0
4	16, 216, 740, 558, 0
5	32, 704, 4704, 10560, 5790, 0
6	64, 2080, 22752, 1 00464, 1 66740, 71370, 0

$$H_n(\tfrac{1}{2},z) = \sum_{k=0}^{n} d_k z^{n-k}, \quad a = 1$$

n	$d_0, d_1, \ldots, d_n$
0	1
1	2, 1
2	4, 12, 3
3	8, 60, 90, 15
4	16, 224, 840, 840, 105
5	32, 720, 5040, 12600, 9450, 945
6	64, 2112, 23760, 1 10880, 2 07900, 1 24740, 10395

TABLE 4.7 *(Continued)*

$V_n(z)$, $z = re^{i\theta}$

r = 1

n/θ	0°	45°	60°	75°	90°
4	0.535(-3)	0.109(-2)	0.185(-2)	0.360(-2)	0.785(-2)
6	0.102(-3)	0.233(-3)	0.436(-3)	0.951(-3)	0.238(-2)
8	0.243(-4)	0.619(-4)	0.125(-3)	0.303(-3)	0.856(-3)
12	0.212(-5)	0.646(-5)	0.150(-4)	0.432(-4)	0.150(-3)
16	0.264(-6)	0.942(-6)	0.247(-5)	0.823(-5)	0.342(-4)
20	0.417(-7)	0.171(-6)	0.497(-6)	0.189(-5)	0.923(-5)

n/θ	105°	120°	135°	150°	165°
4	0.189(-1)	0.490(-1)	0.134(0)	0.370(0)	0.940(0)
6	0.671(-2)	0.208(-1)	0.691(-1)	0.237(0)	0.783(0)
8	0.277(-2)	0.101(-1)	0.403(-1)	0.176(0)	0.880(0)
12	0.619(-3)	0.294(-2)	0.157(-1)	0.917(-1)	0.554(0)
16	0.173(-3)	0.103(-2)	0.699(-2)	0.519(-1)	0.384(0)
20	0.558(-4)	0.406(-3)	0.343(-2)	0.326(-1)	0.355(0)

r = 2

n/θ	0°	45°	60°	75°	90°
4	0.126(-4)	0.366(-4)	0.810(-4)	0.215(-3)	0.670(-3)
6	0.122(-5)	0.417(-5)	0.105(-4)	0.326(-4)	0.123(-3)
8	0.163(-6)	0.642(-6)	0.180(-5)	0.644(-5)	0.287(-4)
12	0.524(-8)	0.264(-7)	0.894(-7)	0.408(-6)	0.243(-5)
16	0.278(-9)	0.174(-8)	0.694(-8)	0.390(-7)	0.298(-6)
20	0.206(-10)	0.156(-9)	0.720(-9)	0.487(-8)	0.465(-7)

n/θ	105°	120°	135°	150°	165°
4	0.236(-2)	0.921(-2)	0.388(-1)	0.176(0)	0.939(0)
6	0.543(-3)	0.273(-2)	0.152(-1)	0.900(-1)	0.530(0)
8	0.154(-3)	0.965(-3)	0.679(-2)	0.515(-1)	0.389(0)
12	0.183(-4)	0.166(-3)	0.176(-2)	0.210(-1)	0.277(0)
16	0.298(-5)	0.373(-4)	0.560(-3)	0.958(-2)	0.175(0)
20	0.600(-6)	0.998(-5)	0.203(-3)	0.485(-2)	0.132(0)

r = 4

n/θ	0°	45°	60°	75°	90°
4	0.519(-7)	0.279(-6)	0.957(-6)	0.428(-5)	0.236(-4)
6	0.201(-8)	0.132(-7)	0.532(-7)	0.293(-6)	0.208(-5)
8	0.120(-9)	0.948(-9)	0.441(-8)	0.292(-7)	0.261(-6)
12	0.960(-12)	0.106(-10)	0.636(-10)	0.584(-9)	0.776(-8)
16	0.155(-13)	0.228(-12)	0.171(-11)	0.209(-10)	0.392(-9)
20	0.395(-15)	0.756(-14)	0.695(-13)	0.110(-11)	0.280(-10)

n/θ	105°	120°	135°	150°	165°
4	0.151(-3)	0.107(-2)	0.795(-2)	0.600(-1)	0.432(0)
6	0.181(-4)	0.182(-3)	0.201(-2)	0.238(-1)	0.294(0)
8	0.296(-5)	0.405(-4)	0.630(-3)	0.107(-1)	0.187(0)
12	0.140(-6)	0.323(-5)	0.898(-4)	0.285(-2)	0.959(-1)
16	0.106(-7)	0.382(-6)	0.174(-4)	0.934(-3)	0.567(-1)
20	0.108(-8)	0.580(-7)	0.408(-5)	0.350(-3)	0.340(-1)

TABLE 4.7 *(Continued)*

$V_n(z), \quad z = re^{i\theta}$

r = 6

n/θ	0°	45°	60°	75°	90°
4	0.637(-9)	0.607(-8)	0.313(-7)	0.225(-6)	0.206(-5)
6	0.124(-10)	0.149(-9)	0.915(-9)	0.830(-8)	0.102(-6)
8	0.407(-12)	0.600(-11)	0.436(-10)	0.490(-9)	0.784(-8)
12	0.115(-14)	0.248(-13)	0.243(-12)	0.401(-11)	0.103(-9)
16	0.753(-17)	0.228(-15)	0.291(-14)	0.674(-13)	0.260(-11)
20	0.860(-19)	0.355(-17)	0.576(-16)	0.181(-14)	0.101(-12)

n/θ	105°	120°	135°	150°	165°
4	0.220(-4)	0.254(-3)	0.294(-2)	0.329(-1)	0.355(0)
6	0.155(-5)	0.272(-4)	0.510(-3)	0.978(-2)	0.194(0)
8	0.164(-6)	0.415(-5)	0.118(-3)	0.358(-2)	0.109(0)
12	0.374(-8)	0.179(-6)	0.103(-4)	0.680(-3)	0.477(-1)
16	0.154(-9)	0.127(-7)	0.134(-5)	0.170(-3)	0.239(-1)
20	0.922(-11)	0.124(-8)	0.224(-6)	0.501(-4)	0.129(-1)

r = 8

n/θ	0°	45°	60°	75°	90°
4	0.134(-10)	0.224(-9)	0.172(-8)	0.195(-7)	0.292(-6)
6	0.150(-12)	0.318(-11)	0.295(-10)	0.429(-9)	0.878(-8)
8	0.300(-14)	0.797(-13)	0.881(-12)	0.162(-10)	0.444(-9)
12	0.359(-17)	0.145(-15)	0.221(-14)	0.622(-13)	0.287(-11)
16	0.111(-19)	0.652(-18)	0.134(-16)	0.551(-15)	0.405(-13)
20	0.649(-22)	0.539(-20)	0.145(-18)	0.841(-17)	0.940(-15)

n/θ	105°	120°	135°	150°	165°
4	0.506(-5)	0.907(-4)	0.152(-2)	0.225(-1)	0.279(0)
6	0.224(-6)	0.639(-5)	0.186(-3)	0.522(-2)	0.134(0)
8	0.161(-7)	0.695(-6)	0.327(-4)	0.158(-2)	0.766(-1)
12	0.195(-9)	0.174(-7)	0.186(-5)	0.220(-3)	0.277(-1)
16	0.474(-11)	0.790(-9)	0.170(-6)	0.431(-4)	0.121(-1)
20	0.180(-12)	0.525(-10)	0.209(-7)	0.104(-4)	0.588(-2)

r = 12

n/θ	0°	45°	60°	75°	90°
4	0.143(-13)	0.740(-12)	0.125(-10)	0.352(-9)	0.138(-7)
6	0.662(-16)	0.434(-14)	0.890(-13)	0.325(-11)	0.177(-9)
8	0.597(-18)	0.498(-16)	0.123(-14)	0.580(-13)	0.435(-11)
12	0.176(-21)	0.231(-19)	0.823(-18)	0.622(-16)	0.845(-14)
16	0.160(-24)	0.321(-22)	0.160(-20)	0.187(-18)	0.435(-16)
20	0.309(-27)	0.924(-25)	0.627(-23)	0.110(-20)	0.419(-18)

n/θ	105°	120°	135°	150°	165°
4	0.611(-6)	0.253(-4)	0.826(-3)	0.188(-1)	0.288(0)
6	0.119(-7)	0.821(-6)	0.506(-4)	0.253(-2)	0.101(0)
8	0.433(-9)	0.484(-7)	0.529(-5)	0.520(-3)	0.453(-1)
12	0.173(-11)	0.457(-9)	0.135(-6)	0.411(-4)	0.122(-1)
16	0.170(-13)	0.952(-11)	0.663(-8)	0.519(-5)	0.425(-2)
20	0.294(-15)	0.324(-12)	0.481(-9)	0.863(-6)	0.171(-2)

TABLE 4.7 *(Concluded)*

$V_n(z), \quad z = re^{i\theta}$

r = 16

n/θ	0°	45°	60°	75°	90°
4	0.306(-16)	0.487(-14)	0.181(-12)	0.128(-10)	0.131(-8)
6	0.695(-19)	0.140(-16)	0.632(-15)	0.574(-13)	0.817(-11)
8	0.331(-21)	0.850(-19)	0.465(-17)	0.546(-15)	0.108(-12)
12	0.312(-25)	0.129(-22)	0.103(-20)	0.200(-18)	0.747(-16)
16	0.104(-28)	0.674(-26)	0.772(-24)	0.240(-21)	0.162(-18)
20	0.806(-32)	0.806(-29)	0.130(-26)	0.627(-24)	0.731(-21)

n/θ	105°	120°	135°	150°	165°
4	0.149(-6)	0.142(-4)	0.887(-3)	0.290(-1)	0.397(0)
6	0.140(-8)	0.224(-6)	0.270(-4)	0.208(-2)	0.103(0)
8	0.280(-10)	0.749(-8)	0.169(-5)	0.285(-3)	0.355(-1)
12	0.423(-13)	0.292(-10)	0.204(-7)	0.129(-4)	0.707(-2)
16	0.187(-15)	0.301(-12)	0.562(-9)	0.107(-5)	0.199(-2)
20	0.163(-17)	0.566(-14)	0.253(-10)	0.128(-6)	0.671(-3)

r = 20

n/θ	0°	45°	60°	75°	90°
4	0.983(-19)	0.487(-16)	0.402(-14)	0.711(-12)	0.194(-9)
6	0.124(-21)	0.770(-19)	0.765(-17)	0.173(-14)	0.644(-12)
8	0.343(-24)	0.271(-21)	0.326(-19)	0.951(-17)	0.493(-14)
12	0.123(-28)	0.157(-25)	0.278(-23)	0.135(-20)	0.135(-17)
16	0.172(-32)	0.352(-29)	0.907(-27)	0.724(-24)	0.135(-20)
20	0.611(-36)	0.197(-32)	0.725(-30)	0.926(-27)	0.309(-23)

n/θ	105°	120°	135°	150°	165°
4	0.574(-7)	0.129(-4)	0.158(-2)	0.781(-1)	1.671(0)
6	0.282(-9)	0.104(-6)	0.238(-4)	0.266(-2)	0.127(0)
8	0.327(-11)	0.202(-8)	0.891(-6)	0.233(-3)	0.344(-1)
12	0.202(-14)	0.341(-11)	0.511(-8)	0.587(-5)	0.502(-2)
16	0.433(-17)	0.183(-13)	0.812(-10)	0.327(-6)	0.113(-2)
20	0.203(-19)	0.199(-15)	0.233(-11)	0.281(-7)	0.323(-3)

4.8.4. Trapezoidal Rule Approximations

Consider

$$\operatorname{Erfc}(az) = \pi^{-\frac{1}{2}} z e^{-a^2z^2} \int_0^\infty \frac{e^{-a^2t^2}}{z^2+t^2} dt, \quad |\arg a| < \pi/4, \quad |\arg z| < \pi/2. \tag{1}$$

The domain of definition of the integral can be extended by rotating the path of integration. If $|\arg a| < \pi/4$, we can have $|\arg z| = \pi/2$ if we interpret the integral as a Cauchy principal value.

Let

$$w = 2\pi r/h, \quad u = az, \quad v = w/2a, \tag{2}$$

$$T(a,z,h) = \pi^{-\frac{1}{2}} z h e^{-a^2z^2} \left[\frac{1}{2z^2} + \sum_{k=1}^{\infty} \frac{e^{-a^2k^2h^2}}{k^2h^2+z^2} \right], \tag{3}$$

$$M(a,z,h) = \pi^{-\frac{1}{2}} z h e^{-a^2z^2} \sum_{k=0}^{\infty} \frac{e^{-a^2(2k+1)^2h^2/4}}{(2k+1)^2h^2/4 + z^2}. \tag{4}$$

Equation (3) is the trapezoidal rule approximation to (1). We call (4) the modified trapezoidal rule approximation. We have

$$\operatorname{Erfc}(az) = T(a,z,h) + P(h), \tag{5}$$

$$\operatorname{Erfc}(az) = M(a,z,h) + Q(h), \tag{6}$$

$$P(h) = -2 \sum_{r=1}^{\infty} G_r, \quad Q(h) = -2 \sum_{r=1}^{\infty} (-)^r G_r, \tag{7}$$

$$G_r = \pi^{-\frac{1}{2}} z e^{-a^2z^2} \int_0^\infty \frac{e^{-a^2t^2} \cos wt}{z^2+t^2} dt, \tag{8}$$

$$G_r = \tfrac{1}{2} \left[e^{zw} \operatorname{Erfc}(u+v) + e^{-zw} \operatorname{Erfc}(u-v) \right], \tag{9}$$

The error terms $P(h)$ and $Q(h)$ depend on G_r. We have the following asymptotic estimates.

$$G_r = \frac{ue^{-(u^2+v^2)}}{2(u^2-v^2)}\left[1 + O(u\pm v)^{-2}\right],$$

$$|u\pm v| \to \infty, \quad |\arg\,(u\pm v)| < 3\pi/4, \tag{10}$$

$$G_r = \frac{\pi^{\frac{1}{2}}e^{-zw}}{2} + \frac{ue^{-(u^2+v^2)}}{2(u^2-v^2)}\left(1 + O\{(u\pm v)^{-2}\}\right),$$

$$|u\pm v| \to \infty, \ |\arg\,(u+v)| < 3\pi/4, \ 3\pi/4 \le |\arg\,(u-v)| \le \pi, \tag{11}$$

$$G_r = \frac{\pi^{\frac{1}{2}}e^{-zw}}{4} - \frac{e^{-zw}(u-v)}{2}\left(1 + O\{(u-v)^2\}\right)$$
$$+ \frac{e^{-(u^2+v^2)}}{4(u+v)}\left(1 + O\{(u+v)^{-2}\}\right),$$

$$|u-v| \to 0, \quad |u+v| \to \infty, \quad |\arg\,(u+v)| < 3\pi/4. \tag{12}$$

In the following discussion, we suppose that if $|u-v| \to 0$ as in (12), then this happens when $r = 1$. Then reliable a priori estimates of the error follow from the appropriate estimates of G_r for $r = 1$ only. To improve the accuracy, the $\pi^{\frac{1}{2}}e^{-zw}$ terms appearing in (11) and (12) should be removed from the respective error terms and incorporated into the evaluation of $T(a,z,h)$ and $M(a,z,h)$ as appropriate. Let

$$T^*(z) = -\sum_{r=1}^{\infty} e^{-zw} = -(e^{2\pi z/h} - 1)^{-1}, \tag{13}$$

$$M^*(z) = -\sum_{r=1}^{\infty} (-)^r e^{-zw} = (e^{2\pi z/h} + 1)^{-1}. \tag{14}$$

Let A, B and C stand for the conditions on $u\pm v$ given in (10), (11) and (12) respectively. Then

$$\text{Erfc}(az) = T(a,z,h) + bT^*(z) + R(h), \tag{15}$$

$$\text{Erfc}(az) = M(a,z,h) + bM^*(z) - R(h), \tag{16}$$

where

$b = 0$, $R(h) = -2G_1$, G_r as in (10), condition A,

$$b = \pi^{\frac{1}{2}},\ R(h) = -\frac{u\exp\left(-(u^2+v^2)\right)}{2(u^2-v^2)}\left[1 + O\{(u\pm v)^{-2}\}\right],$$

$r = 1$, condition B,

$$b = \tfrac{1}{2}\pi^{\frac{1}{2}},\ R(h) = e^{-zw}(u-v)\left(1 + O\{(u-v)^{-2}\}\right)$$

$$-\frac{\exp\left(-(u^2+v^2)\right)}{2(u+v)}\left[1 + O\{(u+v)^{-2}\}\right],\ r = 1,\ \text{condition C.} \tag{17}$$

Notice that if $r = 1$ and a and z are real, conditions A, B and C are equivalent to $z > \pi/ha^2$, $z < \pi/ha^2$ and $z = \pi/ha^2$ respectively.

There is little difference in the accuracy of (15) and (16) with $R(h)$ omitted and both are quite effective even for relatively large values of h for nearly all values of az in the right half plane. Indeed, the accuracy is for the most part good even when $z = 0$ in which event (3) breaks down. However, the approximation cannot be used if $z = ikh$ and is numerically unstable if z is near one of these values. Similar remarks pertain to (4) which breaks down for $z = i(2k+1)h/2$. The difficulty can often be avoided by a judicious choice of h, though even this will be ineffectual if z is close to zero. A better strategy is to use (15) if $1/4 \leq f(y/h) \leq 3/4$ and to use (16) otherwise, where $z = x+iy$ and $f(y/h)$ denotes the fractional part of y/h, i.e., $f(y/h) = y/h - [y/h]$. In particular, use (16) when z is real and small. Indeed, if $z = 0$, (16) with $R(h)$ omitted gives $\text{Erfc}(0) = \frac{1}{2}\pi^{\frac{1}{2}}$ exactly whereas the right hand side of (3) becomes infinite.

The above discussion is based on developments given by Luke (1969) and incorporates later suggestions made by

Chiarella and Reichel (1968), Matta and Reichel (1971) and Hunter and Regan (1972). Matta and Reichel give explicitly the trapezoidal rule approximations for $w(z)$, the Voigt functions and other forms of the error function, see 4.8.1(1-21). For related work on integrals of $w(z)$, see Reichel (1967).

4.8.5. Inequalities

From 4.4.2(2,3) with $\nu = \frac{1}{2}$, we have

$$\frac{z^2}{2z^2+1} < ze^{z^2}\int_z^\infty e^{-t^2}\,dt < \frac{z^2+1}{2z^2+3}, \qquad 0 < z < \infty, \qquad (1)$$

$$\frac{z^2(2z^2+5)}{4z^4+12z^2+3} < ze^{z^2}\int_z^\infty e^{-t^2}\,dt < \frac{2z^4+9z^2+4}{4z^4+20z^2+15}, \qquad 0 < z < \infty. \qquad (2)$$

If $z = 3$, (2) gives

$$0.45217 < 0.45268 < 0.45283. \qquad (3)$$

A further inequality follows from 4.4.2(7) with $p = 2$. The following sharp inequalities are due to Boyd (1959).

$$\frac{\pi/2}{(z^2+\pi)^{\frac{1}{2}} + (\pi-1)z} \le e^{z^2}\int_z^\infty e^{-t^2}\,dt \le \frac{\pi/2}{\left[(\pi-2)^2z^2+\pi\right]^{\frac{1}{2}} + 2z},$$

$$0 \le z \le \infty. \qquad (4)$$

Strand (1965) has shown that

$$|\mathrm{erfc}(z)| < \frac{e^{y^2-x^2}}{\pi^{\frac{1}{2}}x}(1 + y^2/x^2), \quad z = x + iy, \quad x > 0. \qquad (5)$$

For other inequalities and approximations for the error function, see Mitrinovic and Vasic (1970).

Let

$$y_n(z) = i^n\mathrm{erfc}(z), \quad r_n(z) = y_n(z)/y_{n-1}(z). \qquad (6)$$

Amos (1973) has shown that for z real and $n \geq 1$,

$$C_n(z) < r_n(z) < D_n(z) \tag{7}$$

where the pair $C_n(z)$ and $D_n(z)$ can be any of the following pairs:

$$C_n(z) = \frac{(z^2+2n+2)^{\frac{1}{2}} - z}{2n+2}, \quad D_n(z) = C_{n-1}(z), \tag{8}$$

$$C_n(z) = \frac{(z^2+2n-2)^{\frac{1}{2}} - z}{2n}, \quad D_n(z) = \frac{(z^2+2n)^{\frac{1}{2}} - z}{2n}, \tag{9}$$

$$C_n(z) = \frac{a_n}{(z^2+2na_n)^{\frac{1}{2}} + z}, \quad D_n(z) = \frac{1}{\left(z^2+2(n+1)a_{n+1}\right)^{\frac{1}{2}} + z}, \quad z \geq 0, \tag{10}$$

$$C_n(z) = \frac{-z + \left(z^2+2(n+1)a_{n+1}\right)^{\frac{1}{2}}}{2(n+1)a_{n+1}}, \quad D_n(z) = \frac{-z + (z^2+2na_n)^{\frac{1}{2}}}{2n}, \quad z < 0, \tag{11}$$

where in (10) and (11)

$$a_n = \frac{n}{2}\left[\frac{\Gamma(n/2 + \frac{1}{2})}{\Gamma(n/2 + 1)}\right]^2 \tag{12}$$

Equations (8), (9) and the combination (10), (11) work best for large positive z, for large positive $-z$ and for z near zero respectively. Also (10) becomes an equality for $z = 0$. Numerical tables illustrating the effectiveness of these inequalities have been given by Amos (1973). Iterative use of (7) produces an inequality for $y_n(z)$ in terms of $y_0(z) = \operatorname{erfc}(z)$ which coupled with (4) gives a direct inequality for $y_n(z)$.

4.9. Fresnel Integrals

4.9.1. Relation to Error Functions and Other Properties

$$C(z) + iS(z) = (2\pi)^{-1/2} \int_0^z t^{-1/2} e^{it}\, dt = (2\pi)^{-1/2} e^{i\pi/4} \gamma(\tfrac{1}{2}, ze^{-i\pi/2})$$
$$= 2(2\pi)^{-1/2} e^{i\pi/4} \operatorname{Erf}(z^{1/2} e^{-i\pi/4}). \tag{1}$$

$$C(z) + iS(z) = (2\pi)^{-1/2} e^{i\pi/4} \{\pi^{1/2} - \Gamma(\tfrac{1}{2}, ze^{-i\pi/2})\}. \tag{2}$$

$$C(z) + iS(z) = (2z/\pi)^{1/2} \left\{ {}_1F_2\left(\begin{matrix} \frac{1}{4} \\ \frac{5}{4}, \frac{1}{2} \end{matrix} \middle| \frac{-z^2}{4} \right) + \tfrac{1}{3}(iz)\, {}_1F_2\left(\begin{matrix} \frac{3}{4} \\ \frac{7}{4}, \frac{3}{2} \end{matrix} \middle| \frac{-z^2}{4} \right) \right\}. \tag{3}$$

$$C(z) + iS(z) = (2z/\pi)^{1/2} e^{iz} \left\{ {}_1F_2\left(\begin{matrix} 1 \\ \frac{3}{4}, \frac{5}{4} \end{matrix} \middle| \frac{-z^2}{4} \right) - \tfrac{2}{3}(iz)\, {}_1F_2\left(\begin{matrix} 1 \\ \frac{5}{4}, \frac{7}{4} \end{matrix} \middle| \frac{-z^2}{4} \right) \right\}. \tag{4}$$

$$C(z) + iS(z) \sim (1+i)/2 - (2\pi z)^{-1/2} e^{iz} \{(2z)^{-1}\, {}_3F_0(1, 3/4, 5/4; -4/z^2)$$
$$+ i\, {}_3F_0(1, 1/4, 3/4; -4/z^2)\}, \quad |z| \to \infty, \quad |\arg z| < \pi. \tag{5}$$

Approximations for the Fresnel integrals follow from Padé approximations for the error functions, see 4.8.3. We have the connecting relations

$$C(z) + iS(z) = e^{iz} (2z/\pi)^{\frac{1}{2}} \left[\frac{C_n(\frac{1}{2}, iz)}{D_n(\frac{1}{2}, iz)} + V_n(\tfrac{1}{2}, ze^{i\pi/2}) \right]. \tag{6}$$

$$C(z) + iS(z) = \frac{-ie^{iz}}{(2\pi z)^{\frac{1}{2}}} \left[\frac{G_n(\frac{1}{2}, -iz)}{H_n(\frac{1}{2}, -iz)} + T_n(\tfrac{1}{2}, ze^{-i\pi/2}) \right],$$

$$z \neq 0, \quad -3\pi/2 < \arg z < \pi/2. \tag{7}$$

The notation $C(z)$ and $S(z)$ are sometimes used to denote functions closely related to those above.

4.9.2. Expansions in Series of Chebyshev Polynomials

TABLE 4.8

Chebyshev Coefficients for Fresnel Integrals

$$\int_0^x t^{-\frac{1}{2}} \cos t\,dt = x^{\frac{1}{2}} \sum_{n=0}^{\infty} a_n T_{2n}(x/8) \qquad \int_0^x t^{-\frac{1}{2}} \sin t\,dt = x^{\frac{1}{2}} \sum_{n=0}^{\infty} b_n T_{2n+1}(x/8)$$

$0 \le x \le 8$

n	a_n	n	b_n
0	0.76435 13866 41860 00189	0	0.63041 40431 45705 39241
1	-0.43135 54754 76601 79313	1	-0.42344 51140 57053 33544
2	0.43288 19997 97266 53054	2	0.37617 17264 33436 56625
3	-0.26973 31033 83871 11029	3	-0.16249 48915 45095 67415
4	0.08416 04532 08769 35378	4	0.03822 25577 86330 08694
5	-0.01546 52448 44613 81958	5	-0.00564 56347 71321 90899
6	0.00187 85542 34398 22018	6	0.00057 45495 19768 97367
7	-0.00016 26497 76188 87547	7	-0.00004 28707 15321 02004
8	0.00001 05739 76563 83260	8	0.00000 24512 07499 23299
9	-0.00000 05360 93398 89243	9	-0.00000 01109 88418 40868
10	0.00000 00218 16584 54933	10	0.00000 00040 82497 31696
11	-0.00000 00007 29016 21186	11	-0.00000 00001 24498 30219
12	0.00000 00000 20373 32548	12	0.00000 00000 03200 48425
13	-0.00000 00000 00483 44033	13	-0.00000 00000 00070 32416
14	0.00000 00000 00009 86533	14	0.00000 00000 00001 33638
15	-0.00000 00000 00000 17502	15	-0.00000 00000 00000 02219
16	0.00000 00000 00000 00272	16	0.00000 00000 00000 00032
17	-0.00000 00000 00000 00004		

TABLE 4.8 *(Continued)*

$$\int_x^{\infty} t^{-\frac{1}{2}} e^{-it} dt = -ix^{-\frac{1}{2}} e^{-ix} \sum_{n=0}^{\infty} c_n T_n^*(5/x) , \quad x \ge 5$$

$$c_n = R(c_n) + iI(c_n)$$

n	$R(c_n)$	n	$I(c_n)$
0	0.99056 04793 73497 54867	0	0.04655 77987 37516 45606
1	-0.01218 35098 31478 99746	1	0.04499 21302 01239 41396
2	-0.00248 27428 23113 06034	2	-0.00175 42871 39651 45324
3	0.00026 60949 52647 24735	3	-0.00014 65340 02581 06784
4	-0.00000 10790 68987 40635	4	0.00003 91330 40863 01585
5	-0.00000 48836 81753 93328	5	-0.00000 34932 28659 77307
6	0.00000 09990 55266 36813	6	-0.00000 03153 53003 23452
7	-0.00000 00750 92717 37211	7	0.00000 01876 58200 85285
8	-0.00000 00190 79487 57288	8	-0.00000 00377 55280 49302
9	0.00000 00090 90797 29266	9	0.00000 00026 65516 50103
10	-0.00000 00019 66236 03267	10	0.00000 00010 88144 81222
11	0.00000 00001 64772 91058	11	-0.00000 00005 35500 76711
12	0.00000 00000 63079 71380	12	0.00000 00001 31576 54466
13	-0.00000 00000 36423 21895	13	-0.00000 00000 15286 08809
14	0.00000 00000 10536 93030	14	-0.00000 00000 03394 76460
15	-0.00000 00000 01716 43801	15	0.00000 00000 02702 02670
16	-0.00000 00000 00107 12365	16	-0.00000 00000 00946 31418
17	0.00000 00000 00204 09885	17	0.00000 00000 00207 15651
18	-0.00000 00000 00090 06395	18	-0.00000 00000 00012 69314
19	0.00000 00000 00025 50616	19	-0.00000 00000 00013 97562
20	-0.00000 00000 00004 03556	20	0.00000 00000 00008 59293
21	-0.00000 00000 00000 56958	21	-0.00000 00000 00003 10695
22	0.00000 00000 00000 76174	22	0.00000 00000 00000 75146
23	-0.00000 00000 00000 36288	23	-0.00000 00000 00000 06478
24	0.00000 00000 00000 11797	24	-0.00000 00000 00000 05224
25	-0.00000 00000 00000 02467	25	0.00000 00000 00000 03863
26	-0.00000 00000 00000 00016	26	-0.00000 00000 00000 01651
27	0.00000 00000 00000 00331	27	0.00000 00000 00000 00504
28	-0.00000 00000 00000 00203	28	-0.00000 00000 00000 00092
29	0.00000 00000 00000 00083	29	-0.00000 00000 00000 00011
30	-0.00000 00000 00000 00025	30	0.00000 00000 00000 00020
31	0.00000 00000 00000 00004	31	-0.00000 00000 00000 00011
32	0.00000 00000 00000 00001	32	0.00000 00000 00000 00005
33	-0.00000 00000 00000 00001	33	-0.00000 00000 00000 00001
34	0.00000 00000 00000 00001		

$$\int_0^{\infty} t^{-\frac{1}{2}} e^{-it} dt = (\pi/2)^{\frac{1}{2}}(1-i)$$

TABLE 4.8 *(Concluded)*

$$\int_x^\infty t^{-\frac{1}{2}} e^{-it} dt = x^{-\frac{1}{2}} e^{-ix}\left[(2x)^{-1}P(x) - iQ(x)\right]$$

$$P(x) = \sum_{n=0}^{\infty} e_n T_{2n}(8/x) \qquad Q(x) = \sum_{n=0}^{\infty} f_n T_{2n}(8/x)$$

$x \geq 8$

n	e_n	n	f_n
0	0.97462 77909 32968 22410	0	0.99461 54517 94079 28910
1	-0.02424 70187 39693 21371	1	-0.00524 27676 60842 97210
2	0.00103 40090 68429 77317	2	0.00013 32586 42298 83909
3	-0.00008 05245 02469 08016	3	-0.00000 77085 64526 42713
4	0.00000 90596 24819 66582	4	0.00000 07084 80770 32045
5	-0.00000 13101 69967 57743	5	-0.00000 00881 25174 11602
6	0.00000 02277 08203 91497	6	0.00000 00135 97847 17148
7	-0.00000 00455 86235 52026	7	-0.00000 00024 68582 95747
8	0.00000 00102 15675 37083	8	0.00000 00005 09257 89921
9	-0.00000 00025 11145 08133	9	-0.00000 00001 16534 00634
10	0.00000 00006 67047 61275	10	0.00000 00000 29065 78309
11	-0.00000 00001 89315 12852	11	-0.00000 00000 07798 47361
12	0.00000 00000 56898 98935	12	0.00000 00000 02228 02542
13	-0.00000 00000 17982 19359	13	-0.00000 00000 00672 39338
14	0.00000 00000 05941 62963	14	0.00000 00000 00212 96411
15	-0.00000 00000 02042 85065	15	-0.00000 00000 00070 41482
16	0.00000 00000 00727 97580	16	0.00000 00000 00024 19805
17	-0.00000 00000 00267 97428	17	-0.00000 00000 00008 61080
18	0.00000 00000 00101 60694	18	0.00000 00000 00003 16287
19	-0.00000 00000 00039 58559	19	-0.00000 00000 00001 19596
20	0.00000 00000 00015 81262	20	0.00000 00000 00000 46444
21	-0.00000 00000 00006 46411	21	-0.00000 00000 00000 18485
22	0.00000 00000 00002 69981	22	0.00000 00000 00000 07527
23	-0.00000 00000 00001 15038	23	-0.00000 00000 00000 03131
24	0.00000 00000 00000 49942	24	0.00000 00000 00000 01328
25	-0.00000 00000 00000 22064	25	-0.00000 00000 00000 00574
26	0.00000 00000 00000 09910	26	0.00000 00000 00000 00252
27	-0.00000 00000 00000 04520	27	-0.00000 00000 00000 00113
28	0.00000 00000 00000 02092	28	0.00000 00000 00000 00051
29	-0.00000 00000 00000 00982	29	-0.00000 00000 00000 00024
30	0.00000 00000 00000 00467	30	0.00000 00000 00000 00011
31	-0.00000 00000 00000 00225	31	-0.00000 00000 00000 00005
32	0.00000 00000 00000 00110	32	0.00000 00000 00000 00002
33	-0.00000 00000 00000 00054	33	-0.00000 00000 00000 00001
34	0.00000 00000 00000 00027	34	0.00000 00000 00000 00001
35	-0.00000 00000 00000 00014		
36	0.00000 00000 00000 00007		
37	-0.00000 00000 00000 00004		
38	0.00000 00000 00000 00002		
39	-0.00000 00000 00000 00001		
40	0.00000 00000 00000 00001		

$$\int_0^\infty t^{-\frac{1}{2}} e^{-it} dt = (\pi/2)^{\frac{1}{2}}(1-i)$$

4.10. Bibliographic and Numerical Data

4.10.1. References

For general references on the incomplete gamma function and related functions, see Erdélyi et al (1953), Luke (1962), Abramowitz and Stegun (1964), Nielsen (1965), Luke (1969), and references given in these sources. See also the references given in 7.11.1. For a study on the zeros of the incomplete gamma function, see Kölbig (1970,1972).

Tables of integrals (in analytic form) involving the exponential integral have been prepared by Corrington (1961) and Geller and Ng (1969). See Ng and Geller (1969) and Geller and Ng (1971) for similar tables involving the error function.

4.10.2. Description of and References to Tables

$\int t^{v-1}e^{-t}dt$ *and Related Functions*

Khamis (1965): $P(n,x) = \left(2^n\Gamma(n)\right)^{-1}\int_0^x t^{n-1}e^{-t/2}dt$, $n =$ 0.05(0.05)10(0.1)20(0.25)70, x = 0.0001(0.0001)0.001 (0.001)0.01(0.01)1(0.05)6(0.1)16(0.5)66(1)166(2)250, 10D.

Pagurova (1963): $\left(x^m\Gamma(m)\right)^{-1}\int_0^x t^{m-1}e^{-t}dt$, m = 0(0.05)3, x = 0(0.05)1, 7D. Some auxiliary functions are also tabulated.

Narasimha (1966): $g(b,x) = be^{-x}\int_0^1 t^{b-1}e^{xt}dt$, $G(b,x) = -be^{x}\int_1^\infty t^{b-1}e^{-xt}dt$. $g(b,x)$: b = 0(0.2)2(0.5)5, x = 0(0.1)2(0.25)3(0.5)5(1)10, 5D. $G(b,x)$, $-b$ = 0(0.2)2(0.5)5, x as above, 5D.

Pagurova (1959): $e^{x}\int_1^\infty u^{-v}e^{-xu}du$, v = 2(1)10, x = 10(0.1)20, 7D. Also same function for v = 0(0.1)1, x = 0.01(0.01)7(0.05) 12(0.1)20, 6 or 7 figures, but with a maximum of 7D.

Harter (1964): Let $I(u,p) = \left(\Gamma(p+1)\right)^{-1}\int_0^{u(p+1)^{1/2}} t^p e^{-t}dt$, $p(x^2,v) = I(u,p)$, $u = x^2/(2v)^{1/2}$, $p = v/2 - 1$. $I(u,p)$, p = -0.5(0.5)74(1)164, u = 0(0.1) to point where $I(u,p)$ = 1 to 9D, max u = 21.5, 9D. Values of u in $p(x^2,v)$ given v = 1(1)150(2)330, $p = 10^{-y}$, $5\cdot 10^{-y}$, y = 2,3,4; p = 0.025, 0.1(0.1)0.9, 0.95, 0.975, 0.99, 0.995, 0.999, 0.9995, 0.9999, 6S.

Ancker and Gafarian (1962): $J(x,y) = \int_0^x u^{-1} g(y,u)du$, where $g(y,u) = \int_0^u t^{y-1}e^{-t}dt$, x, y = 0.1(0.1)10, 6S.

Frevel (1973): $\left(\Gamma(1 + 1/v)\right)^{-1}\int_0^x \exp\left(-t^v\right)dt$, x = 0(0.01)2 where $1/v$ is the abscissa of the main minimum of $\Gamma(1 + x)$, 5S.

Breig and Crosbie (1974): $A_1(x,y) = \int_1^\infty u^{-1}e^{-ux}dt$, $A_2(x,y) = \int_1^\infty t^{-2}e^{-ux}dt$, $u = (t^2 + y^2)^{\frac{1}{2}}$, $A_3(x,y) = x\int_1^\infty A_2(xt, y/t)dt$, x = 0(0.001)0.01(0.005)0.1(0.1)1.0(0.25)2(1)5; y = 0, 10^u, $2 \cdot 10^u$, $5 \cdot 10^u$, 1000, u = -1(1)2, mostly 5S for all entries whose absolute ln is less than about 44.

$\int t^n e^{-t}dt$, n an integer, and Related Functions

J. Miller, Gerhouser and Matsen (1959): $\int_1^\infty e^{-at}t^n dt$, a = 1/8(1/8)25, n = 0(1)16, 14S.

B. K. Gupta (1963) : $A_n(a) = \int_1^\infty e^{-at}t^n dt$, $F_n = \int_1^\infty e^{-at}t^n Q(t)dt$, where $Q(t) = \frac{1}{2}\ln\frac{t+1}{t-1}$, a = 3.1(0.1)10, n = 0(1)10, 10S.

Krugliak and Whitman (1965): $\int_1^\infty e^{-ax}x^n dx$, $\int_{-1}^1 e^{-ax}x^n dx$, n = 0(1)15, a = 0(0.01)50, 6S; n = 0(1)17, a = 0(0.125)25, 10, 12 or 14S.

Tooper and Mark (1968): $\int_0^x t^{-1}\sinh t\, dt$, x = 1(1)30, 13S.

S. S. Kumar (1962): Let $E_n(x) = \int_1^\infty u^{-n}e^{-xu}du$, $A\left(f(x)\right) = \frac{1}{2}\int_0^\infty E_1(|t-x|)f(t)dt$, $B\left(f(x)\right) = 2\int_x^\infty E_2(t-x)f(t)dt - 2\int_0^\infty E_2(x-t)f(t)dt$. All tables are to 6D.

$E_1(x-kx)$, $A(e^{-kx})$, $B(e^{-kx})$, x = 0.001, 0.005, 0.05, 0.1(0.1)1.5(0.5)9.5, k = 0.01, 0.05, 0.1(0.1)1(1)10(5)30.

$A(x^m)$, $B(x^m)$, x = 0(0.01)0.1(0.02)0.2(0.04)0.64, 0.72, 0.8(0.1)2(1)9, 9.5, m = 0(1)5.

$\frac{1}{2}\int_0^\infty e^{-kt}E_1(t)dt$, $2\int_0^\infty e^{-kt}\{E_2(t) - E_2(-t)\}dt$, k = 0.01, 0.05, 0.1, 0.2(0.2)1(1)10(5)30.

Murnaghan and Wrench (1963): Converging factors $C_n(n+1)$ and $C_n(n+\frac{1}{2})$ and reduced derivatives to facilitate the evaluation of $\mathrm{Ei}(x)$ are tabulated for $n=5(1)20$, 45D; and similarly for $-\mathrm{Ei}(-x)$, $n=4(1)20$, 48D. They also tabulate $\mathrm{Ei}(x)$ and $-\mathrm{Ei}(-x)$, $x=6(1)20$ and $x=21$ for the latter function, to 44S and 45D respectively. Also $R_j=\int_0^1 t^j e^{-t}dt$, $j=0(1)50$, 45D.

Chipman (1972): $F(x)=\int_0^x t^{-1}H(t)dt$, $G(x)=\int_0^x t^{-1}e^{-t}H(t)dt$, $H(t)=\mathrm{Ei}(t)-\gamma-\ln|t|$. $F(\pm x)$, $G(\pm x)$, $x=0.1(0.1)2(0.2)5(0.5)20(1)40(2)80$, 12S.

$\int e^{-t^2}dt$ *and Related Functions*

Akad. Nauk SSSR (1960): $A(x)=\int_0^x e^{-u^2/2}du$, $x=0(0.001)2.5(0.002)3.4(0.005)4(0.01)4.5(0.1)6$, 10D; $\log_{10}(\frac{1}{2}-A(x))$, $x=5(1)50(10)100(50)500$, 10D; $(-)^{s-1}(s!)^{-\frac{1}{2}}(d/dx)^{s-1}\left(e^{-x^2/2}\right)$, $s=2(1)21$, $x=0(0.002)4$, 7D.

Smirnov (1960): $F(x)=\int_0^x f(u)du$, $f(u)=e^{-u^2/2}$, $x=0(0.001)2.5(0.002)3.4(0.005)4(0.01)4.5$, 7D and $x=4.5(0.01)6$, 10D; $-\log[\frac{1}{2}-F(x)]$ $x=5(1)50(10)100(50)500$, 5D; $(-)^{s-1}(s!)^{-\frac{1}{2}}D^{s-1}f(x)$, $D=d/dx$, $s=2(1)21$, $x=0(0.002)4$, 7D. The same data is given in Akad. Nauk SSSR (1960), above.

Stegun and Zucker (1970): $\mathrm{erf}(x)$, $\mathrm{erfc}(x)$, $x=0(0.05)1(0.5)27$, 18S.

Pimbley and Nelson (1964): $x^{\frac{1}{2}}e^{-x}\int_0^x t^{-\frac{1}{2}}e^t dt$, $x=0(0.1)10(1)100$, 12D.

Smirnov and Bol'sev (1962): $T(h,A)=(2\pi)^{-1}\int_0^a u^{-1}exp\left[-h^2u/2\right]dx$, $u=1+x^2$, $h=0(0.01)3$, $a=0(0.01)1$; $h=3(0.05)4$, $a=0.05(0.05)1$; 7S. $h=4(0.1)5.2$, $a=0.1(0.1)1$; 7S. $exp(-\frac{1}{2}h^2)$, $F(h)=(2\pi)^{-\frac{1}{2}}\int_0^h exp(-t^2)dt$, $T(h,1)=\{1-4F^2(h)\}/8$, $h=0(0.001)3(0.005)4(0.01)5(0.1)6$, 7S.

Russel and Lal (1967): $\left[2^{\frac{1}{2}n}\Gamma(\frac{1}{2}n)\right]^{-1}\int_y^\infty t^{\frac{1}{2}n-1}e^{-t^2}dt$, $n=1(1)50$, $y=0(0.001)0.1(0.01)0.1(0.1)107$, 5D.

Britney and Winkler (1970): $(2\pi)^{-\frac{1}{2}}\int_{-\infty}^{z} t^n e^{-t^2/2}dt$, $n=1(1)6$, $z=0(0.01)5$, 11S.

Ihm (1961): $(2\pi)^{-\frac{1}{2}}\int_{-\infty}^{x} e^{-\frac{1}{2}(t-iy)^2}dt$, $x=0(0.1)3.9$, $y=0(0.1)2.5$, 4D.

Martz (1964): $\int_0^z e^{i\pi u^2}du$, $z=x+iy$, $y=-2.60(0.02)1.82$, x is variable and depends on y, 5S.

Faddeeva and Terent'ev (1954): $e^{-z^2}(1+2i\pi^{-\frac{1}{2}}\int_0^z e^{t^2}dt)$, $z=x+iy$, $x=0(0.02)3$, $y=0(0.02)3$; $x=3(0.1)5$, $y=0(0.1)2.9$; $x=0(0.1)5$, $y=3(0.1)5$, 6D.

Karpov (1958): $F(z)=\int_0^z e^{t^2}dt$, $z=\rho e^{i\theta}$, $\theta=45°(0.3125°)48.75°(0.625°)55°(1.25°)65°(2.5°)90°$, $\rho'\leq\rho\leq\rho''$, ρ' depends on θ and takes on one of the values $\frac{1}{2}$, 1, 3/2, ..., 4, ρ'' also depends on θ and takes on the value 3, 4, or 5, the spacing in ρ is 0.01, 5D.

Karpov (1954): $e^{-z^2}\int_0^z e^{t^2}dt$, $z=\rho e^{i\theta}$, $\theta=0°(2.5°)30°(1.25°)35°(0.625°)45°, 90°$, $\rho=0(0.01)5$, finer mesh in some cases depending on θ, $\rho=0(0.01)10$ for $\theta=0$, 5D.

Fried and Conte (1961): $2ie^{-z^2}\int_{-\infty}^{iz} e^{-t^2}dt$, $z=x+iy$. $x=0(0.1)10$, $y=-10(0.1)10$, 6S mostly. This source contains inaccuracies, particularly for $y<0$, according to the authors of the next entry.

Fettis, Caslin and Cramer (1972): $z(p)=\pi^{-\frac{1}{2}}\int_{-\infty}^{\infty}(t-p)^{-1}e^{-t^2}dt = 2ie^{-p^2}\int_{-\infty}^{ip}e^{-t^2}dt$, $z'(p)$, $p=x+iy$ $x=0(0.1)20$, $y=\pm 0(0.1)10$, 11S. First 200 zeros of $z(p)$ and $z'(p)$, 10D.

Cook and Elliott (1960): $U(p,t)$, $V(p,t)$, see 4.8.1(19,20), $p/(1+p)=0(0.05)1$, $t=0(0.025)0.2(0.05)1(0.1)2$, 5D; $\int_{-\infty}^{\infty}\{U(p,t)\}^n dt$ $n=1(1)10$, t as above, 5D.

Hetnarski (1964): Let $U(t,p,a)\pm V(t,p,a)=a^{-\frac{1}{2}}\exp(at\pm pa^{\frac{1}{2}}) \times \operatorname{erfc}\{\pm(at)^{\frac{1}{2}}+p/2t^{\frac{1}{2}}\}$. $U(t,p,a)$, $V(p,t,a)$, $a=1$, $t=0.2$, $p=0.25, 0.3, 0.5, 1.0$; $t=1(1)4$, about 9 values of p from 0.1 to about 4 (depending on t), 5D.

Finn and Mugglestone (1965): $u(x,y)$, see 4.8.1(14), y = 0(0.01)0.2(0.04)0.48, 0.55(0.05)0.7(0.1)1, x = 0(0.25)6(0.5)10(1)22, 6S. Also coefficients are given for a least square quartic polynomial in y, $u(x,y) \sim p+qa+ra^2+sa^3+ta^4$, x = 0(0.25)6(0.5)10,11,12, $0.001 \leq a \leq 0.2$.

Hummer (1965): $r(x,y) = \pi^{-\frac{1}{2}}u(x,y)$, $u(x,y)$ as in the above entry, y = 0, $1\cdot 10^{-n}$, $2\cdot 10^{-n}$, $3\cdot 10^{-n}$, $5\cdot 10^{-n}$, $7\cdot 10^{-n}$ for n = 2, 3, 4, and y = 0.1(0.05)0.5, x = 0(0.05) 5(0.1)10, 8S.

Reichel (1969): Let $U(p,t)$ be defined as in 4.8.1(19). $\int_0^\infty \{U(p,t)\}^n dt$ is tabulated for n = 1(1)25 and various t to 9S.

K. L. Miller, Molmud and Meecham (1972): $\int_0^\infty (t-z)^{-1}t^m e^{-t^2}dt$, $z = x+iy$, x = -10(0.2)10, y = 0(0.2)10, m=0,4, 6S.

Ascari (1968): $A_n = 2^n\Gamma(n/2+1)$, A_n^{-1}, n = 1(1)24, 12S. $A_n i^n \operatorname{erfc}(x)$, n = 1(1)24, x = 0(0.01)5.20, 10S.

Berlyand, Gavrilova and Prudnikov (1961): $2^n\Gamma(n/2+1) \times i^n\operatorname{erfc}(x)$, n = 1(1)30, x = 0(0.01)N, N depends on n and varies monotonically from 3.50 when n = 1 and 2 to 1.0 when n = 26-30. The accuracy goes from 6S initially to 2S near the end of the table. Also erfc(x), x = 0.01 (0.01)3.50, 6S. This table should be used with caution. See Math. Comp. 16(1963), 470-471; 22(1968), 898-899.

Fettis, Caslin and Cramer (1973a): First hundred zeros of erf(z) and of e^{-z^2} erfc$(-iz)$ to 11S. These authors write Erfc(z) instead of our erfc(z).

Fettis, Caslin and Cramer (1973b): First hundred zeros of the derivative of e^{-z^2} erfc$(-iz)$ to 11S. These authors write Erfc(z) instead of our erfc(z).

Spicer (1963): Values of b in $P = (2/\pi^{\frac{1}{2}})\int_0^b e^{-u^2}du$ for P = 0(0.0001)0.9(0.00001)0.99997, 9D.

Krishnan (1965,1966): Let $\phi(x) = (2\pi)^{-\frac{1}{2}} \int_{-\infty}^b e^{-u^2/2}\,du$, $p + (1-p)\phi(k) = \phi(x)$. Values of x, for p = 0.01(0.01) 0.99, $k + 5$ = 0.4(0.1)7.3, 4D.

Murnaghan(1965): Converging factors $C_n(n+1)$ and $C_n(n+\frac{1}{2})$ and reduced derivatives to facilitate the evaluation of $\frac{1}{2}\int_x^\infty t^{-\frac{1}{2}}e^{-t}dt$ are calculated to 63D for $n = 2(1)64$ and $n = 0(1)64$ respectively.

Wrench (1971): Converging factors $C(n+\frac{1}{2})$ and $C(n)$ and reduced derivatives to facilitate the evaluation of $\exp(-x^2)\int_0^x \exp(t^2)dt$ are tabulated for $n = 1(1)40$, 30D.

$\int t^{-1}e^{it}dt$ *and Related Functions*

Wrench and Alley (1972): Converging factors $P_s(s)$, $P_{s+1}(s)$, $O_s(s)$, $O_{s+1}(s)$ and reduced derivatives to facilitate the evaluation of $si(z)$ and $Ci(z)$, $s = 1(1)70$, 33D. They also tabulate $si(x)$, $Si(x)$ and $Ci(x)$, $x = 1(1)70$, 28D.

Ling and Lin (1971): $(2/\pi)Si(n\pi/2)$, $n = 1(1)200$, 25D.

Strömgren(1962): $\int_0^\infty (1+u)^{-1}\ln(1+ux/y)e^{-iux}du$, $x,y = 0(0.1)2(1)10$, 6D.

Luke, Fair, Coombs and Moran (1965): Value of $b, 0<b<1$ such that $\int_0^a u^{-b}(\cos u)du = 0$, $a = 3\pi/2$ is given to 15D.

Halbgewacks and Shah (1967): See entry on p.151.

$\int t^{-\frac{1}{2}}e^{it}dt$ *and Related Functions*

Syrett and Wilson (1966): $\int_0^x \exp(i\pi t^2/2)dt$, $x = 0(0.001)2(0.01)10$, 28S.

Wrench and Alley (1973): Converging of various kinds and reduced derivatives to facilitate evaluation of Fresnel integrals for $x = 2(1)71$, 35D. They also tabulate $\int_0^x \exp(\frac{1}{2}i\pi t^2)dt$, $x = 1(1)6$, 28D; $(2\pi)^{-\frac{1}{2}}\int_0^x t^{-\frac{1}{2}}\exp(it)dt$, $x = 1(1)70$, 28D; and $\exp(-ix)\int_x^\infty t^{-\frac{1}{2}}\exp(it)dt$ $x = 1(1)70$, 28D.

Kreyszig (1957a): z_n and $z_n{}^*$, complex zeros of $\int_0^z \cos t^2 dt$ and $\int_0^z \sin t^2 dt$, respectively, $n=1(1)50$, 2D.

Kreyszig (1957b): Let $c(z)=\int_z^\infty \cos t^2 dt$, $s(z)=\int_z^\infty \sin t^2 dt$. x_n and $x_n{}^*$, real zeros of $c(z)$ and $s(z)$, respectively, $n=0(1)25$, 2D; $c(z), s(z)$, $z=0(0.2)4(0.5)20$, 3D; $c(z), s(z)$, $z=x+iy$, $x=0(1)20$, $y=0(1)5$, 2D.

Walters and Wait (1964): $\int_{-\infty}^{\infty} f(z)e^{-i\pi z^2/2}dz$, 5D where in one case $f(z)=\{1+\exp -2A(z-z_0)\}$ and two other cases where in each instance $(-\infty,\infty)$ is broken up into three subintervals. See Math. Comp. 19(1965), 510,511 for a complete description.

4.10.3. Description of and References to Other Approximations and Expansions

Khovanskii (1963): Continued fraction representations for the incomplete gamma functions.

Barakat (1961): From 4.1(1) and 3.2.2(4), we have $\gamma(v,ix) = (ix)^v \sum_{n=0}^{\infty} \epsilon_n(-i)^n B_n(v)J_n(x)$, $B_n(v)=\int_0^1 t^{v-1}T_n(t)dt$. This paper tabulates $B_n(v)$ to 8D for $v=0.2(0.2)0.8$ and $v=\frac{1}{2}$.

Németh (1963): This gives some interesting coefficients for the evaluation of $F_a(y)=\int_0^x \exp(-y^a)dy$ valid for $0\le t\le 1$, $t=x/4^b$, $ab=1$, 11D; and for the evaluation of $G_a(x)=\Gamma(b+1)-F_a(x)$, valid for $0\le c\le 1$, $c=4/x^a$, $a\ge 1$, 11D.

Verbeeck (1970): Let $E_v(x)=\int_1^\infty t^{-v}e^{-xt}dt$, $x>0$, $v>0$, $1-xe^x E_v(x) = \{\mu_2(z)-\mu_1(z)\}/\mu_2(z)=zw(z)/\mu_2(z)$, $zx=1$. Using the differential equation for $E_v(x)$, one can develop a differential equation involving $w(z)$, $\mu_2(z)$ and their derivatives. This is approximately solved by the tau method assuming that both $\mu_1(z)$ and $\mu_2(z)$ are polynomials in z of degree m such that $\mu_1(0)=\mu_2(0)=1$. For $v+\frac{1}{2}=1(1)6$, $1\le x<\infty$, coefficients in the polynomials for $m=2,3,4$ are given to 9D. Approximations to $E_v(x)$ range from about $6\cdot 10^{-6}$ to about 10^{-9}. The case $v=1(1)5$, $1\le x<\infty$ is treated in a similar fashion. The tau method is also used to get polynomial approximations for the entire function related to $E_v(x)$,

v as above with $0 \le x \le 1$.

Gray, Thompson and McWilliams (1969): A sequence of approximations, μ_n, for $\int_a^\infty e^{-x} x^{z-1} dx$, $z>0$. $\mu_0 = e^{-a} a^z / p$, $\mu_1 = \mu_0 \{1-(z-1)/(p^2+2a)\}$. μ_2 is also given. Here $p = a+1-z$.

Clenshaw (1962): $-E_1(x) - \ln|x|$, $CT(x)$, $a=4$, 20D; $e^x E_1(x)$, $CT^*(y)$, $a=4$, 20D; $\operatorname{erf}(x)$, $CT(x)$, $a=4$, 20D and $CT(y)$, $a=4$, 20D.

Clenshaw, Miller and Woodger (1963): $Ei(x) - \ln|x|)$, $CT(x)$, $a=4$, 15D; $-e^x Ei(-x)$, $CT^*(y)$, $a=4$, 15D; $\operatorname{erf}(x)$, $CT(x)$, $a=4$, 20D and $CT(y)$, $a=4$, 8D.

Luke and Wimp (1963): $e^x \int_x^\infty t^{-1} e^{-t} dt$, $CT^*(y)$, $a=4$, 20d; $Ci(x) + i(\pi/2 - Si(x)) = x^{-1} e^{-ix}[-B(x) + iA(x)]$, $A(x)$, $B(x)$, $CT^*(y)$, $a=4$, 20d (the notation $Ci(x)$ and $Si(x)$ used in this reference differs from that employed here); $e^{x^2} \int_x^\infty e^{-t^2} dt$, $CT(y)$, $a=2$, 20d.

Németh (1964a): $\int_x^\infty t^{-1} e^{-t} dt$, truncated $CT^*(x)$, $a=4$, and truncated $CT^*(y)$, $a=4$, each expressed in powers of x and $1/x$, respectively, 10d. This report also gives a polynomial approximation for $\int_0^1 t^{-v-2}(1-e^{-xt}) dt$, $0<x<4$, 10D. The latter is based on a Chebyshev expansion for the exponential function.

Cody and Thacher (1968): Rational Chebychev approximations are presented for the exponential integral $E_1(x) = -Ei(-x)$ in the intervals $(0,1]$, $[1,4]$ and $[4,\infty)$ with maximal relative errors ranging down to 10^{-21}. The first 40 coefficients in a continued fraction representation of $E_1(x) + \gamma + \ln x$ are given to 25S.

Cody and Thacher (1969): Rational Chebyshev approximations are presented for $Ei(x)$ in the intervals $(0,6]$, $[6,12]$, $[12,24]$ and $[24,\infty)$. Maximal relative errors range from $8 \cdot 10^{-19}$ to $2 \cdot 10^{-21}$. $Ei(y) = 0$, $y = 0.37250\ldots$ is given to 30D.

Luke (1969): (a) Coefficients in the main diagonal Padé approximations to $z^{-1} \int_0^z t^{-1}(1-e^{-t}) dt$ of order $n=1(1)10$, 20S. The absolute value of the errors in these approximations are given for n as above, $z = re^{i\theta}$, $r=1(1)10$, $\theta=0$, $\pi/2$, π, 3S. The relative error is also tabulated for $\theta=\pi$. (b) Coefficients in the main diagonal Padé approximations to $4z^{-2} \int_0^z t^{-1}(1-\cos t) dt$ and $z^{-1} \int_0^z t^{-1} \sin t \, dt$ of order $n=2(2)10$, 20S. The absolute values of the errors in these approximations are given for n as above, $z=1(1)10$, 3S.

Chipman (1972): First 50 coefficients of a continued fraction expansion of $A(-x)=-\int_x^\infty t^{-1}Ei(-t)dt$, $x>0$ and $B(-x)=$ -P.V. $\int_{-\infty}^{x} t^{-1}Ei(-t)e^t dt$, $x>0$ to 18S.

Wimp (1961): $Ci(x)$, $Si(x)$, $CT(x)$, $a=2,5,10$D.

Németh (1965b): $Ci(x)+i(\pi/2-Si(x))=x^{-1}e^{-ix}[-B(x)+iA(x)]$, $A(x)$, $B(x)$, $CT(y)$, $a=8,12$D. There the coefficient for $C_0(a)$ is incorrect. See our f_0 in Table 4.4.

Bulirsch (1967): $Ci(x)$, $Si(x)$, $CT(x)$, $a=16$, 15D; $\int_0^x e^{\frac{1}{2}i\pi t^2}dt$, $CT(x)$, $a=3$, 15D; $ixe^{ix}\int_x^\infty t^{-1}e^{-it}dt$, $CT(y)$, $a=16$, 15D. Also a form related to the Fresnel integrals, $CT(y)$, $a=3$, 15D.

Briones (1964): $Si(\pi x/4)$, $CT(x)$, $a=1$, 10D.

Halbgewachs and Shah (1967): $C(x,\beta)+iS(x,\beta)=(2\pi)^{-\frac{1}{2}}\int_0^a t^{-\beta}e^{it}dt$. $CT(x)$, $a=3\pi/2$. $\beta=0.3084$, 0.3085, 0.9999, 10S. Also $C(x,\beta)$, $S(x,\beta)$, $\theta=\arctan S(x,\beta)/C(x,\beta)$ if $\beta_0<\beta<1$, $\theta=-\arctan S(x,\beta)/S(x,\beta)$ if $0<\beta<\beta_0$, $C(x,\beta_0)=0$, $x=3\pi/2$, $\beta=0(0.1)$ 0.9, 0.306, 0.308, 0.3084, 0.3085, 0.99, 0.999, 0.9999, 0.99999, 10S. β_0 is given to 10D.

Németh (1964b): $\int_0^x e^{-t^2}dt$, truncated $CT(x)$, $a=2$, and truncated $CT(y)$, $a=2$, each expressed in powers of x^2 and $1/x^2$, respectively, 10D.

Németh (1966): Forms related to Erfi(x), $CT(x)$ and $CT(y)$, both for $a=4$, 13D.

Anolik and Mil'ner (1970): $F(x)=\exp(-x^2)\mathrm{Erfi}(x)$, $F'(x)$, $CT(x)$, $a=5$. 16D and 15D for F and F', respectively.

Ray and Pitman (1963): $e^{x^2/2}\int_x^\infty e^{-t^2/2}dt$, $CT(x)$, $a=1$, 7D; $CT(y)$, $a=1$, 7D; but two of the coefficients, namely, b_0 and b_{26} appear to be in error in the sixth decimal place.

Hummer (1964): $e^{-x^2}\int_0^x e^{t^2}dt$, $CT(x)$, $a=5$, 16D.

Wood (1967): $i\,\mathrm{erfc}(x)+x^{1/2}$, $x^{-1}\{4i^2\,\mathrm{erfc}(x)-1-2x^2\}$, $CT(x)$, $a=1$, 7D. Also, $x^{m+1}e^{x^2}i^m\,\mathrm{erfc}(x)$, $CT(y)$, $m=1,2$, $a=1$, 7D.

Cody (1969): Nearly best rational Chebyshev approximations are presented for erf(x) in the interval $|x|<0.5$ and erfc(x) in the intervals $[0.46875,4]$, $[4,\infty)$ with maximal relative errors ranging down to between $6\cdot10^{-19}$ and $3\cdot10^{-20}$.

Hart et al (1969): Best Chebyshev rational approximations for erfc(x), various ranges, with accuracy up to 10^{-23}.

Cody, Paciorek and Thacher (1970): Nearly best rational Chebyshev approximations are presented for $e^{-x^2}\int_0^x e^{t^2}dt$ for $|x|\leq 2.5$, and the intervals $[2.5,3.5]$, $[3.5,5.0]$, $[5.0,\infty)$ with maximal relative errors ranging down to between $2\cdot 10^{-20}$ and $7\cdot 10^{-22}$.

Mosier and Shillady (1972): Let $F(z)=\int_0^1 e^{-zu^2}du$. Approximations for $0<z<22.5$ by a quartic polynomial in $z-s_i$ where the interval i and corresponding s_i are evaluated from a given z by simple formulas given in the paper. Coefficients in the polynomials are given for $i=1(1)19$ to 16D. Error is everywhere less than $4\cdot 10^{-12}$.

Finn and Mugglestone (1965): See entry on p. 147.

Oldham (1968): $F(x)=\frac{1}{2}\pi^{\frac{1}{2}}xe^{x^2}\mathrm{erfc}(x)=\{1+[1+2x^{-2}A(x)]^{\frac{1}{2}}\}^{-1}$, $A(x)=1-(1-2/\pi)e^{-xp(x)}$, $p(x)=a[1-bx^2(1-ax/\pi^2)]$, $a=0.8577$, $b=0.024$. This approximation is nearly best in the Chebyshev sense with a maximal relative error in $F(x)$ of one part in 7000. Approximation becomes exact as $x\to 0$ or $x\to\infty$. For some other heuristic fits to erfc(x), somewhat like that above, see Hart (1966) and Schucaney and Gray (1968).

Strecok (1968): Formulas and coefficients are developed for computing the inverse of the error function to at least 18S for all possible arguments up to $1-10^{-300}$ in magnitude. A formula to evaluate erf(x) to at least 22D for $|x| \leq 5\pi/2$ is also presented.

Boersma (1960): $e^{ix}(4/x)^{1/2}[C(x) - iS(x)]$, $\tau CP(x)$, $a = 4$, 9D. $C(x) - iS(x) = \frac{1}{2}(1 - i) + e^{-ix}(4/x)^{1/2}A(x)$, $A(x)$, $\tau CP(y)$, $a = 4$, 9D.

Németh (1965a): $x^{-1/2}[C(x) + iS(x)]$, $CT(x)$, $a = 8$, 12D. $C(x) + iS(x) = \frac{1}{2}(1 + i) - (2\pi x)^{-1/2}e^{ix}[A(x) + iB(x)]$, $A(x)$, $B(x)$, $CT(y)$, $a = 8$, 15D.

Hangelbroek (1967): $(5/x)^{1/2}[C(x) - iS(x)]$, $CT^*(x)$, $a = 5$, 14D. $C(x) - iS(x) = \frac{1}{2}(1 - i) - (5/x)^{1/2}e^{-ix}[E(x) - iF(x)]$, $E(x)$, $F(x)$, $CT^*(y)$, $a = 5$, 14D.

Syrett and Wilson (1966): Polynomial approximations of the real and imaginary parts of $\int_0^x \exp(i\pi t^2/2)dt$, $x>0$, expressed as a sum of Chebyshev polynomials. These are deduced by truncation of higher order *BCP*. Two sets of approximations are given to cover each of three different ranges. The coefficients in the two sets are given to 8S and 20S respectively.

Cody (1968): Best rational Chebyshev approximations for the real and imaginary parts of $\int_0^x \exp(-i\pi t^2/2)dt$ in the intervals $|x| \leq 1.2$, $[1.2,1.6]$, $[1.6,1.9]$, $[1.9,2.4]$, $[2.4,\infty)$. Maximal relative errors range down to $2\cdot 10^{-19}$.

CHAPTER V THE GENERALIZED HYPERGEOMETRIC FUNCTION ${}_pF_Q$ AND THE G-FUNCTION

5.1. Introduction

This rather lengthy chapter deals with the ${}_pF_q$ and its generalization, the G-function. It is arranged to provide quick access to results useful for the applications. The G-function is significant since numerous special functions appearing in applied mathematics are special cases or are closely related. Thus a result involving a G-function becomes a master or key formula from which a very large number of results can be deduced for Bessel functions, Legendre functions, confluent hypergeometric functions, etc., their combinations and other related functions.

The ${}_pF_q$ has a long history which dates from the times of Euler. We make no attempt to trace its historical development. For this the reader is referred to the volume by Slater (1966). The G-function is of recent origin having been introduced by Meijer in 1936. For material on the G-function, one must consult original research documents except for the volumes by Erdélyi et al (1953,1954) and Luke (1969). Accordingly, in the development of this chapter, we give references for data on the G-function as well as rather recent references on the ${}_pF_q$ as the situation warrants.

Some general references on the special functions are Whittaker and Watson (1927), Bailey (1935), Watson (1945), Erdélyi et al (1953), Hobson (1955), Kratzer and Franz (1960), Rainville (1960), Hochstadt (1961,1971), Sneddon (1961,1972), Luke (1962,1969), Schäfke (1963), Abramowitz and Stegun (1964), Mitrinovic and Djokovic (1964), Mitrinovic, Dokovic and Dragomir (1964), Wang and Kuo (1965), Kuznecov (1965), N.N. Lebedev (1965), Slater (1966), Inui (1967), Arsenin (1968), Spain and M.G. Smith (1970), Ayant and Borg (1971), Mitrinovic (1972), Mitrinovic, Tosic and Janic (1972). See also references in 1.6.1, 6.12.1, 7.11.1, 9.13.1, 10.5.1 and 11.1.

Some important volumes on asymptotics which include applications to special functions are Erdélyi (1956), de Bruijn (1958), Ford (1960), Egrafov (1961), Jeffreys (1962), Heading (1962), Fröman and Fröman (1965), Copson (1965), Wasow (1965), Lauwerier (1966), Berg (1968), Sirovich (1971), Dingle (1973), Nayfek (1973), Olver (1974) and Murray (1974).

For nonnumerical tables of integrals, integral transforms, and series relating to the entire spectrum of special functions, see Magnus and Oberhettinger (1948), Erdélyi et al (1954), Oberhettinger (1957), Ryshik and Gradstein (1957), Gröbner and Hofreiter (1961), Oberhettinger and Higgins (1963), Gradstehyn and Ryzhik (1965), Mangulis (1965), G.E. Roberts and Kaufman (1966), Magnus, Oberhettinger and Soni (1966), Colombo and Lavoine (1972), Oberhettinger (1972,1973a,1973b,1975), Oberhettinger and Badii (1973) and Hansen (1975).

Indices for all types of numerical tables have been prepared by A.V. Lebedev and Fedorova (1956), Burunova (1959), Fletcher, J.C.P. Miller, Rosenhead and Comrie (1962), Greenwood and Hartley (1962), and Schütte (1966). A cumulative index with subject classification system for the journal Mathematics of Computation by Luke, Wimp and Fair (1972) is valuable. See this source for errata on many of the references noted above.

In this volume, we present two sided inequalities for the $_pF_q$, numerous special cases of the $_pF_q$ and a certain class of G-functions. In particular, see 5.14. Further sources of inequalities bearing on the special functions are Erber (1960), Carlson (1965,1966,1970b), Carlson and Tobey (1968) and Mitrinovic and Vasic (1970).

5.2. The $_pF_q$

5.2.1. Power Series

The generalized hypergeometric series is defined by

$$_pF_q(\alpha_1, \alpha_2, \ldots, \alpha_p; \rho_1, \rho_2, \ldots, \rho_q; z) = {}_pF_q\left(\begin{matrix}\alpha_1, \alpha_2, \ldots, \alpha_p \\ \rho_1, \rho_2, \ldots, \rho_q\end{matrix}\middle| z\right)$$

$$= \sum_{k=0}^{\infty}\left[\prod_{h=1}^{p}(\alpha_h)_k z^k \Big/ \prod_{h=1}^{q}(\rho_h)_k k!\right],$$

$$(a)_k = \Gamma(a+k)/\Gamma(a), \text{etc.} \qquad (1)$$

We suppose that none of the ρ_j's is a negative integer or zero. Where no confusion can arise, we simply refer to (1) as a $_pF_q$. It is often convenient to employ a contracted notation and write (1) in the abbreviated form

$$_pF_q(\alpha_p; \rho_q; z) = {}_pF_q\left(\begin{matrix}\alpha_p \\ \rho_q\end{matrix}\middle| z\right) = \sum_{k=0}^{\infty}[(\alpha_p)_k z^k/(\rho_q)_k k!]. \qquad (2)$$

Thus $\Gamma(\alpha_p + k)$ is interpreted as $\Pi_{j=1}^{p} \Gamma(\alpha_j + k)$; $(\alpha_p)_k$ as $\Pi_{j=1}^{p} (\alpha_j)_k$; α_p as $\Pi_{j=1}^{p} \alpha_j$, etc. An empty term is treated as unity so that, for example, if $p = 2$, $(\alpha_h)_k = 1$ for $h > 2$. The α_h's and ρ_h's are called numerator and denominator parameters, respectively, and z is called the variable. The ${}_pF_q$ is symmetric in its numerator parameters, and likewise in its denominator parameters. If a numerator parameter and a denomnominator parameter coalesce, then omit the parameter, whence the ${}_pF_q$ becomes a ${}_{p-1}F_{q-1}$. The ${}_pF_q$ series terminates and, therefore, is a polynomial if a numerator parameter is a negative integer or zero, provided that no denominator parameter is a negative integer or zero.

The ${}_pF_q$ series

$$\begin{aligned}&\text{converges for all finite } z \text{ if } p \leq q,\\ &\text{converges for } |z| < 1 \text{ if } p = q + 1,\\ &\text{diverges for all } z,\ z \neq 0 \text{ if } p > q + 1.\end{aligned} \tag{3}$$

The ${}_pF_q$ series

$$\begin{aligned}&\text{absolutely converges for } |z| = 1 \text{ if } R(\eta) < 0,\\ &\text{conditionally converges for } |z| = 1,\ z \neq 1 \text{ if } 0 \leq R(\eta) < 1,\\ &\text{diverges for } |z| = 1, \text{ if } 1 \leq R(\eta),\end{aligned} \tag{4}$$

$$\eta = \sum_{h=1}^{p} \alpha_h - \sum_{h=1}^{q} \rho_h .$$

If m is a positive integer or zero, then

$${}_{p+1}F_{q+1}\left(\begin{matrix}-m,\ \alpha_p\\ c,\ \rho_q\end{matrix}\,\middle|\, z\right) = \frac{(\alpha_p)_m(-)^m z^m}{(c)_m(\rho_q)_m} \times {}_{q+2}F_p\left(\begin{matrix}-m,\ 1-m-c,\ 1-m-\rho_q\\ 1-m-\alpha_p\end{matrix}\,\middle|\, \frac{(-)^{p+q+1}}{z}\right), \tag{5}$$

where neither c nor any ρ_h is a negative integer or zero. If no ρ_h is a negative integer or zero, if, for convenience, to simplify the discussion, no α_h is a negative integer or zero, and if c is a negative integer or zero, $c = -n$, then

$$ {}_{p+1}F_{q+1}\left(\begin{matrix}-m, \alpha_p\\ -n, \rho_q\end{matrix}\middle| z\right) \quad \text{is not defined if} \quad n < m, \tag{6}$$

$$ {}_{p+1}F_{q+1}\left(\begin{matrix}-m, \alpha_p\\ -n, \rho_q\end{matrix}\middle| z\right) = {}_pF_q\left(\begin{matrix}\alpha_p\\ \rho_q\end{matrix}\middle| z\right) \quad \text{if} \quad n = m, \tag{7}$$

$$ {}_{p+1}F_{q+1}\left(\begin{matrix}-m, \alpha_p\\ -n, \rho_q\end{matrix}\middle| z\right) = \frac{(n-m)!\,(\alpha_p)_m z^m}{n!(\rho_q)_m} \times {}_{q+2}F_p\left(\begin{matrix}-m, n+1-m, 1-m-\rho_q\\ 1-m-\alpha_p\end{matrix}\middle| \frac{(-)^{p+q+1}}{z}\right) + \frac{(n-m)!\,m!(\alpha_\rho)_{n+1}(-)^{m+n}z^{n+1}}{n!(n+1)!\,(\rho_q)_{n+1}} \times {}_{p+1}F_{q+1}\left(\begin{matrix}n+1-m, n+1+\alpha_p\\ n+2, n+1+\rho_q\end{matrix}\middle| z\right) \quad \text{if } n > m. \tag{8}$$

We often have need for a truncated hypergeometric series. Thus,

$$ {}_pF_q^m\left(\begin{matrix}\alpha_p\\ \rho_q\end{matrix}\middle| z\right) = \sum_{k=0}^{m} \frac{(\alpha_p)_k z^k}{(\rho_q)_k k!} = \frac{(\alpha_p)_m z^m}{(\rho_q)_m m!}\, {}_{q+2}F_p\left(\begin{matrix}-m, 1-m-\rho_q, 1\\ 1-m-\alpha_p\end{matrix}\middle| \frac{(-)^{p+q+1}}{z}\right), \tag{9}$$

where no ρ_h is a negative integer or zero.

Let

$$A = {}_pF_q(u_p;\ v_q;\ gx), \quad B = {}_rF_s(a_r;\ b_s;\ wx), \quad AB = \sum_{k=0}^{\infty} c_k x^k. \tag{10}$$

If neither A nor B is a polynomial, then

$$c_k = \frac{w^k (a_r)_k}{(v_q)_k k!}\ {}_{p+s+1}F_{p+r}\left(\begin{matrix}-k, 1-b_s-k, u_p\\ 1-a_r-k, v_q\end{matrix}\middle| \frac{(-)^{r+s+1}g}{w}\right) \tag{11}$$

or

$$c_k = \frac{g^k (u_p)_k}{(v_q)_k k!}\ {}_{q+r+1}F_{p+s}\left(\begin{matrix}-k,\ 1-v_q-k, a_r\\ 1-u_p-k, b_s\end{matrix}\middle| \frac{(-)^{p+q+1}w}{g}\right) \tag{12}$$

If A only is a polynomial of degree m (let $u_h=-m, h=p$), then the above forms for c_k are valid for $k=0,1,\ldots,m$ and

$$c_k = \frac{w^k (a_r)_k}{(b_s)_k k!} \; {}_{p+s+1}F_{q+r}\left(\begin{matrix} -m, u_{p-1}-k, 1-b_s-k \\ v_q, 1-a_r-k \end{matrix} \middle| \frac{(-)^{r+s+1} g}{w} \right) \quad (13)$$

$$k=m+1, m+2, \ldots \; .$$

If both A and B are polynomials (let $a_h=-n$ for $h=r, m\leq n$), then (11) and (12) are valid for $k=1,2,\ldots,m$, (13) is valid for $k=m+1, m+2, \ldots, n$ and

$$c_{n+k} = \frac{(-m)_k (u_{p-1})_k g^k (-w)^n (a_{r-1})_n}{(v_q)_k k! \, (b_s)_n}$$

$$\times \; {}_{p+s+1}F_{q+r}\left(\begin{matrix} k-m, u_{p-1}+k, -n, 1-b_s-n \\ v_q+k, 1+k, 1-a_{r-1}-n \end{matrix} \middle| \frac{(-)^{r+s+1} g}{w} \right),$$

$$k=1, 2, \ldots, m. \quad (14)$$

In the applications, it is often desirable to split a hypergeometric series into its even and odd parts which are again hypergeometric series. Thus

$${}_pF_q\left(\begin{matrix} \alpha_p \\ \rho_q \end{matrix} \middle| z \right) = A(z) + \frac{\alpha_p z}{\rho_q} B(z),$$

$$A(z) = {}_{2p}F_{2q+1}\left(\begin{matrix} \tfrac{1}{2}\alpha_p, \tfrac{1}{2}\alpha_p+\tfrac{1}{2} \\ \tfrac{1}{2}, \tfrac{1}{2}\rho_q, \tfrac{1}{2}\rho_q+\tfrac{1}{2} \end{matrix} \middle| 4^{p-q-1} z^2 \right)$$

$$B(z) = {}_{2p}F_{2q+1}\left(\begin{matrix} \tfrac{1}{2}\alpha_p+\tfrac{1}{2}, \tfrac{1}{2}\alpha_p+1 \\ 3/2, \tfrac{1}{2}\rho_q+\tfrac{1}{2}, \tfrac{1}{2}\rho_q+1 \end{matrix} \middle| 4^{p-q-1} z^2 \right) \quad (15)$$

Clearly

$$2A(z) = {}_pF_q(z) + {}_pF_q(-z),$$

$$\frac{2\alpha_p z}{\rho_q} B(z) = {}_pF_q(z) - {}_pF_q(-z). \quad (16)$$

This process (15) can be generalized so that we can write

$$ {}_pF_q(z) = \sum_{k=0}^{n} \frac{(\alpha_p)_k z^k A_k(z)}{(\rho_q)_k k!}, \tag{17}$$

where each A_k is a ${}_{(n+1)p}F_{(n+1)q+n}$. We omit details as all of this can be viewed as a special case of 5.3.2 (5). Also the latter equation can be used to derive expressions analogous to (16). A number of equations of the kind just described relating to 5.3.2 (5) for $k=2$ and various values of m,n,p and q have been given by Carlson (1970a). Osler (1975) has proved a general result for power series from which a generalization of (16) is readily deduced. We have the following result.

Let n and k be positive integers, $k<n$, $w=\exp(2\pi i/n)$ and

$$F(z) = \sum_{r=0}^{\infty} F_r z^r, \quad |z|<R. \tag{18}$$

Then

$$\sum_{m=0}^{\infty} F_{nm+k} z^{nm+k} = (1/n) \sum_{s=0}^{n-1} w^{-ks} F(zw^s). \tag{19}$$

For hypergeometric functions, we have

$$ {}_{np+1}F_{nq+n}\left(\begin{matrix}(\alpha_p+k)/n, (\alpha_p+k+1)/n, \ldots (\alpha_p+k+n-1)/n, 1 \\ (\rho_{q+1}+k)/n, (\rho_{q+1}+k+1)/n, \ldots (\rho_{q+1}+k+n-1)/n\end{matrix} \middle| x^n\right)$$

$$= \frac{k!(\rho_q)_k x^{-k} n^{(p-q-1)k}}{n(\alpha_p)_k} \sum_{s=0}^{n-1} w^{-ks} {}_pF_q\left(\begin{matrix}\alpha_p \\ \rho_q\end{matrix} \middle| w^s n^{(q-p+1)} x\right) \tag{20}$$

where $\rho_h=1$ if $h=q+1$. In view of this last definition, the ${}_{np+1}F_{nq+n}$ is really an ${}_{np}F_{nq+n-1}$.

5.2.2. Derivatives and Contiguous Relations

$$\frac{d^n}{dz^n} {}_pF_q\left(\begin{matrix}\alpha_p \\ \rho_q\end{matrix} \middle| z\right) = \frac{(\alpha_p)_n}{(\rho_q)_n} {}_pF_q\left(\begin{matrix}\alpha_p + n \\ \rho_q + n\end{matrix} \middle| z\right). \tag{1}$$

$$\frac{d^n}{dz^n}\left[z^\delta {}_pF_q\left(\begin{matrix}\alpha_p \\ \rho_q\end{matrix} \middle| z\right)\right] = (\delta - n + 1)_n z^{\delta-n} {}_{p+1}F_{q+1}\left(\begin{matrix}\delta+1, \alpha_p \\ \delta+1-n, \rho_q\end{matrix} \middle| z\right). \tag{2}$$

If $(\delta + 1 - n)$ is a negative integer or zero, a more convenient form of (2) is

$$\frac{d^n}{dz^n}\left[z^\delta\, {}_pF_q\left(\begin{matrix}\alpha_p\\ \rho_q\end{matrix}\,\middle|\, z\right)\right]$$

$$= \frac{(\alpha_p)_{n-\delta}\, n!}{(\rho_q)_{n-\delta}(n-\delta)!}\, {}_{p+1}F_{q+1}\left(\begin{matrix}\alpha_p + n - \delta,\, n+1\\ \rho_q + n - \delta,\, n + 1 - \delta\end{matrix}\,\middle|\, z\right). \tag{3}$$

$$\frac{d^n}{dz^n}\left[z^{\sigma+n-1}\, {}_{p+1}F_q\left(\begin{matrix}\sigma,\, \alpha_p\\ \rho_q\end{matrix}\,\middle|\, z\right)\right]$$

$$= (\sigma)_n z^{\sigma-1}\, {}_{p+1}F_q\left(\begin{matrix}\sigma + n,\, \alpha_p\\ \rho_q\end{matrix}\,\middle|\, z\right). \tag{4}$$

$$\frac{d^n}{dz^n}\left[z^{\sigma-1}\, {}_pF_{q+1}\left(\begin{matrix}\alpha_p\\ \sigma,\, \rho_q\end{matrix}\,\middle|\, z\right)\right]$$

$$= (\sigma - n)_n z^{\sigma-1-n}\, {}_pF_{q+1}\left(\begin{matrix}\alpha_p\\ \sigma - n,\, \rho_q\end{matrix}\,\middle|\, z\right). \tag{5}$$

Let

$$F = {}_pF_q \tag{6}$$

$$F(\alpha_1+) = {}_pF_q\left(\begin{matrix}\alpha_1+1, \alpha_2, \ldots, \alpha_p\\ \rho_q\end{matrix}\,\middle|\, x\right) \tag{7}$$

$$F(\rho_1-) = {}_pF_q\left(\begin{matrix}\alpha_p\\ \rho_1-1, \rho_2, \ldots, \rho_q\end{matrix}\,\middle|\, x\right) \tag{8}$$

Then if $n=1$, (4) and (5) take the respective forms

$$(\delta + \alpha_1)F = \alpha_1 F(\alpha_1+), \tag{9}$$

$$(\delta + \rho_1 - 1)F = (\rho_1 - 1)F(\rho_1-), \tag{10}$$

where $\delta = z\,d/dz$.

For other contiguous relations, see Rainville (1960).

5.2.3. Integral Representations and Integrals Involving the $_pF_q$

Beta Transforms

$${}_{p+1}F_{q+1}\left(\begin{matrix}\beta,\, \alpha_p\\ \beta + \sigma,\, \rho_q\end{matrix}\,\middle|\, z\right)$$

$$= \frac{\Gamma(\beta+\sigma)z^{1-\beta-\sigma}}{\Gamma(\beta)\Gamma(\sigma)} \int_0^z t^{\beta-1}(z-t)^{\sigma-1} {}_pF_q\left(\begin{matrix}\alpha_p\\ \rho_q\end{matrix}\middle| t\right) dt \tag{1}$$

$$= \frac{\Gamma(\beta+\sigma)}{\Gamma(\beta)\Gamma(\sigma)} \int_0^1 t^{\beta-1}(1-t)^{\sigma-1} {}_pF_q\left(\begin{matrix}\alpha_p\\ \rho_q\end{matrix}\middle| zt\right) dt \tag{2}$$

$$= \frac{2\Gamma(\beta+\sigma)}{\Gamma(\beta)\Gamma(\sigma)} \int_0^{\pi/2} (\sin\theta)^{2\beta-1}(\cos\theta)^{2\sigma-1} {}_pF_q\left(\begin{matrix}\alpha_p\\ \rho_q\end{matrix}\middle| z\sin^2\theta\right) d\theta \tag{3}$$

$$= \frac{2\Gamma(\beta+\sigma)}{\Gamma(\beta)\Gamma(\sigma)} \int_0^{\pi/2} (\cos\theta)^{2\beta-1}(\sin\theta)^{2\sigma-1} {}_pF_q\left(\begin{matrix}\alpha_p\\ \rho_q\end{matrix}\middle| z\cos^2\theta\right) d\theta, \tag{4}$$

under the conditions

$$p \leqslant q+1, \quad R(\beta) > 0, \quad R(\sigma) > 0; \qquad |z| < 1 \quad \text{if} \quad p = q+1.$$

Since the ${}_2F_1(z)$ series in the unit disc D_1 can be analytically continued in the cut complex plane $|\arg(1-z)|<\pi$, D_2, it is clear that (2) serves to analytically continue the ${}_{p+1}F_p$ from D_1 to D_2. We use the same notation for the analytically continued function as for the series. Thus if $p=q+1$, (1)-(4) are valid for $|\arg(1-z)|<\pi$. The process of analytic continuation can be carried out by means of other integral representations as well.

Laplace and Inverse Laplace Transforms

$${}_{p+1}F_q\left(\begin{matrix}\sigma, \alpha_p\\ \rho_q\end{matrix}\middle| \omega/z\right) = \frac{z^\sigma}{\Gamma(\sigma)} \int_0^\infty e^{-zt} t^{\sigma-1} {}_pF_q\left(\begin{matrix}\alpha_p\\ \rho_q\end{matrix}\middle| \omega t\right) dt \tag{5}$$

is valid under the five cases listed below. In each case $R(\sigma) > 0$ and $z \neq 0$. It is convenient to set down the following conditions:

$$R(\sigma - \alpha_j) < 1, \qquad j = 1, 2, \ldots, p, \tag{6}$$

$$\nu = \sum_{h=1}^{p} \alpha_h - \sum_{h=1}^{q} \rho_h . \tag{7}$$

CASE 1. $p < q$, $|\arg z| < \pi/2$.

CASE 2. $p = q-1$, $|\arg z| = \pi/2$, $\arg\omega = \pi$, (14), $R(4\sigma + 2\nu) < 3$.

CASE 3. $p = q$, $|\arg z| < \pi/2$, $R(z) > R(\omega)$ or $R(z) = R(\omega)$, $z \neq \omega$ and $R(\sigma + \nu) < 1$.

CASE 4. $p = q$, $|\arg z| = \pi/2$, $\pi/2 < |\arg \omega| < 3\pi/2$, (6).

CASE 5. $p = q$, $|\arg z| = |\arg \omega| = \pi/2$, $z \neq \omega$, (6), $R(\sigma + \nu) < 1$.

$$\int_0^\infty e^{-zt} t^{\sigma-1} {}_pF_{p-1}\left(\begin{matrix}\alpha_p \\ \rho_{p-1}\end{matrix}\middle| -yt\right) dt = \frac{\Gamma(\rho_{p-1}) z^{-\sigma}}{\Gamma(\alpha_p)} G^{p+1,1}_{p,p+1}\left(\frac{z}{y}\middle|\begin{matrix}1, \rho_{p-1} \\ \sigma, \alpha_p\end{matrix}\right),$$

$$R(\sigma) > 0, \quad |\arg y| < \pi, \quad |\arg z| < \pi/2 \quad \text{or}$$

$$|\arg z| = \pi/2 \quad \text{and} \quad R(\sigma - \alpha_j) < 1, \quad j = 1, 2, \ldots, p. \tag{8}$$

where the G-function is defined by 5.3.1(1).

We also have for $R(\sigma) > 0$ and $z \neq 0$,

$${}_{p+2}F_q\left(\begin{matrix}\sigma, \sigma + \frac{1}{2}, \alpha_p \\ \rho_q\end{matrix}\middle| 4\omega^2/z^2\right) = \frac{z^\sigma}{\Gamma(\sigma)} \int_0^\infty e^{-zt} t^{\sigma-1} {}_pF_q\left(\begin{matrix}\alpha_p \\ \rho_q\end{matrix}\middle| \omega^2 t^2\right) dt,$$

$$\begin{aligned}
&p \leqslant q - 2, && && |\arg z| < \pi/2; \\
&p = q - 1, && \text{(a)} && R(z) > 2|R(\omega)| \geqslant 0 \quad \text{or} \\
& && \text{(b)} && R(z) = 2|R(\omega)| > 0, \\
& && && R(\sigma + \nu) < -\tfrac{1}{2} \quad \text{or} \\
& && \text{(c)} && R(z) = R(\omega) = 0, \\
& && && R(\sigma + \nu) < -\tfrac{1}{2}, \\
& && && R(\sigma - 2\alpha_h) < 1, \quad h = 1, 2, \ldots, p,
\end{aligned} \tag{9}$$

where v is as in (7).

$$\omega^{\beta-1} {}_pF_{q+1}\left(\begin{matrix}\alpha_p \\ \beta, \rho_q\end{matrix}\middle| \omega z\right) = \frac{\Gamma(\beta)}{2\pi i} \int_{c-i\infty}^{c+i\infty} e^{\omega t} t^{-\beta} {}_pF_q\left(\begin{matrix}\alpha_p \\ \rho_q\end{matrix}\middle| z/t\right) dt,$$

$$\omega \text{ real}, \quad \omega \neq 0, \quad R(\beta) > 0, \quad c > 0, \quad p \leqslant q. \tag{10}$$

$\arg(1 - z/c)| < \pi$ if $p = q + 1$.

Mellin Transform

$$\int_0^\infty t^{\sigma-1}\,{}_pF_q\left(\begin{matrix}\alpha_p\\ \rho_q\end{matrix}\middle|\,-\eta t\right)dt=\frac{\eta^{-\sigma}\Gamma(\sigma)\Gamma(\alpha_p-\sigma)\Gamma(\rho_q)}{\Gamma(\alpha_p)\Gamma(\rho_q-\sigma)},$$

$$0<R(\sigma)<R(\alpha_j),\qquad j=1,2,\ldots,p,\qquad \eta\neq 0,\tag{11}$$

$$q=p+1,\qquad R(\nu+2\sigma)<\tfrac{1}{2},\qquad \eta>0,$$

$$q=p,\qquad |\arg\eta|<\pi/2\quad\text{or}\quad|\arg\eta|=\pi/2\quad\text{if}\quad R(\nu+\sigma)<1,$$

$$q=p-1,\qquad |\arg\eta|<\pi,$$

where ν is given by (7). This is a limiting case of (5).

An Indefinite Integral

$$\int_z^\infty t^{\sigma-1}\,{}_pF_q\left(\begin{matrix}\alpha_p\\ \rho_q\end{matrix}\middle|\,-\eta t\right)dt=\frac{(-)^m\eta^m(\alpha_p)_m}{(\rho_q)_m m!}\left[\Psi(1+m)+\sum_{j=1}^{q}\Psi(\rho_j+m)\right.$$

$$\left.-\sum_{j=1}^{p}\Psi(\alpha_j+m)-\ln(\eta z)\right]-\eta^m\sum_{\substack{k=0\\k\neq m}}^{\infty}\frac{(\alpha_p)_k(-)^k(\eta z)^{k-m}}{(k-m)(\rho_q)_k k!},\tag{12}$$

where σ is a negative integer or zero, say $\sigma=-m$. The conditions of validity for (12) are the same as for (11) without $R(\sigma)>0$.

A Mellin-Barnes integral representation for $_pF_q$ follows from 5.3.1(1-4,12).

5.2.4. Evaluation for Special Values of the Variable and Parameters

Throughout this section n is a positive integer or zero. Saalschütz's formula is

$$_3F_2\left(\begin{matrix}-n,\,a,\,b\\ c,\,1+a+b-c-n\end{matrix}\middle|\,1\right)=\frac{(c-a)_n(c-b)_n}{(c)_n(c-a-b)_n}\tag{1}$$

Also

$$ {}_3F_2\left(\begin{matrix}-n, n+a, b\\ c, a+b+1-c\end{matrix}\,\middle|\,1\right) = \frac{(c-b)_n(a-c+1)_n}{(c)_n(a+b+1-c)_n}. \tag{2} $$

Equations (3), (4), and (5) below go by the names Dixon, Watson and Whipple respectively.

$$ {}_3F_2\left(\begin{matrix}a, b, c\\ 1+a-b, 1+a-c\end{matrix}\,\middle|\,1\right) = \frac{\Gamma(1+\frac{1}{2}a)\,\Gamma(1+a-b)}{\Gamma(1+a)\,\Gamma(1+\frac{1}{2}a-b)} \times \frac{\Gamma(1+a-c)\,\Gamma(1+\frac{1}{2}a-b-c)}{\Gamma(1+\frac{1}{2}a-c)\,\Gamma(1+a-b-c)}, \quad R(a-2b-2c) > -2. \tag{3} $$

$$ {}_3F_2\left(\begin{matrix}a, b, c\\ \frac{1}{2}(a+b+1), 2c\end{matrix}\,\middle|\,1\right) = \frac{\Gamma(\frac{1}{2})\,\Gamma(\frac{1}{2}+c)\,\Gamma(\frac{1}{2}+\frac{1}{2}a+\frac{1}{2}b)}{\Gamma(\frac{1}{2}+\frac{1}{2}a)\,\Gamma(\frac{1}{2}+\frac{1}{2}b)} \times \frac{\Gamma(\frac{1}{2}-\frac{1}{2}a-\frac{1}{2}b+c)}{\Gamma(\frac{1}{2}-\frac{1}{2}a+c)\,\Gamma(\frac{1}{2}-\frac{1}{2}b+c)}, \quad R(2c-a-b) > -1. \tag{4} $$

$$ {}_3F_2\left(\begin{matrix}a, 1-a, c\\ f, 2c+1-f\end{matrix}\,\middle|\,1\right) = \frac{\pi\Gamma(f)\,\Gamma(2c+1-f)}{2^{2c-1}\Gamma[c+\frac{1}{2}(a+1-f)]\,\Gamma[\frac{1}{2}(a+f)]\,\Gamma[1+c-\frac{1}{2}(a+f)]\,\Gamma[\frac{1}{2}(1-a+f)]}, \quad R(c) > 0. \tag{5} $$

The left hand side of (5) is meaningful if c is a negative integer or zero, say $c=-n$. In this event its value is not given by the right hand side of (5) unless a is an integer or zero. We do have

$$ {}_3F_2\left(\begin{matrix}-n, a, 1-a\\ f, -2n-f+1\end{matrix}\,\middle|\,1\right) = \frac{2^{2n}(\frac{1}{2}a+\frac{1}{2}f)_n(\frac{1}{2}a-\frac{1}{2}f+\frac{1}{2}-n)_n}{(f)_n(1-2n-f)_n}. \tag{6} $$

$$ {}_3F_2\left(\begin{matrix}a, b, c\\ e, f\end{matrix}\,\middle|\,1\right) = \frac{\Gamma(e)\,\Gamma(f)\,\Gamma(s)}{\Gamma(a)\,\Gamma(s+b)\,\Gamma(s+c)}\,{}_3F_2\left(\begin{matrix}e-a, f-a, s\\ s+b, s+c\end{matrix}\,\middle|\,1\right), \tag{7} $$

$$ s = e+f-a-b-c, \qquad s \neq 0. $$

$$ {}_3F_2\left(\begin{matrix}a, b, c\\ e, f\end{matrix}\,\middle|\,1\right) = \frac{\Gamma(1-a)\,\Gamma(e)\,\Gamma(f)\,\Gamma(c-b)}{\Gamma(e-b)\,\Gamma(f-b)\,\Gamma(1+b-a)\,\Gamma(c)} \times {}_3F_2\left(\begin{matrix}b, b-e+1, b-f+1\\ 1+b-c, 1+b-a\end{matrix}\,\middle|\,1\right) \tag{8} $$

+ a similar expression with b and c interchanged.

Equations (7) and (8) are but two of many relations known as transformation formulas for ${}_3F_2$ series of unit argument. For a detailed description of these, see Bailey (1935), Slater (1966) and Luke (1969).

If in a ${}_3F_2$ form, a numerator parameter exceeds a denominator parameter by a positive integer, say m, the ${}_3F_2$ may be expressed as the sum of $(m+1)$ ${}_2F_1$'s. If the latter can be summed, then so can the ${}_3F_2$. In illustration, since

$$ {}_3F_2\left(\begin{matrix}a, b, c+1\\ d, c\end{matrix}\,\middle|\,z\right) = \sum_{k=0}^{\infty} \frac{(a)_k(b)_k(c+k)z^k}{(d)_k k!\,c} = {}_2F_1\left(\begin{matrix}a, b\\ d\end{matrix}\,\middle|\,z\right) + \frac{abz}{cd}\,{}_2F_1\left(\begin{matrix}a+1, b+1\\ d+1\end{matrix}\,\middle|\,z\right) \tag{9} $$

$$ {}_3F_2\left(\begin{matrix}a, b, c+1\\ d, c\end{matrix}\,\middle|\,1\right) = \frac{\Gamma(d)\Gamma(d-a-b)}{\Gamma(d-a)\Gamma(d-b)}\left[1 - \frac{ab}{c(a+b+1-d)}\right], \quad R(d-a-b)>1. \tag{10} $$

This is a special case of 5.10.2(4).

The formula

$$ {}_3F_2\left(\begin{matrix}a, b, c\\ d+1, c+1\end{matrix}\,\middle|\,z\right) = \frac{c}{c-d}\,{}_2F_1\left(\begin{matrix}a, b\\ d+1\end{matrix}\,\middle|\,z\right) - \frac{d}{c-d}\,{}_3F_2\left(\begin{matrix}a, b, c\\ d, c+1\end{matrix}\,\middle|\,z\right), \tag{11} $$

can be used to prove

$$ {}_3F_2\left(\begin{matrix}-n, n+a, c\\ a+1, c+1\end{matrix}\,\middle|\,1\right) = \frac{a(a-c)_n}{(a-c)(a)_n(c+1)_n}, \quad n>0. \tag{12} $$

$$ {}_3F_2\left(\begin{matrix}-n, n+a+1, c\\ a, c+1\end{matrix}\,\middle|\,1\right) = \frac{n!}{a(a+1)_n}\left(c(-)^n - \frac{(c-a)(a+1-c)_n}{(c+1)_n}\right) \tag{13} $$

Equation (11) can be readily generalized to a $_{p+1}F_p$. For $p=3$,

$$ {}_4F_3\left({a,b,c,d \atop g,c+1,d+1}\middle| z\right) = \frac{d(c-g)}{(c-d)g}\,{}_3F_2\left({a,b,c \atop g+1,c+1}\middle| z\right) $$

$$ \frac{c(d-g)}{(c-d)g}\,{}_3F_2\left({a,b,d \atop g+1,d+1}\middle| z\right), \tag{14} $$

and if $z=1$ and $a+b=g+1$, each $_3F_2$ is Saalschützian and so may be summed, whence the $_4F_3$ can also be summed.

From the result,

$$ z\,{}_{p+1}F_{q+1}\left({\alpha_p, 1 \atop \rho_q, 2}\middle| z\right) = \frac{(\rho_q-1)}{(\alpha_p-1)}\left({}_pF_q\left({\alpha_p-1 \atop \rho_q-1}\middle| z\right)-1\right), \tag{15} $$

we get

$$ {}_3F_2\left({a, b, 1 \atop c, 2}\middle| 1\right) = \frac{(c-1)}{(a-1)(b-1)}\left[\frac{\Gamma(c-1)\,\Gamma(c-a-b+1)}{\Gamma(c-a)\,\Gamma(c-b)} - 1\right], $$
$$ a \neq 1, \quad b \neq 1, \quad R(c-a-b) > -1. \tag{16} $$

By L'Hospital's theorem,

$$ {}_3F_2\left({a, 1, 1 \atop c, 2}\middle| 1\right) = \frac{(c-1)}{(a-1)}\,[\psi(c-1) - \psi(c-a)],\ a \neq 1,\ R(c-a) > 0, \tag{17} $$

$$ {}_3F_2\left({1, 1, 1 \atop c, 2}\middle| 1\right) = (c-1)\,\psi'(c-1), \qquad R(c) > 1. \tag{18} $$

Again,

$$ z^2\,{}_3F_2\left({a, b, 1 \atop c, 3}\middle| z\right) = \frac{2(c-2)_2}{(a-2)_2(b-2)_2}\left[{}_2F_1\left({a-2, b-2 \atop c-2}\middle| z\right) - 1\right] \tag{19} $$
$$ - \frac{2z(c-1)}{(a-1)(b-1)}, $$

leads to

$$ {}_3F_2\left({a, b, 1 \atop c, 3}\middle| 1\right) = \frac{2(c-2)_2}{(a-2)_2(b-2)_2}\left[\frac{\Gamma(c-2)\,\Gamma(c-a-b+2)}{\Gamma(c-a)\,\Gamma(c-b)} - 1\right] $$
$$ - \frac{2(c-1)}{(a-1)(b-1)}, \tag{20} $$
$$ a \neq 1, 2; \quad b \neq 1, 2; \quad R(c-a-b) > -2. $$

$$ {}_4F_3\left(\begin{matrix}-n, n+\lambda, 1, 1\\ d, e, 2\end{matrix}\middle|\, 1\right) = \frac{(d-1)(e-1)}{(n+1)(n+\lambda-1)}\Big[\psi(d+n)+\psi(e+n)-\psi(d-1)-\psi(e-1)\Big], \quad \lambda+1=d+e. \tag{21} $$

$$ {}_4F_3\left(\begin{matrix}1+a, 1+b, 1+c, 1\\ 2+a-b, 2+a-c, 2\end{matrix}\middle|\, 1\right) = \frac{(1+a-b)(1+a-c)}{abc} \times\left[\frac{\Gamma(1+\frac{1}{2}a)\Gamma(1+a-b)\Gamma(1+a-c)\,\Gamma(1+\frac{1}{2}a-b-c)}{\Gamma(1+a)\Gamma(1+\frac{1}{2}a-b)\Gamma(1+\frac{1}{2}a-c)\,\Gamma(1+a-b-c)} -1\right], \quad R(a-2b-2c)>-2. \tag{22} $$

$$ {}_4F_3\left(\begin{matrix}1+a, 1+b, 1, 1\\ 2+a-b, 2+a-c, 2\end{matrix}\middle|\, 1\right) = \frac{(1+a-b)(1+a)}{ab}\Big[\psi(1+\tfrac{1}{2}a)-\psi(1+a)+\Psi(1+a-b)-\Psi(1+\tfrac{1}{2}a-b)\Big], \quad R(a-2b)>-2. \tag{23} $$

$$ {}_4F_3\left(\begin{matrix}b, 2v+1, v, v+3/2\\ 2v+2-b, v+2, v+\frac{1}{2}\end{matrix}\middle|\, g\right) = \frac{(v+2-b)_v\Gamma(v+2)}{2^{2v}(3/2)_v\left(1-\frac{b(g+1)}{2}\right)_v}, \quad g=\pm 1, \tag{24} $$

provided $R(b)<(1-g)/4$. This result is due to Rice (1944).

$$ {}_4F_3\left(\begin{matrix}a+h, a+\frac{1}{2}, b+h, b+\frac{1}{2}\\ \frac{1}{2}+h, a-b+1, a-b+\frac{1}{2}+h\end{matrix}\middle|\, 1\right) = \frac{2^{-2a-1-2h}\Gamma(2a-2b+1+h)}{(1-h+hab)\Gamma(a+1-2b)} \times\left[\frac{\Gamma(\frac{1}{2}-2b)}{\Gamma(\frac{1}{2}+a-2b)} + \frac{(1-2h)\Gamma(\frac{1}{2})}{\Gamma(\frac{1}{2}+a)}\right], \quad h=0,1, R(b)<\tfrac{1}{4}. \tag{25} $$

$$ {}_4F_3\left(\begin{matrix}a, 1+\frac{1}{2}a, b, c\\ \frac{1}{2}a, 1+a-b, 1+a-c\end{matrix}\middle|\, 1\right) = \frac{(1+a-b-c)_b}{(1+a-b)_b}, \quad R(2b+2c-a)<1. \tag{26} $$

$$ {}_4F_3\left(\begin{matrix}-n, n+a, \mu, \mu+\frac{1}{2}\\ \frac{1}{2}a+\frac{1}{2}, \frac{1}{2}a+1, 2\mu\end{matrix}\middle|\, 1\right) = \frac{a(a+1-2\mu)_n}{(a+2n)(a)_n}. \tag{27} $$

The latter is due to Carlitz (1963).

$$ {}_4F_3\left(\begin{matrix}-n, v, \frac{1}{2}v+1, \mu+v+1\\ n+1+v, \frac{1}{2}v, -\mu\end{matrix}\middle|\, -1\right) = \frac{(-)^n\Gamma(\mu+1-n)(v+1)_n}{\Gamma(\mu+1)}. \tag{28} $$

This result is also valid if μ is a positive integer $\geq n$ provided the ${}_4F_3$ is interpreted as a sum of its first $(n+1)$ terms.

$$ {}_3F_2\left({2a,2b,a+b \atop 2a+2b,a+b+\frac{1}{2}}\middle|1\right) = \left[\frac{(b+\frac{1}{2})_a}{(\frac{1}{2})_a}\right]^2. \tag{29}$$

$$ {}_3F_2\left({2a,2b,a+b \atop 2a+2b-1,a+b+\frac{1}{2}}\middle|1\right) = \frac{-(2b-1)(2a-1)}{(2a+2b-1)}\left[\frac{(b+\frac{1}{2})_a}{(\frac{1}{2})_a}\right]^2. \tag{30}$$

$$ {}_3F_2\left({2a,2b-1,a+b-1 \atop 2a+2b-2,a+b-\frac{1}{2}}\middle|1\right) = \frac{[(b-\frac{1}{2})_a]^2}{(\frac{1}{2})_a(-\frac{1}{2})_a}. \tag{31}$$

$$ {}_3F_2\left({1,1,2-a \atop 2,1+a}\middle|-1\right) = \frac{a}{2(a-1)}\left(\gamma+\psi(a)\right). \tag{32}$$

$$ {}_3F_2\left({1,1,2a+1 \atop 2,2a+2}\middle|-1\right) = \frac{2a+1}{4a}\left(\psi(a+\tfrac{1}{2})-\psi(a+1)+2\ln 2\right). \tag{33}$$

$$ {}_3F_2\left({a,1-a,3a-1 \atop 2a,a+\frac{1}{2}}\middle|-1/8\right) = (2^{3a-3}/\pi)\left[\frac{\Gamma(a/2)\Gamma(a+\frac{1}{2})}{\Gamma(3a/2)}\right]^2. \tag{34}$$

The latter is due to Champion, Danielson and Miksell (1969).

In numerous applications of quantum theory, certain coefficients which go by the names of Clebsch-Gordan, Racah, 3-j and 6-j symbols are tabulated. These are related to certain terminating ${}_3F_2$'s and ${}_4F_3$'s of unit argument. Some pertinent references are J. Miller, Gerhouser and Matsen (1959), Shimpuku (1960), Ishidzu (1960), Rotenberg, Bivins, Metropolis and Wooten (1960), Chi (1962), Biedenharn and van Dam (1965), Nikiforov, Uvarov and Levitan (1965), Minton (1970) and W. Miller (1972).

Let

$$ B_m = {}_{p+1}F_p^m\left({\alpha_p, 1 \atop 1+\rho_p}\middle|1\right). \tag{35}$$

Then

$$ B_m - B_{m-1} = \frac{(\alpha_p)_m}{(1+\rho_p)_m}. \tag{36}$$

If

$$ B_m = R+T(\alpha_p)_{m+1}\big/(1+\rho_p)_m, \tag{37}$$

where R and T are independent of m, then

$$B_m = \frac{\rho_p}{(\rho_p-\alpha_p)} - \frac{(\alpha_p)_{m+1}}{(\rho_p-\alpha_p)(1+\rho_p)_m}, \tag{38}$$

provided

$$S_t(\alpha_p) = S_t(\rho_p), \quad t=0,1,\ldots,p-1, \tag{39}$$

where $S_r(A_p)$ is defined by

$$\prod_{i=1}^{p} (x+A_i) = \sum_{r=0}^{p} S_r(A_p)x^{p-r} . \tag{40}$$

Thus under the conditions (39),

$${}_{p+1}F_p\left(\begin{matrix}\alpha_p, 1\\ 1+\rho_p\end{matrix}\,\middle|\,1\right) = \frac{\rho_p}{\rho_p-\alpha_p}\left(1-\frac{\Gamma(\rho_p)}{\Gamma(\alpha_p)}\right) \tag{41}$$

and

$${}_pF_{p-1}\left(\begin{matrix}-n, \alpha_{p-1}\\ 1+\rho_{p-1}\end{matrix}\,\middle|\,1\right) = 0. \tag{42}$$

In the condition $S_t(\alpha_p)=S_t(\rho_p)$ for the latter equation, $\alpha_h=-n$ for $h=p$ and $\rho_h=0$ for $h=p$. The above developments (35)-(40) are due to Slater (1966). In illustration of these results we have

$$ {}_3F_2^{\,m}\left(\begin{matrix}a, b, c\\ e, f\end{matrix}\,\middle|\,1\right) = \frac{(a+1)_m(b+1)_m(c+1)_m}{(e)_m(f)_m m!},$$

$$a+b+c+2=e+f, \quad ab+ac+bc=(e-1)(f-1). \tag{43}$$

Carlitz (1965) showed that

$${}_{p+2}F_{p+1}^{\,m}\left(\begin{matrix}\alpha_p, 1+g, 1\\ 1+\rho_p, g\end{matrix}\,\middle|\,1\right) = 1+\frac{\alpha_p}{(\rho_p-\alpha_p)(1+\rho_p)} - \frac{(\alpha_p)_{m+1}}{(\rho_p-\alpha_p)(1+\rho_p)_m} \tag{44}$$

provided (39) holds for $t=0,1,\ldots,p-2$ and

$$g^{-1} = S_{p-1}(\alpha_p)-S_{p-1}(\rho_p). \tag{45}$$

Verma (1967) has applied the results (35-45) to sum series of the form

$$\sum_{k=-s}^{m} (\alpha_p)_k/(\rho_p)_k .$$

$$_{p+1}F_p\left(\begin{matrix}-n, 1+\alpha_p\\ \alpha_p\end{matrix}\middle| 1\right) = 0 \text{ for } n>p. \tag{46}$$

For evaluation of the above for $n\leq p$, see Luke (1969).

$$\frac{(\alpha_p)}{\Gamma(a+p+1)} \; {}_{p+2}F_{p+1}\left(\begin{matrix}-n, n+a, 1+\alpha_p\\ p+1+a, \alpha_p\end{matrix}\middle| 1\right)$$
$$= \frac{(-)^p(n+a-\alpha_p)n!}{(p-n)!\Gamma(2n+a+1)} \; {}_{p+2}F_{p+1}\left(\begin{matrix}n-p, n+a, 1+n+a-\alpha_p\\ 2n+a+1, n+a-\alpha_p\end{matrix}\middle| 1\right), \tag{47}$$

and the latter is nil if $0\leq p\leq n-1$.

$$_2F_0(-n, 2\,|-1/m) = m = n+1. \tag{48}$$

5.3. The G-Function

5.3.1. Definition and Relation to the $_pF_q$

The G-function is defined by the Mellin-Barnes integral

$$G_{p,q}^{m,n}\left(z\,\middle|\,\begin{matrix}a_1,\ldots,a_p\\ b_1,\ldots,b_q\end{matrix}\right) = G_{p,q}^{m,n}\left(z\,\middle|\,\begin{matrix}a_p\\ b_q\end{matrix}\right)$$
$$= (2\pi i)^{-1}\int_L \frac{\prod_{j=1}^m \Gamma(b_j - s)\prod_{j=1}^n \Gamma(1-a_j+s)}{\prod_{j=m+1}^q \Gamma(1-b_j+s)\prod_{j=n+1}^p \Gamma(a_j-s)}\, z^s\, ds. \tag{1}$$

Where no confusion can result, we often refer to the latter as $G_{p,q}^{m,n}(z)$. Here an empty product is interpreted as unity, $0 \leqslant m \leqslant q$, $0 \leqslant n \leqslant p$, and the parameters a_h and b_h are such that no pole of $\Gamma(b_j - s)$, $j = 1, 2,\ldots, m$, coincides with any pole of $\Gamma(1 - a_k + s)$, $k = 1, 2,\ldots, n$. Thus $(a_k - b_j)$ is not a positive integer. We retain these assumptions throughout. Also $z \neq 0$.

There are three different paths L of integration:

L goes from $\sigma - i\infty$ to $\sigma + i\infty$ so that all poles of $\Gamma(b_j - s)$, $j = 1, 2, \ldots, m$, lie to the right of the path, and all poles of $\Gamma(1 - a_k + s)$, $k = 1, 2, \ldots, n$, lie to the left of the path. For the integral to converge, we need $\delta = m + n - \frac{1}{2}(p + q) > 0$, $|\arg z| < \delta\pi$. If $|\arg z| = \delta\pi$, $\delta \geqslant 0$, the integral converges absolutely when $p = q$ if $R(\nu) < -1$; and when $p \neq q$, if with $s = \sigma + i\tau$, σ and τ real, σ is chosen so that for $\tau \to \pm\infty$, (2)

$$(q - p)\sigma > R(\nu) + 1 - \tfrac{1}{2}(q - p),$$
$$\nu = \sum_{j=1}^{q} b_j - \sum_{j=1}^{p} a_j .$$

L is a loop beginning and ending at $+\infty$ and encircling all poles of $\Gamma(b_j - s)$, $j = 1, 2, \ldots, m$, once in the negative direction, but none of the poles of $\Gamma(1 - a_k + s)$, $k = 1, 2, \ldots, n$. The integral converges if $q \geqslant 1$ and either $p < q$ or $p = q$ and $|z| < 1$. (3)

L is a loop beginning and ending at $-\infty$ and encircling all poles of $\Gamma(1 - a_k + s)$, $k = 1, 2, \ldots, n$, once in the positive direction, but none of the poles of $\Gamma(b_j - s)$, $j = 1, 2, \ldots, m$. The integral converges if $p \geqslant 1$ and either $p > q$ or $p = q$ and $|z| > 1$. (4)

It is supposed that the parameters a_h, b_h, and the variable z are such that at least one of the definitions (2)–(4) makes sense. Where more than one of the definitions has meaning, they lead to the same result so that no confusion arises.

If we use (3), the integral can be evaluated as a sum of residues. If no two of the b_h terms, $h = 1, 2, \ldots, m$, differ by an integer or zero, all poles are simple, and

$$G^{m,n}_{p,q}\left(z \,\middle|\, \begin{matrix} a_p \\ b_q \end{matrix}\right) = \sum_{h=1}^{m} \frac{\prod_{j=1}^{m} \Gamma(b_j - b_h)^* \prod_{j=1}^{n} \Gamma(1 + b_h - a_j)\, z^{b_h}}{\prod_{j=m+1}^{q} \Gamma(1 + b_h - b_j) \prod_{j=n+1}^{p} \Gamma(a_j - b_h)}$$
$$\times {}_pF_{q-1}\left(\begin{matrix} 1 + b_h - a_p \\ 1 + b_h - b_q^* \end{matrix} \,\middle|\, (-)^{p-m-n} z\right),$$
$$p < q \quad \text{or} \quad p = q \quad \text{and} \quad |z| < 1. \tag{5}$$

In the ${}_pF_{q-1}$, $1+b_h-b_q^*$ stands for the set of q-1 parameters $1+b_h-b_1, 1+b_h-b_2, \ldots, 1+b_h-b_{h-1}, 1+b_h-b_{h+1}, \ldots, 1+b_h-b_q$. Thus omit 1 in the set of q parameters $1+b_h-b_q$. Also $\Gamma(b_j-b_h)^*$ means to omit this term when $j=h$. This notational scheme is used throughout.

If $m = 0$, and we use (3), the integrand is analytic on and within the contour and so

$$G_{p,q}^{0,n}\left(z \,\middle|\, \begin{matrix} a_p \\ b_q \end{matrix}\right) = 0. \tag{6}$$

It is clear that the G-function is a many valued function of z with a branch point at the origin. For further discussion of the singularities of the G-function, see 5.3.3 and 5.7.2.

If only of the two b_h's differ by an integer or zero, we can use L'Hospital's theorem on the pertinent portion of the right side of (5) to get a series representation for the G-function on the left. Suppose that $b_2-b_1=s$, s a positive integer or zero. Write (5) in the form

$$G_{p,q}^{m,n}\left(z \,\middle|\, \begin{matrix} a_p \\ b_q \end{matrix}\right) = \sum_{h=1}^{m} C_h^{(m)}, \qquad m \geqslant 2, \tag{7}$$

where the meaning of $C_h^{(m)}$ is evident. Then

$$\begin{aligned}
G_{p,q}^{m,n}\left(z \,\middle|\, \begin{matrix} a_p \\ b_q \end{matrix}\right) &= \sum_{h=3}^{m} C_h^{(m)} + T_{p,q}^{m,n}\left(z \,\middle|\, \begin{matrix} 1, 1+b_q-b_2 \\ 1+a_p-b_2 \end{matrix}\right) \\
&+ \frac{(-)^{s+1} z^{b_2} \prod_{j=3}^{m} \Gamma(b_j-b_2) \prod_{j=1}^{n} \Gamma(1+b_2-a_j)}{s! \prod_{j=m+1}^{q} \Gamma(1+b_2-b_j) \prod_{j=n+1}^{p} \Gamma(a_j-b_2)} \\
&\times \Bigg[\Bigg\{(\gamma + \ln z) - \sum_{j=3}^{m} \psi(b_j-b_2) + \sum_{j=1}^{n} \psi(1+b_2-a_j) \\
&- \sum_{j=m+1}^{q} \psi(1+b_2-b_j) + \sum_{j=n+1}^{p} \psi(a_j-b_2) - \psi(1+s)\Bigg\} \\
&\times {}_pF_{q-1}\left(\begin{matrix} 1+b_2-a_p \\ 1+b_2-b_q^* \end{matrix} \,\middle|\, (-)^{p-m-n} z\right) \\
&+ {}_pF^q_{p\;q-1}\left(\begin{matrix} 1+b_2-a_p \\ 1+b_2-b_q^* \end{matrix} \,\middle\|\, \begin{matrix} 1+b_2-a_p \\ 1+b_2-b_q \end{matrix} \,\middle|\, (-)^{p-m-n} z\right)\Bigg],
\end{aligned}$$

$$b_2 = b_1 + s, \tag{8}$$

where

$$T_{p,q}^{m,n}\left(z\middle|\begin{matrix}1,1+b_q-b_2\\1+a_p-b_2\end{matrix}\right) = T_{p,q}^{m,n}\left(z\middle|\begin{matrix}1,1-s,1+b_3-b_2,\ldots,1+b_q-b_2\\1+a_1-b_2,\ldots,1+a_p-b_2\end{matrix}\right)$$

$$= \frac{z^{b_1}\prod_{j=3}^{m}(\Gamma(b_j-b_1)\prod_{j=1}^{n}\Gamma(1+b_1-a_j)}{\prod_{j=m+1}^{q}\Gamma(1+b_1-b_j)\prod_{j=n+1}^{p}\Gamma(a_j-b_1)}$$

$$\times \sum_{k=0}^{s-1}\frac{(1+b_1-a_p)_k(\ s-1-k)!\,(-)^{k(m+n+1-p)}z^k}{\prod_{j=3}^{q}(1+b_1-b_j)_k k!}$$

$$= \frac{z^{b_1}\{z\exp[i\pi(m+n+1-p)]\}^{s-1}\,\pi^{m+n-p-2}\prod_{j=1}^{p}\Gamma(b_2-a_j)\prod_{j=n+1}^{p}\sin\pi(a_j-b_1)}{(s-1)!\prod_{j=3}^{q}\Gamma(b_2-b_j)\prod_{j=3}^{m}\sin\pi(b_j-b_1)} \tag{9}$$

$$\times\ {}_{q+1}F_p\left(\begin{matrix}1,1+b_q-b_2\\1+a_p-b_2\end{matrix}\middle|\frac{(-)^{q-m-n}}{z}\right), b_2-b_1=s,$$

and

$${}_pF_q^{uv}\left(\begin{matrix}\sigma_p\\\tau_q\end{matrix}\middle\|\begin{matrix}\gamma_u\\\delta_v\end{matrix}\middle| z\right) = \sum_{k=0}^{\infty}\frac{(\sigma_p)_k z^k}{(\tau_q)_k k!}\{\psi(\gamma_u+k)-\psi(\gamma_u)-\psi(\delta_v+k)+\psi(\delta_v)\}. \tag{10}$$

In the latter, $\psi(\gamma_u)$ is short for $\sum_{j=1}^{u}\psi(\gamma_j)$, etc.

If another pair of the b_j's differ by a positive integer or zero, say $b_2-b_3=r$, but b_2 and b_1 do not differ by an integer or zero, then a function independent of $C_3^{(m)}$ follows on application of the above analysis to $C_3^{(m)}+C_4^{(m)}$. If three or more of the b_j's differ by an integer or zero, the previous analysis can be readily extended.

Alternatively, the result (8-10) and extensions thereto can be deduced from (1) by residue theory upon recognizing that there are multiple poles in addition to simple poles. In this connection, see the work of MacRobert (1961, 1962) who developed forms for a special case of the G-function when two or more of the b_j's differ by an integer or zero.

If no two of the a_h terms, $h = 1,\ldots, n$, differ by an integer or zero, then use of (4) leads to

$$G_{p,q}^{m,n}\left(z \,\middle|\, \begin{matrix} a_p \\ b_q \end{matrix}\right) = \sum_{h=1}^{n} \frac{\prod_{j=1}^{n} \Gamma(a_h - a_j)^* \prod_{j=1}^{m} \Gamma(b_j - a_h + 1)\, z^{a_h - 1}}{\prod_{j=n+1}^{p} \Gamma(1 + a_j - a_h) \prod_{j=m+1}^{q} \Gamma(a_h - b_j)}$$

$$\times\, {}_qF_{p-1}\left(\begin{matrix} 1 + b_q - a_h \\ 1 + a_p - a_h^* \end{matrix} \,\middle|\, \frac{(-)^{q-m-n}}{z}\right),$$

$$q < p \quad \text{or} \quad q = p \quad \text{and} \quad |z| > 1. \tag{11}$$

If two or more of the a_h terms differ by an integer or zero, the forms for the G-function in (11) can be deduced from the forms for the G-function in (7) when two or more of the b_h terms differ by an integer or zero in view of the relation 5.3.2(3).

It is clear from (5) and (11) that the $_pF_q$ can be expressed as a G-function. We have the following relations:

$$G_{p,q}^{1,n}\left(z \,\middle|\, \begin{matrix} a_p \\ b_q \end{matrix}\right) = \frac{\prod_{j=1}^{n} \Gamma(1 + b_1 - a_j)\, z^{b_1}}{\prod_{j=2}^{q} \Gamma(1 + b_1 - b_j) \prod_{j=n+1}^{p} \Gamma(a_j - b_1)}$$

$$\times\, {}_pF_{q-1}\left(\begin{matrix} 1 + b_1 - a_p \\ 1 + b_1 - b_q^* \end{matrix} \,\middle|\, (-)^{p-1-n} z\right),$$

$$p < q \quad \text{or} \quad p = q \quad \text{and} \quad |z| < 1; \tag{12}$$

$${}_pF_q\left(\begin{matrix} 1 + \alpha_p - \rho_0 \\ 1 + \rho_q - \rho_0^* \end{matrix} \,\middle|\, z\right) = \frac{\left[\begin{matrix} \prod_{j=1}^{q} \Gamma(1 + \rho_j - \rho_0) \prod_{j=n+1}^{p} \Gamma(\rho_0 - \alpha_j) \\ \times \{z \exp[-i\pi(n + 1 - p)]\}^{(\rho_0 - 1)} \end{matrix}\right]}{\prod_{j=1}^{n} \Gamma(1 + \alpha_j - \rho_0)}$$

$$\times\, G_{p,q+1}^{1,n}\left(z \exp[-i\pi(n+1-p)] \,\middle|\, \begin{matrix} 1 - \alpha_p \\ 1 - \rho_0, 1 - \rho_1, \ldots, 1 - \rho_q \end{matrix}\right),$$

$$p \leqslant q \quad \text{or} \quad p = q + 1 \quad \text{and} \quad |z| < 1; \tag{13}$$

$${}_pF_q\left(\begin{matrix} \alpha_p \\ \rho_q \end{matrix} \,\middle|\, z\right) = \frac{\Gamma(\rho_q)}{\Gamma(\alpha_p)}\, G_{p,q+1}^{1,p}\left(-z \,\middle|\, \begin{matrix} 1 - \alpha_p \\ 0, 1 - \rho_q \end{matrix}\right),$$

$$p \leqslant q \quad \text{or} \quad p = q + 1 \quad \text{and} \quad |z| < 1. \tag{14}$$

$$G_{p,q}^{m,1}\left(z \,\middle|\, \begin{matrix} a_p \\ b_q \end{matrix}\right) = \frac{\prod_{j=1}^{m} \Gamma(b_j - a_1 + 1)\, z^{a_1 - 1}}{\prod_{j=2}^{p} \Gamma(1 + a_j - a_1) \prod_{j=m+1}^{q} \Gamma(a_1 - b_j)}$$

$$\times\, {}_qF_{p-1}\left(\begin{matrix} 1 + b_q - a_1 \\ 1 + a_p - a_1^* \end{matrix} \,\middle|\, \frac{(-)^{q-m-1}}{z}\right),$$

$$q < p \quad \text{or} \quad q = p \quad \text{and} \quad |z| > 1. \tag{15}$$

$${}_pF_q\left(\begin{matrix} 1 + \alpha_p - \rho_0 \\ 1 + \rho_q - \rho_0^* \end{matrix} \,\middle|\, z\right) = \frac{\left[\begin{matrix} \prod_{j=1}^{q} \Gamma(1 + \rho_j - \rho_0) \prod_{j=n+1}^{p} \Gamma(\rho_0 - \alpha_j) \\ \times \{z \exp[-i\pi(n + 1 - p)]\}^{(\rho_0 - 1)} \end{matrix}\right]}{\prod_{j=1}^{n} \Gamma(1 + \alpha_j - \rho_0)}$$

$$\times\, G_{q+1,p}^{n,1}\left(z^{-1} \exp[i\pi(n + 1 - p)] \,\middle|\, \begin{matrix} \rho_0, \rho_1, \ldots, \rho_q \\ \alpha_p \end{matrix}\right),$$

$$p \leqslant q \quad \text{or} \quad p = q + 1 \quad \text{and} \quad |z| < 1. \tag{16}$$

$${}_pF_q\left(\begin{matrix} \alpha_p \\ \rho_q \end{matrix} \,\middle|\, z\right) = \frac{\Gamma(\rho_q)}{\Gamma(\alpha_p)}\, G_{q+1,p}^{p,1}\left(-z^{-1} \,\middle|\, \begin{matrix} 1, \rho_q \\ \alpha_p \end{matrix}\right),$$

$$p \leqslant q \quad \text{or} \quad p = q + 1 \quad \text{and} \quad |z| < 1. \tag{17}$$

The combination (5) and (12) yields the expansion formula

$$G_{p,q}^{m,n}\left(z \,\middle|\, \begin{matrix} a_p \\ b_q \end{matrix}\right) = \sum_{h=1}^{m} \frac{\left[\begin{matrix} \prod_{j=1}^{m} \{\Gamma(b_j - b_h)^* \,\Gamma(1 + b_h - b_j)^*\} \\ \times \exp[-i\pi b_h(p + 1 - m - n)] \end{matrix}\right]}{\prod_{j=n+1}^{p} \{\Gamma(a_j - b_h)\, \Gamma(1 + b_h - a_j)\}}$$

$$\times\, G_{p,q}^{1,p}\left((-)^{p+1-m-n} z \,\middle|\, \begin{matrix} a_p \\ b_h, b_q^* \end{matrix}\right). \tag{18}$$

Also

$$G_{p,q}^{m,n}\left(z \,\middle|\, \begin{matrix} a_p \\ b_q \end{matrix}\right) = \sum_{h=1}^{n} \frac{\left[\begin{matrix} \prod_{j=1}^{n} \{\Gamma(a_h - a_j)^* \,\Gamma(1 + a_j - a_h)^* \\ \times \exp[-i\pi(a_h - 1)(q + 1 - m - n)] \end{matrix}\right]}{\prod_{j=m+1}^{q} \{\Gamma(a_h - b_j)\, \Gamma(1 + b_j - a_h)\}}$$

$$\times\, G_{p,q}^{q,1}\left((-)^{q+1-m-n} z \,\middle|\, \begin{matrix} a_h, a_p^* \\ b_q \end{matrix}\right). \tag{19}$$

Originally the G-function was defined by Meijer (1936) by the series (5). Later, see Meijer (1941a, p. 83; 1946), the series definition was replaced by one in terms of the Mellin–Barnes integral (1) where the path L is as defined in (2) and (3), respectively. The complete definition (1)–(4), except for the discussion involving $|\arg z| = \delta\pi$, $\delta \geqslant 0$, in (2), is due to Erdélyi *et al.* (1953, Vol. 1, p. 207).

The E-function introduced by MacRobert (1937, 1938) can be defined by

$$E(a_1, a_2, ..., a_p; b_1, b_2, ..., b_q; z) = E(a_p; b_q; z) = G_{q+1,p}^{p,1}\left(z \,\middle|\, \begin{matrix} 1, b_q \\ a_p \end{matrix}\right). \tag{20}$$

Both MacRobert's E-function and Meijer's G-function arose from an attempt to give meaning to the $_pF_q$ symbol when $p > q + 1$.

5.3.2. Elementary Properties

The G-function is symmetric in the parameters $a_1, ..., a_n$; $a_{n+1}, ..., a_p$; $b_1, ..., b_m$; and $b_{m+1}, ..., b_q$. Thus, if one of the a_h's, $h = 1, 2, ..., n$, is equal to one of the b_j's, $j = m + 1, ..., q$, the G-function reduces to one of lower order. For example,

$$G_{p,q}^{m,n}\left(z \,\middle|\, \begin{matrix} a_1, ..., a_p \\ b_1, ..., b_{q-1}, a_1 \end{matrix}\right) = G_{p-1,q-1}^{m,n-1}\left(z \,\middle|\, \begin{matrix} a_2, ..., a_p \\ b_1, ..., b_{q-1} \end{matrix}\right), \qquad n, p, q \geqslant 1. \tag{1}$$

Similarly, if one of the a_h's, $h = n + 1, ..., p$, is equal to one of the b_j's, $j = 1, 2, ..., m$, then the G-function reduces to one of lower order. For example,

$$G_{p,q}^{m,n}\left(z \,\middle|\, \begin{matrix} a_1, ..., a_{p-1}, b_1 \\ b_1, b_2, ..., b_q \end{matrix}\right) = G_{p-1,q-1}^{m-1,n}\left(z \,\middle|\, \begin{matrix} a_1, ..., a_{p-1} \\ b_2, ..., b_q \end{matrix}\right), \qquad m, p, q \geqslant 1. \tag{2}$$

In the discussion of the G-function, we can without loss of generality suppose that $p \leq q$ in view of the important relation

$$G_{p,q}^{m,n}\left(z \,\middle|\, \begin{matrix} a_p \\ b_q \end{matrix}\right) = G_{q,p}^{n,m}\left(z^{-1} \,\middle|\, \begin{matrix} 1 - b_q \\ 1 - a_p \end{matrix}\right), \qquad \arg\left(\frac{1}{z}\right) = -\arg z, \tag{3}$$

Another significant relation is

$$z^\sigma G_{p,q}^{m,n}\left(z \,\middle|\, \begin{matrix} a_p \\ b_q \end{matrix}\right) = G_{p,q}^{m,n}\left(z \,\middle|\, \begin{matrix} a_p + \sigma \\ b_q + \sigma \end{matrix}\right). \tag{4}$$

The above equations as well as those which follow are readily proved from 5.3.1(1).

$$G_{p,q}^{m,n}\left(z \,\middle|\, \begin{matrix} a_p \\ b_q \end{matrix}\right) = (2\pi)^u k^v \, G_{pk,qk}^{mk,nk}\left(\left\{\frac{z^k}{k^{k(q-p)}}\right\} \,\middle|\, \begin{matrix} c_{1k}, c_{2k}, \dots, c_{pk} \\ d_{1k}, d_{2k}, \dots, d_{qk} \end{matrix}\right),$$

$$u = (k-1)[\tfrac{1}{2}(p+q) - m - n], \qquad v = \sum_{j=1}^{q} b_j - \sum_{j=1}^{p} a_j + \tfrac{1}{2}(p-q) + 1, \quad (5)$$

where c_{hk} and d_{hk} stand respectively for the set of parameters

$$\frac{a_h}{k}, \quad \frac{a_h+1}{k}, \dots, \quad \frac{a_h+k-1}{k}, \qquad h = 1, 2, \dots, p,$$

$$\frac{b_h}{k}, \quad \frac{b_h+1}{k}, \dots, \quad \frac{b_h+k-1}{k}, \qquad h = 1, 2, \dots, q.$$

$$G_{p+1,q+1}^{m+1,k}\left(z \,\middle|\, \begin{matrix} a, a_p \\ a, b_q \end{matrix}\right) = (-)^r \, G_{p,q}^{m,k}\left(z \,\middle|\, \begin{matrix} a_p \\ b_q \end{matrix}\right), \qquad a_k = a + r, \quad r \text{ an integer.} \quad (6)$$

$$G_{p+1,q+1}^{m,n+1}\left(z \,\middle|\, \begin{matrix} a, a_p \\ b_q, b \end{matrix}\right) = (-)^r \, G_{p+1,q+1}^{m+1,n}\left(z \,\middle|\, \begin{matrix} a_p, a \\ b, b_q \end{matrix}\right), \qquad q \geqslant m, \quad a - b = r,$$

$$r \text{ an integer or zero.} \quad (7)$$

$$(1 - a_1 + b_1)\, G_{p,q}^{m,n}\left(z \,\middle|\, \begin{matrix} a_p \\ b_q \end{matrix}\right) = G_{p,q}^{m,n}\left(z \,\middle|\, \begin{matrix} a_1 - 1, a_2, \dots, a_p \\ b_q \end{matrix}\right) + G_{p,q}^{m,n}\left(z \,\middle|\, \begin{matrix} a_p \\ b_1 + 1, b_2, \dots, b_q \end{matrix}\right),$$

$$m, n \geqslant 1. \quad (8)$$

$$(a_h - a_1)\, G_{p,q}^{m,n}\left(z \,\middle|\, \begin{matrix} a_p \\ b_q \end{matrix}\right) = G_{p,q}^{m,n}\left(z \,\middle|\, \begin{matrix} a_1 - 1, a_2, \dots, a_p \\ b_q \end{matrix}\right) + G_{p,q}^{m,n}\left(z \,\middle|\, \begin{matrix} a_1, \dots, a_{p-1}, a_p - 1 \\ b_q \end{matrix}\right),$$

$$1 \leqslant n \leqslant p - 1, \qquad h = p. \quad (9)$$

$$G_{p,q}^{m,n}\left(z \,\middle|\, \begin{matrix} a_p \\ b_q \end{matrix}\right) = (2\pi i)^{-1} \left\{\exp(i\pi b_{m+1})\, G_{p,q}^{m+1,n}\left(ze^{-i\pi} \,\middle|\, \begin{matrix} a_p \\ b_q \end{matrix}\right) - \exp(-i\pi b_{m+1})\, G_{p,q}^{m+1,n}\left(ze^{i\pi} \,\middle|\, \begin{matrix} a_p \\ b_q \end{matrix}\right)\right\},$$

$$m \leqslant q - 1. \quad (10)$$

$$G_{p,q}^{m,n}\left(z \middle| \begin{matrix} a_p \\ b_q \end{matrix}\right) = (2\pi i)^{-1} \Big\{ \exp(i\pi a_{n+1})\, G_{p,q}^{m,n+1}\left(ze^{-i\pi} \middle| \begin{matrix} a_p \\ b_q \end{matrix}\right)$$

$$- \exp(-i\pi a_{n+1})\, G_{p,q}^{m,n+1}\left(ze^{i\pi} \middle| \begin{matrix} a_p \\ b_q \end{matrix}\right)\Big\},$$

$$n \leqslant p-1. \qquad (11)$$

$$G_{p,q+2}^{m+2,n}\left(z \middle| \begin{matrix} a_p \\ 0, \tfrac{1}{2}, b_q \end{matrix}\right) = \pi \left\{ G_{p,q+2}^{m+1,n}\left(ze^{i\pi} \middle| \begin{matrix} a_p \\ 0, b_q, \tfrac{1}{2} \end{matrix}\right) + i G_{p,q+2}^{m+1,n}\left(ze^{i\pi} \middle| \begin{matrix} a_p \\ \tfrac{1}{2}, b_q, 0 \end{matrix}\right)\right\} \qquad (12)$$

$$\frac{d}{dz}\left\{ z^{-b_1} G_{p,q}^{m,n}\left(z \middle| \begin{matrix} a_p \\ b_q \end{matrix}\right)\right\} = -\, z^{-1-b_1} G_{p,q}^{m,n}\left(z \middle| \begin{matrix} a_p \\ 1+b_1, b_2, \ldots, b_q \end{matrix}\right). \qquad (13)$$

$$\frac{d}{dz}\left\{ z^{-b_h} G_{p,q}^{m,n}\left(z \middle| \begin{matrix} a_p \\ b_q \end{matrix}\right)\right\} = z^{-1-b_h} G_{p,q}^{m,n}\left(z \middle| \begin{matrix} a_p \\ b_1, \ldots, b_{q-1}, 1+b_q \end{matrix}\right)$$

$$m < q, \quad h = q. \qquad (14)$$

$$\frac{d}{dz}\left\{ z^{1-a_1} G_{p,q}^{m,n}\left(z \middle| \begin{matrix} a_p \\ b_q \end{matrix}\right)\right\} = z^{-a_1} G_{p,q}^{m,n}\left(z \middle| \begin{matrix} a_1 - 1, a_2, \ldots, a_p \\ b_q \end{matrix}\right), \qquad n \geqslant 1. \quad (15)$$

$$\frac{d}{dz}\left\{ z^{1-a_h} G_{p,q}^{m,n}\left(z \middle| \begin{matrix} a_p \\ b_q \end{matrix}\right)\right\} = -z^{-a_h} G_{p,q}^{m,n}\left(z \middle| \begin{matrix} a_1, \ldots, a_{p-1}, a_p - 1 \\ b_q \end{matrix}\right),$$

$$n < p, \quad h = p. \qquad (16)$$

$$z^k \frac{d^k}{dz^k}\left\{ G_{p,q}^{m,n}\left(z \middle| \begin{matrix} a_p \\ b_q \end{matrix}\right)\right\} = G_{p+1,q+1}^{m,n+1}\left(z \middle| \begin{matrix} 0, a_p \\ b_q, k \end{matrix}\right), \qquad (17)$$

$$z^k \frac{d^k}{dz^k}\left\{ G_{p,q}^{m,n}\left(z^{-1} \middle| \begin{matrix} a_p \\ b_q \end{matrix}\right)\right\} = (-)^k G_{p+1,q+1}^{m,n+1}\left(z^{-1} \middle| \begin{matrix} 1-k, a_p \\ b_q, 1 \end{matrix}\right). \qquad (18)$$

5.3.3. Analytic Continuation of $G_{p,p}^{m,n}(z)$

If $p=q$ in 5.3.1(1) and L is defined as in 5.3.1(3), then we require $0<|z|<1$, whereas if L is given by 5.3.1(4), $|z|>1$. When L is the path in 5.3.1(2), we have a representation valid for all z, $z\neq 0$, provided $m+n\geq p+1$ and $|\arg z|<(m+n-p)\pi$. When $|z|<1$, we can without altering the value of $G(z)\equiv G_{p,p}^{m,n}(z)$, bend the contour in 5.3.1(2) around so that it coincides with the contour in 5.3.1(3). From 5.7.2(1), $G(z)$ satisfies a linear differential equation of order p and every finite point in the z-plane is an ordinary point save $z=0$ and $z=(-)^{m+n-p}$ which are regular singularities. To define $G(z)$ outside the unit circle, we introduce a cross cut in the z-plane along the straight line from

$$(-)^{m+n-p} \quad \text{to} \quad (-)^{m+n-p}(1 + \infty e^{i\mu}), \qquad -\pi/2 \leqslant \mu \leqslant \pi/2. \tag{1}$$

In (1), we most always take $\mu = 0$. Then in the cut plane $G(z)$ has no singularity except for the branch point at $z = 0$. If $m + n \geqslant p + 2$, the sector $|\arg z| < (m + n - p)\pi$ contains the point $(-)^{m+n-p}$. But in this situation, as noted above, $G(z)$ is regular at $z = (-)^{m+n-p}$. Thus, if $m + n \geqslant p + 2$ and $|\arg z| < (m + n - p)\pi$, the cross cut is superfluous. We have, therefore, proved that if $m + n \geqslant p + 1$ and $|\arg z| < (m + n - p)\pi$, then $G(z)$ can be continued analytically from inside the unit disc with center at the origin to the outside of this disc by means of the expansion 5.3.1(11) with $q=p$. It readily follows that

$$
\begin{aligned}
{}_{p+1}F_p\left(\begin{matrix} a_{p+1} \\ b_p \end{matrix}\,\middle|\, z\right) &= \frac{\Gamma(b_p)}{\Gamma(a_{p+1})} \sum_{h=1}^{p+1} \Gamma(a_h) \frac{\prod_{j=1}^{p+1} \Gamma(a_j - a_h)^*(z^{-1}e^{i\pi})^{a_h}}{\prod_{j=1}^{p} \Gamma(b_j - a_h)} \\
&\quad \times {}_{p+1}F_p\left(\begin{matrix} 1 + a_h - b_p, a_h \\ 1 + a_h - a_{p+1}^* \end{matrix}\,\middle|\, z^{-1}\right), \\
&\qquad 0 < \arg z < 2\pi,
\end{aligned} \tag{2}
$$

and this furnishes the analytic continuation of the ${}_{p+1}F_p$ on the left from inside the unit circle to the outside of this circle. We have therefore completely generalized the ${}_2F_1$ results of Chapter VI with respect to solutions of 5.7.1(1) about the singular points $z=0$ and $z=\infty$, and the connection between these solutions. In the ${}_{p+1}F_p$ situation for $p > 1$, the solutions about the singular point $z = 1$ are not of hypergeometric type and so are rather complicated. We shall not deal with this aspect of the problem. Instead, we refer the reader to the excellent memoir by Nörlund (1955).

For further discussion on the analytic continuation of the function $G_{p,p}^{m,n}(z)$, see 5.9.1(10,11) and 5.9.2(21-23).

5.4. The Confluence Principle

To introduce the concept of confluence, notice that if z is bounded,

$$
\begin{aligned}
\lim_{|\sigma|\to\infty} (1 - z/\sigma)^{-\sigma} &= \lim_{|\sigma|\to\infty} \exp[-\sigma \ln(1 - z/\sigma)] \\
&= e^z \lim_{|\sigma|\to\infty} \exp\{(z^2/2\sigma)[1 + O(z/\sigma)]\} = e^z,
\end{aligned} \tag{1}
$$

and in our hypergeometric notation

$$\lim_{|\sigma|\to\infty} {}_1F_0(\sigma, z/\sigma) = {}_0F_0(z) = e^z. \tag{2}$$

We can get this result from another point of view. Now the general term in the expansion of $(1 - z/\sigma)^{-\sigma}$ is

$$v_k z^k/k! = \Gamma(\sigma + k)z^k/\Gamma(\sigma)\sigma^k k! \tag{3}$$

and from 1.3(13), with k fixed,

$$\lim_{|\sigma|\to\infty} v_k = 1. \tag{4}$$

Thus with z fixed,

$$\lim_{|\sigma|\to\infty} (1 - z/\sigma)^{-\sigma} = \lim_{|\sigma|\to\infty} \sum_{k=0}^{\infty} (v_k z^k/k!) = \sum_{k=0}^{\infty} (z^k/k!) \lim_{|\sigma|\to\infty} v_k$$

$$= \sum_{k=0}^{\infty} z^k/k! = e^z. \tag{5}$$

Again, if in ${}_2F_1(a, b; c; z)$, z is replaced by z/b, we get a power series in z whose radius of convergence is $|b|$. The latter defines an analytic function with singularities at $z = 0, b$, and ∞. Let $|b| \to \infty$. The limiting form defines an entire function with a singularity at $z = \infty$ which is a confluence of the two singularities b and ∞ of ${}_2F_1(a, b; c; z/b)$. Thus,

$$\lim_{|b|\to\infty} {}_2F_1(a, b; c; z/b) = {}_1F_1(a; c; z) \tag{6}$$

and the ${}_1F_1$ is a confluent form of the ${}_2F_1$. For this reason, the ${}_1F_1$ is called a confluent hypergeometric function. A natural generalization of (6) is the statement

$$\lim_{|\sigma|\to\infty} {}_{p+1}F_q\left(\begin{matrix}\alpha_p, \sigma\\ \rho_q\end{matrix}\,\middle|\, z/\sigma\right) = {}_pF_q\left(\begin{matrix}\alpha_p\\ \rho_q\end{matrix}\,\middle|\, z\right), \qquad q \geqslant p. \tag{7}$$

This limit process is called a confluence with respect to σ and the resulting limit of such a process is called a confluent limit. The importance of the result (7) lies in the fact that known representations for ${}_{p+1}F_q$ may be used to deduce similar type representations for ${}_pF_q$ which do not follow from the former by an obvious suppression of a numerator parameter.

In (7), the confluence is with respect to a numerator parameter of the ${}_{p+1}F_q$. However, the limit process can be invoked with respect to a denominator parameter, and

$$\lim_{|\beta|\to\infty} {}_pF_{q+1}\left(\begin{matrix}\alpha_p \\ \rho_q, \beta\end{matrix}\,\middle|\,\beta z\right) = {}_pF_q\left(\begin{matrix}\alpha_p \\ \rho_q\end{matrix}\,\middle|\,z\right), \qquad p \leqslant q+1,$$

$$\begin{aligned} |\arg\beta| &\leqslant \pi - \delta, \qquad 0 < \delta \leqslant \pi/2, \\ |z| &< R \qquad \text{if} \quad |\arg\beta| \leqslant \pi/2, \\ |z| &< |\sin(\arg\beta)|\, R \qquad \text{if} \quad \pi/2 \leqslant |\arg\beta| \leqslant \pi - \delta, \end{aligned} \tag{8}$$

where $R = 1$ if $p = q + 1$ and $R = \infty$ if $p \leqslant q$.

These results are special cases of some general theorems due to Fields(1967). We now present these theorems along with applications to ${}_pF_q$'s.

Theorem 1. *Let*

$$\sum_{k=0}^{\infty} a_k z^k < \infty, \qquad |z| < R. \tag{9}$$

Then

$$F(z, \sigma) = \sum_{k=0}^{\infty} [a_k(\sigma)_k/k!](z/\sigma)^k, \qquad |z/\sigma| < R, \tag{10}$$

and $F(z, \sigma)$ can be rearranged to read

$$F(z, \sigma) = \sum_{j=0}^{\infty} g_j(z)\sigma^{-j}, \qquad |z/\sigma| < R. \tag{11}$$

That is, $F(z, \sigma)$ converges for $|z/\sigma| < R$, where the $g_j(z)$ are entire functions of z given explicitly by (13) *below. Further, for $j \geqslant 1$, $g_j(z)$ can be expressed in terms of the derivatives of $g_0(z) = g(z)$.*

$$g_j(z) = \sum_{k=0}^{\infty} \frac{a_k \alpha_{j,k} z^k}{k!} = \sum_{k=0}^{\infty} \frac{a_k(1-k)_j B_j^{(k)} z^k}{j!\,k!}. \tag{12}$$

where $B_j^{(k)}$ is a generalized Bernoulli number, see 14.2 . If Δ is the forward difference operator with respect to k, then

$$g_j(z) = \sum_{r=0}^{j-1} \frac{(-)^j z^{1+j+r}}{j!\,r!} g_0^{(1+j+r)}(z) \left[\Delta^r \left(\frac{B_j^{(k)}}{k}\right)\right]_{k=j+1} \tag{13}$$

Since the coefficients of $z^{1+j+r} g_0^{(1+j+r)}(z)$ in (13) are independent of $g_0(z)$, the coefficients can be deduced from the special case

$$F(z, \sigma) = (1 - z/\sigma)^{-\sigma} = e^z \exp\left\{\sum_{j=2}^{\infty} (\sigma/j)(z/\sigma)^j\right\}, \qquad g_0(z) = e^z. \tag{14}$$

The first few $g_j(z)$ are as follows:

$$g_0(z) = g(z) = \sum_{k=0}^{\infty} a_k z^k/k!, \qquad g_1(z) = (z^2/2)g^{(2)}(z),$$
$$g_2(z) = (z^3/3)g^{(3)}(z) + (z^4/8)g^{(4)}(z), \tag{15}$$
$$g_3(z) = (z^4/4)g^{(4)}(z) + (z^5/6)g^{(5)}(z) + (z^6/48)g^{(6)}(z),$$
$$g_4(z) = (z^5/5)g^{(5)}(z) + (13z^6/72)g^{(6)}(z) + (z^7/24)g^{(7)}(z) + (z^8/384)g^{(8)}(z).$$

Theorem 2.

$$\begin{aligned}
{}_{p+1}F_q\left(\begin{matrix}\alpha_p, \sigma \\ \rho_q\end{matrix}\,\middle|\, z/\sigma\right) &= {}_pF_q\left(\begin{matrix}\alpha_p \\ \rho_q\end{matrix}\,\middle|\, z\right) + \frac{A_2(z)}{2\sigma} + \frac{1}{24\sigma^2}(8A_3(z) + 3A_4(z)) \\
&\quad + \frac{1}{48\sigma^3}(12A_4(z) + 8A_5(z) + A_6(z)) + \frac{1}{5760\sigma^4} \\
&\quad \times (1152A_5(z) + 1040A_6(z) + 240A_7(z) + 15A_8(z)) + \cdots, \\
A_r(z) &= z^r \frac{d^r}{dz^r}\, {}_pF_q\left(\begin{matrix}\alpha_p \\ \rho_q\end{matrix}\,\middle|\, z\right) = \frac{(\alpha_p)_r z^r}{(\rho_q)_r}\, {}_pF_q\left(\begin{matrix}\alpha_p + r \\ \rho_q + r\end{matrix}\,\middle|\, z\right), \\
& p = q, \quad |z| < |\sigma|; \qquad p < q, \quad |z| < \infty.
\end{aligned} \tag{16}$$

Theorem 3. *Let*

$$f(z) = \sum_{k=0}^{\infty} c_k z^k < \infty, \qquad |z| < R. \tag{17}$$

Then

$$T(z, \beta) = \sum_{k=0}^{\infty} [c_k/(\beta)_k](\beta z)^k \tag{18}$$

converges for all z, $\beta \neq 0, -1, -2, \ldots,$ *and*

$$\begin{aligned}
T(z, \beta) &\sim \sum_{j=0}^{\infty} f_j(z)(-\beta)^{-j}, \qquad |\beta| \to \infty, \quad |\arg\beta| \leqslant \pi - \delta, \quad 0 < \delta \leqslant \pi/2, \\
&|z| < R \quad \text{if} \quad |\arg\beta| \leqslant \pi/2; \\
&|z| < |\sin(\arg\beta)|\, R \quad \text{if} \quad \pi/2 \leqslant |\arg\beta| \leqslant \pi - \delta,
\end{aligned} \tag{19}$$

where the $f_j(z)$ *are analytic functions in* $|z| < R$ *and are given by* (21) *below.*

For $j \geqslant 1$, $f_j(z)$ can be expressed in terms of the derivatives of $f_0(z) = f(z)$.

$$f_j(z) = \sum_{k=0}^{\infty} \frac{c_k(k)_j B_j^{(1-k)}}{j!} z^k. \tag{20}$$

$$f_j(z) = \sum_{r=0}^{2j-2} \frac{z^{r+2}}{j!\, r!} f_0^{(r+2)}(z) \left[\Delta^r \left(\frac{(k)_j B_j^{(1-k)}}{k(k-1)} \right) \right]_{k=2}, \qquad j \geqslant 1. \tag{21}$$

The first few $f_j(z)$ are as follows:

$$\begin{aligned}
f_0(z) &= f(z) = \sum_{k=0}^{\infty} c_k z^k, \qquad f_1(z) = (z^2/2) f^{(2)}(z), \\
f_2(z) &= (z^2/2) f^{(2)}(z) + (2z^3/3) f^{(3)}(z) + (z^4/8) f^{(4)}(z), \\
f_3(z) &= (z^2/2) f^{(2)}(z) + 2z^3 f^{(3)}(z) + (3z^4/2) f^{(4)}(z) \\
&\quad + (z^5/3) f^{(5)}(z) + (z^6/48) f^{(6)}(z), \\
f_4(z) &= (z^2/2) f^{(2)}(z) + (14z^3/3) f^{(3)}(z) + (61z^4/8) f^{(4)}(z) + (62z^5/15) f^{(5)})(z) \\
&\quad + (131z^6/144) f^{(6)}(z) + (z^7/12) f^{(7)}(z) + (z^8/384) f^{(8)}(z).
\end{aligned} \tag{22}$$

Theorem 4.

$$\begin{aligned}
{}_pF_{q+1}\left(\begin{matrix} \alpha_p \\ \rho_q, \beta \end{matrix} \middle| \beta z \right) &= {}_pF_q\left(\begin{matrix} \alpha_p \\ \rho_q \end{matrix} \middle| z \right) - \frac{A_2(z)}{2\beta} + \frac{1}{24\beta^2}\{12A_2(z) + 16A_3(z) + 3A_4(z)\} \\
&\quad - \frac{1}{48\beta^3}\{24A_2(z) + 96A_3(z) + 72A_4(z) + 16A_5(z) + A_6(z)\} \\
&\quad + \frac{1}{5760\beta^4}\{2880A_2(z) + 26880A_3(z) \\
&\quad + 43920A_4(z) + 23808A_5(z) + 5240A_6(z) \\
&\quad + 480A_7(z) + 15A_8(z)\} + O(\beta^{-5}),
\end{aligned} \tag{23}$$

$$|\beta| \to \infty, \quad p \leqslant q, \quad |\arg \beta| \leqslant \pi - \delta, \quad 0 < \delta \leqslant \pi/2,$$

$$p = q + 1, \quad |z| < 1 \quad \text{if} \quad R(\beta) \geqslant 0$$

$$|z| < \sin \delta, \quad 0 < \delta \leqslant \pi/2 \quad \text{if} \quad R(\beta) \leqslant 0,$$

where $A_r(z)$ is defined as in (16).

5.5. Multiplication Theorems

Throughout this section $z \neq 0$ and $m, n, p,$ and q are integers with

$$q \geqslant 1, \qquad 0 \leqslant n \leqslant p \leqslant q, \qquad 0 \leqslant m \leqslant q.$$

Then

$$G_{p,q}^{m,n}\left(zw \,\middle|\, \begin{matrix} a_p \\ b_q \end{matrix}\right) = \sum_{k=0}^{\infty} \frac{(w-1)^k}{k!} G_{p+1,q+1}^{m,n+1}\left(z \,\middle|\, \begin{matrix} 0, a_p \\ b_q, k \end{matrix}\right), \tag{1}$$

valid under the five cases enumerated below.

CASE 1. $p < q$, $|w-1| < 1$.

CASE 2. $p = q$, $(-)^{m+n-p}z \neq 1$, $|w-1| < 1$ if $(-)^{m+n-p}R(z) \leqslant \frac{1}{2}$.

$$|w-1| < \left|1 - \frac{(-)^{m+n-p}}{z}\right| \qquad \text{if} \quad \tfrac{1}{2} \leqslant (-)^{m+n-p}R(z) \leqslant 1.$$

$$|w-1| < |I(z)|/|z| \qquad \text{if} \quad (-)^{m+n-p}R(z) \geqslant 1.$$

CASE 3. $p = q$, $m+n-p \geqslant 1$, $|\arg z| \leqslant (m+n-p-\frac{1}{2})\pi$, $|w-1| < 1$.

CASE 4. $m = 1$, $p < q$, b_1 is a nonnegative integer. That is, under these restrictions (1) is valid for all z and w.

CASE 5. $m = 1$, $p = q$, b_1 is a nonnegative integer,

$$(-)^{m+n-p}z \neq 1, \qquad |w-1| < \left|1 - \frac{(-)^{1+n-p}}{z}\right| \qquad \text{if} \qquad (-)^{1+n-p}R(z) \leqslant 1;$$

$$|w-1| < \frac{|I(z)|}{|z|} \qquad \text{if} \qquad (-)^{1+n-p}R(z) \geqslant 1. \tag{2}$$

With $m, n, p,$ and q as in (1), then

$$G_{p,q}^{m,n}\left(zw \,\middle|\, \begin{matrix} a_p \\ b_q \end{matrix}\right) = \sum_{k=0}^{\infty} \frac{(1-1/w)^k}{k!} G_{p+1,q+1}^{m,n+1}\left(z \,\middle|\, \begin{matrix} 1-k, a_p \\ b_q, 1 \end{matrix}\right), \qquad z \neq 0, \tag{3}$$

is valid under the following three cases.

CASE 1. $p < q$, $R(w) > \frac{1}{2}$.

CASE 2. $p = q$, $(-)^{m+n-p}z \neq 1$, $R(w) > \frac{1}{2}$ if $|1-(-)^{m+n-p}z| \geqslant 1$;

$$|1/w - 1| < |1-(-)^{m+n-p}z| \qquad \text{if} \qquad |1-(-)^{m+n-p}z| \leqslant 1. \tag{4}$$

CASE 3. $p = q$, $m+n-p \geqslant 1$, $|\arg z| \leqslant (m+n-p-\frac{1}{2})\pi$, $R(w) > \frac{1}{2}$.

The following results readily follow from the above equations. In (5)-(7), $m<q$ and in (9)-(11), $n>0$.

$$G_{p,q}^{m,n}\left(zw\,\middle|\,\begin{matrix}a_p\\ b_{q-1},0\end{matrix}\right)=\sum_{k=0}^{\infty}\frac{(w-1)^k}{k!}G_{p,q}^{m,n}\left(z\,\middle|\,\begin{matrix}a_p\\ b_{q-1},k\end{matrix}\right). \tag{5}$$

$$G_{p,q}^{m,n}\left(zw\,\middle|\,\begin{matrix}a_p\\ b_q\end{matrix}\right)=w^{b_h}\sum_{k=0}^{\infty}\frac{(w-1)^k}{k!}G_{p,q}^{m,n}\left(z\,\middle|\,\begin{matrix}a_p\\ b_{q-1},k+b_q\end{matrix}\right),\qquad h=q. \tag{6}$$

$$G_{p,q}^{m,n}\left(zw\,\middle|\,\begin{matrix}a_p\\ b_q\end{matrix}\right)=w^{b_1}\sum_{k=0}^{\infty}\frac{(1-w)^k}{k!}G_{p,q}^{m,n}\left(z\,\middle|\,\begin{matrix}a_p\\ k+b_1,b_2,\ldots,b_q\end{matrix}\right). \tag{7}$$

$$G_{p,q}^{m,n}\left(zw\,\middle|\,\begin{matrix}a_p\\ b_q\end{matrix}\right)=\sum_{k=0}^{\infty}\frac{(1-w)^k}{k!}G_{p+1,q+1}^{m+1,n}\left(z\,\middle|\,\begin{matrix}a_p,0\\ k,b_q\end{matrix}\right), \tag{8}$$

$$G_{p,q}^{m,n}\left(zw\,\middle|\,\begin{matrix}1,a_2,\ldots,a_p\\ b_q\end{matrix}\right)=\sum_{k=0}^{\infty}\frac{(1-1/w)^k}{k!}G_{p,q}^{m,n}\left(z\,\middle|\,\begin{matrix}1-k,a_2,\ldots,a_p\\ b_q\end{matrix}\right), \tag{9}$$

$$G_{p,q}^{m,n}\left(zw\,\middle|\,\begin{matrix}a_p\\ b_q\end{matrix}\right)=w^{a_1-1}\sum_{k=0}^{\infty}\frac{(1-1/w)^k}{k!}G_{p,q}^{m,n}\left(z\,\middle|\,\begin{matrix}a_1-k,a_2,\ldots,a_p\\ b_q\end{matrix}\right). \tag{10}$$

$$G_{p,q}^{m,n}\left(zw\,\middle|\,\begin{matrix}a_p\\ b_q\end{matrix}\right)=w^{a_h-1}\sum_{k=0}^{\infty}\frac{(1/w-1)^k}{k!}G_{p,q}^{m,n}\left(z\,\middle|\,\begin{matrix}a_1,\ldots,a_{p-1},a_p-k\\ b_q\end{matrix}\right),\quad n<p,\ h=p. \tag{11}$$

For generalizations of the above expansions, see 5.10.1(1). Equation (1) was first given by Meijer (1941b). See also Meijer (1952-1956, p.376) and Knottnerus (1960). It is a generalization of well known results for Bessel functions, confluent hypergeometric functions and hypergeometric functions. The Cases 2 and 5 conditions in (2) are different from those given by Meijer. The reasons for this are given by Luke (1969).

For an illustration use (5) with $m=1$, $n=p=0$, $q=2$, and z and w replaced by $z^2/4$ and w^2, respectively. Then with the aid of 8.3(8), we get

$$J_\nu(wz)=w^\nu\sum_{k=0}^{\infty}\frac{(1-w^2)^k(z/2)^k}{k!}J_{\nu+k}(z),\quad |w|<\infty. \tag{12}$$

5.6. Integrals Involving G-Functions

The conditions given for validity of the integrals of this section are sufficient. A detailed description of conditions is given by Luke (1969).

Mellin Transform

Let

$$1 \le n \le p < q, \quad 1 \le m \le q,$$
$$\delta = m + n - \tfrac{1}{2}(p+q), \tag{1}$$

$$-\min_{1 \leqslant h \leqslant m} R(b_h) < R(s) < 1 - \max_{1 \leqslant j \leqslant n} R(a_j).$$

Then

$$\int_0^\infty y^{s-1} G_{p,q}^{m,n}\left(\eta y \,\middle|\, \begin{matrix} a_p \\ b_q \end{matrix}\right) dy = \frac{\eta^{-s} \prod_{j=1}^m \Gamma(b_j + s) \prod_{j=1}^n \Gamma(1 - a_j - s)}{\prod_{j=m+1}^q \Gamma(1 - b_j - s) \prod_{j=n+1}^p \Gamma(a_j + s)},$$

$$\delta > 0, \quad \eta \neq 0, \quad |\arg \eta| < \delta\pi. \tag{2}$$

Integral of the Product of Two G-Functions

Let

$$1 \le n \le p < q, \quad 1 \le m \le q,$$
$$1 \le \nu \le \sigma < \tau, \quad 1 \le \mu \le \tau, \tag{3}$$
$$\delta = m + n - \tfrac{1}{2}(p+q), \qquad \rho = \mu + \nu - \tfrac{1}{2}(\sigma + \tau).$$

We have need for the following conditions:

$$R(b_j + d_h) > -1, \qquad j = 1, 2, \ldots, m, \quad h = 1, 2, \ldots, \mu.$$

$$R(a_j + c_h) < 1, \qquad j = 1, 2, \ldots, n, \quad h = 1, 2, \ldots, \nu.$$

$a_j - b_h$ is not a positive integer, $\quad j = 1, 2, \ldots, n, \quad h = 1, 2, \ldots, m.$ (4)

$c_j - d_h$ is not a positive integer, $\quad j = 1, 2, \ldots, \nu, \quad h = 1, 2, \ldots, \mu.$

$$\eta \neq 0, \qquad \omega \neq 0.$$

Then

$$\int_0^\infty G_{p,q}^{m,n}\left(\eta x \,\middle|\, \begin{matrix} a_p \\ b_q \end{matrix}\right) G_{\sigma,\tau}^{\mu,\nu}\left(\omega x \,\middle|\, \begin{matrix} c_\sigma \\ d_\tau \end{matrix}\right) dx$$

$$= \frac{1}{\eta} G_{q+\sigma,p+\tau}^{n+\mu,m+\nu}\left(\frac{\omega}{\eta} \,\middle|\, \begin{matrix} -b_1, \ldots, -b_m, c_\sigma, -b_{m+1}, \ldots, -b_q \\ -a_1, \ldots, -a_n, d\tau, -a_{n+1}, \ldots, -a_p \end{matrix}\right)$$

$$= \frac{1}{\omega} G_{p+\tau,q+\sigma}^{m+\nu,n+\mu}\left(\frac{\eta}{\omega} \,\middle|\, \begin{matrix} a_1, \ldots, a_n, -d_\tau, a_{n+1}, \ldots, a_p \\ b_1, \ldots, b_m, -c_\sigma, b_{m+1}, \ldots, b_q \end{matrix}\right), \tag{5}$$

$$\delta > 0, \ |\arg \eta| < \delta\pi, \ \rho > 0, \ |\arg \omega| < \rho\pi.$$

Equation (5), due to Meijer (1941a), is a very important master formula. Some special cases are listed below. Conditions of validity are omitted. See (5) and the opening remarks of this section.

$$\int_0^\infty y^{d_1} {}_\sigma F_{\tau-1}\left(\begin{matrix} 1 + d_1 - c_\sigma \\ 1 + d_1 - d_\tau^* \end{matrix} \,\middle|\, -\omega y\right) G_{p,q}^{m,n}\left(zy \,\middle|\, \begin{matrix} a_p \\ b_q \end{matrix}\right) dy$$

$$= \frac{\Gamma(1 + d_1 - d_\tau)}{\Gamma(1 + d_1 - c_\sigma)} \omega^{-d_1-1} G_{p+\tau,q+\sigma}^{m+\sigma,n+1}\left(\frac{z}{\omega} \,\middle|\, \begin{matrix} a_1, \ldots, a_n, -d_\tau, a_{n+1}, \ldots, a_p \\ b_1, \ldots, b_m, -c_\sigma, b_{m+1}, \ldots, b_q \end{matrix}\right), \tag{6}$$

$$\int_0^\infty y^{\alpha-1}(y + \beta)^{-\sigma} G_{p,q}^{m,n}\left(zy \,\middle|\, \begin{matrix} a_p \\ b_q \end{matrix}\right) dy = \frac{\beta^{\alpha-\sigma}}{\Gamma(\sigma)} G_{p+1,q+1}^{m+1,n+1}\left(\beta z \,\middle|\, \begin{matrix} 1 - \alpha, a_p \\ \sigma - \alpha, b_q \end{matrix}\right). \tag{7}$$

The Hankel-, Y-, K-, and $\mathbf{H}$ transforms are

$$\int_0^\infty x^{-a} J_\nu(2[\omega x]^{1/2}) G_{p,q}^{m,n}\left(\eta x \,\middle|\, \begin{matrix} a_p \\ b_q \end{matrix}\right) dx$$

$$= \omega^{a-1} G_{p+2,q}^{m,n+1}\left(\frac{\eta}{\omega} \,\middle|\, \begin{matrix} a - \tfrac{1}{2}\nu, a_p, a + \tfrac{1}{2}\nu \\ b_q \end{matrix}\right), \tag{8}$$

$$\int_0^\infty x^{-a} Y_\nu(2[\omega x]^{1/2}) G_{p,q}^{m,n}\left(\eta x \,\middle|\, \begin{matrix} a_p \\ b_q \end{matrix}\right) dx$$

$$= \omega^{a-1} G_{p+3,q+1}^{m,n+2}\left(\frac{\eta}{\omega} \,\middle|\, \begin{matrix} a + \tfrac{1}{2}\nu, a - \tfrac{1}{2}\nu, a_p, a + \tfrac{1}{2}\nu + \tfrac{1}{2} \\ b_q, a + \tfrac{1}{2}\nu + \tfrac{1}{2} \end{matrix}\right), \tag{9}$$

$$\int_0^\infty x^{-a} K_\nu(2[\omega x]^{1/2}) G_{p,q}^{m,n}\left(\eta x \,\middle|\, \begin{matrix} a_p \\ b_q \end{matrix}\right) dx$$

$$= \frac{\omega^{a-1}}{2} G_{p+2,q}^{m,n+2}\left(\frac{\eta}{\omega} \,\middle|\, \begin{matrix} a + \tfrac{1}{2}\nu, a - \tfrac{1}{2}\nu, a_p \\ b_q \end{matrix}\right), \tag{10}$$

$$\int_0^\infty x^{-a}\mathbf{H}_\nu(2[\omega x]^{1/2})\, G_{p,q}^{m,n}\left(\eta x \,\middle|\, \begin{matrix} a_p \\ b_q \end{matrix}\right) dx$$

$$= \omega^{a-1}\, G_{p+3,q+1}^{m+1,n+1}\left(\frac{\eta}{\omega} \,\middle|\, \begin{matrix} a-\frac{1}{2}-\frac{1}{2}\nu,\, a_p,\, a-\frac{1}{2}\nu,\, a+\frac{1}{2}\nu \\ a-\frac{1}{2}-\frac{1}{2}\nu,\, b_q \end{matrix}\right), \tag{11}$$

respectively.

$$\int_0^\infty (z+y)^{-1}\, G_{p,q}^{m,n}\left(\eta y^2 \,\middle|\, \begin{matrix} a_p \\ b_q \end{matrix}\right) dy = (2\pi)^{-1}\, G_{p+2,q+2}^{m+2,n+2}\left(\eta z^2 \,\middle|\, \begin{matrix} 0,\, \frac{1}{2},\, a_p \\ 0,\, \frac{1}{2},\, b_q \end{matrix}\right),$$

$$|\arg z| < \pi. \tag{12}$$

Laplace Transform and Inverse Laplace Transform

$$\int_0^\infty e^{-\omega y} y^{-\alpha}\, G_{p,q}^{m,n}\left(zy \,\middle|\, \begin{matrix} a_p \\ b_q \end{matrix}\right) dy = \omega^{\alpha-1}\, G_{p+1,q}^{m,n+1}\left(\frac{z}{\omega} \,\middle|\, \begin{matrix} \alpha,\, a_p \\ b_q \end{matrix}\right). \tag{13}$$

This is a special case of (5). (13) is also valid for $|\arg \omega| = \pi/2$ provided further that $R(\alpha - a_j) > -1, \quad j = 1, 2, \ldots, n$.

$$\int_0^\infty e^{-\beta x}\, G_{p,q}^{m,n}\left(\alpha x^2 \,\middle|\, \begin{matrix} a_p \\ b_q \end{matrix}\right) dx = \pi^{-1/2}\beta^{-1}\, G_{p+2,q}^{m,n+2}\left(\frac{4\alpha}{\beta^2} \,\middle|\, \begin{matrix} 0,\, \frac{1}{2},\, a_p \\ b_q \end{matrix}\right),$$

$$\delta > 0, \quad |\arg \alpha| < \delta\pi, \quad R(\beta) > 0, \quad R(b_j) > -\tfrac{1}{2}, \quad j = 1, 2, \ldots, m, \tag{14}$$

δ as in (3) provided further that $a_j - b_h$ is not a positive integer as in (4). (14) is also valid for $|\arg \beta| = \pi/2$ provided $R(a_j) < 1$, $j = 1, 2, \ldots, n$.

Suppose m, n, p, q and δ are as in (3), and $a_j - b_h$ is as in (4). Let

$$c > 0, \quad z \text{ real}, \quad z \neq 0, \quad y \neq 0,$$

$$R(\alpha - b_h) < 1, \qquad h = 1, 2, \ldots, m. \tag{15}$$

Then

$$z^{-\alpha}\, G_{p\, q+1}^{m,n+1}\left(zy \,\middle|\, \begin{matrix} a_p \\ b_q,\, \alpha \end{matrix}\right) = (2\pi i)^{-1} \int_{c-i\infty}^{c+i\infty} e^{\omega z}\omega^{\alpha-1}\, G_{p,q}^{m,n}\left(\frac{y}{\omega} \,\middle|\, \begin{matrix} a_p \\ b_q \end{matrix}\right) d\omega,$$

$$\delta > 0, \; |\arg y| < \delta\pi, \; c > \max\left(0, R\left[(-)^{m+n-p}\, y\right]\right) \text{ if } p = q. \tag{16}$$

$$\int_{c-i\infty}^{c+i\infty} t^{\sigma-1}\, {}_rF_s\left(\begin{matrix}1+\sigma-c_r\\1+\sigma-d_s\end{matrix}\middle|\,\eta t\right) G_{p,q}^{m,n}\left(\frac{z}{t}\,\middle|\,\begin{matrix}a_p\\b_q\end{matrix}\right) dt$$

$$= \frac{2\pi i\Gamma(1+\sigma-d_s)}{\Gamma(1+\sigma-c_r)}\,\eta^{-\sigma}\, G_{p+r,q+s+1}^{m,n+r}\left(\eta z\,\middle|\,\begin{matrix}c_r\,,\,a_p\\b_q\,,\,d_s\,,\,\sigma\end{matrix}\right),$$

$$c > 0,\quad q \geqslant 1,\quad 0 \leqslant n \leqslant p \leqslant q,\quad 0 \leqslant m \leqslant q;$$

$$\eta \neq 0,\quad z \neq 0,\quad c > |\,z\,| \quad \text{if} \quad p = q;$$

$$R(c_j - b_h) < 1,\qquad j = 1, 2,\ldots, r,\qquad h = 1, 2,\ldots, m;$$

$$r = s,\qquad \eta > 0,\qquad R\left(\sigma + \sum_{j=1}^{r} d_j - \sum_{j=1}^{r} c_j - b_h\right) < 1,\qquad h = 1, 2,\ldots, m;$$

$$r = s+1,\qquad |\arg\eta| < \pi/2,\qquad c < R(1/\eta). \tag{17}$$

Fourier Transforms

$$\int_0^\infty \cos\gamma x\, G_{p,q}^{m,n}\left(\alpha x^2\,\middle|\,\begin{matrix}a_p\\b_q\end{matrix}\right) dx = \pi^{1/2}\gamma^{-1}\, G_{p+2,q}^{m,n+1}\left(\frac{4\alpha}{\gamma^2}\,\middle|\,\begin{matrix}\tfrac{1}{2},\,a_p\,,\,0\\b_q\end{matrix}\right),\qquad \gamma > 0.$$

$$\delta > 0,\quad |\arg\alpha| < \delta\pi,\quad R(b_j) > -\tfrac{1}{2},\quad j = 1, 2,\ldots, m,\quad R(a_j) < 1,$$

$$j = 1, 2,\ldots, n, \tag{18}$$

$$\int_0^\infty \sin\gamma x\, G_{p,q}^{m,n}\left(\alpha x^2\,\middle|\,\begin{matrix}a_p\\b_q\end{matrix}\right) dx = \pi^{1/2}\gamma^{-1}\, G_{p+2,q}^{m,n+1}\left(\frac{4\alpha}{\gamma^2}\,\middle|\,\begin{matrix}0,\,a_p\,,\,\tfrac{1}{2}\\b_q\end{matrix}\right),\qquad \gamma > 0,$$

$$\delta > 0,\quad |\arg\alpha| < \delta\pi,\quad R(b_j) > -1,\quad j = 1, 2,\ldots, m,\quad R(a_j) < 1,$$

$$j = 1, 2,\ldots, n, \tag{19}$$

again with δ as in (1) and $a_j - b_h$ as in (4). See also (13) and (14) with ω and β replaced by $i\omega$ and $i\beta$ respectively.

Euler Transforms

Let m, n, p, q and δ be as in (1) save that $m \geq 2$. Let $z \neq 0$ and

$$R(b_1 + b_2) < R(\alpha + \beta),\quad R(a_j) < 1 + R(b_k),\quad k = 1, 2,\quad j = 1, 2,\ldots, n.$$

$$a_j - b_h \text{ is not a positive integer},\quad j = 1, 2,\ldots, n,\quad h = 1, 2,\ldots, m. \tag{20}$$

Then

$$\int_1^\infty y^{-\beta}(y-1)^{\alpha+\beta-b_1-b_2-1}\,{}_2F_1\left(\begin{matrix}\alpha-b_1\,,\,\alpha-b_2\\ \alpha+\beta-b_1-b_2\end{matrix}\middle|\,1-y\right)$$
$$\times\, G_{p,q}^{m,n}\left(zy\,\middle|\,\begin{matrix}a_p\\ \alpha,\beta,b_3,\ldots,b_q\end{matrix}\right)dy=\Gamma(\alpha+\beta-b_1-b_2)\,G_{p,q}^{m,n}\left(z\,\middle|\,\begin{matrix}a_p\\ b_q\end{matrix}\right),$$
$$\delta>0,\ |\arg z|<\delta\pi. \tag{21}$$

$$\int_0^1 y^{-\alpha}(1-y)^{\alpha-\beta-1}\,G_{p\,q}^{m,n}\left(zy\,\middle|\,\begin{matrix}a_p\\ b_q\end{matrix}\right)dy=\Gamma(\alpha-\beta)\,G_{p+1,q+1}^{m,n+1}\left(z\,\middle|\,\begin{matrix}\alpha,a_p\\ b_q\,,\beta\end{matrix}\right),$$
$$0\leqslant n\leqslant p<q,\qquad 1\leqslant m\leqslant q,\qquad R(\beta)<R(\alpha)<R(b_j)+1$$
$$\text{for}\quad j=1,2,\ldots,m, \tag{22}$$

provided further that a_j-b_h is as in (4).

5.7. Differential Equations

5.7.1. THE $_pF_Q$

Consider the differential equation

$$[\delta(\delta+\rho_q-1)-z(\delta+\alpha_p)]\,U(z)=0,\qquad \delta=zD,\quad D=d/dz, \tag{1}$$

where the notation is compact as in 5.2.1. That is, $(\delta+\alpha_p)$ stands for $\prod_{j=1}^{p}(\delta+\alpha_j)$, etc. For connections between the δ and D operators and other related data, see 14.3.

The order of (1) is $\max(p, q+1)$. If $p<q+1$, the singular points of (1) are at $z=0$ and $z=\infty$; $z=0$ is a regular singularity, $z=\infty$ an irregular singularity. If $p=q+1$, $z=0$. $z=1$, and $z=\infty$ are regular singularities.

The fundamental solutions of (1) near $z=0$ are proportional to

$$U_h(z)=z^{1-\rho_h}\,{}_pF_q\left(\begin{matrix}1+\alpha_p-\rho_h\\ 1+\rho_q-\rho_h^{*}\end{matrix}\middle|\,z\right),\qquad h=0,1,\ldots,q, \tag{2}$$

provided that no two of the ρ_h's differ by an integer or zero. Here and elsewhere, the asterisk sign (*) indicates that the term $1+\rho_j-\rho_h$ is omitted when $j=h$. Also $\rho_0=1$.

$$U_0=F(z)={}_pF_q\left(\begin{matrix}\alpha_p\\ \rho_q\end{matrix}\middle|\,z\right), \tag{3}$$

$$\left[(\delta + \rho_q - 1) - z \sum_{k=0}^{p-1} f_k g_{k+2}\right] F(z) - f_p \int_0^z F(t)\, dt = (\rho_q - 1),$$

$$(\rho_q - 1) = \prod_{h=1}^{q} (\rho_h - 1), \quad \text{etc.}, \quad g_k = (\delta + \alpha_k)(\delta + \alpha_{k+1}) \cdots (\delta + \alpha_p),\ g_{p+1} = 1,$$

$$f_0 = 1, \qquad f_k = (\alpha_1 - 1)(\alpha_2 - 1) \cdots (\alpha_k - 1). \tag{4}$$

If a numerator parameter is unity, say $\alpha_1 = 1$, then $F(z)$ satisfies a nonhomogeneous differential equation of order $\max(p - 1, q)$. Thus,

$$[(\delta + \rho_q - 1) - z(\delta + \alpha_2) \cdots (\delta + \alpha_p)]\, F(z) = (\rho_q - 1),$$

$$F(z) = {}_pF_q\left(\begin{matrix}1, \alpha_2, \ldots, \alpha_p \\ \rho_q\end{matrix} \middle| z\right). \tag{5}$$

We also have

$$[\delta(\delta + \rho_q - 1) - z(\delta + \alpha_p)]\, A(z) = \mu(\mu + \rho_q - 1)\, z^{\mu},$$

$$A(z) = z^{\mu}\, {}_{p+1}F_{q+1}\left(\begin{matrix}1, \mu + \alpha_p \\ \mu + 1, \mu + \rho_q\end{matrix} \middle| z\right), \tag{6}$$

$$[(\delta + \mu)(\delta + \mu + \rho_q - 1) - z(\delta + \mu + \alpha_p)]\, B(z) = \mu(\mu + \rho_q - 1),$$

$$B(z) = z^{-\mu} A(z). \tag{7}$$

If $p \geqslant q + 1$ and no two of the α_j's differ by an integer or zero, there are p fundamental solutions of (8) near $z = \infty$ proportional to

$$V_h(z) = z^{-\alpha_h}\, {}_{q+1}F_{p-1}\left(\begin{matrix}1 + \alpha_h - \rho_q \\ 1 + \alpha_h - \alpha_p^*\end{matrix} \middle| \frac{(-)^{q+1-p}}{z}\right), \qquad h = 1, 2, \ldots, p. \tag{8}$$

Here again, $\rho_0 = 1$.

As previously remarked, the point $z = 1$ is a regular singularity of (1) when $p = q + 1$. For fundamental solutions near this singularity, we refer the reader to the excellent work of Nörlund (1955).

If some of the ρ_h's differ by an integer, fundamental solutions near $z=0$ can be obtained upon relating the ${}_pF_q$ to the G-function and applying the developments in 5.3.1. For more complete details, see Luke (1969). Similar remarks pertain to fundamental solutions of (1) near $z=\infty$ if some of the α_j's differ by an integer or zero.

5.7.2. THE G-FUNCTION

The differential equation

$$[(-)^\tau z(\delta - a_p + 1) - (\delta - b_q)]\, y(z) = 0, \qquad \delta = z\, d/dz, \tag{1}$$

$$\tau = m+n-p,$$

is satisfied by $G^{m,n}_{p,q}(z)$. This equation is of order $\max(p, q)$ and in from 5.3.2(3),we can suppose that $p \leqslant q$. If $p < q$, the only singularities of (1) are at $z = 0$, a regular singularity, and at $z = \infty$, an irregular singularity. If $p = q$, $z = 0$ and $z = \infty$ are regular singularities, and in addition there is a regular singularity at $z = (-)^\tau$. The solutions of (1) in the neighborhood of the singularities $z = 0$ and $z = \infty$ have been fully investigated by Meijer (1946, pp. 344–356). No fundamental system for the neighborhood of $z = (-)^\tau$ has been given in the literature. In this connection, see Nörlund (1955).

In the neighborhood of $z = 0$, the q functions

$$y_h(z) = \exp[i\pi(\tau + 1)\, b_h]\, G^{1,p}_{p,q}\left(z \exp[-i\pi(\tau + 1)] \,\middle|\, \begin{matrix} a_p \\ b_h, b_1, \ldots, b_{h-1}, b_{h+1}, \ldots, b_q \end{matrix}\right),$$
$$h = 1, 2, \ldots, q, \tag{2}$$

form a fundamental system of solutions for (1) provided that no two of the b_h terms $h = 1, 2, \ldots, m$, differ by an integer or zero. This condition is really not essential, see the discussion surrounding 5.3.1(7). Clearly

$$y_h(z) = \frac{\Gamma(1 + b_h - a_p)}{\Gamma(1 + b_h - b_q)}\, z^{b_h}\, {}_pF_{q-1}\left(\begin{matrix} 1 + b_h - a_p \\ 1 + b_h - b_q^* \end{matrix} \,\middle|\, (-)^\tau z\right),$$
$$p \leqslant q - 1, \quad \text{or} \quad p = q \quad \text{and} \quad |z| < 1. \tag{3}$$

It follows that $G^{m,n}_{p,q}(z)$ is a linear combination of the functions (2). The pertinent expression is given by 5.3.1(18).

Next we consider solutions of (1) in the vicinity of the irregular singularity $z = \infty$. We distinguish two cases.

CASE 1. $p < q$. For every value of $\arg z$, it is possible to find integers λ and ω such that

$$|\arg z + (\nu - 2\lambda + 1)\,\pi| < (\tfrac{1}{2}\sigma + 1)\,\pi, \tag{4}$$

$$|\arg z + (\nu - 2\psi)\,\pi| < (\sigma + \epsilon)\,\pi, \qquad \psi = \omega, \omega + 1, \ldots, \omega + \sigma - 1, \tag{5}$$

where ν, σ, and ϵ are defined in 5.9.1(1,2). Then the p functions
If condition (C) [see 5.7(5)] holds, then the p functions

$$G^{q,1}_{p,q}(z \exp[i\pi(\nu - 2\lambda + 1)] \| a_t), \qquad t = 1, 2, \ldots, p \tag{6}$$

are linearly independent and satisfy (1) provided no two of the a_h's differ by an integer or zero. Further from 5.9.1(7,8), these functions tend algebraically to zero or infinity. The condition on the a_h's is really not essential in view of the comments following 5.3.1(11). Throughout our discussion, when the condition on the a_h's does not hold, it is presumed that linearly independent solutions are to be found by the methods previously noted. In this event the algebraic character of the solutions is not altered. Thus (6) gives rise to p solutions in the region of $z=\infty$. To get the remaining $\sigma=q-p$ solutions, consider

$$G_{p,q}^{q,0}\left(z \exp[i\pi(\nu - 2\psi)] \,\middle|\, \begin{matrix} a_p \\ b_q \end{matrix}\right), \qquad \psi = \omega, \omega + 1, \ldots, \omega + \sigma - 1. \tag{7}$$

These functions also satisfy (1) and, in virtue of 5.9.1(12) and (5), tend exponentially to zero or to infinity as $|z| \to \infty$. Furthermore, the solutions (7) are linearly independent. Thus a fundamental system of (1) valid near $z=\infty$ is formed by the p functions (6) and the $\sigma=q-p$ functions (7).

CASE 2. $p=q$. We assume that

$$\arg z - \tau\pi \neq \pm 2k\pi, \qquad k = 0, 1, 2, \ldots . \tag{8}$$

Then from (4) we can find an integer λ such that

$$|\arg z - (\tau + 2\lambda - 1)\pi| < \pi. \tag{9}$$

In this case, (6) with $q = p$ satisfies (1). Further, because of 5.9.1(7,11), these p solutions are linearly independent.

There may occur in a formula not the p functions (6), or the same set of functions with $q = p$, but only the first n of these, namely,

$$G_{p,q}^{q,1}(z \exp[i\pi(\nu - 2\lambda + 1)] \| a_t), \qquad t = 1, 2, \ldots, n, \tag{10}$$

or the latter set with $q = p$.

If we have to do with less than the σ functions $G_{p,q}^{q,0}(z)$, namely, the $\alpha(\alpha < \sigma)$ functions

$$G_{p,q}^{q,0}(z \exp[i\pi(\nu - 2\psi)]), \qquad \psi = \omega, \omega + 1, \ldots, \omega + \alpha - 1, \tag{11}$$

then the condition (5) need not be satisfied for $\psi = \omega, \omega + 1, \ldots, \omega + \sigma - 1$, but only for $\psi = \omega, \omega + 1, \ldots, \omega + \alpha - 1$.

5.8. Series of G-Functions

5.8.1. Introduction

The multiplication theorems of 5.5. permit for the expansion of G-functions in series of G-functions. Another group of formulas of this kind is taken up in 5.10.1. In 5.8.3 we delineate four important expansions which express $G_{p,q}^{m,n}(z)$ as a finite sum of G-functions with the same p, q but with $m = q$ and $n = 0$ or $n = 1$.

Corresponding to each expansion theorem there is a companion theorem (in one instance there is more than one companion theorem) which gives the conditions under which the expansion theorem expresses $G_{p,q}^{m,n}(z)$ in terms of fundamental solutions of 5.7.2(1) valid near $z=\infty$. These data coupled with the asymptotic expansions of $G_{p,q}^{q,1}(z)$ and $G_{p,q}^{q,0}(z)$ given by 5.9.1 lead to the asymptotic expansions for $G_{p,q}^{m,n}(z)$ which are developed in 5.9.2. The asymptotic expansion for the $_pF_q$ is treated in 5.9.3.

Our presentation of the expansion theorems and asymptotic expansions is extracted from the work of Meijer (1946). He gives generalizations of the expansion theorems in 5.8.3, but the need for these and related results do not seem to arise in practice. His developments are clouded with a mass of special notation and data which are only of secondary interest. The reader will undoubtedly prefer to first read a simplified treatment of the subject given by Luke (1969) and Fields (1972a).

For an excellent treatment of asymptotic expansions for the H-function which includes the G-function as a special case, see the work of Braaksma (1963).

5.8.2. Notation

We freely use the notation of 5.9.1(1-5). Other notation follows. Let

$$\Delta^{m,n}{}_{q}(t) = (-)^{\nu+1} \frac{\prod_{j=1}^{n} \Gamma(a_t - a_j)^* \Gamma(1 + a_j - a_t)}{\prod_{j=m+1}^{q} \Gamma(a_t - b_j)\, \Gamma(1 + b_j - a_t)}. \tag{1}$$

Where no confusion can result, we simply write $\Delta(t)$. This practice of omitting super and subscripts when appropriate is usually adopted in the definitions which follow. Thus we define

$$A = A^{m,n}{}_{q} = (-)^{\nu} (2\pi i)^{-\nu} \exp\left\{ i\pi \left(\sum_{j=1}^{n} a_j - \sum_{j=m+1}^{q} b_j \right) \right\}. \tag{2}$$

If in the right-hand side of (2), i is replaced by $-i$, the resulting expression is notated as $\bar{A}^{m,n}_{q}$ or simply as $\bar{A}$ when appropriate. The bar over a letter convention also applies to the quantities defined in (3)–(5), (13)–(16) below. Thus,

$$B = B^{m,n}_{p} = (-)^{\tau} (2\pi i)^{\tau} \exp \left\{ i\pi \left(\sum_{j=1}^{m} b_j - \sum_{j=n+1}^{p} a_j \right) \right\}, \tag{3}$$

$$\bar{B} = \bar{B}^{m,n}_{p} = (2\pi i)^{\tau} \exp \left\{ -i\pi \left(\sum_{j=1}^{m} b_j - \sum_{j=n+1}^{p} a_j \right) \right\}. \tag{4}$$

In these and all other definitions, it is supposed that the parameters are such that the definitions make sense. Let

$$\frac{\prod_{j=m+1}^{q} [1 - x \exp(2i\pi b_j)]}{\prod_{j=1}^{n} [1 - x \exp(2i\pi a_j)]} = \sum_{k=0}^{\infty} \Omega^{m,n}_{q}(k)\, x^k,$$

$$| x \exp(2i\pi a_j)| < 1, \qquad \Omega^{m,n}_{q}(k) = \Omega(k). \tag{5}$$

We next present some developments to facilitate the computation of $\Omega(k)$. Put

$$\xi_0 = 1, \qquad \xi_1 = \sum_{j=1}^{n} \exp(2i\pi a_j), \qquad \xi_2 = \sum_{1 \leqslant j < k \leqslant n} \exp[2i\pi(a_j + a_k)], \ldots,$$

$$\xi_r = \sum_{1 \leqslant h_1 < h_2 < \cdots < h_{r-1} \leqslant h_r}^{n} \exp\left(2i\pi \sum_{j=1}^{r} a_{h_j}\right), \qquad r \geqslant 1, \tag{6}$$

$$\zeta_0 = 1, \qquad \zeta_r = \sum_{1 \leqslant h_1 < h_2 < \cdots < h_{r-1} \leqslant h_r}^{q} \exp\left(2i\pi \sum_{j=1}^{r} b_{h_j}\right), \qquad r \geqslant 1. \tag{7}$$

Then

$$\sum_{r=0}^{q-m} (-)^r \zeta_r x^r = \left(\sum_{r=0}^{n} (-)^r \xi_r x^r \right) \left(\sum_{k=0}^{\infty} \Omega(k)\, x^k \right), \tag{8}$$

and so with $\Omega(0) = 1$, we have the recursion formula

$$\Omega(k) = (-)^k (\zeta_k - \xi_k) + \sum_{r=1}^{k-1} (-)^{r-1} \xi_r \Omega(k - r), \qquad k \geqslant 1. \tag{9}$$

In particular,

$$\Omega(1) = \xi_1 - \zeta_1, \qquad \Omega(2) = \zeta_2 - \xi_2 + \xi_1\Omega(1),$$
$$\Omega(3) = \xi_3 - \zeta_3 + \xi_1\Omega(2) - \xi_2\Omega(1). \tag{10}$$

Also

$$\Omega(k+1) = [1/(k+1)] \sum_{r=0}^{k} (S_{r+1} - T_{r+1})\, \Omega(k-r), \tag{11}$$

$$S_r = \sum_{j=1}^{n} \exp(2i\pi a_j r), \qquad T_r = \sum_{j=m+1}^{q} \exp(2i\pi b_j r), \qquad r \geqslant 1. \tag{12}$$

Let

$$\frac{\prod_{j=n+1}^{p} [1 - x\exp(2i\pi a_j)]}{\prod_{j=1}^{m} [1 - x\exp(2i\pi b_j)]} = \sum_{\lambda=0}^{\infty} E_p^{m,n}(\lambda)\, x^\lambda,$$

$$| x\exp(2i\pi b_j | < 1, \qquad E_p^{m,n}(\lambda) = E(\lambda), \tag{13}$$

$$\frac{\prod_{j=n+1}^{p} [1 - x\exp(2i\pi a_j)]}{[1 - x\exp(2i\pi a_l)] \prod_{j=1}^{m} [1 - x\exp(2i\pi b_j)]} = \sum_{r=0}^{\infty} \theta_p^{m,n}(l, r)\, x^r,$$

$$| x\exp(2i\pi b_j)| < 1, \qquad | x\exp(2i\pi a_l)| < 1,$$

$$\theta_p^{m,n}(l, r) = \theta(l, r). \tag{14}$$

By appropriate change of notation, recursion formulas for the evaluation of $E(\lambda)$ and $\theta(l, r)$ follow from (6)–(12).

Let

$$\Phi_{p,q}^{m,n}(h, \lambda) = \sum_{r=0}^{h+\lambda-1} E(r)\, \Omega^{0,p}_{\ q}(h + \lambda - r - 1), \qquad \Phi_{p,q}^{m,n}(h, \lambda) = \Phi(h, \lambda), \tag{15}$$

$$\psi_{p,q}^{m,n}(h, \lambda) = \sum_{r=1}^{h} E(\lambda - r)\, \bar{\Omega}^{0,p}_{\ q}(h - r), \qquad \psi_{p,q}^{m,n}(h, \lambda) = \psi(h, \lambda), \tag{16}$$

$$R_{p,q}^{m,n}(h, \lambda) = A^{m,n}_{\ q}\Phi(h, \lambda) - A^{0,p}_{\ q}\bar{B}\bar{\psi}(h, -\tau - \lambda + 1), \qquad R_{p,q}^{m,n}(h, \lambda) = R(h, \lambda), \tag{17}$$

$$T_{p,q}^{m,n}(l, \lambda) = -\{\exp(i\pi a_l)\, B\theta(l, \lambda - 1) + \exp(-i\pi a_l)\, \bar{B}\bar{\theta}(l, -\tau - \lambda)\}$$
$$\times \Delta^{0,p}_{\ q}(l) + \exp[i\pi a_l(\tau + 2\lambda - 1)]\, \Delta^{m,n}_{\ q}(l),$$

$$T_{p,q}^{m,n}(l, \lambda) = T(l, \lambda). \tag{18}$$

$$D_{p,q}^{m,n}(\lambda) = (-2\pi i)^{p-q} \exp\left\{i\pi\left(\sum_{h=1}^{p} a_h - \sum_{h=1}^{q} b_h\right)\right\}$$
$$\times \{BE(\lambda) - \bar{B}\bar{E}(-\tau - \lambda)\}, \qquad D_{p,q}^{m,n}(\lambda) = D(\lambda). \tag{19}$$

5.8.3. Expansion Theorems

Theorem 1.1. *Under conditions* (A), (B), $1 \leqslant n \leqslant p \leqslant q$, $1 \leqslant m \leqslant q$, $\nu + 1 \leqslant 0$, λ *an arbitrary integer such that* $0 \leqslant \lambda \leqslant -\nu - 1$, *then*

$$G_{p,q}^{m,n}(z) = \sum_{t=1}^{n} \exp[-i\pi a_t(\nu + 2\lambda + 1)] \, \Delta(t) \, G_{p,q}^{q,1}(z \exp[i\pi(\nu + 2\lambda + 1)] \| a_t). \quad (1)$$

Theorem 1.2. *Under conditions* (A), (B), $1 \leqslant n \leqslant p < q$, $2 \leqslant m \leqslant q$, $\nu + 1 \leqslant 0$, $| \arg z | < \rho\pi$, λ *as in* (1), $-\eta\pi < \arg z + 2\lambda\pi < \rho\pi$, *then* (1) *expresses* $G_{p,q}^{m,n}(z)$ *in terms of fundamental solutions valid near* $z = \infty$. (2)

Theorem 2.1. *Under conditions* (A), (B), $q \geqslant 1$, $0 \leqslant n \leqslant p \leqslant q$, $0 \leqslant m \leqslant q$, r *an arbitrary integer such that* $r \geqslant \max(0, \nu + 1)$, *then*

$$G_{p,q}^{m,n}(z) = A \sum_{k=0}^{r-1} \Omega(k) \, G_{p,q}^{q,0}(z \exp[i\pi(\nu - 2k)])$$

$$+ \sum_{t=1}^{n} \exp[i\pi a_t(2r - \nu - 1)] \, \Delta(t) \, G_{p,q}^{q,1}(z \exp[i\pi(\nu - 2r + 1)] \| a_t). \quad (3)$$

This expansion is also valid if i is replaced by $-i$.

Theorem 2.2. *Under conditions* (A), (B), $1 \leqslant n \leqslant p < q$, $2 \leqslant m \leqslant q$, $\nu + 1 \leqslant 0$, $-\eta\pi < \arg z < (\tau + \epsilon)\pi$, r *an arbitrary integer such that* $r \geqslant 0$ *and* $-\rho\pi < 2r\pi - \arg z < \eta\pi$, *then* (3) *expresses* $G_{p,q}^{m,n}(z)$ *in terms of fundamental solutions valid near* $z = \infty$. (4)

Theorem 2.3. *Under conditions* (A), (B), $0 \leqslant n \leqslant p < q$, $1 \leqslant m \leqslant q$, $\frac{3}{4}p + \frac{1}{4}q - \frac{1}{2}\epsilon < m + n \leqslant q + 1$, $-\rho\pi < \arg z < (\tau + \epsilon)\pi$, r *an arbitrary integer such that* $r \geqslant \nu + 1$, $-\rho\pi < 2r\pi - \arg z < \eta\pi$, *then* (3) *expresses* $G_{p,q}^{m,n}(z)$ *in terms of fundamental solutions valid near* $z = \infty$. (5)

Theorem 2.4. *Under conditions* (A), (B), $1 \leqslant n \leqslant p < q$, $2 \leqslant m \leqslant q$, $\nu + 1 \leqslant 0$, $-(\tau + \epsilon)\pi < \arg z < -\eta\pi$, r *an arbitrary integer* $\geqslant 0$ *such that* $-\rho\pi < 2r\pi + \arg z < \eta\pi$, *then* (3) *with* i *replaced by* $-i$ *expresses* $G_{p,q}^{m,n}(z)$ *in terms of fundamental solutions valid near* $z = \infty$. (6)

Theorem 2.5. *Under conditions* (A), (B), $0 \leqslant n \leqslant p < q$, $1 \leqslant m \leqslant q$, $\frac{3}{4}p + \frac{1}{4}q - \frac{1}{2}\epsilon < m + n \leqslant q + 1$, $-(\tau + \epsilon)\pi < \arg z < \rho\pi$, r *an arbitrary integer such that* $r \geqslant \nu + 1$, $-\rho\pi < 2r\pi + \arg z < \eta\pi$, *then* (3) *with i replaced by $-i$ expresses $G^{m,n}_{p,q}(z)$ in terms of fundamental solutions valid near $z = \infty$.* (7)

Theorem 3.1. *Under conditions* (A), (B), $q \geqslant 1$, $0 \leqslant n \leqslant p \leqslant q$, $0 \leqslant m \leqslant q$, $\nu + 1 \geqslant 0$, *r an arbitrary integer such that* $0 \leqslant r \leqslant \nu + 1$, *then*

$$G^{m,n}_{p,q}(z) = A \sum_{k=0}^{r-1} \Omega(k)\, G^{q,0}_{p,q}(z e^{i\pi(\nu-2k)}) + \bar{A} \sum_{s=0}^{\nu-r} \bar{\Omega}(k)\, G^{q,0}_{p,q}(z e^{-i\pi(\nu-2s)})$$

$$+ \sum_{t=1}^{n} \exp[i\pi a_t(2r - \nu - 1)]\, \Delta(t)\, G^{q,1}_{p,q}(z \exp[i\pi(\nu - 2r + 1)] \| a_t). \qquad (8)$$

This expansion formula is also valid if i is replaced by $-i$ provided that $r \geqslant \max(0, \nu + 1)$.

Theorem 3.2. *Under conditions* (A), (B), $0 \leqslant n \leqslant p < q$, $1 \leqslant m \leqslant q$, $p + 1 \leqslant m + n \leqslant \frac{3}{4}q + \frac{1}{4}p - \frac{1}{2}\epsilon + 1$, $|\arg z| < (\tau + \epsilon)\pi$, r *an arbitrary integer such that* $0 \leqslant r \leqslant \nu + 1$, $-\rho\pi < 2r\pi - \arg z < \eta\pi$, *then* (8) *expresses $G^{m,n}_{p,q}(z)$ in terms of fundamental solutions valid near $z = \infty$.* (9)

Theorem 4.1. *Under conditions* (A), (C), $q \geqslant 1$, $0 \leqslant n \leqslant p \leqslant q$, $0 \leqslant m \leqslant q$, *$\lambda$ and μ arbitrary integers with* $0 \leqslant \mu \leqslant q - p$, *then*

$$G^{m,n}_{p,q}(z) = \sum_{h=1}^{q-p-\mu} R(h, \lambda)\, G^{q,0}_{p,q}(z \exp[i\pi(\nu - 2h - 2\lambda + 2)])$$

$$+ \sum_{k=1}^{\mu} \bar{R}(k, -\tau - \lambda + 1)\, G^{q,0}_{p,q}(z \exp[i\pi(2p - q - m - n + 2k - 2\lambda)])$$

$$+ \sum_{k=1}^{p} \exp[i\pi a_k(q - p - 2\mu)]\, T(k, \lambda)$$

$$\times\, G^{q,1}_{p,q}(z \exp[i\pi(2p - q - m - n - 2\lambda + 2\mu + 1)] \| a_k). \qquad (10)$$

Theorem 4.2. *Under conditions* (A), (C), $0 \leqslant n \leqslant p < q$, $0 \leqslant m \leqslant q$, *$\lambda$ and μ arbitrary integers such that* $(\tau + \epsilon + 2\lambda - 2)\pi \leqslant \arg z < (\tau + \epsilon + 2\lambda)\pi$, $(m + n - \frac{3}{2}p + \frac{1}{2}q + 2\lambda - 2)\pi < 2\mu\pi + \arg z < (m + n - \frac{5}{2}p + \frac{3}{2}q + 2\lambda)\pi$, *then* (10) *expresses $G^{m,n}_{p,q}(z)$ in terms of fundamental solutions valid near $z = \infty$.* (11)

5.9. Asymptotic Expansions

5.9.1. $G_{P,Q}^{Q,N}(z)$, N=0,1

To simplify the exposition in this section as well as in Sections 5.8.2 and 5.9.2, it is convenient to introduce some notation. We let

$$m + n - p = \tau, \qquad q - m - n = \nu, \tag{1}$$

$$m + n - \tfrac{1}{2}(p + q) = \tfrac{1}{2}(\tau - \nu) = \rho, \qquad \tfrac{3}{2}q - \tfrac{1}{2}p - m - n + 2 = \eta, \qquad q - p = \sigma,$$
$$\epsilon = \tfrac{1}{2} \quad \text{if} \quad \sigma = 1, \qquad \epsilon = 1 \quad \text{if} \quad \sigma > 1. \tag{2}$$

Certain restrictions on the parameters a_h and b_h of the G-function enter the hypotheses of the theorems. These are called (A), (B), and (C) and are as follows:

$$\text{(A)} \qquad a_j - b_h \neq 1, 2, 3, \ldots \qquad \text{for} \qquad j = 1, 2, \ldots, n, \quad h = 1, 2, \ldots, m. \tag{3}$$

$$\text{(B)} \qquad a_j - a_t \neq 0, \pm 1, \pm 2, \ldots \qquad \text{for} \quad j, t = 1, 2, \ldots, n, \quad j \neq t. \tag{4}$$

$$\text{(C)} \qquad a_j - a_h \neq 0, \pm 1, \pm 2, \ldots \qquad \text{for} \quad j, h = 1, 2, \ldots, p, \quad j \neq h. \tag{5}$$

Actually (B) can always be deleted from our assumptions provided we understand that when (B) does not hold, a passage to the limit is required as in the discussion surrounding 5.3.1(7). A similar remark also holds for (C).

Define

$$G_{p,q}^{q,1}(z \parallel a_t) = G_{p,q}^{q,1}\left(z \,\middle|\, \begin{matrix} a_t, a_1, \ldots, a_{t-1}, a_{t+1}, \ldots, a_p \\ b_q \end{matrix}\right). \tag{6}$$

Further, with $1 \leqslant t \leqslant p < q$, let us formally write

$$E_{p,q}(z \parallel a_t) = \frac{z^{a_t - 1}\Gamma(1 + b_q - a_t)}{\Gamma(1 + a_p - a_t)}\, {}_qF_{p-1}\left(\begin{matrix} 1 + b_q - a_t \\ 1 + a_p - a_t^* \end{matrix} \,\middle|\, -\frac{1}{z}\right), \tag{7}$$

provided condition (A) [see (3)] holds with $j = t$ and $m = q$.

Theorem 1. *If* $1 \leqslant t \leqslant p < q$, (A), *then*

$$G_{p,q}^{q,1}(z \parallel a_t) \sim E_{p,q}(z \parallel a_t),$$
$$|z| \to \infty, \qquad |\arg z| \leqslant (\tfrac{1}{2}\sigma + 1)\pi - \delta, \qquad \delta \geqslant 0. \tag{8}$$

Theorem 2. *If* $1 \leqslant n \leqslant p < q$, $1 \leqslant m \leqslant q$, (A) *and* (B), *then with* $\rho > 0$,

$$G^{m,n}_{p,q}(z) \sim \sum_{t=1}^{n} \exp[-i\pi(\nu + 1)\, a_t]\, \Delta^{m,n}_{\ \ \ q}(t)\, E_{p,q}(z \exp[i\pi(\nu + 1)] \,\|\, a_t),$$

$$|\, z \,| \to \infty, \qquad |\arg z\,| \leqslant \rho\pi - \delta, \qquad \delta \geqslant 0, \tag{9}$$

where $\Delta^{m,n}_{\ \ \ q}(t)$ *is defined by* 5.8.2(1).

Theorem 3. *If* $m + n \geqslant p + 1$, (A) *and* (B), *and* $|\arg z\,| < (m + n - p)\pi$, *then* $G^{m,n}_{p,p}(z)$ *can be continued analytically outside the unit circle by the expansion*

$$G^{m,n}_{p,p}(z) = \sum_{t=1}^{n} \exp[-i\pi(p - m - n + 1)\, a_t]\, \Delta^{m,n}_{\ \ \ p}(t)$$

$$\times\, E_{p,p}(z \exp[i\pi(p - m - n + 1)] \,\|\, a_t). \tag{10}$$

Theorem 4. *If* (A) *holds for* $n = 1$ *and* $m = p$, *and* $|\arg z\,| < \pi$, *then* $G^{p,1}_{p,p}(z \,\|\, a_t)$ *can be continued analytically outside the unit circle by the expression*

$$G^{p,1}_{p,p}(z \,\|\, a_t) = E_{p,p}(z \,\|\, a_t). \tag{11}$$

Theorem 3 is the statement 5.3.3(2). Theorem 4 is the special case n=1 of Theorem 3.

Theorem 5.

$$G^{q,0}_{p,q}\left(z \,\middle|\, \begin{matrix} a_p \\ b_q \end{matrix}\right) \sim H_{p,q}(z),$$

$$|\, z \,| \to \infty, \qquad |\arg z\,| \leqslant (\sigma + \epsilon)\,\pi - \delta, \qquad \delta > 0, \tag{12}$$

where σ *and* ϵ *are defined in* (1), (2) *and where for brevity we write*

$$H_{p,q}(z) = \frac{(2\pi)^{(\sigma-1)/2}}{\sigma^{1/2}} \exp\{-\sigma(z)^{1/\sigma}\}\, z^{\theta} \sum_{k=0}^{\infty} M_k(z)^{-k/\sigma},$$

$$M_0 = 1, \qquad \sigma\theta = \{\tfrac{1}{2}(1 - \sigma) + \Xi_1 - \Lambda_1\},$$

$$\Xi_1 = \sum_{h=1}^{q} b_h\,, \qquad \Lambda_1 = \sum_{h=1}^{p} a_h\,. \tag{13}$$

$$M_1 = (\Lambda_2 - \Xi_2) - \frac{(\Lambda_1 - \Xi_1)}{2\sigma}\,[\sigma(\Lambda_1 + \Xi_1) + (\Lambda_1 - \Xi_1)] - \frac{(\sigma^2 - 1)}{24\sigma}, \tag{14}$$

where

$$\Xi_2 = \sum_{h=2}^{q} \sum_{k=1}^{h-1} b_h b_k , \qquad \Lambda_2 = \sum_{h=2}^{p} \sum_{k=1}^{h-1} a_h a_k . \tag{15}$$

The M_k's are independent of z and can be found by use of a recurrence formula, see Luke (1969). One way to derive the recurrence formula is to substitute the form for $H_{p,q}(z)$ in the differential equation for $G_{p,q}^{q,0}(z)$ and equate like powers of z. Some special cases are treated in 5.9.3.

5.9.2. $G_{P,Q}^{M,N}(Z)$

As previously noted, asymptotic expansions of $G_{p,q}^{m,n}(z)$ for all values of m, n, p, q (with $p < q$), and $\arg z$ for $|z| \to \infty$ can be derived from the theorems of 5.8.3 and Theorems 1 and 5 of 5.9.1. Indeed, in his work Meijer (1946) investigates all of the G-functions which appear on the right-side of the expansion formulas in 5.8.3, and from these he determines those which are dominant.

It is helpful to clarify the concept of dominance. We say that $A(z)$ is dominant compared with $B(z)$ or $B(z)$ is subdominant with respect to $A(z)$ if the order of the lead term of the asymptotic expansion of $B(z)$ is less than the order of the error term in the asymptotic expansion of $A(z)$. To illustrate, consider the asymptotic expansions

$$A_1(z) \sim e^z \sum_{k=0}^{\infty} a_{1k} z^{-k}, \qquad A_2(z) \sim z^6 \sum_{k=0}^{\infty} a_{2k} z^{-k},$$

$$A_3(z) \sim z^{1/2} \sum_{k=0}^{\infty} a_{3k} z^{-k}, \qquad A_4(z) \sim e^{iz} z^{-8} \sum_{k=0}^{\infty} a_{4k} z^{-k},$$

$$A_5(z) \sim e^{-z} \sum_{k=0}^{\infty} a_{5k} z^{-k}, \qquad A_6(z) \sim e^{-2z} \sum_{k=0}^{\infty} a_{6k} z^{-k},$$

where it is supposed that no a_{i0} , $i = 1, 2, \ldots, 6$, is zero. Assume $z > 0$. Then clearly $A_1(z)$ is dominant compared with $A_2(z), \ldots, A_6(z)$. Compared with $A_5(z)$ and $A_6(z)$, $A_2(z)$, $A_3(z)$, and $A_4(z)$ are dominant, and $A_5(z)$ is dominant compared with $A_6(z)$. However, $A_2(z)$ is not dominant compared with $A_3(z)$ and $A_4(z)$. Thus among the functions $A_2(z), \ldots, A_6(z)$, there are three dominants. They are $A_2(z)$, $A_3(z)$, and $A_4(z)$. Qualitatively, we say that $A_1(z)$ is an exponential growth, $A_5(z)$ and $A_6(z)$ are exponential decays, $A_2(z)$ and $A_3(z)$ are algebraic, and $A_4(z)$ is algebraic and sinusoidal.

To get asymptotic expansions for $G_{p,q}^{m,n}(z)$ it is only necessary to retain the dominant term or terms in the right-hand sides of the expansion formulas in 5.8.3, unless, of course, the coefficients of these dominants vanish. The coefficients depend on the parameters a_j and b_j and in general do not vanish. We suppose, therefore, that if there is only one dominant function, the coefficient of this function is not zero. If there are two or more dominants, then at least one of them has a nonvanishing coefficient. If all the coefficients of the dominant functions vanish, it is necessary to further explore the nature of the above expansion formulas. We omit such details.

We follow Meijer and give only the dominant term or terms in the asymptotic expansions. However, for the benefit of the reader, we tell from which expansion theorems the results follow so that the complete expansion may be deduced if desired. This is illustrated later when we record the complete asymptotic expansion for the $_pF_q$, $p \leqslant q$. In the theorems which follow, we make free use of the notation introduced in 5.8.2 and 5.9.1.

Theorem 1. *Under conditions* (A), (B), $1 \leqslant n \leqslant p < q$, $1 \leqslant m \leqslant q$, $\rho > 0$, *then*

$$G_{p,q}^{m,n}(z) \sim \sum_{t=1}^{n} \exp[-i\pi a_t(\nu+1)]\, \Delta(t)\, E_{p,q}(z \exp[i\pi(\nu+1)] \| a_t),$$

$$|z| \to \infty, \qquad |\arg z| \leqslant \rho\pi - \delta, \qquad \delta > 0. \tag{1}$$

This is proved from 5.8.3(1, 2) if $\nu + 1 \leqslant 0$ and from 5.8.3(3, 5) if $\rho > 0$ and $\nu + 1 \geqslant 0$.

Theorem 2. *If* $1 \leqslant p + 1 \leqslant m \leqslant q$, *then:*

$$G_{p,q}^{m,0}(z) \sim A^{m,0}_{q} H_{p,q}(ze^{i\pi(q-m)}) \qquad \textit{if} \quad m \leqslant q-1,$$

$$|z| \to \infty, \qquad \delta \leqslant \arg z \leqslant (m-p+1)\pi - \delta, \qquad \delta > 0; \tag{2}$$

$$G_{p,q}^{m,0}(z) \sim \bar{A}^{m,0}_{q} H_{p,q}(ze^{-i\pi(q-m)}) \qquad \textit{if} \quad m \leqslant q-1,$$

$$|z| \to \infty, \qquad \delta - (m-p+1)\pi \leqslant \arg z \leqslant -\delta, \qquad \delta > 0; \tag{3}$$

$$G_{p,q}^{m,0}(z) \sim A^{m,0}_{q} H_{p,q}(ze^{i\pi(q-m)}) + \bar{A}^{m,0}_{q} H_{p,q}(ze^{-i\pi(q-m)}),$$

$$\textit{if} \quad m \leqslant (q-1), \quad z \to +\infty, \qquad \textit{that is} \quad \arg z = 0; \tag{4}$$

$$G_{p,q}^{q,0}(z) \sim H_{p,q}(z),$$

$$|z| \to \infty, \qquad |\arg z| \leqslant (\sigma + \epsilon)\pi - \delta, \qquad \delta > 0. \tag{5}$$

Equations (2)–(4) are proved from 5.8.3(3). Equation (5) is the same as 5.9.1(12).

Theorem 3. *Under conditions* (A), $0 \leqslant n \leqslant p \leqslant q-2$, $1 \leqslant \tau \leqslant \frac{1}{2}\sigma$ [*or* $p+1 \leqslant m+n \leqslant \frac{1}{2}(p+q)$], *then*

$$G_{p,q}^{m,n}(z) \sim AH_{p,q}(ze^{i\pi\nu}),$$
$$|z| \to \infty, \qquad \delta \leqslant \arg z \leqslant (\tau+1)\pi - \delta, \qquad \delta > 0; \tag{6}$$

$$G_{p,q}^{m,n}(z) \sim \bar{A}H_{p,q}(ze^{-i\pi\nu}),$$
$$|z| \to \infty, \qquad \delta - (\tau+1)\pi \leqslant \arg z \leqslant -\delta, \qquad \delta > 0; \tag{7}$$

$$G_{p,q}^{m,n}(z) \sim AH_{p,q}(ze^{i\pi\nu}) + \bar{A}H_{p,q}(ze^{-i\pi\nu}) \qquad \textit{with} \quad 1 \leqslant \tau < \tfrac{1}{2}\sigma,$$
$$z \to +\infty, \qquad \textit{that is} \quad \arg z = 0; \tag{8}$$

$$G_{p,q}^{m,n}(z) \sim AH_{p,q}(ze^{i\pi\nu}) + \bar{A}H_{p,q}(ze^{-i\pi\nu})$$
$$+ \sum_{t=1}^{n} \exp[-i\pi a_t(\nu+1)]\, \Delta(t)\, E_{p,q}(ze^{i\pi(\nu+1)} \,\|\, a_t) \tag{9}$$

with $\rho = 0$, *condition* (B) *and* $z \to +\infty$, *that is*, $\arg z = 0$.

This is proved from 5.8.3(8,9).

Theorem 4. *Under conditions* (A), $0 \leqslant n \leqslant p < q$, $1 \leqslant m \leqslant q$, $\rho > 0$, *then*

$$G_{p,q}^{m,n}(z) \sim AH_{p,q}(ze^{i\pi\nu}),$$
$$|z| \to \infty, \qquad \delta + \rho\pi \leqslant \arg z \leqslant (\tau+\epsilon)\pi - \delta, \qquad \delta > 0; \tag{10}$$

$$G_{p,q}^{m,n}(z) \sim \bar{A}H_{p,q}(ze^{-i\pi\nu}),$$
$$|z| \to \infty, \qquad \delta - (\tau+\epsilon)\pi < \arg z \leqslant -\rho\pi + \delta, \qquad \delta > 0; \tag{11}$$

$$G_{p,q}^{m,n}(z) \sim AH_{p,q}(ze^{i\pi\nu}) + \sum_{t=1}^{n} \exp[-i\pi a_t(\nu+1)]\, \Delta(t)\, E_{p,q}(ze^{i\pi(\nu+1)} \,\|\, a_t), \tag{12}$$

provided also that condition (B) *holds*, $|z| \to \infty$, $\arg z = \rho\pi$;

$$G_{p,q}^{m,n}(z) \sim \bar{A}H_{p,q}(ze^{-i\pi\nu}) + \sum_{t=1}^{n} \exp[i\pi a_t(\nu+1)]\, \Delta(t)\, E_{p,q}(ze^{-i\pi(\nu+1)} \,\|\, a_t), \tag{13}$$

provided also that condition (B) *holds*, $|z| \to \infty$, $\arg z = -\rho\pi$.

Equations (10) and (12) follow from 5.8.3(3, 4, 5), respectively, while (11) and (13) follow from 5.8.3(3, 6, 7), respectively.

Theorem 5. *Under conditions* (A), $0 \leqslant n \leqslant p \leqslant q-2$, $1 \leqslant m \leqslant q$; *if* $\tau \leqslant 1$, λ *is an arbitrary integer; if* $\tau \geqslant 2$, λ *is either an arbitrary integer* $\geqslant 0$ *or an arbitrary integer* $\leqslant -\tau$; *then*

$$G_{p,q}^{m,n}(z) \sim D(\lambda)\, H_{p,q}(ze^{i\pi(\nu-2\lambda)}),$$

$$|z| \to \infty, \qquad \delta_1 + (\tau + 2\lambda - 1)\pi \leqslant \arg z \leqslant (\tau + 2\lambda + 1)\pi + \delta_2, \qquad (14)$$

where δ_1 *and* δ_2 *are arbitrary small quantities whose signs are chosen so that the closed interval* $[\delta_1 + a, \delta_2 + b]$ *is contained in the open interval* (a, b);

$$G_{p,q}^{m,n}(z) \sim D(\lambda)\, H_{p,q}(z \exp[i\pi(\nu - 2\lambda)]) + D(\lambda - 1)\, H_{p,q}(z \exp[i\pi(\nu - 2\lambda + 2)]),$$

$$q > p + 2, \qquad |z| \to \infty, \qquad \arg z = (\tau + 2\lambda - 1)\pi; \qquad (15)$$

$$G_{p,p+2}^{m,n}(z) \sim D_{p,p+2}^{m,n}(\lambda)\, H_{p,p+2}(z \exp[-i\pi(\tau + 2\lambda - 2)])$$

$$+ D_{p,p+2}^{m,n}(\lambda - 1)\, H_{p,p+2}(z \exp[-i\pi(\tau + 2\lambda - 4)])$$

$$+ \sum_{t=1}^{p} \exp[-2i\pi(\lambda + 1)\, a_t]\, T_{p,p+2}^{m,n}(t, \lambda)\, E_{p,p+2}(z \exp[-i\pi(\tau - 3)] \| a_t), \qquad (16)$$

provided condition (C) *holds*, $|z| \to \infty$, $\arg z = (\tau + 2\lambda - 1)\pi$.

This is proved from 5.8.3(10,11). Note that the case $m=0$ need not be considered in view of 5.3.1(6). For special values of λ and τ, Theorem 5 reduces to previous statements. These are enumerated as follows:

If $\tau \geqslant 1$ and $\lambda = 0$, then (14) reduces to (6).
If $\tau \geqslant 1$ and $\lambda = -\tau$, then (14) reduces to (7).
If $\tau > 1$ and $\lambda = 0$, then (15) reduces to (6).
If $\tau = 1$ and $\lambda = 0$, then (15) reduces to (8) with $\tau = 1$.
If $\tau > 1$ and $\lambda = 0$, then (16) reduces to (12) with $q = p + 2$.
If $\tau = 1$ and $\lambda = 0$, then (16) reduces to (9) with $q = p + 2$.

Theorem 6.1. *Under the conditions* (A), (C), $p \geqslant 1$, $0 \leqslant n \leqslant p$, $1 \leqslant m \leqslant p + 1$, λ *an arbitrary integer, then*

$$G_{p,p+1}^{m,n}(z) \sim \sum_{t=1}^{p} \exp[-i\pi a_t(2\lambda + 1)]\, T_{p,p+1}^{m,n}(t, \lambda)\, E_{p,p+1}(z \exp[-i\pi(\tau - 2)] \| a_t),$$

$$|z| \to \infty, \qquad \delta_1 + (\tau + 2\lambda - \tfrac{3}{2})\pi \leqslant \arg z \leqslant (\tau + 2\lambda - \tfrac{1}{2})\pi + \delta_2. \qquad (17)$$

This expansion is not valid when $n = \lambda = 0$ and $m = p + 1$. For $n = 0$ and $m = p + 1$, see (2)–(5). The expansion (17) is also valid if

$$\tau \geqslant 2, \qquad -\tau < \lambda < 0, \qquad (\tau + 2\lambda - \tfrac{1}{2})\,\pi \leqslant \arg z \leqslant (\tau + 2\lambda + \tfrac{1}{2})\,\pi + \delta_2$$

or

$$\tau \geqslant 2, \qquad 1 - \tau < \lambda < 1, \qquad \arg z = (\tau + 2\lambda - \tfrac{3}{2})\,\pi.$$

REMARK. If $1 \leqslant t \leqslant n$ and $1 - \tau \leqslant \lambda \leqslant 0$, then $T^{m,n}_{p,p+1}(t, \lambda) = \exp[i\pi a_t(\tau + 2\lambda - 1)]\, \Delta^{m,n}_{p+1}(t)$, and if $n + 1 \leqslant t \leqslant p$ and $1 - \tau \leqslant \lambda \leqslant 0$, then $T^{m,n}_{p,p+1}(t, \lambda) = 0$, so that (17) with $n \geqslant 1$, $\tau \geqslant 1$, and $1 - \tau \leqslant \lambda \leqslant 0$ is the same as (1) with $q = p + 1$.

Theorem 6.2. *Under the conditions* (A), $p \geqslant 1$, $0 \leqslant n \leqslant p$, $1 \leqslant m \leqslant p + 1$, $\tau \geqslant 2$, *λ an arbitrary integer $\geqslant 0$ or $\leqslant -\tau$,*

$$G^{m,n}_{p,p+1}(z) \sim D^{m,n}_{p,p+1}(\lambda)\, H_{p,p+1}(z \exp[-i\pi(\tau + 2\lambda - 1)]),$$

$$\delta_1 + (\tau + 2\lambda - \tfrac{1}{2})\,\pi \leqslant \arg z \leqslant (\tau + 2\lambda + \tfrac{1}{2})\,\pi + \delta_2 . \qquad (18)$$

This is also valid if $\tau \leqslant 1$ with λ an arbitrary integer.

Theorem 6.3. *Under conditions* (A), (C), $p \geqslant 1$, $0 \leqslant n \leqslant p$, $1 \leqslant m \leqslant p + 1$, $\tau \geqslant 2$, *λ an arbitrary integer $\geqslant 0$ or $\leqslant -\tau$,*

$$G^{m,n}_{p,p+1}(z) \sim D^{m,n}_{p,p+1}(\lambda)\, H_{p,p+1}(z \exp[-i\pi(\tau + 2\lambda - 1)])$$

$$+ \sum_{t=1}^{p} \exp[-i\pi a_t(2\lambda + 1)]\, T^{m,n}_{p,p+1}(t, \lambda)\, E_{p,p+1}(z \exp[-i\pi(\tau - 2)] \,\|\, a_t),$$

$$|\, z \,| \to \infty, \qquad \arg z = (\tau + 2\lambda - \tfrac{1}{2})\,\pi. \qquad (19)$$

This is also valid if $\tau \leqslant 1$ and λ is an arbitrary integer.

Theorem 6.4. *Under conditions* (A), (C), $p \geqslant 1$, $0 \leqslant n \leqslant p$, $1 \leqslant m \leqslant p + 1$, $\tau \geqslant 2$, *λ an arbitrary integer $\geqslant 1$ or $\leqslant -\tau + 1$,*

$$G^{m,n}_{p,p+1}(z) \sim D^{m,n}_{p,p+1}(\lambda - 1)\, H_{p,p+1}(z \exp[-i\pi(\tau + 2\lambda - 3)])$$

$$+ \sum_{t=1}^{p} \exp[-i\pi a_t(2\lambda + 1)]\, T^{m,n}_{p,p+1}(t, \lambda)\, E_{p,p+1}(z \exp[-i\pi(\tau - 2)] \,\|\, a_t),$$

$$|\, z \,| \to \infty, \qquad \arg z = (\tau + 2\lambda - \tfrac{3}{2})\,\pi. \qquad (20)$$

This is also valid if $\tau \leqslant 1$ and λ is an arbitrary integer.

Equations (17)–(20) are proved from 5.8.3(10, 11).

The following and last theorem of this section concerns the analytic continuation of $G_{p,p}^{m,n}(z)$ in the general case.

Theorem 7. *Under conditions* (A), (C), $p \geqslant 1$, $0 \leqslant n \leqslant p$, $0 \leqslant m \leqslant p$, λ *an arbitrary integer such that* $(\tau + 2\lambda - 2)\pi < \arg z < (\tau + 2\lambda)\pi$, *then* $G_{p,p}^{m,n}(z)$ *can be expressed in terms of fundamental solutions valid near* $z = \infty$ *by the formula*

$$G_{p,p}^{m,n}(z) = \sum_{t=1}^{p} T_{p,p}^{m,n}(t, \lambda)\, G_{p,p}^{p,1}(z \exp[-i\pi(\tau + 2\lambda - 1)] \| a_t), \tag{21}$$

and can be continued analytically outside the unit circle $|z| = 1$ *by the expansion*

$$G_{p,p}^{m,n}(z) = \sum_{t=1}^{p} \exp(-2i\pi\lambda a_t)\, T_{p,p}^{m,n}(t, \lambda)\, E_{p,p}(z \exp[-i\pi(\tau - 1)] \| a_t). \tag{22}$$

If $\tau = m + n - p \geqslant 1$ *and* $1 - \tau \leqslant \lambda \leqslant 0$, *and we use the properties of* $T_{p,p}^{m,n}(t, \lambda)$ *as in the remark following* (17), *then*

$$G_{p,p}^{m,n}(z) = \sum_{t=1}^{n} \exp[i\pi a_t(\tau - 1)]\, \Delta^{m,n}_{p}(t)\, E_{p,p}(z \exp[-i\pi(\tau - 1)] \| a_t),$$

$$|\arg z| < \tau\pi, \qquad \arg z \neq (\tau - 2k)\,\pi, \qquad k = 1, 2, \ldots, [\tau/2]. \tag{23}$$

Observe that if $p = q$, $-\tau = \nu = \rho$ *and the expansion in* (23) *is the same as that in* (1) *with* $p = q$.

Theorem 7 follows from 5.8.3(10, 11). It is essentially the same as the Theorem given by 5.9.1(10) though slightly less general for we now have excluded those values of z for which $\arg z = (\tau - 2k)\pi$, $k = 1, 2, \ldots, [\tau/2]$.

5.9.3. $_pF_Q(z)$

Let

$$L_{p,q}(z) = \sum_{t=1}^{p} L_{p,q}^{(t)}(z),$$

$$L_{p,q}^{(t)}(z) = \frac{z^{-\alpha_t}\Gamma(\alpha_t)\,\Gamma(\alpha_p - \alpha_t)^*}{\Gamma(\rho_q - \alpha_t)}\, {}_{q+1}F_{p-1}\left(\begin{matrix}\alpha_t\,, 1 + \alpha_t - \rho_q \\ 1 + \alpha_t - \alpha_p^*\end{matrix} \,\middle|\, \frac{(-)^{q-p}}{z}\right), \tag{1}$$

or formally from 5.3.1(11) without the restriction on p, q, and z,

$$L_{p,q}(z) = G^{p,1}_{q+1,p}\left(z^{-1}\,\middle|\,\begin{matrix}1,\rho_q\\ \alpha_p\end{matrix}\right). \tag{2}$$

Observe that if two or more of the α_h's differ by an integer or zero, we can define $L_{p,q}(z)$ by a limit process as previously noted. In our present notation

$$A^{1,p}_{q+1}H_{p,q+1}(z\exp[i\pi(q-p-2k)]) = K_{p,q}(z\exp[-i\pi(2k+1)]), \tag{3}$$

$$K_{p,q}(z) = \frac{(2\pi)^{(1-\beta)/2}}{\beta^{1/2}}[\exp\{\beta z^{1/\beta}\}]\,z^{\gamma}\sum_{r=0}^{\infty} N_r z^{-r/\beta}, \qquad N_0 = 1,$$

$$\beta = q+1-p,\quad \beta\gamma = (\beta-1)/2 + B_1 - C_1,\quad B_1 = \sum_{h=1}^{p}\alpha_h,\quad C_1 = \sum_{h=1}^{q}\rho_h. \tag{4}$$

For computation of the N_r's, see the remarks following 5.9.1(15). For $r=1$, we have

$$\begin{aligned} N_1 = {} & C_2 - B_2 + (2\beta)^{-1}(B_1 - C_1)[\beta(B_1+C_1) + B_1 - C_1 - 2] \\ & + (24\beta)^{-1}(\beta-1)(\beta-11), \end{aligned} \tag{5}$$

$$B_2 = \sum_{s=2}^{p}\sum_{t=1}^{s-1}\alpha_s\alpha_t, \qquad C_2 = \sum_{s=2}^{q}\sum_{t=1}^{s-1}\rho_s\rho_t.$$

For convenience in the applications, we record

$$\begin{aligned} K_{p,q}(ze^{i\pi}) + K_{p,q}(ze^{-i\pi}) = {} & \frac{2(2\pi)^{(1-\beta)/2}}{\beta^{1/2}}[\exp\{\beta z^{1/\beta}\cos\pi/\beta\}]\,z^{\gamma} \\ & \times \sum_{r=0}^{\infty} N_r z^{-r/\beta}\cos(\pi r/\beta - \pi\gamma - \beta z^{1/\beta}\sin\pi/\beta). \end{aligned} \tag{6}$$

The asymptotic expansions for ${}_pF_q(z)$ are divided into two cases. Here we consider the case $0 \leqslant p \leqslant q-1$, that is, $\beta \geqslant 2$. See (33) below for the case $p=q$.

$$\begin{aligned} {}_pF_q\left(\begin{matrix}\alpha_p\\ \rho_q\end{matrix}\,\middle|\,-z\right) \sim {} & \frac{\Gamma(\rho_q)}{\Gamma(\alpha_p)}\left\{\sum_{k=0}^{r-1}\Gamma^{1,p}_{q+1}(k)\,K_{p,q}(z\exp[-i\pi(2k+1)])\right. \\ & \left. + \sum_{s=0}^{\beta-r-1}\bar{\Gamma}^{1,p}_{q+1}(k)\,K_{p,q}(z\exp[i\pi(2s+1)]) + L_{p,q}(z)\right\}, \\ & 0 \leqslant p \leqslant q-1, \end{aligned} \tag{7}$$

where $|z| \to \infty$, r an arbitrary integer such that $0 \leqslant r \leqslant \beta$,

$$|\arg z| \leqslant 2\pi - \delta, \qquad \delta > 0,$$
$$\delta_1 + (4r - 3\beta - 2)\,\pi/2 \leqslant \arg z \leqslant (4r - \beta + 2)\,\pi/2 + \delta_2\,,$$

where δ_1 and δ_2 have the same meaning as in 5.9.2(14) and $\bar{\Gamma}^{1,p}_{\ q+1}(k)$ is $\Gamma^{1,p}_{\ q+1}(k)$ (defined by (10) below) with i replaced by $-i$. If we take $r = [(\beta + 1)/2]$, then the restrictions on $\arg z$ are

$$\begin{aligned} \delta - 3\pi/2 \leqslant \arg z \leqslant 2\pi - \delta \qquad & \text{if} \quad \beta = 3, \\ |\arg z| \leqslant 2\pi - \delta \qquad & \text{if} \quad \beta \neq 3, \quad \delta > 0, \end{aligned} \tag{8}$$

while if $r = [\beta/2]$,

$$\begin{aligned} \delta - 2\pi \leqslant \arg z \leqslant 3\pi/2 - \delta \qquad & \text{if} \quad \beta = 3, \\ |\arg z| \leqslant 2\pi - \delta \qquad & \text{if} \quad \beta \neq 3, \quad \delta > 0. \end{aligned} \tag{9}$$

By definition

$$\frac{\prod_{j=1}^{q} [1 - x \exp(2i\pi\rho_j)]}{\prod_{j=1}^{p} [1 - x \exp(-2i\pi\alpha_j)]} = \sum_{k=0}^{\infty} \Gamma^{1,p}_{\ q+1}(k)\, x^k, \qquad |x \exp(-2i\pi\alpha_j)| < 1, \tag{10}$$

Let

$$\xi_0 = 1, \quad \xi_1 = \sum_{j=1}^{p} \exp(-2i\pi\alpha_j), \quad \xi_2 = \sum_{1 \leqslant j < k \leqslant p}^{p} \exp[-2i\pi(\alpha_j + \alpha_k)], \ldots,$$
$$\xi_r = \sum_{1 \leqslant h_1 < h_2 < \cdots h_{r-1} \leqslant h_r}^{p} \exp\left(-2i\pi \sum_{j=1}^{r} \alpha_{h_j}\right), \qquad r \geqslant 1, \tag{11}$$

$$\zeta_0 = 1, \quad \zeta_r = \sum_{1 \leqslant h_1 < h_2 < \cdots h_{r-1} \leqslant h_r}^{q} \exp\left(-2i\pi \sum_{j=1}^{r} \rho_{h_j}\right), \qquad r \geqslant 1. \tag{12}$$

Then

$$\sum_{r=0}^{q} (-)^r \zeta_r x^r = \left(\sum_{r=0}^{p} (-)^r \xi_r x^r\right)\left(\sum_{k=0}^{\infty} \Gamma^{1,p}_{\ q+1}(k)\, x^k\right), \tag{13}$$

and so with $\Gamma^{1,p}_{\ q+1}(0) = 1$,

$$\Gamma^{1,p}_{\ q+1}(k) = (-)^k(\zeta_k - \xi_k) + \sum_{r=1}^{k-1} (-)^{r-1} \xi_r \Gamma^{1,p}_{\ q+1}(k - r), \qquad k \geqslant 1, \tag{14}$$

$$\Gamma^{1,p}_{q+1}(1) = \xi_1 - \zeta_1, \qquad \Gamma^{1,p}_{q+1}(2) = \zeta_2 - \xi_2 + \xi_1\Gamma^{1,p}_{q+1}(1),$$
$$\Gamma^{1,p}_{q+1}(3) = \xi_3 - \zeta_3 + \xi_1\Gamma^{1,p}_{q+1}(2) - \xi_2\Gamma^{1,p}_{q+1}(1), \text{ etc.} \tag{15}$$

If only the dominant terms are of interest, we get the following results

$${}_pF_q\left(\begin{matrix}\alpha_p\\ \rho_q\end{matrix}\middle| z\right) \sim \frac{\Gamma(\rho_q)}{\Gamma(\alpha_p)} K_{p,q}(z),$$
$$0 \leqslant p \leqslant q-1, \qquad |z| \to \infty, \qquad |\arg z| \leqslant \pi - \delta, \qquad \delta > 0. \tag{16}$$

$${}_pF_q\left(\begin{matrix}\alpha_p\\ \rho_q\end{matrix}\middle| -z\right) \sim \frac{\Gamma(\rho_q)}{\Gamma(\alpha_p)} \{K_{p,q}(ze^{i\pi}) + K_{p,q}(ze^{-i\pi})\},$$
$$0 \leqslant p < q-1, \qquad \text{that is,} \qquad \beta \geqslant 3;$$
$$z \to +\infty, \qquad \text{that is,} \quad \arg z = 0. \tag{17}$$

The case $\beta = 2$ has many important applications. We record the asymptotic expansion for this situation in full:

$${}_pF_{p+1}\left(\begin{matrix}\alpha_p\\ \rho_{p+1}\end{matrix}\middle| -z\right) \sim \frac{\Gamma(\rho_{p+1})}{\Gamma(\alpha_p)} \{K_{p,p+1}(ze^{i\pi}) + K_{p,p+1}(ze^{-i\pi}) + L_{p,p+1}(z)\},$$
$$|z| \to \infty, \qquad |\arg z| \leqslant 2\pi - \delta, \qquad \delta > 0. \tag{18}$$

To facilitate computations in (17),(18), see (6). For numerous applications it is convenient to replace z by $z^2/4$ and write

$${}_pF_{p+1}\left(\begin{matrix}\alpha_p\\ \rho_{p+1}\end{matrix}\middle| \frac{-z^2}{4}\right) \sim \frac{\Gamma(\rho_{p+1})}{\Gamma(\alpha_p)} \{K_{p,p+1}([\tfrac{1}{2}ze^{i\pi/2}]^2) + K_{p,p+1}([\tfrac{1}{2}ze^{-i\pi/2}]^2)$$
$$+ L_{p,p+1}(z^2/4)\},$$
$$|z| \to \infty, \qquad |\arg z| \leqslant \pi - \delta, \qquad \delta > 0. \tag{19}$$

Also

$${}_pF_{p+1}\left(\begin{matrix}\alpha_p\\ \rho_{p+1}\end{matrix}\middle| \frac{z^2}{4}\right) \sim \frac{\Gamma(\rho_{p+1})}{\Gamma(\alpha_p)} \{K_{p,p+1}([\tfrac{1}{2}z]^2) + K_{p,p+1}([\tfrac{1}{2}ze^{i\epsilon\pi}]^2)$$
$$+ L_{p,p+1}([\tfrac{1}{2}ze^{i\epsilon\pi}]^2)\},$$
$$|z| \to \infty, \quad \delta - (2+\epsilon)\pi/2 \leqslant \arg z \leqslant (2-\epsilon)\pi/2 - \delta, \quad \epsilon = \pm 1, \quad \delta > 0. \tag{20}$$

The apparent discrepancy in (8) when $|\arg z| < \pi/2$ is a case of Stoke's phenomenon.

Let us write

$$K_{p,p+1}([\tfrac{1}{2}z]^2) = \frac{e^z z^{2\gamma}}{2^{2\gamma+1}\pi^{1/2}} \sum_{k=0}^{\infty} d_k z^{-k}, \qquad d_0 = 1,$$

$$\gamma = \tfrac{1}{2}\left\{\tfrac{1}{2} + \sum_{h=1}^{p} \alpha_h - \sum_{h=1}^{p+1} \rho_h\right\} = \tfrac{1}{2}\{\tfrac{1}{2} + B_1 - C_1\}. \tag{21}$$

We record below recursion formulas for the d_k's in the cases $p = 0$, 1, and 2. In this situation observe that

$$d_k = 2^k N_k \tag{22}$$

N_k as in (4,5) with $q=p+1$. When $q=p+1$, it can be shown that

$$N_2 = \tfrac{1}{2}N_1^2 - U_2, \quad N_3 = N_1^3/6 - N_1U_2 - U_3,$$

$$\begin{aligned}
U_2 &= (16)^{-1}(B_1 - C_1)(8B_2 - 5B_1^2 - C_1^2 - 2B_1C_1 + 6B_1 + 2C_1 - 3/4) \\
&\quad + (4)^{-1}(B_2 - C_2)(B_1 + C_1 - 2) - (2)^{-1}(B_3 - C_3), \\
U_3 &= (192)^{-1}(B_1 - C_1)(C_1^3 + 5B_1C_1^2 + 35B_1^2C_1 - 105B_1^3 + 236B_1^2 \\
&\quad + 160B_1B_2 - 24B_2C_1 - 8C_1C_2 - 40B_1C_1 - 4C_1^2 - 192B_2 \\
&\quad - 64B_3 - 291B_1/2 - C_1/2 + 9) + (24)^{-1}(B_2 - C_2)(2C_2 - 10B_2 \\
&\quad + 6C_1 - 30B_1 - 7B_1C_1 + 15B_1^2 + 73/4) \\
&\quad + (6)^{-1}(B_3 - C_3)(C_1 - 3B_1 + 6) + (3)^{-1}(B_4 - C_4) + 21/1024,
\end{aligned} \tag{23}$$

As remarked N_1 is given by (4,5) with $q=p+1$. Also B_i and C_i, $i=1,2$, are as in (4,5). To complete the definition of U_2 and U_3, we need the definitions

$$B_3 = \sum_{r=3}^{p} \sum_{s=2}^{r-1} \sum_{t=1}^{s-1} \alpha_r\alpha_s\alpha_t, \qquad C_3 = \sum_{r=3}^{p+1} \sum_{s=2}^{r-1} \sum_{t=1}^{s-1} \rho_r\rho_s\rho_t. \tag{24}$$

CASE 1. $p = 0$. If $\rho_1 = \nu + 1$,

$$d_k = \frac{(\tfrac{1}{2} + \nu)_k (\tfrac{1}{2} - \nu)_k}{2^k k!}. \tag{25}$$

See the comment following (37).

CASE 2. $p = 1$.

$$\begin{aligned}
2(k+1)\, d_{k+1} &= [3k^2 + 2k(1 + C_1 - 3B_1) + 4N_1]\, d_k \\
&\quad - (k - 2\gamma - 1)(k - 2\gamma + 1 - 2\rho_1)(k - 2\gamma + 1 - 2\rho_2)\, d_{k-1},
\end{aligned} \tag{26}$$

CASE 3. $p = 2$.

$$\begin{aligned}2(k+1)\,d_{k+1} = {} & [5k^2 + 2k(3 + B_1 - 3C_1 - 10\gamma) + 4N_1]\,d_k \\ & - [4k^3 - 6k^2(C_1 + 4\gamma) + 2k(24\gamma^2 + 12\gamma C_1 + C_1 + 4C_2 - 1) \\ & - 32\gamma^3 - 24\gamma^2 C_1 - 4\gamma(C_1 + 4C_2 - 1) + 2C_1 - 4C_2 - 8C_3 - 1]\,d_{k-1} \\ & + (k - 2\gamma - 2)(k - 2\gamma - 2\rho_1)(k - 2\gamma - 2\rho_2)(k - 2\gamma - 2\rho_3)\,d_{k-2},\end{aligned} \tag{27}$$

When β=3, the complete asymptotic expansion is

$${}_pF_{p+2}\left(\begin{matrix}\alpha_p \\ \rho_{p+2}\end{matrix}\middle|\, -z\right) \sim \frac{\Gamma(\rho_{p+2})}{\Gamma(\alpha_p)}\{U(z) + V(z) + L_{p,p-2}(z)\}, \qquad |z| \to \infty, \tag{28}$$

$$U(z) = K_{p,p+2}(ze^{i\pi}) + K_{p,p+2}(ze^{-i\pi}),$$

and either

$$V(z) = V_1(z) = \Gamma^{1,p}_{p+3}(1)\,K_{p,p+2}(ze^{-3i\pi}),$$

or (29)

$$V(z) = V_2(z) = \bar{\Gamma}^{1,p}_{p+3}(1)\,K_{p,p+2}(ze^{3i\pi}),$$

according as $\delta - 3\pi/2 \leqslant \arg z \leqslant 2\pi - \delta$ or $\delta - 2\pi \leqslant \arg z \leqslant 3\pi/2 - \delta$, respectively, $\delta > 0$. The apparent discrepancy in (15) when $|\arg z| < 3\pi/2$ is a case of Stoke's phenomenon. It is readily shown that

$$V_1(z) = \frac{[\sum_{j=1}^{p} \exp(-2i\pi\alpha_j) - \sum_{j=1}^{p+2} \exp(-2i\pi\rho_j)]}{(2\pi)\,3^{1/2}} \exp\{-3z^{1/3}\}\, z^{\gamma} e^{-3i\pi\gamma} \times \sum_{r=0}^{\infty} (-)^r N_r z^{-r/3}. \tag{30}$$

When $\beta = 4$,

$${}_pF_{p+3}\left(\begin{matrix}\alpha_p \\ \rho_{p+3}\end{matrix}\middle|\, -z\right) \sim \frac{\Gamma(\rho_{p+3})}{\Gamma(\alpha_p)}\{A(z) + B(z) + L_{p,p+3}(z)\},$$

$$|z| \to \infty, \qquad |\arg z| \leqslant 2\pi - \delta, \qquad \delta > 0, \tag{31}$$

$$A(z) = K_{p,p+3}(ze^{i\pi}) + K_{p,p+3}(ze^{-i\pi}),$$

$$\begin{aligned}B(z) &= \Gamma^{1,p}_{p+4}(1)\,K_{p,p+3}(ze^{-3i\pi}) + \bar{\Gamma}^{1,p}_{p+4}(1)\,K_{p,p+3}(ze^{3i\pi}) \\ &= \frac{\exp\{-4z^{1/4}\cos\pi/4\}\,z^{\gamma}}{(2\pi)^{3/2}} \sum_{r=0}^{\infty} (-)^r N_r z^{-r/4} b_r(z),\end{aligned}$$

$$\begin{aligned}b_r(z) = {} & \left\{\sum_{j=1}^{p} \cos 2\pi\alpha_j - \sum_{j=1}^{p+3} \cos 2\pi\rho_j\right\} \cos((\pi r/4) + 3\pi\gamma + 4z^{1/4}\sin\pi/4) \\ & - \left\{\sum_{j=1}^{p} \sin 2\pi\alpha_j - \sum_{j=1}^{p+3} \cos 2\pi\rho_j\right\} \sin((\pi r/4) + 3\pi\gamma + 4z^{1/4}\sin\pi/4).\end{aligned} \tag{32}$$

This concludes our discussion for the case $0 \le p \le q-1$. We now turn to the case $p=q$.

$$ {}_pF_p\left(\begin{matrix}\alpha_p\\ \rho_p\end{matrix}\middle| -z\right) \sim \frac{\Gamma(\rho_p)}{\Gamma(\alpha_p)}\{K_{p,p}(ze^{-i\pi}) + L_{p,p}(z)\}, $$

$$ |z| \to \infty, \qquad \delta - \pi/2 \leqslant \arg z \leqslant 3\pi/2 - \delta, \qquad \delta > 0. \tag{33} $$

$$ {}_pF_p\left(\begin{matrix}\alpha_p\\ \rho_p\end{matrix}\middle| -z\right) \sim \frac{\Gamma(\rho_p)}{\Gamma(\alpha_p)}\{K_{p,p}(ze^{i\pi}) + L_{p,p}(z)\}, $$

$$ |z| \to \infty, \qquad \delta - 3\pi/2 \leqslant \arg z \leqslant \pi/2 - \delta, \qquad \delta > 0. \tag{34} $$

We may also write

$$ {}_pF_p\left(\begin{matrix}\alpha_p\\ \rho_p\end{matrix}\middle| z\right) \sim \frac{\Gamma(\rho_p)}{\Gamma(\alpha_p)}\{K_{p,p}(z) + L_{p,p}(ze^{i\epsilon\pi})\}, \qquad |z| \to \infty, $$

$$ \delta - (2+\epsilon)\,\pi/2 \leqslant \arg z \leqslant (2-\epsilon)\,\pi/2 - \delta, \qquad \epsilon = \pm 1, \qquad \delta > 0. \tag{35} $$

Let

$$ K_{p,p}(2z) = e^{2z}(2z)^{\gamma} \sum_{k=0}^{\infty} d_k z^{-k}, \qquad d_0 = 1, \qquad \gamma = \sum_{h=1}^{p} (\alpha_h - \rho_h) = B_1 - C_1. \tag{36} $$

In the sequel, we give the recursion formulas for d_k for the cases $p = 1$ and $p = 2$.

CASE 1. $p = q = 1$.

$$ d_k = \frac{(\rho_1 - \alpha_1)_k\,(1-\alpha_1)_k}{2^k k!}. \tag{37} $$

Note that if $\rho_1 = 2\alpha_1 = 2\nu + 1$, then (37) and (25) are the same and the d_k's are coefficients associated with the asymptotic expansion of $I_\nu(z)$, see 9.5(7).

CASE 2. $p = q = 2$.

$$ \begin{aligned} 4(k+1)\,d_{k+1} &= 2[2k^2 - k(2\gamma + B_1 - 1) + N_1]\,d_k \\ &\quad - (k-1-\gamma)(k-\rho_1-\gamma)(k-\rho_2-\gamma)\,d_{k-1}, \end{aligned} \tag{38} $$

where N_1 is given by (5), with $p = q = 2$ and $\beta = 1$. If $\alpha_2 = 1$, $\rho_1 = \alpha_1 + \nu + \frac{1}{2}$, $\rho_2 = \alpha_1 - \nu + \frac{1}{2}$, then d_k is given by (25).

If one of the numerator terms is unity, say $\alpha_1=1$, then Kim (1972) has shown that with $\alpha_2=\alpha$,

$$ K_{2,2}(2z) = e^{2z}(2z)^{\gamma}\ {}_2F_0(\rho_1-\alpha, \rho_2-\alpha; 1/2z). \tag{39} $$

A closed form expression for the d_k's for general p has been given by Joshi and McDonald (1972).

5.10. Expansions in Series of Generalized Jacobi, Generalized Laguerre and Chebyshev Polynomials

5.10.1. Expansions for G-Functions

In this section we delineate important expansion theorems for G-functions in series of other G-functions whose coefficients are extended Jacobi or extended Laguerre polynomials. The corresponding expansions for ${}_pF_q$'s are given in 5.10.2. The results include so many special cases and works of other authors that we make no attempt to supply much of a bibliography. In this connection, see Luke (1969) and the references given there.

Theorem 1. *Assumptions*:

(1) *Let* $t \geqslant 1, 0 \leqslant m \leqslant q, 0 \leqslant k \leqslant p, q + s \geqslant 1, z \neq 0$.

(2) *Let none of the following quantities be negative integers*: b_h ; $\alpha_j + b_h$, $j = 1, 2,\ldots, t$; $b_h - a_j$, $j = 1, 2,\ldots, k$. *Here and in what follows* $h = 1, 2,\ldots, m$.

(3) *Let* $-R(b_h) < \min(\alpha_1, \frac{1}{2}\alpha_1 + \frac{1}{4})$, $\alpha_1 = \beta + 1$, $\lambda = \alpha + \beta + 1$, $\alpha > -1$, $\beta > -1$.

(4) *Let the following inequalities be satisfied.*

$$R(1 + b_h - c_j) > 0, \quad j = 1, 2,\ldots, r; \qquad R(\beta_j + b_h) > 0, \quad j = 1, 2,\ldots, u;$$

$$R\left\{\sum_{j=1}^{s} d_j - \sum_{j=1}^{r} c_j + \sum_{j=1}^{u} \beta_j - \sum_{j=1}^{t} \alpha_j - 2b_h\right\} < s - r + \tfrac{1}{2}.$$

(5) *Let*

(a) $p + r \leqslant q + s - 1$ *and* $r + u + 1 = s + t$ (*which implies* $p + t \leqslant q + u$), $|z| < \infty$, $0 < \omega < 1$; *or*

(b) $p + r = q + s$ *and* $r + u + 1 = s + t$ (*which implies* $p + t = q + u + 1$), $z \neq (-)^{m+k-p}$, $|\arg(1 - (-)^{m+k-p}z)| < \pi$, $0 < \omega < 1$.

Then

$$G_{p+r,q+s}^{m,k+r}\left(z\omega \,\middle|\, \begin{matrix} c_r, a_p \\ b_q, d_s \end{matrix}\right) = \frac{\Gamma(1-c_r)\Gamma(\beta_u)}{\Gamma(\alpha_t)\Gamma(1-d_s)} \sum_{n=0}^{\infty} \frac{(-)^n(2n+\lambda)\Gamma(n+\lambda)}{n!}$$
$$\times\, G_{p+t+1,q+u+2}^{m,k+t+1}\left(z \,\middle|\, \begin{matrix} 0, 1-\alpha_t, a_p \\ b_q, 1-\beta_u, -n-\lambda, n \end{matrix}\right)$$
$$\times\, {}_{r+u+2}F_{s+t}\left(\begin{matrix} -n, n+\lambda, 1-c_r, \beta_u \\ \alpha_t, 1-d_s \end{matrix} \,\middle|\, \omega\right). \tag{1}$$

REMARK 1. The inequality involving b_h in assumption (3) is omitted unless the ${}_{r+u+2}F_{s+t}$ polynomial in (2) reduces to a ${}_2F_1$. If such a reduction takes place, assumption (4) is omitted.

REMARK 2. In the above and all similar type theorems the given conditions are only sufficient.

When $r=s=u=0$, $t=1$, and $\alpha_1=\beta+1$, Theorem 1 reads as follows.

Theorem 2. *Assumptions:*

(1) *Let* $0\leq m\leq q$, $0\leq k\leq p$, $q\geq 1$, $z\neq 0$.

(2) *Let none of the following quantities be a negative integer:* b_h, $\beta+1+b_h$, b_h-a_j, $j=1,2,\ldots,k$ *where here and in the sequel* $h=1,2,\ldots,m$.

(3) *Let* $-R(b_h)<\min(\beta+1,\ \frac{1}{2}\beta+3/4)$, $\beta>-1$.

(4) (a) $p\leq q-1$, $|z|<\infty$, $0<\omega<1$ *or*

(b) $p=q$, $z\neq(-)^{m+k-p}$, $\left|\arg\left(1-(-)^{m+k-p}z\right)\right|<\pi$, $0<\omega<1$.

Then

$$G_{p,q}^{m,k}\left(z\omega\,\middle|\,\begin{matrix}a_p\\ b_q\end{matrix}\right)=\sum_{n=0}^{\infty}\frac{(2n+\lambda)\Gamma(n+\lambda)}{\Gamma(n+\beta+1)}G_{p+2,q+2}^{m,k+2}\left(z\,\middle|\,\begin{matrix}0,\,-\beta,\,a_p\\ b_q,\,n,\,-n-\lambda\end{matrix}\right)R_n^{(\alpha,\beta)}(\omega)\,. \tag{2}$$

Theorem 3. *Assumptions:*

(1) *Let* $0\leqslant m\leqslant q$, $0\leqslant k\leqslant p$, $\delta=m+k-\frac{1}{2}(p+q)$.

(2) *Let* $\alpha>-1$, $\beta>-1$, $\lambda=\alpha+\beta+1$.

(3) *Suppose* b_j-a_h *is not a negative integer,* $j=1,2,\ldots,m$, $h=1,2,\ldots,k$.

(4) *Let* $|\arg z|<\delta\pi$, $\delta>0$, $0\leqslant\omega\leqslant 1$, *and* $|\arg(1-(-)^{\delta}z)|<\pi$ *if* $p=q$.

Then the expansion (2) *is valid. The result is also true for* $|\arg z|=\delta\pi$, $\delta\geq 0$ *provided* $p\leq q$ *and*

$$R\{\nu-(q-p)b_j\}<\tfrac{1}{2}(q-p)-g-2,\qquad j=1,2,\ldots,m,$$

$$g=\max(\alpha,\beta,-\tfrac{1}{2}),\qquad \nu=\sum_{j=1}^{q}b_j-\sum_{j=1}^{p}a_j\,. \tag{3}$$

Theorem 4. *Assumptions:*

(1) *Let* $0 \leqslant m \leqslant q$, $0 \leqslant k \leqslant p$, $q + s \geqslant 1$, $z \neq 0$.

(2) *Let none of the following quantities be negative integers*: b_h ; $\alpha_j + b_h$, $j = 1, 2,..., t$; $b_h - a_j$, $j = 1, 2,..., k$. *Here and in what follows*, $h = 1, 2,..., m$.

(3) *Let the following inequalities be satisfied.*

$$R(1 + b_h - c_j) > 0, \quad j = 1, 2,..., r; \qquad R(\beta_j + b_h) > 0, \quad j = 1, 2,..., u,$$

$$R\left\{\sum_{j=1}^{s} d_j - \sum_{j=1}^{r} c_j + \sum_{j=1}^{u} \beta_j - \sum_{j=1}^{t} \alpha_j - 2b_h\right\} < s - r + \tfrac{1}{2}.$$

(4) *Let*

(a) $p + r < q + s - 1$ *and* $r + u + 1 = s + t$ (*which implies* $p + t < q + u$), $|z| < \infty$, $0 < \omega < \infty$; *or*

(b) $p + r = q + s - 1$ *and* $r + u + 1 = s + t$ (*which implies* $p + t = q + u$), $(-)^{m+k-p} R(z) < \frac{1}{2}$, $0 < \omega < \infty$; *or*

(c) $p + r = q + s - 1$ *and* $r + u + 1 = s + t$ (*which implies* $p + t = q + u$), $m + k - p \geqslant 1$, $z \neq 0$, $|\arg z| \leqslant (m + k - p - \frac{1}{2})\pi$, $0 < \omega < \infty$; *or*

(d) $p + r < q + s$ *and* $r + u = s + t$ (*which implies* $p + t < q + u$), $|z| < \infty$, $|\omega - 1| < 1$; *or*

(e) $p + r = q + s$ *and* $r + u = s + t$ (*which implies* $p + t = q + u$), $(-)^{m+k-p} R(z) < \frac{1}{2}$, $|\omega - 1| < 1$; *or*

(f) $p = q$, $r = s = t = u = 0$, $(-)^{m+k-p} z \neq 1$, $|\omega - 1| < 1$ *if* $(-)^{m+k-p} R(z) \leqslant \frac{1}{2}$, $|\omega - 1| < |1 - (-)^{m+k-p}/z|$ *if* $\frac{1}{2} \leqslant (-)^{m+k-p} R(z) \leqslant 1$, $|\omega - 1| < |I(z)|/|z|$ *if* $(-)^{m+k-p} R(z) \geqslant 1$; *or*

(g) $p = q$, $r = s = t = u = 0$, $m + k - p \geqslant 1$, $|\arg z| \leqslant (m + k - p - \frac{1}{2})\pi$, $|\omega - 1| < 1$.

In the case of conditions 4(d)–(g), *the third inequality in assumption* (3) *may be omitted. Then*

$$G_{p+r,q+s}^{m,k+r}\left(z\omega \,\middle|\, \begin{matrix} c_r, a_p \\ b_q, d_s \end{matrix}\right) = \frac{\Gamma(1 - c_r)\Gamma(\beta_u)}{\Gamma(1 - d_s)\Gamma(\alpha_t)} \sum_{n=0}^{\infty} \frac{(-)^n}{n!}$$

$$\times\, G_{p+t+1,q+u+1}^{m,k+t+1}\left(z \,\middle|\, \begin{matrix} 0, 1 - \alpha_t, a_p \\ b_q, 1 - \beta_u, n \end{matrix}\right)$$

$$\times\, {}_{r+u+1}F_{s+t}\left(\begin{matrix} -n, 1 - c_r, \beta_u \\ \alpha_t, 1 - d_s \end{matrix} \,\middle|\, \omega\right). \tag{4}$$

REMARK. Theorem 4 is a confluent form of Theorem 1. The polynomial in (4) is, except for a factor, the Laguerre polynomial when $r=u=s=0$ and $t=1$.

Theorem 5.

$$G_{p,q}^{m,k}\left(z\omega \,\middle|\, \begin{matrix} a_p \\ b_q \end{matrix}\right) = \sum_{n=0}^{\infty} \frac{(2n+\lambda)\Gamma(n+\lambda)}{\Gamma(n+\beta+1)}$$

$$\times\, G_{p+2,q+2}^{m+2,k}\left(z \,\middle|\, \begin{matrix} a_p\,,\, 1-n,\, n+\lambda+1 \\ 1,\, \beta+1,\, b_q \end{matrix}\right) R_n^{(\alpha,\beta)}(1/\omega)$$

$$= \sum_{n=0}^{\infty} \frac{(-)^n(2n+\lambda)\Gamma(n+\lambda)}{\Gamma(n+\beta+1)}$$

$$\times\, G_{p+2,q+2}^{m+1,k+1}\left(z \,\middle|\, \begin{matrix} 1-n,\, a_p\,,\, n+\lambda+1 \\ \beta+1,\, b_q\,,\, 1 \end{matrix}\right) R_n^{(\alpha,\beta)}(1/\omega), \quad (5)$$

valid under the same conditions as for Theorem 3 except that $1 \leqslant \omega \leqslant \infty$. *If* $|\arg z| = \delta\pi$, $\delta \geqslant 0$, (5) *is also valid provided further that* $q \geqslant p$ *and* (3) *is replaced by*

$$R\{\nu - (q-p)a_h\} < \tfrac{1}{2}(p-q) - g - 2, \qquad h = 1, 2, \ldots, k,$$

$$g = \max(\alpha, \beta, -\tfrac{1}{2}). \quad (6)$$

When $k = 1$, (5) can be proved under a different set of conditions.

Theorem 6. *Assumptions:*

(1) *Let* $0 \leqslant m \leqslant q$, $p \geqslant 0$, $q > p+1$.
(2) *Let* $b_j\,,\, \beta+1 \neq 0, -1, -2, \ldots, j = 1, 2, \ldots, q$.
(3) *Let* $m > \tfrac{1}{2}(p+q-1)$, $|\arg z| < \pi(m + \tfrac{1}{2}(1-p-q))$, $z \neq 0$.
(4) *Let* $1 \leqslant \omega \leqslant \infty$.

Then

$$G_{p+1,q}^{m,1}\left(z\omega \,\middle|\, \begin{matrix} 1, a_p \\ b_q \end{matrix}\right) = \sum_{n=0}^{\infty} \frac{(2n+\lambda)\Gamma(n+\lambda)}{\Gamma(n+\beta+1)} \Phi_n^{(m)}(z) R_n^{(\alpha,\beta)}(1/\omega), \quad (7)$$

$$\Phi_n^{(m)}(z) = G_{p+3,q+2}^{m+2,1}\left(z \,\middle|\, \begin{matrix} 1, a_p\,,\, 1-n,\, n+\lambda+1 \\ 1,\, \beta+1,\, b_q \end{matrix}\right). \quad (8)$$

Theorem 6 is valuable for the applications as will be seen from the discussion surrounding (16) below. But first it is informative to quote a result which depicts the asymptotic behaviour of (8) when n is large. For further details in connection with the following theorem, see Wimp (1967), Luke (1969) and Fields (1973).

Theorem 7. *Let*

(1) $$d_1 = \frac{(2\beta + 2\nu + 2 + p - q)}{2(q-p+1)},$$

(2) $$d_2 = \frac{2[\beta + \nu - \lambda - 1/2 - (\lambda/2 + 1)(q - p - 1)]}{(q - p + 1)},$$

(3) $$\nu = \sum_{j=1}^{q} b_j - \sum_{j=1}^{p} a_j ,$$

(4) $$\delta = \min\left(\frac{2}{q - p + 1}, \frac{q - p - 1}{q - p + 1}\right),$$

(5) $$V_n(x) = \exp\left\{-n\left[(q - p + 1)x^{-\rho/2} - \tfrac{1}{3}x^{-3\rho/2} + \frac{(4 + p - q)}{10(q - p + 1)}x^{-5\rho/2} + \dots\right]\right\},$$

(6) $$\rho/2 = (q - p - 1)/(q - p + 1),$$

(7) $\mu_k^{(m)}$ *be the coefficient of* x^k *in the product*

$$\prod_{j=m+1}^{q} [1 - x\exp(2i\pi b_j)],$$

(8) *the assumptions of Theorem* 6 *be satisfied. Then*

$$\Phi_n^{(m)}(z) = \frac{(-)^n(2\pi)^{\frac{1}{2}(q-p)}(-2\pi i)^{m-q} z^{d_1}n^{d_2}\exp(-i\pi[\sum_{j=m+1}^{q} b_j - d_1(q - m)])}{(q - p + 1)^{1/2}}$$

$$\times \sum_{k=0}^{q-m} \mu_k^{(m)} \exp(-2i\pi d_1 k)$$

$$\times V_n(n\{z\exp[i\pi(q - m - 2k)]\}^{-1/(q-p-1)})[1 + O(n^{-\delta})]. \quad (9)$$

Theorem 6 is very important for the applications. Note that $G_{p+1,q}^{m,1}(z\omega)$ has an asymptotic representation in descending powers of $z\omega$. Thus (7) may be viewed as a summation process which connects a generally divergent sequence into a convergent one. As special cases of (7), we get expansions for confluent hypergeometric functions (which include Bessel functions) and Lommel functions about $z = \infty$ in series of descending Jacobi polynomials.

In illustration, from 8.3(3) and Theorem 6, we have

$$(\omega z)^a U(a; c; \omega z) = [\Gamma(a)\Gamma(1 + a - c)]^{-1} G_{1,2}^{2,1}\left(\omega z \,\middle|\, \begin{matrix} 1 \\ a, 1 + a - c \end{matrix}\right), \quad (10)$$

$$(\omega z)^a U(a; c; \omega z) = \sum_{n=0}^{\infty} C_n(z) T_n^*(1/\omega),$$

$$a, 1 + a - c \neq 0, -1, -2,\ldots,$$

$$z \neq 0, \qquad |\arg z| < 3\pi/2, \qquad 1 \leqslant \omega \leqslant \infty, \tag{11}$$

where

$$\begin{aligned} C_n(z) &= \frac{\epsilon_n}{\Gamma(\frac{1}{2})\Gamma(a)\Gamma(1+a-c)} G_{3,4}^{4,1}\left(z \,\middle|\, \begin{matrix} 1, 1-n, n+1 \\ 1, \frac{1}{2}, a, 1+a-c \end{matrix}\right) \\ &= \frac{\epsilon_n(-)^n}{\Gamma(\frac{1}{2})\Gamma(a)\Gamma(1+a-c)} G_{2,3}^{3,1}\left(z \,\middle|\, \begin{matrix} 1-n, n+1 \\ \frac{1}{2}, a, 1+a-c \end{matrix}\right) \\ &= \frac{\epsilon_n(-)^n}{\Gamma(\frac{1}{2})\Gamma(a)\Gamma(1+a-c)} G_{3,2}^{1,3}\left(\frac{1}{z} \,\middle|\, \begin{matrix} \frac{1}{2}, 1-a, c-a \\ n, -n \end{matrix}\right). \end{aligned} \tag{12}$$

Since (11) converges, for z fixed and z as in (11),

$$\lim_{n\to\infty} C_n(z) = 0, \tag{13}$$

and for later use we note that

$$\lim_{\omega\to\infty} (\omega z)^a U(a; c; \omega z) = 1, \qquad |\arg z| < 3\pi/2, \tag{14}$$

which implies that

$$\sum_{n=0}^{\infty} (-)^n C_n(z) = 1, \qquad |\arg z| < 3\pi/2. \tag{15}$$

From 5.13.4, $C_n(z)$ satisfies the recurrence formula

$$\begin{aligned} \frac{2C_n(z)}{\epsilon_n} = 2(n+1) \left\{1 - \frac{(2n+3)(n+a+1)(n+b+1)}{2(n+2)(n+a)(n+b)} \right. \\ \left. - \frac{2z}{(n+a)(n+b)}\right\} C_{n+1}(z) \\ + \left\{1 - \frac{2(n+1)(2n+3-2z)}{(n+a)(n+b)}\right\} C_{n+2}(z) \\ - \frac{(n+1)(n+3-a)(n+3-b)}{(n+2)(n+a)(n+b)} C_{n+3}(z), \\ b = 1 + a - c. \end{aligned} \tag{16}$$

Application of (9) gives

$$C_n(z) = \frac{4(-)^n \pi^{1/2} n^{2(2a-c-1)/3} z^{(4a+1-2c)/6}}{3^{1/2}\Gamma(a)\Gamma(a+1-c)} \exp(-3n^{2/3}z^{1/3})[1 + O(n^{-1/3})],$$
$$|\arg z| < 3\pi/2. \tag{17}$$

Now (16) is a third order difference equation, and according to the Birkhoff–Trjitzinsky (1932) theory of singular difference equations, there exist two other linearly independent solutions of this difference equation which have the behavior

$$\begin{aligned} y_r(n) &= \delta_r(-)^n n^s \exp(-3\omega_r n^{2/3} z^{1/3})[1 + O(n^{-1/3})], \qquad r = 1, 2, \\ s &= 2(2a-c-1)/3, \qquad \omega_r = e^{2i\pi r/3}, \qquad |\arg z| < 3\pi/2, \end{aligned} \tag{18}$$

where δ_r is independent of n. Note that (18) can be used to characterize $C_n(z)$ if we take $r = 0$. In (18), the notation $y_r(n)$ is that of 12.2 with $h = r$. The backward recurrence scheme, 12.2(3-8), applied to (15,16) gives the desired values of $C_n(z)$ for $|\arg z| < \pi$. If $|\arg z| = \pi$, the procedure converges if it is modified by 12.2(16-18). For expansion of $U(a;\, c;\, \omega z)$ in series of $T_{2n}(1/\omega)$, see G. F. Miller (1966a)

For another example of Theorem 6, we consider a G-function which includes as special cases Lommel functions, cosine and sine integrals, Fresnel integrals, and the product of Hankel functions [see 8.3(17,21, 22,31)]. We have

$$G_{1,3}^{3,1}\left(\frac{\omega^2 z^2}{4}\,\middle|\,\begin{matrix}1\\ a, b, c\end{matrix}\right) = \sum_{n=0}^{\infty} A_n(z) T_{2n}(1/\omega),$$

$$\begin{aligned} A_n(z) &= \epsilon_n \pi^{-1/2}\, G_{3,5}^{5,1}\left(\frac{z^2}{4}\,\middle|\,\begin{matrix}1, 1-n, n+1\\ 1, \tfrac{1}{2}, a, b, c\end{matrix}\right) \\ &= \epsilon_n(-)^n \pi^{-1/2}\, G_{2,4}^{4,1}\left(\frac{z^2}{4}\,\middle|\,\begin{matrix}1-n, n+1\\ \tfrac{1}{2}, a, b, c\end{matrix}\right) \\ &= \epsilon_n(-)^n \pi^{-1/2}\, G_{4,2}^{1,4}\left(\frac{4}{z^2}\,\middle|\,\begin{matrix}\tfrac{1}{2}, 1-a, 1-b, 1-c\\ n, -n\end{matrix}\right), \end{aligned}$$

$$z \neq 0, \qquad |\arg z| < \pi, \qquad 1 \leqslant \omega \leqslant \infty, \tag{19}$$

provided further that none of the numbers a, b, c is a negative integer or zero. We also have the normalization relation

$$\sum_{n=0}^{\infty} (-)^n A_n(z) = \Gamma(a)\Gamma(b)\Gamma(c), \qquad |\arg z| < \pi. \tag{20}$$

Under the same conditions as for (19), from (9)

$$A_n(z) = \frac{(-)^n 2^{3/2}\pi}{n}\left(\frac{nz}{2}\right)^{\frac{1}{2}(a+b+c-1)} \exp\left[-4\left(\frac{nz}{2}\right)^{1/2}\right][1 + O(n^{-1/2})]. \tag{21}$$

From the developments in 5.13.4, $A_n(z)$ satisfies a fourth order difference equation. Again, by the Birkhoff–Trjitzinsky (1932) theory, there exist three other linearly independent solutions of this difference equation which have the behavior

$$\eta_n^{(r)}(z) = \delta_r(-)^n n^s \exp[-4\omega_r(nz/2)^{1/2}[1 + O(n^{-1/2})], \qquad r = 1, 2, 3,$$
$$s = \tfrac{1}{2}(a + b + c - 1), \qquad \omega_r = e^{i\pi r/2} \qquad |\arg z| < \pi, \tag{22}$$

where δ_r is independent of n. Note that with $r = 0$, (22) characterizes $A_n(z)$. Just as in the developments surrounding (19) we can show that the backward recurrence scheme described by 12.2(3-8) when applied to the difference equation for $A_n(z)$ and (32) produces the desired values of $A_n(z)$ for $|\arg z| < \pi/2$. If $|\arg z| = \pi/2$, then the procedure also converges provided it is modified according to the provisions of 12.2(16-18).

5.10.2. EXPANSIONS FOR ${}_pF_q$

Throughout this section

$$\lambda = \alpha + \beta + 1, \qquad \alpha > -1, \qquad \beta > -1, \qquad \lambda \text{ not a negative integer}, \tag{1}$$

$$\epsilon_0 = 1, \qquad \epsilon_n = 2 \qquad \text{for} \quad n \geqslant 1. \tag{2}$$

All the results in this section are special cases of those given in 5.10.1.

Theorem 1. *Assumptions:*

(1) *Let* $t \geqslant 1$, $\alpha_1 = \beta + 1$, $\lambda = \alpha + \beta + 1$, *where* λ *is not a negative integer.*

(2) *Let*

(a) $p + r \leqslant q + s$ *and* $r + u + 1 = s + t$ (*which implies* $p + t \leqslant q + u + 1$), $|z| < \infty$, $|\omega| < \infty$; *or*

(b) $p + r = q + s + 1$ *and* $r + u + 1 = s + t$ (*which implies* $p + t = q + u + 2$), $z \neq 1$, $|\arg(1 - z)| < \pi$, $0 \leqslant \omega \leqslant 1$; *or*

(c) $p + r \leqslant q + s$ *and* $s + t \leqslant r + u$ (*which implies* $p + t \leqslant q + u$), $|z| < \infty$, $|\omega| < \infty$; *or*

(d) $p + r = q + s + 1$ *and* $s + t \leqslant r + u$ (*which implies* $p + t \leqslant q + u + 1$), $z\omega \neq 1$, $|\arg(1 - z\omega)| < \pi$; *or*

(e) $r + u + 2 \leqslant s + t$ *and* $p + t \leqslant q + u + 1$ (*which implies* $p + r + 1 \leqslant q + s$), $|z| < \infty$, $|\omega| < \infty$; *or*

(f) $r + u + 2 \leqslant s + t$ *and* $p + t = q + u + 2$ (*which implies* $p + r \leqslant q + s$), $z \neq 1$, $|\arg(1 - z)| < \pi$, $\omega \geqslant 0$.

Then

$$
{}_{p+r}F_{q+s}\left(\begin{matrix}a_p\,,\,c_r\\ b_q\,,\,d_s\end{matrix}\middle|\, z\omega\right) = \sum_{n=0}^{\infty} \frac{(-)^n(\alpha_t)_n(a_p)_n z^n}{n!\,(\beta_u)_n(b_q)_n(n+\lambda)_n}
$$

$$
\times\ {}_{p+t}F_{q+u+1}\left(\begin{matrix}\alpha_t+n,\, a_p+n\\ \lambda+1+2n,\, \beta_u+n,\, b_q+n\end{matrix}\middle|\, z\right)
$$

$$
\times\ {}_{r+u+2}F_{s+t}\left(\begin{matrix}-n,\, n+\lambda,\, c_r\,,\, \beta_u\\ \alpha_t\,,\, d_s\end{matrix}\middle|\, \omega\right). \tag{3}
$$

Theorem 2. *Assumptions: Let*

(a) $p+r \leqslant q+s$ *and* $r+u+1 = s+t$ (*which implies* $p+t \leqslant q+u$), $|z| < \infty$, $|\omega| < \infty$; *or*

(b) $p+r = q+s$ *and* $r+u+1 = s+t$ (*which implies* $p+t = q+u+1$), $R(z) < \frac{1}{2}$, $|\omega| < \infty$; *or*

(c) $p+r \leqslant q+s$ *and* $r+u = s+t$ (*which implies* $p+t \leqslant q+u$), $|z| < \infty$, $|\omega| < \infty$; *or*

(d) $p+r = q+s+1$ *and* $r+u = s+t$ (*which implies* $p+t = q+u+1$), $R(z) < \frac{1}{2}$, $|1-\omega| < |1-1/z|$; *or*

(e) $u = t = 0$, $p = q+1$ *and* $r+1 < s$ (*which implies* $p+r < q+s$), $R(z) < \frac{1}{2}$, $|\omega| < \infty$; *or*

(f) $r = s = t = u = 0, p = q+1, z \neq 1$, $|1-\omega| < |1-1/z|$ *if* $R(z) \leqslant 1$, $|\omega - 1| < |I(z)|/|z|$ *if* $R(z) \geqslant 1$; *or*

(g) $p+r \leqslant q+s$ *and* $s+t \leqslant r+u-1$ (*which implies* $p+t \leqslant q+u$), $|z| < \infty$, $|\omega| < \infty$; *or*

(h) $p+r = q+s+1$ *and* $s+t \leqslant r+u-1$ (*which implies* $p+t \leqslant q+u$), $|z\omega| < 1$.

(i) $p + t \leq q + u$, $r + u < s + t$ (*which implies* $p + r < q + s$) $|z| < \infty$, $|\omega| < \infty$, $0 < 1 + Min(R\ a_h, c_i, \alpha_j, \beta_k)$, $h = 2,3,\ldots p$; i, j, *and* $k = 1,2,\ldots,r$, t *and* u *respectively.*

Then

$$
{}_{p+r}F_{q+s}\left(\begin{matrix}a_p\,,\,c_r\\ b_q\,,\,d_s\end{matrix}\middle|\, z\omega\right) = \sum_{n=0}^{\infty} \frac{(-)^n(\alpha_t)_n(a_p)_n z^n}{n!\,(\beta_u)_n(b_q)_n}\ {}_{p+t}F_{q+u}\left(\begin{matrix}\alpha_t+n,\, a_p+n\\ \beta_u+n,\, b_q+n\end{matrix}\middle|\, z\right)
$$

$$
\times\ {}_{r+u+1}F_{s+t}\left(\begin{matrix}-n,\, c_r\,,\, \beta_u\\ \alpha_t\,,\, d_s\end{matrix}\middle|\, \omega\right). \tag{4}
$$

Conditions (a)-(h) are given by Luke (1969) while conditions (i) are due to Diaz and Osler (1974).

The following equations are special cases of Theorem 1.

$$
{}_pF_q\left(\begin{matrix}a_p\\b_q\end{matrix}\middle| z\omega\right) = \sum_{n=0}^{\infty} C_n R_n^{(\alpha,\beta)}(\omega),
$$

$$
C_n = \frac{(a_p)_n z^n}{(b_q)_n(n+\lambda)_n}\, {}_{p+1}F_{q+1}\left(\begin{matrix}\beta+1+n,\, a_p+n\\ \lambda+1+2n,\, b_q+n\end{matrix}\middle| z\right),
$$

$$
p \leqslant q, \quad |z| < \infty, \quad |\omega| < \infty;
$$

$$
p = q+1, \quad z \neq 1, \quad |\arg(1-z)| < \pi, \quad 0 \leqslant \omega \leqslant 1. \tag{5}
$$

$$
{}_pF_q\left(\begin{matrix}a_p\\b_q\end{matrix}\middle| z(1+\epsilon\omega)\right) = \sum_{n=0}^{\infty} C_n P_n^{(\alpha,\beta)}(\omega), \qquad \epsilon = \pm 1,
$$

$$
C_n = \frac{(a_p)_n (2z)^n \epsilon^n}{(b_q)_n(n+\lambda)_n}\, {}_{p+1}F_{q+1}\left(\begin{matrix}\dfrac{\lambda+1}{2}+\dfrac{\epsilon(\beta-\alpha)}{2}+n,\, a_p+n\\ \lambda+1+2n,\, b_q+n\end{matrix}\middle| 2z\right),
$$

$$
p \leqslant q, \quad |z| < \infty, \quad |\omega| < \infty;
$$

$$
p = q+1, \quad z \neq \tfrac{1}{2}, \quad |\arg(1-2z)| < \pi, \quad -1 \leqslant \omega \leqslant 1. \tag{6}
$$

$$
x^{\epsilon}\, {}_pF_q\left(\begin{matrix}a_p\\b_q\end{matrix}\middle| zx^2\right) = \sum_{n=0}^{\infty} C_n P_{2n+\epsilon}^{(\alpha,\alpha)}(x), \qquad \epsilon = 0, 1,
$$

$$
C_n = \frac{(a_p)_n (2n+\epsilon)!\,(\alpha+1)_n z^n}{(b_q)_n n!\,(n+\alpha+\frac{1}{2}+\epsilon)_n(\alpha+1)_{2n+\epsilon}}
$$

$$
\times\, {}_{p+1}F_{q+1}\left(\begin{matrix}\frac{1}{2}+\epsilon+n,\, a_p+n\\ \frac{3}{2}+\epsilon+\alpha+2n,\, b_q+n\end{matrix}\middle| z\right),
$$

$$
p \leqslant q, \quad |z| < \infty, \quad |x| < \infty;
$$

$$
p = q+1, \quad z \neq 1, \quad |\arg(1-z)| < \pi, \quad -1 \leqslant x \leqslant 1. \tag{7}
$$

$$
{}_pF_q\left(\begin{matrix}a_p\\b_q\end{matrix}\middle| z\omega\right) = \sum_{n=0}^{\infty} C_n P_n^{(\alpha,\alpha)}(\omega),
$$

$$
C_n = \frac{(a_p)_n (2z)^n}{(b_q)_n(n+2\alpha+1)_n}\, {}_{2p}F_{2q+1}\left(\begin{matrix}\dfrac{a_p+n}{2},\, \dfrac{a_p+1+n}{2}\\ \frac{3}{2}+\alpha+n,\, \dfrac{b_q+n}{2},\, \dfrac{b_q+1+n}{2}\end{matrix}\middle| \frac{z^2}{4^{q+1-p}}\right),
$$

$$
p \leqslant q, \quad |z| < \infty, \quad |\omega| < \infty;
$$

$$
p = q+1, \quad z^2 \neq 1, \quad |\arg(1-z^2)| < \pi, \quad -1 \leqslant \omega \leqslant 1. \tag{8}
$$

$$x^{\delta}\,{}_pF_q\left(\begin{matrix}a_p\\b_q\end{matrix}\,\middle|\, zx^2\right) = \sum_{n=0}^{\infty} B_n T_{2n+\delta}(x), \qquad \delta = 0, 1,$$

$$B_n = \frac{[\epsilon_n(1-\delta)+\delta](a_p)_n(z/4)^n}{(b_q)_n n!}\,{}_{p+1}F_{q+1}\left(\begin{matrix}\frac{1}{2}+\delta+n,\, a_p+n\\1+\delta+2n,\, b_q+n\end{matrix}\,\middle|\, z\right),$$

$$p \leqslant q, \quad |z| < \infty, \quad |x| < \infty;$$

$$p = q+1, \quad z \neq 1, \quad |\arg(1-z)| < \pi, \quad -1 \leqslant x \leqslant 1. \tag{9}$$

$${}_pF_q\left(\begin{matrix}a_p\\b_q\end{matrix}\,\middle|\, z\omega\right) = \sum_{n=0}^{\infty} B_n T_n^*(\omega),$$

$$B_n = \frac{\epsilon_n(a_p)_n(z/4)^n}{(b_q)_n n!}\,{}_{p+1}F_{q+1}\left(\begin{matrix}\frac{1}{2}+n,\, a_p+n\\1+2n,\, b_q+n\end{matrix}\,\middle|\, z\right),$$

$$p \leqslant q, \quad |z| < \infty, \quad |\omega| < \infty;$$

$$p = q+1, \quad z \neq 1, \quad |\arg(1-z)| < \pi, \quad 0 \leqslant \omega \leqslant 1. \tag{10}$$

$${}_pF_q\left(\begin{matrix}a_p\\b_q\end{matrix}\,\middle|\, z\omega\right) = \sum_{n=0}^{\infty} B_n T_n(\omega),$$

$$B_n = \frac{\epsilon_n(a_p)_n(z/2)^n}{(b_q)_n n!}\,{}_{2p}F_{2q+1}\left(\begin{matrix}\dfrac{a_p+n}{2},\, \dfrac{a_p+1+n}{2}\\1+n,\, \dfrac{b_q+n}{2},\, \dfrac{b_q+1+n}{2}\end{matrix}\,\middle|\, \frac{z^2}{4^{q+1-p}}\right),$$

$$p \leqslant q, \quad |z| < \infty, \quad |\omega| < \infty;$$

$$p = q+1, \quad z^2 \neq 1, \quad |\arg(1-z^2)| < \pi, \quad -1 \leqslant \omega \leqslant 1. \tag{11}$$

5.11. Expansions in Series of Bessel Functions

$${}_rF_s\left(\begin{matrix}c_r\\d_s\end{matrix}\,\middle|\, \frac{z^2\omega^2}{4}\right) = \left(\frac{2}{z}\right)^{\lambda} \sum_{n=0}^{\infty} \frac{(-)^n(2n+\lambda)\Gamma(n+\lambda)}{n!} I_{2n+\lambda}(z)$$

$$\times\, {}_{r+2}F_s\left(\begin{matrix}-n,\, n+\lambda,\, c_r\\d_s\end{matrix}\,\middle|\, \omega^2\right),$$

λ not a negative integer, $r \leqslant s$ or $r = s+1$ and $(z\omega)^2 \neq 1$,

$$|\arg(1 - z^2\omega^2)| < \pi. \tag{1}$$

$$ {}_rF_s\left(\begin{matrix} c_r \\ d_s \end{matrix}\middle| \frac{z^2\omega^2}{4}\right) = \left(\frac{2}{z}\right)^\nu \Gamma(1+\nu) \sum_{n=0}^{\infty} \frac{(-)^n(z/2)^n}{n!} I_{n+\nu}(z) $$

$$ \times {}_{r+2}F_s\left(\begin{matrix} -n, c_r, 1+\nu \\ d_s \end{matrix}\middle| \omega^2\right), $$

$$ r \leqslant s \quad \text{or} \quad r = s+1 \quad \text{and} \quad |z\omega| < 2. \tag{2} $$

$$ {}_rF_s\left(\begin{matrix} c_r \\ d_s \end{matrix}\middle| z\omega\right) = \frac{e^{z/2}}{z^\lambda(\frac{1}{2})_\lambda} \left\{ 2^{2\lambda}(1)_\lambda(\tfrac{1}{2})_\lambda I_\lambda\left(\frac{z}{2}\right) + 2\sum_{n=1}^{\infty} \frac{(-)^n(n+\lambda)(n)_{2\lambda}}{n} I_{n+\lambda}\left(\frac{z}{2}\right) \right. $$

$$ \left. \times {}_{r+2}F_{s+1}\left(\begin{matrix} -n, n+2\lambda, c_r \\ \frac{1}{2}+\lambda, d_s \end{matrix}\middle| \omega\right)\right\}, $$

λ not a negative integer, $\quad r \leqslant s$ or $r = s+1$ and $(z\omega) \neq 1$,

$$ |\arg(1 - z\omega)| < \pi. \tag{3} $$

$$ {}_rF_s\left(\begin{matrix} c_r \\ d_s \end{matrix}\middle| z^2\omega^2\right) = \frac{2\Gamma(\mu+1)\Gamma(\nu+1)(2/z)^{\mu+\nu}}{\Gamma(\mu+\nu+1)} $$

$$ \times \sum_{n=0}^{\infty} \frac{(-)^n[n+(\mu+\nu)/2]\Gamma(n+\mu+\nu)}{n!} I_{\mu+n}(z) I_{\nu+n}(z) $$

$$ \times {}_{r+4}F_{s+2}\left(\begin{matrix} -n, n+\mu+\nu, \mu+1, \nu+1, c_r \\ \dfrac{\mu+\nu+1}{2}, \dfrac{\mu+\nu+2}{2}, d_s \end{matrix}\middle| \omega^2\right), $$

$\mu+\nu$ is not a negative integer, $\quad r \leqslant s$ or $r = s+1$ and $(z\omega)^2 \neq 1$,

$$ |\arg(1 - z^2\omega^2)| < \pi. \tag{4} $$

5.12. Polynomial and Rational Approximations

In this section we present polynomial and rational approximations for the $_pF_q$ and a certain G-function which includes the $_pF_q$ series as a special case. The data can be developed along the lines suggested in 13.4 and 13.5. There are several points of view. For details, see Luke (1969), Fields and Wimp (1970) and Fields (1972b).

Let

$$ H(z) = {}_pF_q(\alpha_p;\rho_q;z). \tag{1} $$

Let $H_n(z)$ be an approximation to $H(z)$ and let $S_n(z)$ be the remainder. We write

$$H(z) = H_n(z) + S_n(z), \tag{2}$$

$$H_n(z) = A_n(z,\gamma)/B_n(\gamma), \tag{3}$$

$$S_n(z) = X_n(z,\gamma)/B_n(\gamma). \tag{4}$$

Then formally at least

$$\begin{aligned} B_n(\gamma) &= \sum_{k=0}^{n} A_{n,k}\gamma^{-k}, \qquad A_{n,k} = \frac{(-n)_k\,(n+\lambda)_k\,(\rho_q - a)_k\,(c_f)_k}{(\beta+1)_k\,(\alpha_p+1-a)_k\,(d_g)_k}, \\ B_n(\gamma) &= {}_{q+f+3}F_{p+g+1}\left(\begin{matrix} -n, n+\lambda, \rho_q - a, c_f, 1 \\ \beta+1, \alpha_p+1-a, d_g \end{matrix}\middle| \frac{1}{\gamma}\right), \end{aligned} \tag{5}$$

$$\begin{aligned} A_n(z,\gamma) &= \gamma^{-a} \sum_{k=0}^{n-a} a_k (z/\gamma)^k \sum_{r=0}^{n-k-a} A_{n,r+k+a}/\gamma^r \\ &= \left[\frac{-n(n+\lambda)(\rho_q-1)\,c_f}{\gamma(\beta+1)\,\alpha_p d_g}\right]^a \sum_{k=0}^{n-a} \frac{(a-n)_k\,(n+\lambda+a)_k\,(\alpha_p)_k\,(c_f+a)_k\,(z/\gamma)^k}{(\beta+1+a)_k\,(\alpha_p+1)_k\,(d_g+a)_k\,k!} \\ &\quad\times {}_{q+f+3}F_{p+g+1}\left(\begin{matrix} -n+a+k, n+\lambda+a+k, \rho_q+k, c_f+a+k, 1 \\ \beta+1+a+k, \alpha_p+1+k, d_g+a+k \end{matrix}\middle| \frac{1}{\gamma}\right) \\ &= \gamma^{-a} \sum_{k=0}^{n-a} \gamma^{-k} \sum_{r=0}^{n-k-a} a_r A_{n,r+k+a} (z/\gamma)^r \\ &= \left[\frac{-n(n+\lambda)(\rho_q-1)\,c_f}{\gamma(\beta+1)\,\alpha_p d_g}\right]^a \sum_{k=0}^{n-a} \frac{(a-n)_k\,(n+\lambda+a)_k\,(\rho_q)_k\,(c_f+a)_k\,\gamma^{-k}}{(\beta+1+a)_k\,(\alpha_p+1)_k\,(d_g+a)_k} \\ &\quad\times {}_{p+q+f+2}F_{p+q+g+1}\left(\begin{matrix} -n+a+k, n+\lambda+a+k, \rho_q+k, c_f+a+k, \alpha_p \\ \beta+1+a+k, \alpha_p+1+k, d_g+a+k, \rho_q \end{matrix}\middle| \frac{z}{\gamma}\right), \end{aligned} \tag{6}$$

$$\begin{aligned} X_n(z,\gamma) &= \sum_{k=0}^{\infty} a_{k+1-a} z^{k+1-a} \\ &\quad\times {}_{p+q+f+3}F_{p+q+g+2}\left(\begin{matrix} -n, n+\lambda, \alpha_p+1-a+k, \rho_q-a, c_f, 1 \\ \beta+1, \rho_q+1-a+k, 2-a+k, \alpha_p+1-a, d_g \end{matrix}\middle| \frac{z}{\gamma}\right), \end{aligned} \tag{7}$$

$$X_n(z, \gamma) = \left(\frac{z\alpha_p}{\rho_q}\right)^{1-a} \sum_{k=0}^{n} \frac{(-n)_k (n+\lambda)_k (\rho_q - a)_k (c_f)_k (z/\gamma)^k}{(\beta+1)_k (\rho_q + 1 - a)_k (d_g)_k (2-a)_k}$$

$$\times {}_{p+1}F_{q+1}\left(\begin{matrix}\alpha_p + 1 - a + k, 1 \\ \rho_q + 1 - a + k, 2 - a + k\end{matrix}\middle| z\right). \qquad (8)$$

In the above equations a_k is the coefficient of z^k in $H(z)$, $a=$ 0 or 1, and α, β and λ are as in 5.10.2(1). In Theorem 6 ahead, λ is further restricted.

Notice that $\gamma^n A_n(z,\gamma)$ is a polynomial in z of degree $n-a$ and $\gamma^n B_n(\gamma)$ is a polynomial in γ of degree n. However, when $\gamma = z$, $z^n A_n(z,z) = z^n A_n(z)$ and $z^n B_n(z)$ are polynomials in z of degree $n-a$ and n respectively. Thus the approximations are quite flexible as they may be polynomials or the ratio of polynomials.

We have the following convergence theorem.

Theorem 1. *If*

(1) $p \leqslant q$ (Case 1), *or*

(2) $p = q + 1$, $z \neq 1$, $|\arg(1 - z)| < \pi$ (Case 2),

(3) α_h *is not a negative integer or zero*, $h = 1, 2, \ldots, p$,

(4) $R(\rho_h) > 0$, $h = 1, 2, \ldots, q$,

(5) $\rho_h \neq a$,

(6) $a = 0$ *or* 1,

(7) $f = g$,

(8) $0 < z/\gamma \leqslant 1$, z/γ *fixed.*

then as $n \to \infty$, *the approximations* $A_n(z, \gamma)/B_n(\gamma)$ [*see*(5) and (6)] *converge uniformly to* ${}_pF_q(\alpha_p ; \rho_q ; z)$ *on compact subsets C of the z-plane which exclude* $z = 0$ [*if* $p = q + 1$, *the points of C must also satisfy hypothesis*(2)].

REMARK. Hypotheses (4) and (5) of the theorem are not overly stringent since

$${}_pF_q\left(\begin{matrix}\alpha_p \\ \rho_q\end{matrix}\middle| z\right) = \sum_{k=0}^{m-1} \frac{(\alpha_p)_k z^k}{(\rho_q)_k k!} + \frac{(\alpha_p)_m z^m}{(\rho_q)_m m!} \, {}_{p+1}F_{q+1}\left(\begin{matrix}\alpha_p + m, 1 \\ \rho_q + m, 1 + m\end{matrix}\middle| z\right),$$

and we can apply our analysis to the latter ${}_{p+1}F_{q+1}$ for m large enough.

The proof of the theorem is constructive and leads to an estimate of the error. Under appropriate conditions, $X_n(z,\gamma)$ is at most of algebraic growth in n while $B_n(\gamma)$ is of exponential growth in n and this assures convergence in view of (4). We have

$$S_n(z) = \frac{O(n^\mu)}{B_n(\gamma)}, \quad n \to \infty, \text{ uniformly for } 0 < z/\gamma \leqslant 1,$$

$$\mu = \max(2a, -2R(c_h), \eta, 2\eta + R(\lambda)), \qquad h = 1, 2, \ldots, f, \tag{9}$$

$$\eta = a - \tfrac{1}{2} + R\left\{\sum_{j=1}^{f} (c_j - d_j) - \beta\right\}.$$

Next, we give asymptotic results for $B_n(\gamma)$ with $f=g$.

$$B_n(\gamma) = \delta_n\{1 - (\gamma/2)\, n^{p-q}[1 + O(n^{-1})]\} \qquad \text{if } p < q$$

$$= \delta_n e^{-\gamma\mu_n}\{1 - (\gamma^2\mu_n/8n) + O(n^{-2})\} \qquad \text{if } p = q, \tag{10}$$

where

$$\delta_n = \frac{(-)^n(n+\lambda)_n(\rho_q - a)_n(c_f)_n(1)_n}{(\beta+1)_n(\alpha_p + 1 - a)_n(d_f)_n\gamma^n}$$

$$= \frac{(-)^n 2^{\lambda-1}E(4/\gamma)^n n^\omega \Gamma(\rho_q - a + n)\, n!}{\Gamma(\alpha_p + 1 - a + n)}[1 + O(n^{-1})],$$

$$E = \frac{\Gamma(\beta+1)\,\Gamma(\alpha_p + 1 - a)\,\Gamma(d_f)}{\Gamma(\rho_q - a)\,\Gamma(c_f)\,\Gamma(\frac{1}{2})}, \quad \omega = \sum_{j=1}^{f} (c_j - d_j) - \tfrac{1}{2} - \beta,$$

$$\mu_n = \frac{(n+\beta)(n+\alpha_p - a)(n + d_f - 1)}{(2n+\lambda-1)(n+\rho_p - 1 - a)(n + c_f - 1)}. \tag{11}$$

When $p = q + 1$,

$$B_n(\gamma) \sim E(\sinh \xi/2)^{2\sigma}(\cosh \xi/2)^{-2\sigma-\lambda}\{1 + O(N^{-2})\}\, N^{2\sigma}\cosh\{N\xi + O(N^{-1})\},$$

$$\cosh \xi = 1 - 2/\gamma, \qquad p = q + 1, \qquad |\arg(-\gamma)| \leqslant \pi - \epsilon, \qquad \epsilon > 0, \tag{12}$$

$$B_n(\gamma) \sim (-)^n E(\cosh \eta/2)^{2\sigma}(\sinh \eta/2)^{-2\sigma-\lambda}\{1 + O(N^{-2})\}\, N^{2\sigma}\cosh\{N\eta + O(N^{-1})\},$$

$$\cosh \eta = 2/\gamma - 1, \quad p = q + 1, \quad \gamma \neq 1, \quad |\arg(1/\gamma - 1)| \leqslant \pi - \epsilon, \quad \epsilon > 0, \tag{13}$$

where

$$N^2 = n(n+\lambda), \qquad 2\sigma = a + \omega - q + \sum_{h=1}^{q} \rho_h - \sum_{h=1}^{q+1} \alpha_h \tag{14}$$

and E and ω are defined in (11).

Corollary. *If in the above theorem,*

(1) $p \leqslant q$,

(2) $H(z) = {}_pF_q(\alpha_p; \rho_q; z) = \sum_{k=0}^{n-1} \frac{(\alpha_p)_k z^k}{(\rho_q)_k k!} + T_n(z)$,

then

$$S_n(z)/T_n(z) = O((4z/\gamma)^{-n} n^{\varphi}),$$

$$\varphi = p + a(q-p) + \max\{a, 3a-\eta, a-\eta-2R(c_h), a+\eta+R(\lambda)\}, \quad (15)$$

with h and η as in (9).

Theorem 2. *If*

$$p = 2, \; q = 1, \; \alpha_1 = 1, \alpha_2 = b, \; \beta_1 = c + 1,$$

$$\gamma = z, \; a = 0, \; \beta = c - a, \; i.e., \; \lambda = c + 1 - a,$$

and z is replaced by $-1/z$, *then the approximations* $H_n(z) = A_n(z)/B_n(z)$ *of Theorem* 1 *to* ${}_2F_1(1, b; c+1; -1/z)$ *occupy the* $(n-a, n)$ *positions of the Padé Table. See* 6.10. *See also the remark following* 4.3.1(1).

REMARKS.

(1) The particular choice of $A_{n,k}$ made in (5) is such that $B_n(\gamma)$ satisfies a linear recursion formula of finite length. Further, $A_n(z,\gamma)$ satisfies a closely related linear recursion formula in that the homogeneous portions of both recursion formulas are identical. All of this is described in 5.13.3.

(2) We note that $H(z) = {}_pF_q(\alpha_p; \rho_q; z)$ has no poles in the finite plane. However, the rational approximations to $H(z)$ have poles at the zeros of $B_n(\gamma)$ as defined by (5), and it is important to know the nature of these zeros. If $p = q + 1$, and if all the numerator and denominator parameters in the definition of $B_n(\gamma)$ are real, then under further suitable restrictions on these parameters, it appears from a result given by Luke(1969, Vol. 1, 7.4.2(8)) that all zeros of $B_n(\gamma)$ lie on the real axis from $\gamma = 1$ to $\gamma = \infty$ for n sufficiently large. Precise information on the zeros is an open question. In any event, it is clear that neighborhoods of these zeros must be excluded from domains of applicability of the polynomial and rational approximations. If $p \leqslant q$, very little concerning the zeros of $B_n(\gamma)$ is known except when

$B_n(\gamma)$ reduces to ${}_2F_0(-n, n+\lambda; 1/\gamma)$. The latter is called a Bessel polynomial which, except for a multiplication factor, becomes the modified Bessel function $K_{n+\frac{1}{2}}(\gamma/2)$, when $\lambda = 1$, whence the nature of the zeros can be deduced from the work of Olver (1960). In general, the zeros of the Bessel polynomial are simple and lie in the left half plane. These polynomials are important in a Gaussian quadrature process for the inversion of Laplace transforms and considerable data on the zeros are available. In this connection, see Kublanovskaja and Smirnova(1959), Salzer (1961), Skoblja(1964a,1964b,1965), Krylov and Skoblja(1968) and Piessens (1971). We have previously conjectured that all zeros lie in a circle about the origin whose radius increases at most linearly with n when $p \leq q$. It is clear from (10) and (11) that to achieve a high degree of accuracy we want $n \gg |\gamma|$. Thus the location of the zeros does not appear critical.

(3) When $p \leqslant q$, the hypergeometric series representation for $H(z)$ converges for all z. But, if $p = q+1$, the ${}_{q+1}F_q$ series converges if $|z| < 1$. However, the function $H(z)$ for which the hypergeometric series representation is valid only in the unit disk, is well defined throughout the complex plane provided $|\arg(1-z)| < \pi$. Thus our approximation process converges in a domain where the hypergeometric series representation diverges. In illustration, $z^{-1}\ln(1+z)$ may be represented by the series ${}_2F_1(1, 1; 2; -z)$, $|z| < 1$. However, the approximation process converges for all z provided we exclude the negative real axis from $z = -1$ to $z = -\infty$, and sufficiently small neighborhoods of the zeros of $B_n(\gamma)$. For the $z^{-1}\ln(1+z)$ case, if $f = 0$ and $\beta = 0$, then $B_n(\gamma)$ is essentially a Jacobi polynomial and all its zeros lie on the real axis from $\gamma = -1$ to $\gamma = -\infty$.

(4) Let $H_k^*(z)$ be the sum of the first $k+1$ terms of the Taylor series expansion for $H(z)$. Now $H_n(z)$ can be conceived as a weighted sum of the polynomials $H_k^*(z)$, $k = 0, 1, \ldots, n$. Thus, the economization process is a *summability technique*. A method of summability is said to be regular if it sums a convergent series to its ordinary sum. It follows that the summability process described by the equations (2)-(8) subject, of course, to the conditions required for convergence, is regular.

The important virtue of our economization process when $p = q+1$ is that, under the liberal conditions stated, it converges to the function $H(z)$ in a domain where the hypergeometric series representation for $H(z)$ is divergent. Thus from the summability point of view, the economization process converts a divergent sequence into a convergent sequence. Under suitable hypotheses, the same is true when $p \geq q+2$ if

$H(z)$ is a G-function which has the $_pF_q$ series as its asymptotic expansion for large z in an appropriate region of the complex plane. We now turn to such approximations.

Consider for arbitrary p and q

$$H(z) = \left(\Gamma(\rho_q)/\Gamma(\alpha_p)\right) G^{1,p}_{p,q+1}\left(-z\left|\begin{matrix}1-\alpha_p\\0,\ 1-\rho_q\end{matrix}\right.\right), \tag{16}$$

$$H(z) = {}_pF_q(\alpha_p;\rho_q;z),$$

$$p \le q \text{ or } p = q+1 \text{ and } |\arg(1-z)| < \pi, \tag{17}$$

$$H(z) \sim {}_pF_q(\alpha_p;\rho_q;z),\ p \ge q+2,$$

$$z \to 0,\ |\arg(-z)| \le (p+1-q)\pi/2 - \varepsilon,\ \varepsilon > 0. \tag{18}$$

Then our approximations for (16), formally at least, can take the same form as in (1)-(7). The theorem of convergence and corollary when $p \le q + 1$ are the same as before. However, the present point of view permits for approximations for the above G-function when $p \ge q + 2$.

Theorem 3. *If either*

(1) $p = q + 2$ *and* $|\arg z| < \pi/2$, *or*
$p = q + 2$, $|\arg z| = \pi/2$, *and* $R(\sigma) > -1 - q$, *or*
$p = q + 3$, $z > 0$, *and* $R(\sigma) > -1 - q/2$,

$$\sigma = \left(\sum_{h=1}^{p} \alpha_h - \sum_{h=1}^{q} \rho_h + (q-p)/2\right)\Big/(p-q-1)$$

hold, and

(2) $z \neq 0$,

(3) α_h *is not a negative integer or zero*, $h = 1, 2, \ldots, p$,

(4) $R(\rho_h) > 0$, $h = 1, 2, \ldots, q$,

(5) $a = 0$ *or* 1,

(6) $\rho_h \neq a$,

(7) $f = g$,

(8) $0 < \gamma/z \le 1$,

then as $n \to \infty$ *the rational approximations*

$$A_n(-1/z, -1/\gamma)/B_n(-1/\gamma)$$

converge uniformly to $H(-1/z)$, *where* $H(z)$ *is given by* (16), *on all compact subsets of the z-plane which exclude* $z = 0$.

Theorem 4. *If*

(1) $p=3$, $q=0$ *and* $w=2z^{\frac{1}{2}}>0$,

(2) $\alpha_1 = 1$, $\alpha_2 = \frac{1}{2}(1+\nu-\mu)$, $\alpha_3 = \frac{1}{2}(1-\nu-\mu)$, $R(\mu) > -3/2$,

(3) $f=g=1$, $c_1=2$, $d_1=1$,

and hypotheses (3), (5) *and* (8) *of Theorem* 3 *hold, then as* $n \to \infty$, *the rational approximations of Theorem* 3 *converge uniformly to* $H(-1/z) = w^{1-\mu}S_{\mu,\nu}(w)$, *see* 8.3(17), *on all compact subsets of the real axis* $0<z<\infty$.

Theorem 5. *If*

(1) $p = 2$, $q = 0$, $|\arg z| < \pi$, $z \neq 0$,

(2) $\alpha_1 = 1$, $\alpha_2 = 1 - \nu$,

(3) ν *is not a positive integer*,

and hypotheses (5), (7) *and* (8) *of Theorem* 3 *hold with* $f = 0$, *then as* $n \to \infty$, *the rational approximations of Theorem* 3 *converge to* $z^{1-\nu}e^{z}\Gamma(\nu,z)$, *see* 4.1(3), *on all compact subsets of the z-plane which exclude* $z = 0$. *Further, if* α *is fixed,* $\beta \to \infty$ *and* $\gamma = z$, *the above approximations occupy the* $(n-\alpha,n)$ *positions of the Padé table. See* 4.3.2.

As in the case of Theorem 1, proof of Theorems 3, 4 and 5 are constructive and in each case it can be shown that the error can be described by

$$S_n(-1/z) = \frac{O(n^{\mu})}{B_n(-1/\gamma)}, \quad n \to \infty, \quad \text{uniformly for } 0<\gamma/z\leq 1, \tag{19}$$

$z \neq 0$ with μ as in (9). For $f = g$, $p \geq q + 2$,

$$B_n(-1/\gamma) = {}_{q+f+3}F_{p+f+1}\left(\begin{matrix}-n, n+\lambda, \rho_q - a, c_f, 1\\ \beta+1, \alpha_p+1-a, d_f\end{matrix}\middle| -\gamma\right), \tag{20}$$

$$B_n(-1/\gamma) = A(N^2\gamma)^{\omega}\exp\left(\delta(N^2\gamma)^{1/\delta} - \theta\gamma/3 + O(N^{-1})\right),$$

$$|\arg\gamma| \leq \pi - \varepsilon, \quad \varepsilon > 0,$$

$$A = \frac{(2\pi)^{(1-\delta)/2}\Gamma(\beta+1)\Gamma(\alpha_p+1-a)\Gamma(d_f)}{\delta^{\frac{1}{2}}\Gamma(\rho_q-a)\Gamma(c_f)},$$

$\delta = p-q+1 \geq 3$, $N^2 = n(n+\lambda)$, $\theta = 1$ if $\delta = 3$, $\theta = 0$ if $\delta > 3$,

$$\omega = \delta^{-1}\left(a(p-q) - \tfrac{1}{2}(p+q) + \sum_{j=1}^{q}\rho_j - \sum_{j=1}^{p}\alpha_j - \beta + \sum_{j=1}^{f}(c_j - d_j)\right). \quad (21)$$

In (21), it is supposed that none of the parameters $\rho_j - a$, c_j is a negative integer or zero. A complete asymptotic description of $B_n(\gamma)$ for arbitrary f,g,p and q and valid for other regions of the complex γ-plane is available in Luke (1969). See this same reference for a confluent form of (21) called the extended Laguerre polynomial.

REMARKS.

(1) In Theorems 3-5, it is more convenient to deal with $H(-1/z)$ rather than $H(z)$ as it is in this form that the $_pF_q$ arises when $p \geq q+2$.

(2) Under rather liberal conditions on the parameters, we conjecture that Theorem 3 holds for $|\arg z| \leq \pi - \varepsilon$, $\varepsilon > 0$ and $p \geq q+2$. This is suggested by Theorem 5 and error coefficients for rational approximations to $K_\nu(z)$ given in 9.3. A partial result in this direction is Theorem 6 below.

(3) Proof of Theorems 3-5 can be developed by showing that the error satisfies a differential equation closely related to that satisfied by $H(z)$ followed by solving this equation. In this connection, see the discussion developed for the τ-method in 13.4. In 5.14 recursion formulas are developed for $H_n(z,\gamma)$ and $B_n(\gamma)$ as defined by (5), (6). Clearly, the error in the approximation process also satisfies a linear difference equation. Here the discussion in 13.5 is pertinent. So the error can be characterized by solving this difference eqequation. This is the approach used by Fields (1972b) to prove the following theorem.

Theorem 6. *If*

(1) $p \geq q+2$,

(2) $\gamma = z \neq 0$ *and* $|\arg z| < \pi$,

(3) $\rho_j - \alpha_k$ *is not a negative integer or zero,* $j = 1, 2, \ldots, q$; $k = 0, 1, \ldots, p$; $\alpha_0 = a$,

(4) *none of the numbers* $\rho_k - \rho_j$, $\alpha_k - \alpha_j$ *is an integer for* $j \neq k$,

(5) $a = 0, 1, \ldots,$ *or* $p - 1$,

(6) $\beta = 0$, $\alpha = 0, 1, \ldots,$ *or* $p - 1 - a$, *that is,* $\lambda = 1, 2, \ldots,$ *or* $p - a$,

(7) $f = g = 0$,

then as $n \to \infty$, *the rational approximations of Theorem* 3 *with* $\gamma = z$ *converge uniformly to* $H(-1/z)$ *on all compact subsets of the* z*-plane which exclude* $z = 0$. *Further, for the error, we have the asymptotic result*

$$S_n(-1/z) = O\left(\exp\left[-\xi(z)|zN^2|^{1/\delta}\right]\right) \tag{22}$$

$$\xi(z) = 2\delta\ (\cos\phi)\{\cos(\phi + |\arg z|/\delta)\},\quad \phi = \tfrac{1}{2}\pi - \pi/\delta,\quad |\arg z| < \pi,$$

where δ *and* N^2 *are as in* (21).

REMARKS.

(1) It appears that condition (4) of Theorem 6 can be weakened, for when it is violated, log terms occur which would not alter the asymptotic of the error. Condition (3) is also suspect.

(2) Aside from the restriction on arg z, Theorem 3 is in the main stronger than Theorem 6 because in the latter we require $\gamma = z$. Also the conditions on β and λ in Theorem 3 are more lenient than the corresponding condition (6) of Theorem 6. However, in practice, we usually elect to have the latter condition fulfilled, because in this instance when $\gamma = z$, both numerator and denominator polynomials of the rational approximation satisfy the same linear homogeneous difference equation.

(3) Fields (1972b,1973) has developed a closed form representation of the error in the rational approximations of Theorem 6, and has determined pertinent asymptotic results. These data are omitted here.

5.13 Recurrence Formulas for Polynomials and Functions Occurring in Approximations to Generalized Hypergeometric Functions

5.13.1. Introduction

From the work in 5.12, we see that the denominator polynomials of the rational approximations for hypergeometric functions are polynomials of hypergeometric type. Further, the numerator polynomials can also be expressed in terms of such polynomials. Again in 5.10, we note that in the expansion of hypergeometric functions in series of extended Jacobi and Laguerre polynomials, the coefficients of these polynomials are of the hypergeometric family.

In the theory and practice of computation, recursion formulas are valuable in obtaining approximations of higher order from those of lower order. In particular, for the case of Padé approximations, it is known classically that the numerator and denominator polynomials must satisfy the same three-term recurrence formula.

In 5.13.2, we give recursion formulas for extended Jacobi and Laguerre polynomials. These expressions are linear, nonhomogeneous and of finite length. Thus in the notation of 5.12(5), we have a recursion formula for $B_n(\gamma)$. In view of 13.5(7), $A_n(z,\gamma)$ satisfies another nonhomogeneous recursion formula which varies from that satisfied by $B_n(\gamma)$ only in the nonhomogeneous term. Data for $B_n(\gamma)$ and $A_n(z,\gamma)$ are described in 5.13.3. Further, if $\gamma = z$, under certain conditions which can always be met in practice, both $B_n(z)$ and $A_n(z)$ satisfy the same recursion formula. This is most useful for the applications. One advantage of a convergent Taylor series expansion is that additional accuracy is obtained by adding more terms. In the case of many polynomial and rational approximations, it happens that the nth approximant is totally unrelated to previous approximants. Thus the schemata employed to get a particular approximant must be used anew to get the approximant next in line. The advantage cited for Taylor series expansions is essentially retained by our rational approximations, and this, coupled with the fact that these rational approximations can converge in domains where the Taylor series fails to converge, enhances the utility of these rational approximations.

In 5.13.3(6), we also show that the hypergeometric polynomial which appears in the third expression of 5.12(6) can be generated by a first order inhomogeneous recursion formula, and this offers an alternative scheme for the evaluation of

$A_n(z,\gamma)$.

As already noted, the coefficients in the expansion of generalized hypergeometric functions in series of extended Jacobi polynomials are functions of hypergeometric type. These functions also satisfy a difference equation which facilitates their computation. This is presented in 5.13.4. Except for the data surrounding 5.13.2(41-50), the results of this section stem from the work of Wimp (1968) and Fields, Luke and Wimp (1968) and Luke (1969).

5.13.2. Recursion Formulas for Extended Jacobi and Laguerre Functions

Consider the extended Jacobi function

$$\mathscr{E}_n(z,\lambda) = {}_{r+3}F_s\left(\begin{matrix}-n, n+\lambda, a_r, 1\\ b_s\end{matrix}\,\middle|\, z\right), \qquad n \text{ arbitrary},$$

$$r+3 \leqslant s \quad \text{or} \quad r+2=s \quad \text{and} \quad |\arg(1-z)| < \pi. \tag{1}$$

Associated with (1) is the function

$$\mathscr{K}_n(z,\lambda) = \frac{(b_s-1)\,z^{-1}}{(n+1)(n+\lambda-1)(a_r-1)}\,{}_{s+1}F_{r+2}\left(\begin{matrix}2-b_s, 1\\ 2+n, 2-n-\lambda, 2-a_r\end{matrix}\,\middle|\,\frac{(-)^{s-r}}{z}\right),$$

$$r+1 \geqslant s \quad \text{or} \quad r+2=s \quad \text{and} \quad |\arg(1-1/z)| < \pi, \tag{2}$$

in the sense that both are particular solutions of the differential equation

$$[(\delta+b_s-1) - z(\delta-n)(\delta+n+\lambda)(\delta+a_r)]\,\mathscr{F}(z) = (b_s-1), \qquad \delta = z\,d/dz. \tag{3}$$

In addition to the fact that (1) and (2) satisfy the same linear nonhomogeneous difference equation, it turns out that a suitably normalized basis of solutions of the related homogeneous differential equation

$$[(\delta+b_s-1) - z(\delta-n)(\delta+n+\lambda)(\delta+a_r)]\,\mathscr{F}(z) = 0 \tag{4}$$

at its singular points $z=0$ and $z=\infty$ also satisfies the related homogeneous difference equation.

To describe these bases of solutions normalized with respect to n, it is convenient to put down the following conditions:

$r+3 \leqslant s$ or $r+2=s$ and $|\arg(1-z)| < \pi$, no two of the parameters b_h, $h=1,2,\ldots,s$ differ by an integer or zero. (5)

$r + 1 \geqslant s$ or $r + 2 = s$ and $|\arg(1 - 1/z)| < \pi$, no two of the parameters $-n$, $n + \lambda$, a_j, $j = 1, 2,\ldots, r$ differ by an integer or zero. (6)

Under conditions (5), we take for our normalized basis of solutions of (3)

$$\mathscr{F}_{n,h}(z, \lambda) = (n + b_h)_{1-b_h}(n + \lambda)_{1-b_h} z^{1-b_h}$$
$$\times {}_{r+2}F_{s-1}\left(\begin{matrix}1 - b_h - n, 1 - b_h + n + \lambda, 1 - b_h + a_r \\ 1 - b_h + b_s^*\end{matrix}\middle| z\right),$$
$$h = 1, 2,\ldots, s. \tag{7}$$

Under conditions (6), we take for our normalized basis of solutions of (3)

$$\mathscr{G}_{n,j}(z, \lambda) = (n + 1 + a_j)_{-a_j}(n + \lambda)_{-a_j} z^{-a_j}$$
$$\times {}_sF_{r+1}\left(\begin{matrix}1 + a_j - b_s \\ 1 + a_j + n, 1 + a_j - n - \lambda, 1 + a_j - a_r^*\end{matrix}\middle| \frac{(-)^{s-r}}{z}\right),$$
$$j = 1, 2,\ldots, r, \tag{8}$$

$$\mathscr{G}_{n,r+1}(z, \lambda) = \frac{\Gamma(n + 1)\,\Gamma(2n + \lambda)\,\Gamma(n + a_r)(ze^{i\varphi})^n}{\Gamma(n + \lambda)\,\Gamma(n + b_s)}$$
$$\times {}_sF_{r+1}\left(\begin{matrix}1 - b_s - n \\ 1 - 2n - \lambda, 1 - a_r - n\end{matrix}\middle| \frac{(-)^{s-r}}{z}\right), \tag{9}$$

$$\mathscr{G}_{n,r+2}(z, \lambda) = \frac{\Gamma(n + 1)\,\Gamma(n + \lambda + 1 - b_s)\exp[i\varphi(s - r - 1)\, n] z^{-n-\lambda}}{\Gamma(n + \lambda)\,\Gamma(2n + \lambda + 1)\,\Gamma(n + \lambda + 1 - a_r)}$$
$$\times {}_sF_{r+1}\left(\begin{matrix}n + \lambda + 1 - b_s \\ 2n + \lambda + 1, n + \lambda + 1 - a_r\end{matrix}\middle| \frac{(-)^{s-r}}{z}\right),$$
$$e^{i\varphi} = -1. \tag{10}$$

We have the following

Theorem 1. *The functions* $\mathscr{E}_n(z, \lambda)$ *and* $\mathscr{K}_n(z, \lambda)$ *under the conditions on* r, s, *and* z *in* (5) *and* (6), *respectively, satisfy the difference equation*

$$\Phi_n(z, \lambda) + \sum_{m=1}^{t} [A_m(n, \lambda) + zB_m(n, \lambda)]\, \Phi_{n-m}(z, \lambda) = \frac{(b_s - 1)(n + \lambda)_n}{(n + b_s - 1)(n + \lambda - t)_n},$$
$$t = \max(r + 2, s), \qquad B_t(n, \lambda) = 0, \tag{11}$$

where

$$A_m(n, \lambda) = \frac{(n+1-m)_m(2n+\lambda-2m)_{2m}(n-m-1+b_s)}{m!\,(n+\lambda-m)_m(2n+\lambda-t-m)_m(n-1+b_s)}$$

$$\times\ {}_{s+2}F_{s+1}\left(\begin{matrix}-m, 2n+\lambda-t-m, n-m+b_s\\ 2n+\lambda+1-2m, n-m-1+b_s\end{matrix}\,\middle|\, 1\right) \tag{12}$$

$$= \frac{(-)^s(n+1-m)_m(2n+\lambda-2m)(2n+\lambda-t+1)_{t-1}(n+\lambda-t+1-b_s)}{(t-m)!\,(n+\lambda-m)_m(2n+\lambda-t-m)_m(n-1+b_s)}$$

$$\times\ {}_{s+2}F_{s+1}\left(\begin{matrix}-t+m, 2n+\lambda-t-m, n+\lambda-t+2-b_s\\ 2n+\lambda+1-t, n+\lambda-t+1-b_s\end{matrix}\,\middle|\, 1\right), \tag{13}$$

$$B_m(n, \lambda) = \frac{(n+1-m)_m(2n+\lambda-2m)_{2m}(n-m+a_r)}{(m-1)!\,(n+\lambda-m)_m(2n+\lambda-t-m+1)_{m-1}(n-1+b_s)}$$

$$\times\ {}_{r+2}F_{r+1}\left(\begin{matrix}1-m, 2n+\lambda-t-m+1, n-m+1+a_r\\ 2n+\lambda+1-2m, n-m+a_r\end{matrix}\,\middle|\, 1\right) \tag{14}$$

$$= \frac{(-)^r(n+1-m)_m(2n+\lambda-2m)(2n+\lambda-t+1)_{t-1}(n+\lambda-t+1-a_r)}{(t-m-1)!\,(n+\lambda-m)_m(2n+\lambda-t-m+1)_{m-1}(n-1+b_s)}$$

$$\times\ {}_{r+2}F_{r+1}\left(\begin{matrix}-t+m+1, 2n+\lambda-t-m+1, n+\lambda-t+2-a_r\\ 2n+\lambda+1-t, n+\lambda-t+1-a_r\end{matrix}\,\middle|\, 1\right). \tag{15}$$

Further, the functions $\mathscr{F}_{n,h}(z, \lambda)$, $h = 1, 2, \ldots, s$, *and* $\mathscr{G}_{n,j}(z, \lambda)$, $j = 1, 2, \ldots, r+2$, *under the conditions* (5) *and* (6), *respectively, satisfy the difference equation*

$$\Phi_n(z, \lambda) + \sum_{m=1}^{t} [A_m(n, \lambda) + zB_m(n, \lambda)]\,\Phi_{n-m}(z, \lambda) = 0. \tag{16}$$

Finally, if no b_h *is equal to any* a_j, *none of the above functions satisfy a nontrivial equation of the form specified of lower order than* t.

Corollary 1. *The function*

$$\mathscr{E}(n, z, \lambda) = \frac{\Gamma(b_s)}{\Gamma(-n)\,\Gamma(n+\lambda)\,\Gamma(a_r)}$$

$$\times\ G^{1,r+3}_{r+3,s+1}\left(-z\,\middle|\,\begin{matrix}0, 1+n, 1-n-\lambda, 1-a_r\\ 0, 1-b_s\end{matrix}\right) \tag{17}$$

satisfies (11), *and satisfies no nontrivial equation of the same form, of lower order than* t, *provided no* b_h *is equal to any* a_j.

REMARK 1. The extended Jacobi function (1) loses its specialized appearance if we put $s = q + 1$ and set $b_h = 1$ for $h = q + 1$.

REMARK 2. As previously noted, under conditions (5) and (6), the functions $\mathscr{F}_{n,h}(z, \lambda)$ and $\mathscr{G}_{n,j}(z, \lambda)$, respectively, form a basis of solutions of the differential equation (4), and hence are linearly independent as functions of z. They are also linearly independent as functions of n.

REMARK 3. If the conditions on the parameters in (5) and (6) are violated, additional solutions of the difference equation (16) can be constructed via the same limit processes used to get additional solutions for the differential equation (4).

REMARK 4. If n is a positive integer or zero, no restrictions on r, s, and z are necessary for $\mathscr{E}_n(z, \lambda)$ to be well defined and for (11) to hold. In this connection, care must be taken in certain limit processes which can arise. For example, suppose $r = 0$, $s = 1$, $b_1 = 1$. Then from (12),

$$A_1(n, \lambda) = -(2n + \lambda - 2)(\lambda - 1)[(n + \lambda - 1)(2n + \lambda - 3)]^{-1}.$$

If we want $A_1(n, \lambda)$ for $n = 1$ and $\lambda = 1$, we must first set $n = 1$ and then let $\lambda \to 1$. Thus, $A_1(n, \lambda) = -1$ for $n = 1$ and $A_1(n, \lambda) = 0$ for $n \neq 1$.

As a corollary to the above results, we give a difference equation satisfied by the extended Laguerre function and the family of functions which satisfy the same differential equation as does the extended Laguerre function. We have need for the following sets of conditions:

$$r + 2 \leqslant s \text{ or } r + 1 = s \text{ and } |\arg(1 - z)| < \pi, \text{ no two of the parameters } b_h,\ h = 1, 2, \ldots, s \text{ differ by an integer or zero;} \tag{18}$$

$$r \geqslant s \text{ or } r + 1 = s \text{ and } |\arg(1 - 1/z)| < \pi, \text{ no two of the parameters } -n,\ a_j,\ j = 1, 2, \ldots, r \text{ differ by an integer or zero.} \tag{19}$$

Also, let

$$\lim_{\lambda \to \infty} \Phi_n(z/\lambda, \lambda) = \Phi_n(z),$$

$$\Phi_n(z, \lambda) = \mathscr{E}_n(z, \lambda), \quad \mathscr{K}_n(z, \lambda), \quad \mathscr{F}_{n,h}(z, \lambda), \quad h = 1, 2, \ldots, s,$$
$$\mathscr{G}_{n,j}(z, \lambda), \quad j = 1, 2, \ldots, r + 1. \tag{20}$$

A limiting form of the above theorem [see (11)–(16)] and corollary [see (17)] is the following result.

Corollary 2. *The functions $\mathscr{E}_n(z)$ and $\mathscr{K}_n(z)$ under the conditions on r, s, and z in* (18) *and* (19), *respectively, satisfy the difference equation*

$$\Phi_n(z) + \sum_{m=1}^{t^*} [A_m(n) + zB_m(n)]\, \Phi_{n-m}(z) = \frac{(b_s - 1)}{(n + b_s - 1)},$$

$$t^* = \max(r + 1, s), \tag{21}$$

where

$$A_m(n) = \frac{(n+1-m)_m(n-m-1+b_s)}{m!\,(n-1+b_s)}\; {}_{s+1}F_s\left(\begin{matrix}-m, n-m+b_s \\ n-m-1+b_s\end{matrix}\middle|\, 1\right) \tag{22}$$

$$= \frac{(n+1-m)_m(-)^m}{m!\,(n-1+b_s)} \sum_{u=0}^{s-m} \frac{(s-u)!}{(s-m-u)!}\, B^{(-m)}_{s-m-u} S_u(n-m-1+b_s), \tag{23}$$

$$B_m(n) = \frac{(n+1-m)_m(n-m+a_r)}{(m-1)!\,(n-1+b_s)}\; {}_{r+1}F_r\left(\begin{matrix}1-m, n-m+1+a_r \\ n-m+a_r\end{matrix}\middle|\, 1\right) \tag{24}$$

$$= \frac{(n+1-m)_m(-)^{m-1}}{(m-1)!\,(n-1+b_s)} \sum_{u=0}^{r+1-m} \frac{(r-u)!}{(r+1-m-u)!}\, B^{(1-m)}_{r+1-m-u} S_u(n-m+a_r), \tag{25}$$

where $B_k^{(a)}$ is defined in 14.2 *and where $S_u(\beta_q)$ are symmetric polynomials implicitly defined by*

$$\prod_{i=1}^{q} (x + \beta_i) = \sum_{u=0}^{q} S_u(\beta_q)\, x^{q-u}. \tag{26}$$

In particular, $A_m(n) = 0$, $m \geqslant s + 1$, and $B_m(n) = 0$, $m \geqslant r + 2$. Also, the functions $\mathscr{F}_{n,h}(z)$, $h = 1, 2,\ldots, s$, and $\mathscr{G}_{n,j}(z)$, $j = 1,2,\ldots, r + 1$, under the conditions (18) *and* (19), *respectively, satisfy the difference equation*

$$\Phi_n(z) + \sum_{m=1}^{t^*} [A_m(n) + zB_m(n)]\, \Phi_{n-m}(z) = 0. \tag{27}$$

Further, the function

$$\mathscr{E}(n, z) = \frac{\Gamma(b_s)}{\Gamma(-n)\,\Gamma(a_r)}\, G^{1,r+2}_{r+2,s+1}\left(-z\,\middle|\,\begin{matrix}0, 1+n, 1-a_r \\ 0, 1-b_s\end{matrix}\right) \tag{28}$$

satisfies (21). *Finally, so long as no b_h is equal to any a_j, the above functions satisfy no nontrivial equation of the form specified, of lower order than t^*.*

To illustrate the principal results of this section, we have for n a positive integer that

$$\mathscr{F}_n(z, \lambda) = {}_4F_3\left(\begin{matrix}-n, n+\lambda, a_1, a_2 \\ b_1, b_2, b_3\end{matrix}\middle|\, z\right) \tag{29}$$

satisfies

$$\sum_{m=0}^{4} [A_m(n, \lambda) + zB_m(n, \lambda)] \mathscr{F}_{n-m}(z, \lambda) = 0,$$

$$A_0(n, \lambda) = 1, \qquad B_0(n, \lambda) = B_4(n, \lambda) = 0, \tag{30}$$

where

$$A_1(n, \lambda) = \frac{(n-1)(2n+\lambda-2)_2(n+b-2)}{(n+\lambda-1)(2n+\lambda-5)(n+b-1)} \times \left[1 - \frac{n(2n+\lambda-5)(n+b-1)}{(n-1)(2n+\lambda-1)(n+b-2)}\right],$$

$$A_2(n, \lambda) = \frac{(n-2)_2(2n+\lambda-4)_4(n+b-3)}{2(n+\lambda-2)_2(2n+\lambda-6)_2(n+b-1)} \times \left\{1 - \frac{2(n-1)(2n+\lambda-6)(n+b-2)}{(n-2)(2n+\lambda-3)(n+b-3)} + \frac{n(2n+\lambda-6)_2(n+b-1)}{(n-2)(2n+\lambda-3)_2(n+b-3)}\right\},$$

$$A_3(n, \lambda) = \frac{(n-2)_2(2n+\lambda-2)_2}{(n+\lambda-2)_2(2n+\lambda-5)(n+b-1)} \times \left\{\frac{(n+\lambda-4)(2n+\lambda-3)(n+\lambda-3-b)}{(n+\lambda-3)(2n+\lambda-7)} - (n+\lambda-2-b)\right\},$$

$$A_4(n, \lambda) = \frac{(n-3)_3(2n+\lambda-3)_3(n+\lambda-3-b)}{(n+\lambda-3)_3(2n+\lambda-7)_3(n+b-1)}, \qquad 1 + \sum_{m=1}^{4} A_m(n, \lambda) = 0,$$

$$B_1(n, \lambda) = \frac{(2n+\lambda-2)_2(n+a-1)}{(n+\lambda-1)(n+b-1)},$$

$$B_2(n, \lambda) = \frac{(n-1)(2n+\lambda-2)_2(2n+\lambda-4)}{(n+\lambda-2)_2(2n+\lambda-5)(n+b-1)} \times \{(2n+\lambda-3)(n+a-2) - (2n+\lambda-5)(n+a-1)\},$$

$$B_3(n, \lambda) = \frac{(n-2)_2(2n+\lambda-3)_3(n+\lambda-3-a)}{(n+\lambda-3)_3(2n+\lambda-5)(n+b-1)}. \tag{31}$$

Here $(n + u + b)$ is short for $\prod_{i=1}^{3} (n + u + b_i)$, and $(n + u + a)$ is short for $\prod_{i=1}^{2} (n + u + a_i)$, etc.

By use of the confluence principle,

$$\mathscr{F}_n^*(z, \lambda) = {}_3F_3\left(\begin{matrix} -n, n+\lambda, a \\ b_1, b_2, b_3 \end{matrix} \middle| z\right) \tag{32}$$

satisfies

$$\sum_{m=0}^{4} [A_m(n, \lambda) + zB_m^*(n, \lambda)] \mathscr{F}_{n-m}^*(z, \lambda) = 0, \qquad B_0^*(n, \lambda) = B_4^*(n, \lambda) = 0, \tag{33}$$

with $A_m(n, \lambda)$ as in (31), and

$$B_1^*(n, \lambda) = \frac{(2n + \lambda - 2)_2(n + a - 1)}{(n + \lambda - 1)(n + b - 1)},$$

$$B_2^*(n, \lambda) = -\frac{(n - 1)(2n + \lambda - 2)_2(2n + \lambda - 4)(\lambda - 1 - 2a)}{(n + \lambda - 2)_2(2n + \lambda - 5)(n + b - 1)}, \tag{34}$$

$$B_3^*(n, \lambda) = -\frac{(n - 2)_2(2n + \lambda - 3)_3(n + \lambda - 3 - a)}{(n + \lambda - 3)_3(2n + \lambda - 5)(n + b - 1)}.$$

For another example,

$$f_n(z, \lambda) = {}_3F_2\left(\begin{matrix}-n, n + \lambda, a \\ b_1, b_2\end{matrix}\middle| z\right) \tag{35}$$

satisfies

$$\sum_{m=0}^{3} [C_m(n, \lambda) + zD_m(n, \lambda)] f_{n-m}(z, \lambda) = 0,$$

$$C_0(n, \lambda) = 1, \qquad D_0(n, \lambda) = D_3(n, \lambda) = 0, \tag{36}$$

$$C_1(n, \lambda) = \frac{(n - 1)(2n + \lambda - 2)_2(n + b - 2)}{(n + \lambda - 1)(2n + \lambda - 4)(n + b - 1)} - \frac{n(2n + \lambda - 2)}{n + \lambda - 1},$$

$$C_2(n, \lambda) = \frac{(n - 1)(2n + \lambda - 1)(n + \lambda - b - 1)}{(n + \lambda - 1)(n + b - 1)} - \frac{(n - 1)(n + \lambda - 3)(2n + \lambda - 2)_2(n + \lambda - b - 2)}{(n + \lambda - 2)_2(2n + \lambda - 5)(n + b - 1)}, \tag{37}$$

$$C_3(n, \lambda) = -\frac{(n - 2)_2(2n + \lambda - 2)_2(n + \lambda - b - 2)}{(2n + \lambda - 5)_2(n + \lambda - 2)_2(n + b - 1)},$$

$$D_1(n, \lambda) = \frac{(2n + \lambda - 2)_2(n + a - 1)}{(n + \lambda - 1)(n + b - 1)},$$

$$D_2(n, \lambda) = -\frac{(n - 1)(2n + \lambda - 2)_2(n + \lambda - a - 2)}{(n + \lambda - 2)_2(n + b - 1)},$$

where $(n + b - 1)$ represents $(n + b_1 - 1)(n + b_2 - 1)$, etc. We can also deduce (35)-(37) from (29)-(31) in the following manner. In (29) put $a_2 = b_3 = n + \lambda - 3$. Then it is easy to see that $A_4(n, \lambda) = B_3(n, \lambda) = 0$, $A_i(n, \lambda) = C_i(n, \lambda)$, $i = 1, 2, 3$, and $B_j(n, \lambda) = D_j(n, \lambda)$, $j = 1, 2$.

Again, application of the confluence principle to (35) with respect to λ shows that

$$f_n(z) = {}_2F_2\left(\begin{matrix}-n, a \\ b_1, b_2\end{matrix}\middle| z\right) \tag{38}$$

satisfies

$$\sum_{m=0}^{3} [C_m(n) + zD_m(n)] f_{n-m}(z) = 0,$$

$$C_0(n) = 1, \qquad D_0(n) = D_3(n) = 0, \tag{39}$$

$$C_1(n) = \frac{(n-1)(n+b-2)}{(n+b-1)} - n, \qquad C_3(n) = -\frac{(n-2)(n-1)}{n+b-1},$$

$$C_2(n) = -[1 + C_1(n) + C_3(n)],$$

$$D_1(n) = \frac{n+a-1}{n+b-1}, \qquad D_2(n) = -\frac{(n-1)}{n+b-1}. \tag{40}$$

The following result due to Wimp (1969) is a generalization of a well known result for Jacobi polynomials.

Theorem 2. *Let*

$$F_n(z,\lambda) = {}_{p+2}F_q(-n, n+\lambda, a_p; b_q; z) \tag{41}$$

where n *is arbitrary if* $p \le q-1$ *(we also require* $|\arg(1-z)| < \pi$ *if* $p = q-1$*) and* n *is a positive integer if* $p \ge q$*. Then*

$$z(u - vz)dF_n(z,\lambda)/dz = \sum_{m=0}^{t} \{P_m(n,\lambda) + zQ_m(n,\lambda)\} F_{n-m}(z,\lambda), \tag{42}$$

where

$$v = 1 \text{ if } p+1 \ge q; \; v = 0 \text{ if } p+1 < q;$$

$$u = 0 \text{ if } p+1 > q, \; u = 1 \text{ if } p+1 \le q;$$

$$t = \max(p+1, q), \tag{43}$$

and no such equation of lower order than t exists provided no $a_h = b_j$, $h = 1, 2, \ldots, p$; $j = 1, 2, \ldots, q$. The $P_m(n,\lambda)$'s and $Q_m(n,\lambda)$'s are unique and

$$P_m(n,\lambda) = \xi\left((-)^{m+1}u + (-)^{t+1}/m!\right)$$

$$\times \sum_{s=0}^{m} \frac{(-m)_s (n-s)(n-s-1+b_q)}{(m+s-2n-\lambda)_{t-1-m}}, \quad m > 0;$$

$$= n\left(u - \frac{(n-1+b_q)}{(2n+\lambda-t)_t}\right), \quad m=0; \tag{44}$$

$$Q_m(n,\lambda) = \xi\left((-)^m|v + (-)^{w+1}/(m-1)!\right)$$

$$\times \sum_{s=0}^{m-1} \frac{(1-m)_s(n-s-1+a_p)}{(m+s+1-2n-\lambda)_{t-m}}, \quad m>0;$$

$$= -vn, \quad m=0,$$

$$\xi = \frac{(2m-2n-\lambda)\Gamma(n+1)\Gamma(n-m+\lambda)}{\Gamma(n+\lambda)\Gamma(n+1-m)}. \tag{45}$$

REMARK. If $p+1=q$ and $z=1$ in (41), we get a recursion formula for $F_n(1,\lambda)$ of order $t=q$, which if of order one less than that of (11) with $r=p$, $s=q+1$ and $b_h=1$ for $h=q+1$.

The following and last theorem of this section is due to Tsai (1970).

Theorem 3. *For arbitrary n, let*

$$G_n(z) = \frac{z^n\Gamma(b_q)}{\Gamma(a_p)} G^{1,p}_{p+2,q+1}\left(\frac{1}{z^2}\middle|\begin{matrix}1-a_p,\ \tfrac{1}{2}n+1,\ \tfrac{1}{2}n+\tfrac{1}{2}\\ 0,\ 1-b_q\end{matrix}\right) \tag{46}$$

$$= \frac{(2z)^n}{\Gamma(n+1)}\ {}_{p+2}F_q(-\tfrac{1}{2}n, \tfrac{1}{2}-\tfrac{1}{2}n, a_p; b_q; -1/z^2),$$

$$q > p+1 \text{ or } q = p+1 \text{ and } |\arg(1+1/z^2)| < \pi. \tag{47}$$

If n is a positive integer, the hypergeometric function reduces to polynomial and the restrictions on p, q and z can be omitted. Then

$$B_0G_n(z) + \sum_{m=1}^{w}\left(A_m(2z)^{m-2} + B_m(2z)^m\right)G_{n-m}(z) = 0,$$

$$w = \max(p+2,\ q+1), \tag{48}$$

where

$$A_m = \frac{(n-m+2a_p)}{2^p\,(m-2)!}\;{}_{p+1}F_p\left(\begin{matrix}2-m,\; n+1-m+2a_p\\ n-m+2a_p\end{matrix}\,\middle|\, 1\right), \tag{49}$$

$$B_m = \frac{(n-m-2+2b_{q+1})}{2^{q+1}m!}\;{}_{q+2}F_{q+1}\left(\begin{matrix}-m,\; n-m-1+2b_{q+1}\\ n-m-2+2b_{q+1}\end{matrix}\,\middle|\, 1\right). \tag{50}$$

Further, if no $a_i = b_j$, $i = 1,2,\ldots,p$; $j = 1,2,\ldots,q$, *then* $G_n(z)$ *does not satisfy a nontrivial recurrence formula of the form specified of order lower than* w.

REMARKS. If $p = 0$ and $q = 1$, (47) is essentially a Jacobi function in view of 11.3.1(3) and 6.4(3). Alternative forms for A_m and B_m which resemble (23) and (25) are given by Tsai (1970)

5.13.3. Recursion Formulas for the Numerator and Denominator Polynomials in the Rational Approximations for the Generalized Hypergeometric Function

In this section, we give the recurrence formulas for the polynomials which enter the rational approximations for ${}_pF_q(z)$ as stated in 5.12. The following theorem summarizes the results in a form convenient for the applications.

Theorem. *Let*

$$H(z) = {}_pF_q(\alpha_p\,;\rho_q\,;z),$$

$$B_n(\gamma) = {}_{q+f+3}F_{p+g+1}\left(\begin{matrix}-n,\, n+\lambda,\, \rho_q - a,\, c_f\,,\, 1\\ \beta+1,\, \alpha_p+1-a,\, d_g\end{matrix}\,\middle|\,\frac{1}{\gamma}\right),$$

$$A_n(z,\gamma) = \left[\frac{-n(n+\lambda)(\rho_q-1)\,c_f}{\gamma(\beta+1)\,\alpha_p\,d_g}\right]^a$$

$$\times \sum_{k=0}^{n-a} \frac{(a-n)_k(n+\lambda+a)_k(\alpha_p)_k(c_f+a)_k(z/\gamma)^k}{(\beta+1+a)_k(\alpha_p+1)_k(d_g+a)_k k!}$$

$$\times\; {}_{q+f+3}F_{p+g+1}\left(\begin{matrix}-n+a+k,\, n+\lambda+a+k,\, \rho_q+k,\, c_f+a+k, 1\\ \beta+1+a+k,\, \alpha_p+1+k,\, d_g+a+k\end{matrix}\,\middle|\,\frac{1}{\gamma}\right), \tag{1}$$

so that $A_n(z,\gamma)/B_n(\gamma)$ *is at least a formal approximation to* $H(z)$.

Let t, K_m *and* L_m *be defined by* (4), see p. 245. *Then*

$$t = \max(q + f + 2, p + g + 1),$$

$$K_m = \frac{(n + 1 - m)_m(2n + \lambda - 2m)_{2m}(n - m + \beta)(n - m + \alpha_p - a)(n - m + d_g - 1)}{m!\,(n + \lambda - m)_m(2n + \lambda - t - m)_m(n + \beta)(n + \alpha_p - a)(n + d_g - 1)}$$

$$\times\ {}_{p+g+3}F_{p+g+2}\left(\begin{matrix}-m, 2n + \lambda - t - m, n - m + \beta + 1, n - m + \alpha_p + 1 - a, n - m + d_g \\ 2n + \lambda + 1 - 2m, n - m + \beta, n - m + \alpha_p - a, n - m + d_g - 1\end{matrix}\,\middle|\, 1\right)$$

$$= \frac{(-)^{p+g+1}(n + 1 - m)_m(2n + \lambda - 2m)(2n + \lambda - t + 1)_{t-1}(n + \lambda - t - \beta)(n + \lambda - t - \alpha_p + a)(n + \lambda - t + 1 - d_g)}{(t - m)!\,(n + \lambda - m)_m(2n + \lambda - t - m)_m(n + \beta)(n + \alpha_p - a)(n + d_g - 1)}$$

$$\times\ {}_{p+g+3}F_{p+g+2}\left(\begin{matrix}-t + m, 2n + \lambda - t - m, n + \lambda - t + 1 - \beta, n + \lambda - t + 1 - \alpha_p + a, n + \lambda - t + 2 - d_g \\ 2n + \lambda + 1 - t, n + \lambda - t - \beta, n + \lambda - t - \alpha_p + a, n + \lambda - t + 1 - d_g\end{matrix}\,\middle|\, 1\right), \qquad (4)$$

$$L_m = \frac{(n + 1 - m)_m(2n + \lambda - 2m)_{2m}(n - m + \rho_q - a)(n - m + c_f)}{(m - 1)!\,(n + \lambda - m)_m(2n + \lambda - t - m + 1)_{m-1}(n + \beta)(n + \alpha_p - a)(n + d_g - 1)}$$

$$\times\ {}_{q+f+2}F_{q+f+1}\left(\begin{matrix}1 - m, 2n + \lambda - t - m + 1, n - m + 1 + \rho_q - a, n - m + 1 + c_f \\ 2n + \lambda + 1 - 2m, n - m + \rho_q - a, n - m + c_f\end{matrix}\,\middle|\, 1\right)$$

$$= \frac{(-)^{q+f}(n + 1 - m)_m(2n + \lambda - 2m)(2n + \lambda - t + 1)_{t-1}(n + \lambda - t + 1 - \rho_q + a)(n + \lambda - t + 1 - c_f)}{(t - m - 1)!\,(n + \lambda - m)_m(2n + \lambda - t - m + 1)_{m-1}(n + \beta)(n + \alpha_p - a)(n + d_g - 1)}$$

$$\times\ {}_{q+f+2}F_{q+f+1}\left(\begin{matrix}-t + m + 1, 2n + \lambda - t - m + 1, n + \lambda - t + 2 - \rho_q + a, n + \lambda - t + 2 - c_f \\ 2n + \lambda + 1 - t, n + \lambda - t + 1 - \rho_q + a, n + \lambda - t + 1 - c_f\end{matrix}\,\middle|\, 1\right).$$

$$\sum_{m=0}^{t}\left(K_m+\frac{1}{\gamma}L_m\right)h_{n-m}(\gamma)=\frac{\beta(\alpha_p-a)(d_g-1)(n+\lambda)_n}{(n+\beta)(n+\alpha_p-a)(n+d_g-1)(n+\lambda-t)_n}, \tag{2}$$

$$\sum_{m=0}^{t}\left(K_m+\frac{1}{\gamma}L_m\right)\psi_{n-m}(z,\gamma)=\left[-\frac{nc_f(\rho_q-1)}{\gamma}\right]^a[\beta(d_g-1)\,\alpha_p]^{1-a}$$
$$\times\frac{(2n+\lambda-t)_t\Gamma(n+\lambda+a-t)}{(n+\beta)(n+\alpha_p-a)(n+d_g-1)\,\Gamma(n+\lambda)}$$
$$\times\,{}_{f+2}F_{g+1}\left(\begin{matrix}a-n,n+\lambda+a-t,c_f+a\\ \beta+a,d_g+a-1\end{matrix}\,\middle|\,\frac{z}{\gamma}\right). \tag{3}$$

The difference equations (2) *and* (3) *are homogeneous if*

$$\beta(\alpha_p-a)(d_g-1)=0 \qquad \text{and} \qquad [c_f(\rho_q-1)]^a=0,$$

respectively. In the special case $\gamma=z$, $\lambda=\alpha+\beta+1$, (3) *is homogeneous when* $f=g=0$ *if* $\alpha+2+a-t$ *is a negative integer or zero.*

Corollary. If $\beta=0$, $f=g=0$ and $\gamma=z$, then $B_n(z)$ and $A_n(z,z)$ satisfy the same homogeneous difference equation provided that the number r is a negative integer or zero where

$$r=q-a-\alpha \text{ if } p\le q+1,$$

$$r=p-a-\alpha-1 \text{ if } p>q+1.$$

REMARK 1. Note that the ${}_{f+2}F_{g+1}$ on the right-hand side of (3) is an extended Jacobi polynomial and thus can be generated by an application of the recursion formula developed in 5.12.3.

REMARK 2. If $a=0$ and $\beta(d_g-1)=0$, a more convenient form of (3) is

$$\sum_{m=0}^{t}\left(K_m+\frac{1}{\gamma}L_m\right)A_{n-m}(z,\gamma)=-\frac{n(z/\gamma)(2n+\lambda-t)_t\Gamma(n+\lambda+1-t)\,\alpha_pc_f}{(n+\beta)(n+\alpha_p)(n+d_g-1)\,\Gamma(n+\lambda)}$$
$$\times\,{}_{f+3}F_{g+2}\left(\begin{matrix}1-n,n+\lambda+1-t,c_f+1,1\\ \beta+1,d_g,2\end{matrix}\,\middle|\,\frac{z}{\gamma}\right). \tag{5}$$

REMARK 3. If in $B_n(\gamma)$ [see (1)], a numerator parameter c and a denominator parameter d coalesce, then (2) and (3) will in general reduce in length only if $c=d=n+\lambda+1-t$. For particular numerical values of $c = d(\neq n + \lambda + 1 - t)$, (2) and (3) though still valid are no longer the desired recursion formulas of shortest length.

REMARK 4. Confluent forms of the theorem and corollary follow upon replacing γ by $\gamma\lambda$, and letting $\lambda \to +\infty$.

REMARK 5. For an alternative technique to evaluate $A_n(z, \gamma)$, consider the form of the third expression in 5.12(11). Let

$$V_{n,k}(\gamma) = {}_{r+3}F_s\left(\begin{matrix} -n + a + k, n + \lambda + a + k, a_r + k, 1 \\ b_s + k \end{matrix}\middle| \frac{1}{\gamma}\right).$$

Then

$$V_{n,k-1}(\gamma) = 1 + \frac{(-n + a - 1 + k)(n + \lambda + a - 1 + k)(a_r - 1 + k)}{(b_s - 1 + k)\gamma} V_{n,k}(\gamma),$$

$$V_{n,n-a}(\gamma) = 1. \tag{6}$$

5.13.4. Recursion Formula for Coefficients in the Expansion of the G-Function in Series of Extended Jacobi Polynomials

From the expansion formula 5.10.1(1), it is sufficient to consider

$$V(n, z, \lambda) = G^{m,k}_{p+1,q+2}\left(z \middle| \begin{matrix} a_{p+1} \\ b_q, n, -n-\lambda \end{matrix}\right), \quad 0 \leqslant m \leqslant q, \quad 0 \leqslant k \leqslant p + 1. \tag{1}$$

From 5.7.2(1), $V(n, z, \lambda)$ satisfies the differential equation

$$[(-)^{p+1-k-m} z(\delta + 1 - a_{p+1}) - (\delta - n)(\delta + n + \lambda)(\delta - b_q)]\, V(z) = 0,$$
$$\delta = z d/dz. \tag{2}$$

As in the developments of 5.13.3, the function $V(n, z, \lambda)$ and the G-functions which comprise the basis of solutions of (2) at its singular points $z = 0$ and $z = \infty$ suitably normalized with respect to n, satisfy the same difference equation. We have need for the following condition.

$(a_i - b_j)$ is not a positive integer, $i = 1, 2, \ldots, k$, $j = 1, 2, \ldots, m$ with $b_h = n$ if $h = q + 1$, and $b_h = -n - \lambda$ if $h = q + 2$. (3)

We have the following theorem.

Theorem. *Under the conditions* (3), $V(n, z, \lambda)$ *satisfies the difference equation*

$$\sum_{r=0}^{t} (H_r(n, \lambda) + z^{-1}J_r(n, \lambda))\, V_{n+r}(z) = 0,$$

$$H_0(n, \lambda) = 1, \qquad J_0(n, \lambda) = J_t(n, \lambda) = 0, \qquad t = \max(q + 2, p + 1), \tag{4}$$

where

$$H_r(n, \lambda) = \frac{(-)^r(2n + \lambda + 2r)(2n + \lambda)_r}{(2n + \lambda)\, r!}$$

$$\times\ {}_{p+3}F_{p+2}\left(\begin{matrix}-r, 2n + \lambda + r, n + 2 - a_{p+1} \\ 2n + \lambda + t + 1, n + 1 - a_{p+1}\end{matrix}\,\middle|\, 1\right) \tag{5}$$

$$= \frac{(-)^{r+p+1}(2n + \lambda + 1)_t(n + \lambda + r - 1 + a_{p+1})}{(t - r)!\,(2n + \lambda + r)_r(n + 1 - a_{p+1})}$$

$$\times\ {}_{p+3}F_{p+2}\left(\begin{matrix}r - t, 2n + \lambda + r, n + \lambda + r + a_{p+1} \\ 2n + \lambda + 2r + 1, n + \lambda + r - 1 + a_{p+1}\end{matrix}\,\middle|\, 1\right), \tag{6}$$

$$J_r(n, \lambda) = \frac{(-)^{r+p+1+k-m}(2n + \lambda + 2r)(2n + \lambda + 1)_r(n + 1 - b_q)}{(r - 1)!\,(n + 1 - a_{p+1})}$$

$$\times\ {}_{q+2}F_{q+1}\left(\begin{matrix}1 - r, 2n + \lambda + r + 1, n + 2 - b_q \\ 2n + \lambda + t + 1, n + 1 - b_q\end{matrix}\,\middle|\, 1\right) \tag{7}$$

$$= \frac{(-)^{r+q+p+1+k-m}(2n + \lambda + r)(2n + \lambda + 1)_t(n + \lambda + r + b_q)}{(t - 1 - r)!\,(2n + \lambda + r)_r(n + 1 - a_{p+1})}$$

$$\times\ {}_{q+2}F_{q+1}\left(\begin{matrix}r + 1 - t, 2n + \lambda + r + 1, n + \lambda + r + 1 + b_q \\ 2n + \lambda + 2r + 1, n + \lambda + r + b_q\end{matrix}\,\middle|\, 1\right). \tag{8}$$

Moreover, the above functions satisfy no nontrivial equation of the form (4) *of lower order than t, provided that no* $a_i = b_j$, $i = 1, 2,..., p + 1$, $j = 1, 2,..., q$.

REMARK 1. With $m = q + 2$, (1) also satisfies (4), and with $m = q + 1$, (1) multiplied by $e^{i\varphi n}$, $e^{i\varphi} = -1$, also satisfies (4).

REMARK 2. The basis of solutions of (2) in the neighborhood of the singular points at $z = 0$ and $z = \infty$ can be readily deduced from the discussion in 5.7.2. Thus if $p > q + 1$, the functions

$$e^{i\varphi n}\, G^{1,p+1}_{p+1,q+2}\left(z \exp[-i\pi(m + k - p)]\,\middle|\,\begin{matrix}a_{p+1} \\ n, b_q, -n - \lambda\end{matrix}\right),$$

$$e^{i\varphi n}\, G^{1,p+1}_{p+1,q+2}\left(z \exp[-i\pi(m+k-p)] \,\middle|\, \begin{matrix} a_{p+1} \\ -n-\lambda, b_q, n \end{matrix}\right), \qquad e^{i\varphi} = -1,$$

$$G^{1,p+1}_{p+1,q+2}\left(z \exp[-i\pi(m+k-p)] \,\middle|\, \begin{matrix} a_{p+1} \\ b_h, b_1, \ldots, b_{h-1}, b_{h+1}, \ldots, b_q, n, -n-\lambda \end{matrix}\right),$$

$$h = 1, 2, \ldots, q, \tag{9}$$

$$G^{0,p+1}_{p+1,q+2}\left(z \exp[i\pi(2r+m+k-p-1)] \,\middle|\, \begin{matrix} a_{p+1} \\ b_q, n, -n-\lambda \end{matrix}\right),$$

$$r = 1, 2, \ldots, p-q-1,$$

satisfy (2). These functions also satisfy (4). Likewise if $q+1 > p$, the functions

$$G^{q+2,1}_{p+1,q+2}\left(z \exp[i\pi(q-m-k+1)] \,\middle|\, \begin{matrix} a_r, a_1, \ldots, a_{r-1}, a_{r+1}, \ldots, a_{p+1} \\ b_q, n, -n-\lambda \end{matrix}\right),$$

$$r = 1, 2, \ldots, p+1, \tag{10}$$

$$G^{q+2,0}_{p+1,q+2}\left(z \exp[-i\pi(2h+m+k-q)] \,\middle|\, \begin{matrix} a_{p+1} \\ b_q, n, -n-\lambda \end{matrix}\right),$$

$$h = 1, 2, \ldots, q+1-p,$$

satisfy both (2) and (4). Furthermore, the bases of solutions of (2) described above are linearly independent qua functions of n.

REMARK 3. If the conditions which must be imposed on the parameters in (1) are violated so that some of the solutions given in 5.7.2 are dependent, then additional solutions can be found via limit processes. Such solutions suitably normalized also satisfy (4).

REMARK 4. There is a close relationship between the recursion formulas for the extended Jacobi function and for $V(n, z, \lambda)$. To see this, note that

$$\mathscr{E}(-n, -z^{-1}, -\lambda) = \frac{\Gamma(b_s)}{\Gamma(n)\,\Gamma(-n-\lambda)\,\Gamma(a_r)}\, G^{r+3,1}_{s+1,r+3}\left(z \,\middle|\, \begin{matrix} 1, b_s \\ a_r, 1, n, -n-\lambda \end{matrix}\right), \tag{11}$$

and except for a multiplicative function, the right-hand side of (11) can be brought into the form (1) by a simple change of notation.

REMARK 5. A recursion formula for

$$V(n, z) = G^{m,k}_{p+1,q+1}\left(z \,\middle|\, \begin{matrix} a_{p+1} \\ b_q, n \end{matrix}\right) \tag{12}$$

and for the functions which satisfy the same differential equation as does $V(n, z)$ at the singular points $z = 0$ and $z = \infty$ follows by confluence from (4).

REMARK 6. If we are concerned with the coefficients in the expansion of a ${}_uF_v$ in series of extended Jacobi polynomials, then from 5.10.2(5) it is sufficient to analyze

$$C_n(z,\lambda) = \frac{(a_{p+1})_n z^n}{(b_q)_n(n+\lambda)_n n!}\,{}_{p+1}F_{q+1}\left(\begin{matrix}a_{p+1}+n\\ b_q+n,\lambda+1+2n\end{matrix}\middle| z\right),$$

$$p \leqslant q \quad \text{or} \quad p = q+1 \quad \text{and} \quad |\arg(1-z)| < \pi. \tag{13}$$

Then for n a positive integer,

$$C_n(z,\lambda) = \frac{\Gamma(b_q)\,\Gamma(n+\lambda)(2n+\lambda)(-)^n}{\Gamma(a_{p+1})\,n!}\,G^{1,p+1}_{p+1,q+2}\left(-z\middle|\begin{matrix}1-a_{p+1}\\ n,1-b_q,-n-\lambda\end{matrix}\right), \tag{14}$$

and except for a multiplicative factor this is the first G-function given in Remark 2 with $k = p+1$ and $m = 0$.

For applications, it is convenient to have available the functions which satisfy the same differential equation and difference equation as does $C_n(z,\lambda)$. When $p \leqslant q$ or when $p = q+1$ and $|\arg(1-z)| < \pi$, they are as follows.

$$P_n^{(h)}(z,\lambda) = \frac{(2n+\lambda)\,\Gamma(n+\lambda)\,z^{1-b_h}}{\Gamma(2-b_h-n)\,\Gamma(n+\lambda+2-b_h)\,n!\,\Gamma(1+b_q-b_h)}$$

$$\times\,{}_{p+1}F_{q+1}\left(\begin{matrix}1+a_{p+1}-b_h\\ 1+b_q-b_h^*,2-b_h-n,n+\lambda+2-b_h\end{matrix}\middle| z\right),$$

$$h = 1,2,\ldots,q, \tag{15}$$

$$Q_n(z,\lambda) = \frac{(2n+\lambda)\,\Gamma(n+\lambda)(a_{p+1})_{-n-\lambda}z^{-n-\lambda}}{\Gamma(1-2n-\lambda)(b_q)_{-n-\lambda}n!}$$

$$\times\,{}_{p+1}F_{q+1}\left(\begin{matrix}a_{p+1}-n-\lambda\\ b_q-n-\lambda,1-\lambda-2n\end{matrix}\middle| z\right). \tag{16}$$

In (13), (15), and (16), we suppose that all parameters are such that the functions make sense. As functions of z or n, these solutions are linearly independent provided that none of the numbers $\lambda+1+2n$, b_h+n, $h = 1,2,\ldots,q$ are integers or zero, and that no two of these numbers differ by an integer or zero. If $\lambda = 0$, and n is a positive integer, $Q_n(z,\lambda) = C_n(z,\lambda)$. In this event, a linearly independent solution can be defined by

$$Q_n^*(z) = \lim_{\lambda\to 0}\frac{C_n(z,\lambda)-Q_n(z,\lambda)}{2\lambda}. \tag{17}$$

Thus,

$$Q_n^*(z) = \frac{(a_{p+1})_n z^n}{(b_q)_n (2n)!} \sum_{k=0}^{\infty} \frac{(a_{p+1}+n)_k z^k}{(b_q+n)_k (2n+1)_k k!} \Big\{ \ln z - \psi(2n+1+k) - \psi(1+k)$$

$$+ \sum_{h=1}^{p+1} \psi(a_h+n+k) - \sum_{h=1}^{q} \psi(b_h+n+k) \Big\}$$

$$- \frac{(-)^{n(p+q+1)}(1-b_q)_n (2n-1)!}{(1-a_{p+1})_n z^n} \sum_{k=0}^{2n-1} \frac{(a_{p+1}-n)_k z^k}{(b_q-n)_k (1-2n)_k k!}. \tag{18}$$

Similarly, in all other situations where the parameters are such that solutions coalesce, we can construct independent solutions after the manner of 5.3.1(7-10).

In illustration of the above results, we have

$$C_n(z, \lambda) = \frac{(a)_n z^n}{(b)_n (n+\lambda)_n n!} \, {}_4F_3\left(\begin{matrix} a_1+n, a_2+n, a_3+n, a_4+n \\ b_1+n, b_2+n, \lambda+1+2n \end{matrix} \middle| \, z \right) \tag{19}$$

satisfies

$$\sum_{m=0}^{4} [E_m(n, \lambda) + z^{-1} F_m(n, \lambda)] \, C_{n+m}(z, \lambda) = 0,$$

$$E_0(n, \lambda) = 1, \qquad F_0(n, \lambda) = F_4(n, \lambda) = 0, \tag{20}$$

where

$$E_1(n, \lambda) = -\frac{(2n+\lambda)(n+1)}{(n+\lambda)} \left[1 - \frac{(2n+\lambda+1)(n+a+1)}{(2n+\lambda+5)(n+a)}\right],$$

$$E_2(n, \lambda) = \frac{(2n+\lambda)_2 (n+1)_2}{2(n+\lambda)_2}$$

$$\times \left[1 - \frac{2(2n+\lambda+2)(n+a+1)}{(2n+\lambda+5)(n+a)} + \frac{(2n+\lambda+2)_2 (n+a+2)}{(2n+\lambda+5)_2 (n+a)}\right],$$

$$E_3(n, \lambda) = -\frac{(2n+\lambda)_3 (n+1)_3 (n+\lambda+3-a)}{(n+\lambda)_3 (2n+\lambda+5)_2 (n+a)}$$

$$\times \left[1 - \frac{(2n+\lambda+3)(n+\lambda+4-a)}{(2n+\lambda+7)(n+\lambda+3-a)}\right], \tag{21}$$

$$E_4(n, \lambda) = \frac{(2n+\lambda)_4 (n+1)_4 (n+\lambda+4-a)}{(n+\lambda)_4 (2n+\lambda+5)_4 (n+a)},$$

$$F_1(n, \lambda) = -\frac{(2n+\lambda)_2 (n+1)(n+b)}{(n+\lambda)(n+a)},$$

$$F_2(n, \lambda) = \frac{(2n+\lambda)_3 (n+1)_2 (n+b)}{(n+\lambda)_2 (n+a)} \left[1 - \frac{(2n+\lambda+3)(n+b+1)}{(2n+\lambda+5)(n+b)}\right],$$

$$F_3(n, \lambda) = -\frac{(2n+\lambda)_4 (n+1)_3 (n+\lambda+4-b)}{(n+\lambda)_3 (2n+\lambda+5)_2 (n+a)}.$$

Here $n + a$ is short for $(n + a_1)(n + a_2)(n + a_3)(n + a_4)$, $n + b$ is short for $(n + b_1)(n + b_2)$, etc.

Similarly,

$$D_n(z, \lambda) = \frac{(a)_n z^n}{(b)_n (n + \lambda)_n n!} \, {}_3F_2\left(\begin{matrix} a_1 + n, a_2 + n, a_3 + n \\ b + n, \lambda + 1 + 2n \end{matrix} \middle| z\right) \tag{22}$$

satisfies

$$\begin{gathered} \sum_{m=0}^{3} [G_m(n, \lambda) + z^{-1} H_m(n, \lambda)] \, D_{n+m}(z, \lambda) = 0, \\ G_0(n, \lambda) = 1, \qquad H_0(n, \lambda) = H_3(n, \lambda) = 0, \end{gathered} \tag{23}$$

where

$$\begin{aligned} G_1(n, \lambda) &= -\frac{(2n + \lambda)(n + 1)}{(n + \lambda)}\left[1 - \frac{(2n + \lambda + 1)(n + a + 1)}{(2n + \lambda + 4)(n + a)}\right], \\ G_2(n, \lambda) &= \frac{(2n + \lambda)_3 (n + 1)_2 (n + \lambda + 3 - a)}{(n + \lambda)_2 (2n + \lambda + 4)_2 (n + a)} \\ &\quad \times \left[1 - \frac{(2n + \lambda + 5)(n + \lambda + 2 - a)}{(2n + \lambda + 2)(n + \lambda + 3 - a)}\right], \\ G_3(n, \lambda) &= \frac{(2n + \lambda)_3 (n + 1)_3 (n + \lambda + 3 - a)}{(n + \lambda)_3 (2n + \lambda + 4)_3 (n + a)}, \\ H_1(n, \lambda) &= -\frac{(2n + \lambda)_2 (n + 1)(n + b)}{(n + \lambda)(n + a)}, \\ H_2(n, \lambda) &= -\frac{(2n + \lambda)_3 (n + 1)_2 (n + \lambda + 3 - b)}{(n + \lambda)_2 (2n + \lambda + 4)(n + a)}. \end{aligned} \tag{24}$$

Equations (22-24) can be deduced directly from (19-21) by putting $a_4 = b = n + \lambda + 4$. Similarly a three term recursion formula for

$$E_n(z, \lambda) = \frac{(a)_n z^n}{(n + \lambda)_n n!} \, {}_2F_1\left(\begin{matrix} a_1 + n, a_2 + n \\ \lambda + 1 + 2n \end{matrix} \middle| z\right). \tag{25}$$

follows from (22-24) by setting $a_3 = b = n + \lambda + 3$.

5.14. Inequalities

For details on the development of the following inequalities, their extension and possible improvement, see Luke (1972a). For other inequalities, see the references in 5.1.

Throughout this section we employ our usual shorthand notation for the ${}_pF_q$ and G-function. Also let

$$\theta = \frac{\alpha_p}{\rho_p}, \qquad \varphi = \frac{\alpha_p + 1}{\rho_p + 1}, \qquad \eta = \frac{\alpha_p + 2}{\rho_p + 2}. \tag{1}$$

As a reminder, note for example that

$$\varphi = \prod_{j=1}^{p} (\alpha_j + 1)/(\rho_j + 1) \tag{2}$$

In Theorems 1-8, the inequalities become equalities if $z \to 0$ or if any of the numerator parameters $\to 0$. In Theorems 9-12, the inequalities become equalities if $z \to \infty$, and likewise for certain values of the parameters as, for example, when $\sigma \to 0$ in (17).

Theorem 1.

$$[1 + \sigma\theta z]^{-1} < {}_{p+1}F_p\left(\begin{matrix}\sigma, \alpha_p \\ \rho_p\end{matrix}\middle| -z\right)$$

$$< 1 - \frac{2\sigma\theta}{(\sigma+1)\varphi} + \frac{2\sigma\theta}{(\sigma+1)\varphi}\left[1 + \frac{(\sigma+1)\varphi z}{2}\right]^{-1}, \tag{3}$$

$$z > 0, \qquad 0 < \sigma \leqslant 1, \qquad \rho_j \geqslant \alpha_j > 0, \qquad j = 1, 2, \ldots, p;$$

Theorem 2.

$$1 - \frac{a(c+1)}{c(a+1)} + \frac{a(c+1)}{c(a+1)}\left[1 + \frac{(a+1)\theta z}{c+1}\right]^{-1} < {}_{p+2}F_{p+1}\left(\begin{matrix}1, a, \alpha_p \\ c, \rho_p\end{matrix}\middle| -z\right)$$

$$< 1 - \frac{\theta}{\varphi} + \frac{\theta}{\varphi}\left[1 + \frac{a\varphi z}{c}\right]^{-1}, \tag{4}$$

$$z > 0, \qquad 0 < c \leqslant a, \qquad \rho_j \geqslant \alpha_j > 0, \qquad j = 1, 2, \ldots, p.$$

Theorem 3.

$$-\frac{1}{\sigma} + \frac{(\sigma+1)}{\sigma}\left[1 + \frac{\theta z}{\sigma+1}\right]^{-1} < {}_{p+1}F_{p+1}\left(\begin{matrix}1, \alpha_p \\ \sigma, \rho_p\end{matrix}\middle| -z\right)$$

$$< 1 - \frac{\theta}{\varphi} + \frac{\theta}{\varphi}\left[1 + \frac{\varphi z}{\sigma}\right]^{-1}, \tag{5}$$

$$z > 0, \qquad \sigma > 0, \qquad \rho_j \geqslant \alpha_j > 0, \qquad j = 1, 2, \ldots, p.$$

Theorem 4.

$$(2-\sigma)^{-1}[1+\theta z]^{-1}-\frac{\sigma-1}{2-\sigma}+\frac{(\sigma-1)\theta}{(2-\sigma)\varphi}[1-\{1+(\sigma-1)\varphi z\}^{-1}]$$

$$<{}_{p+1}F_p\left(\begin{matrix}\sigma,\alpha_p\\ \rho_p\end{matrix}\middle|-z\right)$$

$$<\frac{\sigma}{2-\sigma}-\frac{2(\sigma-1)}{2-\sigma}\left[1+\frac{\sigma\theta z}{2}\right]^{-1}-\frac{\sigma\theta}{(2-\sigma)\varphi}[1-\{1+\varphi z\}^{-1}],$$

$$z>0,\qquad 1\leqslant\sigma\leqslant 2,\qquad \rho_j\geqslant\alpha_j>0,\qquad j=1,2,\ldots,p \tag{6}$$

Theorem 5.

$$1-\frac{\sigma\theta}{(\sigma+1)\varphi}[1-\{1+(\sigma+1)\varphi z\}^{-1}]<{}_{p+1}F_p\left(\begin{matrix}\sigma,\alpha_p\\ \rho_p\end{matrix}\middle|-z\right)$$

$$<1-\frac{\sigma\theta z}{\sigma+2}-\frac{2\sigma(\sigma+1)}{(\sigma+2)^2}\left[1-\left\{1+\frac{(\sigma+2)\theta z}{2}\right\}^{-1}\right], \tag{7}$$

$$z>0,\qquad -1<\sigma<0,\qquad \rho_j\geqslant\alpha_j>0,\qquad j=1,2,\ldots,p.$$

Theorem 6.

$$1-\frac{2\sigma\theta}{(\sigma+1)\varphi}+\frac{2\sigma\theta}{(\sigma+1)\varphi}\left[1+\frac{(\sigma+1)\varphi z}{2}\right]^{-1}<{}_{p+1}F_p\left(\begin{matrix}\sigma,\alpha_p\\ \rho_p\end{matrix}\middle|-z\right)$$

$$<1-\sigma\theta z+\frac{3\sigma(\sigma+1)\theta\varphi z}{2(\sigma+2)\eta}-\frac{9\sigma(\sigma+1)\theta\varphi}{2(\sigma+2)^2\eta^2}$$

$$\times\left[1-\left(1+\frac{(\sigma+2)\eta z}{3}\right)^{-1}\right], \tag{8}$$

$$z>0,\qquad -1<\sigma<0,\qquad \rho_j\geqslant\alpha_j>0,\qquad j=1,2,\ldots,\rho.$$

Theorem 7.

$$(1+\theta z)^{-\sigma}<{}_{p+1}F_p\left(\begin{matrix}\sigma,\alpha_p\\ \rho_p\end{matrix}\middle|-z\right)<1-\theta+\theta(1+z)^{-\sigma},$$

$$z>0,\qquad \sigma>0,\qquad \rho_j\geqslant\alpha_j>0,\qquad j=1,2,\ldots,p; \tag{9}$$

$$(1-\theta z)^{-\sigma}<{}_{p+1}F_p\left(\begin{matrix}\sigma,\alpha_p\\ \rho_p\end{matrix}\middle|z\right)<1-\theta+\theta(1-z)^{-\sigma},$$

$$0<z<1,\qquad \sigma>0,\qquad \rho_j\geqslant\alpha_j>0,\qquad j=1,2,\ldots,p; \tag{10}$$

$$1-\sigma\theta\left(1-\frac{\varphi}{2}\right)z-\frac{\sigma\theta\varphi z}{2(1+z)^{\sigma+1}}$$

$$<{}_{p+1}F_p\left(\begin{matrix}\sigma,\alpha_p\\ \rho_p\end{matrix}\middle|-z\right)<1-\frac{\sigma\theta z}{(1+(\varphi z/2))^{\sigma+1}}, \tag{11}$$

$$z>0,\qquad \sigma>0,\qquad \rho_j\geqslant\alpha_j>0,\qquad j=1,2,\ldots,p;$$

$$1+\frac{\sigma\theta z}{(1-(\varphi z/2))^{\sigma+1}}<{}_{p+1}F_p\left(\begin{matrix}\sigma,\alpha_p\\ \rho_p\end{matrix}\middle| z\right)$$

$$<1+\sigma\theta\left(1-\frac{\varphi}{2}\right)z+\frac{\sigma\theta\varphi z}{2(1-z)^{\sigma+1}}, \tag{12}$$

$$0<z<1,\qquad \sigma>0,\qquad \rho_j\geqslant\alpha_j>0,\qquad j=1,2,\ldots,p.$$

Theorem 8.

$$e^{-\theta z}<{}_pF_p\left(\begin{matrix}\alpha_p\\ \rho_p\end{matrix}\middle| -z\right)<1-\theta+\theta e^{-z}, \tag{13}$$

$$e^{\theta z}<{}_pF_p\left(\begin{matrix}\alpha_p\\ \rho_p\end{matrix}\middle| z\right)<1-\theta+\theta e^{z}, \tag{14}$$

$$1-\theta z\left(1-\frac{\varphi}{2}+\frac{\varphi}{2}e^{-z}\right)<{}_pF_p\left(\begin{matrix}\alpha_p\\ \rho_p\end{matrix}\middle| -z\right)<1-\theta z e^{-\varphi z/2}, \tag{15}$$

$$1+\theta z e^{\varphi z/2}<{}_pF_p\left(\begin{matrix}\alpha_p\\ \rho_p\end{matrix}\middle| z\right)<1+\theta z\left(1-\frac{\varphi}{2}+\frac{\varphi}{2}e^{z}\right),$$

$$z>0,\qquad \rho_j\geqslant\alpha_j>0,\qquad j=1,2,\ldots,p. \tag{16}$$

Theorem 9.

$$1-\frac{\sigma\epsilon\theta}{z+\sigma\epsilon\theta}<\frac{\Gamma(\rho_p)}{\Gamma(\sigma)\Gamma(\epsilon)\Gamma(\alpha_p)}\,G^{p+2,1}_{p+1,p+2}\left(z\middle|\begin{matrix}1,\rho_p\\ \sigma,\epsilon,\alpha_p\end{matrix}\right)$$

$$<1-\frac{\sigma\epsilon\theta}{z+[(\sigma+1)(\epsilon+1)\varphi/2]}, \tag{17}$$

$$z>0,\qquad 0<\sigma\leqslant 1,\qquad \epsilon>0,\qquad \rho_j\geqslant\alpha_j>0,\qquad j=1,2,\ldots,p.$$

Theorem 10.

$$1-\frac{\epsilon a\theta/c}{z+[\epsilon(a+1)\theta/(c+1)]}<\frac{\Gamma(c)\,\Gamma(\rho_p)}{\Gamma(\epsilon)\,\Gamma(a)\,\Gamma(\alpha_p)}\,G^{p+3,1}_{p+2,p+3}\left(z\middle|\begin{matrix}1,c,\rho_p\\ \epsilon,1,a,\alpha_p\end{matrix}\right)$$

$$<1-\frac{\epsilon a\theta/c}{z+[(\epsilon+1)\alpha\varphi/c]} \tag{18}$$

$$z>0,\qquad 0<c\leqslant a,\qquad \epsilon>0,\qquad \rho_j\geqslant\alpha_j>0,\qquad j=1,2,\ldots,p;$$

$$1-\frac{\epsilon\theta}{(2-\sigma)}[z+\epsilon\theta]^{-1}+\frac{\epsilon(\sigma-1)^2\theta}{(2-\sigma)}[z+(\epsilon+1)(\sigma-1)\varphi]^{-1}$$

$$<\frac{\Gamma(\rho_p)}{\Gamma(\epsilon)\,\Gamma(\sigma)\,\Gamma(\alpha_p)}\,G^{p+2,1}_{p+1,p+2}\left(z\middle|\begin{matrix}1,\rho_p\\ \epsilon,\sigma,\alpha_p\end{matrix}\right) \tag{19}$$

$$<1+\frac{\epsilon\sigma(\sigma-1)\theta}{(2-\sigma)}\left[z+\frac{\epsilon\sigma\theta}{2}\right]^{-1}-\frac{\epsilon\sigma\theta}{(2-\sigma)}[z+(\epsilon+1)\varphi]^{-1},$$

$$z>0,\qquad 1\leqslant\sigma\leqslant 2,\qquad \epsilon>0,\qquad \rho_j\geqslant\alpha_j>0,\qquad j=1,2,\ldots,p;$$

Theorem 11.

$$\frac{1}{\Gamma(\epsilon)\,\Gamma(\sigma)}\, G_{1,2}^{2,1}\left(\frac{z}{\theta}\,\middle|\,\begin{matrix}1\\ \epsilon,\ \sigma\end{matrix}\right) < \frac{\Gamma(\rho_p)}{\Gamma(\epsilon)\,\Gamma(\sigma)\,\Gamma(\alpha_p)}\, G_{p+1,p+2}^{p+2,1}\left(z\,\middle|\,\begin{matrix}1,\ \rho_p\\ \epsilon,\ \sigma,\ \alpha_p\end{matrix}\right)$$
$$< 1 - \theta + \frac{\theta}{\Gamma(\epsilon)\,\Gamma(\sigma)}\, G_{1,2}^{2,1}\left(z\,\middle|\,\begin{matrix}1\\ \epsilon,\ \sigma\end{matrix}\right),$$
$$z > 0, \qquad \sigma > 0, \qquad \epsilon > 0, \qquad \rho_j \geqslant \alpha_j > 0, \qquad j = 1, 2,\ldots, p; \tag{20}$$

$$1 - \frac{\epsilon\sigma\theta(1-\varphi/2)}{z} - \frac{\theta\varphi}{2\Gamma(\epsilon)\,\Gamma(\sigma)z}\, G_{1,2}^{2,1}\left(z\,\middle|\,\begin{matrix}1\\ \epsilon+1,\ \sigma+1\end{matrix}\right)$$
$$< \frac{\Gamma(\rho_p)}{\Gamma(\epsilon)\,\Gamma(\sigma)\,\Gamma(\alpha_p)}\, G_{p+1,p+2}^{p+2,1}\left(z\,\middle|\,\begin{matrix}1,\ \rho_p\\ \epsilon,\ \sigma,\ \alpha_p\end{matrix}\right)$$
$$< 1 - \frac{\theta}{\Gamma(\epsilon)\,\Gamma(\sigma)z}\, G_{1,2}^{2,1}\left(\frac{\varphi z}{2}\,\middle|\,\begin{matrix}1\\ \epsilon+1,\ \sigma+1\end{matrix}\right),$$
$$z > 0, \qquad \sigma > 0, \qquad \epsilon > 0, \qquad \rho_j \geqslant \alpha_j > 0, \qquad j = 1, 2,\ldots, p. \tag{21}$$

CHAPTER VI THE GAUSSIAN HYPERGEOMETRIC FUNCTION $_2F_1$

6.1. Introduction

In this chapter we delineate for the most part only those properties of the $_2F_1$ which are not special cases of the material in Chapter 5.

References pertinent to this chapter will be found in 5.1. Except for some work by Runckel, (1971), there seems to be very little in the literature on the zeros of the general $_2F_1$. Döring (1966a) develops a representation for the zeros of $w = f(z)$ when $a(z)w'' + b(z)w' + c(z)w = 0$.

6.2. Elementary Properties

6.2.1. DERIVATIVES

$$\frac{d^n}{dz^n}\left[z^\delta(1-z)^{a+b-c}\,{}_2F_1\left(\begin{matrix}a, b\\ c\end{matrix}\middle| z\right)\right]$$
$$= (\delta-n+1)_n z^{\delta-n}\,{}_3F_2\left(\begin{matrix}c-a, c-b, \delta+1\\ c, \delta+1-n\end{matrix}\middle| z\right). \qquad (1)$$

$$\frac{d^n}{dz^n}\left[z^{n+c-1}(1-z)^{n+a+b-c}\,{}_2F_1\left(\begin{matrix}a+n, b+n\\ c+n\end{matrix}\middle| z\right)\right]$$
$$= (c)_n z^{c-1}(1-z)^{a+b-c}\,{}_2F_1\left(\begin{matrix}a, b\\ c\end{matrix}\middle| z\right). \qquad (2)$$

$$\frac{d^n}{dz^n}\left[z^{c-a+n-1}(1-z)^{a+b-c}\,{}_2F_1\left(\begin{matrix}a, b\\ c\end{matrix}\middle| z\right)\right]$$
$$= (c-a)_n z^{c-a-1}(1-z)^{a+b-c-n}\,{}_2F_1\left(\begin{matrix}a-n, b\\ c\end{matrix}\middle| z\right). \qquad (3)$$

$$\frac{d^n}{dz^n}\left[z^{c-1}(1-z)^{a+b-c}\,{}_2F_1\left(\begin{matrix}a, b\\ c\end{matrix}\middle| z\right)\right]$$
$$= (c-n)_n z^{c-n-1}(1-z)^{a+b-c-n}\,{}_2F_1\left(\begin{matrix}a-n, b-n\\ c-n\end{matrix}\middle| z\right). \qquad (4)$$

$$\frac{d^n}{dz^n}\left[(1-z)^{a+b-c}\,{}_2F_1\left(\begin{matrix}a, b\\ c\end{matrix}\middle| z\right)\right]$$
$$= \frac{(c-a)_n(c-b)_n}{(c)_n}(1-z)^{a+b-c-n}\,{}_2F_1\left(\begin{matrix}a, b\\ c+n\end{matrix}\middle| z\right). \qquad (5)$$

$$\frac{d^n}{dz^n}\left[z^\alpha(1-z)^\beta\right] = (\alpha-n+1)_n z^{\alpha-n}(1-z)^{\beta-n}\,{}_2F_1\left(\begin{matrix}-n,\ \alpha+1+\beta-n\\ \alpha+1-n\end{matrix}\,\middle|\,z\right),$$

α not an integer $< n$, (6)

$$\frac{d^n}{dz^n}\left[z^s(1-z)^\beta\right] = \frac{n!(-\beta)_{n-s}}{(n-s)!}(1-z)^{\beta-n}\,{}_2F_1\left(\begin{matrix}-s,\ \beta+1\\ n-s+1\end{matrix}\,\middle|\,z\right),$$

s an integer, $s \leqslant n$. (7)

$$\frac{d^n}{dz^n}\left[z^{c-1}(1-z)^{b-c+n}\,{}_2F_1\left(\begin{matrix}a,\ b\\ c\end{matrix}\,\middle|\,z\right)\right] = (c-n)_n z^{c-1-n}(1-z)^{b-c}\,{}_2F_1\left(\begin{matrix}a-n,\ b\\ c-n\end{matrix}\,\middle|\,z\right). \tag{8}$$

$$\frac{d^n}{dz^n}\left[(1-z)^{a+n-1}\,{}_2F_1\left(\begin{matrix}a,\ b\\ c\end{matrix}\,\middle|\,z\right)\right] = \frac{(-)^n(a)_n(c-b)_n}{(c)_n}(1-z)^{a-1}\,{}_2F_1\left(\begin{matrix}a+n,\ b\\ c+n\end{matrix}\,\middle|\,z\right). \tag{9}$$

$$A(z) = {}_2F_1\left(\begin{matrix}\nu+1,\ \nu+\mu+1\\ \nu+\mu+\lambda+2\end{matrix}\,\middle|\,z\right) = \frac{(-)^{\mu+1}(\nu+\mu+\lambda+1)!}{\lambda!\,\nu!(\nu+\mu)!\,(\mu+\lambda)!}\,\frac{d^{\nu+\mu}}{dz^{\nu+\mu}}\left[(1-z)^{\mu+\lambda}\frac{d^\lambda}{dz^\lambda}\{z^{-1}\ln(1-z)\}\right], \tag{10}$$

where ν, μ, λ are positive integers or zero. Note that the right-hand side of (10) gives the analytic continuation of the $_2F_1$ on the left throughout the entire complex plane, the positive real axis from 1 to ∞ excluded.

6.2.2. Contiguous Relations

The six functions

$$_2F_1(a\pm 1, b; c; z), \qquad {}_2F_1(a, b\pm 1, c; z), \qquad {}_2F_1(a, b; c\pm 1; z) \tag{1}$$

are called *contiguous* to $_2F_1(a, b; c; z)$. We use the notation $_2F_1(a+)$ and $_2F_1(a-)$ to designate the $_2F_1$ with a replaced by $(a+1)$ and $(a-1)$, respectively, etc. Gauss proved that between the $_2F_1$ and any two functions contiguous to it, there exists a linear relation with coefficients which are linear in z. There are fifteen relations of this kind. Only four of the fifteen are really independent, as all others may be obtained by elimination and use of the fact that the $_2F_1$ is symmetric in a and b. In view of symmetry, it is sufficient to record only nine of the fifteen relations. They are as follows:

$$(c-a)\,{}_2F_1(a-) + (2a-c-az+bz)\,{}_2F_1 + a(z-1)\,{}_2F_1(a+) = 0. \quad (2)$$

$$c(c-1)(z-1)\,{}_2F_1(c-)$$
$$+\,c[c-1-(2c-a-b-1)z]\,{}_2F_1 + (c-a)(c-b)z\,{}_2F_1(c+) = 0. \quad (3)$$

$$c[a+(b-c)z]\,{}_2F_1 - ac(1-z)\,{}_2F_1(a+) + (c-a)(c-b)z\,F(c+) = 0. \quad (4)$$

$$c(1-z)\,{}_2F_1 - c\,{}_2F_1(a-) + (c-b)z\,{}_2F_1(c+) = 0. \quad (5)$$

$$(b-a)\,{}_2F_1 + a\,{}_2F_1(a+) - b\,{}_2F_1(b+) = 0. \quad (6)$$

$$(c-a-b)\,{}_2F_1 + a(1-z)\,{}_2F_1(a+) - (c-b)\,F(b-) = 0. \quad (7)$$

$$(c-a-1)\,{}_2F_1 + a\,{}_2F_1(a+) - (c-1)\,{}_2F_1(c-) = 0. \quad (8)$$

$$(b-a)(1-z)\,{}_2F_1 - (c-a)\,{}_2F_1(a-) + (c-b)\,{}_2F_1(b-) = 0. \quad (9)$$

$$[a-1+(b+1-c)z]\,{}_2F_1$$
$$+\,(c-a)\,{}_2F_1(a-) - (c-1)(1-z)\,{}_2F_1(c-) = 0. \quad (10)$$

Equation (2) can be viewed as a second order difference equation in the parameter a. A second solution of this difference equation can be taken in the form

$$\left(\Gamma(1+a-c)/\Gamma(a)\right)w_2 \quad (11)$$

where w_2 is given by 6.3(6). Thus the two solutions of the differential equation 6.3(1) also satisfy the same difference equation. Similarly, a second solution of (3) can be taken in the form

$$\left(\Gamma(c-1)\Gamma(1+a-c)\Gamma(1+b-c)/\Gamma(1-c)\right)w_2. \quad (12)$$

6.2.3. Integral Representations

$${}_2F_1\left({a,b \atop c}\,\middle|\,z\right) = \frac{\Gamma(c)}{\Gamma(c-a)\Gamma(a)}\int_0^1 t^{a-1}(1-t)^{c-a-1}(1-zt)^{-b}\,dt,$$
$$R(c) > R(a) > 0, \qquad |\arg(1-z)| < \pi. \quad (1)$$

The series on the left converges if $|z| < 1$, but the integral on the right is single valued and analytic if $|\arg(1-z)| < \pi$, and so the integral gives the analytic continuation of the ${}_2F_1$. We use the same notation for the analytically continued function as for the series. Also

$${}_2F_1\left({a,b \atop c}\,\middle|\,z\right) = \frac{\Gamma(c)}{\Gamma(a)\Gamma(c-a)}\int_0^\infty t^{a-1}(1+t)^{b-c}(1+t-zt)^{-b}\,dt,$$
$$R(c) > R(a) > 0, \qquad |\arg(1-z)| < \pi, \quad (2)$$

or

$${}_2F_1\left(\begin{matrix}a, b\\ a+b+1-c\end{matrix}\middle| 1-z\right)$$
$$= \frac{\Gamma(a+b+1-c)}{\Gamma(a)\Gamma(b+1-c)} \int_0^\infty t^{a-1}(1+t)^{c-a-1}(1+zt)^{-b}\, dt,$$
$$R(a) > 0, \qquad R(b+1-c) > 0, \qquad |\arg z| < \pi. \tag{3}$$

6.3. Differential Equations

The unique second order linear differential equation with three isolated regular singularities at $z = 0, 1$, and ∞ can be written in the form

$$[z(1-z)\, D^2 + \{c - (a+b+1)\, z\}\, D - ab]\, w = 0, \qquad D = d/dz \tag{1}$$

or

$$[\delta(\delta + c - 1) - z(\delta + a)(\delta + b)]\, w = 0, \qquad \delta = zD. \tag{2}$$

Integration of (2) leads to

$$[(\delta + c - 1) - z(\delta + b) - z(a-1)]\, w - (a-1)(b-1) \int_0^z w(t)\, dt = c - 1 \tag{3}$$

so that if $b = 1$, w satisfies the first-order nonhomogeneous differential equation

$$[(\delta + c - 1) - z(\delta + 1) - z(a-1)]\, w = c - 1. \tag{4}$$

The linearly independent solutions of (1) near the origin are proportional to

$$w_1 = {}_2F_1(a, b; c; z), \tag{5}$$

and

$$w_2 = z^{1-c}\, {}_2F_1(1 + a - c, 1 + b - c; 2 - c; z), \tag{6}$$

provided that c is not an integer or zero.

It is clear that we must enlarge the definition of the solutions of (1) when c is an integer or zero. First we note that w_1 is a polynomial of degree m, a positive integer or zero if either $a = -m$ or $b = -m$. Throughout our discussion, since w_1 is symmetric in a and b, if one of these is specialized, we will let it be a. If $a = -m$ and c is a negative integer, $m + c < 1$, then a well-defined solution of (1) is

$$f(a, b; c; z) = \sum_{k=0}^{-a} \frac{(a)_k (b)_k z^k}{(c)_k k!} = \sum_{k=0}^{m} \frac{(-m)_k (b)_k z^k}{(1-s)_k k!}$$

$$= \frac{(s-1-m)!\,(b)_m z^m}{(s-1)!}\,{}_2F_1\left(\begin{matrix}-m,\, s-m\\ 1-m-b\end{matrix}\,\middle|\, z^{-1}\right),$$

$$c = 1-s, \qquad s \text{ a positive integer}, \qquad m < s, \tag{7}$$

and this is independent of w_2. Likewise, if a and c are positive integers, $a < c$, then

$$z^{1-c} f(1+a-c,\, 1+b-c;\, 2-c;\, z)$$

$$= z^{-s} \sum_{k=0}^{s-m} \frac{(m-s)_k (b-s)_k z^k}{(1-s)_k k!}$$

$$= \frac{(m-1)!\,\Gamma(b-m)\, z^{-m}}{(s-1)!\,\Gamma(b-s)}\,{}_2F_1\left(\begin{matrix}m-s,\, m\\ 1+m-b\end{matrix}\,\middle|\, z^{-1}\right), \tag{8}$$

$a = m$ a positive integer, $c = 1+s$, s a positive integer or zero, $m < s+1$, is a well-defined solution of (1) and is independent of w_1. In (7) and (8), the ${}_2F_1$ notation applies only when $(1-m-b)$ and $(1+m-b)$, respectively, are not negative integers.

If $a = c$, $w_1(z) = (1-z)^{-b}$ and a second solution is given by (6) provided that a is not a positive integer. When $a = c$, it can be shown that a solution of (1) is

$$u = (1-a)(1-z)^{-b} \int^z t^{-a}(1-t)^{b-1}\, dt. \tag{9}$$

In particular, if m and n are positive integers or zero, and the constant of integration in (9) is ignored,

$$u = z^{-n-1}(1-z)^{-1} \sum_{k=0}^{n-m} \frac{(m-n)_k}{(-n)_k} \left(\frac{z}{z-1}\right)^k,$$

$$c = a = n+2, \qquad b = n-m+1, \qquad n \geqslant m, \tag{10}$$

$$u = (1-m)(1-z)^{-m-n-1} \left\{ \sum_{\substack{k=0\\ k\neq m-1}}^{m+n} \frac{(-)^k \binom{m+n}{k} z^{k+1-n}}{k+1-m} + (-)^{m-1}\binom{m+n}{m-1} \ln z \right\},$$

$$c = a = m, \qquad b = m+n+1. \tag{11}$$

If c is a negative integer or zero, say $c = -m$, w_1 is undefined. In this event, a solution can be taken proportional to

$$\lim_{c\to -m} \frac{w_1}{\Gamma(c)} = \frac{(a)_{m+1}(b)_{m+1} z^{m+1}}{(m+1)!}\,{}_2F_1(m+a+1,\, m+b+1;\, m+2;\, z). \tag{12}$$

However, this is proportional to w_2 with $c = -m$.

Delineation of the complete solution of (1) is taken up in 6.6. See the next section for solutions of (1) in the neighborhood of the singularities $z = 1$ and $z = \infty$.

The differential equation

$$\frac{h(h-1)}{(h')^2} y'' + \left\{ \frac{h(h-1)}{(h')^3} \left(\frac{2\alpha h'}{z} + 2f'h' - h'' \right) + \frac{(a+b+1)\, h - c}{h'} \right\} y' + \left\{ \left(\frac{\alpha}{z} + f' \right) \left(\frac{(a+b+1)\, h - c}{h'} \right) + \frac{h(h-1)}{(h')^3} \left[\frac{\alpha(\alpha-1)\, h'}{z^2} + \frac{2\alpha f' h'}{z} + f''h' + (f')^2 h' - \frac{\alpha h''}{z} - f'h'' \right] + ab \right\} y = 0, \tag{13}$$

with $h \equiv h(z)$, $f \equiv f(z)$, and $y \equiv y(z)$, is satisfied by

$$y(z) = z^{-\alpha} e^{-f(z)} {}_2F_1 \left(\begin{matrix} a, b \\ c \end{matrix} \middle| h(z) \right). \tag{14}$$

6.4. Kummer Solutions and Transformation Formulae

Kummer's 24 solutions of 6.3(1) are as follows:

$$w_1 = {}_2F_1(a, b; c; z) \tag{1}$$

$$= (1-z)^{c-a-b} {}_2F_1(c-a, c-b; c; z) \tag{2}$$

$$= (1-z)^{-a} {}_2F_1(a, c-b; c; z/(z-1)) \tag{3}$$

$$= (1-z)^{-b} {}_2F_1(c-a, b; c; z/(z-1)). \tag{4}$$

$$w_2 = z^{1-c} {}_2F_1(1+a-c, 1+b-c; 2-c; z) \tag{5}$$

$$= z^{1-c}(1-z)^{c-a-b} {}_2F_1(1-a, 1-b; 2-c; z) \tag{6}$$

$$= z^{1-c}(1-z)^{c-a-1} {}_2F_1(1+a-c, 1-b; 2-c; z/(z-1)) \tag{7}$$

$$= z^{1-c}(1-z)^{c-b-1} {}_2F_1(1+b-c, 1-a; 2-c; z/(z-1)). \tag{8}$$

$$w_3 = {}_2F_1(a, b; a+b+1-c; 1-z) \tag{9}$$

$$= z^{1-c} {}_2F_1(a+1-c, b+1-c; a+b+1-c; 1-z) \tag{10}$$

$$= z^{-a} {}_2F_1(a, a+1-c; a+b+1-c; 1-z^{-1}) \tag{11}$$

$$= z^{-b} {}_2F_1(b, b+1-c; a+b+1-c; 1-z^{-1}). \tag{12}$$

$$w_4 = (1-z)^{c-a-b}\, {}_2F_1(c-a, c-b; c+1-a-b; 1-z) \tag{13}$$
$$= z^{1-c}(1-z)^{c-a-b}\, {}_2F_1(1-a, 1-b; c+1-a-b; 1-z) \tag{14}$$
$$= z^{a-c}(1-z)^{c-a-b}\, {}_2F_1(c-a, 1-a; c+1-a-b; 1-z^{-1}) \tag{15}$$
$$= z^{b-c}(1-z)^{c-a-b}\, {}_2F_1(c-b, 1-b; c+1-a-b; 1-z^{-1}). \tag{16}$$

$$w_5 = (z^{-1}e^{i\pi})^{a}\, {}_2F_1(a, a+1-c; a+1-b; z^{-1}) \tag{17}$$
$$= (z^{-1}e^{i\pi})^{c-b}(1-z)^{c-a-b}\, {}_2F_1(1-b, c-b; a+1-b; z^{-1}) \tag{18}$$
$$= (1-z)^{-a}\, {}_2F_1(a, c-b; a+1-b; (1-z)^{-1}) \tag{19}$$
$$= (z^{-1}e^{i\pi})^{c-1}(1-z)^{c-1-a}\, {}_2F_1(1-b, a+1-c; a+1-b; (1-z)^{-1}). \tag{20}$$

$$w_6 = (z^{-1}e^{i\pi})^{b}\, {}_2F_1(b, b+1-c; b+1-a; z^{-1}) \tag{21}$$
$$= (z^{-1}e^{i\pi})^{c-a}(1-z)^{c-a-b}\, {}_2F_1(1-a, c-a; b+1-a; z^{-1}) \tag{22}$$
$$= (1-z)^{-b}\, {}_2F_1(b, c-a; b+1-a; (1-z)^{-1}) \tag{23}$$
$$= (z^{-1}e^{i\pi})^{c-1}(1-z)^{c-1-b}\, {}_2F_1(1-a, b+1-c; b+1-a; (1-z)^{-1}). \tag{24}$$

Conditions for convergence of the above series can be deduced from 5.2.1(3,4).

The pairs (3), (4) and (7), (8) furnish the analytic continuation of w_1 and w_2, respectively, from the interior of the unit circle with center at the origin to the half-plane $R(z) < \frac{1}{2}$. Similarly, the pairs (11), (12) and (15), (16) give the analytic continuation of w_3 and w_4, respectively, from the interior of the unit circle with center at $z = 1$ to the half-plane $R(z) > \frac{1}{2}$, and the pairs (19), (20) and (23), (24) provide the analytic continuation of w_5 and w_6, respectively, from the exterior of the unit circle with center at the origin to the interior of this circle which lies in the left half-plane.

It follows from Kummer's solutions that if c is not an integer or zero, one of the fundamental solutions of 6.3(1) is composed of only a finite number of terms whenever at least one of the numbers $a, b, c-a, c-b$ is an integer or zero. This is the same as saying that at least one of the eight numbers $\pm(c-1) \pm (a-b) \pm (a+b-c)$ is an odd integer. Such solutions are called degenerate.

6.5. Analytic Continuation

The integral 6.2.3(1) defines a single valued analytic function of z in the domain $|\arg(1-z)| < \pi$, and so serves for the analytic continuation of the $_2F_1$ hypergeometric series into this domain. It is convenient to denote the analytic continuation of the $_2F_1$ series by $_2F_1$, and this means the principal branch of the analytic function generated by the hypergeometric series. Similarly, the Mellin-Barnes integral representation(see 5.3.1 and the final comment in 5.2.3) serves for the analytic continuation of the $_2F_1$ series into the domain $|\arg(-z)| < \pi$.

We suppose for the moment that w_1 and w_2, w_3 and w_4, and w_5 and w_6, see 6.4, are the fundamental solutions of 6.3(1) about $z = 0$, 1 and ∞. Clearly, all six of these quantities cannot be independent, and any three must be linearly related. This gives rise to twenty relations. Let D_1, D_2 and D_3 represent the domains $|z| < 1$, $|1-z| < 1$, $|z| > 1$. Then (1)-(4), (5)-(8) and (9)-(12) below give the relations connecting the functions in D_1 and D_3, D_1 and D_2 and D_2 and D_3, respectively. We omit the eight equations which involve one function from each domain.

$$w_1 = \frac{\Gamma(b-a)\,\Gamma(c)}{\Gamma(b)\,\Gamma(c-a)}\,w_5 + \frac{\Gamma(a-b)\,\Gamma(c)}{\Gamma(a)\,\Gamma(c-b)}\,w_6\,. \tag{1}$$

$$w_2 = \frac{\Gamma(b-a)\,\Gamma(2-c)\,e^{i\pi(1-c)}}{\Gamma(1-a)\,\Gamma(1+b-c)}\,w_5 + \frac{\Gamma(a-b)\,\Gamma(2-c)\,e^{i\pi(1-c)}}{\Gamma(1-b)\,\Gamma(1+a-c)}\,w_6\,. \tag{2}$$

$$w_5 = \frac{\Gamma(1+a-b)\,\Gamma(1-c)}{\Gamma(1-b)\,\Gamma(1+a-c)}\,w_1 + \frac{\Gamma(1+a-b)\,\Gamma(c-1)\,e^{i\pi(c-1)}}{\Gamma(a)\,\Gamma(c-b)}\,w_2\,. \tag{3}$$

$$w_6 = \frac{\Gamma(1+b-a)\,\Gamma(1-c)}{\Gamma(1-a)\,\Gamma(1+b-c)}\,w_1 + \frac{\Gamma(1+b-a)\,\Gamma(c-1)\,e^{i\pi(c-1)}}{\Gamma(b)\,\Gamma(c-a)}\,w_2\,. \tag{4}$$

$$w_3 = \frac{\Gamma(a+b+1-c)\,\Gamma(1-c)}{\Gamma(b+1-c)\,\Gamma(a+1-c)}\,w_1 + \frac{\Gamma(a+b+1-c)\,\Gamma(c-1)}{\Gamma(a)\,\Gamma(b)}\,w_2\,. \tag{5}$$

$$w_4 = \frac{\Gamma(c+1-a-b)\,\Gamma(1-c)}{\Gamma(1-a)\,\Gamma(1-b)}\,w_1 + \frac{\Gamma(c+1-a-b)\,\Gamma(c-1)}{\Gamma(c-a)\,\Gamma(c-b)}\,w_2\,. \tag{6}$$

$$w_1 = \frac{\Gamma(c-a-b)\,\Gamma(c)}{\Gamma(c-a)\,\Gamma(c-b)}\,w_3 + \frac{\Gamma(a+b-c)\,\Gamma(c)}{\Gamma(a)\,\Gamma(b)}\,w_4\,. \tag{7}$$

$$w_2 = \frac{\Gamma(c-a-b)\,\Gamma(2-c)}{\Gamma(1-a)\,\Gamma(1-b)}\,w_3 + \frac{\Gamma(a+b-c)\,\Gamma(2-c)}{\Gamma(a+1-c)\,\Gamma(b+1-c)}\,w_4\,. \tag{8}$$

$$w_3 = \frac{\Gamma(b-a)\,\Gamma(a+b+1-c)\,e^{-i\pi a}}{\Gamma(1+b-c)\,\Gamma(b)}\,w_5 + \frac{\Gamma(a-b)\,\Gamma(a+b+1-c)\,e^{-i\pi b}}{\Gamma(1+a-c)\,\Gamma(a)}\,w_6\,. \tag{9}$$

$$w_4 = \frac{\Gamma(b-a)\,\Gamma(c+1-a-b)\,e^{i\pi(b-c)}}{\Gamma(1-a)\,\Gamma(c-a)} w_5 + \frac{\Gamma(a-b)\,\Gamma(c+1-a-b)\,e^{i\pi(a-c)}}{\Gamma(1-b)\,\Gamma(c-b)} w_6 . \tag{10}$$

$$w_5 = \frac{\Gamma(c-a-b)\,\Gamma(1+a-b)\,e^{i\pi a}}{\Gamma(1-b)\,\Gamma(c-b)} w_3 + \frac{\Gamma(a+b-c)\,\Gamma(1+a-b)\,e^{i\pi(c-b)}}{\Gamma(1+a-c)\,\Gamma(a)} w_4 . \tag{11}$$

$$w_6 = \frac{\Gamma(c-a-b)\,\Gamma(1+b-a)\,e^{i\pi b}}{\Gamma(1-a)\,\Gamma(c-a)} w_3 + \frac{\Gamma(a+b-c)\,\Gamma(1+b-a)\,e^{i\pi(c-a)}}{\Gamma(1+b-c)\,\Gamma(b)} w_4 . \tag{12}$$

6.6. The Complete Solution and Wronskians

We now take up the question of obtaining the complete solution to the hypergeometric equation 6.3(1) in the singular cases previously mentioned. Recall that w_1 and w_2 are not independent or one of these is not defined if c is an integer or zero, and none of the numbers a, b, $c - a$, $c - b$ is an integer. The same may be said for the pairs w_3, w_4 and w_5, w_6 if $(a + b + 1 - c)$ and $(a + 1 - b)$, respectively, are integers or zero. In the degenerate cases where any of the numbers a, b, $c - a$, $c - b$ is an integer, fundamental solutions are easily picked out from Kummer's set of 24 solutions; see the concluding remarks of 6.4. In this section we develop expansions which are solutions of 6.3(1) when the denominator parameters of the ${}_2F_1$'s appearing in w_1, w_2,..., w_6 are integers. Thus, for example, we develop expansions which satisfy 6.3(1) and which are independent of w_1 when c is a positive integer or zero. Furthermore, the analytic continuation of these solutions is automatically provided. Except in some of the degenerate cases, the solutions involve logarithms and are called logarithmic solutions. In this connection, the degenerate solutions are a simple by-product of the logarithmic solutions. Though the key results are a special case of 5.3.1(8-10), it is worthwhile to present them in the setting of the ${}_2F_1$ in view of the applications.

To simplify the analysis, consider

$$\begin{aligned} V(z) &= (ze^{-i\pi})^{-a_1}\,{}_2F_1\left(\begin{matrix}1+a_1-b_0,\, 1+a_1-b_1\\ 1+a_1-a_2\end{matrix}\middle| z^{-1}\right) \\ &= \frac{\Gamma(b_0-b_1)\,\Gamma(1+a_1-a_2)}{\Gamma(b_0-a_2)\,\Gamma(1+a_1-b_1)}(ze^{-i\pi})^{(1-b_0)}\,{}_2F_1\left(\begin{matrix}1+a_1-b_0,\, 1+a_2-b_0\\ 1+b_1-b_0\end{matrix}\middle| z\right) \\ &\quad + \frac{\Gamma(b_1-b_0)\,\Gamma(1+a_1-a_2)}{\Gamma(b_1-a_2)\,\Gamma(1+a_1-b_0)}(ze^{-i\pi})^{(1-b_1)}\,{}_2F_1\left(\begin{matrix}1+a_1-b_1,\, 1+a_2-b_1\\ 1+b_0-b_1\end{matrix}\middle| z\right), \\ &\quad 0 < \arg z < 2\pi. \end{aligned} \tag{1}$$

This is advantageous as it can represent any of the formulas 6.5(1–4) by making the substitutions in the accompanying tabulation.

$V(z)$	Equation	z	a_1	a_2	b_0	b_1
w_1	6.5(1)	$z^{-1}e^{2\pi i}$	0	$1-c$	$1-a$	$1-b$
$e^{i\pi(c-1)}w_2$	6.5(2)	$z^{-1}e^{2\pi i}$	$1-c$	0	$1-a$	$1-b$
w_5	6.5(3)	z	a	b	1	c
w_6	6.5(4)	z	b	a	1	c

(2)

Similarly Eqs. 6.5(5–8) are given by

$$W(z) = {}_2F_1\left(\begin{matrix}1+a_1-b_0,\, b_1-a_2\\ 1+a_1-a_2\end{matrix}\,\middle|\, z\right)$$

$$= \frac{\Gamma(b_0-b_1)\,\Gamma(1+a_1-a_2)}{\Gamma(b_0-a_2)\,\Gamma(1+a_1-b_1)}\,{}_2F_1\left(\begin{matrix}1+a_1-b_0,\, b_1-a_2\\ 1+b_1-b_0\end{matrix}\,\middle|\, 1-z\right),$$

$$+ (1-z)^{b_0-b_1}\frac{\Gamma(b_1-b_0)\,\Gamma(1+a_1-a_2)}{\Gamma(b_1-a_2)\,\Gamma(1+a_1-b_0)}\,{}_2F_1\left(\begin{matrix}1+a_1-b_1,\, b_0-a_2\\ 1+b_0-b_1\end{matrix}\,\middle|\, 1-z\right),$$

$$|\arg(1-z)| < \pi. \tag{3}$$

$W(z)$	Equation	z	a_1	a_2	b_0	b_1
w_3	6.5(5)	$1-z$	a	$c-b$	1	c
$(1-z)^{a+b-c}w_4$	6.5(6)	$1-z$	$c-a$	b	1	c
w_1	6.5(7)	z	a	$1+a-c$	1	$a+b+1-c$
$z^{c-1}w_2$	6.5(8)	z	$a+1-c$	a	1	$a+b+1-c$

(4)

Equations 6.5(9-10) and 6.5(11-12) can be represented in terms of W and V, respectively, since

$$z^{1+a_1-b_0}\,{}_2F_1\left(\begin{matrix}1+a_1-b_0,\, 1+a_1-b_1\\ 1+a_1-a_2\end{matrix}\,\middle|\, 1-z\right) = W(1-z^{-1}) \tag{5}$$

$$e^{-i\pi a_1}(z^{-1}e^{i\pi})^{1+a_1-b_0}(1-z)^{1-b_0}\;{}_2F_1\left(\begin{matrix}1+a_1-b_0,\, b_1-a_2\\ 1+a_1-a_2\end{matrix}\,\middle|\, z^{-1}\right)$$

$$= V(\{1-z\}\,e^{2\pi i}). \tag{6}$$

Also $W(z)$ is readily related to $V(1-z^{-1})$ and so actually only developments based on (1),(2) are required. Nonetheless, we find it convenient to give data for both $V(z)$ and $W(z)$.

Let

$$V_1(z) = \frac{(ze^{-i\pi})^{(1-b_1)}\Gamma(1 + a_1 - a_2)}{\Gamma(b_1 - a_2)\,\Gamma(1 + a_1 - b_0)}$$

$$\times \sum_{k=0}^{s-1} \frac{(1 + a_1 - b_1)_k(1 + a_2 - b_1)_k(s - 1 - k)!\,(-)^k z^k}{k!}$$

$$= \frac{(z^{-1}e^{i\pi})^{b_0}\Gamma(1 + a_1 - a_2)(1 + a_1 - b_1)_{s-1}(1 + a_2 - b_1)_{s-1}}{\Gamma(b_1 - a_2)\,\Gamma(1 + a_1 - b_0)(s - 1)!}$$

$$\times\ {}_3F_2\left(\begin{matrix}1 - s,\ 1,\ 1\\ 1 + b_0 - a_1\,,\ 1 + b_0 - a_2\end{matrix}\middle|\, z^{-1}\right), \tag{7}$$

It is understood that if $s = 0$, $V_1(z)$ is nil, i.e., $\sum_{k=0}^{s-1}$ is nil if $s = 0$. This convention is retained throughout. Then

$$V(z) = V_1(z) + \frac{(-)^{s+1}(ze^{-i\pi})^{(1-b_0)}\Gamma(1 + a_1 - a_2)}{s!\,\Gamma(b_0 - a_2)\,\Gamma(1 + a_1 - b_1)}$$

$$\times \Big\{[\gamma + \ln(ze^{-i\pi}) + \psi(1 + a_1 - b_0) + \psi(b_0 - a_2) - \psi(1 + s)]$$

$$\times\ {}_2F_1\left(\begin{matrix}1 + a_1 - b_0\,,\ 1 + a_2 - b_0\\ 1 + s\end{matrix}\middle|\, z\right)$$

$$+ \sum_{k=0}^{\infty} \frac{(1 + a_1 - b_0)_k(1 + a_2 - b_0)_k z^k}{(1 + s)_k k!}$$

$$\times (\psi(1 + a_1 - b_0 + k) - \psi(1 + a_1 - b_0)$$

$$+ \psi(1 + a_2 - b_0 + k) - \psi(1 + a_2 - b_0)$$

$$- \psi(1 + s + k) + \psi(1 + s) - \psi(1 + k) + \psi(1))\Big\}, \tag{8}$$

$b_1 = b_0 + s$, s is a positive integer or zero,

$1 + a_1 - b_0$ is not a negative integer or zero,

$1 + a_2 - b_0$ is not an integer or zero,

$$0 < \arg z < 2\pi.$$

We now consider the restrictions imposed on $1 + a_1 - b_0$ and $1 + a_2 - b_0$ in (13). If $1 + a_1 - b_0$ is a negative integer or zero, then $V_1(z) = 0$ and (8) reduces to

$$V(z) = \frac{(-)^m(m + s)!\,\Gamma(1 + a_1 - a_2)}{s!\,\Gamma(b_0 - a_2)}(ze^{-i\pi})^{(1-b_0)}\ {}_2F_1\left(\begin{matrix}-m,\ 1 + a_2 - b_0\\ 1 + s\end{matrix}\middle|\, z\right),$$

$1 + a_1 - b_0 = -m$, m a positive integer or zero, (9)

which is the statement (1).

If $(b_0 - a_2)$ is a negative integer or zero, a similar analysis gives

$$V(z) = V_1(z) + \frac{(-)^{m+s+1}(m-1)!\,\Gamma(1+a_1-a_2)}{s!\,\Gamma(1+a_1-b_1)}(ze^{-i\pi})^{(1-b_0)} \times {}_2F_1\left(\begin{matrix} m, 1+a_1-b_0 \\ 1+s \end{matrix}\middle| z\right), \tag{10}$$

$$V_1(z) = \frac{(s-1)!\,\Gamma(1+a_1-a_2)(ze^{-i\pi})^{(1-b_1)}}{(s-m)!\,\Gamma(1+a_1-b_0)} \times \sum_{k=0}^{s-m} \frac{(m-s)_k(1+a_1-b_1)_k z^k}{(1-s)_k k!}$$

$$= \frac{(m-1)!\,\Gamma(1+a_1-a_2)\,\Gamma(a_1-a_2)}{(s-m)!\,\Gamma(1+a_1-b_0)\,\Gamma(1+a_1-b_1)}(ze^{-i\pi})^{(1-b_1)}z^{s-m} \times {}_2F_1\left(\begin{matrix} m-s, m \\ m+b_0-a_1 \end{matrix}\middle| z^{-1}\right), \tag{11}$$

$$b_0 - a_2 = 1 - m, \qquad m \text{ a positive integer},$$

and this too can be directly deduced from (1).

Now suppose that $(b_0 - a_2)$ is a positive integer. Then

$$\begin{aligned} V(z) &= V_1(z) + \frac{(-)^{s-1}(ze^{-i\pi})^{(1-b_0)}\Gamma(1+a_1-a_2)}{s!\,m!\,\Gamma(1+a_1-b_1)} \\ &\times \Big\{[\gamma + \ln(ze^{-i\pi}) + \psi(1+a_1-b_0) \\ &+ \psi(1+m) - \psi(1+s)]\,{}_2F_1\left(\begin{matrix} -m, 1+a_1-b_0 \\ 1+s \end{matrix}\middle| z\right) \\ &+ \sum_{k=0}^{m} \frac{(-m)_k(1+a_1-b_0)_k z^k}{(1+s)_k k!} \\ &\times (\psi(1+a_1-b_0+k) - \psi(1+a_1-b_0) + \psi(1+m-k) - \psi(1+m) \\ &- \psi(1+s+k) + \psi(1+s) - \psi(1+k) + \psi(1)) + (-)^m m! \\ &\times \sum_{k=m+1}^{\infty} \frac{(1+a_1-b_0)_k(k-m-1)!\,z^k}{(1+s)_k k!}\Big\}, \end{aligned}$$

$$b_1 = b_0 + s, \qquad 1 + a_2 - b_0 = -m, \qquad 0 < \arg z < 2\pi, \tag{12}$$

where m and s are positive integers or zero. Note that the last infinite series in (12) can be expressed in terms of a ${}_3F_2$.

Thus (11) is valid without the restrictions on $(1 + a_1 - b_0)$ and $(1 + a_2 - b_0)$. The point is that without these conditions, the expansion is free of logarithms unless $(b_0 - a_2)$ is a positive integer.

If $a_1 = b_0 + \nu$, $a_2 = b_0 - \mu - \lambda - 1$, and $b_1 = b_0 - \mu$, or if $a_1 = b_0 + \nu + \mu$, $a_2 = b_0 - \lambda - 1$, and $b_1 = b_0 + \mu$, where μ, ν, and λ are positive integers or zero, then the logarithmic solution for $V(z)$ can be put in the form 6.2.1(10). If $b_0 = a_2$ or $b_1 = a_2$, see 6.3(9).

For $W(z)$ we have

$$W_1(z) = \frac{(1-z)^{-s}\,\Gamma(1+a_1-a_2)}{\Gamma(b_1-a_2)\,\Gamma(1+a_1-b_0)} \times \sum_{k=0}^{s-1} \frac{(1+a_1-b_1)_k(b_0-a_2)_k(s-1-k)!\,(-)^k(1-z)^k}{k!}$$

$$= \frac{(-)^{s+1}\Gamma(1+a_1-a_2)(1+a_1-b_1)_{s-1}(b_0-a_2)_{s-1}}{(1-z)\Gamma(b_1-a_2)\Gamma(1+a_1-b_0)(s-1)!} \times {}_3F_2\left(\begin{matrix}1-s,1,1\\1+b_0-a_1,\,2-b_1+a_2\end{matrix}\middle|\frac{1}{1-z}\right), \tag{13}$$

$$W(z) = W_1(z) + \frac{(-)^{s+1}\Gamma(1+a_1-a_2)}{s!\,\Gamma(b_0-a_2)\,\Gamma(1+a_1-b_1)}$$
$$\times \Big\{[\gamma + \ln(1-z) + \psi(1+a_1-b_0) + \psi(b_1-a_2) - \psi(1+s)]$$
$$\times {}_2F_1\left(\begin{matrix}1+a_1-b_0,\,b_1-a_2\\1+s\end{matrix}\middle|1-z\right)$$
$$+ \sum_{k=0}^{\infty} \frac{(1+a_1-b_0)_k(b_1-a_2)_k(1-z)^k}{(1+s)_k k!}$$
$$\times (\psi(1+a_1-b_0+k) - \psi(1+a_1-b_0) + \psi(b_1-a_2+k) - \psi(b_1-a_2)$$
$$- \psi(1+s+k) + \psi(1+s) - \psi(1+k) + \psi(1))\Big\}, \tag{14}$$

where $b_1 = b_0 + s$, s is a positive integer or zero, neither of the numbers $1 + a_1 - b_0$ nor $b_1 - a_2$ is a negative integer or zero, and $|\arg(1-z)| < \pi$.

If $(1 + a_1 - b_0)$ is a negative integer or zero, then directly from (3), and without the restriction that b_1 is a positive integer, we have

$$W(z) = {}_2F_1\left(\begin{matrix}-m,\,b_1-a_2\\1-m-a_2\end{matrix}\middle|z\right) = \frac{(b_1)_m}{(a_2)_m}\,{}_2F_1\left(\begin{matrix}-m,\,b_1-a_2\\b_1\end{matrix}\middle|1-z\right),$$
$$a_1 = -m, \qquad m \text{ a positive integer or zero.} \tag{15}$$

If $(b_1 - a_2)$ is a negative integer or zero, we again arrive at the form (15) since $W(z)$ is symmetric in its numerator parameters.

Suppose now that $b_0 = b_1 + s$, s a positive integer or zero. Then $W(z)$ is given by the right-hand side of (11) if there we interchange the roles of b_0 and b_1 and multiply throughout by $(1-z)^s$.

If $a_1 = \nu + 1$, $a_2 = -\mu - \lambda$, and $b_1 = \nu - \lambda + 1$, or if $a_1 = \nu + \mu + 1$, $a_2 = -\lambda$, and $b_1 = \nu - \lambda + 1$, where μ, ν, and λ are positive integers or zero, then the logarithmic solution is also given by 6.2.1(10). If $b_0 = a_2$ or $b_1 = 1 + a_1$, see 6.3(9).

Corresponding results for $T(z)$ and $U(z)$ follow from (5) and (6). For this and a complete table of solutions to 6.3(1) in the degenerate cases, see Luke (1969). Kummer type relations for the logarithmic solutions have been given by Nörlund (1963). See also Luke (1969).

Wronskians

Let

$$W_{ij} = W(w_i, w_j) = w_i(z)\, w_j'(z) - w_i'(z)\, w_j(z), \tag{16}$$

$$A = z^{-c}(1-z)^{c-a-b-1}. \tag{17}$$

Then

$$W_{12} = (1-c)\, A, \qquad W_{13} = -\frac{\Gamma(a+b+1-c)\,\Gamma(c)}{\Gamma(a)\,\Gamma(b)} A, \tag{18}$$

$$W_{14} = -\frac{\Gamma(c+1-a-b)\,\Gamma(c)}{\Gamma(c-a)\,\Gamma(c-b)} A, \qquad W_{15} = \frac{\Gamma(1+a-b)\,\Gamma(c)\, e^{i\pi c}}{\Gamma(a)\,\Gamma(c-b)} A, \tag{19}$$

$$W_{23} = -\frac{\Gamma(a+b+1-c)\,\Gamma(2-c)}{\Gamma(a+1-c)\,\Gamma(b+1-c)} A, \quad W_{24} = -\frac{\Gamma(c+1-a-b)\,\Gamma(2-c)}{\Gamma(1-a)\,\Gamma(1-b)} A, \tag{20}$$

$$W_{25} = -\frac{\Gamma(1+a-b)\,\Gamma(2-c)}{\Gamma(1-b)\,\Gamma(1+a-c)} A, \qquad W_{34} = -(c-a-b)\, A, \tag{21}$$

$$\begin{aligned} W_{35} &= \frac{\Gamma(a+b+1-c)\,\Gamma(1+a-b)\, e^{i\pi(c-b)}}{\Gamma(a+1-c)\,\Gamma(a)} A, \\ W_{45} &= \frac{\Gamma(c+1-a-b)\,\Gamma(1+a-b)\, e^{i\pi a}}{\Gamma(1-b)\,\Gamma(c-b)} A, \end{aligned} \tag{22}$$

$$W_{56} = (b-a)\, e^{i\pi c} A. \tag{23}$$

To get W_{i6} for $i = 1, 2, 3, 4$ simply interchange the roles of a and b in W_{i5}.

6.7. Quadratic Transformations

$${}_2F_1(2a, 2a-c+1; c; z) = (1+z)^{-2a}\, {}_2F_1\left(a, a+\tfrac{1}{2}; c; \frac{4z}{(1+z)^2}\right), \tag{1}$$

$$_2F_1(2a, 2a - c + 1; c; z) = (1 - z)^{-2a}\,{}_2F_1\left(a, c - a - \tfrac{1}{2}; c; -\frac{4z}{(1 - z)^2}\right), \tag{2}$$

$$_2F_1(a, a + \tfrac{1}{2}(1 - c); \tfrac{1}{2}(c + 1); z^2) = (1 + z)^{-2a}\,{}_2F_1\left(a, \tfrac{1}{2}c; c; \frac{4z}{(1 + z)^2}\right), \tag{3}$$

$$_2F_1(2a, 2b; a + b + \tfrac{1}{2}; z) = {}_2F_1(a, b; a + b + \tfrac{1}{2}; 4z(1 - z)), \tag{4}$$

$$_2F_1(2a - 1, 2b - 1; a + b - \tfrac{1}{2}; z) = (1 - 2z)\,{}_2F_1(a, b; a + b - \tfrac{1}{2}; 4z(1 - z)), \tag{5}$$

$$_2F_1(a, 1 - a; c; z) = (1 - z)^{c-1}\,{}_2F_1(\tfrac{1}{2}(c - a), \tfrac{1}{2}(c + a - 1); c; 4z(1 - z)), \tag{6}$$

$$_2F_1(a, b; 2a; z) = \left(\frac{2}{1 + y}\right)^{2b}{}_2F_1\left(b, b + \tfrac{1}{2} - a; a + \tfrac{1}{2}; \left(\frac{1 - y}{1 + y}\right)^2\right), \quad y = (1 - z)^{1/2}, \tag{7}$$

$$_2F_1(a, b; 2a; 2z) = (1 - z)^{-b}\,{}_2F_1\left(\tfrac{1}{2}b, \tfrac{1}{2}(b + 1); a + \tfrac{1}{2}; \left(\frac{z}{1 - z}\right)^2\right). \tag{8}$$

The above formulas are valid near $z = 0$ and $(1 - z)^{\delta} > 0$ if δ is real and $z < 1$. Many other forms can be derived from (1)-(8) by use of Kummer's formulas 6.4.

Quadratic transformation formulas for the logarithmic solutions of the hypergeometric differential equation have been given by Nörlund (1963). See also Luke (1969).

6.8. The $_2F_1$ for Special Values of the Argument

$$_2F_1(a, b; c; 1) = \frac{\Gamma(c)\,\Gamma(c - a - b)}{\Gamma(c - a)\,\Gamma(c - b)},$$

c not a negative integer or zero, $\quad R(c - a - b) > 0. \qquad (1)$

$$\lim_{z \to 1}\,(1 - z)^{a+b-c}\,{}_2F_1(a, b; c; z) = \frac{\Gamma(c)\,\Gamma(a + b - c)}{\Gamma(a)\,\Gamma(b)},$$

c not a negative integer or zero, $\quad R(a + b - c) > 0. \qquad (2)$

$$_2F_1(-n, b; c; 1) = \frac{(c - b)_n}{(c)_n}, \tag{3}$$

$$_2F_1(-n, n + \lambda; c; 1) = \frac{(-)^n(1 + \lambda - c)_n}{(c)_n}, \tag{4}$$

where n is a positive integer or zero and c is not a negative integer or zero. If c is a negative integer or zero, say $c = -m$, and $m > n$, then

$$\sum_{k=0}^{n} \frac{(-n)_k(b)_k}{(-m)_k k!} = \frac{(m-n)!\,(m+b+1-n)_n}{m!}, \tag{5}$$

which is known as Vandermonde's theorem.

$${}_2F_1(2a, 2a-c+1; c; -1) = \frac{2^{-2a}(c-a)_a}{(\frac{1}{2})_a}. \tag{6}$$

$${}_2F_1(2a, 2b; a+b+\tfrac{1}{2}; \tfrac{1}{2}) = \frac{\pi^{1/2}\Gamma(a+b+\frac{1}{2})}{\Gamma(a+\frac{1}{2})\,\Gamma(b+\frac{1}{2})}. \tag{7}$$

$${}_2F_1(a, 1-a; c; \tfrac{1}{2}) = \frac{2^{1-c}\pi^{1/2}\Gamma(c)}{\Gamma[(c+a)/2]\,\Gamma[(c-a+1)/2]}. \tag{8}$$

$$\begin{aligned} {}_2F_1(2a, 2b; a+b+1; \tfrac{1}{2}) &= \frac{\pi^{\frac{1}{2}}\Gamma(a+b+1)}{b-a} \\ &\times \left[\frac{1}{\Gamma(b)\Gamma(a+\frac{1}{2})} - \frac{1}{\Gamma(a)\Gamma(b+\frac{1}{2})}\right]. \end{aligned} \tag{9}$$

$$\begin{aligned} {}_2F_1(2a, 2b; a+b; \tfrac{1}{2}) &= \pi^{\frac{1}{2}}\Gamma(a+b) \\ &\times \left[\frac{1}{\Gamma(b)\Gamma(a+\frac{1}{2})} + \frac{1}{\Gamma(a)\Gamma(b+\frac{1}{2})}\right]. \end{aligned} \tag{10}$$

$$\begin{aligned} {}_2F_1(2a, 2b; a+b-\tfrac{1}{2}; \tfrac{1}{2}) &= \pi^{\frac{1}{2}}\Gamma(a+b-\tfrac{1}{2}) \\ &\times \left[\frac{a+b-\frac{1}{2}}{\Gamma(a+\frac{1}{2})\Gamma(b+\frac{1}{2})} + \frac{2}{\Gamma(a)\Gamma(b)}\right]. \end{aligned} \tag{11}$$

Equations (9)-(11) are due to Mitra (1943).

$${}_2F_1\left(\begin{matrix}-n, c\\ 2c\end{matrix}\,\middle|\, 2\right) = \frac{(-)^n 2^n (c)_n}{(2c)_n}\,{}_2F_1\left(\begin{matrix}-n, 1-n-2c\\ 1-n-c\end{matrix}\,\middle|\,\frac{1}{2}\right) = 0,\ n \text{ odd}, \tag{12}$$

$${}_2F_1\left(\begin{matrix}-2n, c\\ 2c\end{matrix}\,\middle|\, 2\right) = \frac{(\frac{1}{2})_n}{(c+\frac{1}{2})_n}. \tag{13}$$

The results (12) and (13) are valid when c is a negative integer provided that in (12), $-2c > n$ and the $_2F_1$ is the sum of its first $n+1$ terms or $1-c$ terms whichever is the smaller; and provided that in (13), $c > n$ and the $_2F_1$ is the sum of its first $2n+1$ or $1-c$ terms whichever is the smaller.

$$f(2a, 2b; c; \tfrac{1}{2}) = \frac{\Gamma(\frac{1}{2} - a)\,\Gamma(\frac{1}{2} - b)}{\pi^{1/2}\Gamma(1 - c)}, \tag{14}$$

where $c = a + b + \frac{1}{2}$ is a negative integer, a or $b = 0, -1, \ldots, -[|c|/2]$;

$$f(2a, 2b; c; \tfrac{1}{2}) = 0, \tag{15}$$

where c is as in (14), a or $b = -\frac{1}{2}, -\frac{3}{2}, \ldots, \frac{1}{2} + [\frac{1}{2}c]$.

$$f(2a, 2a - c + 1; c; -1) = \frac{2^{-2a-\epsilon}(1 - c)_a}{(\frac{1}{2} - a)_a}, \tag{16}$$

where c is a negative integer, a is a nonpositive integer $\geqslant c/2$, $\epsilon = 0$; $2a = c - 1, c - 2, \ldots, 2c - 1$, $\epsilon = 1$;

$$f(a, 1 - a; c; \tfrac{1}{2}) = \frac{\Gamma(1 - (a + c)/2)\,\Gamma((1 + a - c)/2)}{2^c\pi^{1/2}\Gamma(1 - c)}, \tag{17}$$

where a is an integer, $c \leqslant a \leqslant 1 - c$, c a negative integer. In (14)-(17), $f(a, b; c; z)$ is defined by 6.3(7).

$$_2F_1(a, \tfrac{1}{2}; 3a; 3/4) = \frac{(16/27)^a\Gamma(a)\Gamma(3a)}{\{\Gamma(2a)\}^2}. \tag{18}$$

$$_2F_1(a, 1 - a; 3a; \tfrac{1}{2}) = \frac{2^{-a}\Gamma(a)\Gamma(3a)}{\{\Gamma(2a)\}^2}. \tag{19}$$

$$_2F_1(1 - a, \tfrac{1}{2}; 2a + \tfrac{1}{2}; \tfrac{1}{4}) = \frac{2\Gamma(a)\Gamma(2a + \frac{1}{2})}{3\Gamma(a + \frac{1}{2})\Gamma(2a)}. \tag{20}$$

Equations (18)-(20) are due to Spiegel (1962).

$$_2F_1(-a, \tfrac{1}{2} - a; 2a + 3/2; -1/3) = \frac{(8/9)^{2a}\Gamma(4/3)\Gamma(2a + 3/2)}{\Gamma(3/2)\Gamma(2a + 4/3)}. \tag{21}$$

$$_2F_1(3a, 3a + \tfrac{1}{2}; 3a + 5/6; 1/9) = \frac{(3/4)^{3a}\Gamma(\frac{1}{2})\Gamma(2a + 5/6)}{\Gamma(a + \frac{1}{2})\Gamma(a + 5/6)}. \tag{22}$$

$$_2F_1(a + 1/3, 3a; 2a + 2/3; e^{i\pi/3}) = \frac{e^{i\pi a/2}\Gamma(a)\Gamma(2a + 2/3)}{\Gamma(3a)\Gamma(2/3)}. \tag{23}$$

6.9. Expansion in Series of Chebyshev Polynomials

In addition to the following, see 5.10.2.

$$ {}_2F_1\left(\begin{matrix}a, b\\ c\end{matrix}\middle| zx\right) = \sum_{n=0}^{\infty} C_n(z) T_n^*(x), $$

$$ 0 \leqslant x \leqslant 1, \qquad z \neq 1, \qquad |\arg(1-z)| < \pi, $$

$$ C_n(z) = \frac{\epsilon_n (a)_n (b)_n z^n}{2^{2n} (c)_n n!}\, {}_3F_2\left(\begin{matrix}a+n, b+n, \frac{1}{2}+n\\ c+n, 1+2n\end{matrix}\middle| z\right), \tag{1} $$

$$ \begin{aligned} \frac{2C_n(z)}{\epsilon_n} &= (n+1)\left\{2 - \frac{(2n+3)(n+a+1)(n+b+1)}{(n+2)(n+a)(n+b)} \right.\\ &\quad \left. + \frac{4(n+c)}{z(n+a)(n+b)}\right\} C_{n+1}(z) + \frac{2}{(n+a)(n+b)} \\ &\quad \times \left\{[(n+2-a)(n+2-b)(n+\tfrac{3}{2}) \right.\\ &\quad \left. - (n+3-a)(n+3-b)(n+1)] + \frac{2(n+1)(n+3-c)}{z}\right\} C_{n+2}(z) \\ &\quad - \frac{(n+1)(n+3-a)(n+3-b)}{(n+2)(n+a)(n+b)} C_{n+3}(z), \end{aligned} \tag{2} $$

$$ \sum_{n=0}^{\infty} (-)^n C_n(z) = 1. \tag{3} $$

With z as in (1), evaluation of $C_n(z)$ by the backward recurrence scheme is convergent.

6.10. Padé Approximations for ${}_2F_1(1, \sigma; \rho + 1; -1/z)$

Let

$$ E(z) = {}_2F_1(1, \sigma; \rho+1; -1/z). \tag{1} $$

Then with $a = 0$ or 1

$$ E(z) = \frac{C_n(z)}{D_n(z)} + R_n(z), \tag{2} $$

$$ D_n(z) = {}_2F_1(-n, n+\rho+1-a; \sigma+1-a; -z), \tag{3} $$

$$ D_n(-1) = \frac{(-)^n (\rho+1-\sigma)_n}{(\sigma+1-a)_n}, \tag{4} $$

$$C_n(z) = \left[\frac{n(n+\rho+1-a)\,z}{\sigma}\right]^a \sum_{k=0}^{n-a} \frac{(-)^k (a-n)_k\,(n+\rho+1)_k\,z^k}{(\sigma+1)_k\,(1+a)_k}$$

$$\times\ {}_4F_3\left(\begin{matrix}-n+a+k, n+\rho+1+k, \sigma, 1\\ \sigma+1+k, 1+a+k, \rho+1\end{matrix}\,\middle|\,1\right), \tag{5}$$

$$R_n(z) = \frac{F_n(z)}{D_n(z)}, \tag{6}$$

$$F_n(z) = \left(\frac{-\sigma}{\rho+1}\right)^{1-a} \frac{n!\,(\rho+1-\sigma)_n}{(\rho+2-a)_{2n}\,z^{n+1-a}}\ {}_2F_1\left(\begin{matrix}n+1, n+1+\sigma-a\\ 2n+\rho+2-a\end{matrix}\,\middle|\,-\frac{1}{z}\right), \tag{7}$$

$$F_n(z) = \left(\frac{-\sigma}{\rho+1}\right)^{1-a} \frac{n!\,(\rho+1-\sigma)_n\,z^a}{(\rho+2-a)_{2n}\,(z+1)^{n+1}}$$

$$\times\ {}_2F_1\left(\begin{matrix}n+1, n+\rho+1-\sigma\\ 2n+\rho+2-a\end{matrix}\,\middle|\,\frac{1}{z+1}\right), \tag{8}$$

$$F_n(z) = \frac{(-\sigma)^{1-a}\,\rho^a(\rho+1-\sigma)_n\,z^\sigma(z+1)^{\rho-\sigma}}{(\rho+1-a)_n} \int_z^\infty \frac{(t-z)^n}{t^{n+\sigma+1-a}(t+1)^{n+\rho+1-\sigma}}\,dt,$$

$$R(\rho) > a-1,\ z \neq 0,\ |\arg z| < \pi. \tag{9}$$

$$C_0(z) = 1, \qquad C_1(z) = 1 - \frac{\sigma(\rho+2)}{(\sigma+1)(\rho+1)} + \frac{(\rho+2)\,z}{(\sigma+1)},$$

$$C_2(z) = 1 - \frac{2\sigma(\rho+3)}{(\sigma+1)(\rho+1)} + \frac{\sigma(\rho+3)(\rho+4)}{(\sigma+2)(\rho+2)(\rho+1)} \tag{10}$$

$$+ \left\{\frac{2(\rho+3)}{\sigma+1} - \frac{\sigma(\rho+3)(\rho+4)}{(\rho+1)(\sigma+1)(\sigma+2)}\right\} z + \frac{(\rho+3)(\rho+4)\,z^2}{(\sigma+1)(\sigma+2)},$$

$$L_0(z) = 0, \qquad L_1(z) = \frac{(\rho+1)\,z}{\sigma},$$

$$L_2(z) = \frac{(\rho+2)(\rho+3)\,z}{\sigma(\sigma+1)}\left\{z + \frac{2+2\rho+\sigma(\rho-1)}{(\rho+1)(\rho+3)}\right\}, \tag{11}$$

$$D_0(z) = 1, \qquad D_1(z) = 1 + \frac{(\rho+2-a)\,z}{(\sigma+1-a)},$$

$$D_2(z) = 1 + \frac{2(\rho+3-a)\,z}{(\sigma+1-a)} + \frac{(\rho+3-a)(\rho+4-a)\,z^2}{(\sigma+1-a)(\sigma+2-a)}. \tag{12}$$

In the above, $C_n(z)$ is $C_n(z)$ for $a=0$ and is $L_n(z)$ for $a=1$. Further polynomials are easily developed since both $D_n(z)$ and $C_n(z)$ satisfy the recurrence formula

$$\frac{(n+\rho+1-a)(n+\sigma+1-a)}{(2n+\rho+2-a)(2n+\rho+1-a)} D_{n+1}(z)$$
$$= \left\{z + \frac{2n^2+2n(\rho+1-a)+(\rho-a)(\sigma+1-a)}{(2n+\rho-a)(2n+\rho+2-a)}\right\} D_n(z)$$
$$- \frac{n(n+\rho-\sigma)}{(2n+\rho-a)(2n+\rho+1-a)} D_{n-1}(z), \tag{13}$$

To discuss the remainder, let

$$e^{-\xi} = 2z + 1 \mp 2(z^2+z)^{\frac{1}{2}} \tag{14}$$

where the sign is chosen so that $|e^{-\xi}| < 1$. This is possible for all z, $|\arg(1+1/z)| < \pi$. If $-1 \le z \le 0$, $|e^{-\xi}| = 1$. For further details on (14), see 2.4.4. Also put

$$(1 - e^{\xi}) = (e^{\xi} - 1)\, e^{\mp i\pi} \tag{15}$$

where the upper (lower) sign is taken when $I(z) > (<)\, 0$. Then

$$R_n(z) = \frac{\begin{bmatrix}(-\sigma)^{1-a}\rho^a \pi z^\sigma (\rho+1-\sigma)_n (\sigma+1-a)_n \\ \times \exp[-(2n+\sigma+1-a)\,\xi](1+e^{-\xi})^{2\rho-2\sigma}[1+O(n^{-1})]\end{bmatrix}}{\begin{bmatrix}2^{2\rho-2\sigma-1}(\rho+1-a)_n\, n! \\ \times \{1+\exp[\pm i\pi(\sigma+\frac{1}{2}-a)]\exp[-(2n+\rho+1-a)\,\xi]\}\end{bmatrix}}, \tag{16}$$

or

$$R_n(z) = \frac{\begin{bmatrix}(-\sigma)^{1-a}\rho^a \pi z^\sigma \Gamma(\rho+1-a)\exp[-(2n+\sigma+1-a)\,\xi] \\ \times (1+e^{-\xi})^{2\rho-2\sigma}[1+O(n^{-1})]\end{bmatrix}}{\begin{bmatrix}\Gamma(\rho+1-\sigma)\,\Gamma(\sigma+1-a)\,2^{2\rho-2\sigma-1} \\ \times \{1+\exp[\pm i\pi(\sigma+\frac{1}{2}-a)]\exp[-(2n+\rho+1-a)\,\xi]\}\end{bmatrix}}. \tag{17}$$

In (16) and (17), take the upper (lower) sign when $I(z) > (<)\, 0$. It follows that for z fixed,

$$\lim_{n\to\infty} R_n(z) = 0, \qquad z \neq 0, \quad z \neq -1, \quad |\arg(1+1/z)| < \pi. \tag{18}$$

In the applications it is convenient to have the Padé approximations for $E(1/z)$. Again with $a = 0$ or 1, we have

$$H(z) = {}_2F_1(1, \sigma; \rho+1; -z), \tag{19}$$

$$H(z) = \frac{A_n(z)}{B_n(z)} + V_n(z) = \frac{C_n(1/z)}{D_n(1/z)} + R_n(1/z), \tag{20}$$

$$B_n(z) = \frac{(\sigma+1-a)_n z^n}{(n+\rho+1-a)_n} D_n\left(\frac{1}{z}\right)$$
$$= {}_2F_1(-n, -n+a-\sigma; -2n+a-\rho; -z), \tag{21}$$

$$A_n(z) = \frac{(\sigma+1-a)_n z^n}{(n+\rho+1-a)_n} C_n(1/z) , \tag{22}$$

where $D_n(z)$ and $C_n(z)$ are given by (3) and (5) respectively.

We let $A_n(z) = A_n(z)$ if $a = 0$ and $A_n(z) = P_n(z)$ if $a = 1$. Thus,

$$\begin{aligned} A_0(z) &= 1, \qquad A_1(z) = 1 + \frac{(\rho+1-\sigma)\, z}{(\rho+1)(\rho+2)}, \\ A_2(z) &= 1 + \left\{\frac{2(\sigma+2)}{\rho+4} - \frac{\sigma}{\rho+1}\right\} z \\ &+ \left\{\frac{(\sigma+1)(\sigma+2)}{(\rho+3)(\rho+4)} - \frac{2\sigma(\sigma+2)}{(\rho+1)(\rho+4)} + \frac{\sigma(\sigma+1)}{(\rho+1)(\rho+2)}\right\} z^2, \end{aligned} \tag{23}$$

$$\begin{aligned} P_0(z) &= 0, \qquad P_1(z) = 1, \\ P_2(z) &= 1 + \left\{\frac{2(\sigma+1)}{\rho+3} - \frac{\sigma}{\rho+1}\right\} z. \end{aligned} \tag{24}$$

Also,

$$\begin{aligned} B_0(z) &= 1, \qquad B_1(z) = 1 + \frac{(\sigma+1-a)\, z}{(\rho+2-a)}, \\ B_2(z) &= 1 + \frac{2(\sigma+2-a)\, z}{(\rho+4-a)} + \frac{(\sigma+2-a)(\sigma+1-a)\, z^2}{(\rho+4-a)(\rho+3-a)}, \end{aligned} \tag{25}$$

Both $A_n(z)$ and $B_n(z)$ satisfy the recurrence formula

$$\begin{aligned} B_{n+1}(z) &= \left\{1 + \frac{[2n^2 + 2n(\rho+1-a) + (\rho-a)(\sigma+1-a)]\, z}{(2n+\rho-a)(2n+\rho+2-a)}\right\} B_n(z) \\ &- \frac{n(n+\rho-\sigma)(n+\rho-a)(n+\sigma-a)\, z^2}{(2n+\rho-1-a)(2n+\rho-a)^2\,(2n+\rho+1-a)} B_{n-1}(z). \end{aligned} \tag{26}$$

For the error, observe that $V_n(z) = R_n(1/z)$. With $V_n(z) = S_n(z)/B_n(z)$, then

$$S_n(z) = \frac{(-\sigma)^{1-a}\, \rho^a (\sigma+1-a)_n\, (\rho+1-\sigma)_n\, (1+z)^{\rho-\sigma}}{(\rho+1-a)_n\, (n+\rho+1-a)_n\, z^\rho} \int_0^z \frac{(z-t)^n\, t^{n+\rho-a}}{(t+1)^{n+\rho+1-\sigma}}\, dt,$$

$$R(\rho) > a - 1, \qquad |\arg z| < \pi. \tag{27}$$

Let

$$e^{-\zeta} = z + 2 \mp 2(z+1)^{\frac{1}{2}}/z , \tag{28}$$

where the sign is chosen so that $|e^{-\zeta}| < 1$. This is possible for all z, $|\arg(1+z)| < \pi$. If $z \leq -1$, $|e^{-\zeta}| = 1$. Note that

(28) is the same as (14) if in the latter we replace z by $1/z$. For further details on (14), see 2.4.4. Also put

$$(1 - e^{\zeta}) = (e^{\zeta} - 1)\, e^{\pm i\pi} \tag{29}$$

where the upper (lower) sign is taken when $I(z) > (<)\, 0$. Then

$$V_n(z) = \frac{\begin{bmatrix}(-\sigma)^{1-a}\, \rho^a \pi z^{-\sigma} (\rho + 1 - \sigma)_n\, (\sigma + 1 - a)_n \\ \times \exp[-(2n + \sigma + 1 - a)\,\zeta](1 + e^{-\zeta})^{2\rho - 2\sigma}\, [1 + O(n^{-1})]\end{bmatrix}}{\begin{bmatrix}2^{2\rho-2\sigma-1}(\rho + 1 - a)_n\, n! \\ \times \{1 + \exp[\mp i\pi(\sigma + \frac{1}{2} - a)] \exp[-(2n + \rho + 1 - a)\,\zeta]\}\end{bmatrix}}, \tag{30}$$

$$V_n(z) = \frac{\begin{bmatrix}(-\sigma)^{1-a}\, \rho^a \pi z^{-\sigma} \Gamma(\rho + 1 - a) \exp[-(2n + \sigma + 1 - a)\,\zeta] \\ \times (1 + e^{-\zeta})^{2\rho - 2\sigma}\, [1 + O(n^{-1})]\end{bmatrix}}{\begin{bmatrix}2^{2\rho-2\sigma-1}\Gamma(\rho + 1 - \sigma)\, \Gamma(\sigma + 1 - a) \\ \times \{1 + \exp[\mp i\pi(\sigma + \frac{1}{2} - a)] \exp[-(2n + \rho + 1 - a)\,\zeta]\}\end{bmatrix}}. \tag{31}$$

In (30) and (31) take the upper (lower) sign when $I(z) > (<)\, 0$. It follows that for z fixed,

$$\lim_{n\to\infty} V_n(z) = 0, \qquad |\arg(1 + z)| < \pi. \tag{32}$$

6.11. Inequalities

It is convenient to rewrite 6.10(20) in the form

$$H(z) = H_n(z,a) + V_n(z), \quad H_n(z,a) = A_n(z)/B_n(z). \tag{1}$$

If

$$z > 0, \quad \sigma + 1 - a > 0, \quad n + \rho + 1 - a > 0,$$

then

$$\operatorname{sign} V_n(z) = \operatorname{sign}\left(\frac{(-\sigma)^{1-a}\rho^a(\rho + 1 - \sigma)_n}{(\rho + 1 - a)_n}\right). \tag{2}$$

Further, if

$$z > 0, \quad \rho \geq 0, \quad \sigma > 0, \quad \rho + 1 - \sigma > 0,$$

then

$$H_m(z,1) < H(z) < H_n(z,0); \quad m,\ n > 0. \tag{3}$$

Also, if the conditions for (3) hold except that $\rho + 1 - \sigma < 0$, and $\rho + 1 - \sigma$ is not a negative integer, then (3) holds provided $(\rho + 1 - \sigma)_r$ is positive, $r = n$ or $r = m$; but if $(\rho + 1 - \sigma)_r$ is negative, then (3) holds with reversed inequality signs. If

$z = 0$ or if $\sigma = 0$, the inequalities become equalities.

The special case $\rho = 0$ has been given in 2.5. These results are the bases for the development of inequalities for ${}_{p+1}F_p$ see 5.13. For a ${}_2F_1$ inequality, see Flett (1972/73).

6.12. Bibliographic and Numerical Data

6.12.1. References

Numerous references are given in 5.1 and throughout Chapter 5 as well. For work and further references on Legendre functions, see Hobson (1955) and Robin (1957,1958,1959). We do not treat elliptic functions and integrals except in so far as $P_{-\frac{1}{2}}(1 - 2k^2) = (2/\pi)K(k)$ is a special case of data relating to $P^m_{n-\frac{1}{2}}(x)$. For data on elliptic functions and integrals, see the journal Mathematics of Computation, and in particular the reviews given therein of the works of Fettis *et al.* See also Luke (1969,1970a) and references in 8.2.6.

6.12.2. Description of and References to Tables

Pearson (1968): Let $B(p,q) = \Gamma(p+q)/\Gamma(p)\Gamma(q)$, $I_x(p,q) = B(p,q)\int_0^x t^{p-1}(1-t)^{q-1}dt$. $B(p,q)$, $q = 0.5(0.5)11(1)50$, $p = q(0.5)11(1)50$, 8S. $I_x(p,q)$ with p and q as above and $x = 0(0.01)$ to point where entry is 1 to 7D. $I_x(p,0.5)$, $p = 11.5(1)14.5$, $x = 0.27(0.01)$, 7D. $I_x(p,q)$, $q = 0.5$, $p = 0.5(0.5)11(0.5)16$, $x = 0.975$, 0.98, 0.985, 0.9880 (0.0005)0.9985, 0.9988(0.0001)0.9999, 8D; $q = 1(0.5)3$, $p = 0(0.5)11(1)16$, $x = 0.975$, 0.98, 0.985, 0.988(0.001) 0.999, 8D.

Osborn and Madey (1968): $B(p,q)$, $I_x(p,q)/B(p,q)$ and $I_x(p,q)$, $p,q = 0.5(0.1)2$, $x = 0.1(0.1)1$, 5D.

Amos (1963): Tables for solution of $I_x(p,q) = R$, $R = 0.001$, 0.0025, $2p = 1(1)30$, 40, 60, 120, $2q = 1(1)10$, 12, 15, 20, 24, 30, 40, 60, 120, 5S.

Harter (1964): Tables as in Amos entry above, $p = 1(1)40$, $q = 1(1)40$, various $R \leq 0.5$, 7S. Also tables for $R > 0.5$.

Mackiernan (1970): Tables involving ${}_2F_1(1/3,5/6;4/3;-1/a^2x^2)$ and ${}_2F_1(1/2,1/6;3/2;-a^2x^2)$, $a = \Gamma(1/3)/\Gamma(1/2)\Gamma(5/6)$, $x = 10^\alpha(10^\beta)10^\gamma$, $\alpha = -3(1)5$, $\beta = \alpha - 1$, $\gamma = \alpha + 1$, 9S. Note that in view of Kummer's transformation formula, the above functions are essentially incomplete beta functions. See 8.2.3, 6.4 and all the above entries.

Smirnov (1961): Let

$$A_\nu = (\pi\nu)^{-\frac{1}{2}}\Gamma(\tfrac{1}{2}+\tfrac{1}{2}\nu)/\Gamma(\tfrac{1}{2}\nu), \quad B(t) = (1+t^2/\nu)^{-\frac{1}{2}-\frac{1}{2}\nu},$$

$$S_\nu(t) = \tfrac{1}{2} + A_\nu tB(t)(1+t^2/\nu)\,{}_2F_1(1-\tfrac{1}{2}\nu,\ 1;\ 3/2;\ -t^2/\nu).$$

$S_\nu(t)$, $A_\nu B(t)$, $t=0(0.01)3(0.02)4.5(0.05)6.5$, $\nu=1(1)12$ and for almost as many values of t with $\nu=13(1)24$, 6D; $S_\nu(t)$, $t=6.5(1)90$, $\nu=1(1)10$ and $t=0(0.01)2.5(0.02)3.5$ $(0.05)5$, $\nu=25(1)35$, 6D; $S_\nu(t)$, $t=0(0.01)2.5(0.02)5$, $1000/\nu=0(2)30$, 6D. Other tables are also provided.

Steidley (1963): A glossary of closed form expressions for ${}_2F_1(a, b; c; z)$ for all possible combinations of $a=0(\frac{1}{2})7$, $b=0(\frac{1}{2})7$, $c=\frac{1}{2}(\frac{1}{2})5$.

Consul (1963): ${}_2F_1(a, b; c; z)$, $a=\frac{1}{2}$, 3/2, 1(1)30, $b=c-1$, c, $c+1$, $c=1$, 2, 3, $z=0.01(0.01)0.1(0.1)0.9$, 0.95, up to 7S.

Carpenter (1962): Consider the expansion

$$(1-2a\cos t+a^2)^{-t} = (1-a^2)^{-t}\{\tfrac{1}{2}G_t^{(0)} + \sum_{k=1}^{\infty} G_t^{(k)}\cos kt\},$$

$$G_t^{(k)} = \{2\Gamma(t+k)/k!\,\Gamma(t+1)\}(1-a^2)^t\,{}_2F_1(t,\ t+k;\ t+1;\ a^2).$$

$\log_{10} G_t^{(k)}$ $k=12(1)24$, $t=0.5$, 1.5, 8D.

Alexander and Vok (1963): Except for a multiplication factor independent of x, this report gives tables of $x^{p-2}(1-x)^{n-p}\,{}_2F_1(n, n; p-1; yx)$ and its integral with respect to x; $p=2(1)10$, $n=p(1)25$, $y=0(0.01)0.69$, $x=0$ $(0.01)1$, 5D; $y=0.70(0.01)1$, $x=0(0.01)0.66(0.005)1$, 5D.

Bark, Zhurina and Karmazina (1962): This is a translation of French texts into Russian together with reproduction of certain tables and recalculation of other tables in these texts. Tables of $P_n^m(\cos\theta)$ are reproduced from "*Tables de fonctions de Legendre associées*," Editions de la Revue d'Optique, Paris, 1952. Also, as many errors were found in tables originally given by G. Prevost("*Tables de fonctions spheriques et de leurs intégrales pour calculer les coefficients du développement en serie de polynome de Laplace*," Gauthier-Villars, Paris, 1933), these entries were recalculated and an abbreviated version put in the present volume. Tables give $P_n^m(x)$ and its integral, $n=0(1)8$, $m=0(1)n$; $n=9$, $m=0,1$ and $m=10$, $n=0$; $x=0(0.01)1$, 5 or 6S.

Centre National d'Études des Télecommunications (1966): $P_n^m(\cos\theta)$, $m=3(1)5$, $n=-0.5(0.1)10$, $\theta=0^\circ(1^\circ)180^\circ$, 9S-13S.

Fettis and Caslan (1970): $P_{n-\frac{1}{2}}^m(s)$, $m=0(1)10$, $n=0(1)N$, $s=1.1(0.1)10$ and $s=\cosh t$, $t=0(0.1)3$, 11S. N is variable, $N=160$ for $s=1.1$ and $N=35$ for $s=10$.

Low (1966): First two ν zeros of $P_{\nu-\frac{1}{2}}^{(-m)}(\cos\theta)$, $m=2$, $\theta=\pi/12$ $(\pi/12)\pi/2$, 2D.

Waterman (1963): ν_i- and μ_i- zeros of $P'_\nu(\cos\theta)$ and $\partial P'_\mu(\cos\theta)/\partial\theta$, $i=1(1)30$, $\theta=165^\circ$, 5D.

Wilcox (1968): Zeros as in the Waterman entry above, but for $i=1(1)50$, 7S.

Hayakawa (1962): First ten non trivial zeros of $P'_\nu(\cos\theta)=0$, $\theta=\pi/6$, 2D. Corresponding values of $P_\nu(\cos\theta)$, $\theta=\pi/120$, $\pi/60(\pi/60)\pi/6$, are given to 6D. This is strange since zeros are needed to more than 2D to produce other 6D values.

Katsura, Inoue, Yamashita and Kilpatrick (1965): $\int_{-1}^{1} P_n^{m_1}(x)P_n^{m_2}(x)P_n^{m_3}(x)dx$ is tabulated for all values of the n_i's from 0 to 8 and all permissible values of the m_i's such that $m_1+m_2+m_3=0$, 11-15D. Integrals of the product of four such functions are similarly tabulated for all values of the n_i's from 0 to 4 and all permissible values of the m_i's such that $m_1+m_2+m_3+m_4=0$.

Albasiny, Bell and Cooper (1963): The same integral as in Katsura above is tabulated for all values of the n_i's such that $n_1\le n_2\le n_3$, $n_1+n_2+n_3$ is even, $|n_1-n_2|\le n_3\le n_1+n_2$; n_1, $n_2\le 12$, $n_3\le 24$, and all permissible values of the m_i's such that $m_1\pm m_2\pm m_3=0$, 12D.

Gutschick and Ludwig (1969): Let $I=\int_{-1}^{1} P_{m_1}^{n_1}(x)P_{m_2}^{n_2}(x)dx$. This table gives exact values of all nonvanishing and nonredundant integrals I, m_i, $n_i=0(1)12$, $i=1,2$.

Zhurina and Karmazina (1960, 1962): $P_{-\frac{1}{2}+i\tau}(x)$, $\tau=0(0.01)50$, $x=-0.9(0.1)0.9$, 7S; τ as before, $x=1.1(0.1)2(0.2)5$ $(10)60$, 7D.

Zhurina and Karmazina (1962): With $\theta=\arccos x$, coefficients for evaluation of $P_{-\frac{1}{2}+i\tau}(\cos\theta)$ and $P'_{-\frac{1}{2}+i\tau}(\cos\theta)$ are given for $x=-0.99(0.01)0.99$, 7D, when the Bessel functions $I_0(\tau\theta)$ and $I_1(\tau\theta)$ are known. With $\eta = \text{arc}\cosh x$, coefficients for evaluation of $P_{-\frac{1}{2}+i\tau}(\cosh\eta)$ and $P'_{-\frac{1}{2}+i\tau}(\cosh\eta)$ are given for $x=1.01(0.01)3(0.05)5(0.1)10(1)60$, 7D when the Bessel functions $J_0(\tau\eta)$ and $J_1(\tau\eta)$ are known. Tables not requiring the Bessel functions give the first eight coefficients in the expansions of $P_{-\frac{1}{2}+i\tau}(\cos\theta)$ and $(1+4\tau^2)^{-1}P'_{-\frac{1}{2}+i\tau}(\cos\theta)$ in powers of τ^2, $x=-0.90(0.01)0.99$, 7D, and the first eight coefficients in expansions of $P_{-\frac{1}{2}+i\tau}(\cosh\eta)$ and $(1+4\tau^2)^{-1}\times P'_{-\frac{1}{2}+i\tau}(\cosh\eta)$ in powers of τ^2, $x=1.01(0.01)3(0.05)5(0.1)10(1)60$, 7D.

Zhurina and Karmazina (1963): $P'_{-\frac{1}{2}+i\tau}(x)$, $\tau = 0(0.01)25$, $x = -0.9(0.1)0.9, 1.1(0.1)2(0.2)5(0.5)10(10)60$. 7S

Murty (1971): $P_{-\frac{1}{2}+i\tau}(x)$, $\tau=0.1(0.1)10$ and $x=1(0.1)10$, 8S. Also given are zeros, corresponding first derivatives, and coordinates of bend points for $\tau=0.9(0.1)10$, 8S.

Ben Daniel and Carr (1960): $f_1(x) = {}_2F_1(-\frac{1}{2}\nu, \frac{1}{2}\nu+\frac{1}{2}; \frac{1}{2}; x^2)$, $f_2(x) = x\,{}_2F_1(\frac{1}{2}-\frac{1}{2}\nu, 1+\frac{1}{2}\nu; 3/2; x^2)$, $\nu=0(0.0625)1(0.125)10(0.25)36$, $x=0(0.01)0.99$, 5S. Zeros of $f_1(x)$ and $f_2(x)$ in the stated x range are given for $\nu=0.4375(0.0625)36$, 4S. Note that the Legendre functions $P_\nu(x)$ and $Q_\nu(x)$ are linear combinations of $f_1(x)$ and $f_2(x)$. Certain integrals of $f_1(x)$ and $f_2(x)$ are given, but some of this data is incorrect. See Math. Comp. 16(1962), 117-119.

Murty and Taylor (1969): $P_\nu(x)$, $\nu=-0.5(0.02)0.5$, $x=-1(0.01)1$, 7D; $\nu=-0.5(0.1)8.5$, $x=-1(0.02)1$, 7D. Values of x_j and $P'_\nu(x_j)$ such that $P_\nu(x_j)=0$ and values of y_j and $P_\nu(y_j)$ such that $P'_\nu(y_j)=0$, $\nu=0.1(0.1)8.5$, 7D.

Richards and Mullineux (1963): $P_\nu(\cos\theta)$, $\theta=40^\circ(10^\circ)140^\circ$, $\nu=0.1(0.1)2$, 9D; $\theta=5^\circ(5^\circ)30^\circ$, $140^\circ(10^\circ)170^\circ$, $\nu=0.1(0.1)2$, 6D. Let ν_r be the r-th positive zero of $P_\nu(\cos\theta)$. For $r=1(1)6$, ν_r is tabulated mostly to 5D for various θ. For example, for $r=1$, $\theta=10^\circ(5^\circ)50^\circ(10^\circ)160^\circ$, 6D, and for $r=6$, $\theta=70^\circ(10^\circ)140^\circ$, 5D.

Fettis and Caslin (1969a): Part I: $Q_{n-\frac{1}{2}}^{m}(s)$, $m=0(1)5$, $n=0(1)N$, $s=1.1(0.1)10$, N varies from 88 at $s=1.1$ to 29 at $s=10$, 11S; m and n as above, $s=\cosh\eta$, $\eta=0.1(0.1)3$, N varies from 400 at $\eta=0.1$ to 28 at $\eta=3$, 11S; $m=0(1)10$, $n=0$ or 1, $s=1.1(0.1)10$ and $s=\cosh\eta$, $\eta=0.1(0.1)3$, 16S. Part II: $Q_{n-\frac{1}{2}}^{m}(s)$, $m=5(1)10$, $n=0(1)N$, $s=1.1(0.1)10$, N varies from 160 at $s=1.1$ to 35 at $s=10$, 11S; m and n as above, $s=\cosh\eta$, $\eta=0.1(0.1)3$, N varies from 450 at $\eta=0.1$ to 34 at $\eta=3$, 11S.

Fettis (1970): $Q_{n-\frac{1}{2}}(s)$, $n=0(1)5$, $s=1.01$, 13S.

CHAPTER VII THE CONFLUENT HYPERGEOMETRIC FUNCTION

7.1. Introduction

As previously noted in 5.4(6) the confluent hypergeometric series follows from that for ${}_2F_1(a, b; c; z/b)$ by use of the confluence principle, that is take the limit as $b \to \infty$. This concept is useful to deduce properties of the ${}_1F_1$ from those for the ${}_2F_1$, especially in those instances where results for the ${}_1F_1$ do not follow from those for the ${}_pF_q$ by setting $p = q = 1$. Thus many of the important results for the ${}_1F_1$ needed for the applications readily follow from those for the ${}_pF_q$ and the ${}_2F_1$, and actually could be omitted here. However, the ${}_1F_1$ is very important for the applications, see 7.10, and so a rather complete picture of the function is presented.

7.2. Integral Representations

$$ {}_1F_1(a; c; z) = \frac{\Gamma(c)}{\Gamma(a)\,\Gamma(c-a)} \int_0^1 e^{zt} t^{a-1} (1-t)^{c-a-1}\, dt, $$

$$ R(c) > R(a) > 0. \qquad (1) $$

Another solution of the confluent hypergeometric differential equation 7.4(1) is

$$ U(a; c; z) = [\Gamma(a)]^{-1} \int_0^\infty e^{-zt} t^{a-1} (1+t)^{c-a-1}\, dt, \qquad R(a) > 0, \qquad R(z) > 0. \quad (2) $$

We also have

$$ U(a; c; z) = (2\pi i)^{-1} \int_{\gamma-i\infty}^{\gamma+i\infty} \frac{\Gamma(a+s)\,\Gamma(-s)\,\Gamma(1-c-s)\, z^s}{\Gamma(a)\,\Gamma(a-c+1)}\, ds, $$

$$ -R(a) < \gamma < \min(0, 1 - R(c)), \qquad |\arg z| < 3\pi/2. \qquad (3) $$

Further (3) is valid for all γ as long as neither a nor $1 + a - c$ is a negative integer or zero, provided that the path of integration separates the poles of $\Gamma(a + s)$ from those of $\Gamma(-s)\Gamma(1 - c - s)$.

In the literature, the notation $\Phi(a; c; z)$ and $\psi(a; c; z)$ is often used in place of ${}_1F_1(a; c; z)$ and $U(a; c; z)$, respectively.

7.3. Elementary Relations

7.3.1. Derivatives

$$\frac{d^n}{dz^n}\,{}_1F_1(a;c;z) = \frac{(a)_n}{(c)_n}\,{}_1F_1\left(\begin{matrix}a+n\\c+n\end{matrix}\middle|\,z\right). \tag{1}$$

$$\frac{d^n}{dz^n}[z^\delta\,{}_1F_1(a;c;z)] = (\delta-n+1)_n\,z^{\delta-n}\,{}_2F_2\left(\begin{matrix}a,\delta+1\\c,\delta+1-n\end{matrix}\middle|\,z\right). \tag{2}$$

$$\frac{d^n}{dz^n}[z^{a+n-1}\,{}_1F_1(a;c;z)] = (a)_n\,z^{a-1}\,{}_1F_1(a+n;c;z). \tag{3}$$

$$\frac{d^n}{dz^n}[z^\delta e^{-z}\,{}_1F_1(a;c;z)] = (\delta-n+1)_n\,z^{\delta-n}\,{}_2F_2\left(\begin{matrix}c-a,\delta+1\\c,\delta+1-n\end{matrix}\middle|\,-z\right). \tag{4}$$

$$\frac{d^n}{dz^n}\left[z^{n+c-1}e^{-z}\,{}_1F_1\left(\begin{matrix}a+n\\c+n\end{matrix}\middle|\,z\right)\right] = (c)_n\,z^{c-1}\,e^{-z}\,{}_1F_1(a;c;z). \tag{5}$$

$$\frac{d^n}{dz^n}[z^{c-a+n-1}e^{-z}\,{}_1F_1(a;c;z)] = (c-a)_n\,z^{c-a-1}\,e^{-z}\,{}_1F_1\left(\begin{matrix}a-n\\c\end{matrix}\middle|\,z\right). \tag{6}$$

$$\frac{d^n}{dz^n}[e^{-z}\,{}_1F_1(a;c;z)] = \frac{(-)^n(c-a)_n}{(c)_n}\,e^{-z}\,{}_1F_1\left(\begin{matrix}a\\c+n\end{matrix}\middle|\,z\right). \tag{7}$$

$$\frac{d^n}{dz^n}[z^\delta e^{-z}] = (\delta-n+1)_n\,z^{\delta-n}\,e^{-z}\,{}_1F_1\left(\begin{matrix}-n\\\delta+1-n\end{matrix}\middle|\,z\right). \tag{8}$$

$$\frac{d^n}{dz^n}U(a;c;z) = (-)^n(a)_n\,U(a+n;c+n;z). \tag{9}$$

$$\frac{d^n}{dz^n}[z^{c-1}U(a;c;z)] = (-)^n(1+a-c)_n\,z^{c-n-1}U(a;c-n;z). \tag{10}$$

$$\frac{d^n}{dz^n}[z^{a+n-1}U(a;c;z)] = (a)_n(1+a-c)_n\,z^{a-1}U(a+n;c;z). \tag{11}$$

$$\frac{d^n}{dz^n}[e^{-z}U(a;c;z)] = (-)^n\,e^{-z}U(a;c+n;z). \tag{12}$$

$$\frac{d^n}{dz^n}[e^{-z}z^{c-a+n-1}U(a;c;z)] = (-)^n\,e^{-z}z^{c-a-1}U(a-n;c;z). \tag{13}$$

$$\frac{d^n}{dz^n}[e^{-z}z^{c-1}U(a;c;z)] = (-)^n\,e^{-z}z^{c-n-1}(1-a)_n\,U(a-n;c-n;z). \tag{14}$$

7.3.2. Contiguous Relations

In the following ${}_1F_1(a\,\pm)$ stands for ${}_1F_1(a \pm 1;\, c;\, z)$, etc., see 6.2.2, and $U(c \pm 1)$ stands for $U(a;\, c \pm 1;\, z)$, etc.

$$(c-a)\,{}_1F_1(a-)+(2a-c+z)\,{}_1F_1-a\,{}_1F_1(a+)=0. \tag{1}$$

$$-c(c-1)\,{}_1F_1(c-)+c(c-1+z)\,{}_1F_1-(c-a)z\,{}_1F_1(c+)=0. \tag{2}$$

$$c(a+z)\,{}_1F_1-ac\,{}_1F_1(a+)-(c-a)z\,{}_1F_1(c+)=0. \tag{3}$$

$$c\,{}_1F_1-c\,{}_1F_1(a-)-z\,{}_1F_1(c+)=0. \tag{4}$$

$$(c-a-1)\,{}_1F_1+a\,{}_1F_1(a+)-(c-1)\,{}_1F_1(c-)=0. \tag{5}$$

$$(a-1+z)\,{}_1F_1+(c-a)\,{}_1F_1(a-)-(c-1)\,{}_1F_1(c-)=0. \tag{6}$$

$$(c-a-1)\,U(c-)-(c-1+z)U+zU(c+)=0. \tag{7}$$

$$(a+z)U+a(c-a-1)\,U(a+)-zU(c+)=0. \tag{8}$$

$$(c-a)\,U+U(a-)-zU(c+)=0. \tag{9}$$

$$U-aU(a+)-U(c-)=0. \tag{10}$$

$$(a-1+z)U-U(a-)+(a-c+1)\,U(c-)=0. \tag{11}$$

$$U(a-)-(2a-c+z)U+a(a-c+1)\,U(a+)=0. \tag{12}$$

With the notation of 7.5, the functions

$$w_1,\ \frac{\Gamma(1+a-c)}{\Gamma(a)}w_2,\ \frac{\Gamma(1+a-c)}{\Gamma(1-c)}w_3,\ \frac{\Gamma(1-a)}{\Gamma(1-c)}w_4 \tag{13}$$

satisfy the same difference equation in a, that is (1). All of the functions in (13) when multiplied by $\Gamma(a)/\Gamma(1+a-c)$, satisfy the difference equation (7). Thus the solutions of the differential equation 7.4(1) satisfy the same difference equation in a. Similarly, the functions

$$w_1,\ \frac{\Gamma(1+a-c)\Gamma(c-1)}{\Gamma(1-c)\Gamma(a)}w_2,\ \frac{\Gamma(1+a-c)}{\Gamma(1-c)}w_3,\ \frac{\Gamma(1-a)}{\Gamma(1-c)}w_4 \tag{14}$$

satisfy the same difference equation in c, that is (2) and all the functions in (14) when multiplied by $\Gamma(1-c)/\Gamma(1+a-c)$ satisfy the difference equation (8).

7.3.3. Products of Confluent Functions

$${}_1F_1\left(\begin{matrix}a\\c\end{matrix}\middle|\,z\right){}_1F_1\left(\begin{matrix}a\\c\end{matrix}\middle|\,-z\right)={}_2F_3\left(\begin{matrix}a,\ c-a\\c,\ \tfrac{1}{2}c,\ \tfrac{1}{2}c+\tfrac{1}{2}\end{matrix}\middle|\,z^2/4\right). \tag{1}$$

$$ {}_1F_1\left(\begin{matrix}a\\c\end{matrix}\middle|z\right){}_1F_1\left(\begin{matrix}1+a-c\\2-c\end{matrix}\middle|-z\right) = {}_2F_3\left(\begin{matrix}\tfrac12+a-\tfrac12c,\ \tfrac12-a+\tfrac12c\\ \tfrac12c+\tfrac12,\ 3/2-\tfrac12c,\ \tfrac12\end{matrix}\middle|z^2/4\right) $$

$$ +\frac{(2a-c)(c-1)z}{c(c-2)}{}_2F_3\left(\begin{matrix}1+a-\tfrac12c,\ 1-a+\tfrac12c\\ \tfrac12c+1,\ 2-\tfrac12c,\ 3/2\end{matrix}\middle|z^2/4\right). \quad (2) $$

$$ {}_1F_1\left(\begin{matrix}a\\2a\end{matrix}\middle|z\right){}_1F_1\left(\begin{matrix}b\\2b\end{matrix}\middle|-z\right) = e^{-z}{}_1F_1\left(\begin{matrix}a\\2a\end{matrix}\middle|z\right){}_1F_1\left(\begin{matrix}b\\2b\end{matrix}\middle|z\right) $$

$$ = {}_2F_3\left(\begin{matrix}\tfrac12(a+b),\ \tfrac12(a+b+1)\\ \tfrac12+a,\ \tfrac12+b,\ a+b\end{matrix}\middle|z^2/4\right) \quad (3) $$

For further relations of this kind, see 8.3 and 8.4.

7.4. Differential Equations

The linearly independent solutions of

$$ [zD^2+(c-z)D-a]\,w(z)=0, \qquad D=d/dz, \quad (1) $$

or

$$ [\delta(\delta+c-1)-z(\delta+a)]\,w(z)=0, \qquad \delta=zD, \quad (2) $$

are proportional to

$$ w_1={}_1F_1\left(\begin{matrix}a\\c\end{matrix}\middle|z\right) \quad\text{and}\quad w_2=z^{1-c}\,{}_1F_1\left(\begin{matrix}1+a-c\\2-c\end{matrix}\middle|z\right), \quad (3) $$

provided c is not an integer or zero. Sometimes the notation

$$ \Phi(a;c;z)={}_1F_1(a;c;z) \quad (4) $$

is used. The differential equation

$$ hy''+\left\{\frac{2\alpha h}{z}+2f'h-\frac{hh''}{h'}-hh'+ch'\right\}y' $$

$$ +\left\{h'\left(\frac{\alpha}{z}+f'\right)(c-h)+h\left[\frac{\alpha(\alpha-1)}{z^2}+\frac{2\alpha f'}{z}+f''+(f')^2\right.\right. $$

$$ \left.\left.-\frac{h''}{h'}\left(\frac{\alpha}{z}+f'\right)\right]-a(h')^2\right\}y=0 $$

is satisfied by

$$ y=z^{-\alpha}e^{-f(z)}\,{}_1F_1\left(\begin{matrix}a\\c\end{matrix}\middle|h(z)\right) \quad (5) $$

where $f = f(z)$, etc.

If $a = -m$, a positive integer or zero, and c is a negative integer such that $m + c < 1$, then a well defined solution of (1) is

$$f(a; c; z) = \sum_{k=0}^{-a} \frac{(a)_k z^k}{(c)_k k!} = \sum_{k=0}^{m} \frac{(-m)_k z^k}{(1-s)_k k!}$$

$$= \frac{(s-1-m)! z^m}{(s-1)!} {}_2F_0(-m, s-m; -z^{-1}),$$

$$c = 1 - s, \qquad s \text{ a positive integer}, \qquad m < s, \tag{6}$$

and this is independent of w_2. Again if a and c are positive integers, $a < c$, then

$$z^{1-c} f(1 + a - c; 2 - c; z) = z^{-s} \sum_{k=0}^{s-m} \frac{(m-s)_k z^k}{(1-s)_k k!}$$

$$= \frac{(m-1)! z^{-m}}{(s-1)!} {}_2F_0(m - s, m; -z^{-1}),$$

$$a = m, \qquad m \text{ a positive integer},$$

$$c = 1 + s, \qquad s \text{ a positive integer or zero}, \qquad m < s + 1, \tag{7}$$

is a well-defined solution of (1) and is independent of w_2. Observe that (6) and (7) are confluent forms of 6.3(7,8) respectively. The complete solution of (1) is treated in the next section.

The confluent forms of 6.4(1,5) are also known as Kummer's relations. Thus

$$w_1 = {}_1F_1(a; c; z) = e^z {}_1F_1(c - a; c; -z), \tag{8}$$

$$w_2 = z^{1-c} {}_1F_1(1 + a - c; 2 - c; z) = z^{1-c} e^z {}_1F_1(1 - a; 2 - c; -z). \tag{9}$$

7.5. The Complete Solution and Wronskians

If c is not an integer, then w_1 and w_2 as defined by 7.4(3) are independent solutions of 7.4(1). Now

$$w_3 = U(a; c; z) = z^{1-c} U(1 + a - c; 2 - c; z), \tag{1}$$

where the U-function is defined by 7.2(3), is also a solution. The relation between these solutions is

$$w_3 = U(a; c; z) = \frac{\Gamma(1-c)}{\Gamma(1+a-c)} w_1 + \frac{\Gamma(c-1)}{\Gamma(a)} w_2 ,$$
$$w_1 = {}_1F_1\left(\begin{matrix} a \\ c \end{matrix} \middle| z\right), \qquad w_2 = z^{1-c} {}_1F_1\left(\begin{matrix} 1+a-c \\ 2-c \end{matrix} \middle| z\right). \tag{2}$$

With $\varepsilon = \pm 1$,

$$w_4 = e^z U(c-a; c; ze^{-i\epsilon\pi}) = \frac{\Gamma(1-c)}{\Gamma(1-a)} w_1 - \frac{\Gamma(c-1)}{\Gamma(c-a)} e^{i\epsilon\pi c} w_2 , \tag{3}$$

is also a solution. By elimination, we have

$$w_1 = \frac{\Gamma(c)}{\Gamma(c-a)} e^{i\epsilon\pi a} w_3 + \frac{\Gamma(c)}{\Gamma(a)} e^{i\epsilon\pi(a-c)} w_4 , \tag{4}$$

$$w_2 = e^{i\epsilon\pi(a-c)} \left\{ -\frac{\Gamma(2-c)}{\Gamma(1-a)} w_3 + \frac{\Gamma(2-c)}{\Gamma(1+a-c)} w_4 \right\}. \tag{5}$$

In (4), (5), the convention

$$\begin{aligned} \epsilon = \operatorname{sign}(I(z)) &= 1 \qquad \text{if } I(z) > 0, \\ &= -1 \qquad \text{if } I(z) < 0, \end{aligned} \tag{6}$$

is often used.

Thus any two of the four quantities w_1, w_2, w_3, and w_4 form a fundamental system if a, c, and $c - a$ are not integers. If a is a negative integer or zero, w_3 and w_1 differ only by a constant multiple. The same is true for w_3 and w_2 if $c - a$ is a positive integer. Again if a is a positive integer w_4 is a constant multiple of w_2, and likewise for w_4 and w_1 if $c - a$ is a negative integer or zero. If c is an integer, zero included, either $w_1 = w_2$ or one of these is not defined. If c is a negative integer or zero, say $c = 1 - n$, $n = 1, 2, 3,\ldots$, then

$$\lim_{c\to 1-n} \frac{w_1}{\Gamma(c)} = \frac{(a)_n}{n!} z^n {}_1F_1(a+n; 1+n; z) \tag{7}$$

satisfies 7.4 (1), but this is a multiple of w_2, and so we get no new solution. The derivation of (2) assumes that c is not an integer, but clearly this is not essential, for by continuity it holds also for integer c. Indeed, as in the ${}_2F_1$ studies, we can use (2) to derive a logarithmic solution of 7.4 (1) when c is an integer. For the present situation, we can use the developments for $w_3 = W(z)$ [see 6.6(3,4,14)] provided we treat all terms involving a_2 as empty. That is, if a term involving a_2 appears as a product, treat it as unity, and if it appears as a sum, treat it as zero. We find

$$w_3 = U(a; 1+s; z) = \frac{z^{-s}}{\Gamma(a)} \sum_{k=0}^{s-1} \frac{(a-s)_k}{k!} (s-1-k)!(-)^k z^k$$
$$+ \frac{(-)^{s+1}}{s!\Gamma(a-s)} \Big\{ [\gamma + \ln z + \psi(a) - \psi(1+s)]\, {}_1F_1\left(\begin{matrix} a \\ s+1 \end{matrix} \middle| z \right)$$
$$+ \sum_{k=0}^{\infty} \frac{(a)_k z^k}{(1+s)_k k!} (\psi(a+k) - \psi(a) - \psi(1+s+k) + \psi(1+s)$$
$$- \psi(1+k) + \psi(1)) \Big\}, \qquad (8)$$

where s is a positive integer or zero. If a is an integer or zero, the logarithmic portion of (8) does not appear. When a is a positive integer or zero, w_3 is proportional to 7.4(7).

Similarly, from 6.6(1,2,8) with $w_6 = V(z)$, we get

$$w_4 = e^z U(1+s-a; 1+s; ze^{-i\epsilon\pi})$$
$$= \frac{(-)^s z^{-s}}{\Gamma(1+s-a)} \sum_{k=0}^{s-1} \frac{(a-s)_k (s-1-k)!(-)^k z^k}{k!}$$
$$+ \frac{(-)^{s+1}}{s!\Gamma(1-a)} \Big\{ [\gamma + \ln(ze^{-i\epsilon\pi}) + \psi(1-a) - \psi(1+s)]\, {}_1F_1\left(\begin{matrix} a \\ 1+s \end{matrix} \middle| z \right)$$
$$+ \sum_{k=0}^{\infty} \frac{(a)_k z^k}{(1+s)_k k!} (\psi(a+k) - \psi(a) - \psi(1+s+k) + \psi(1+s)$$
$$- \psi(1+k) + \psi(1)) \Big\}, \qquad (9)$$

where a is not an integer or zero and s is a positive integer or zero.

If a is a positive integer, then w_4 is a linear combination of both w_1 and w_2 unless $a > s$, in which case w_4 is a multiple of w_1. This follows from (2) and may also be deduced from (9). If a is a negative integer or zero, then from (9) or from 6.6(12), we have

$$e^z U(1+s+m; 1+s; ze^{-i\epsilon\pi}) = \frac{(-)^s z^{-s}}{(m+s)!} \sum_{k=0}^{s-1} \frac{(-m-s)_k (s-1-k)!(-)^k z^k}{k!}$$
$$+ \frac{(-)^{s+1}}{s!m!} \Big\{ [\gamma + \ln(ze^{-i\epsilon\pi}) + \psi(1+m)$$
$$- \psi(1+s)]\, {}_1F_1\left(\begin{matrix} -m \\ 1+s \end{matrix} \middle| z \right)$$
$$+ \sum_{k=0}^{m} \frac{(-m)_k z^k}{(1+s)_k k!} (\psi(1+m-k) - \psi(1+m)$$
$$- \psi(1+s+k) + \psi(1+s) - \psi(1+k) + \psi(1))$$
$$+ (-)^m m! \sum_{k=m+1}^{\infty} \frac{(k-m-1)!\, z^k}{(1+s)_k k!} \Big\}, \qquad (10)$$

where m and s are positive integers or zero.

Similar expansions for $U(a;\ 1-\varepsilon;\ z)$ and $e^{z}U(m-a;\ m;\ y)$, $m=1-\varepsilon$, $y=z\exp(-i\varepsilon\pi)$, follow from (8) and (9-10), respectively, in view of (1).

Wronskians

Let W_{ij} denote the Wronskian of the solutions w_i and w_j. Thus,

$$W_{ij} = W(w_i\,,\, w_j) = w_i(z)\, w_j'(z) - w_i'(z)\, w_j(z). \tag{11}$$

Also let

$$B = z^{-c}e^{z}. \tag{12}$$

Then, with ε as in (6),

$$W_{12} = (1-c)\,B, \qquad W_{13} = -\frac{\Gamma(c)}{\Gamma(a)}\,B, \tag{13}$$

$$W_{14} = \frac{\Gamma(c)}{\Gamma(a)}\,e^{i\epsilon\pi c}B, \qquad W_{23} = -\frac{\Gamma(c)}{\Gamma(1+a-c)}\,B, \tag{14}$$

$$W_{24} = -\frac{\Gamma(c)}{\Gamma(1-a)}\,B, \qquad W_{34} = \frac{B}{\sin \pi c}\,[e^{i\epsilon\pi c}\sin \pi(c-a) + \sin \pi a], \tag{15}$$

7.6. Asymptotic Expansions

$$U(a;\, c;\, z) \sim z^{-a}\,{}_2F_0(a,\, 1+a-c;\, -z^{-1}),$$
$$|\,z\,| \to \infty, \qquad |\arg z\,| \leqslant 3\pi/2 - \epsilon, \qquad \epsilon > 0. \tag{1}$$

If $R_n(z)$ is the error when the ${}_2F_0$ is truncated after n terms, then

$$|R_n(z)| \leqslant \left|\frac{(1+a-c)_n(a)_n}{n!}\right| x^{-a_1-n}, \qquad x = R(z) > 0, \quad a_1 = R(a). \tag{2}$$

Thus if a, c, and z are real, the error does not exceed the first term neglected and is of the same sign as the first term neglected.

The asymptotic expansion for ${}_1F_1(a;\, c;\, z)$ follows from 7.5(4).

$$\begin{aligned}{}_1F_1(a;\, c;\, z) \sim \frac{\Gamma(c)}{\Gamma(c-a)}\,(z^{-1}e^{i\epsilon\pi})^{a}\,{}_2F_0(a,\, 1+a-c;\, -z^{-1})& \\ + \frac{\Gamma(c)}{\Gamma(a)}\,e^{z}z^{a-c}\,{}_2F_0(c-a,\, 1-a;\, z^{-1}),&\end{aligned}$$

$$\epsilon = 1 \text{ if } I(z) > 0, \quad \epsilon = -1 \text{ if } I(z) < 0, \quad |\,z\,| \to \infty, \quad |\arg z\,| < \pi, \tag{3}$$

$$ {}_1F_1(a; c; z) \sim \frac{\Gamma(c)}{\Gamma(a)} e^z z^{a-c} \qquad \text{as} \quad R(z) \to \infty, \tag{4} $$

$$ {}_1F_1(a; c; z) \sim \frac{\Gamma(c)}{\Gamma(c-a)} (-z)^{-a} \quad \text{as} \quad R(z) \to -\infty. \tag{5} $$

We can also write

$$ {}_1F_1(a; c; z) \sim \frac{\Gamma(c)}{\Gamma(a)} e^z z^{a-c} \, {}_2F_0(c-a, 1-a; z^{-1}) + \frac{\Gamma(c)}{\Gamma(c-a)} z^{-a} e^{-i\delta a\pi} \, {}_2F_0(a, 1+a-c; -z^{-1}), $$

$$ |z| \to \infty, \qquad \delta = \pm 1, \qquad -(2+\delta)\pi/2 < \arg z < (2-\delta)\pi/2. \tag{6} $$

We now present a convergent series expansion in z for a function simply related to ${}_1F_1$ which under certain conditions on the parameters is a useful expansion when these parameters are large. We have

$$ {}_1F_1(a; c; z) = e^{az/c} W(z), $$

$$ W(z) = \sum_{k=0}^{\infty} v_k z^k, \quad |z| < \infty, \tag{7} $$

$$ v_0 = 1, \; v_1 = 0, \; v_2 = \frac{a(c-a)}{2c^2(c+1)}, \; v_3 = \frac{a(c-a)(c-2a)}{3c^3(c+1)(c+2)}, $$

$$ v_4 = \frac{a(c-a)}{8c^4(c+1)(c+2)(c+3)} \{2(c-2a)^2 + a(c-a)(c+2)\}, \ldots, $$

$$ v_{k+1} = \frac{\{c(c-2a)kv_k + a(c-a)v_{k-1}\}}{(k+1)(k+c)c^2}. \tag{8} $$

In the following discussion, let z be fixed. If both a and c are $O(n)$, then $v_2 = O(n^{-1})$, $v_3 = O(n^{-2})$ and $v_k = O(n^{2-k})$, $k \geq 4$. If both a and c are $O(n)$, but $c - a$ is independent of n, then $v_k = O(n^{-k})$, $k \geq 2$. Under the same conditions on a and c but $c - 2a$ is independent of n, then $v_{2k} = O(n^{-k})$, $k \geq 1$, and $v_{2k+1} = O(n^{-k-2})$, $k \geq 1$.

Next, we give a uniform asymptotic representation for the ${}_1F_1$ when c is large and $\tfrac{1}{2}c - a$ is bounded.

$$e^{-z/2}\,{}_1F_1(a;c;z) \sim \sum_{n=0}^{\infty} d_n(z)/u^n, \qquad d_0(z) = 1,$$

$$u = \tfrac{1}{2}(c-1), \quad k = c/2 - a,$$

$$d_{n+1}(z) = -\tfrac{1}{2}z\,d_n'(z) + \tfrac{1}{2}\int_0^z (t/4 - k)\,d_n(t)\,dt. \tag{9}$$

$$d_1(z) = \frac{z^2}{16} - \frac{kz}{2}, \qquad d_2(z) = \frac{z^4}{512} - \frac{kz^3}{32} + \frac{z^2(2k^2-1)}{16} + \frac{kz}{4},$$

$$d_3(z) = \frac{z^6}{24576} - \frac{kz^5}{1024} + \frac{z^4(4k^2-3)}{512} - \frac{k(4k^2-13)z^3}{192} - \frac{z^2(3k^2-1)}{16} - \frac{kz}{8}. \tag{10}$$

If the expansion (9) is truncated after n terms, the remainder is $O(u^{-n})$ uniformly in z for $|R(z)| \leq b$, b fixed but arbitrary.

7.7. Expansions in Series of Chebyshev Polynomials

In addition to the following, see 5.10.1(11) and 5.10.2.

$${}_1F_1\left(\begin{matrix}a\\c\end{matrix}\,\middle|\, zx\right) = \sum_{n=0}^{\infty} C_n(z)T_n^*(x), \qquad 0 \leqslant x \leqslant 1,$$

$$C_n(z) = \frac{\epsilon_n(a)_n z^n}{2^{2n}(c)_n n!}\,{}_2F_2\left(\begin{matrix}a+n,\ \frac{1}{2}+n\\ c+n,\ 1+2n\end{matrix}\,\middle|\, z\right), \tag{1}$$

$$\begin{aligned}\frac{2C_n(z)}{\epsilon_n} &= (n+1)\left\{2 - \frac{(2n+3)(n+a+1)}{(n+2)(n+a)} + \frac{4(n+c)}{z(n+a)}\right\} C_{n+1}(z)\\ &\quad + \frac{2}{(n+a)}\left\{(n+3-a)(n+1) - (n+2-a)(n+\tfrac{3}{2})\right.\\ &\quad \left. + \frac{2(n+1)(n+3-c)}{z}\right\} C_{n+2}(z) + \frac{(n+1)(n+3-a)}{(n+2)(n+a)}\,C_{n+3}(z).\end{aligned} \tag{2}$$

$$\sum_{n=0}^{\infty} (-)^n C_n(z) = 1. \tag{3}$$

The coefficients $C_n(z)$ are easily generated by use of (2) in the backward direction. See 12.2(3-7).

7.8. Expansions in Series of Bessel Functions

$$ {}_1F_1(a; c; z\omega) = \Gamma(\lambda + 1)e^{z/2}\left(\frac{4}{z}\right)^{\lambda} \sum_{n=0}^{\infty} \frac{(-)^n(n + \lambda)(2\lambda)_n}{n!\,\lambda} I_{n+\lambda}\left(\frac{z}{2}\right) $$

$$ \times\, {}_3F_2\left(\begin{matrix}-n, n + 2\lambda, a \\ \lambda + \frac{1}{2}, c\end{matrix}\,\middle|\, \omega\right). \tag{1} $$

If we designate the ${}_3F_2$ in (1) evaluated at $\omega = 1$ by $R_n(a, c, \lambda)$, then

$$ (n + c)(n + 2\lambda)R_{n+1}(a, c, \lambda) = (c - 2a)(2n + 2\lambda)R_n(a, c, \lambda) + n(n - c + 2\lambda)R_{n-1}(a, c, \lambda), \tag{2} $$

$$ R_n(a, c, \lambda) = (-)^n R_n(c - a, c, \lambda), \tag{3} $$

$$ R_n(a, c, a - \tfrac{1}{2}) = \frac{(-)^n(2a - c)_n}{(c)_n}, \tag{4} $$

$$ R_n(a, c, c - a - \tfrac{1}{2}) = \frac{(c - 2a)_n}{(c)_n}, \tag{5} $$

$$ R_{2n+1}(a, 2a, \lambda) = 0, \qquad R_{2n}(a, 2a, \lambda) = \frac{(\frac{1}{2})_n(\lambda + \frac{1}{2} - a)_n}{(a + \frac{1}{2})_n(\lambda + \frac{1}{2})_n}. \tag{6} $$

$$ {}_1F_1(a; c; z) = \Gamma(a + \tfrac{1}{2})e^{z/2}\left(\frac{4}{z}\right)^{a-\frac{1}{2}} \sum_{n=0}^{\infty} \frac{(n + a - \frac{1}{2})(2a - 1)_n(2a - c)_n}{n!\,(c)_n(a - \frac{1}{2})} I_{n+a-\frac{1}{2}}\left(\frac{z}{2}\right). \tag{7} $$

$$ {}_1F_1(a; c; z) = \Gamma(c - a + \tfrac{1}{2})e^{z/2}\left(\frac{4}{z}\right)^{c-a-\frac{1}{2}} $$

$$ \times \sum_{n=0}^{\infty} \frac{(-)^n(n + c - a - \frac{1}{2})(2c - 2a - 1)_n(c - 2a)_n}{n!\,(c - a - \frac{1}{2})(c)_n} I_{n+c-a-\frac{1}{2}}\left(\frac{z}{2}\right). \tag{8} $$

$$ {}_1F_1(1; c; z) = e^z\left(\frac{\pi}{2z}\right)^{1/2}\Big[\sum_{n=0}^{\infty} \frac{(-)^n(4n + 1)\{(2 - c)/2\}_n}{\{(1 + c)/2\}_n} I_{2n+\frac{1}{2}}(z) $$

$$ - \left(\frac{c - 1}{c}\right) \sum_{n=0}^{\infty} \frac{(-)^n(4n + 3)\{(3 - c)/2\}_n}{\{(2 + c)/2\}_n} I_{2n+\frac{3}{2}}(z)\Big]. \tag{9} $$

7.9. Inequalities

$$-1 + 2\,{}_2F_1\left(\begin{matrix}1, a\\ c\end{matrix}\middle| -\frac{z}{2}\right) < {}_1F_1\left(\begin{matrix}a\\ c\end{matrix}\middle| -z\right) < {}_2F_1\left(\begin{matrix}1, a\\ c\end{matrix}\middle| -z\right), \tag{1}$$

$$-1 + 2\left[1 + \frac{az}{2c}\right]^{-1} < {}_1F_1\left(\begin{matrix}a\\ c\end{matrix}\middle| -z\right)$$

$$< 1 - \frac{a(c+1)}{c(a+1)} + \frac{a(c+1)}{c(a+1)}\left[1 + \frac{(a+1)z}{c+1}\right]^{-1},$$

$$z > 0, \qquad c \geqslant a > 0; \tag{2}$$

$$-\frac{1}{c} + \frac{(c+1)}{2}\left[1 + \frac{az}{c+1}\right]^{-1} < {}_1F_1\left(\begin{matrix}a\\ c\end{matrix}\middle| -z\right)$$

$$< \frac{1-a}{1+a} + \frac{2a}{a+1}\left[1 + \frac{(a+1)z}{2c}\right]^{-1},$$

$$z > 0, \qquad a < 1, \qquad c > 0. \tag{3}$$

For further inequalities, see 5.13. See also Lorch (1967).

7.10. Other Notations and Related Functions

In the literature, a notation introduced by Whittaker is often used to designate solutions of the confluent hypergeometric differential equation. These functions are called Whittaker functions. The connection between these functions and the ${}_1F_1$ and U functions are outlined below.

$$M_{k,m}(z) = e^{-z/2}z^{c/2}\,{}_1F_1(a; c; z), \tag{1}$$

$$W_{k,m}(z) = W_{k,-m}(z) = \frac{\Gamma(-2m)}{\Gamma(\frac{1}{2} - m - k)} M_{k,m}(z) + \frac{\Gamma(2m)}{\Gamma(\frac{1}{2} + m - k)} M_{k,-m}(z). \tag{2}$$

$$k = c/2 - a, \qquad m = (c-1)/2. \tag{3}$$

$$W_{k,m}(z) = e^{-z/2}z^{c/2}U(a; c; z). \tag{4}$$

$$M_{k,m}(ze^{i\pi/2}) = \exp[i\pi(m + \tfrac{1}{2})]\, M_{-k,m}(ze^{-i\pi/2}). \tag{5}$$

$$W_{k,m}(ze^{i\pi/2}) = \frac{\pi}{\sin 2m\pi}\left[-\frac{\exp[i\pi(\frac{1}{2} + m)]}{\Gamma(\frac{1}{2} - m - k)\,\Gamma(1 + 2m)} M_{-k,m}(ze^{-i\pi/2}) + \frac{\exp[i\pi(\frac{1}{2} - m)]}{\Gamma(\frac{1}{2} + m - k)\,\Gamma(1 - 2m)} M_{-k,-m}(ze^{-i\pi/2})\right]. \tag{6}$$

$$M_{-k,m}(ze^{\pm i\pi}) = \exp[\pm i\pi(\tfrac{1}{2} + m)]\, M_{k,m}(z). \tag{7}$$

$$W_{-k,m}(ze^{\pm i\pi}) = \frac{\pi}{\sin 2m\pi}\left[-\frac{\exp[\pm i\pi(\frac{1}{2} + m)]}{\Gamma(\frac{1}{2} - m + k)\,\Gamma(1 + 2m)} M_{k,m}(z) + \frac{\exp[\pm i\pi(\frac{1}{2} - m)]}{\Gamma(\frac{1}{2} + m + k)\,\Gamma(1 - 2m)} M_{k,-m}(z)\right]. \tag{8}$$

7.11. Bibliographic and Numerical Data

7.11.1. References

General references on confluent hypergeometric functions are Tricomi (1952, 1954, 1960), Buchholtz (1953), Slater (1960), and Jankovic (1961). See also the references given in 5.1.

Ng and Geller (1970) list indefinite integrals of confluent hypergeometric functions. Zeros of confluent hypergeometric functions have been studied by Dyson (1960), Lambert-Goffart (1962) and Johnston (1964).

7.11.2. Description of and References to Tables and Other Approximations

Heatley (1964): $e^{-x}{}_1F_1(a;\ c;\ x)$, $a = \frac{1}{4}(\frac{1}{4})1$, $c = -\frac{1}{2}$, $\frac{1}{2}(\frac{1}{2})3$; $a = 2\frac{1}{4}(\frac{1}{4})3$, $c = 3$; $a = 1\frac{1}{4}(\frac{1}{4})2$, $c = 2$; $x^{\frac{1}{2}} = 0(0.2)4,5$, 7S except that if $a = 1$, $c = -\frac{1}{2}$, $\frac{1}{2}(\frac{1}{2})3$, $x^{\frac{1}{2}} = 0(0.1)3$. $T(m,n,r) = r^{2n-m+1}e^{-r^2}\Gamma(\frac{1}{2}m+\frac{1}{2})\,{}_1F_1(\frac{1}{2}m+\frac{1}{2};\ n+1;\ r^2)/n!$, $m = -\frac{1}{2}$, $\frac{1}{2}$, $n = -2(\frac{1}{2})2$, $r = 0(0.2)4$, 5, 6, 10, 25; $m = 1$, $n = -2(\frac{1}{2})3$, $r = 0(0.1)3$, 9S.

Heatley (1965): $e^{-x}{}_1F_1(a;\ c;\ x)$, $a = 0.25$, 0.75, $c = 1$, 2, 3 and $a = 1.25$, 2.25, $c = 2,3$, $x^{\frac{1}{2}} = 2(0.2)4$, 5D; $T(m,n,r)$ as in the above entry, $m = -0.5$, 0.5, $n = -2(1)2$, $r = 2$ (0.2)4, 5D; $m = 2$, $n = 0$, $r = 0(0.1)5$ as well as $2r\pi^{-\frac{1}{2}}\times T(2,0,r)$, 7D; $m = 3(2)9$, $m+n = 5.5(2)17.5$, $r = 0(0.1)$ 5, 7D.

Johnston (1964): Evaluation of ${}_1F_1(a;\ c;\ x)$ and its zeros to 7D. For a complete description, see Math. Comp. 19 (1965), 343.

Zhurina and Osipova (1964): ${}_1F_1(a;\ c;\ x)$, $U(a;\ c;\ x)$, $a = -0.98(0.02)1.10$, $c = 2$, $x = 0(0.01)4$, 7S. Use tables with caution as identifying notation is omitted on page headings.

Korobochkin and Filippov (1965): Tables relating to the solutions of 7.4(1) with $c = 2$ are given. They are much too long to detail here, but see Math. Rev. 32(1966), #5926 and Math. Comp. 21(1967), 265-267.

Kireeva and Karpov (1961): $D_\nu(z)$, see 8.2.7(4), $z = x(1+i)$, $\nu = 0(0.1)2$, $\pm x = 0(0.01)5$; $\nu = 0(0.05)2$, $\pm x = 5(0.01)10$, generally to 6D.

Karpov and Cistova (1964): $D_\nu(z)$ as above, $\nu = -2(0.1)0$, $\pm x = 0(0.01)5$; $\nu = -2(0.05)0$, $\pm x = 5(0.01)10$, generally to 6D.

Karpov and Cistova (1968): $D_\nu(z)$ as above, $0 \le z < \infty$; $\exp(-z^2/4)D_\nu(z)$, $-\infty < z \le 0$, $\exp(-z^2/4)D_\nu(iz)$, $0 \le z < \infty$. The tabular interval in z is 0.01 for $|z| \le 5$ and 0.001 or 0.0001 in $w = 1/z$ for $|z| > 5$. The range in ν is -1 (0.1)1. Tabular entries are 7D if less than unity in magnitude, otherwise 8S.

Murzewski and Sowa (1972): $D_{-n}(x)$, $n = 1(1)20$, $x = 0(0.05)3$, 7S.

Luk'ianov, Teplov and Akimova (1961): $F_L(\eta,\rho)$, $G_L(\eta,\rho)$, $G_L'(\eta,\rho)$, see 8.2.7(1-3), $L = 0(1)15$, $\rho = 1(0.2)20$, $\ln \eta = -\infty$, $-0.8(0.1)0.8$, 5S, except that $G_L'(\eta,\rho)$ is given for $L = 0,1$ only.

Lutz and Karvelis (1963): $F_L(\eta,\rho)$, $G_L(\eta,\rho)$, their first derivatives and phase of $F_L(\eta,\rho) + iG_L(\eta,\rho)$, $L = 10$, $\eta = 1$, 2,3, $\rho = 1,5,10$, 6S.

Curtis (1964): Tables are closely related to $F_L(\eta,\rho)$ and $G_L(\eta,\rho)$ as above and their derivatives, and are too long to describe here. See Math. Rev. 29(1965), #4915 and Math. Comp. 19(1965), 341-342.

Strecok and Gregory (1972): Best Chebyshev rational approximations to $G_0(\eta,\rho)$ and $G_0'(\eta,\rho)$ along segments in the (η,ρ) plane: $\rho = 2\eta$, $3.5 < \eta \le 15$, $2 < \eta \le 3.5$, $1 \le \eta \le 2$; $\rho = 1$, $\eta > 1$; $\rho = 30$, $15 \le \eta \le 18.5$, $18.5 < \eta \le 22$. Also coefficients for rational approximations to $\ln (U)$ and $\ln (-U)$, $U = G_0(\eta,30)$ for $22 < \eta \le 30$. Accuracy is at least 13S.

CHAPTER VIII IDENTIFICATION OF THE $_pF_q$ AND G-FUNCTIONS WITH THE SPECIAL FUNCTIONS

8.1. Introduction

In Chapters V-VII, numerous results for the $_pF_q$ and its generalization, the G-function, are given. A vast number of special functions are particular cases of this material or are closely related to same. To facilitate use of our work, we identify the more important special functions with the $_pF_q$ and $G^{m,n}_{p,q}$ notation.

The exponential function e^z, the Bessel function $J_\nu(z)$, and the sine integral $Si(z)$ are examples of named functions. Named functions are identified in terms of the $_pF_q$ notation in 8.2 or references are given where the material is located in this volume. In 8.3 named functions are expressed in terms of the G-function while in 8.4 the G-function is expressed in terms of named functions.

8.2. Named Special Functions Expressed as $_pF_q$'s

8.2.1. Elementary Functions

The binomial function is treated in Chapter II. The exponential function, logs, circular and hyperbolic functions and their inverses are discussed in Chapter III.

8.2.2. The Incomplete Gamma Function and Related Functions

See Chapter IV.

8.2.3. The Gaussian Hypergeometric Function

See Chapter VI. Also

$$aB(a, b)\, x^{-a}(1-x)^{-b}I_x(a, b) = {}_2F_1\left(\begin{matrix}1, a+b\\ a+1\end{matrix}\,\middle|\, x\right), \qquad |x| < 1, \tag{1}$$

$$B(a, b) = \int_0^1 t^{a-1}(1-t)^{b-1}\,dt = \Gamma(a)\,\Gamma(b)/\Gamma(a+b), \tag{2}$$

$$B(a, b)\, I_x(a, b) = \int_0^x t^{a-1}(1-t)^{b-1}\, dt. \tag{3}$$

$I_x(a, b)$ is essentially the incomplete beta function.

8.2.4. Legendre Functions

$$P_\nu^\mu(z) = [\Gamma(1-\mu)]^{-1}[(z+1)/(z-1)]^{\mu/2} \times {}_2F_1(-\nu, \nu+1; 1-\mu; (1-z)/2), \tag{1}$$

$$Q_\nu^\mu(z) = \frac{e^{i\pi\mu}\pi^{1/2}\Gamma(\mu+\nu+1)}{2^{\nu+1}\Gamma(\nu+\frac{3}{2})} z^{-\mu-\nu-1}(z^2-1)^{\mu/2} \times {}_2F_1((\mu+\nu+1)/2, (\mu+\nu+2)/2; \nu+\tfrac{3}{2}; z^{-2}), \tag{2}$$

where the complex z-plane is cut along the real axis from -1 to 1.

$$P_\nu^\mu(x) = [\Gamma(1-\mu)]^{-1}[(1+x)/(1-x)]^{\mu/2} \times {}_2F_1(-\nu, \nu+1; 1-\mu; (1-x)/2), \qquad -1 < x < 1. \tag{3}$$

$$Q_\nu^\mu(x) = \tfrac{1}{2}e^{-i\pi\mu}[e^{-i\pi\mu/2}Q_\nu^\mu(x+i0) + e^{i\pi\mu/2}Q_\nu^\mu(x-i0)], \qquad -1 < x < 1. \tag{4}$$

For references on Legendre functions, see Abramowitz and Stegun (1964), Erdélyi *et al.* (1953, Vol. 1, Chapter 3), Hobson (1955), Robin (1957), and Snow (1952).

8.2.5. Orthogonal Polynomials

See Chapter XI.

8.2.6. Complete Elliptic Integrals

$$\mathbf{K}(k) = \int_0^{\pi/2} (1-k^2\sin^2\theta)^{-1/2}\, d\theta = \tfrac{1}{2}\pi\, {}_2F_1(\tfrac{1}{2}, \tfrac{1}{2}; 1; k^2), \qquad |k^2| < 1. \tag{1}$$

$$\mathbf{E}(k) = \int_0^{\pi/2} (1-k^2\sin^2\theta)^{1/2}\, d\theta = \tfrac{1}{2}\pi\, {}_2F_1(-\tfrac{1}{2}, \tfrac{1}{2}; 1; k^2), \qquad |k^2| < 1. \tag{2}$$

For references on complete and incomplete elliptic integrals and related topics, see Abramowitz and Stegun (1964), Byrd and Friedman (1954), Cayley (1961), Erdélyi *et al.* (1953, Vol. 2), Milne–Thomson (1950), Neville (1951), Oberhettinger and Magnus (1949), and Tricomi (1948). See also the index by Luke, Wimp and Fair (1972) for location of other references.

8.2.7. Confluent Hypergeometric Functions, Whittaker Functions and Bessel Functions

See Chapters VII, IX and X.

The so-called Coulomb wave functions are essentially confluent hypergeometric functions. Thus,

$$F_L(\eta, \rho) = C_L(\eta)\, \rho^{L+1} e^{-i\rho}\, {}_1F_1\left(\begin{matrix} L+1-i\eta \\ 2L+2 \end{matrix} \middle| 2i\rho\right),$$
$$C_L(\eta) = \frac{2^L e^{-\pi\eta/2} \,|\, \Gamma(L+1+i\eta)|}{(2L+1)!}. \tag{1}$$

In view of 7.4(8), $F_L(\eta, \rho)$ is real if ρ, η, and L are real. In the applications L is usually a positive integer or zero.

$$F_L(\eta, \rho) + iG_L(\eta, \rho) = \frac{(-)^{L+1}\Gamma(L+1-i\eta)\, e^{\pi\eta/2}(2\rho)^{L+1} e^{-i\rho}}{|\,\Gamma(L+1-i\eta)|} \times U(L+1-i\eta; 2L+2; 2i\rho). \tag{2}$$

$$F_L(\eta, \rho) + iG_L(\eta, \rho) = \frac{\Gamma(L+1-i\eta)\, e^{\pi\eta/2} i^{L+1}}{|\,\Gamma(L+1-i\eta)|} W_{i\eta, L+\frac{1}{2}}(2i\rho). \tag{3}$$

For further material on Coulomb wave functions, see Abramowitz and Stegun (1964) and Curtis (1964). We have followed the notation of the former.

The parabolic cylinder function is defined by

$$\begin{aligned} D_\nu(z) &= 2^{(\nu-1)/2} e^{-z^2/4} z U((1-\nu)/2; 3/2; z^2/2) \\ &= 2^{(\nu+1/2)/2} z^{-1/2} W_{(\nu+1/2)/2, -1/4}(z^2/2) \\ &= 2^{\nu/2} e^{-z^2/4} \left[\frac{\Gamma(1/2)}{\Gamma[(1-\nu)/2]}\, {}_1F_1\left(\begin{matrix} -\nu/2 \\ 1/2 \end{matrix} \middle| z^2/2\right) \right. \\ &\quad \left. + \frac{z\Gamma(-1/2)}{2^{1/2}\Gamma(-\nu/2)}\, {}_1F_1\left(\begin{matrix} (1-\nu)/2 \\ 3/2 \end{matrix} \middle| z^2/2\right) \right]. \end{aligned} \tag{4}$$

For further material on (4), see Erdélyi *et al.* (1953, Vol. 2, Chapter 8) J. C. P. Miller (1955), and Abramowitz and Stegun (1964).

8.3. Named Functions Expressed in Terms of the G-Function

$${}_pF_q\left(\begin{matrix} \alpha_p \\ \rho_q \end{matrix} \middle| z\right) = \frac{\Gamma(\rho_q)}{\Gamma(\alpha_p)} G^{1,p}_{p,q+1}\left(-z \middle| \begin{matrix} 1-\alpha_p \\ 0, 1-\rho_q \end{matrix}\right),$$
$$p \leqslant q \quad \text{or} \quad p = q+1 \quad \text{and} \quad |z| < 1. \tag{1}$$

$${}_pF_q\left(\begin{matrix}\alpha_p\\ \rho_q\end{matrix}\,\middle|\, z\right) = \frac{\Gamma(\rho_q)}{\Gamma(\alpha_p)}\, G^{p,1}_{q+1,p}\left(-z^{-1}\,\middle|\,\begin{matrix}1,\rho_q\\ \alpha_p\end{matrix}\right),$$

$$p \leqslant q \quad \text{or} \quad p = q+1 \quad \text{and} \quad |z| < 1. \tag{2}$$

$$z^a e^{z/2} M_{k,m}(z) = z^{a+c/2}\,{}_1F_1(a;\; c;\; z)$$
$$= z^{1/2}\Gamma(2m+1)\,\Gamma(k+\tfrac{1}{2}-m)\,G^{1,0}_{1,2}\left(z\,\middle|\,\begin{matrix}a+\frac{1}{2}+k\\ a+m,\, a-m\end{matrix}\right)$$
$$= \frac{(2\pi z)^{1/2}\,\Gamma(2m+1)\,\Gamma(k+\frac{1}{2}-m)}{2^{k-a}}$$
$$\times\, G^{2,0}_{2,4}\left(\frac{z^2}{4}\,\middle|\,\begin{matrix}\frac{a}{2}+\frac{1}{4}+\frac{k}{2},\,\frac{a}{2}+\frac{3}{4}+\frac{k}{2}\\ \frac{a+m}{2},\,\frac{a+m+1}{2},\,\frac{a-m}{2},\,\frac{a-m+1}{2}\end{matrix}\right). \tag{3}$$

$$z^a e^{-z/2} M_{k,m}(z) = z^{a+c/2}\,{}_1F_1(c-a;\; c;\; -z)$$
$$= \frac{z^{1/2}\Gamma(2m+1)}{\Gamma(k+\frac{1}{2}+m)}\,G^{1,1}_{1,2}\left(z\,\middle|\,\begin{matrix}a+\frac{1}{2}-k\\ a+m,\, a-m\end{matrix}\right)$$
$$= \frac{(z/2\pi)^{1/2}\,2^{k+a}\,\Gamma(2m+1)}{\Gamma(k+\frac{1}{2}+m)}$$
$$\times\, G^{2,2}_{2,4}\left(\frac{z^2}{4}\,\middle|\,\begin{matrix}\frac{a}{2}+\frac{1}{4}-\frac{k}{2},\,\frac{a}{2}+\frac{3}{4}-\frac{k}{2}\\ \frac{a+m}{2},\,\frac{a+m+1}{2},\,\frac{a-m}{2},\,\frac{a-m+1}{2}\end{matrix}\right). \tag{4}$$

$$z^a e^{z/2} W_{k,m}(z) = z^{a+c/2}\,U(a;\; c;\; z)$$
$$= [\Gamma(\tfrac{1}{2}-k+m)\,\Gamma(\tfrac{1}{2}-k-m)]^{-1}\,G^{2,1}_{1,2}\left(z\,\middle|\,\begin{matrix}a+k+1\\ a+\frac{1}{2}+m,\, a+\frac{1}{2}-m\end{matrix}\right)$$
$$= \frac{z^{1/2}2^{a-k}}{(2\pi)^{3/2}\,\Gamma(\frac{1}{2}-k+m)\,\Gamma(\frac{1}{2}-k-m)}$$
$$\times\, G^{4,2}_{2,4}\left(\frac{z^2}{4}\,\middle|\,\begin{matrix}\frac{a}{2}+\frac{1}{4}+\frac{k}{2},\,\frac{a}{2}+\frac{3}{4}+\frac{k}{2}\\ \frac{a+m}{2},\,\frac{a+m+1}{2},\,\frac{a-m}{2},\,\frac{a-m+1}{2}\end{matrix}\right). \tag{5}$$

$$z^a e^{-z/2} W_{k,m}(z) = z^{a+c/2}e^{-z}\,U(a;\; c;\; z)$$
$$= z^{1/2}\,G^{2,0}_{1,2}\left(z\,\middle|\,\begin{matrix}a+\frac{1}{2}-k\\ a+m,\, a-m\end{matrix}\right)$$
$$= (z/2\pi)^{1/2}\,2^{a+k}$$
$$\times\, G^{4,0}_{2,4}\left(\frac{z^2}{4}\,\middle|\,\begin{matrix}\frac{a}{2}+\frac{1}{4}-\frac{k}{2},\,\frac{a}{2}+\frac{3}{4}-\frac{k}{2}\\ \frac{a+m}{2},\,\frac{a+m+1}{2},\,\frac{a-m}{2},\,\frac{a-m+1}{2}\end{matrix}\right). \tag{6}$$

$$z^\mu J_\nu(z) = 2^\mu\, G^{1,0}_{0,2}\left(\frac{z^2}{4}\,\middle|\, \frac{\mu+\nu}{2}, \frac{\mu-\nu}{2}\right). \tag{7}$$

$$z^\mu J_\nu(z) = 2^{2\mu}\, G^{2,0}_{0,4}\left(\frac{z^4}{256}\,\middle|\, \frac{\mu+\nu}{4}, \frac{\mu+\nu+2}{4}, \frac{\mu-\nu}{4}, \frac{\mu-\nu+2}{4}\right). \tag{8}$$

$$\pi^{1/2} e^{-z}(2z)^\mu I_\nu(z) = G^{1,1}_{1,2}\left(2z\,\middle|\, {\tfrac{1}{2}+\mu \atop \mu+\nu, \mu-\nu}\right). \tag{9}$$

$$z^\mu Y_\nu(z) = (-)^m\, 2^\mu\, G^{2,0}_{1,3}\left(\frac{z^2}{4}\,\middle|\, {\frac{\mu-\nu-1}{2}-m \atop \frac{\mu+\nu}{2}, \frac{\mu-\nu}{2}, \frac{\mu-\nu-1}{2}-m}\right),$$

$$m = 0, \pm 1, \pm 2, \ldots . \tag{10}$$

$$z^\mu K_\nu(z) = 2^{\mu-1}\, G^{2,0}_{0,2}\left(\frac{z^2}{4}\,\middle|\, \frac{\mu+\nu}{2}, \frac{\mu-\nu}{2}\right). \tag{11}$$

$$e^{-z}(2z)^\mu K_\nu(z) = \pi^{1/2}\, G^{2,0}_{1,2}\left(2z\,\middle|\, {\tfrac{1}{2}+\mu \atop \mu+\nu, \mu-\nu}\right). \tag{12}$$

$$e^{z}(2z)^\mu K_\nu(z) = \pi^{-1/2}\cos\nu\pi\, G^{2,1}_{1,2}\left(2z\,\middle|\, {\tfrac{1}{2}+\mu \atop \mu+\nu, \mu-\nu}\right). \tag{13}$$

$$(z/2)^\omega s_{\mu,\nu}(z) = 2^{\mu-1}\Gamma[(\mu+\nu+1)/2]\,\Gamma[(\mu-\nu+1)/2]$$

$$\times\, G^{1,1}_{1,3}\left(\frac{z^2}{4}\,\middle|\, {\frac{\mu+\omega+1}{2} \atop \frac{\mu+\omega+1}{2}, \frac{\omega+\nu}{2}, \frac{\omega-\nu}{2}}\right). \tag{14}$$

$$(z/2)^\omega \mathbf{H}_\nu(z) = G^{1,1}_{1,3}\left(\frac{z^2}{4}\,\middle|\, {\frac{\nu+\omega+1}{2} \atop \frac{\nu+\omega+1}{2}, \frac{\omega+\nu}{2}, \frac{\omega-\nu}{2}}\right). \tag{15}$$

$$(z/2)^\omega S_{\mu,\nu}(z) = \frac{2^{\mu-1}}{\Gamma[(1-\mu+\nu)/2]\,\Gamma[(1-\mu-\nu)/2]}$$

$$\times\, G^{3,1}_{1,3}\left(\frac{z^2}{4}\,\middle|\, {\frac{\mu+\omega+1}{2} \atop \frac{\mu+\omega+1}{2}, \frac{\omega+\nu}{2}, \frac{\omega-\nu}{2}}\right). \tag{16}$$

$$(z/2)^\omega[\mathbf{H}_\nu(z) - Y_\nu(z)] = \frac{\cos\nu\pi}{\pi^2}\, G^{3,1}_{1,3}\left(\frac{z^2}{4}\,\middle|\, {\frac{\nu+\omega+1}{2} \atop \frac{\nu+\omega+1}{2}, \frac{\omega+\nu}{2}, \frac{\omega-\nu}{2}}\right). \tag{17}$$

$$(z/2)^{\omega}[I_{\nu}(z) - \mathbf{L}_{\nu}(z)] = \pi^{-1} G_{1,3}^{2,1}\left(\frac{z^2}{4}\,\middle|\, \begin{matrix} \dfrac{\nu+\omega+1}{2} \\ \dfrac{\nu+\omega+1}{2}, \dfrac{\omega+\nu}{2}, \dfrac{\omega-\nu}{2} \end{matrix}\right). \tag{18}$$

$$(z/2)^{\omega}[I_{-\nu}(z) - \mathbf{L}_{\nu}(z)] = \pi^{-1} \cos \nu\pi\, G_{1,3}^{2,1}\left(\frac{z^2}{4}\,\middle|\, \begin{matrix} \dfrac{\nu+\omega+1}{2} \\ \dfrac{\nu+\omega+1}{2}, \dfrac{\omega-\nu}{2}, \dfrac{\omega+\nu}{2} \end{matrix}\right). \tag{19}$$

$$\begin{aligned} &\mathrm{Ci}(z) + i\,\mathrm{si}(z) \\ &= -\frac{z^{-1}e^{iz}}{\pi^{1/2}}\left\{2z^{-1} G_{1,3}^{3,1}\left(\frac{z^2}{4}\,\middle|\, \begin{matrix} 1 \\ \frac{3}{2}, 1, 1 \end{matrix}\right) + i\, G_{1,3}^{3,1}\left(\frac{z^2}{4}\,\middle|\, \begin{matrix} 1 \\ \frac{1}{2}, 1, 1 \end{matrix}\right)\right\}. \end{aligned} \tag{20}$$

$$\begin{aligned} &C(z) + iS(z) \\ &= \tfrac{1}{2}(1+i) - \frac{(2\pi z)^{-1/2} e^{iz}}{2^{1/2}\pi}\left\{\left(\frac{2}{z}\right) G_{1,3}^{3,1}\left(\frac{z^2}{4}\,\middle|\, \begin{matrix} 1 \\ \frac{3}{4}, \frac{5}{4}, 1 \end{matrix}\right)\right. \\ &\quad \left. + i\, G_{1,3}^{3,1}\left(\frac{z^2}{4}\,\middle|\, \begin{matrix} 1 \\ \frac{1}{4}, \frac{3}{4}, 1 \end{matrix}\right)\right\}. \end{aligned} \tag{21}$$

$$\begin{aligned} &z^a M_{k,m}(iz)\, M_{k,m}(-iz) = z^{a+c}\, {}_1F_1(a;\ c;\ iz)\, {}_1F_1(a;\ c;\ -iz) \\ &= \frac{2^{a+1}\pi^{1/2}[\Gamma(2m+1)]^2}{\Gamma(\frac{1}{2}+k+m)\,\Gamma(\frac{1}{2}-k+m)} \\ &\quad \times G_{2,4}^{1,2}\left(\frac{z^2}{4}\,\middle|\, \begin{matrix} \dfrac{a}{2}+1+k, \dfrac{a}{2}+1-k \\ \dfrac{a+1}{2}+m, \dfrac{a+1}{2}-m, \dfrac{a}{2}+1, \dfrac{a+1}{2} \end{matrix}\right). \end{aligned} \tag{22}$$

$$\begin{aligned} &z^a M_{k,m}(z)\, W_{-k,m}(z) = z^{a+1} e^{-z}\, {}_1F_1(a;\ c;\ z) U(1-a;\ 2-c;\ z) \\ &= \frac{2^a \Gamma(2m+1)}{\pi^{1/2}\Gamma(\frac{1}{2}+k+m)} \\ &\quad \times G_{2,4}^{3,1}\left(\frac{z^2}{4}\,\middle|\, \begin{matrix} \dfrac{a}{2}+1-k, \dfrac{a}{2}+1+k \\ \dfrac{a+1}{2}+m, \dfrac{a}{2}+1, \dfrac{a+1}{2}, \dfrac{a+1}{2}-m \end{matrix}\right). \end{aligned} \tag{23}$$

$$\begin{aligned} &z^a W_{k,m}(z)\, W_{-k,m}(z) = z^{a+1} e^{-z} U(a;\ c;\ z) U(1-a;\ 2-c;\ z) \\ &= \frac{2^a}{\pi^{1/2}} G_{2,4}^{4,0}\left(\frac{z^2}{4}\,\middle|\, \begin{matrix} \dfrac{a}{2}+1+k, \dfrac{a}{2}+1-k \\ \dfrac{a+1}{2}+m, \dfrac{a+1}{2}-m, \dfrac{a}{2}+1, \dfrac{a+1}{2} \end{matrix}\right). \end{aligned} \tag{24}$$

$$z^a W_{k,m}(iz)\, W_{k,m}(-iz) = z^{a+c} U(a;\ c;\ ze^{i\pi/2}) U(a;\ c;\ ze^{-i\pi/2})$$

$$= \frac{2^a}{\pi^{1/2}\Gamma(\frac{1}{2}-k+m)\,\Gamma(\frac{1}{2}-k-m)}$$

$$\times\ G_{2,4}^{4,1}\left(\frac{z^2}{4}\,\middle|\,\begin{matrix}\frac{a}{2}+1+k, \frac{a}{2}+1-k\\ \frac{a+1}{2}+m, \frac{a+1}{2}-m, \frac{a}{2}+1, \frac{a+1}{2}\end{matrix}\right). \tag{25}$$

$$z^\omega J_\mu(z)\, J_\nu(z)$$

$$= \pi^{-1/2}\, G_{2,4}^{1,2}\left(z^2\,\middle|\,\begin{matrix}\frac{\omega+1}{2}, \frac{\omega}{2}\\ \frac{\omega+\mu+\nu}{2}, \frac{\omega-\mu+\nu}{2}, \frac{\omega+\mu-\nu}{2}, \frac{\omega-\mu-\nu}{2}\end{matrix}\right). \tag{26}$$

$$z^\omega J_\nu^2(z) = \pi^{-1/2}\, G_{1,3}^{1,1}\left(z^2\,\middle|\,\begin{matrix}\frac{\omega+1}{2}\\ \frac{\omega}{2}+\nu, \frac{\omega}{2}-\nu, \frac{\omega}{2}\end{matrix}\right). \tag{27}$$

$$z^\omega J_{-\nu}(z)\, J_\nu(z) = \pi^{-1/2}\, G_{1,3}^{1,1}\left(z^2\,\middle|\,\begin{matrix}\frac{\omega+1}{2}\\ \frac{\omega}{2}, \frac{\omega}{2}+\nu, \frac{\omega}{2}-\nu\end{matrix}\right). \tag{28}$$

$$z^\omega J_\nu(z)\, I_\nu(z) = \pi^{1/2} 2^{3\omega/2}\, G_{0,4}^{1,0}\left(\frac{z^4}{64}\,\middle|\,\frac{\omega+2\nu}{4}, \frac{\omega-2\nu}{4}, \frac{\omega}{4}, \frac{\omega+2}{4}\right). \tag{29}$$

$$z^\omega J_\nu(z)\, Y_\nu(z) = -\pi^{-1/2}\, G_{1,3}^{2,0}\left(z^2\,\middle|\,\begin{matrix}\frac{\omega+1}{2}\\ \frac{\omega}{2}, \frac{\omega}{2}+\nu, \frac{\omega}{2}-\nu\end{matrix}\right). \tag{30}$$

$$z^\omega I_\nu(z)\, K_\mu(z)$$

$$= (4\pi)^{-1/2}\, G_{2,4}^{2,2}\left(z^2\,\middle|\,\begin{matrix}\frac{\omega}{2}, \frac{\omega+1}{2}\\ \frac{\omega+\mu+\nu}{2}, \frac{\omega-\mu+\nu}{2}, \frac{\omega+\mu-\nu}{2}, \frac{\omega-\mu-\nu}{2}\end{matrix}\right). \tag{31}$$

$$z^\omega J_\nu(z)\, K_\nu(z) = \pi^{-1/2} 2^{3\omega/2-1/2}\, G_{0,4}^{3,0}\left(\frac{z^4}{64}\,\middle|\,\frac{\omega+2\nu}{4}, \frac{\omega+2}{4}, \frac{\omega}{4}, \frac{\omega-2\nu}{4}\right). \tag{32}$$

$$z^{\omega} K_{\mu}(z)\, K_{\nu}(z) = 2^{-1}\pi^{1/2}\, G_{2,4}^{4,0}\left(z^2 \,\middle|\, \begin{matrix} \frac{\omega}{2}, \frac{\omega+1}{2} \\ \frac{\omega+\mu+\nu}{2}, \frac{\omega-\mu+\nu}{2}, \frac{\omega+\mu-\nu}{2}, \frac{\omega-\mu-\nu}{2} \end{matrix}\right). \tag{33}$$

$$z^{\omega} H_{\nu}^{(1)}(z)\, H_{\nu}^{(2)}(z) = \pi^{-5/2} 2 \cos \nu\pi\, G_{1,3}^{3,1}\left(z^2 \,\middle|\, \begin{matrix} \frac{\omega+1}{2} \\ \frac{\omega}{2}+\nu, \frac{\omega}{2}-\nu, \frac{\omega}{2} \end{matrix}\right). \tag{34}$$

$$z^{\omega}[J_{\mu}(z)\, J_{\nu}(z) + J_{-\mu}(z)\, J_{-\nu}(z)] = \{[2\cos(\mu+\nu)\,\pi/2]/\pi^{1/2}\} \times G_{2,4}^{2,1}\left(z^2 \,\middle|\, \begin{matrix} \frac{\omega+1}{2}, \frac{\omega}{2} \\ \frac{\omega+\mu+\nu}{2}, \frac{\omega-\mu-\nu}{2}, \frac{\omega-\mu+\nu}{2}, \frac{\omega+\mu-\nu}{2} \end{matrix}\right). \tag{35}$$

$$z^{\omega}[J_{\mu}(z)\, J_{\nu}(z) - J_{-\mu}(z)\, J_{-\nu}(z)] = -\{[2\sin(\mu+\nu)\,\pi/2]/\pi^{1/2}\} \times G_{2,4}^{2,1}\left(z^2 \,\middle|\, \begin{matrix} \frac{\omega}{2}, \frac{\omega+1}{2} \\ \frac{\omega+\mu+\nu}{2}, \frac{\omega-\mu-\nu}{2}, \frac{\omega-\mu+\nu}{2}, \frac{\omega+\mu-\nu}{2} \end{matrix}\right). \tag{36}$$

$$z^{\omega}[H_{\mu}^{(1)}(z)\, H_{\nu}^{(1)}(z) - H_{\mu}^{(2)}(z)\, H_{\nu}^{(2)}(z)] = \frac{4}{i\pi^{1/2}}\, G_{2,4}^{3,0}\left(z^2 \,\middle|\, \begin{matrix} \frac{\omega}{2}, \frac{\omega+1}{2} \\ \frac{\omega+\mu+\nu}{2}, \frac{\omega-\mu+\nu}{2}, \frac{\omega+\mu-\nu}{2}, \frac{\omega-\mu-\nu}{2} \end{matrix}\right). \tag{37}$$

$$z^{\omega}[I_{\mu}(z)\, I_{\nu}(z) - I_{-\mu}(z)\, I_{-\nu}(z)] = -\frac{\sin(\mu+\nu)\,\pi}{\pi^{3/2}} \times G_{2,4}^{2,2}\left(z^2 \,\middle|\, \begin{matrix} \frac{\omega}{2}, \frac{\omega+1}{2} \\ \frac{\omega+\mu+\nu}{2}, \frac{\omega-\mu-\nu}{2}, \frac{\omega-\mu+\nu}{2}, \frac{\omega+\mu-\nu}{2} \end{matrix}\right). \tag{38}$$

$$(z^\omega/2i)[e^{i\pi(\nu-\mu)/2}H_\nu^{(1)}(z)\,H_\mu^{(2)}(z) - e^{i\pi(\mu-\nu)/2}H_\mu^{(1)}(z)\,H_\nu^{(2)}(z)]$$
$$= \pi^{-5/2}(\cos\mu\pi - \cos\nu\pi)$$
$$\times G_{2,4}^{4,1}\left(z^2 \,\middle|\, \begin{matrix} \frac{\omega}{2}, \frac{\omega+1}{2} \\ \frac{\omega+\mu+\nu}{2}, \frac{\omega-\mu+\nu}{2}, \frac{\omega+\mu-\nu}{2}, \frac{\omega-\mu-\nu}{2} \end{matrix}\right). \tag{39}$$

$$\tfrac{1}{2}(z^\omega)[e^{i\pi(\nu-\mu)/2}H_\nu^{(1)}(z)\,H_\mu^{(2)}(z) + e^{i\pi(\mu-\nu)/2}H_\mu^{(1)}(z)\,H_\nu^{(2)}(z)]$$
$$= \pi^{-5/2}(\cos\mu\pi + \cos\nu\pi)$$
$$\times G_{2,4}^{4,1}\left(z^2 \,\middle|\, \begin{matrix} \frac{\omega+1}{2}, \frac{\omega}{2} \\ \frac{\omega+\mu+\nu}{2}, \frac{\omega-\mu+\nu}{2}, \frac{\omega+\mu-\nu}{2}, \frac{\omega-\mu-\nu}{2} \end{matrix}\right). \tag{40}$$

8.4. The G-Function Expressed as a Named Function

$$G_{p,q}^{m,n}\left(z\,\middle|\,\begin{matrix}a_p\\b_q\end{matrix}\right) = \sum_{h=1}^{m} \frac{\prod_{j=1}^{m}\Gamma(b_j-b_h)^*\prod_{j=1}^{n}\Gamma(1+b_h-a_j)\,z^{b_h}}{\prod_{j=m+1}^{q}\Gamma(1+b_h-b_j)\prod_{j=n+1}^{p}\Gamma(a_j-b_h)}$$
$$\times\, {}_pF_{q-1}\left(\begin{matrix}1+b_h-a_p\\1+b_h-b_q^*\end{matrix}\,\middle|\,(-)^{p-m-n}\,z\right),$$
$$p<q \quad\text{or}\quad p=q \quad\text{and}\quad |z|<1. \tag{1}$$

$$G_{p,q}^{m,n}\left(z\,\middle|\,\begin{matrix}a_p\\b_q\end{matrix}\right) = \sum_{h=1}^{n} \frac{\prod_{j=1}^{n}\Gamma(a_h-a_j)^*\prod_{j=1}^{m}\Gamma(b_j-a_h+1)\,z^{a_h-1}}{\prod_{j=n+1}^{p}\Gamma(1+a_j-a_h)\prod_{j=m+1}^{q}\Gamma(a_h-b_j)}$$
$$\times\, {}_qF_{p-1}\left(\begin{matrix}1+b_q-a_h\\1+a_p-a_h^*\end{matrix}\,\middle|\,\frac{(-)^{q-m-n}}{z}\right),$$
$$q<p \quad\text{or}\quad q=p \quad\text{and}\quad |z|>1. \tag{2}$$

$$G_{p,q}^{1,n}\left(z\,\middle|\,\begin{matrix}a_p\\b_q\end{matrix}\right) = \frac{\prod_{j=1}^{n}\Gamma(1+b_1-a_j)\,z^{b_1}}{\prod_{j=2}^{q}\Gamma(1+b_1-b_j)\prod_{j=n+1}^{p}\Gamma(a_j-b_1)}$$
$$\times\, {}_pF_{q-1}\left(\begin{matrix}1+b_1-a_p\\1+b_1-b_q^*\end{matrix}\,\middle|\,(-)^{p-1-n}\,z\right),$$
$$p<q \quad\text{or}\quad p=q \quad\text{and}\quad |z|<1. \tag{3}$$

$$G_{p,q}^{m,1}\left(z\,\middle|\,\begin{matrix}a_p\\b_q\end{matrix}\right) = \frac{\prod_{j=1}^{m}\Gamma(b_j-a_1+1)\,z^{a_1-1}}{\prod_{j=2}^{p}\Gamma(1+a_j-a_1)\prod_{j=m+1}^{q}\Gamma(a_1-b_j)}$$
$$\times\, {}_qF_{p-1}\left(\begin{matrix}1+b_q-a_1\\1+a_p-a_1^*\end{matrix}\,\middle|\,\frac{(-)^{q-m-1}}{z}\right),$$
$$q<p \quad\text{or}\quad q=p \quad\text{and}\quad |z|>1. \tag{4}$$

$$G_{1,2}^{1,1}\left(z\,\middle|\,\begin{matrix}1-k\\ \frac{1}{2}+m,\ \frac{1}{2}-m\end{matrix}\right)=\frac{e^{-z/2}\Gamma(\frac{1}{2}+k+m)}{\Gamma(2m+1)}\,M_{k,m}(z). \tag{5}$$

$$G_{0,2}^{1,0}(z\mid a,b)=z^{\frac{1}{2}(a+b)}J_{a-b}(2z^{1/2}). \tag{6}$$

$$G_{1,2}^{1,0}\left(z\,\middle|\,\begin{matrix}\frac{1}{2}\\ a,\ -a\end{matrix}\right)=\pi^{-1/2}(\cos a\pi)\,e^{z/2}I_a(z/2). \tag{7}$$

$$G_{0,2}^{2,0}(z\mid a,b)=2z^{\frac{1}{2}(a+b)}K_{a-b}(2z^{1/2}). \tag{8}$$

$$G_{1,2}^{2,0}\left(z\,\middle|\,\begin{matrix}\frac{1}{2}\\ b,\ -b\end{matrix}\right)=\pi^{-1/2}e^{-z/2}K_b(z/2). \tag{9}$$

$$G_{1,2}^{2,1}\left(z\,\middle|\,\begin{matrix}\frac{1}{2}\\ b,\ -b\end{matrix}\right)=\frac{\pi^{1/2}e^{z/2}}{\cos b\pi}\,K_b(z/2). \tag{10}$$

$$G_{1,2}^{2,0}\left(z\,\middle|\,\begin{matrix}a\\ b,\ c\end{matrix}\right)=z^be^{-z}U(a-c;\ 1+b-c;\ z). \tag{11}$$

$$G_{1,2}^{2,1}\left(z\,\middle|\,\begin{matrix}a\\ b,\ c\end{matrix}\right)=\Gamma(b-a+1)\,\Gamma(c-a+1)\times z^bU(b+1-a;\ b+1-c;\ z). \tag{12}$$

$$G_{0,4}^{1,0}(z\mid a+b/2,a,a-b/2,a+1/2)=\pi^{-1/2}z^aJ_b(x)\,I_b(x),\quad x=2^{3/2}z^{1/4}. \tag{13}$$

$$G_{0,4}^{1,0}(z\mid a+1/2,a,a+b/2,a-b/2)=\frac{\pi^{-1/2}z^a}{2\sin b\pi/2}\times[J_b(x)\,I_{-b}(x)-J_{-b}(x)\,I_b(x)],\quad x=2^{3/2}z^{1/4}. \tag{14}$$

$$G_{0,4}^{2,0}(z\mid a,a+\tfrac{1}{2},b,b+\tfrac{1}{2})=z^{\frac{1}{2}(a+b)}J_{2(a-b)}(4z^{1/4}). \tag{15}$$

$$G_{0,4}^{2,0}(z\mid a,-a,0,\tfrac{1}{2})=\frac{-\pi^{1/2}}{\sin 2\pi a}\,[J_{2a}(xe^{i\pi/4})\,J_{2a}(xe^{-i\pi/4})-J_{-2a}(xe^{i\pi/4})\,J_{-2a}(xe^{-i\pi/4})],\quad x=2^{3/2}z^{1/4}. \tag{16}$$

$$G_{0,4}^{2,0}(z \mid 0, \tfrac{1}{2}, a, -a) = \frac{\pi^{1/2}}{i \sin 2\pi a} \times [e^{2i\pi a} J_{2a}(xe^{-i\pi/4})\, J_{-2a}(xe^{i\pi/4}) - e^{-2i\pi a} J_{2a}(xe^{i\pi/4})\, J_{-2a}(xe^{-i\pi/4})],$$
$$x = 2^{3/2} z^{1/4}. \tag{17}$$

$$G_{0,4}^{3,0}(z \mid 3a - \tfrac{1}{2}, a, -a - \tfrac{1}{2}, a - \tfrac{1}{2}) = \frac{2\pi^{1/2} z^{a-1/2}}{\cos 2\pi a} K_{4a}(x)[J_{4a}(x) + J_{-4a}(x)],$$
$$x = 2^{3/2} z^{1/4}. \tag{18}$$

$$G_{0,4}^{3,0}(z \mid 0, a - \tfrac{1}{2}, -a - \tfrac{1}{2}, -\tfrac{1}{2}) = 4(\pi/z)^{1/2} K_{2a}(x) \times [\cos a\pi\, J_{2a}(x) - \sin a\pi\, Y_{2a}(x)],$$
$$x = 2^{3/2} z^{1/4}. \tag{19}$$

$$G_{0,4}^{3,0}(z \mid -\tfrac{1}{2}, a - \tfrac{1}{2}, -a - \tfrac{1}{2}, 0) = -4(\pi/z)^{1/2} K_{2a}(x) \times [\sin \pi a\, J_{2a}(x) + \cos \pi a\, Y_{2a}(x)],$$
$$x = 2^{3/2} z^{1/4}. \tag{20}$$

$$G_{0,4}^{3,0}(z \mid a, b + \tfrac{1}{2}, b, 2b - a) = (2\pi)^{1/2} z^b K_{2(a-b)}(x)\, J_{2(a-b)}(x),$$
$$x = 2^{3/2} z^{1/4}. \tag{21}$$

$$G_{0,4}^{4,0}(z \mid a, a + \tfrac{1}{2}, b, b + \tfrac{1}{2}) = 4\pi z^{\frac{1}{2}(a+b)} K_{2(a-b)}(4z^{1/4}). \tag{22}$$

$$G_{0,4}^{4,0}(z \mid a, a + \tfrac{1}{2}, b, 2a - b) = 8\pi^{1/2} z^a K_{2(b-a)}(xe^{i\pi/4})\, K_{2(b-a)}(xe^{-i\pi/4}),$$
$$x = 2^{3/2} z^{1/4}. \tag{23}$$

$$G_{2,4}^{4,0}\left(z^2 \,\middle|\, \begin{matrix} \frac{1}{2} + k, \frac{1}{2} - k \\ 0, \frac{1}{2}, m, -m \end{matrix}\right) = \pi^{1/2} z^{-1} W_{k,m}(2z)\, W_{-k,m}(2z). \tag{24}$$

$$G_{2,4}^{4,1}\left(z^2 \,\middle|\, \begin{matrix} \frac{1}{2} + k, \frac{1}{2} - k \\ 0, \frac{1}{2}, m, -m \end{matrix}\right) = \pi^{1/2} z^{-1} \Gamma(\tfrac{1}{2} + m - k)\, \Gamma(\tfrac{1}{2} - m - k) \times W_{k,m}(2iz)\, W_{k,m}(-2iz). \tag{25}$$

$$G_{1,3}^{1,1}\left(z \,\middle|\, \begin{matrix} \frac{1}{2} \\ 0, a, -a \end{matrix}\right) = \pi^{1/2} J_a(z^{1/2})\, J_{-a}(z^{1/2}). \tag{26}$$

$$G_{1,3}^{1,1}\left(z \,\middle|\, \begin{matrix} \frac{1}{2} \\ a, 0, -a \end{matrix}\right) = \pi^{1/2} J_a^{\,2}(z^{1/2}). \tag{27}$$

$$G_{1,3}^{1,1}\left(z\,\middle|\,\begin{matrix} a \\ a, b, a-\frac{1}{2}\end{matrix}\right) = z^{(2a+2b-1)/4}\mathbf{H}_{a-b-1/2}(2z^{1/2}). \tag{28}$$

$$G_{1,3}^{2,0}\left(z\,\middle|\,\begin{matrix} a-\frac{1}{2} \\ a, b, a-\frac{1}{2}\end{matrix}\right) = z^{(a+b)/2}Y_{b-a}(2z^{1/2}). \tag{29}$$

$$G_{1,3}^{2,0}\left(z\,\middle|\,\begin{matrix} \frac{1}{2} \\ -a, a, 0\end{matrix}\right) = \frac{\pi^{1/2}}{2\sin a\pi}\,[J_{-a}^2(z^{1/2}) - J_a^2(z^{1/2})]. \tag{30}$$

$$G_{1,3}^{2,0}\left(z\,\middle|\,\begin{matrix} \frac{1}{2} \\ 0, a, -a\end{matrix}\right) = -\pi^{1/2}J_a(z^{1/2})\,Y_a(z^{1/2}). \tag{31}$$

$$G_{1,3}^{2,1}\left(z\,\middle|\,\begin{matrix} \frac{1}{2} \\ 0, a, -a\end{matrix}\right) = 2\pi^{1/2}I_a(z^{1/2})\,K_a(z^{1/2}). \tag{32}$$

$$G_{1,3}^{2,1}\left(z\,\middle|\,\begin{matrix} \frac{1}{2} \\ a, -a, 0\end{matrix}\right) = \frac{\pi^{3/2}}{\sin 2\pi a}\,[I_{-a}^2(z^{1/2}) - I_a^2(z^{1/2})]. \tag{33}$$

$$G_{1,3}^{2,1}\left(z\,\middle|\,\begin{matrix} a+\frac{1}{2} \\ a+\frac{1}{2}, b, a\end{matrix}\right) = \frac{\pi z^{(a+b)/2}}{\cos(b-a)\,\pi}\,[I_{b-a}(2z^{1/2}) - \mathbf{L}_{a-b}(2z^{1/2})]. \tag{34}$$

$$G_{1,3}^{2,1}\left(z\,\middle|\,\begin{matrix} a+\frac{1}{2} \\ a, a+\frac{1}{2}, b\end{matrix}\right) = \pi z^{(a+b)/2}[I_{a-b}(2z^{1/2}) - \mathbf{L}_{a-b}(2z^{1/2})]. \tag{35}$$

$$G_{1,3}^{3,0}\left(z\,\middle|\,\begin{matrix} a+\frac{1}{2} \\ a+b, a-b, a\end{matrix}\right) = 2\pi^{-1/2}z^aK_b^2(z^{1/2}). \tag{36}$$

$$G_{1,3}^{3,1}\left(z\,\middle|\,\begin{matrix} a+\frac{1}{2} \\ a+\frac{1}{2}, -a, a\end{matrix}\right) = \frac{\pi^2}{\cos 2\pi a}\,[\mathbf{H}_{2a}(2z^{1/2}) - Y_{2a}(2z^{1/2})]. \tag{37}$$

$$G_{1,3}^{3,1}\left(z\,\middle|\,\begin{matrix} a \\ a, b, -b\end{matrix}\right) = 2^{2-2a}\Gamma(1-a-b)\,\Gamma(1-a+b)S_{2a-1,2b}(2z^{1/2}). \tag{38}$$

$$G_{1,3}^{3,1}\left(z\,\middle|\,\begin{matrix} a+\frac{1}{2} \\ b, 2a-b, a\end{matrix}\right) = \frac{\pi^{5/2}z^a}{2\cos(b-a)\,\pi}\,H_{b-a}^{(1)}(z^{1/2})\,H_{b-a}^{(2)}(z^{1/2}). \tag{39}$$

See 8.3(20, 21) for G-functions with $m = q = 3$, $n = p = 1$.

$$G_{2,4}^{1,2}\left(z\,\middle|\,\begin{matrix} a+\frac{1}{2}, a \\ a+b, a-c, a+c, a-b\end{matrix}\right) = \pi^{1/2}z^aJ_{b+c}(z^{1/2})\,J_{b-c}(z^{1/2}). \tag{40}$$

$$G_{2,4}^{2,2}\left(z\,\middle|\,\begin{matrix} a, a+\frac{1}{2} \\ b, c, 2a-c, 2a-b\end{matrix}\right) = 2\pi^{1/2}z^aI_{b+c-2a}(z^{1/2})\,K_{b-c}(z^{1/2}). \tag{41}$$

$$G_{2,4}^{3,0}\left(z\,\middle|\,\begin{matrix} 0, \frac{1}{2} \\ a, b, -b, -a\end{matrix}\right) = (i\pi^{1/2}/4)[H_{a-b}^{(1)}(z^{1/2})\,H_{a+b}^{(1)}(z^{1/2}) - H_{a-b}^{(2)}(z^{1/2})\,H_{a+b}^{(2)}(z^{1/2})]. \tag{42}$$

$$G_{2,4}^{3,1}\left(z \,\middle|\, \begin{matrix} \frac{1}{4}+k, \frac{1}{4}-k \\ m-\frac{1}{4}, \frac{1}{4}, -\frac{1}{4}, -m-\frac{1}{4} \end{matrix}\right) = \frac{\pi^{1/2}\Gamma(m+\frac{1}{2}-k)}{z^{3/4}\Gamma(2m+1)} \times M_{-k,m}(2z^{1/2})\, W_{k,m}(2z^{1/2}). \quad (43)$$

$$G_{2,4}^{4,0}\left(z \,\middle|\, \begin{matrix} \frac{1}{2}+a, \frac{1}{2}-a \\ 0, \frac{1}{2}, b, -b \end{matrix}\right) = (\pi/z)^{1/2}\, W_{a,b}(2z^{1/2})\, W_{-a,b}(2z^{1/2}). \quad (44)$$

$$G_{2,4}^{4,0}\left(z \,\middle|\, \begin{matrix} a, a+\frac{1}{2} \\ a+b, a+c, a-c, a-b \end{matrix}\right) = 2\pi^{-1/2}z^{a}K_{b+c}(z^{1/2})\, K_{b-c}(z^{1/2}). \quad (45)$$

$$G_{2,4}^{4,1}\left(z \,\middle|\, \begin{matrix} 0, \frac{1}{2} \\ a, b, -b, -a \end{matrix}\right) = \frac{i\pi^{5/2}}{4\sin\pi a \sin\pi b} \times [e^{-i\pi b}H_{a-b}^{(1)}(z^{1/2})\, H_{a+b}^{(2)}(z^{1/2}) - e^{i\pi b}H_{a+b}^{(1)}(z^{1/2})\, H_{a-b}^{(2)}(z^{1/2})]. \quad (46)$$

$$G_{2,4}^{4,1}\left(z \,\middle|\, \begin{matrix} \frac{1}{2}, 0 \\ a, b, -b, -a \end{matrix}\right) = \frac{\pi^{5/2}}{4\cos\pi a \cos\pi b} \times [e^{-i\pi b}H_{a-b}^{(1)}(z^{1/2})\, H_{a+b}^{(2)}(z^{1/2}) + e^{i\pi b}H_{a+b}^{(1)}(z^{1/2})\, H_{a-b}^{(2)}(z^{1/2})]. \quad (47)$$

$$G_{2,4}^{4,1}\left(z \,\middle|\, \begin{matrix} \frac{1}{2}+a, \frac{1}{2}-a \\ 0, \frac{1}{2}, b, -b \end{matrix}\right) = (\pi/z)^{1/2}\,\Gamma(\tfrac{1}{2}+b-a)\,\Gamma(\tfrac{1}{2}-b-a) \times W_{a,b}(2iz^{1/2})\, W_{a,b}(-2iz^{1/2}). \quad (48)$$

CHAPTER IX BESSEL FUNCTIONS AND THEIR INTEGRALS

9.1. Introduction

In the hypergeometric notation, Bessel functions of the first kind can be expressed in terms of a ${}_0F_1$ or a ${}_1F_1$. Thus numerous properties follow from those for the ${}_pF_q$ and from the developments in Chapter VII. The analytical material in this chapter is kept to a minimum. Our principal emphases are on coefficients for expansions in infinite series of Chebyshev polynomials of the first kind and rational approximations.

9.2. Definitions, Connecting Relations and Power Series

$$J_\nu(z) = \frac{(z/2)^\nu}{\Gamma(\nu+1)}\, {}_0F_1(\,;1+\nu;-z^2/4). \tag{1}$$

$$I_\nu(z) = \frac{(z/2)^\nu}{\Gamma(\nu+1)}\, {}_0F_1(\,;1+\nu;z^2/4) = e^{-i\nu\pi/2} J_\nu(ze^{i\pi/2}),\quad -\pi < \arg z \leqslant \pi/2. \tag{2}$$

$$J_\nu(z) = \frac{(z/2)^\nu e^{\mp iz}}{\Gamma(\nu+1)}\, {}_1F_1(\tfrac{1}{2}+\nu;1+2\nu;\pm 2iz). \tag{3}$$

$$I_\nu(z) = \frac{(z/2)^\nu e^{\mp z}}{\Gamma(\nu+1)}\, {}_1F_1(\tfrac{1}{2}+\nu;1+2\nu;\pm 2z) \tag{4}$$

$$Y_\nu(z) = (\csc \nu\pi)[(\cos \nu\pi)\, J_\nu(z) - J_{-\nu}(z)]. \tag{5}$$

$$\pi Y_n(z) = \left[\frac{\partial J_\nu(z)}{\partial\nu} - (-)^n \frac{\partial J_{-\nu}(z)}{\partial\nu}\right]_{\nu=n},\qquad (\pi/2)\, Y_0(z) = \left[\frac{\partial J_\nu(z)}{\partial\nu}\right]_{\nu=0}. \tag{6}$$

$$C_\nu(z) = AJ_\nu(z) + BY_\nu(z), \tag{7}$$

where A and B are independent of z.

$$\begin{aligned}K_\nu(z) &= (\pi/2)(\csc \nu\pi)[I_{-\nu}(z) - I_\nu(z)]\\ &= \pi^{1/2}e^{-z}(2z)^\nu\, U(\tfrac{1}{2}+\nu;1+2\nu;2z) = (\pi/2z)^{1/2}\, W_{0,\nu}(2z).\end{aligned} \tag{8}$$

$$2K_n(z) = (-)^n \left[\frac{\partial I_{-\nu}(z)}{\partial\nu} - \frac{\partial I_\nu(z)}{\partial\nu}\right]_{\nu=n},\qquad K_0(z) = -\left[\frac{\partial I_\nu(z)}{\partial\nu}\right]_{\nu=0}. \tag{9}$$

$$H_\nu^{(1)}(z) = J_\nu(z) + iY_\nu(z),\qquad H_\nu^{(2)}(z) = J_\nu(z) - iY_\nu(z). \tag{10}$$

$$K_\nu(z) = \tfrac{1}{2}\pi i\epsilon e^{i\epsilon\nu\pi/2} H_\nu^{(3-\epsilon)/2}(ze^{i\epsilon\pi/2}),$$
$$-(3+\epsilon)\,\pi/4 < \arg z \leqslant (3-\epsilon)\,\pi/4, \quad \epsilon = \pm 1. \tag{11}$$

$$Y_\nu(ze^{im\pi}) = e^{-im\nu\pi} Y_\nu(z) + 2i(\sin m\nu\pi)(\cot \nu\pi)\, J_\nu(z), \tag{12}$$

$$K_\nu(ze^{im\pi}) = e^{-im\nu\pi} K_\nu(z) - \frac{i\pi(\sin m\nu\pi)}{\sin \nu\pi}\, I_\nu(z), \tag{13}$$

where m is an integer or zero.

$$\begin{aligned} Y_n(z) &= (2/\pi)[\gamma + \ln(z/2)]\, J_n(z) \\ &\quad - \frac{(z/2)^{-n}}{\pi} \sum_{k=0}^{n-1} \frac{(n-k-1)!}{k!} \left(\frac{z^2}{4}\right)^k \\ &\quad - \frac{(z/2)^n}{\pi} \sum_{k=0}^{\infty} (-)^k \frac{[\psi(k+1) + \psi(n+k+1) - 2\psi(1)]}{k!\,(n+k)!} \left(\frac{z^2}{4}\right)^k. \end{aligned} \tag{14}$$

$$\begin{aligned} K_n(z) &= (-)^{n+1}[\gamma + \ln(z/2)]\, I_n(z) \\ &\quad + (1/2)(z/2)^{-n} \sum_{k=0}^{n-1} \frac{(-)^k(n-k-1)!}{k!} \left(\frac{z^2}{4}\right)^k \\ &\quad + \frac{(-)^n}{2} (z/2)^n \sum_{k=0}^{\infty} \frac{[\psi(k+1) + \psi(n+k+1) - 2\psi(1)]}{k!\,(n+k)!} \left(\frac{z^2}{4}\right)^k. \end{aligned} \tag{15}$$

Kelvin Functions

$$\mathrm{ber}_\nu(z) + i\,\mathrm{bei}_\nu(z) = J_\nu(ze^{3i\pi/4}) = e^{i\nu\pi} J_\nu(ze^{-i\pi/4}). \tag{16}$$

$$\mathrm{ker}_\nu(z) + i\,\mathrm{kei}_\nu(z) = \exp(-\tfrac{1}{2}i\pi\nu)\, K_\nu(ze^{i\pi/4}) = \tfrac{1}{2}i\pi H_\nu^{(1)}(ze^{3i\pi/4}). \tag{17}$$

Airy Functions

Let

$$\xi = (\tfrac{2}{3})\, z^{3/2}. \tag{18}$$

Then

$$\mathrm{Ai}(z) = (z^{1/2}/3)\{I_{-1/3}(\xi) - I_{1/3}(\xi)\} = \pi^{-1}(z/3)^{1/2}\, K_{1/3}(\xi). \tag{19}$$

$$\mathrm{Bi}(z) = (z/3)^{1/2}\{I_{-1/3}(\xi) + I_{1/3}(\xi)\}. \tag{20}$$

$$\mathrm{Ai}(-z) = (z^{1/2}/3)\{J_{-1/3}(\xi) + J_{1/3}(\xi)\}. \tag{21}$$

$$\mathrm{Bi}(-z) = (z/3)^{1/2}\{J_{-1/3}(\xi) - J_{1/3}(\xi)\}. \tag{22}$$

$$\mathrm{Ai}'(z) = -(z/3)\{I_{-2/3}(\xi) - I_{2/3}(\xi)\} = -(z/3^{1/2}\pi)\, K_{2/3}(\xi). \tag{23}$$

$$\mathrm{Bi}'(z) = 3^{-1/2}z\{I_{-2/3}(\xi) + I_{2/3}(\xi)\}. \tag{24}$$

$$\mathrm{Ai}'(-z) = -(z/3)\{J_{-2/3}(\xi) - J_{2/3}(\xi)\}. \tag{25}$$

$$\mathrm{Bi}'(-z) = 3^{-1/2}z\{J_{-2/3}(\xi) + J_{2/3}(\xi)\}. \tag{26}$$

9.3. Difference-Differential Formulas

Let $W_\nu(z)$ represent any of the Bessel functions of the first three kinds or the modified Bessel functions. With each $W_\nu(z)$ we associate two parameters a and b as outlined in the following tabulation:

$W_\nu(z)$	a	b
$J_\nu(z)$, $Y_\nu(z)$, $H_\nu^{(1)}(z)$, $H_\nu^{(2)}(z)$	1	1
$I_\nu(z)$	-1	1
$K_\nu(z)$	1	-1

(1)

$$aW_{\nu+1}(z) + bW_{\nu-1}(z) = (2\nu/z)\, W_\nu(z). \tag{2}$$

$$-aW_{\nu+1}(z) + bW_{\nu-1}(z) = 2W_\nu'(z). \tag{3}$$

$$zW_\nu'(z) + \nu W_\nu(z) = bzW_{\nu-1}(z). \tag{4}$$

$$zW_\nu'(z) - \nu W_\nu(z) = -azW_{\nu+1}(z). \tag{5}$$

$$(z^{-1}d/dz)^m\{z^\nu W_\nu(z)\} = b^m z^{\nu-m} W_{\nu-m}(z). \tag{6}$$

$$(z^{-1}d/dz)^m\{z^{-\nu} W_\nu(z)\} = (-a)^m z^{-\nu-m} W_{\nu+m}(z). \tag{7}$$

$$W_\nu^{(m)}(z) = 2^{-m} \sum_{k=0}^{m} (-)^k \binom{m}{k} a^k b^{m-k} W_{\nu-m+2k}(z). \tag{8}$$

$$[z^2D^2 + zD + (abz^2 - \nu^2)]\, W_\nu(z) = 0, \qquad D = d/dz. \tag{9}$$

If

$$y(z) = z^{-\alpha} e^{-f(z)} C_\nu[h(z)],$$

then with $y = y(z)$, $h = h(z)$, and $f = f(z)$,

$$\begin{aligned} z^2y'' + z\{2\alpha + z[2f' - (h''/h') + (h'/h)]\}\, y' + \{&[\alpha(\alpha-1) + 2\alpha f'z + f''z^2 \\ &+ (f')^2z^2 - (h''/h')(\alpha z + f'z^2)] \\ &+ (h'/h)(\alpha z + f'z^2) \\ &+ (h')^2(z^2 - \nu^2z^2/h^2)\}\, y = 0. \end{aligned} \tag{10}$$

A generalization of (2) follows.

$$K_{\nu+m}(z) = T_{m,\nu}K_\nu(z) + T_{m-1,\nu+1}(z)\, K_{\nu-1}(z), \tag{11}$$

$$T_{2m+\epsilon,\nu}(z) = \left\{\frac{2(m+1)(m+\nu)}{z}\right\}^\epsilon \times {}_4F_1\left(\begin{matrix}-m, m+1+\epsilon, m+\nu+\epsilon, 1-m-\nu \\ \frac{1}{2}+\epsilon\end{matrix}\,\middle|\, \frac{1}{z^2}\right), \quad \epsilon = 0, 1, \tag{12}$$

$$T_{m+1,\nu}(z) = (2/z)(\nu+m)\, T_{m,\nu}(z) + T_{m-1,\nu}(z). \tag{13}$$

Wronskians

We define

$$W\{u(z), v(z)\} = u(z)\, v'(z) - u'(z)\, v(z). \tag{14}$$

Then

$$W\{J_\nu(z), J_{-\nu}(z)\} = -(2/\pi z) \sin \nu\pi, \tag{15}$$

$$W\{J_\nu(z), Y_\nu(z)\} = 2/\pi z, \tag{16}$$

$$W\{H_\nu^{(1)}(z), H_\nu^{(2)}(z)\} = -4i/\pi z, \tag{17}$$

$$W\{I_\nu(z), I_{-\nu}(z)\} = -(2/\pi z) \sin \nu\pi, \tag{18}$$

$$W\{I_\nu(z), K_\nu(z)\} = -1/z. \tag{19}$$

9.4. Products of Bessel Functions

$$J_\mu(z)\, J_\nu(z) = \frac{(z/2)^{\mu+\nu}}{\Gamma(\mu+1)\,\Gamma(\nu+1)}\, {}_2F_3\left(\begin{matrix}(\mu+\nu+1)/2, (\mu+\nu+2)/2 \\ \mu+1, \nu+1, \mu+\nu+1\end{matrix}\,\middle|\, -z^2\right), \tag{1}$$

$$J_\nu(z)\, J_{\nu+1}(z) = \frac{(z/2)^{2\nu+1}}{\Gamma(\nu+1)\,\Gamma(\nu+2)}\, {}_1F_2\left(\begin{matrix}\nu+3/2 \\ \nu+2, 2\nu+2\end{matrix}\,\middle|\, -z^2\right), \tag{2}$$

$$J_\nu^2(z) = \frac{(z/2)^{2\nu}}{[\Gamma(\nu+1)]^2}\, {}_1F_2(\nu+\tfrac{1}{2}; \nu+1, 2\nu+1; -z^2). \tag{3}$$

$$J_{-\nu}(z)\, J_\nu(z) = \frac{(\sin \nu\pi)}{\nu\pi}\, {}_1F_2(\tfrac{1}{2}; 1+\nu, 1-\nu; -z^2). \tag{4}$$

$$J_\nu(z)\, I_\nu(z) = \frac{(z/2)^{2\nu}}{[\Gamma(\nu+1)]^2}\, {}_0F_3\left(; \frac{\nu+1}{2}, \frac{\nu+2}{2}, \nu+1; \frac{-z^4}{64}\right). \tag{5}$$

$$J_{-\nu}(z)\, I_\nu(z) = \frac{(\sin \nu\pi)}{\nu\pi}\, {}_0F_3\left(; 1+\tfrac{1}{2}\nu, 1-\tfrac{1}{2}\nu, \tfrac{1}{2}; \frac{-z^4}{64}\right) - \frac{2(\sin \nu\pi)(z/2)^2}{\pi(1-\nu^2)}\, {}_0F_3\left(; \frac{3+\nu}{2}, \frac{3-\nu}{2}, \frac{3}{2}; \frac{-z^4}{64}\right). \tag{6}$$

9.5. Asymptotic Expansions for Large Variable

$$H_\nu^{(1)}(z) \sim (2/\pi z)^{1/2} \exp[i(z - \tfrac{1}{2}\nu\pi - \tfrac{1}{4}\pi)]\, {}_2F_0\left(\tfrac{1}{2}+\nu, \tfrac{1}{2}-\nu; \frac{1}{2iz}\right),$$

$$|z| \to \infty, \quad \delta - \pi \leqslant \arg z \leqslant 2\pi - \delta, \quad \delta > 0. \tag{1}$$

$$H_\nu^{(2)}(z) \sim (2/\pi z)^{1/2} \exp[-i(z - \tfrac{1}{2}\nu\pi - \tfrac{1}{4}\pi)]\, {}_2F_0\left(\tfrac{1}{2}+\nu, \tfrac{1}{2}-\nu; -\frac{1}{2iz}\right),$$

$$|z| \to \infty, \quad \delta - 2\pi \leqslant \arg z \leqslant \pi - \delta, \quad \delta > 0. \tag{2}$$

$$J_\nu(z) = (2/\pi z)^{1/2}[A(z)\cos\theta + B(z)\sin\theta], \tag{3}$$

$$Y_\nu(z) = (2/\pi z)^{1/2}[A(z)\sin\theta - B(z)\cos\theta],$$

$$\theta = z - \tfrac{1}{2}\nu\pi - \tfrac{1}{4}\pi, \tag{4}$$

$$A(z) \sim {}_4F_1\left(\begin{matrix}\tfrac{1}{4}+\tfrac{1}{2}\nu, \tfrac{1}{4}-\tfrac{1}{2}\nu, \tfrac{3}{4}+\tfrac{1}{2}\nu, \tfrac{3}{4}-\tfrac{1}{2}\nu \\ \tfrac{1}{2}\end{matrix}\,\middle|\, -\frac{1}{z^2}\right), \tag{5}$$

$$B(z) \sim \frac{(\tfrac{1}{4}-\nu^2)}{2z}\, {}_4F_1\left(\begin{matrix}\tfrac{3}{4}+\tfrac{1}{2}\nu, \tfrac{3}{4}-\tfrac{1}{2}\nu, \tfrac{5}{4}+\tfrac{1}{2}\nu, \tfrac{5}{4}-\tfrac{1}{2}\nu \\ \tfrac{3}{2}\end{matrix}\,\middle|\, -\frac{1}{z^2}\right),$$

$$|z| \to \infty, \quad |\arg z| \leqslant \pi - \delta, \quad \delta > 0. \tag{6}$$

$$I_\nu(z) \sim \frac{e^z}{(2\pi z)^{1/2}}\, {}_2F_0\left(\tfrac{1}{2}+\nu, \tfrac{1}{2}-\nu; \frac{1}{2z}\right)$$

$$+ \frac{\exp[-z - \epsilon(\tfrac{1}{2}+\nu)\, i\pi]}{(2\pi z)^{1/2}}\, {}_2F_0\left(\tfrac{1}{2}+\nu, \tfrac{1}{2}-\nu; -\frac{1}{2z}\right),$$

$$|z| \to \infty, \ \delta - (2+\epsilon)\pi \leqslant \arg z \leqslant (2-\epsilon)\pi - \delta,$$

$$\delta > 0, \ \epsilon = \pm 1. \tag{7}$$

$$K_\nu(z) \sim (\pi/2z)^{1/2} e^{-z}\, {}_2F_0\left(\tfrac{1}{2}+\nu, \tfrac{1}{2}-\nu; -\frac{1}{2z}\right),$$

$$|z| \to \infty, \quad |\arg z| \leqslant 3\pi/2 - \delta, \quad \delta > 0. \tag{8}$$

If ν is half an odd integer, the above hypergeometric series terminate in which case we can replace asymptotic equality by equality and omit the restrictions on arg z.

For uniform asymptotic expansions, see the work of Olver (1954,1974).

9.6. Integrals of Bessel Functions

$$\int_0^z t^\mu J_\nu(t)\,dt = \frac{z^{\mu+\nu+1}}{2^\nu(\mu+\nu+1)\,\Gamma(\nu+1)}\, {}_1F_2\left(\begin{matrix}\tfrac{1}{2}(\mu+\nu+1) \\ \tfrac{1}{2}(\mu+\nu+3), \nu+1\end{matrix}\,\middle|\, \frac{-z^2}{4}\right),$$

$$R(\mu+\nu) > -1. \tag{1}$$

$$\int_0^z t^\mu H_\nu^{(1)}(t)\,dt = (2^\mu/\pi)\,e^{\frac{1}{2}i(\mu-\nu)\pi}\,\Gamma[\tfrac{1}{2}(\mu-\nu+1)]\,\Gamma[\tfrac{1}{2}(\mu+\nu+1)] + (\mu+\nu-1)\,zH_\nu^{(1)}(z)\,S_{\mu-1,\nu-1}(z) - zH_{\nu-1}^{(1)}(z)\,S_{\mu,\nu}(z),$$
$$R(\mu \pm \nu) > -1. \qquad (2)$$

See 10.1(2) for $S_{\mu,\nu}(z)$.

$$\int_0^z e^{it}t^\mu J_\nu(t)\,dt = \frac{z^{\mu+1}(z/2)^\nu}{(\mu+\nu+1)\,\Gamma(\nu+1)}\,{}_2F_2\left(\begin{matrix}\frac{1}{2}+\nu,\ \mu+\nu+1\\ 2\nu+1,\ \mu+\nu+2\end{matrix}\,\middle|\,2iz\right),$$
$$R(\mu+\nu) > -1. \qquad (3)$$

$$\int_0^z e^{it}t^\mu H_\nu^{(1)}(t)\,dt = \frac{2}{\pi}\,\frac{e^{\frac{1}{2}i(\mu-\nu)\pi}\Gamma(\mu+\nu+1)\,\Gamma(\mu-\nu+1)}{2^\mu(\frac{3}{2})_\mu} + ze^{\frac{1}{2}i\mu\pi}[(\mu-\nu)(\mu+\nu+1)^{-1}H_\nu^{(1)}(z)\,H_{\mu,\nu+1}(ze^{-\frac{1}{2}i\pi}) + iH_{\nu+1}^{(1)}(z)\,H_{\mu,\nu}(ze^{-\frac{1}{2}i\pi})] + (\mu+\nu+1)^{-1}\,e^{iz}z^{\mu+1}H_\nu^{(1)}(z),$$
$$R(\mu \pm \nu) > -1. \qquad (4)$$

See 10.1(11) for $H_{\mu,\nu}(z)$. With

$$j_{\alpha,\nu}(z) = [\Gamma(\alpha)]^{-1}\int_0^z (z-t)^{\alpha-1}\,J_\nu(t)\,dt, \qquad R(\alpha) > 0, \qquad R(\nu) > -1, \qquad (5)$$

$$j_{\alpha,\nu}(z) = \frac{2^\alpha(z/2)^{\nu+\alpha}}{\Gamma(\nu+\alpha+1)} \times {}_2F_3\left(\begin{matrix}(\nu+1)/2,\ (\nu+2)/2\\ \nu+1,\ (\nu+\alpha+1)/2,\ (\nu+\alpha+2)/2\end{matrix}\,\middle|\,-z^2/4\right). \qquad (6)$$

If ν is a positive integer r, then $j_{\alpha,r}(z)$ is the rth repeated integral of $J_\nu(z)$.

For the development of the above and many other results on integrals of Bessel functions, hypergeometric functions, and related topics, see Luke (1962a). See also the discussion in 10.1.

9.7. Expansions in Series of Chebyshev Polynomials

In this section we give analytical formulas for the expansion of Bessel functions in series of Chebyshev polynomials of the first kind. Formulas for analogous expansions for integrals of Bessel functions follow from the developments in 11.6. Expansions also follow by use of the appropriate theorems in 5.10. Following the analytical portion of this section we present extensive tables of coefficients for many particular Bessel functions and integrals involving these functions.

$$J_\nu(ax) = (ax/2)^\nu \sum_{n=0}^{\infty} B_n T_{2n}(x), \qquad -1 \leqslant x \leqslant 1, \tag{1}$$

$$B_n = \frac{\epsilon_n(-)^n(a/4)^{2n}}{n!\,\Gamma(n+\nu+1)}\, {}_1F_2\left(\begin{matrix}\frac{1}{2}+n \\ 1+2n,\ \nu+1+n\end{matrix}\,\middle|\, -\frac{a^2}{4}\right),$$

$$\frac{2B_n}{\epsilon_n} = -\left[1 - \frac{a^2}{16(n+2)(n+\nu+1)}\right]\frac{16(n+1)(n+\nu+1)}{a^2}B_{n+1}$$

$$-\left[(n+2-\nu) - \frac{a^2}{16(n+1)}\right]\frac{16(n+1)}{a^2}B_{n+2} - \frac{(n+1)}{(n+2)}B_{n+3}, \tag{2}$$

$$\sum_{n=0}^{\infty} B_n = \{\Gamma(\nu+1)\}^{-1}. \tag{3}$$

$$J_{2k}(ax) = \sum_{n=0}^{\infty} \epsilon_n J_{k+n}(a/2) J_{k-n}(a/2) T_{2n}(x), \qquad -1 \leqslant x \leqslant 1. \tag{4}$$

$$J_{2k+1}(ax) = 2\sum_{n=0}^{\infty} J_{k+n+1}(a/2) J_{k-n}(a/2) T_{2n+1}(x), \qquad -1 \leqslant x \leqslant 1. \tag{5}$$

In (4), (5), k is a positive integer or zero.

$$J_0(ax) = \sum_{n=0}^{\infty} \epsilon_n (-)^n J_n^2(a/2) T_{2n}(x), \qquad -1 \leqslant x \leqslant 1. \tag{6}$$

$$J_1(ax) = 2\sum_{n=0}^{\infty} (-)^n J_n(a/2) J_{n+1}(a/2) T_{2n+1}(x), \qquad -1 \leqslant x \leqslant 1. \tag{7}$$

$$I_\nu(ax) = (ax/2)^\nu e^{ax} \sum_{n=0}^{\infty} C_n(a) T_n^*(x), \quad 0 < x \leq 1,$$

$$C_n(a) = \frac{\varepsilon_n (-)^n 2^{2\nu}\Gamma(n+\nu+\frac{1}{2})(a/2)^n}{\pi^{\frac{1}{2}}\Gamma(n+2\nu+1)}\, {}_2F_2\left(\begin{matrix}\frac{1}{2}+\nu+n,\ \frac{1}{2}+n \\ 1+2\nu+n,\ 1+n\end{matrix}\,\middle|\, -2a\right), \tag{8}$$

$$\frac{2C_n(a)}{\varepsilon} = -\frac{(n+1)}{a(n+2)(2n+1+2\nu)}\{a(2n+5-2\nu)$$

$$+ 4(n+2)(n+1+2\nu)\}C_{n+1}(a)$$

$$+ \{a(2n+1+2\nu)\}^{-1}\{a(2n+1+2\nu) - 4(n+1)(n+2-2\nu)\}C_{n+2}(a)$$

$$+ \frac{(n+1)(2n+5-2\nu)}{(n+2)(2n+1+2\nu)}C_{n+3}(a), \tag{9}$$

$$\sum_{n=0}^{\infty} (-)^n C_n(a) = \{\Gamma(\nu+1)\}^{-1} . \tag{10}$$

$$\begin{aligned} Y_m(ax) &= (2/\pi)[\gamma + \ln(ax/2)]J_m(ax) + N_{m-1}(ax) - (1/\pi)W_m(ax), \\ K_m(ax) &= (-)^{m+1}[\gamma + \ln(ax/2)]I_m(ax) - (\pi/2)i^m N_{m-1}(iax) + (i^m/2)W_m(iax), \\ N_{m-1}(ax) &= -\frac{1}{\pi}\sum_{n=0}^{m-1}\frac{(m-n-1)!}{n!}\left(\frac{ax}{2}\right)^{2n-m}, \qquad m>0, \\ &= 0, \qquad m=0, \end{aligned} \tag{11}$$

$$\begin{aligned} W_m(ax) &= \sum_{n=0}^{\infty} A_n T_n(x), \qquad -1 \leqslant x \leqslant 1, \\ A_n &= \frac{\varepsilon_n\{(-)^m + (-)^n\}}{\pi^{\frac{1}{2}}2^{2n+1}} \\ &\quad \times \sum_{k=0}^{\infty} \frac{(-)^k(-m-2k)_n[h_{k+m}+h_k]}{[\frac{1}{2}(k+m-n+1)]_{n+1/2}\,k!\,(m+k)!}\left(\frac{a}{2}\right)^{m+2k}, \\ h_0 &= 0, \qquad h_k = \sum_{r=1}^{k} 1/r = \psi(k+1) - \psi(1). \end{aligned} \tag{12}$$

$$\begin{aligned} Y_0(ax) &= (2/\pi)[\gamma + \ln \tfrac{1}{2}(ax)]J_0(ax) + \sum_{n=0}^{\infty} B_n T_{2n}(x), \qquad 0 < x \leqslant 1, \\ B_n &= -(2/\pi)\epsilon_n(-)^n(\tfrac{1}{2})_n \sum_{k=0}^{\infty}\frac{(\frac{1}{2})_k(\frac{1}{2}a)^k}{k!\,(n+k)!} \\ &\quad \times\{\psi(n+1+k) - \psi(1) + \psi(\tfrac{1}{2}) - \psi(k+\tfrac{1}{2})\}J_{k+2n}(a), \\ B_n &= \frac{2\epsilon_n(\frac{1}{4}a)^{2n}(-)^{n+1}}{\pi(n!)^2}\sum_{k=0}^{\infty}\frac{(-)^k(\frac{1}{2}a)^{2k}(n+\frac{1}{2})_k h_{n+k}}{(n+1)_k(2n+1)_k k!} . \end{aligned} \tag{13}$$

$$\begin{aligned} Y_1(ax) &= (2/\pi)[\gamma + \ln \tfrac{1}{2}(ax)]J_1(ax) - (2/\pi ax) + \sum_{n=0}^{\infty} C_n T_{2n+1}(x), \quad 0 < x \leqslant 1, \\ C_n &= \frac{2(-)^{n+1}(\frac{1}{4}a)^{2n+1}}{\pi n!\,(n+1)!}\sum_{k=0}^{\infty}\frac{(-)^k(\frac{1}{2}a)^{2k}(n+\frac{3}{2})_k[h_{n+k+1}+h_{n+k}]}{(n+2)_k(2n+2)_k k!} . \end{aligned} \tag{14}$$

$$K_0(ax) = -[\gamma + \ln \tfrac{1}{2}(ax)]I_0(ax) + \sum_{n=0}^{\infty} B_n T_{2n}(x), \qquad 0 < x \leqslant 1,$$

$$B_n = \epsilon_n(\tfrac{1}{2})_n \sum_{k=0}^{\infty} \frac{(-)^k(\frac{1}{2})_k(\frac{1}{2}a)^k}{k!\,(n+k)!}$$
$$\times \{\psi(n+1+k) - \psi(1) + \psi(\tfrac{1}{2}) - \psi(\tfrac{1}{2}+k)\} I_{k+2n}(a),$$
$$B_n = \frac{\epsilon_n(\frac{1}{4}a)^{2n}}{(n!)^2} \sum_{k=0}^{\infty} \frac{(\frac{1}{2}a)^{2k}(n+\frac{1}{2})_k h_{n+k}}{(n+1)_k(2n+1)_k k!}. \tag{15}$$

$$K_1(ax) = [\gamma + \ln \tfrac{1}{2}(ax)]I_1(ax) + (1/ax) + \sum_{n=0}^{\infty} C_n T_{2n+1}(x), \qquad 0 < x \leqslant 1,$$

$$C_n = -\frac{(\frac{1}{4}a)^{2n+1}}{n!\,(n+1)!} \sum_{k=0}^{\infty} \frac{(\frac{1}{2}a)^{2k}(n+\frac{3}{2})_k[h_{n+k+1} + h_{n+k}]}{(n+2)_k(2n+2)_k k!}. \tag{16}$$

Next we present expansions in descending series of the variable. To this end we make use of 9.2(8) and 5.10.1(16-23). To get from the notation employed in the latter set of equations to that preferred here, put

$$a = \tfrac{1}{2} + \nu, \quad c = 1 + 2\nu, \quad z = 2\lambda \tag{17}$$

and $\omega = z/\lambda$ where z is now a new variable. Thus

$$K_\nu(z) = (\pi/2z)^{\frac{1}{2}} e^{-z} \sum_{n=0}^{\infty} C_n(\lambda) T_n^*(\lambda/z),$$
$$\lambda/z \le 1, \quad |\arg z| < 3\pi/2 . \tag{18}$$

Clearly

$$\lim_{n\to\infty} C_n(\lambda) = 0, \tag{19}$$

the recursion formula for $C_n(\lambda)$ follows from 5.10.1(16) and (17), and

$$\sum_{n=0}^{\infty} (-)^n C_n(\lambda) = 1. \tag{20}$$

From (18) and 9.2(11),

$$H_\nu^{(1)}(z) = (2/\pi z)^{\frac{1}{2}} e^{i\theta} \sum_{n=0}^{\infty} H_n(\lambda) T_n^*(\lambda/z), \quad H_n(\lambda) = C_n(\lambda e^{-i\pi/2}),$$

$$\theta = z - \tfrac{1}{2}\nu\pi - \tfrac{1}{4}\pi, \quad \lambda/z \le 1, \quad -\pi < \arg z < 2\pi. \tag{21}$$

The expansion for the Hankel function $H_\nu^{(2)}(z)$ follows from (21) upon replacing i by $-i$. So we have expansions for $J_\nu(z)$ and $Y_\nu(z)$ in descending series of Chebyshev polynomials in view of 9.2(10).

We now consider a descending type expansion for $I_\nu(z)$ where z is positive. From 8.3(9) and 5.10.1(5),

$$I_\nu(z) = (2\pi z)^{-\frac{1}{2}}e^z F_\nu(z),\ F_\nu(z) = G_{1,2}^{1,1}\left(2z\,\middle|\,\begin{matrix}1\\ \frac{1}{2}+\nu,\ \frac{1}{2}-\nu\end{matrix}\right), \quad (22)$$

$$F_\nu(z) = \sum_{n=0}^{\infty} M_n(\lambda)T_n^*(\lambda/z),\ \lambda/z \geq 1,\ z>0,$$

$$M_n(\lambda) = \pi^{-\frac{1}{2}}\epsilon_n(-)^n G_{2,3}^{2,1}\left(2\lambda\,\middle|\,\begin{matrix}1-n,\ n+1\\ \frac{1}{2},\ \frac{1}{2}+\nu,\ \frac{1}{2}-\nu\end{matrix}\right). \quad (23)$$

$M_n(\lambda)$ satisfies the same recurrence formula as does $C_n(\lambda)$, see the remark after (19), provided that λ is replaced by $-\lambda$. Again the expansion formula (23) converges for ν and λ fixed. So

$$\lim_{n\to\infty} M_n(\lambda) = 0. \quad (24)$$

Also

$$\sum_{n=0}^{\infty} (-)^n M_n(\lambda) = 1. \quad (25)$$

It can be shown from the discussion surrounding 5.10.1(17, 18) that values for $M_n(\lambda)$ cannot be deduced by the usual backward recurrence procedure unless modified as described by 12.2(16-19). For application of this technique for the construction of coefficients $M_n(\lambda)$ for various ν, see Luke (1971-1972,II).

Actually the recursion formula 5.10.1(16) is not valid if either a or b is a negative integer unless $n+a>0$ or $n+b>0$. If, for example, $a=-m$, m a positive integer, then 5.10.1(16) is only valid for $n>m$. However, a further relation can be obtained if first we multiply through by $n+a$ and then set $n+a=0$. When a and c are defined by (17) and $\nu=\frac{1}{2}$ (in which event $F_{\frac{1}{2}}(z)=1-e^{-2z}$), Luke (1971-1972,II) shows how to get an expansion for the descending exponential in series of Chebyshev polynomials of the reciprocal variable as illustrated by 3.2.2(11).

An examination of the tables of coefficients clearly manifests the superiority of the Chebyshev expansions over their at least formal series counterparts in powers of the variable. Indeed an expansion like (18) is convergent and yet it can be rearranged to give the divergent asymptotic expansion of $K_\nu(z)$. The expansion (18) is still deficient in the sense that a different set of coefficients is required for each choice of λ and ν. To improve matters, one can develop coefficients for the expansion of $C_n(\lambda)$ in series of Chebyshev polynomials involving the parameter ν. We illustrate. Suppose λ is given and $0 \leq \nu \leq 1$. Then for any specific ν we can use the backward recursion technique to compute $C_n(\lambda)$. Our idea is to write

$$C_n(\lambda) = \sum_{r=0}^{\infty} L_{r,n}(\lambda) T_r^*(\nu), \quad 0 \leq \nu \leq 1 \tag{26}$$

and to compute approximations to $L_{r,n}(\lambda)$ by use of either Theorem 1 or Theorem 3 of 11.7, where for the application at hand, a slight change of notation is required. In practice, we found it convenient to compute coefficients using both Theorems 1 and 3. A comparison of the results affords some appreciation of the accuracy of the coefficients when compared with those for expansion in an infinite series of Chebyshev polynomials, and the average value of the two coefficients for each r is more accurate than either by themselves in view of the comment following 11.7(11). Simple criteria for an *a priori* determination of the number of coefficients required to achieve a certain level of accuracy might not be available and some experimentation might be necessary.

Construction of tables of coefficients based on these ideas for expansion of $J_\nu(z)$, $Y_\nu(z)$, $I_\nu(z)$ and $K_\nu(z)$ in both ascending and descending series of Chebyshev polynomials can be found in Luke (1971-1972). It is remarkable that only 205 coefficients are required to compute $(2z/\pi)^{\frac{1}{2}} e^z K_\nu(z)$ with an accuracy of about 20D for all $z \geq 5$ and all ν, $0 \leq \nu \leq 1$. Since $K_\nu(z) = K_{-\nu}(z)$ and use of the recurrence formula for $K_\nu(z)$ is stable in the forward direction, we can readily evaluate $K_\nu(z)$ for $z \geq 5$ and all $\nu \geq 0$.

TABLE 9.1

CHEBYSHEV COEFFICIENTS FOR $J_0(x)$ AND $Y_0(x)$

$$J_0(x) = \sum_{n=0}^{\infty} a_n T_{2n}(x/8)$$

$$-8 \le x \le 8$$

$$Y_0(x) = (2/\pi)(\gamma + \ln x/2) J_0(x) + \sum_{n=0}^{\infty} b_n T_{2n}(x/8)$$

$$0 < x \le 8$$

n	a_n	n	b_n
0	0.15772 79714 74890 11956	0	-0.02150 51114 49657 55061
1	-0.00872 34423 52852 22129	1	-0.27511 81330 43518 79146
2	0.26517 86132 03336 80987	2	0.19860 56347 02554 15556
3	-0.37009 49938 72649 77903	3	0.23425 27461 09021 80210
4	0.15806 71023 32097 26128	4	-0.16563 59817 13650 41312
5	-0.03489 37694 11408 88516	5	0.04462 13795 40669 28217
6	0.00481 91800 69467 60450	6	-0.00693 22862 91523 18829
7	-0.00046 06261 66206 27505	7	0.00071 91174 03752 30309
8	0.00003 24603 28821 00508	8	-0.00005 39250 79722 93939
9	-0.00000 17619 46907 76215	9	0.00000 30764 93288 10848
10	0.00000 00760 81635 92419	10	-0.00000 01384 57181 23009
11	-0.00000 00026 79253 53056	11	0.00000 00050 51054 36909
12	0.00000 00000 78486 96314	12	-0.00000 00001 52582 85043
13	-0.00000 00000 01943 83469	13	0.00000 00000 03882 86747
14	0.00000 00000 00041 25321	14	-0.00000 00000 00084 42875
15	-0.00000 00000 00000 75885	15	0.00000 00000 00001 58748
16	0.00000 00000 00000 01222	16	-0.00000 00000 00000 02608
17	-0.00000 00000 00000 00017	17	0.00000 00000 00000 00038

$$J_0(x) + iY_0(x) = (2/\pi x)^{\frac{1}{2}} e^{i(x-\pi/4)} \sum_{n=0}^{\infty} c_n T_n^*(5/x), \quad x \ge 5$$

$$c_n = R(c_n) + iI(c_n)$$

n	$R(c_n)$	n	$I(c_n)$
0	0.99898 80898 58965 15390	0	-0.01233 15205 78544 14382
1	-0.00133 84285 49971 85578	1	-0.01224 94962 81259 47486
2	-0.00031 87898 78061 89289	2	0.00009 64941 84993 42287
3	0.00000 85112 32210 65665	3	0.00001 36555 70490 35682
4	0.00000 06915 42349 13894	4	-0.00000 08518 06644 42635
5	-0.00000 00907 70101 53734	5	-0.00000 00272 44053 41355
6	0.00000 00014 54928 07929	6	0.00000 00096 46421 33771
7	0.00000 00009 26762 48672	7	-0.00000 00006 83347 51799
8	-0.00000 00001 39166 19797	8	-0.00000 00000 60627 38000
9	0.00000 00000 03237 97518	9	0.00000 00000 21695 71634
10	0.00000 00000 02535 35729	10	-0.00000 00000 02304 89890
11	-0.00000 00000 00559 09032	11	-0.00000 00000 00122 55390
12	0.00000 00000 00041 91896	12	0.00000 00000 00092 31372
13	0.00000 00000 00008 73316	13	-0.00000 00000 00016 77838
14	-0.00000 00000 00003 61861	14	0.00000 00000 00000 75375
15	0.00000 00000 00000 59438	15	0.00000 00000 00000 46244
16	-0.00000 00000 00000 00964	16	-0.00000 00000 00000 15906
17	-0.00000 00000 00000 02436	17	0.00000 00000 00000 02500
18	0.00000 00000 00000 00789	18	0.00000 00000 00000 00015
19	-0.00000 00000 00000 00125	19	-0.00000 00000 00000 00135
20	-0.00000 00000 00000 00002	20	0.00000 00000 00000 00044
21	0.00000 00000 00000 00008	21	-0.00000 00000 00000 00007
22	-0.00000 00000 00000 00003		

TABLE 9.2

Chebyshev Coefficients for $J_1(x)$ and $Y_1(x)$

$$J_1(x) = \sum_{n=0}^{\infty} a_n T_{2n+1}(x/8) \qquad J_1(x) = (x/8) \sum_{n=0}^{\infty} b_n T_{2n}(x/8)$$

$-8 \le x \le 8$

n	a_n	n	b_n
0	0.05245 81903 34656 48458	0	0.64835 87706 05264 92084
1	0.04809 64691 58230 37394	1	-1.19180 11605 41216 87251
2	0.31327 50823 61567 18380	2	1.28799 40988 57677 62038
3	-0.24186 74084 47407 48475	3	-0.66144 39341 34543 25277
4	0.07426 67962 16787 03781	4	0.17770 91172 39728 28328
5	-0.01296 76273 11735 17510	5	-0.02917 55248 06154 20766
6	0.00148 99128 96667 63839	6	0.00324 02701 82683 85747
7	-0.00012 22786 85054 32427	7	-0.00026 04443 89348 58068
8	0.00000 75626 30229 69605	8	0.00001 58870 19239 93213
9	-0.00000 03661 30855 23363	9	-0.00000 07617 58780 54003
10	0.00000 00142 77324 38731	10	0.00000 00294 97070 07278
11	-0.00000 00004 58570 03076	11	-0.00000 00009 42421 29816
12	0.00000 00000 12351 74811	12	0.00000 00000 25281 23664
13	-0.00000 00000 00283 17735	13	-0.00000 00000 00577 74042
14	0.00000 00000 00005 59509	14	0.00000 00000 00011 38572
15	-0.00000 00000 00000 09629	15	-0.00000 00000 00000 19554
16	0.00000 00000 00000 00146	16	0.00000 00000 00000 00295
17	-0.00000 00000 00000 00002	17	-0.00000 00000 00000 00004

$$Y_1(x) = (2/\pi)(\gamma + \ln x/2)J_1(x) - (2/\pi x) + \sum_{n=0}^{\infty} c_n T_{2n+1}(x/8)$$

$0 < x \le 8$

n	c_n
0	-0.04017 29465 44414 07579
1	-0.44444 71476 30558 06261
2	-0.02271 92444 28417 73587
3	0.20664 45410 17490 51976
4	-0.08667 16970 56948 52366
5	0.01763 67030 03163 13441
6	-0.00223 56192 94485 09524
7	0.00019 70623 02701 54078
8	-0.00001 28858 53299 24086
9	0.00000 06528 47952 35852
10	-0.00000 00264 50737 17479
11	0.00000 00008 78030 11712
12	-0.00000 00000 24343 27870
13	0.00000 00000 00572 61216
14	-0.00000 00000 00011 57794
15	0.00000 00000 00000 20347
16	-0.00000 00000 00000 00314
17	0.00000 00000 00000 00004

TABLE 9.2 *(Concluded)*

$$J_1(x)+iY_1(x) = (2/\pi x)^{\frac{1}{2}} e^{i(x-3\pi/4)} \sum_{n=0}^{\infty} d_n T_n^*(5/x), \quad x \ge 5$$

$$d_n = R(d_n)+iI(d_n)$$

n	$R(d_n)$	n	$I(d_n)$
0	1.00170 22348 53820 99565	0	0.03726 17150 00537 65365
1	0.00225 55728 46561 17976	1	0.03714 53224 79807 68994
2	0.00054 32164 87508 01325	2	-0.00013 72632 38201 90679
3	-0.00001 11794 61895 40836	3	-0.00001 98512 94687 59687
4	-0.00000 09469 01382 39192	4	0.00000 10700 14057 38568
5	0.00000 01110 32677 12082	5	0.00000 00383 05261 71449
6	-0.00000 00012 94398 92684	6	-0.00000 00116 28723 27663
7	-0.00000 00011 14905 94420	7	0.00000 00007 59733 09244
8	0.00000 00001 57637 23196	8	0.00000 00000 75476 07460
9	-0.00000 00000 02830 45747	9	-0.00000 00000 24752 78088
10	-0.00000 00000 02932 16857	10	0.00000 00000 02493 89256
11	0.00000 00000 00617 80854	11	0.00000 00000 00156 19784
12	-0.00000 00000 00043 16153	12	-0.00000 00000 00103 38521
13	-0.00000 00000 00010 13289	13	0.00000 00000 00018 12876
14	0.00000 00000 00003 97343	14	-0.00000 00000 00000 70876
15	-0.00000 00000 00000 63161	15	-0.00000 00000 00000 52042
16	0.00000 00000 00000 00575	16	0.00000 00000 00000 17235
17	0.00000 00000 00000 02696	17	-0.00000 00000 00000 02625
18	-0.00000 00000 00000 00847	18	-0.00000 00000 00000 00038
19	0.00000 00000 00000 00130	19	0.00000 00000 00000 00148
20	0.00000 00000 00000 00003	20	-0.00000 00000 00000 00047
21	-0.00000 00000 00000 00009	21	0.00000 00000 00000 00008
22	0.00000 00000 00000 00003	22	0.00000 00000 00000 00000
23	-0.00000 00000 00000 00001	23	-0.00000 00000 00000 00001

TABLE 9.3

CHEBYSHEV COEFFICIENTS FOR $\int^x t^{-m} J_0(t)\,dt$ AND $\int^x t^{-m} Y_0(t)\,dt$, $m = 0, 1$

$$\int_0^x J_0(t)dt = \sum_{n=0}^{\infty} a_n T_{2n+1}(x/8) \qquad -8 \le x \le 8$$

$$\int_0^x Y_0(t)dt = (2/\pi)(\gamma + \ln x/2)\int_0^x J_0(t)dt - \sum_{n=1}^{\infty} b_n T_{2n+1}(x/8) \qquad 0 \le x \le 8$$

n	a_n	n	b_n
0	1.29671 75412 10529 84167	0	1.52325 89274 53589 03192
1	-0.36520 27407 41585 37488	1	0.16707 19381 81103 39620
2	0.50821 88856 60789 27112	2	0.19604 60450 17129 95275
3	-0.30180 69121 16998 30875	3	-0.27450 26073 93900 63315
4	0.08576 03874 41558 28731	4	0.10180 66421 62423 09366
5	-0.01444 10725 38500 54169	5	-0.01978 67970 11808 59820
6	0.00162 45557 64822 73217	6	0.00244 75401 49909 44840
7	-0.00013 14897 32007 27470	7	-0.00021 24429 21144 18655
8	0.00000 80523 00171 47464	8	0.00001 37438 21090 86322
9	-0.00000 03869 53377 61818	9	-0.00000 06908 35405 49799
10	0.00000 00150 02074 18186	10	0.00000 00278 19570 53702
11	-0.00000 00004 79607 04238	11	-0.00000 00009 18984 49486
12	0.00000 00000 12868 92765	12	0.00000 00000 25377 49742
13	-0.00000 00000 00294 08710	13	-0.00000 00000 00594 95975
14	0.00000 00000 00005 79477	14	0.00000 00000 00011 99595
15	-0.00000 00000 00000 09949	15	-0.00000 00000 00000 21031
16	0.00000 00000 00000 00150	16	0.00000 00000 00000 00324
17	-0.00000 00000 00000 00002	17	-0.00000 00000 00000 00004

$$\int_x^{\infty} t^{-1}J_0(t)dt + (\gamma + \ln x/2) = \int_0^x t^{-1}\left[1 - J_0(t)\right]dt = \sum_{n=0}^{\infty} a_n T_{2n}(x/8), \quad 0 < x \le 8$$

$$\int_x^{\infty} t^{-1}Y_0(t)dt = (2/\pi)(\gamma + \ln x/2)\int_0^x t^{-1}\left[1 - J_0(t)\right]dt + \pi/6 - \pi^{-1}(\gamma + \ln x/2)^2 - \sum_{n=0}^{\infty} b_n T_{2n}(x/8), \quad 0 < x \le 8$$

n	a_n	n	b_n
0	1.35105 09191 81876 36388	0	1.17024 76315 91773 78917
1	0.83791 03073 48683 76979	1	0.70944 45491 34294 92278
2	-0.35047 96397 85294 62711	2	-0.28401 50820 96576 85981
3	0.12777 41586 77531 98659	3	0.13265 78987 78083 99830
4	-0.02981 03569 82555 60990	4	-0.03696 71557 94706 40644
5	0.00455 21984 11693 87328	5	0.00636 04002 91027 60459
6	-0.00048 40862 19671 85359	6	-0.00073 70115 58540 43150
7	0.00003 78020 28599 16883	7	0.00006 14636 16385 29090
8	-0.00000 22588 69085 06771	8	-0.00000 38715 36647 14065
9	0.00000 01066 46090 68423	9	0.00000 01909 48867 24907
10	-0.00000 00040 80054 43149	10	-0.00000 00075 81934 39230
11	0.00000 00001 29099 96251	11	0.00000 00002 47764 60612
12	-0.00000 00000 03435 77839	12	-0.00000 00000 06783 71842
13	0.00000 00000 00077 99552	13	0.00000 00000 00157 94562
14	-0.00000 00000 00001 52842	14	-0.00000 00000 00003 16655
15	0.00000 00000 00000 02612	15	0.00000 00000 00000 05525
16	-0.00000 00000 00000 00039	16	-0.00000 00000 00000 00085
17	0.00000 00000 00000 00001	17	0.00000 00000 00000 00001

TABLE 9.3 *(Continued)*

$$\int_x^\infty J_o(t)dt+i\int_x^\infty Y_o(t)dt = (2/\pi x)^{\frac{1}{2}}e^{i(x+\pi/4)} \sum_{n=0}^{\infty} c_n T_n^*(5/x) \text{ , } x \geq 5$$

$$c_n = R(c_n)+iI(c_n)$$

n	$R(c_n)$	n	$I(c_n)$
0	0.98740 76158 14884 26270	0	-0.05776 66747 40994 51444
1	-0.01622 95522 38987 83538	1	-0.05561 79374 24115 22950
2	-0.00327 41117 97339 24011	2	0.00240 40410 70872 61157
3	0.00036 92769 92655 13937	3	0.00019 64777 76330 32259
4	-0.00000 20837 13476 09414	4	-0.00005 46215 76498 13484
5	-0.00000 68286 10172 02808	5	0.00000 49615 33956 28297
6	0.00000 14114 08894 67207	6	0.00000 04376 23929 01943
7	-0.00000 01069 99598 18439	7	-0.00000 02647 66396 96766
8	-0.00000 00268 57064 68353	8	0.00000 00534 45098 22653
9	0.00000 00128 52949 03326	9	-0.00000 00037 89065 39485
10	-0.00000 00027 82927 64282	10	-0.00000 00015 36724 96861
11	0.00000 00002 33536 92269	11	0.00000 00007 57412 49246
12	0.00000 00000 89167 79341	12	-0.00000 00001 86161 17165
13	-0.00000 00000 51517 91688	13	0.00000 00000 21633 44301
14	0.00000 00000 14904 80479	14	0.00000 00000 04800 15078
15	-0.00000 00000 02427 97151	15	-0.00000 00000 03821 66065
16	-0.00000 00000 00151 48359	16	0.00000 00000 01338 43845
17	0.00000 00000 00288 66165	17	-0.00000 00000 00292 98739
18	-0.00000 00000 00127 37706	18	0.00000 00000 00017 95082
19	0.00000 00000 00036 07236	19	0.00000 00000 00019 76575
20	-0.00000 00000 00005 70713	20	-0.00000 00000 00012 15265
21	-0.00000 00000 00000 80559	21	0.00000 00000 00004 39396
22	0.00000 00000 00001 07730	22	-0.00000 00000 00001 06272
23	-0.00000 00000 00000 51320	23	0.00000 00000 00000 09161
24	0.00000 00000 00000 16683	24	0.00000 00000 00000 07387
25	-0.00000 00000 00000 03489	25	-0.00000 00000 00000 05463
26	-0.00000 00000 00000 00022	26	0.00000 00000 00000 02336
27	0.00000 00000 00000 00468	27	-0.00000 00000 00000 00713
28	-0.00000 00000 00000 00287	28	0.00000 00000 00000 00131
29	0.00000 00000 00000 00117	29	0.00000 00000 00000 00015
30	-0.00000 00000 00000 00035	30	-0.00000 00000 00000 00028
31	0.00000 00000 00000 00006	31	0.00000 00000 00000 00016
32	0.00000 00000 00000 00001	32	-0.00000 00000 00000 00006
33	-0.00000 00000 00000 00002	33	0.00000 00000 00000 00002
34	0.00000 00000 00000 00001		

$$\int_0^\infty J_o(t)dt = 1 \text{ , } \int_0^\infty Y_o(t)dt = 0$$

TABLE 9.3 *(Concluded)*

$$\int_x^\infty t^{-1}J_0(t)dt+i\int_x^\infty t^{-1}Y_0(t)dt = (2/\pi x)^{\frac{1}{2}}x^{-1}e^{i(x+\pi/4)}\sum_{n=0}^{\infty} c_n T_n^*(5/x)\ , \quad x \geq 5$$

$$c_n = R(c_n)+iI(c_n)$$

n	$R(c_n)$	n	$I(c_n)$
0	0.95360 15080 97385 58095	0	-0.13917 92593 02000 01236
1	-0.05860 83885 38723 31670	1	-0.12902 06572 61350 67062
2	-0.01020 28357 56598 56676	2	0.01103 00434 81095 35741
3	0.00196 01270 40436 22581	3	0.00051 81718 08568 80364
4	-0.00009 57497 76977 56219	4	-0.00030 92821 01739 75681
5	-0.00003 57047 94770 43714	5	0.00004 64709 84430 47525
6	0.00001 16967 79604 30223	6	-0.00000 00819 88453 40928
7	-0.00000 16438 62464 52682	7	-0.00000 19188 83810 06925
8	-0.00000 00741 58457 51760	8	0.00000 05781 36677 61104
9	0.00000 01143 43875 27717	9	-0.00000 00844 89977 73317
10	-0.00000 00360 09032 14141	10	-0.00000 00052 56121 61520
11	0.00000 00060 12573 86446	11	0.00000 00076 32577 90924
12	0.00000 00001 91246 56215	12	-0.00000 00026 86439 63177
13	-0.00000 00005 48920 28385	13	0.00000 00005 42799 49860
14	0.00000 00002 27404 45656	14	-0.00000 00000 17443 65343
15	-0.00000 00000 56714 90865	15	-0.00000 00000 39756 92920
16	0.00000 00000 06075 10983	16	0.00000 00000 20696 83990
17	0.00000 00000 02520 60520	17	-0.00000 00000 06396 23674
18	-0.00000 00000 01912 55246	18	0.00000 00000 01163 59235
19	0.00000 00000 00740 56501	19	0.00000 00000 00067 59603
20	-0.00000 00000 00189 50214	20	-0.00000 00000 00165 57337
21	0.00000 00000 00020 21389	21	0.00000 00000 00084 25597
22	0.00000 00000 00011 03617	22	-0.00000 00000 00028 24474
23	-0.00000 00000 00008 89993	23	0.00000 00000 00006 07698
24	0.00000 00000 00003 88558	24	-0.00000 00000 00000 03171
25	-0.00000 00000 00001 19200	25	-0.00000 00000 00000 77237
26	0.00000 00000 00000 21456	26	0.00000 00000 00000 48022
27	0.00000 00000 00000 02915	27	-0.00000 00000 00000 19502
28	-0.00000 00000 00000 04877	28	0.00000 00000 00000 05671
29	0.00000 00000 00000 02737	29	-0.00000 00000 00000 00862
30	-0.00000 00000 00000 01080	30	-0.00000 00000 00000 00269
31	0.00000 00000 00000 00308	31	0.00000 00000 00000 00309
32	-0.00000 00000 00000 00042	32	-0.00000 00000 00000 00167
33	-0.00000 00000 00000 00020	33	0.00000 00000 00000 00066
34	0.00000 00000 00000 00020	34	-0.00000 00000 00000 00019
35	-0.00000 00000 00000 00011	35	0.00000 00000 00000 00003
36	0.00000 00000 00000 00004	36	0.00000 00000 00000 00001
37	-0.00000 00000 00000 00001	37	-0.00000 00000 00000 00001
		38	0.00000 00000 00000 00001

TABLE 9.4

CHEBYSHEV COEFFICIENTS FOR $\int_0^x u^{-1} \int_0^u J_0(t)\, dt\, du$ AND $\int_0^x u^{-1} \int_0^u t^{-1}[1 - J_0(t)]\, dt\, du$

$$\int_0^x u^{-1} \int_0^u J_o(t)dtdu = \sum_{n=0}^{\infty} a_n T_{2n+1}(x/8) \qquad \int_0^x u^{-1} \int_0^u t^{-1}\left[1-J_o(t)\right]dtdu = \sum_{n=0}^{\infty} b_n T_{2n}(x/8)$$

$-8 \le x \le 8$

n	a_n	n	b_n
0	3.85110 65386 99851 87808	0	1.13957 65086 78636 96373
1	-0.72972 87522 49245 55384	1	0.93209 57655 07534 47899
2	0.26315 29260 69072 40311	2	-0.16895 03959 70272 64027
3	-0.07224 84046 52510 63367	3	0.03292 46309 03043 99123
4	0.01313 01703 34335 31551	4	-0.00499 54087 19772 04386
5	-0.00163 36430 02632 63733	5	0.00056 00714 36392 68677
6	0.00014 64957 48125 82590	6	-0.00004 70358 11046 76174
7	-0.00000 98932 81920 38199	7	0.00000 30386 77449 29417
8	0.00000 05210 11566 29297	8	-0.00000 01550 36662 86762
9	-0.00000 00219 97004 09935	9	0.00000 00063 92857 22944
10	0.00000 00007 61315 52798	10	-0.00000 00002 17264 25071
11	-0.00000 00000 21997 60276	11	0.00000 00000 06187 75546
12	0.00000 00000 00538 75574	12	-0.00000 00000 00149 78664
13	-0.00000 00000 00011 32884	13	0.00000 00000 00003 11944
14	0.00000 00000 00000 20679	14	-0.00000 00000 00000 05648
15	-0.00000 00000 00000 00331	15	0.00000 00000 00000 00090
16	0.00000 00000 00000 00005	16	-0.00000 00000 00000 00001

TABLE 9.5

CHEBYSHEV COEFFICIENTS FOR $I_0(x)$ AND $K_0(x)$

$$I_o(x) = \sum_{n=0}^{\infty} a_n T_{2n}(x/8) \qquad K_o(x) = -(\gamma + \ln x/2) I_o(x) + \sum_{n=0}^{\infty} b_n T_{2n}(x/8)$$

$-8 \le x \le 8$ $\qquad$ $0 < x \le 8$

n	a_n	n	b_n
0	127.73343 98121 81083 56301	0	240.27705 96407 20389 10102
1	190.49432 01727 42844 19322	1	369.47407 39728 67282 63764
2	82.48903 27440 24099 61321	2	169.97341 16984 01148 04378
3	22.27481 92424 62230 87742	3	49.02046 37772 63439 39371
4	4.01167 37601 79348 53351	4	9.38849 73252 68442 32756
5	0.50949 33654 39982 87079	5	1.25947 97636 67703 58618
6	0.04771 87487 98174 13524	6	0.12377 69641 14924 54118
7	0.00341 63317 66012 34095	7	0.00924 43098 62866 90621
8	0.00019 24693 59688 11366	8	0.00054 06238 96492 55807
9	0.00000 87383 15496 62236	9	0.00002 53737 96028 08704
10	0.00000 03260 91050 57896	10	0.00000 09754 78302 83898
11	0.00000 00101 69726 72769	11	0.00000 00312 49571 77932
12	0.00000 00002 68828 12895	12	0.00000 00008 46434 70610
13	0.00000 00000 06096 89280	13	0.00000 00000 19628 88451
14	0.00000 00000 00119 89083	14	0.00000 00000 00393 96098
15	0.00000 00000 00002 06305	15	0.00000 00000 00006 90835
16	0.00000 00000 00000 03132	16	0.00000 00000 00000 10673
17	0.00000 00000 00000 00042	17	0.00000 00000 00000 00146
18	0.00000 00000 00000 00001	18	0.00000 00000 00000 00002

TABLE 9.5 *(Concluded)*

$$I_0(x) = (2\pi x)^{-\frac{1}{2}} e^x \sum_{n=0}^{\infty} c_n T_n^*(8/x)$$

$x \geq 8$

n	c_n
0	1.00827 92054 58740 03188
1	0.00844 51226 24920 94320
2	0.00017 27006 30777 56653
3	0.00000 72475 91099 95896
4	0.00000 05135 87726 87802
5	0.00000 00568 16965 80812
6	0.00000 00085 13091 22285
7	0.00000 00012 38425 36400
8	0.00000 00000 29801 67230
9	-0.00000 00000 78956 69832
10	-0.00000 00000 33127 12763
11	-0.00000 00000 04497 33864
12	0.00000 00000 01799 79030
13	0.00000 00000 00965 74832
14	0.00000 00000 00038 60424
15	-0.00000 00000 00104 03934
16	-0.00000 00000 00023 95045
17	0.00000 00000 00009 55447
18	0.00000 00000 00004 44315
19	-0.00000 00000 00000 85864
20	-0.00000 00000 00000 70878
21	0.00000 00000 00000 08676
22	0.00000 00000 00000 11194
23	-0.00000 00000 00000 01211
24	-0.00000 00000 00000 01813
25	0.00000 00000 00000 00249
26	0.00000 00000 00000 00299
27	-0.00000 00000 00000 00062
28	-0.00000 00000 00000 00049
29	0.00000 00000 00000 00016
30	0.00000 00000 00000 00007
31	-0.00000 00000 00000 00004
32	-0.00000 00000 00000 00001
33	0.00000 00000 00000 00001

$$K_0(x) = (\pi/2x)^{\frac{1}{2}} e^{-x} \sum_{n=0}^{\infty} d_n T_n^*(5/x)$$

$x \geq 5$

n	d_n
0	0.98840 81742 30825 80035
1	-0.01131 05046 46928 28069
2	0.00026 95326 12762 72369
3	-0.00001 11066 85196 66535
4	0.00000 06325 75108 50049
5	-0.00000 00450 47337 64110
6	0.00000 00037 92996 45568
7	-0.00000 00003 64547 17921
8	0.00000 00000 39043 75576
9	-0.00000 00000 04579 93622
10	0.00000 00000 00580 81063
11	-0.00000 00000 00078 83236
12	0.00000 00000 00011 36042
13	-0.00000 00000 00001 72697
14	0.00000 00000 00000 27545
15	-0.00000 00000 00000 04589
16	0.00000 00000 00000 00796
17	-0.00000 00000 00000 00143
18	0.00000 00000 00000 00027
19	-0.00000 00000 00000 00005
20	0.00000 00000 00000 00001

TABLE 9.6

CHEBYSHEV COEFFICIENTS FOR $I_1(x)$ AND $K_1(x)$

$$I_1(x) = \sum_{n=0}^{\infty} a_n T_{2n+1}(x/8) \qquad I_1(x) = (x/8) \sum_{n=0}^{\infty} b_n T_{2n}(x/8)$$

$-8 \le x \le 8$

n	a_n	n	b_n
0	220.60142 69235 23778 56112	0	129.94511 89032 38645 86560
1	125.35426 68371 52356 46451	1	181.31261 60405 70265 39103
2	42.86523 40931 28256 85130	2	69.39591 76337 34447 53798
3	9.45300 52294 34910 53517	3	16.33455 05525 22066 16461
4	1.42965 77090 76213 46814	4	2.57145 99063 47754 90572
5	0.15592 42954 76256 29116	5	0.28785 55118 04672 03057
6	0.01276 80490 81733 88545	6	0.02399 30791 47840 55176
7	0.00081 08879 00690 69214	7	0.00154 30190 15627 21914
8	0.00004 10104 61938 23750	8	0.00007 87567 85754 16515
9	0.00000 16880 42203 43687	9	0.00000 32641 38122 30986
10	0.00000 00575 86950 54206	10	0.00000 01119 46284 56389
11	0.00000 00016 53453 53976	11	0.00000 00032 27616 52023
12	0.00000 00000 40484 76606	12	0.00000 00000 79290 55929
13	0.00000 00000 00854 96289	13	0.00000 00000 01678 97282
14	0.00000 00000 00015 72708	14	0.00000 00000 00030 95296
15	0.00000 00000 00000 25419	15	0.00000 00000 00000 50120
16	0.00000 00000 00000 00364	16	0.00000 00000 00000 00718
17	0.00000 00000 00000 00005	17	0.00000 00000 00000 00009

$$K_1(x) = (\gamma + \ln x/2) I_1(x) + x^{-1} - \sum_{n=0}^{\infty} c_n T_{2n+1}(x/8)$$

$0 < x \le 8$

n	c_n
0	418.88944 61663 96890 97522
1	249.89554 90428 68080 38961
2	91.18031 93387 41787 75763
3	21.44499 50539 62240 43921
4	3.43841 53928 80464 59793
5	0.39484 60929 40938 23432
6	0.03382 87455 26884 19281
7	0.00223 57203 34170 88760
8	0.00011 71310 22460 84561
9	0.00000 49754 27122 13645
10	0.00000 01746 04931 76984
11	0.00000 00051 43294 11806
12	0.00000 00001 28903 39664
13	0.00000 00000 02780 94119
14	0.00000 00000 00052 17097
15	0.00000 00000 00000 85869
16	0.00000 00000 00000 01250
17	0.00000 00000 00000 00016

TABLE 9.6 *(Concluded)*

$$I_1(x) = (2\pi x)^{-\frac{1}{2}} e^x \sum_{n=0}^{\infty} d_n T_n^*(8/x)$$

$x \geq 8$

n	d_n
0	0.97580 06023 26285 92615
1	-0.02446 74429 63276 38489
2	-0.00027 72053 60763 82886
3	-0.00000 97321 46728 02013
4	-0.00000 06297 24238 63981
5	-0.00000 00659 61142 15424
6	-0.00000 00096 13872 91942
7	-0.00000 00014 01140 90103
8	-0.00000 00000 47563 16654
9	0.00000 00000 81530 68107
10	0.00000 00000 35408 14832
11	0.00000 00000 05102 56407
12	-0.00000 00000 01804 40934
13	-0.00000 00000 01023 59447
14	-0.00000 00000 00052 67784
15	0.00000 00000 00107 09419
16	0.00000 00000 00026 11976
17	-0.00000 00000 00009 56129
18	-0.00000 00000 00004 71335
19	0.00000 00000 00000 82924
20	0.00000 00000 00000 74262
21	-0.00000 00000 00000 08045
22	-0.00000 00000 00000 11657
23	0.00000 00000 00000 01107
24	0.00000 00000 00000 01884
25	-0.00000 00000 00000 00233
26	-0.00000 00000 00000 00311
27	0.00000 00000 00000 00061
28	0.00000 00000 00000 00051
29	-0.00000 00000 00000 00016
30	-0.00000 00000 00000 00008
31	0.00000 00000 00000 00004
32	0.00000 00000 00000 00001
33	-0.00000 00000 00000 00001

$$K_1(x) = (\pi/2x)^{\frac{1}{2}} e^{-x} \sum_{n=0}^{\infty} e_n T_n^*(5/x)$$

$x \geq 5$

n	e_n
0	1.03595 08587 72358 33071
1	0.03546 52912 43331 11380
2	-0.00046 84750 28166 88856
3	0.00001 61850 63810 05343
4	-0.00000 08451 72048 12368
5	0.00000 00571 32218 10284
6	-0.00000 00046 45554 60661
7	0.00000 00004 35417 33857
8	-0.00000 00000 45757 29704
9	0.00000 00000 05288 13281
10	-0.00000 00000 00662 61293
11	0.00000 00000 00089 04792
12	-0.00000 00000 00012 72607
13	0.00000 00000 00001 92086
14	-0.00000 00000 00000 30450
15	0.00000 00000 00000 05046
16	-0.00000 00000 00000 00871
17	0.00000 00000 00000 00156
18	-0.00000 00000 00000 00029
19	0.00000 00000 00000 00006
20	-0.00000 00000 00000 00001

TABLE 9.7

CHEBYSHEV COEFFICIENTS FOR THE INTEGRALS OF $I_0(x)$ AND $K_0(x)$

$$\int_0^x I_0(t)\,dt = \sum_{n=0}^{\infty} a_n T_{2n+1}(x/8)$$

$-8 \le x \le 8$

n	a_n
0	259.89023 78064 77291 73120
1	144.00704 99049 58326 10668
2	48.17137 08012 49494 98863
3	10.43608 31327 33075 62509
4	1.55652 46198 84162 51676
5	0.16791 80424 15203 17657
6	0.01363 15129 32972 85978
7	0.00085 96966 41686 46061
8	0.00004 32308 33927 40972
9	0.00000 17709 94620 21966
10	0.00000 00601 75490 25739
11	0.00000 00017 21895 40848
12	0.00000 00000 42036 99778
13	0.00000 00000 00885 48177
14	0.00000 00000 00016 25211
15	0.00000 00000 00000 26216
16	0.00000 00000 00000 00375
17	0.00000 00000 00000 00005

$$\int_0^x K_0(t)\,dt = -(\gamma + \ln x/2)\int_0^x I_0(t)\,dt + \sum_{n=0}^{\infty} b_n T_{2n+1}(x/8)$$

$0 < x \le 8$

n	b_n
0	494.46973 43920 76091 07855
1	287.91209 46138 67131 72634
2	102.78276 72089 04561 78295
3	23.73730 13023 60633 33053
4	3.75137 61611 89713 66941
5	0.42591 84155 79600 67827
6	0.03616 35210 25074 58274
7	0.00237 27602 55936 31706
8	0.00012 35768 37415 85715
9	0.00000 52235 40841 92078
10	0.00000 01825 58521 84278
11	0.00000 00053 58787 33400
12	0.00000 00001 33900 85232
13	0.00000 00000 02881 22921
14	0.00000 00000 00053 92916
15	0.00000 00000 00000 88585
16	0.00000 00000 00000 01287
17	0.00000 00000 00000 00017

TABLE 9.7 *(Concluded)*

$$\int_0^x I_0(t)dt = (2\pi x)^{-\frac{1}{2}}e^x \sum_{n=0}^{\infty} c_n T_n^*(8/x)$$

$x \geq 8$

n	c_n
0	1.04790 61157 24010
1	0.05153 72954 31712
2	0.00412 80137 15044
3	0.00055 55143 81221
4	0.00004 73076 79123
5	-0.00002 09628 57759
6	-0.00001 12080 67263
7	-0.00000 08489 39008
8	0.00000 11097 75766
9	0.00000 02714 52545
10	-0.00000 01255 88772
11	-0.00000 00463 22049
12	0.00000 00192 61390
13	0.00000 00070 57828
14	-0.00000 00038 20465
15	-0.00000 00008 91488
16	0.00000 00008 40548
17	0.00000 00000 37019
18	-0.00000 00001 77371
19	0.00000 00000 31033
20	0.00000 00000 30584
21	-0.00000 00000 15045
22	-0.00000 00000 02560
23	0.00000 00000 04289
24	-0.00000 00000 00820
25	-0.00000 00000 00750
26	0.00000 00000 00481
27	0.00000 00000 00001
28	-0.00000 00000 00124
29	0.00000 00000 00052
30	0.00000 00000 00010
31	-0.00000 00000 00019
32	0.00000 00000 00006
33	0.00000 00000 00002
34	-0.00000 00000 00003
35	0.00000 00000 00001

$$\int_x^{\infty} K_0(t)dt = (\pi/2x)^{\frac{1}{2}}e^{-x} \sum_{n=0}^{\infty} d_n T_n^*(5/x)$$

$x \geq 5$

n	d_n
0	0.94845 03112 70840 01200
1	-0.04843 36883 20144 26839
2	0.00284 52735 36373 53655
3	-0.00024 11658 65541 40493
4	0.00002 57606 49079 80233
5	-0.00000 32467 76974 97624
6	0.00000 04644 74520 32803
7	-0.00000 00735 32096 70299
8	0.00000 00126 56166 77192
9	-0.00000 00023 37737 80696
10	0.00000 00004 58849 84282
11	-0.00000 00000 94967 63567
12	0.00000 00000 20598 29645
13	-0.00000 00000 04658 58335
14	0.00000 00000 01094 04452
15	-0.00000 00000 00265 86219
16	0.00000 00000 00066 65442
17	-0.00000 00000 00017 19670
18	0.00000 00000 00004 55561
19	-0.00000 00000 00001 23679
20	0.00000 00000 00000 34352
21	-0.00000 00000 00000 09747
22	0.00000 00000 00000 02821
23	-0.00000 00000 00000 00832
24	0.00000 00000 00000 00250
25	-0.00000 00000 00000 00076
26	0.00000 00000 00000 00024
27	-0.00000 00000 00000 00007
28	0.00000 00000 00000 00002
29	-0.00000 00000 00000 00001

$$\int_0^{\infty} K_0(t)dt = \pi/2$$

TABLE 9.8

Chebyshev Coefficients for $\int_0^x t^{-1}[I_0(t) - 1]\,dt$ and $\int_x^\infty t^{-1}K_0(t)\,dt$

$$\int_0^x t^{-1}\left[I_0(t)-1\right]dt = \sum_{n=0}^{\infty} a_n T_{2n}(x/8)\ , \quad -8 \le x \le 8$$

$$\int_x^\infty t^{-1}K_0(t)dt = (\gamma+\ln x/2)\int_0^x t^{-1}\left[I_0(t)-1\right]dt+\pi^2/24+\tfrac{1}{2}(\gamma+\ln x/2)^2 - \sum_{n=0}^{\infty} b_n T_{2n}(x/8)$$

$0 < x \le 8$

n	a_n	n	b_n
0	22.41195 92097 11540 88911	0	43.28142 98168 47953 30024
1	31.48627 97258 09661 46640	1	62.20884 20010 93457 93811
2	11.25818 19942 74855 41180	2	23.82776 90009 53506 19306
3	2.53024 75874 10407 84810	3	5.72830 10540 31355 24026
4	0.38520 74947 27554 40692	4	0.92590 00275 61214 88322
5	0.04205 20436 91893 04074	5	0.10649 72792 14668 10418
6	0.00343 78483 10239 86068	6	0.00911 20168 98979 26527
7	0.00021 77312 36377 39044	7	0.00060 06120 38440 63728
8	0.00001 09765 68565 04757	8	0.00003 13670 06001 11057
9	0.00000 04503 30397 26279	9	0.00000 13280 13478 58260
10	0.00000 00153 13864 76566	10	0.00000 00464 54582 16292
11	0.00000 00004 38364 93355	11	0.00000 00013 64208 52879
12	0.00000 00000 10702 91920	12	0.00000 00000 34092 14393
13	0.00000 00000 00225 42982	13	0.00000 00000 00733 53295
14	0.00000 00000 00004 13666	14	0.00000 00000 00013 72714
15	0.00000 00000 00000 06671	15	0.00000 00000 00000 22542
16	0.00000 00000 00000 00095	16	0.00000 00000 00000 00327
17	0.00000 00000 00000 00001	17	0.00000 00000 00000 00004

TABLE 9.8 *(Concluded)*

$$\int_0^x t^{-1}[I_0(t)-1]\,dt = (2\pi x)^{-\frac{1}{2}} x^{-1} e^{x} \sum_{n=0}^{\infty} c_n T_n^*(8/x) + \ln(8/x)\ , \quad x \ge 8$$

$$\int_x^{\infty} t^{-1} K_0(t)\,dt = (\pi/2x)^{\frac{1}{2}} x^{-1} e^{-x} \sum_{n=0}^{\infty} d_n T_n^*(5/x)\ , \quad x \ge 5$$

n	c_n	n	d_n
0	1.13730 59729 57956	0	0.87823 74557 33144 87755
1	0.15028 68683 04482	1	-0.11074 68238 59845 47294
2	0.01257 12054 42504	2	0.00976 72723 38298 76020
3	-0.00173 03619 76412	3	-0.00108 21377 22273 23340
4	-0.00186 71346 48509	4	0.00014 12539 10868 05608
5	-0.00058 97905 96544	5	-0.00002 08930 59305 37890
6	0.00001 45540 83296	6	0.00000 34136 22130 56458
7	0.00007 63759 19373	7	-0.00000 06052 68064 17063
8	0.00001 26111 20267	8	0.00000 01149 73564 34993
9	-0.00000 94572 16158	9	-0.00000 00231 70530 87425
10	-0.00000 27572 45188	10	0.00000 00049 16677 78545
11	0.00000 14640 94877	11	-0.00000 00010 91904 21437
12	0.00000 04717 17955	12	0.00000 00002 52548 52893
13	-0.00000 02887 79135	13	-0.00000 00000 60588 90216
14	-0.00000 00660 40921	14	0.00000 00000 15026 38416
15	0.00000 00647 67951	15	-0.00000 00000 03841 29509
16	0.00000 00039 16081	16	0.00000 00000 01009 69303
17	-0.00000 00143 81343	17	-0.00000 00000 00272 30795
18	0.00000 00021 13812	18	0.00000 00000 00075 21065
19	0.00000 00027 23932	19	-0.00000 00000 00021 23894
20	-0.00000 00011 87240	20	0.00000 00000 00006 12335
21	-0.00000 00003 05554	21	-0.00000 00000 00001 80005
22	0.00000 00003 78160	22	0.00000 00000 00000 53891
23	-0.00000 00000 48200	23	-0.00000 00000 00000 16415
24	-0.00000 00000 77950	24	0.00000 00000 00000 05082
25	0.00000 00000 40800	25	-0.00000 00000 00000 01598
26	0.00000 00000 04548	26	0.00000 00000 00000 00510
27	-0.00000 00000 12601	27	-0.00000 00000 00000 00165
28	0.00000 00000 04067	28	0.00000 00000 00000 00054
29	0.00000 00000 01659	29	-0.00000 00000 00000 00018
30	-0.00000 00000 01874	30	0.00000 00000 00000 00006
31	0.00000 00000 00410	31	-0.00000 00000 00000 00002
32	0.00000 00000 00335	32	0.00000 00000 00000 00001
33	-0.00000 00000 00284		
34	0.00000 00000 00047		
35	0.00000 00000 00059		
36	-0.00000 00000 00046		
37	0.00000 00000 00007		
38	0.00000 00000 00010		
39	-0.00000 00000 00008		
40	0.00000 00000 00002		

TABLE 9.9

CHEBYSHEV COEFFICIENTS FOR $\int_0^x u^{-1} \int_0^u I_0(t)\,dt\,du$ AND $\int_0^x u^{-1} \int_0^u t^{-1}[I_0(t) - 1]\,dt\,du$

$$\int_0^x u^{-1}\int_0^u I_0(t)\,dt\,du = \sum_{n=0}^{\infty} a_n T_{2n+1}(x/8) \qquad \int_0^x u^{-1}\int_0^u t^{-1}[I_0(t)-1]\,dt\,du = \sum_{n=0}^{\infty} b_n T_{2n}(x/8)$$

$-8 \le x \le 8$

n	a_n	n	b_n
0	50.14955 31577 82108 82091	0	5.21732 54461 49819 20862
1	21.91121 15812 45618 93454	1	6.66881 93468 06710 15591
2	6.02040 88719 94394 86289	2	1.72261 47594 80346 43570
3	1.09046 33297 92063 57844	3	0.30624 58948 23843 63682
4	0.13847 94671 44940 89547	4	0.03844 55904 67473 95253
5	0.01293 55793 78590 11645	5	0.00355 90727 29586 97459
6	0.00092 27043 32133 77199	6	0.00025 19556 73815 28618
7	0.00005 17773 31569 82422	7	0.00001 40463 56291 35972
8	0.00000 23415 19659 51103	8	0.00000 06316 04983 33167
9	0.00000 00870 52899 76329	9	0.00000 00233 64357 47100
10	0.00000 00027 05430 21237	10	0.00000 00007 22904 90096
11	0.00000 00000 71285 23431	11	0.00000 00000 18972 80968
12	0.00000 00000 01611 92086	12	0.00000 00000 00427 50839
13	0.00000 00000 00031 61091	13	0.00000 00000 00008 35723
14	0.00000 00000 00000 54259	14	0.00000 00000 00000 14304
15	0.00000 00000 00000 00822	15	0.00000 00000 00000 00216
16	0.00000 00000 00000 00011	16	0.00000 00000 00000 00003

TABLE 9.10

Chebyshev Coefficients for ber(x), bei(x), ker(x), kei(x), and Their Derivatives

$$\mathrm{ber}(x) = \sum_{n=0}^{\infty} a_n T_{2n}(x^2/64) \qquad \mathrm{bei}(x) = \sum_{n=0}^{\infty} b_n T_{2n+1}(x^2/64)$$

$$\mathrm{ber}'(x) = (x/8) \sum_{n=0}^{\infty} c_n T_{2n+1}(x^2/64) \qquad \mathrm{bei}'(x) = (x/8) \sum_{n=0}^{\infty} d_n T_{2n}(x^2/64)$$

$$-8 \le x \le 8$$

n	a_n	n	b_n
0	2.25521 15482 79523 90138	0	-29.34949 10970 21269 22722
1	10.84058 01738 13068 20665	1	-8.98868 87413 38202 57684
2	8.71271 74101 86675 55916	2	3.46690 09758 41511 39894
3	-0.85344 63696 95052 22986	3	-0.14735 80153 21209 28048
4	0.01904 82639 34734 39291	4	0.00192 21031 54268 04953
5	-0.00015 59976 15956 17446	5	-0.00001 04178 99277 03635
6	0.00000 05829 62923 95910	6	0.00000 00277 43180 21356
7	-0.00000 00011 36930 89629	7	-0.00000 00000 40549 17690
8	0.00000 00000 01270 22191	8	0.00000 00000 00035 22916
9	-0.00000 00000 00000 87119	9	-0.00000 00000 00000 01933
10	0.00000 00000 00000 00039	10	0.00000 00000 00000 00001

n	c_n	n	d_n
0	25.78109 24425 89600 75371	0	-9.99884 34643 81679 05729
1	14.94051 22687 76532 54706	1	-5.32294 13802 52723 50097
2	-2.48492 25515 96818 57127	2	8.16009 17317 54580 36429
3	0.07541 65574 88338 11832	3	-0.50716 07078 49198 13307
4	-0.00077 64982 50599 45331	4	0.00859 23457 75034 34863
5	0.00000 34898 29181 41896	5	-0.00005 71184 19171 87424
6	-0.00000 00079 48362 33566	6	0.00000 01800 26851 82568
7	0.00000 00000 10153 93838	7	-0.00000 00003 03819 56248
8	-0.00000 00000 00007 83687	8	0.00000 00000 00299 26429
9	0.00000 00000 00000 00387	9	-0.00000 00000 00000 18356
		10	0.00000 00000 00000 00007

$$\mathrm{ber}(x) + i\,\mathrm{bei}(x) = J_0(xe^{3i\pi/4})$$

$$\mathrm{ber}'(x) + i\,\mathrm{bei}'(x) = e^{-i\pi/4}\left[\mathrm{ber}_1(x) + i\,\mathrm{bei}_1(x)\right] = e^{-i\pi/4} J_1(xe^{3i\pi/4})$$

TABLE 9.10 *(Continued)*

$$\ker(x) = -(\gamma+\ln x/2)\,\mathrm{ber}(x)+(\pi/4)\,\mathrm{bei}(x)-(x/8)^4 \sum_{n=0}^{\infty} e_n T_{2n}(x^2/64)$$

$$\mathrm{kei}(x) = -(\gamma+\ln x/2)\,\mathrm{bei}(x)-(\pi/4)\,\mathrm{ber}(x)+(x/8)^2 \sum_{n=0}^{\infty} f_n T_{2n}(x^2/64)$$

$$\ker'(x) = -(\gamma+\ln x/2)\,\mathrm{ber}'(x)-x^{-1}\mathrm{ber}(x)+(\pi/4)\,\mathrm{bei}'(x)-(x/8)^3 \sum_{n=0}^{\infty} g_n T_{2n}(x^2/64)$$

$$\mathrm{kei}'(x) = -(\gamma+\ln x/2)\,\mathrm{bei}'(x)-x^{-1}\mathrm{bei}(x)-(\pi/4)\,\mathrm{ber}'(x)+(x/8) \sum_{n=0}^{\infty} h_n T_{2n}(x^2/64)$$

$0 < x \le 8$

n	e_n	n	f_n
0	5.03749 13279 40243 09626	0	-34.11314 87924 14490 76243
1	-82.13362 54977 30465 74995	1	-33.37426 03178 96596 83482
2	8.61760 65894 62441 70191	2	15.96104 66759 83989 93950
3	-0.20943 20427 43605 42250	3	-0.76688 42692 52450 82998
4	0.00183 72709 07812 58606	4	0.01089 43699 35866 72405
5	-0.00000 72563 87142 14307	5	-0.00006 29825 20880 94100
6	0.00000 00148 11082 80775	6	0.00000 01765 55262 95937
7	-0.00000 00000 17194 29932	7	-0.00000 00002 69199 99368
8	0.00000 00000 00012 18762	8	0.00000 00000 00242 40280
9	-0.00000 00000 00000 00557	9	-0.00000 00000 00000 13718
10	0.00000 00000 00000 00000	10	0.00000 00000 00000 00005

n	g_n	n	h_n
0	-30.24095 21143 27190 84320	0	-10.38306 52726 52518 42353
1	-65.51939 55565 94624 78265	1	4.63400 89303 76357 74285
2	12.30542 66464 04620 11634	2	17.69391 91618 10812 91434
3	-0.41155 11950 45437 01403	3	-1.29878 38466 03610 97255
4	0.00455 69840 80054 95188	4	0.02419 84772 00411 21282
5	-0.00002 16804 15192 22712	5	-0.00017 21444 83306 79904
6	0.00000 00517 18527 18270	6	0.00000 05719 22142 65045
7	-0.00000 00000 68679 74648	7	-0.00000 00010 07561 98801
8	0.00000 00000 00054 79415	8	0.00000 00000 01028 97780
9	-0.00000 00000 00000 02784	9	-0.00000 00000 00000 65109
10	0.00000 00000 00000 00001	10	0.00000 00000 00000 00027

$$\ker(x)-i\,\mathrm{kei}(x) = K_0(xe^{-i\pi/4})$$

$$\ker'(x)-i\,\mathrm{kei}'(x) = e^{i\pi/4}\left[\ker_1(x)-i\,\mathrm{kei}_1(x)\right] = e^{3i\pi/4}K_1(xe^{-i\pi/4})$$

$$\mathrm{ber}(x)-i\,\mathrm{bei}(x) = (i/\pi)\left[\overline{K_0(xe^{-3i\pi/4})}-K_0(xe^{-i\pi/4})\right]$$

$$\mathrm{ber}'(x)-i\,\mathrm{bei}'(x) = \frac{e^{i\pi/4}}{\pi}\left[\overline{K_1(xe^{-3i\pi/4})}+K_1(xe^{-i\pi/4})\right]$$

TABLE 9.10 *(Continued)*

$$K_0(xe^{-i\pi/4}) = (\pi/2x)^{\frac{1}{2}} e^{i(\pi/8+u)} e^{-u} \sum_{n=0}^{\infty} p_n T_n^*(5/x)$$

$$K_1(xe^{-i\pi/4}) = (\pi/2x)^{\frac{1}{2}} e^{i(\pi/8+u)} e^{-u} \sum_{n=0}^{\infty} q_n T_n^*(5/x)$$

$$u = 2^{-\frac{1}{2}}x, \quad x \geq 5$$

$$p_n = R(p_n)+iI(p_n)$$

n	$R(p_n)$	n	$I(p_n)$
0	0.99125 27590 13757 71500	0	-0.00790 65568 61206 00284
1	-0.00870 45775 92248 60627	1	-0.00761 52111 06332 59121
2	0.00004 88558 00279 13133	2	0.00028 01428 53466 02342
3	0.00000 54968 42797 59179	3	-0.00001 08679 91455 73912
4	-0.00000 06378 02857 72398	4	0.00000 03430 71791 35686
5	0.00000 00537 88925 26933	5	0.00000 00047 78727 85761
6	-0.00000 00038 03665 36911	6	-0.00000 00028 77466 56739
7	0.00000 00001 80405 87493	7	0.00000 00004 46278 05555
8	0.00000 00000 07409 37643	8	-0.00000 00000 53663 29543
9	-0.00000 00000 04004 46940	9	0.00000 00000 05345 74325
10	0.00000 00000 00807 04345	10	-0.00000 00000 00376 17314
11	-0.00000 00000 00126 91627	11	-0.00000 00000 00005 44569
12	0.00000 00000 00016 78693	12	0.00000 00000 00009 39965
13	-0.00000 00000 00001 74350	13	-0.00000 00000 00002 53051
14	0.00000 00000 00000 07994	14	0.00000 00000 00000 50860
15	0.00000 00000 00000 02512	15	-0.00000 00000 00000 08650
16	-0.00000 00000 00000 01072	16	0.00000 00000 00000 01240
17	0.00000 00000 00000 00280	17	-0.00000 00000 00000 00130
18	-0.00000 00000 00000 00060	18	0.00000 00000 00000 00001
19	0.00000 00000 00000 00011	19	0.00000 00000 00000 00005
20	-0.00000 00000 00000 00002	20	-0.00000 00000 00000 00002

$$q_n = R(q_n)+iI(q_n)$$

n	$R(q_n)$	n	$I(q_n)$
0	1.02638 45771 27877 23444	0	0.02493 11563 84580 59657
1	0.02632 27379 40162 54838	1	0.02443 02125 32635 26335
2	-0.00007 11145 12850 82471	2	-0.00048 50015 54807 11200
3	-0.00000 83456 68118 73449	3	0.00001 55186 22052 84493
4	0.00000 08572 29001 62198	4	-0.00000 04354 08068 17209
5	-0.00000 00676 79783 56927	5	-0.00000 00075 27108 70232
6	0.00000 00045 60014 86316	6	0.00000 00036 02315 70219
7	-0.00000 00002 03852 67136	7	-0.00000 00005 34668 13801
8	-0.00000 00000 09832 00251	8	0.00000 00000 62439 08190
9	0.00000 00000 04715 61149	9	-0.00000 00000 06062 51798
10	-0.00000 00000 00924 49058	10	0.00000 00000 00411 47493
11	0.00000 00000 00142 72392	11	0.00000 00000 00008 52051
12	-0.00000 00000 00018 56911	12	-0.00000 00000 00010 79488
13	0.00000 00000 00001 88902	13	0.00000 00000 00002 83532
14	-0.00000 00000 00000 07963	14	-0.00000 00000 00000 56177
15	-0.00000 00000 00000 02892	15	0.00000 00000 00000 09441
16	0.00000 00000 00000 01189	16	-0.00000 00000 00000 01336
17	-0.00000 00000 00000 00307	17	0.00000 00000 00000 00138
18	0.00000 00000 00000 00065	18	0.00000 00000 00000 00000
19	-0.00000 00000 00000 00012	19	-0.00000 00000 00000 00005
20	0.00000 00000 00000 00002	20	0.00000 00000 00000 00002
		21	-0.00000 00000 00000 00001

TABLE 9.10 *(Continued)*

$$K_0(xe^{-3i\pi/4}) = (\pi/2x)^{\frac{1}{2}} e^{i(3\pi/8+u)} e^{u} \sum_{n=0}^{\infty} r_n T_n^*(5/x)$$

$$u = 2^{-\frac{1}{2}}x \ , \ x \geq 5$$

$$r_n = R(r_n) + iI(r_n)$$

n	$R(r_n)$	n	$I(r_n)$
0	1.00865 35386 89731 55708	0	-0.01000 16524 38338 83257
1	0.00855 53633 96689 45433	1	-0.01040 34198 75862 28268
2	-0.00012 20782 59092 27154	2	-0.00040 81988 35331 06358
3	-0.00002 52264 01820 40156	3	-0.00000 46809 27085 12980
4	-0.00000 11699 68331 07769	4	0.00000 19525 36816 30629
5	0.00000 01850 46432 39172	5	0.00000 01858 59426 18316
6	0.00000 00294 31144 49948	6	-0.00000 00215 16287 69336
7	-0.00000 00031 11401 06759	7	-0.00000 00049 37472 78376
8	-0.00000 00008 81098 29884	8	0.00000 00005 58401 43948
9	0.00000 00001 19962 57531	9	0.00000 00001 64741 32092
10	0.00000 00000 31414 27816	10	-0.00000 00000 29146 34909
11	-0.00000 00000 07594 76163	11	-0.00000 00000 05854 28040
12	-0.00000 00000 00981 30065	12	0.00000 00000 02039 93683
13	0.00000 00000 00548 50142	13	0.00000 00000 00112 49107
14	-0.00000 00000 00011 22097	14	-0.00000 00000 00143 88766
15	-0.00000 00000 00035 71498	15	0.00000 00000 00014 63097
16	0.00000 00000 00007 09864	16	0.00000 00000 00007 96481
17	0.00000 00000 00001 39760	17	-0.00000 00000 00002 69885
18	-0.00000 00000 00000 88635	18	-0.00000 00000 00000 07907
19	0.00000 00000 00000 08455	19	0.00000 00000 00000 25347
20	0.00000 00000 00000 05989	20	-0.00000 00000 00000 05618
21	-0.00000 00000 00000 02407	21	-0.00000 00000 00000 00925
22	0.00000 00000 00000 00084	22	0.00000 00000 00000 00822
23	0.00000 00000 00000 00225	23	-0.00000 00000 00000 00149
24	-0.00000 00000 00000 00079	24	-0.00000 00000 00000 00042
25	0.00000 00000 00000 00001	25	0.00000 00000 00000 00030
26	0.00000 00000 00000 00009	26	-0.00000 00000 00000 00006
27	-0.00000 00000 00000 00003	27	-0.00000 00000 00000 00002
		28	0.00000 00000 00000 00001

TABLE 9.10 *(Concluded)*

$$K_1(xe^{-3i\pi/4}) = (\pi/2x)^{\frac{1}{2}} e^{i(3\pi/8+u)} e^{u} \sum_{n=0}^{\infty} s_n T_n^*(5/x)$$

$$u = 2^{-\frac{1}{2}}x \ , \ x \geq 5$$

$$s_n = R(s_n)+iI(s_n)$$

n	$R(s_n)$	n	$I(s_n)$
0	0.97373 65178 99961 22053	0	0.02842 97638 80044 02130
1	-0.02612 98016 58743 37580	1	0.02908 87508 37542 66347
2	0.00016 59892 49190 36549	2	0.00066 92210 59207 47506
3	0.00003 40793 65284 60986	3	0.00000 81050 90636 93453
4	0.00000 15975 29404 14505	4	-0.00000 23807 46811 43423
5	-0.00000 02107 64430 98560	5	-0.00000 02345 76227 48777
6	-0.00000 00355 67189 76407	6	0.00000 00233 39348 00016
7	0.00000 00032 72161 51934	7	0.00000 00058 11070 54672
8	0.00000 00010 20552 39677	8	-0.00000 00005 79873 34960
9	-0.00000 00001 24683 08692	9	-0.00000 00001 89227 72015
10	-0.00000 00000 36026 08278	10	0.00000 00000 30503 17990
11	0.00000 00000 08012 42285	11	0.00000 00000 06758 81745
12	0.00000 00000 01159 54210	12	-0.00000 00000 02168 21058
13	-0.00000 00000 00587 01806	13	-0.00000 00000 00145 99461
14	0.00000 00000 00005 82342	14	0.00000 00000 00155 08873
15	0.00000 00000 00038 83581	15	-0.00000 00000 00014 15973
16	-0.00000 00000 00007 24642	16	-0.00000 00000 00008 78294
17	-0.00000 00000 00001 59248	17	0.00000 00000 00002 81413
18	0.00000 00000 00000 93671	18	0.00000 00000 00000 11786
19	-0.00000 00000 00000 07995	19	-0.00000 00000 00000 27123
20	-0.00000 00000 00000 06525	20	0.00000 00000 00000 05711
21	0.00000 00000 00000 02502	21	0.00000 00000 00000 01059
22	-0.00000 00000 00000 00060	22	-0.00000 00000 00000 00867
23	-0.00000 00000 00000 00241	23	0.00000 00000 00000 00149
24	0.00000 00000 00000 00082	24	0.00000 00000 00000 00046
25	0.00000 00000 00000 00000	25	-0.00000 00000 00000 00032
26	-0.00000 00000 00000 00009	26	0.00000 00000 00000 00006
27	0.00000 00000 00000 00003	27	0.00000 00000 00000 00002
		28	-0.00000 00000 00000 00001

TABLE 9.11. Chebyshev Coefficients for $x^{-\nu}J_\nu(x)$, $J_\nu(x)$, and $Y_\nu(x)$, $\nu = \pm\frac{1}{4}$

$$x^{-\nu}J_\nu(x) = \sum_{n=0}^{\infty} A_n(\nu)T_{2n}(x/8)\ , \quad -8 \le x \le 8$$

$$J_{\frac{1}{4}}(x)+iY_{\frac{1}{4}}(x) = (2/\pi x)^{\frac{1}{2}}e^{i(x-3\pi/8)} \sum_{n=0}^{\infty} b_n T_n^*(5/x)\ , \quad x \ge 5$$

n	$A_n(\nu)$, $\nu = -\frac{1}{4}$	n	$A_n(\nu)$, $\nu = +\frac{1}{4}$
0	0.16292 58079 76064 15480	0	0.14095 04424 34505 80450
1	0.10511 50518 33156 12343	1	-0.09889 25831 80256 71358
2	0.13242 24755 27353 25249	2	0.29237 16573 91930 84943
3	-0.47013 66584 22099 94613	3	-0.27196 83738 64697 27999
4	0.24169 85159 18686 62294	4	0.10031 55706 12324 83139
5	-0.05877 81532 27298 74785	5	-0.02032 93309 73457 26178
6	0.00866 64845 97492 60851	6	0.00264 33080 03301 67923
7	-0.00087 05421 69721 19155	7	-0.00024 10962 16786 09119
8	0.00006 38666 80734 66882	8	0.00001 63496 60758 29721
9	-0.00000 35864 44044 78645	9	-0.00000 08589 20905 24090
10	0.00000 01594 93491 95509	10	0.00000 00360 46120 29190
11	-0.00000 00057 64899 44734	11	-0.00000 00012 37640 33287
12	0.00000 00001 72875 43422	12	0.00000 00000 35438 30324
13	-0.00000 00000 04373 41020	13	-0.00000 00000 00859 63902
14	0.00000 00000 00094 63995	14	0.00000 00000 00017 89909
15	-0.00000 00000 00001 77249	15	-0.00000 00000 00000 32349
16	0.00000 00000 00000 02902	16	0.00000 00000 00000 00512
17	-0.00000 00000 00000 00042	17	-0.00000 00000 00000 00007
18	0.00000 00000 00000 00001		

$$b_n = R(b_n)+iI(b_n)$$

n	$R(b_n)$	n	$I(b_n)$
0	0.99926 16997 83419 16006	0	-0.00925 32994 85605 33084
1	-0.00097 66473 08281 74196	1	-0.00919 40369 61713 80152
2	-0.00023 27900 19041 64303	2	0.00006 97280 65099 73685
3	0.00000 61210 11436 83637	3	0.00000 98816 25563 33149
4	0.00000 04986 81089 81470	4	-0.00000 06109 30822 84656
5	-0.00000 00649 96102 30523	5	-0.00000 00196 87937 84503
6	0.00000 00010 23765 43683	6	0.00000 00069 01194 27121
7	0.00000 00006 62955 27042	7	-0.00000 00004 86456 16615
8	-0.00000 00000 99177 51648	8	-0.00000 00000 43469 67802
9	0.00000 00000 02274 72135	9	0.00000 00000 15468 44093
10	0.00000 00000 01809 21934	10	-0.00000 00000 01637 96188
11	-0.00000 00000 00397 82148	11	-0.00000 00000 00088 03327
12	0.00000 00000 00029 70029	12	0.00000 00000 00065 74118
13	0.00000 00000 00006 23361	13	-0.00000 00000 00011 92199
14	-0.00000 00000 00002 57380	14	0.00000 00000 00000 53128
15	0.00000 00000 00000 42190	15	0.00000 00000 00000 32944
16	-0.00000 00000 00000 00666	16	-0.00000 00000 00000 11304
17	-0.00000 00000 00000 01733	17	0.00000 00000 00000 01773
18	0.00000 00000 00000 00560	18	0.00000 00000 00000 00012
19	-0.00000 00000 00000 00089	19	-0.00000 00000 00000 00096
20	-0.00000 00000 00000 00001	20	0.00000 00000 00000 00031
21	0.00000 00000 00000 00006	21	-0.00000 00000 00000 00005
22	-0.00000 00000 00000 00002		

$$Y_{\frac{1}{4}}(x) = J_{\frac{1}{4}}(x)-2^{\frac{1}{2}}J_{-\frac{1}{4}}(x)\ , \quad Y_{-\frac{1}{4}}(x) = 2^{-\frac{1}{2}}\left[Y_{\frac{1}{4}}(x)+J_{\frac{1}{4}}(x)\right]$$

TABLE 9.12

Chebyshev Coefficients for $x^{-\nu}J_\nu(x)$, $J_\nu(x)$, and $Y_\nu(x)$, $\nu = \pm\frac{1}{3}$

$$x^{-\nu}J_\nu(x) = \sum_{n=0}^{\infty} A_n(\nu)T_{2n}(x/8) , \quad -8 \le x \le 8$$

$$J_{\frac{1}{3}}(x)+iY_{\frac{1}{3}}(x) = (2/\pi x)^{\frac{1}{2}}e^{i(x-5\pi/12)} \sum_{n=0}^{\infty} b_nT_n^*(5/x) , \quad x \ge 5$$

n	$A_n(\nu)$, $\nu = -\frac{1}{3}$	n	$A_n(\nu)$, $\nu = +\frac{1}{3}$
0	0.15921 18677 83189 60054	0	0.13437 97942 27756 44764
1	0.13935 64007 14896 26069	1	-0.11984 86974 10732 73261
2	0.05406 07319 77580 98869	2	0.28710 68687 28687 47721
3	-0.49889 25940 08176 73779	3	-0.24251 07291 97801 31136
4	0.27634 28099 41437 09958	4	0.08569 82591 68484 14668
5	-0.06962 17491 14124 72070	5	-0.01691 33949 03176 58537
6	0.01050 51479 21580 21916	6	0.00215 74934 76006 00765
7	-0.00107 36049 94092 01433	7	-0.00019 38490 54788 73423
8	0.00007 98632 63462 05385	8	0.00001 29832 45927 14860
9	-0.00000 45371 68415 38583	9	-0.00000 06748 56479 41628
10	0.00000 02038 09350 93453	10	0.00000 00280 59239 71246
11	-0.00000 00074 32215 74519	11	-0.00000 00009 55461 01654
12	0.00000 00002 24648 92145	12	0.00000 00000 27154 60828
13	-0.00000 00000 05724 18891	13	-0.00000 00000 00654 22144
14	0.00000 00000 00124 68853	14	0.00000 00000 00013 53678
15	-0.00000 00000 00002 34948	15	-0.00000 00000 00000 24324
16	0.00000 00000 00000 03868	16	0.00000 00000 00000 00383
17	-0.00000 00000 00000 00056	17	-0.00000 00000 00000 00005
18	0.00000 00000 00000 00001		

$$b_n = R(b_n)+iI(b_n)$$

n	$R(b_n)$	n	$I(b_n)$
0	0.99946 50123 35293 52860	0	-0.00685 69474 27212 69534
1	-0.00070 77638 36250 14260	1	-0.00681 43334 21755 53208
2	-0.00016 87952 26919 56340	2	0.00005 01456 42019 67809
3	0.00000 43855 75455 07034	3	0.00000 71142 29548 86394
4	0.00000 03580 42553 40564	4	-0.00000 04367 92366 55846
5	-0.00000 00464 10843 62258	5	-0.00000 00141 58283 43520
6	0.00000 00007 21026 33596	6	0.00000 00049 24417 94772
7	0.00000 00004 73021 33710	7	-0.00000 00003 45775 27763
8	-0.00000 00000 70556 54580	8	-0.00000 00000 31070 98997
9	0.00000 00000 01600 09976	9	0.00000 00000 11008 31850
10	0.00000 00000 01288 42696	10	-0.00000 00000 01162 71069
11	-0.00000 00000 00282 67591	11	-0.00000 00000 00063 01191
12	0.00000 00000 00021 03354	12	0.00000 00000 00046 74357
13	0.00000 00000 00004 44013	13	-0.00000 00000 00008 46207
14	-0.00000 00000 00001 82827	14	0.00000 00000 00000 37472
15	0.00000 00000 00000 29922	15	0.00000 00000 00000 23430
16	-0.00000 00000 00000 00462	16	-0.00000 00000 00000 08025
17	-0.00000 00000 00000 01232	17	0.00000 00000 00000 01257
18	0.00000 00000 00000 00398	18	0.00000 00000 00000 00009
19	-0.00000 00000 00000 00063	19	-0.00000 00000 00000 00068
20	-0.00000 00000 00000 00001	20	0.00000 00000 00000 00022
21	0.00000 00000 00000 00004	21	-0.00000 00000 00000 00004
22	-0.00000 00000 00000 00001		

$$Y_{\frac{1}{3}}(x) = 3^{-\frac{1}{2}}\left[J_{\frac{1}{3}}(x)-2J_{-\frac{1}{3}}(x)\right] , \quad Y_{-\frac{1}{3}}(x) = \tfrac{1}{2}\left[Y_{\frac{1}{3}}(x)+3^{\frac{1}{2}}J_{\frac{1}{3}}(x)\right]$$

TABLE 9.13

CHEBYSHEV COEFFICIENTS FOR $x^{-\nu}J_\nu(x)$ AND $Y_\nu(x)$, $\nu = \pm\frac{1}{2}$

$$x^{-\nu}J_\nu(x) = \sum_{n=0}^{\infty} A_n(\nu)T_{2n}(x/8), \quad -8 \le x \le 8$$

$$x^{\frac{1}{2}}J_{-\frac{1}{2}}(x) = (2/\pi)^{\frac{1}{2}}\cos x, \quad x^{-\frac{1}{2}}J_{\frac{1}{2}}(x) = (2/\pi)^{\frac{1}{2}}x^{-1}\sin x$$

n	$A_n(\nu)$, $\nu = -\frac{1}{2}$	n	$A_n(\nu)$, $\nu = \frac{1}{2}$
0	0.13695 75288 64404 54420	0	0.12075 45258 19019 13667
1	0.18030 86984 49846 86416	1	-0.14790 26923 59076 04910
2	-0.16812 61413 05732 38261	2	0.26404 76389 44269 13135
3	-0.53869 31975 99532 31563	3	-0.18993 42276 85509 14475
4	0.35658 25672 87920 10667	4	0.06203 76898 44444 51299
5	-0.09697 01449 38138 68753	5	-0.01164 29440 41549 09142
6	0.01535 73976 79715 15188	6	0.00143 13492 58107 83329
7	-0.00162 64975 12771 22330	7	-0.00012 48957 81762 72750
8	0.00012 44801 97174 49578	8	0.00000 81639 34433 01290
9	-0.00000 72417 50186 78414	9	-0.00000 04155 84588 23173
10	0.00000 03320 13004 88450	10	0.00000 00169 65472 88075
11	-0.00000 00123 26950 96296	11	-0.00000 00005 68332 12611
12	0.00000 00003 78635 88392	12	0.00000 00000 15915 30668
13	-0.00000 00000 09789 27737	13	-0.00000 00000 00378 30022
14	0.00000 00000 00216 09309	14	0.00000 00000 00007 73095
15	-0.00000 00000 00004 12203	15	-0.00000 00000 00000 13732
16	0.00000 00000 00000 06865	16	0.00000 00000 00000 00214
17	-0.00000 00000 00000 00101	17	-0.00000 00000 00000 00003
18	0.00000 00000 00000 00001		

$$Y_{\frac{1}{2}}(x) = -J_{-\frac{1}{2}}(x), \quad Y_{-\frac{1}{2}}(x) = J_{\frac{1}{2}}(x)$$

Further Chebyshev coefficients for cos x and x^{-1} sinx are given in Table 3.6.

TABLE 9.14. Chebyshev Coefficients for $x^{-\nu}J_\nu(x)$, $J_\nu(x)$, and $Y_\nu(x)$, $\nu = \pm\frac{2}{3}$

$$x^{-\nu}J_\nu(x) = \sum_{n=0}^{\infty} A_n(\nu)T_{2n}(x/8), \quad -8 \le x \le 8$$

$$J_{\frac{2}{3}}(x)+iY_{\frac{2}{3}}(x) = (2/\pi x)^{\frac{1}{2}}e^{i(x-7\pi/12)} \sum_{n=0}^{\infty} b_n T_n^*(5/x), \quad x \ge 5$$

n	$A_n(\nu)$, $\nu = -\frac{2}{3}$	n	$A_n(\nu)$, $\nu = +\frac{2}{3}$
0	0.08307 65189 70714 13774	0	0.10706 16423 86466 43961
1	0.14677 82421 85396 65543	1	-0.15964 55672 68641 34784
2	-0.49064 81271 48274 31500	2	0.23154 28348 62294 91180
3	-0.53708 96254 93875 54809	3	-0.14613 03842 74246 10067
4	0.45119 46685 89249 45197	4	0.04445 81121 09228 63513
5	-0.13366 52251 07965 02245	5	-0.00795 89793 03527 55087
6	0.02229 27277 97850 46819	6	0.00094 45189 12900 68469
7	-0.00245 08688 34375 83891	7	-0.00008 01183 62342 38651
8	0.00019 31741 98309 31795	8	0.00000 51145 43070 17215
9	-0.00001 15157 21499 94324	9	-0.00000 02550 98880 33266
10	0.00000 05391 20814 71979	10	0.00000 00102 28602 86309
11	-0.00000 00203 86970 98485	11	-0.00000 00003 37191 66450
12	0.00000 00006 36537 86340	12	0.00000 00000 09306 21345
13	-0.00000 00000 16702 19785	13	-0.00000 00000 00218 28231
14	0.00000 00000 00373 70183	14	0.00000 00000 00004 40647
15	-0.00000 00000 00007 21757	15	-0.00000 00000 00000 07739
16	0.00000 00000 00000 12159	16	0.00000 00000 00000 00119
17	-0.00000 00000 00000 00180	17	-0.00000 00000 00000 00002
18	0.00000 00000 00000 00002		

$$b_n = R(b_n)+iI(b_n)$$

n	$R(b_n)$	n	$I(b_n)$
0	1.00063 42623 22219 31510	0	0.00962 40948 68137 48329
1	0.00083 96170 22157 46942	1	0.00957 62502 42782 57395
2	0.00020 10007 10698 37296	2	-0.00005 63476 40295 43109
3	-0.00000 48012 97392 84074	3	-0.00000 80532 21333 51369
4	-0.00000 03975 54267 85494	4	0.00000 04712 25026 88650
5	0.00000 00496 32219 05089	5	0.00000 00158 78185 97396
6	-0.00000 00006 98219 51174	6	-0.00000 00052 40952 29525
7	-0.00000 00005 03120 14746	7	0.00000 00003 58294 24769
8	0.00000 00000 73554 41231	8	0.00000 00000 33441 00636
9	-0.00000 00000 01538 03494	9	-0.00000 00000 11503 44855
10	-0.00000 00000 01352 57091	10	0.00000 00000 01193 85552
11	0.00000 00000 00292 26229	11	0.00000 00000 00068 41269
12	-0.00000 00000 00021 24855	12	-0.00000 00000 00048 54549
13	-0.00000 00000 00004 66698	13	0.00000 00000 00008 68390
14	0.00000 00000 00001 88628	14	-0.00000 00000 00000 36775
15	-0.00000 00000 00000 30537	15	-0.00000 0000(00000 24374
16	0.00000 00000 00000 00400	16	0.00000 00000 00000 08242
17	0.00000 00000 00000 01274	17	-0.00000 00000 00000 01278
18	-0.00000 00000 00000 00407	18	-0.00000 00000 00000 00013
19	0.00000 00000 00000 00064	19	0.00000 00000 00000 00070
20	0.00000 00000 00000 00001	20	-0.00000 00000 00000 00023
21	-0.00000 00000 00000 00004	21	0.00000 00000 00000 00004
22	0.00000 00000 00000 00001		

$$Y_{\frac{2}{3}}(x) = -3^{-\frac{1}{2}}\left[J_{\frac{2}{3}}(x)+2J_{-\frac{2}{3}}(x)\right], \quad Y_{-\frac{2}{3}}(x) = \tfrac{1}{2}\left[-Y_{\frac{2}{3}}(x)+3^{\frac{1}{2}}J_{\frac{2}{3}}(x)\right]$$

TABLE 9.15. CHEBYSHEV COEFFICIENTS FOR $x^{-\nu}J_\nu(x)$, $J_\nu(x)$, AND $Y_\nu(x)$, $\nu = \pm\frac{3}{4}$

$$x^{-\nu}J_\nu(x) = \sum_{n=0}^{\infty} A_n(\nu)T_{2n}(x/8)\ , \quad -8 \le x \le 8$$

$$J_{\frac{3}{4}}(x)+iY_{\frac{3}{4}}(x) = (2/\pi x)^{\frac{1}{2}}e^{i(x-5\pi/8)} \sum_{n=0}^{\infty} b_n T_n^*(5/x)\ , \quad x \ge 5$$

n	$A_n(\nu)$, $\nu = -\frac{3}{4}$	n	$A_n(\nu)$, $\nu = +\frac{3}{4}$
0	0.03818 88739 23737 32877	0	0.10033 19891 39053 98255
1	0.08376 61801 83353 92342	1	-0.16051 88065 96379 52297
2	-0.69230 31627 37762 26562	2	0.21380 33597 57252 48053
3	-0.51214 95647 45616 04811	3	-0.12741 31877 14541 20984
4	0.50327 80857 82480 03641	4	0.03750 21128 48965 00073
5	-0.15627 56201 74437 04345	5	-0.00656 39297 55197 89837
6	0.02678 43130 08894 89412	6	0.00076 57670 81903 28333
7	-0.00300 22381 05763 86516	7	-0.00006 40663 07078 44199
8	0.00024 02372 95880 85029	8	0.00000 40426 83397 24072
9	-0.00001 45009 90789 33805	9	-0.00000 01996 26343 77317
10	0.00000 06861 46611 92642	10	0.00000 00079 33846 85596
11	-0.00000 00261 89586 90084	11	-0.00000 00002 59478 82930
12	0.00000 00008 24520 46257	12	0.00000 00000 07110 11587
13	-0.00000 00000 21797 17179	13	-0.00000 00000 00165 67797
14	0.00000 00000 00491 03745	14	0.00000 00000 00003 32430
15	-0.00000 00000 00009 54345	15	-0.00000 00000 00000 05805
16	0.00000 00000 00000 16171	16	0.00000 00000 00000 00089
17	-0.00000 00000 00000 00241	17	-0.00000 00000 00000 00001
18	0.00000 00000 00000 00003		

$$b_n = R(b_n)+iI(b_n)$$

n	$R(b_n)$	n	$I(b_n)$
0	1.00095 37439 52710 45606	0	0.01548 04303 74290 44633
1	0.00126 28060 52685 56358	1	0.01540 99141 20986 59178
2	0.00030 27053 90738 16892	2	-0.00008 30726 67787 65841
3	-0.00000 70122 06892 12271	3	-0.00001 19032 14813 89763
4	-0.00000 05834 75836 04373	4	0.00000 06845 99586 60370
5	0.00000 00718 80236 07645	5	0.00000 00233 76238 49484
6	-0.00000 00009 74117 68360	6	-0.00000 00075 77172 19576
7	-0.00000 00007 27214 53447	7	0.00000 00005 13083 45546
8	0.00000 00001 05563 41210	8	0.00000 00000 48531 37999
9	-0.00000 00000 02141 16158	9	-0.00000 00000 16523 48209
10	-0.00000 00000 01945 94119	10	0.00000 00000 01704 14810
11	0.00000 00000 00418 22305	11	0.00000 00000 00099 55759
12	-0.00000 00000 00030 15384	12	-0.00000 00000 00069 57786
13	-0.00000 00000 00006 71694	13	0.00000 00000 00012 39364
14	0.00000 00000 00002 69722	14	-0.00000 00000 00000 51632
15	-0.00000 00000 00000 43496	15	-0.00000 00000 00000 34954
16	0.00000 00000 00000 00533	16	0.00000 00000 00000 11768
17	0.00000 00000 00000 01824	17	-0.00000 00000 00000 01817
18	-0.00000 00000 00000 00581	18	-0.00000 00000 00000 00020
19	0.00000 00000 00000 00090	19	0.00000 00000 00000 00100
20	0.00000 00000 00000 00002	20	-0.00000 00000 00000 00032
21	-0.00000 00000 00000 00006	21	0.00000 00000 00000 00005
22	0.00000 00000 00000 00002		

$$Y_{\frac{3}{4}}(x) = -J_{\frac{3}{4}}(x)-2^{\frac{1}{2}}J_{-\frac{3}{4}}(x)\ , \quad Y_{-\frac{3}{4}}(x) = 2^{-\frac{1}{2}}\left[-Y_{\frac{3}{4}}(x)+J_{\frac{3}{4}}(x)\right]$$

TABLE 9.16

CHEBYSHEV COEFFICIENTS FOR INTEGRALS OF $J_\nu(x)$ AND $Y_\nu(x)$, $\nu = \pm\frac{1}{3}$

$$x^{-\nu-1}\int_0^x J_\nu(t)dt = \sum_{n=0}^{\infty} A_n(\nu)T_{2n}(x/8), \quad -8 \le x \le 8$$

n	$A_n(\nu)$, $\nu = -\frac{1}{3}$	n	$A_n(\nu)$, $\nu = +\frac{1}{3}$
0	0.49379 43365 93762 93657	0	0.20596 67464 51231 80193
1	-0.35949 38354 55150 73127	1	-0.24095 45570 05558 76587
2	0.31318 37689 85315 16662	2	0.14858 48589 54665 87865
3	-0.17035 70492 39821 19708	3	-0.05632 17821 06799 12091
4	0.04915 61260 43071 92861	4	0.01272 26120 62536 21429
5	-0.00857 73428 55470 88154	5	-0.00186 14682 89847 37708
6	0.00100 27729 49057 45264	6	0.00018 99142 72414 47005
7	-0.00008 42278 32929 16249	7	-0.00001 42774 55373 54148
8	0.00000 53395 62526 54229	8	0.00000 08242 28296 25590
9	-0.00000 02649 39051 82298	9	-0.00000 00377 11283 71224
10	0.00000 00105 80234 53666	10	0.00000 00014 01916 03435
11	-0.00000 00003 47653 76927	11	-0.00000 00000 43202 00019
12	0.00000 00000 09569 46158	12	0.00000 00000 01121 95792
13	-0.00000 00000 00223 95824	13	-0.00000 00000 00024 89612
14	0.00000 00000 00004 51252	14	0.00000 00000 00000 47759
15	-0.00000 00000 00000 07912	15	-0.00000 00000 00000 00800
16	0.00000 00000 00000 00122	16	0.00000 00000 00000 00012
17	-0.00000 00000 00000 00002		

$$\int_0^x Y_{\frac{1}{3}}(t)dt = 3^{-\frac{1}{2}}\left[\int_0^x J_{\frac{1}{3}}(t)dt - 2\int_0^x J_{-\frac{1}{3}}(t)dt\right]$$

$$\int_0^x Y_{-\frac{1}{3}}(t)dt = \frac{1}{2}\left[\int_0^x Y_{\frac{1}{3}}(t)dt + 3^{\frac{1}{2}}\int_0^x J_{\frac{1}{3}}(t)dt\right]$$

TABLE 9.16 *(Concluded)*

$$\int_x^\infty J_{\frac{1}{3}}(t)dt + i\int_x^\infty Y_{\frac{1}{3}}(t)dt = (2/\pi x)^{\frac{1}{2}} e^{i(x+\pi/12)} \sum_{n=0}^{\infty} b_n T_n^*(5/x)$$

$$x \geq 5$$

$$b_n = R(b_n) + iI(b_n)$$

n	$R(b_n)$	n	$I(b_n)$
0	0.98882 97080 37670 12116	0	-0.05279 76450 18858 06940
1	-0.01440 53104 12239 58488	1	-0.05091 36487 30090 05506
2	-0.00291 83264 76044 29914	2	0.00210 91347 29260 48518
3	0.00032 23047 73967 46914	3	0.00017 39472 10378 58509
4	-0.00000 16071 41227 96707	4	-0.00004 75608 24576 78148
5	-0.00000 59416 22057 21583	5	0.00000 42892 90391 25461
6	0.00000 12228 53452 96953	6	0.00000 03819 83428 35272
7	-0.00000 00923 77123 07165	7	-0.00000 02295 18964 55980
8	-0.00000 00233 04071 46982	8	0.00000 00462 66615 71955
9	0.00000 00111 32223 66035	9	-0.00000 00032 74424 16012
10	-0.00000 00024 09277 54885	10	-0.00000 00013 31618 98732
11	0.00000 00002 02063 67281	11	0.00000 00006 55903 01345
12	0.00000 00000 77236 27307	12	-0.00000 00001 61190 06899
13	-0.00000 00000 44613 05827	13	0.00000 00000 18729 44392
14	0.00000 00000 12906 73409	14	0.00000 00000 04157 33564
15	-0.00000 00000 02102 48123	15	-0.00000 00000 03309 50464
16	-0.00000 00000 00131 19321	16	0.00000 00000 01159 06808
17	0.00000 00000 00249 98027	17	-0.00000 00000 00253 72586
18	-0.00000 00000 00110 30909	18	0.00000 00000 00015 54586
19	0.00000 00000 00031 23915	19	0.00000 00000 00017 11719
20	-0.00000 00000 00004 94252	20	-0.00000 00000 00010 52435
21	-0.00000 00000 00000 69763	21	0.00000 00000 00003 80525
22	0.00000 00000 00000 93296	22	-0.00000 00000 00000 92034
23	-0.00000 00000 00000 44444	23	0.00000 00000 00000 07934
24	0.00000 00000 00000 14448	24	0.00000 00000 00000 06398
25	-0.00000 00000 00000 03022	25	-0.00000 00000 00000 04731
26	-0.00000 00000 00000 00019	26	0.00000 00000 00000 02023
27	0.00000 00000 00000 00405	27	-0.00000 00000 00000 00618
28	-0.00000 00000 00000 00249	28	0.00000 00000 00000 00113
29	0.00000 00000 00000 00101	29	0.00000 00000 00000 00013
30	-0.00000 00000 00000 00030	30	-0.00000 00000 00000 00024
31	0.00000 00000 00000 00005	31	0.00000 00000 00000 00014
32	0.00000 00000 00000 00001	32	-0.00000 00000 00000 00006
33	-0.00000 00000 00000 00001	33	0.00000 00000 00000 00002
34	0.00000 00000 00000 00001		

$$\int_0^\infty J_{\frac{1}{3}}(t)dt = 1, \quad \int_0^\infty Y_{\frac{1}{3}}(t)dt = -3^{-\frac{1}{2}}$$

TABLE 9.17

CHEBYSHEV COEFFICIENTS FOR $x^{-\nu}I_\nu(x)$ AND $K_\nu(x)$, $\nu = \pm\frac{1}{4}$

$$x^{-\nu}I_\nu(x) = \sum_{n=0}^{\infty} A_n(\nu)T_{2n}(x/8)\ ,\quad -8 \le x \le 8$$

n	$A_n(\nu)$, $\nu = -\frac{1}{4}$	n	$A_n(\nu)$, $\nu = +\frac{1}{4}$
0	209.59062 15375 52424 77370	0	77.21811 30087 15195 08688
1	317.00792 25021 81224 06653	1	113.42825 38420 28391 70394
2	141.45293 77868 21139 93217	2	47.64374 69803 57541 86212
3	39.56383 78741 05546 28468	3	12.42423 30269 63289 02498
4	7.38064 65080 42638 48929	4	2.16151 94052 25023 35004
5	0.96908 46487 05676 88986	5	0.26571 80340 11124 56938
6	0.09360 34898 73674 25397	6	0.02414 83312 33806 45829
7	0.00689 38664 96563 77869	7	0.00168 16199 50850 98591
8	0.00039 86253 25375 74699	8	0.00009 23567 02672 68623
9	0.00001 85368 70819 17986	9	0.00000 40958 08321 17864
10	0.00000 07072 31576 84242	10	0.00000 01495 62011 59572
11	0.00000 00225 13813 21018	11	0.00000 00045 71268 75757
12	0.00000 00006 06620 55308	12	0.00000 00001 18588 43261
13	0.00000 00000 14005 88481	13	0.00000 00000 02642 67000
14	0.00000 00000 00280 06939	14	0.00000 00000 00051 11578
15	0.00000 00000 00004 89596	15	0.00000 00000 00000 86603
16	0.00000 00000 00000 07544	16	0.00000 00000 00000 01296
17	0.00000 00000 00000 00103	17	0.00000 00000 00000 00017
18	0.00000 00000 00000 00001		

$$K_{\frac{1}{4}}(x) = 2^{-\frac{1}{2}}\pi\left[I_{-\frac{1}{4}}(x) - I_{\frac{1}{4}}(x)\right]$$

$$K_{-\frac{1}{4}}(x) = K_{\frac{1}{4}}(x)$$

TABLE 9.17 *(Concluded)*

$$I_{\frac{1}{4}}(x) = (2\pi x)^{-\frac{1}{2}} e^{x} \sum_{n=0}^{\infty} b_n T_n^*(8/x) \ , \ x \geq 8$$

$$K_{\frac{1}{4}}(x) = (\pi/2x)^{\frac{1}{2}} e^{-x} \sum_{n=0}^{\infty} c_n T_n^*(5/x) \ , \ x \geq 5$$

n	b_n	n	c_n
0	1.00619 92270 14122 57068	0	0.99128 81656 75147 07489
1	0.00631 99620 31140 71764	1	-0.00850 62567 20022 24415
2	0.00012 56131 27965 63848	2	0.00019 70491 57408 35126
3	0.00000 52052 40761 57340	3	-0.00000 80377 10166 53940
4	0.00000 03591 84411 38806	4	0.00000 04554 01498 42688
5	0.00000 00355 85362 88796	5	-0.00000 00323 27352 81652
6	0.00000 00036 05011 66436	6	0.00000 00027 16130 28432
7	-0.00000 00001 26294 10401	7	-0.00000 00002 60644 07112
8	-0.00000 00002 96595 12439	8	0.00000 00000 27882 69016
9	-0.00000 00001 18337 70164	9	-0.00000 00000 03267 69068
10	-0.00000 00000 21655 68249	10	0.00000 00000 00414 08700
11	0.00000 00000 03032 03737	11	-0.00000 00000 00056 16859
12	0.00000 00000 03041 10001	12	0.00000 00000 00008 09018
13	0.00000 00000 00530 77173	13	-0.00000 00000 00001 22930
14	-0.00000 00000 00204 52639	14	0.00000 00000 00000 19600
15	-0.00000 00000 00105 49334	15	-0.00000 00000 00000 03264
16	0.00000 00000 00005 50090	16	0.00000 00000 00000 00566
17	0.00000 00000 00014 35975	17	-0.00000 00000 00000 00102
18	0.00000 00000 00001 13866	18	0.00000 00000 00000 00019
19	-0.00000 00000 00001 87253	19	-0.00000 00000 00000 00004
20	-0.00000 00000 00000 31536	20	0.00000 00000 00000 00001
21	0.00000 00000 00000 26035		
22	0.00000 00000 00000 05770		
23	-0.00000 00000 00000 04034		
24	-0.00000 00000 00000 00914		
25	0.00000 00000 00000 00697		
26	0.00000 00000 00000 00125		
27	-0.00000 00000 00000 00130		
28	-0.00000 00000 00000 00012		
29	0.00000 00000 00000 00025		
30	-0.00000 00000 00000 00001		
31	-0.00000 00000 00000 00005		
32	0.00000 00000 00000 00001		
33	0.00000 00000 00000 00001		

TABLE 9.18

CHEBYSHEV COEFFICIENTS FOR $x^{-\nu}I_\nu(x)$ AND $K_\nu(x)$, $\nu = \pm\frac{1}{3}$

$$x^{-\nu}I_\nu(x) = \sum_{n=0}^{\infty} A_n(\nu)T_{2n}(x/8)\ ,\quad -8 \le x \le 8$$

n	$A_n(\nu)$, $\nu = -\frac{1}{3}$	n	$A_n(\nu)$, $\nu = +\frac{1}{3}$
0	246.76446 77610 21950 14476	0	65.17366 54884 16526 95482
1	374.90419 63977 61772 38356	1	95.23190 27040 92163 97860
2	168.94690 54665 51934 32342	2	39.59331 03435 21063 45067
3	47.81292 47970 09021 28986	3	10.20622 18597 97178 58861
4	9.02594 18911 35278 81180	4	1.75552 58978 34803 58155
5	1.19852 10167 30412 23176	5	0.21351 04236 00373 85749
6	0.11697 76116 93126 92166	6	0.01921 26868 62831 23642
7	0.00869 83095 97421 34176	7	0.00132 57975 71063 41284
8	0.00050 74091 61720 07577	8	0.00007 22075 54088 40446
9	0.00002 37873 65337 43276	9	0.00000 31776 13642 90919
10	0.00000 09143 62383 07318	10	0.00000 01152 07184 65074
11	0.00000 00293 09947 86363	11	0.00000 00034 97935 67902
12	0.00000 00007 94846 32582	12	0.00000 00000 90183 77386
13	0.00000 00000 18462 55797	13	0.00000 00000 01998 07607
14	0.00000 00000 00371 27729	14	0.00000 00000 00038 43793
15	0.00000 00000 00006 52493	15	0.00000 00000 00000 64790
16	0.00000 00000 00000 10105	16	0.00000 00000 00000 00965
17	0.00000 00000 00000 00139	17	0.00000 00000 00000 00013
18	0.00000 00000 00000 00002		

$$K_{\frac{1}{3}}(x) = 3^{-\frac{1}{2}}\pi\left[I_{-\frac{1}{3}}(x) - I_{\frac{1}{3}}(x)\right]$$

$$K_{-\frac{1}{3}}(x) = K_{\frac{1}{3}}(x)$$

TABLE 9.18 *(Concluded)*

$$I_{\frac{1}{3}}(x) = (2\pi x)^{-\frac{1}{2}} e^{x} \sum_{n=0}^{\infty} b_n T_n^*(8/x) \, , \quad x \geq 8$$

$$K_{\frac{1}{3}}(x) = (\pi/2x)^{\frac{1}{2}} e^{-x} \sum_{n=0}^{\infty} c_n T_n^*(5/x) \, , \quad x \geq 5$$

n	b_n	n	c_n
0	1.00458 61710 93207 34833	0	0.99353 64122 76093 38920
1	0.00467 34791 99873 59910	1	-0.00631 44392 60798 63137
2	0.00009 08034 04815 03519	2	0.00014 30095 80961 13131
3	0.00000 37262 16110 59392	3	-0.00000 57870 60592 02472
4	0.00000 02520 73237 89921	4	0.00000 03265 50333 19976
5	0.00000 00227 82110 77259	5	-0.00000 00231 23231 95077
6	0.00000 00012 91332 27669	6	0.00000 00019 39555 14434
7	-0.00000 00006 11915 15648	7	-0.00000 00001 85897 88507
8	-0.00000 00003 75616 85308	8	0.00000 00000 19868 42439
9	-0.00000 00001 16415 45893	9	-0.00000 00000 02326 78966
10	-0.00000 00000 14443 25071	10	0.00000 00000 00294 68313
11	0.00000 00000 05373 68740	11	-0.00000 00000 00039 95293
12	0.00000 00000 03074 27194	12	0.00000 00000 00005 75225
13	0.00000 00000 00297 65801	13	-0.00000 00000 00000 87375
14	-0.00000 00000 00265 19963	14	0.00000 00000 00000 13927
15	-0.00000 00000 00091 36476	15	-0.00000 00000 00000 02319
16	0.00000 00000 00015 52212	16	0.00000 00000 00000 00402
17	0.00000 00000 00014 12177	17	-0.00000 00000 00000 00072
18	-0.00000 00000 00000 22965	18	0.00000 00000 00000 00013
19	-0.00000 00000 00001 98300	19	-0.00000 00000 00000 00003
20	-0.00000 00000 00000 12744	20	0.00000 00000 00000 00001
21	0.00000 00000 00000 28761		
22	0.00000 00000 00000 02981		
23	-0.00000 00000 00000 04505		
24	-0.00000 00000 00000 00457		
25	0.00000 00000 00000 00764		
26	0.00000 00000 00000 00044		
27	-0.00000 00000 00000 00137		
28	0.00000 00000 00000 00003		
29	0.00000 00000 00000 00025		
30	-0.00000 00000 00000 00003		
31	-0.00000 00000 00000 00004		
32	0.00000 00000 00000 00001		
33	0.00000 00000 00000 00001		

TABLE 9.19

CHEBYSHEV COEFFICIENTS FOR $x^{-\nu}I_\nu(x)$ AND $K_\nu(x)$, $\nu = \pm\frac{1}{2}$

$$x^{-\nu}I_\nu(x) = \sum_{n=0}^{\infty} A_n(\nu)T_{2n}(x/8), \quad -8 \le x \le 8$$

$$x^{\frac{1}{2}}I_{-\frac{1}{2}}(x) = (2/\pi)^{\frac{1}{2}}\cosh x, \quad x^{-\frac{1}{2}}I_{\frac{1}{2}}(x) = (2/\pi)^{\frac{1}{2}}\, x^{-1}\sinh x$$

n	$A_n(\nu)$, $\nu = -\frac{1}{2}$	n	$A_n(\nu)$, $\nu = \frac{1}{2}$
0	341.14680 66877 57709 35762	0	46.30161 21376 63508 58533
1	522.76731 23162 06884 75983	1	66.92307 67839 81516 78476
2	240.22615 12568 58864 67852	2	27.25731 02358 01156 57567
3	69.60690 52290 19078 72448	3	6.86653 89697 66800 61513
4	13.45808 70162 18460 04060	4	1.15472 07749 19002 05399
5	1.82814 08870 29306 20115	5	0.13749 54616 57570 59483
6	0.18223 08830 13487 19697	6	0.01213 27205 25685 67827
7	0.01381 56240 12655 81706	7	0.00082 22993 97455 19710
8	0.00082 03903 31650 28533	8	0.00004 40495 14611 83834
9	0.00003 90944 64565 53339	9	0.00000 19090 65804 91177
10	0.00000 15256 16735 55748	10	0.00000 00682 41975 61328
11	0.00000 00495 92652 91520	11	0.00000 00020 44885 46493
12	0.00000 00013 62496 55968	12	0.00000 00000 52077 85488
13	0.00000 00000 32034 55257	13	0.00000 00000 01140 62540
14	0.00000 00000 00651 57899	14	0.00000 00000 00021 70695
15	0.00000 00000 00011 57419	15	0.00000 00000 00000 36218
16	0.00000 00000 00000 18106	16	0.00000 00000 00000 00534
17	0.00000 00000 00000 00251	17	0.00000 00000 00000 00007
18	0.00000 00000 00000 00003		

$$K_{\frac{1}{2}}(x) = (\pi/2)\left[I_{-\frac{1}{2}}(x) - I_{\frac{1}{2}}(x)\right] = (\pi/2x)^{\frac{1}{2}}e^{-x}$$

$$K_{-\frac{1}{2}}(x) = K_{\frac{1}{2}}(x)$$

Further Chebyshev coefficients for e^x and e^{-x} are given in Table 3.3.

TABLE 9.20

CHEBYSHEV COEFFICIENTS FOR $x^{-\nu}I_\nu(x)$ AND $K_\nu(x)$, $\nu = \pm\frac{2}{3}$

$$x^{-\nu}I_\nu(x) = \sum_{n=0}^{\infty} A_n(\nu)T_{2n}(x/8)\ , \quad -8 \le x \le 8$$

n	$A_n(\nu)$, $\nu = -\frac{2}{3}$	n	$A_n(\nu)$, $\nu = +\frac{2}{3}$
0	469.95312 79949 20422 45819	0	32.77479 24030 35026 33440
1	726.03216 50664 95923 67181	1	46.83793 90742 83086 52136
2	340.10272 33771 79941 55999	2	18.68656 02563 49831 12085
3	100.90075 85218 30026 31764	3	4.60121 27076 45102 94050
4	19.98542 28723 93609 46223	4	0.75671 34574 71448 13715
5	2.77810 11375 55155 75221	5	0.08824 17620 77463 53461
6	0.28291 28347 05747 25953	6	0.00763 78220 14940 49230
7	0.02187 48334 95706 09367	7	0.00050 85533 07618 34336
8	0.00132 26199 57420 50365	8	0.00002 68011 17196 81826
9	0.00006 40820 86136 66998	9	0.00000 11441 45898 26759
10	0.00000 25392 92565 78414	10	0.00000 00403 31192 20640
11	0.00000 00837 21455 18318	11	0.00000 00011 92919 42611
12	0.00000 00023 30625 06863	12	0.00000 00000 30013 89391
13	0.00000 00000 55474 05288	13	0.00000 00000 00649 93739
14	0.00000 00000 01141 38753	14	0.00000 00000 00012 23720
15	0.00000 00000 00020 49509	15	0.00000 00000 00000 20213
16	0.00000 00000 00000 32389	16	0.00000 00000 00000 00295
17	0.00000 00000 00000 00454	17	0.00000 00000 00000 00004
18	0.00000 00000 00000 00006		

$$K_{\frac{2}{3}}(x) = 3^{-\frac{1}{2}}\pi\left[I_{-\frac{2}{3}}(x) - I_{\frac{2}{3}}(x)\right]$$

$$K_{-\frac{2}{3}}(x) = K_{\frac{2}{3}}(x)$$

TABLE 9.20 *(Concluded)*

$$I_{\frac{2}{3}}(x) = (2\pi x)^{-\frac{1}{2}} e^{x} \sum_{n=0}^{\infty} b_n T_n^*(8/x)\ ,\quad x \geq 8$$

$$K_{\frac{2}{3}}(x) = (\pi/2x)^{\frac{1}{2}} e^{-x} \sum_{n=0}^{\infty} c_n T_n^*(5/x)\ ,\quad x \geq 5$$

n	b_n	n	c_n
0	0.99363 49867 16925 14075	0	1.00914 95380 72789 40218
1	-0.00646 71526 00616 03301	1	0.00897 12068 42483 59755
2	-0.00010 60188 22351 54487	2	-0.00017 13895 98261 53943
3	-0.00000 41406 57716 23469	3	0.00000 65547 92549 82352
4	-0.00000 02916 95418 20778	4	-0.00000 03595 19190 48499
5	-0.00000 00365 71574 33200	5	0.00000 00250 24412 18493
6	-0.00000 00075 81590 37399	6	-0.00000 00020 74924 13355
7	-0.00000 00019 23008 52343	7	0.00000 00001 97223 66561
8	-0.00000 00004 20438 79538	8	-0.00000 00000 20946 47303
9	-0.00000 00000 39372 03510	9	0.00000 00000 02440 93253
10	0.00000 00000 19007 44203	10	-0.00000 00000 00307 90652
11	0.00000 00000 10137 63568	11	0.00000 00000 00041 60827
12	0.00000 00000 01331 29494	12	-0.00000 00000 00005 97399
13	-0.00000 00000 00676 92205	13	0.00000 00000 00000 90528
14	-0.00000 00000 00311 72156	14	-0.00000 00000 00000 14400
15	0.00000 00000 00011 86909	15	0.00000 00000 00000 02393
16	0.00000 00000 00040 21108	16	-0.00000 00000 00000 00414
17	0.00000 00000 00004 78382	17	0.00000 00000 00000 00074
18	-0.00000 00000 00004 73482	18	-0.00000 00000 00000 00014
19	-0.00000 00000 00001 15940	19	0.00000 00000 00000 00003
20	0.00000 00000 00000 58663	20	-0.00000 00000 00000 00001
21	0.00000 00000 00000 20610		
22	-0.00000 00000 00000 08266		
23	-0.00000 00000 00000 03374		
24	0.00000 00000 00000 01364		
25	0.00000 00000 00000 00527		
26	-0.00000 00000 00000 00257		
27	-0.00000 00000 00000 00076		
28	0.00000 00000 00000 00052		
29	0.00000 00000 00000 00009		
30	-0.00000 00000 00000 00011		
31	0.00000 00000 00000 00000		
32	0.00000 00000 00000 00002		

TABLE 9.21

CHEBYSHEV COEFFICIENTS FOR $x^{-\nu}I_\nu(x)$ AND $K_\nu(x)$, $\nu = \pm\frac{3}{4}$

$$x^{-\nu}I_\nu(x) = \sum_{n=0}^{\infty} A_n(\nu)T_{2n}(x/8)\ ,\quad -8 \le x \le 8$$

n	$A_n(\nu)$, $\nu = -\frac{3}{4}$	n	$A_n(\nu)$, $\nu = +\frac{3}{4}$
0	550.85016 30341 70284 56795	0	27.53708 02195 73353 85694
1	854.33983 22965 82786 49311	1	39.12415 98284 07127 82013
2	404.01608 32602 60410 71583	2	15.44817 01263 74054 70556
3	121.28594 02003 53382 69889	3	3.76092 53817 01842 83709
4	24.31702 99859 04328 91121	4	0.61173 09235 84474 84880
5	3.41979 37183 85947 02679	5	0.07060 21276 80523 59245
6	0.35204 85520 80364 77796	6	0.00605 30181 43426 79264
7	0.02749 25423 21295 73950	7	0.00039 95102 75267 90197
8	0.00167 75241 42529 80727	8	0.00002 08849 58933 48629
9	0.00008 19619 02809 85823	9	0.00000 08849 49892 15498
10	0.00000 32729 89641 93685	10	0.00000 00309 79261 90894
11	0.00000 01086 86125 93681	11	0.00000 00009 10421 10195
12	0.00000 00030 45750 72505	12	0.00000 00000 22768 74494
13	0.00000 00000 72946 04113	13	0.00000 00000 00490 27234
14	0.00000 00000 01509 60227	14	0.00000 00000 00009 18212
15	0.00000 00000 00027 25484	15	0.00000 00000 00000 15091
16	0.00000 00000 00000 43293	16	0.00000 00000 00000 00219
17	0.00000 00000 00000 00610	17	0.00000 00000 00000 00003
18	0.00000 00000 00000 00008		

$$K_{\frac{3}{4}}(x) = 2^{-\frac{1}{2}}\pi\left[I_{-\frac{3}{4}}(x) - I_{\frac{3}{4}}(x)\right]$$

$$K_{-\frac{3}{4}}(x) = K_{\frac{3}{4}}(x)$$

TABLE 9.21 *(Concluded)*

$$I_{\frac{3}{4}}(x) = (2\pi x)^{-\frac{1}{2}} e^{x} \sum_{n=0}^{\infty} b_n T_n^*(8/x) , \quad x \geq 8$$

$$K_{\frac{3}{4}}(x) = (\pi/2x)^{\frac{1}{2}} e^{-x} \sum_{n=0}^{\infty} c_n T_n^*(5/x) , \quad x \geq 5$$

n	b_n	n	c_n
0	0.98980 19115 24008 91053	0	1.01476 24350 64637 87104
1	-0.01035 09365 14827 02366	1	0.01449 34617 87809 66495
2	-0.00015 85263 84973 08076	2	-0.00025 87162 07241 80365
3	-0.00000 60527 21962 69398	3	0.00000 96912 18911 49213
4	-0.00000 04158 38597 31055	4	-0.00000 05261 29313 98850
5	-0.00000 00487 99346 56591	5	0.00000 00363 96854 28973
6	-0.00000 00089 86835 43794	6	-0.00000 00030 05472 75589
7	-0.00000 00019 83283 58282	7	0.00000 00002 84827 79992
8	-0.00000 00003 58969 60092	8	-0.00000 00000 30182 90699
9	-0.00000 00000 08766 61846	9	0.00000 00000 03511 09500
10	0.00000 00000 25819 44847	10	-0.00000 00000 00442 27227
11	0.00000 00000 09780 23878	11	0.00000 00000 00059 69557
12	0.00000 00000 00565 05071	12	-0.00000 00000 00008 56248
13	-0.00000 00000 00851 65936	13	0.00000 00000 00001 29645
14	-0.00000 00000 00270 24995	14	-0.00000 00000 00000 20607
15	0.00000 00000 00040 96038	15	0.00000 00000 00000 03423
16	0.00000 00000 00040 49626	16	-0.00000 00000 00000 00592
17	0.00000 00000 00001 11252	17	0.00000 00000 00000 00106
18	-0.00000 00000 00005 24821	18	-0.00000 00000 00000 00020
19	-0.00000 00000 00000 70416	19	0.00000 00000 00000 00004
20	0.00000 00000 00000 69690	20	-0.00000 00000 00000 00001
21	0.00000 00000 00000 14472		
22	-0.00000 00000 00000 10173		
23	-0.00000 00000 00000 02430		
24	0.00000 00000 00000 01667		
25	0.00000 00000 00000 00362		
26	-0.00000 00000 00000 00301		
27	-0.00000 00000 00000 00044		
28	0.00000 00000 00000 00058		
29	0.00000 00000 00000 00002		
30	-0.00000 00000 00000 00011		
31	0.00000 00000 00000 00001		
32	0.00000 00000 00000 00002		
33	-0.00000 00000 00000 00001		

TABLE 9.22

Chebyshev Coefficients for Integrals of $I_\nu(x)$ and $K_\nu(x)$, $\nu = \pm\frac{1}{3}$

$$x^{-\nu-1}\int_0^x I_\nu(t)dt = \sum_{n=0}^{\infty} A_n(\nu)T_{2n}(x/8)\ ,\quad -8 \le x \le 8$$

n	$A_n(\nu)$, $\nu = -\frac{1}{3}$	n	$A_n(\nu)$, $\nu = +\frac{1}{3}$
0	37.13171 10624 42210 88716	0	9.91967 05951 33311 97436
1	51.83684 32807 69385 04232	1	12.99446 00285 78086 99911
2	20.31771 26547 47443 38419	2	4.62139 70994 91553 94909
3	4.93462 28026 35283 23263	3	1.01563 65613 79056 52862
4	0.80311 33029 50707 39916	4	0.15040 41767 77394 06957
5	0.09290 41695 17721 69149	5	0.01595 11335 74471 35477
6	0.00799 10820 74845 93537	6	0.00126 73333 96268 80628
7	0.00052 94196 86074 27989	7	0.00007 80876 84856 61104
8	0.00002 77877 19996 42354	8	0.00000 38349 21534 56630
9	0.00000 11822 95948 59139	9	0.00000 01534 78030 78076
10	0.00000 00415 59236 93036	10	0.00000 00050 98099 81234
11	0.00000 00012 26322 12416	11	0.00000 00001 42731 59557
12	0.00000 00000 30791 43101	12	0.00000 00000 03412 41214
13	0.00000 00000 00665 59749	13	0.00000 00000 00070 45636
14	0.00000 00000 00012 51272	14	0.00000 00000 00001 26866
15	0.00000 00000 00000 20640	15	0.00000 00000 00000 02009
16	0.00000 00000 00000 00301	16	0.00000 00000 00000 00028
17	0.00000 00000 00000 00004		

$$\int_0^x K_{\frac{1}{3}}(t)dt = 3^{-\frac{1}{2}}\pi\left[\int_0^x I_{-\frac{1}{3}}(t)dt - \int_0^x I_{\frac{1}{3}}(t)dt\right]$$

$$\int_0^x K_{-\frac{1}{3}}(t)dt = \int_0^x K_{\frac{1}{3}}(t)dt$$

TABLE 9.22 *(Concluded)*

$$\int_0^x I_{\frac{1}{3}}(t)dt = (2\pi x)^{-\frac{1}{2}} e^x \sum_{n=0}^{\infty} b_n T_n^*(8/x) , \quad x \geq 8$$

$$\int_x^{\infty} K_{\frac{1}{3}}(t)dt = (\pi/2x)^{\frac{1}{2}} e^{-x} \sum_{n=0}^{\infty} c_n T_n^*(5/x) , \quad x \geq 5$$

n	b_n	n	c_n
0	1.04314 26375 97623	0	0.95277 51681 62171 87270
1	0.04614 81441 17153	1	-0.04445 34999 08360 31283
2	0.00332 37687 31804	2	0.00253 35888 40117 18913
3	0.00031 70067 42678	3	-0.00021 19435 87680 21300
4	-0.00002 61018 94776	4	0.00002 24934 77764 46526
5	-0.00003 38911 70760	5	-0.00000 28253 15829 60541
6	-0.00000 90742 58771	6	0.00000 04034 11595 60099
7	0.00000 10372 78957	7	-0.00000 00637 94575 69250
8	0.00000 12507 86213	8	0.00000 00109 72930 75016
9	0.00000 00544 75035	9	-0.00000 00020 26007 52732
10	-0.00000 01660 20071	10	0.00000 00003 97563 42317
11	-0.00000 00164 80732	11	-0.00000 00000 82270 07660
12	0.00000 00258 34647	12	0.00000 00000 17842 37550
13	0.00000 00019 21568	13	-0.00000 00000 04035 02151
14	-0.00000 00046 13844	14	0.00000 00000 00947 56173
15	0.00000 00001 21014	15	-0.00000 00000 00230 25871
16	0.00000 00008 54979	16	0.00000 00000 00057 72709
17	-0.00000 00001 63658	17	-0.00000 00000 00014 89326
18	-0.00000 00001 41781	18	0.00000 00000 00003 94536
19	0.00000 00000 65190	19	-0.00000 00000 00001 07110
20	0.00000 00000 14602	20	0.00000 00000 00000 29750
21	-0.00000 00000 18313	21	-0.00000 00000 00000 08441
22	0.00000 00000 02011	22	0.00000 00000 00000 02443
23	0.00000 00000 03621	23	-0.00000 00000 00000 00721
24	-0.00000 00000 01682	24	0.00000 00000 00000 00216
25	-0.00000 00000 00284	25	-0.00000 00000 00000 00066
26	0.00000 00000 00521	26	0.00000 00000 00000 00020
27	-0.00000 00000 00131	27	-0.00000 00000 00000 00006
28	-0.00000 00000 00080	28	0.00000 00000 00000 00002
29	0.00000 00000 00068	29	-0.00000 00000 00000 00001
30	-0.00000 00000 00009		
31	-0.00000 00000 00014		
32	0.00000 00000 00009		
33	0.00000 00000 00000		
34	-0.00000 00000 00002		
35	0.00000 00000 00001		

$$\int_0^{\infty} K_{\frac{1}{3}}(t)dt = 3^{-\frac{1}{2}}\pi$$

9.8. Expansions in Series of Bessel Functions

In addition to the following, see 9.7(4-7).

$$J_\nu(\lambda z) = \lambda^\nu \sum_{n=0}^{\infty} \frac{(1-\lambda^2)^n (z/2)^n}{n!} J_{\nu+n}(z). \tag{1}$$

$$Y_\nu(\lambda z) = \lambda^\nu \sum_{n=0}^{\infty} \frac{(1-\lambda^2)^n (z/2)^n}{n!} Y_{\nu+n}(z), \qquad |1-\lambda^2| < 1. \tag{2}$$

$$J_\nu(z) = \frac{\Gamma(a)}{\Gamma(\nu+1)} \left(\frac{z}{2}\right)^{\nu+1-a} \sum_{n=0}^{\infty} \frac{(\nu+1-a)_n}{n!\,(\nu+1)_n} \left(\frac{z}{2}\right)^n J_{n+a-1}(z),$$

$$(a-1) \text{ is not a negative integer.} \tag{3}$$

$$J_\nu(z) = \frac{\Gamma(a)}{\Gamma(\nu+1)} \left(\frac{z}{2}\right)^{\nu-a} \sum_{n=0}^{\infty} \frac{(-)^n (2n+a)(a-\nu)_n (a)_n}{n!\,(\nu+1)_n} J_{2n+a}(z),$$

$$a \text{ is not a negative integer.} \tag{4}$$

If $a = 0$, (4) becomes

$$J_\nu(z) = \frac{(z/2)^\nu}{\Gamma(\nu+1)} \sum_{n=0}^{\infty} \frac{(-)^n \epsilon_n (-\nu)_n}{(\nu+1)_n} J_{2n}(z). \tag{5}$$

$$\frac{\partial J_\nu(z)}{\partial \nu} = [\ln z/2 - \psi(\nu+1)] J_\nu(z) + \frac{\Gamma(a)(z/2)^{\nu+1-a}}{\Gamma(\nu+1)} \sum_{n=1}^{\infty} \frac{(\nu+1-a)_n}{n!\,(\nu+1)_n}$$

$$\times \{\psi(\nu+1+n-a) - \psi(\nu+1-a) - \psi(n+\nu+1)$$

$$+ \psi(\nu+1)\} (z/2)^n J_{n+a-1}(z),$$

$$(a-1) \text{ not a negative integer.} \tag{6}$$

$$\frac{\partial J_\nu(z)}{\partial \nu} = [\ln z/2 - \psi(\nu+1)] J_\nu(z) + \frac{(z/2)^{\nu+1}}{\Gamma(\nu+1)} \sum_{n=0}^{\infty} \frac{(z/2)^n J_{n+1}(z)}{n!\,(\nu+n+1)^2}. \tag{7}$$

$$\frac{\pi}{2} Y_m(z) = [\ln z/2 - \psi(m+1)] J_m(z) - \frac{m!}{2} (z/2)^{-m} \sum_{n=0}^{m-1} \frac{(z/2)^n J_n(z)}{n!\,(m-n)}$$

$$+ m! \sum_{n=1}^{\infty} \frac{(z/2)^n J_{m+n}(z)}{(n+m)!\,n}. \tag{8}$$

$$K_m(z) = (-)^{m+1}[\ln z/2 - \psi(m+1)]I_m(z) + (m!/2)(z/2)^{-m}$$
$$\times \sum_{n=0}^{m-1} \frac{(-)^n(z/2)^n I_n(z)}{n!(m-n)} + (-)^{m+1}m! \sum_{n=1}^{\infty} \frac{(-)^n(z/2)^n I_{n+m}(z)}{(n+m)!n}. \tag{9}$$

An alternative expansion for the third expression in (8) and (9) follows from

$$m! \sum_{n=1}^{\infty} \frac{(z/2)^n J_{n+m}(z)}{(n+m)!n} = -\sum_{n=1}^{\infty} \frac{(-)^n(2n+m)}{n(n+m)} J_{2n+m}(z). \tag{10}$$

Further expansions are given by Luke (1969). See this same reference for expansions of some integrals involving Bessel functions in series of Bessel functions.

9.9. Rational Approximations

9.9.1. Introduction

In the following, we consider rational approximations for the Bessel functions for both small and large z. By z small, we mean $|z| < r$, r fixed and finite. By z large, we mean $R < |z| < \infty$, R fixed and finite with z restricted to some sector of the complex plane. The approximations are special cases or extensions of the results in 5.12 and apply to the integral portions of the respective functions. Certain rational approximations for $I_\nu(z)$ and $I_{\nu+1}(z)/I_\nu(z)$ are taken up in 9.9.2. These approximations are intimately connected with certain procedures for evaluation of the Bessel functions by use of the backward recurrence formula. The latter schema are presented in 9.10. The development of rational approximations for $K_m(z)$ and $Y_m(z)$, z small and m an integer along with coefficients for $m = 0,1$ are presented in a recent paper by Luke (1973) and will not be given here.

9.9.2. $I_\nu(z)$, z Small

The modified Bessel functions can be represented in terms of hypergeometric functions in two different ways. Thus

$$I_\nu(z) = U_\nu(z)\,{}_0F_1(\nu+1;\ z^2/4), \tag{1}$$

$$I_\nu(z) = U_\nu(z)e^{z}\,{}_1F_1(\nu+\tfrac{1}{2};\ 2\nu+1;\ -2z),$$

$$U_\nu(z) = (z/2)^\nu/\Gamma(\nu+1), \tag{2}$$

which we refer to as Cases I and II respectively. Results for the Bessel function $J_\nu(z)$ are easily achieved since

$$J_\nu(z) = e^{i\nu\pi/2} I_\nu(ze^{-i\pi/2}). \tag{3}$$

Thus in the approximations for ${}_0F_1$ given ahead, to get the corresponding forms for $J_\nu(z)$, we need only replace z^2 by $-z^2$, but in the approximations for ${}_1F_1$ given ahead, we must replace z by $-iz$. If z in $J_\nu(z)$ is real and if ν is real, $J_\nu(z)$ is also real. In this event, the $J_\nu(z)$ representation in terms of the ${}_1F_1$ is not convenient as complex arithmetic is required.

CASE I. In 5.12(1-11), let

$$\begin{gathered} p = 0,\ q = 1,\ \rho_1 = \nu + 1,\ f = g = 0,\ a = 0, \\ \alpha = 1 - \delta,\ \beta = \nu,\ \lambda = \nu + 2 - \delta,\ \delta = 0 \text{ or } \delta = 1, \\ \gamma = z \text{ and replace } z \text{ by } z^2/4. \end{gathered} \tag{4}$$

Then

$${}_0F_1(\nu + 1;\ z^2/4) = \{A_n(z)/B_n(z)\} + R_n(z), \tag{5}$$

$$B_n(z) = {}_0F_1^n(-2n + 1 - \lambda;\ z^2/4), \tag{6}$$

$$\begin{gathered} A_n(z) = {}_2F_3^r\left(\begin{matrix} -m,\ \tfrac{1}{2} - m \\ -2n - \lambda + 1,\ \nu + 1,\ -2m \end{matrix}\middle| z^2\right), \\ r = [m],\ m = n + \tfrac{1}{2} - \tfrac{1}{2}\delta, \end{gathered} \tag{7}$$

$$\begin{gathered} B_0(z) = 1,\ B_1(z) = 1 - \frac{z^2}{4(\lambda + 1)}, \\ B_2(z) = 1 - \frac{z^2}{4(\lambda + 3)} + \frac{z^4}{32(\lambda + 2)(\lambda + 3)}, \\ B_3(z) = 1 - \frac{z^2}{4(\lambda + 5)} + \frac{z^4}{32(\lambda + 4)(\lambda + 5)} - \frac{z^6}{384(\lambda + 3)(\lambda + 4)(\lambda + 5)}, \end{gathered} \tag{8}$$

$$\begin{gathered} A_0(z) = 1,\ A_1(z) = 1 + \frac{(2 - \delta)z^2}{4(\lambda + 1)(\nu + 1)}, \\ A_2(z) = 1 + \frac{(4 - \delta)z^2}{4(\lambda + 3)(\nu + 1)} + \frac{(2 - \delta)(3 - \delta)z^4}{32(\lambda + 2)(\lambda + 3)(\nu + 1)(\nu + 2)}, \end{gathered} \tag{9}$$

$$A_3(z) = 1 + \frac{(6-\delta)z^2}{4(\lambda+5)(\nu+1)} + \frac{(4-\delta)(5-\delta)z^4}{32(\lambda+4)(\lambda+5)(\nu+1)(\nu+2)}$$

$$+ \frac{(2-\delta)(3-\delta)(4-\delta)z^6}{384(\lambda+3)(\lambda+4)(\lambda+5)(\nu+1)(\nu+2)(\nu+3)} .$$

Both $A_n(z)$ and $B_n(z)$ satisfy the recurrence formula

$$B_n(z) = (1+F_1y)B_{n-1}(z) + (E+F_2y)yB_{n-2}(z) + F_3y^3B_{n-3}(z) ,$$

$$n \geq 3, \; y = z^2/4 ,$$

$$F_1 = \frac{(n-\lambda+1)}{n(2n+\lambda-1)(2n+\lambda-4)} , \quad E = \frac{(n+\lambda-3)}{n(2n+\lambda-4)(2n+\lambda-3)} ,$$

$$F_2 = \frac{E(n+2\lambda-4)}{(n+\lambda-3)(2n+\lambda-5)(2n+\lambda-2)} , \qquad (10)$$

$$F_3 = \frac{-E}{(2n+\lambda-6)(2n+\lambda-5)^2(2n+\lambda-4)} .$$

This formula is stable when used in the forward direction.
For the error, we have

$$\frac{(z/2)^\nu R_n(z)}{\Gamma(\nu+1)} = \frac{(-)^n(z/2)^{2n+\nu}O(n^\mu)\left[1+O(n^{-1})\right]}{\Gamma(\nu+1)n!(n+\lambda)_n} , \qquad (11)$$

$$\mu = 1-\delta-R(\nu) \text{ if } -1 < R(\nu) \leq 1-\delta; \; \mu = 0 \text{ if } R(\nu) \geq 1-\delta,$$

and so for z and ν fixed, $R(\nu) > -1$, the approximation process is convergent. Another formulation of the error is given by $E_{m,\nu}(z)$ of 9.10.3(10-13) with $m = 0$. There the restriction on ν is relaxed to $-\pi < \arg \nu \leq \pi$, ν not a negative integer.

CASE II. In 5.12(1-11), let

$$p = q = 1, \; \alpha_1 = \nu + \tfrac{1}{2}, \; \rho_1 = 2\nu + 1, \; f = g = 0, \qquad (12)$$

$a = 0, \alpha = 1$, $\beta = 2\nu$, $\lambda = 2\nu + 2$, $\gamma = z$ and replace z by $2z$.

Then

$${}_1F_1(\nu+\tfrac{1}{2};\; 2\nu+1;\; -2z) = \{G_n(z)/H_n(z)\} + S_n(z) , \qquad (13)$$

$$H_n(z) = {}_1F_1^n(-n-\nu-\tfrac{1}{2};\; -2n-2\nu-1;\; 2z) , \qquad (14)$$

$$G_n(z) = {}_2F_3^r(-\tfrac{1}{2}n,\ \tfrac{1}{2}-\tfrac{1}{2}n;\ -n-\nu,\ \nu+1,\ -n;\ z^2),\ r = [n/2], \tag{15}$$

$$H_0(z) = 1,\ H_1(z) = 1+z,\ H_2(z) = 1+z+(2\nu+3)z^2/2(2\nu+4),$$

$$H_3(z) = 1+z+\frac{(2\nu+5)z^2}{2(2\nu+6)}+\frac{(2\nu+3)z^3}{6(2\nu+6)}, \tag{16}$$

$$G_0(z) = G_1(z) = 1,\ G_2(z) = 1+z^2/4(\nu+1)(\nu+2),$$

$$G_3(z) = 1+z^2/2(\nu+1)(\nu+3). \tag{17}$$

If for $A_n(z)$ as given by (7) we write $A_n(z,\delta)$, then $G_{2n}(z) = A_n(z,1)$. Both $G_n(z)$ and $H_n(z)$ satisfy the same recurrence formula

$$H_n(z) = (1+Q_1z)H_{n-1}(z)+(P+Q_2z)zH_{n-2}(z)+Q_3z^3H_{n-3}(z),\ n\geq 3,$$

$$Q_1 = (n+2\nu-1)/2n(n+\nu-1),\ P = -Q_1, \tag{18}$$

$$Q_2 = 1/4(n+\nu)(n+\nu-1),\ Q_3 = P/4(n+\nu-1)(n+\nu-2).$$

This formula is stable when used in the forward direction.
For the error, we have

$$\frac{(z/2)^\nu e^z S_n(z)}{\Gamma(\nu+1)} = \frac{(z/2)^\nu(2z)^n(\nu+3/2)_n O(n^\omega)}{\Gamma(\nu+1)(n+2\nu+2)_n n!},\ R(\nu) > -1,$$

$$\omega = 1-2R(\nu)\ \text{if}\ -1<R(\nu)\leq\tfrac{1}{2},\ \omega = 0\ \text{if}\ R(\nu)\geq\tfrac{1}{2}. \tag{19}$$

Another formulation of the error is given by $F_{m,\nu}(z)$ of 9.10.3(18-21) with $m = 0$ if there we set $\delta = 1$ and replace n by $n/2$. There the restriction on ν is relaxed to $-\pi<\arg\nu\leq\pi$, ν not a negative integer.

We now turn to rational approximations for

$$S_\nu(z) = 2(\nu+1)I_{\nu+1}(z)/zI_\nu(z). \tag{20}$$

We have

$$S_\nu(z) = \{2(\nu+1)/z\}[C_n(z,a)/D_n(z,a)] + R_n(z), \tag{21}$$

$$C_n(z,a) = {}_2F_3^{n-a}(a-n,\ \tfrac{1}{2}-n;\ a-2n-\nu-1,\ \nu+2,\ a-2n;\ z^2), \tag{22}$$

$$D_n(z,a) = {}_2F_3^n(-n,\ a-n-\tfrac{1}{2};\ a-2n-\nu-1,\ \nu+2,\ a-2n-1;\ z^2), \tag{23}$$

where $a=0$ or $a=1$. This approximation to $S_\nu(z)$ is given by $2(\nu+1)/z$ times the ratio $W_{1,\nu}(z)/W_{0,\nu}(z)$ where $W_{m,\nu}(z)$ is found by the backward recursion process described in 9.10.2, and explicitly depicted by 9.10.3(1-5). Indeed, in (7), let

$$A_n(z) \equiv A_n(z,\delta,\nu). \tag{24}$$

Then

$$C_n(z,a) = A_{n-a}(z,1-a,\nu+1),\quad D_n(z,a) = A_n(z,a,\nu). \tag{25}$$

Both $C_n(z,a)$ and $D_n(z,a)$ satisfy the recurrence formula

$$D_{n+1}(z,a) = (1+z^2/E)D_n(z,a) - (z^4/F)D_{n-1}(z,a),$$

$$E = 2(2n+\nu+1-a)(2n+\nu+3-a), \tag{26}$$

$$F = 16(2n+\nu-a)(2n+\nu+1-a)^2(2n+\nu+2-a).$$

This formula is stable when used in the forward direction.
For the error, we have

$$R_n(z) = \frac{(-)^{1-a}(\nu+1)(z/2)^{2n-a}I_{2n+2+\nu-a}(z)}{(\nu+1)_{2n+1-a}D_n(z,a)I_\nu(z)}, \tag{27}$$

and so for z and ν fixed,

$$\lim_{n\to\infty} R_n(z) = 0, \tag{28}$$

unless z is a zero of $I_\nu(z)$ in which case we know that z is pure imaginary if ν is real. Also for z and ν fixed,

$$\lim_{n\to\infty} D_n(z,a) = (z/2)^{-\nu}\Gamma(\nu+1)I_\nu(z), \tag{29}$$

so that approximations for zeros of $I_\nu(z)$ follow by finding zeros of $D_n(z,a)$.

The approximations to $S_\nu(z)$ occupy the $(n-a,n)$ positions of the Padé matrix table. In view of (27), we have the inequality

$$\frac{C_n(z,1)}{D_n(z,1)} < S_\nu(z) < \frac{C_n(z,0)}{D_n(z,0)}, \quad z > 0, \quad \nu > -1, \tag{30}$$

with equality if $z = 0$ provided $n > 0$.

9.9.3. $K_\nu(z)$, z Large

Use 8.3(13) with $\mu = \frac{1}{2}$ and Theorem 6 of 5.12 with

$$\begin{gathered} p = 2,\ q = 0,\ \alpha_1 = \tfrac{1}{2} + \nu,\ \alpha_2 = \tfrac{1}{2} - \nu,\ a = 0, \\ \alpha = \beta = 0,\ \lambda = 1, \text{ and } z \text{ replaced by } 2z. \end{gathered} \tag{1}$$

Then

$$K_\nu(z) = (\pi/2z)^{1/2} e^{-z}\{E_n(z) + R_n(z)\}, \tag{2}$$

$$E_n(z) = \varphi_n(z)/f_n(z),$$

$$\varphi_n(z) = \sum_{k=0}^{n} \frac{(\frac{1}{2}-\nu)_k (\frac{1}{2}+\nu)_k (-n)_k (n+1)_k}{(\frac{3}{2}-\nu)_k (\frac{3}{2}+\nu)_k (1)_k\, k!} \times {}_3F_3\left(\begin{matrix} -n+k, n+1+k, 1 \\ 1+k, \frac{3}{2}-\nu+k, \frac{3}{2}+\nu+k \end{matrix} \middle| -2z\right),$$

$$f_n(z) = {}_2F_2\left(\begin{matrix} -n, n+1 \\ \frac{3}{2}-\nu, \frac{3}{2}+\nu \end{matrix} \middle| -2z\right), \tag{3}$$

where $R_n(z)$ is the remainder term and

$$\lim_{n\to\infty} R_n(z) = 0, \qquad z \text{ fixed}, \quad |\arg z| \leqslant \pi. \tag{4}$$

In particular we have

$$f_0(z) = 1, \qquad f_1(z) = a_1^{-1}(16z + a_1),$$

$$f_2(z) = (a_1a_2)^{-1}(768z^2 + 48a_2z + a_1a_2),$$

$$f_3(z) = (a_1a_2a_3)^{-1}(61440z^3 + 3840a_3z^2 + 96a_2a_3z + a_1a_2a_3),$$

$$\varphi_0(z) = 1, \qquad \varphi_1(z) = a_1^{-1}[16z + (4\nu^2 + 7)],$$

$$\varphi_2(z) = (a_1a_2)^{-1}[768z^2 + 48(4\nu^2 + 23)z + (16\nu^4 + 248\nu^2 + 129)],$$

$$\varphi_3(z) = (a_1a_2a_3)^{-1}[61440z^3 + 3840z^2(4\nu^2 + 47) + 96z(16\nu^4 + 504\nu^2 + 1025) + (64\nu^6 + 4048\nu^4 + 15628\nu^2 + 5055)],$$

$$a_n = (2n+1)^2 - 4\nu^2. \tag{5}$$

From 5.13.3, both $\varphi_n(z)$ and $f_n(z)$ satisfy the recurrence formula

$$f_n(z) + (P_1 - Q_1 z) f_{n-1}(z) + (P_2 - Q_2 z) f_{n-2}(z) + P_3 f_{n-3}(z) = 0, \qquad n > 2,$$

$$Q_1 = Q_2 = \frac{16(2n-1)}{a_n},$$

$$P_1 = -\frac{(2n-1)(12n^2 - 20n - a_0)}{(2n-3)\, a_n}, \qquad P_2 = \frac{(12n^2 - 28n + 8 - a_0)}{a_n},$$

$$P_3 = -\frac{(2n-1)\, a_{n-3}}{(2n-3)\, a_n}, \qquad P_1 + P_2 + P_3 + 1 = 0, \tag{6}$$

where a_n is defined in (5).

Approximations for $J_\nu(z)$ and $Y_\nu(z)$ follow from the connecting relations 9.2(10,11), and each is valid for $|\arg z| < \pi/2$.

In the following tables, we present coefficients for the polynomials $\varphi_n(z)$ and $f_n(z)$ for $n = 0(1)6$ and for each $\nu = 0$, 1/4, 1/3, 2/3, 3/4, 1. Further polynomials can be generated by use of (6). Following the coefficients for $\nu = 0$, 1/3, 2/3, 1, we record $|R_n(z)|$ for $z = re^{i\theta}$, $r = 1,2(2)10$, $\theta = 0(\pi/4)3\pi/4$ and $n = 2, 4, 6, 10, 15, 20$. For $\nu = 0,1$, we also give values of $|R_n(z)|$ for $r = 12, 16, 20$ with θ and n as just described. A missing entry means that $|R_n(z)| < 0.5w$, $w = 10^{-20}$. Approximate error coefficients for $\nu = 1/4$ and $\nu = 3/4$ can be inferred from the corresponding data for $\nu = 0$, 1/3 and $\nu = 2/3$, 1, respectively. Observe that $R_n(z) = 0$ if $\nu = N + \frac{1}{2}$, N a positive integer or zero and $n \geq N$.

If n and z are fixed, the rational approximations weaken as ν increases. For $0 \leq \nu \leq 1$, rational approximations for $K_{\nu+m}(z)$, m a positive integer, can be developed with the aid of 9.3(11-13). Notice that $K_{-\nu}(z) = K_\nu(z)$.

TABLE 9.23

COEFFICIENTS FOR $K_0(z)$

$$K_o(z) = (\pi/2z)^{\frac{1}{2}} e^{-z} \left[\varphi_n(z)/f_n(z) + R_n(z)\right]$$

$$f_n(z) = b_o^{-1} \sum_{k=0}^{n} b_k z^k$$

n	$b_o, b_1, \ldots, b_n$
0	1
1	9, 16
2	75, 400, 256
3	735, 7840, 12544, 4096
4	8505, 1 51200, 4 35456, 3 31776, 65536
5	1 14345, 30 49200, 136 60416, 178 42176, 79 29856, 10 48576
6	17 56755, 655 85520, 4197 47328, 8223 62112, 6091 57120, 1772 09344, 167 77216

$$\varphi_n(z) = b_o^{-1} \sum_{k=0}^{n} c_k z^k$$

n	$c_o, c_1, \ldots, c_n$
0	1
1	7, 16
2	43, 368, 256
3	337, 6560, 12032, 4096
4	3273, 1 15296, 3 98592, 3 23584, 65536
5	38103, 21 32976, 119 10912, 169 24672, 77 98784, 10 48576
6	5 18019, 423 87600, 3486 85824, 7574 48704, 5881 85600, 1751 12192, 167 77216

TABLE 9.23 *(Continued)*

$\|R_n(z)\|$, $z = re^{i\theta}$				
		r = 1		
n/θ	0	π/4	π/2	3π/4
2	0.701(-3)	0.105(-2)	0.307(-2)	0.172(-1)
4	0.489(-5)	0.190(-4)	0.197(-3)	0.409(-2)
6	0.513(-7)	0.612(-6)	0.176(-4)	0.117(-2)
10	0.396(-10)	0.154(-8)	0.255(-6)	0.129(-3)
15	0.619(-14)	0.225(-11)	0.252(-8)	0.117(-4)
20	0.109(-16)	0.638(-14)	0.397(-10)	0.147(-5)
		r = 2		
2	0.161(-3)	0.215(-3)	0.510(-3)	0.231(-2)
4	0.744(-7)	0.188(-5)	0.175(-4)	0.360(-3)
6	0.668(-9)	0.273(-7)	0.858(-6)	0.716(-4)
10	0.887(-13)	0.155(-10)	0.421(-8)	0.431(-5)
15	0.622(-17)	0.434(-14)	0.126(-10)	0.206(-6)
20	0.907(-21)	0.274(-17)	0.680(-13)	0.134(-7)
		r = 4		
2	0.232(-4)	0.286(-4)	0.530(-4)	0.144(-3)
4	0.729(-7)	0.158(-6)	0.846(-6)	0.100(-4)
6	0.227(-10)	0.980(-9)	0.198(-7)	0.112(-5)
10	0.851(-15)	0.986(-13)	0.250(-10)	0.279(-7)
15	0.206(-20)	0.370(-17)	0.166(-13)	0.547(-9)
20		0.386(-21)	0.229(-16)	0.166(-10)
		r = 6		
2	0.638(-5)	0.753(-5)	0.121(-4)	0.248(-4)
4	0.176(-7)	0.295(-7)	0.109(-6)	0.741(-6)
6	0.316(-10)	0.128(-9)	0.152(-8)	0.466(-7)
10	0.478(-16)	0.461(-14)	0.740(-12)	0.522(-9)
15		0.466(-19)	0.170(-15)	0.496(-11)
20			0.896(-19)	0.846(-13)
		r = 8		
2	0.241(-5)	0.277(-5)	0.408(-5)	0.704(-5)
4	0.529(-8)	0.794(-8)	0.232(-7)	0.105(-6)
6	0.104(-10)	0.268(-10)	0.212(-9)	0.363(-8)
10	0.501(-18)	0.510(-15)	0.483(-13)	0.185(-10)
15		0.199(-20)	0.474(-17)	0.888(-13)
20			0.113(-20)	0.880(-15)

TABLE 9.23 *(Concluded)*

$\lvert R_n(z) \rvert$, $z = re^{i\theta}$				
		r = 10		
n/θ	0	π/4	π/2	3π/4
2	0.111(-5)	0.124(-5)	0.172(-5)	0.266(-5)
4	0.192(-8)	0.270(-8)	0.667(-8)	0.229(-7)
6	0.348(-11)	0.733(-11)	0.428(-10)	0.470(-9)
10	0.509(-17)	0.891(-16)	0.513(-14)	0.105(-11)
15		0.183(-21)	0.245(-18)	0.257(-14)
20				0.153(-16)
		r = 12		
2	0.577(-6)	0.639(-6)	0.846(-6)	0.121(-5)
4	0.803(-9)	0.108(-8)	0.236(-8)	0.663(-8)
6	0.129(-11)	0.240(-11)	0.112(-10)	0.859(-10)
10	0.261(-17)	0.204(-16)	0.764(-15)	0.875(-13)
15		0.243(-22)	0.195(-19)	0.108(-15)
20				0.395(-18)
		r = 16		
2	0.202(-6)	0.219(-6)	0.273(-6)	0.356(-6)
4	0.190(-9)	0.240(-9)	0.445(-9)	0.963(-9)
6	0.231(-12)	0.373(-12)	0.127(-11)	0.597(-11)
10	0.441(-18)	0.176(-17)	0.337(-16)	0.150(-14)
15			0.324(-21)	0.474(-18)
20				0.793(-21)
		r = 20		
2	0.879(-7)	0.941(-7)	0.113(-6)	0.139(-6)
4	0.595(-10)	0.722(-10)	0.120(-9)	0.221(-9)
6	0.556(-13)	0.823(-13)	0.226(-12)	0.780(-12)
10	0.792(-19)	0.232(-18)	0.276(-17)	0.617(-16)
15				0.537(-20)

TABLE 9.24

COEFFICIENTS FOR $K_{1/4}(z)$

$$K_{\frac{1}{4}}(z) = (\pi/2z)^{\frac{1}{2}} e^{-z}\left[\varphi_n(z)/f_n(z) + R_n(z)\right]$$

$$f_n(z) = b_0^{-1} \sum_{k=0}^{n} b_k z^k$$

n	$b_0, b_1, \ldots, b_n$
0	1
1	35, 64
2	1155, 6336, 4096
3	45045, 4 94208, 7 98720, 2 62144
4	145 49535, 2660 48640, 7739 59680, 5927 07584, 1174 40512
5	3346 39305, 91786 78080, 4 15358 36160, 5 45290 97728, 2 43101 85984, 32212 25472
6	7 52938 43625, 289 12835 95200, 1869 11262 72000, 3680 71409 66400, 2734 89592 32000, 797 25330 43200, 75 59142 44096

$$\varphi_n(z) = b_0^{-1} \sum_{k=0}^{n} c_k z^k$$

n	$c_0, c_1, \ldots, c_n$
0	1
1	29, 64
2	771, 5952, 4096
3	25803, 4 32768, 7 74144, 2 62144
4	74 10015, 2176 68480, 7244 14464, 5816 97536, 1174 40512
5	1549 37655, 70545 12960, 3 74997 93408, 5 24151 68512, 2 40081 96096, 32212 25472
6	3 21775 91145, 209 93363 39520, 1628 19901 44000, 3461 19575 96160, 2664 02896 28160, 790 16660 82816, 75 59142 44096

TABLE 9.25

COEFFICIENTS FOR $K_{1/3}(z)$

$$K_{\frac{1}{3}}(z) = (\pi/2z)^{\frac{1}{2}} e^{-z}\left[\varphi_n(z)/f_n(z) + R_n(z)\right]$$

$$f_n(z) = b_o^{-1} \sum_{k=0}^{n} b_k z^k$$

n	$b_o, b_1, \ldots, b_n$
0	1
1	77, 144
2	17017, 95472, 62208
3	74 36429, 834 42528, 1359 24480, 447 89760
4	10782 82205, 2 01652 77600, 5 91271 48800, 4 54616 06400, 90296 15616
5	16 71337 41775, 468 84270 42000, 2138 43188 16000, 2818 61959 68000, 1259 63137 84320, 167 17688 34048
6	25354 18862 72675, 9 95728 13517 99600, 64 88002 32877 44000, 128 27537 78503 68000, 95 54304 00540 67200, 27 89680 65337 58976, 2 64808 18331 32032

$$\varphi_n(z) = b_o^{-1} \sum_{k=0}^{n} c_k z^k$$

n	$c_o, c_1, \ldots, c_n$
0	1
1	67, 144
2	12697, 91152, 62208
3	49 87459, 756 66528, 1328 14080, 447 89760
4	6668 37485, 1 74043 10880, 5 63053 93920, 4 48345 49760, 90296 15616
5	9 68195 54225, 386 78602 87440, 1983 11735 46240, 2737 35305 62560, 1248 02187 26400, 167 17688 34048
6	13905 52108 89875, 7 88585 24582 56080, 58 61263 19890 40640, 122 57574 15046 34880, 93 70409 43366 14400, 27 71291 19620 13696, 2 64808 18331 32032

TABLE 9.25 *(Concluded)*

$|R_n(z)|$, $z = re^{i\theta}$

r = 1

n/θ	0	π/4	π/2	3π/4
2	0.358(-3)	0.551(-3)	0.164(-2)	0.921(-2)
4	0.241(-5)	0.979(-5)	0.103(-3)	0.216(-2)
6	0.243(-7)	0.311(-6)	0.918(-5)	0.611(-3)
10	0.194(-10)	0.775(-9)	0.132(-6)	0.672(-4)
15	0.273(-14)	0.113(-11)	0.130(-8)	0.612(-5)
20	0.514(-17)	0.318(-14)	0.204(-10)	0.762(-6)

r = 2

n/θ	0	π/4	π/2	3π/4
2	0.836(-4)	0.113(-3)	0.268(-3)	0.120(-2)
4	0.393(-7)	0.963(-6)	0.905(-5)	0.186(-3)
6	0.339(-9)	0.139(-7)	0.441(-6)	0.369(-4)
10	0.435(-13)	0.780(-11)	0.215(-8)	0.221(-5)
15	0.305(-17)	0.217(-14)	0.641(-11)	0.105(-6)
20	0.443(-21)	0.137(-17)	0.345(-13)	0.685(-8)

r = 4

n/θ	0	π/4	π/2	3π/4
2	0.121(-4)	0.149(-4)	0.276(-4)	0.744(-4)
4	0.374(-7)	0.811(-7)	0.434(-6)	0.514(-5)
6	0.120(-10)	0.498(-9)	0.101(-7)	0.573(-6)
10	0.429(-15)	0.497(-13)	0.127(-10)	0.142(-7)
15	0.102(-20)	0.186(-17)	0.838(-14)	0.277(-9)
20		0.193(-21)	0.115(-16)	0.844(-11)

r = 6

n/θ	0	π/4	π/2	3π/4
2	0.332(-5)	0.392(-5)	0.630(-5)	0.128(-4)
4	0.901(-8)	0.151(-7)	0.560(-7)	0.379(-6)
6	0.161(-10)	0.649(-10)	0.772(-9)	0.237(-7)
10	0.242(-16)	0.233(-14)	0.374(-12)	0.264(-9)
15		0.234(-19)	0.857(-16)	0.251(-11)
20			0.453(-19)	0.428(-13)

r = 8

n/θ	0	π/4	π/2	3π/4
2	0.126(-5)	0.144(-5)	0.212(-5)	0.364(-5)
4	0.271(-8)	0.407(-8)	0.119(-7)	0.538(-7)
6	0.529(-11)	0.137(-10)	0.108(-9)	0.187(-8)
10	0.230(-18)	0.258(-15)	0.244(-13)	0.934(-11)
15		0.100(-20)	0.239(-17)	0.448(-13)
20			0.694(-21)	0.444(-15)

r = 10

n/θ	0	π/4	π/2	3π/4
2	0.576(-6)	0.647(-6)	0.894(-6)	0.138(-5)
4	0.982(-9)	0.138(-8)	0.341(-8)	0.117(-7)
6	0.177(-11)	0.373(-11)	0.218(-10)	0.239(-9)
10	0.258(-17)	0.451(-16)	0.260(-14)	0.530(-12)
15		0.923(-22)	0.124(-18)	0.129(-14)
20				0.773(-17)

TABLE 9.26

Coefficients for $K_{2/3}(z)$

$$K_{\frac{2}{3}}(z) = (\pi/2z)^{\frac{1}{2}} e^{-z}\left[\varphi_n(z)/f_n(z)+R_n(z)\right]$$

$$f_n(z) = b_o^{-1} \sum_{k=0}^{n} b_k z^k$$

n	$b_o, b_1, \ldots, b_n$
0	1
1	65, 144
2	13585, 90288, 62208
3	11 54725, 153 48960, 264 38400, 89 57952
4	8233 18925, 1 82396 80800, 5 65517 37600, 4 47091 38432, 90296 15616
5	88 34212 06525, 2935 67662 47600, 14158 67003 71200, 19189 16221 50144, 8719 89980 03712, 1170 23818 38336
6	18993 55594 02875, 8 83638 66405 27600, 60 88228 11596 16000, 123 77009 62868 42880, 93 73892 28539 90400, 27 67613 30476 64640, 2 64808 18331 32032

$$\varphi_n(z) = b_o^{-1} \sum_{k=0}^{n} c_k z^k$$

n	$c_o, c_1, \ldots, c_n$
0	1
1	79, 144
2	19633, 96336, 62208
3	18 67115, 175 26240, 273 09312, 89 57952
4	14475 07805, 2 21591 27520, 6 05021 94432, 4 55870 17728, 90296 15616
5	166 05327 38435, 3767 03898 67440, 15691 30158 43584, 19985 57431 23456, 8833 67295 71328, 1170 23818 38336
6	37744 29588 92075, 11 89865 90275 81200, 69 81102 18604 10880, 131 78147 97732 24960, 96 31344 68584 24320, 27 93358 54481 08032, 2 64808 18331 32032

TABLE 9.26 *(Concluded)*

$|R_n(z)|$, $z = re^{i\theta}$

n/θ	0	π/4	π/2	3π/4
		r = 1		
2	0.377(-3)	0.640(-3)	0.203(-2)	0.112(-1)
4	0.224(-5)	0.107(-4)	0.120(-3)	0.252(-2)
6	0.201(-7)	0.329(-6)	0.104(-4)	0.704(-3)
10	0.176(-10)	0.793(-9)	0.146(-6)	0.766(-4)
15	0.173(-14)	0.113(-11)	0.142(-8)	0.693(-5)
20	0.425(-17)	0.316(-14)	0.221(-10)	0.859(-6)
		r = 2		
2	0.936(-4)	0.129(-3)	0.311(-3)	0.137(-2)
4	0.466(-7)	0.104(-5)	0.100(-4)	0.206(-3)
6	0.358(-9)	0.145(-7)	0.478(-6)	0.403(-4)
10	0.405(-13)	0.792(-11)	0.228(-8)	0.239(-5)
15	0.284(-17)	0.217(-14)	0.674(-11)	0.113(-6)
20	0.412(-21)	0.136(-17)	0.360(-13)	0.734(-8)
		r = 4		
2	0.136(-4)	0.169(-4)	0.313(-4)	0.828(-4)
4	0.403(-7)	0.876(-7)	0.471(-6)	0.552(-5)
6	0.138(-10)	0.524(-9)	0.107(-7)	0.608(-6)
10	0.441(-15)	0.510(-13)	0.132(-10)	0.149(-7)
15	0.100(-20)	0.188(-17)	0.866(-14)	0.290(-9)
20		0.194(-21)	0.119(-16)	0.879(-11)
		r = 6		
2	0.374(-5)	0.442(-5)	0.709(-5)	0.142(-4)
4	0.971(-8)	0.163(-7)	0.603(-7)	0.404(-6)
6	0.171(-10)	0.685(-10)	0.816(-9)	0.249(-7)
10	0.253(-16)	0.240(-14)	0.389(-12)	0.275(-9)
15		0.238(-19)	0.881(-16)	0.259(-11)
20			0.464(-19)	0.441(-13)
		r = 8		
2	0.141(-5)	0.162(-5)	0.238(-5)	0.405(-5)
4	0.291(-8)	0.438(-8)	0.127(-7)	0.573(-7)
6	0.558(-11)	0.144(-10)	0.114(-9)	0.195(-8)
10	0.166(-18)	0.266(-15)	0.253(-13)	0.966(-11)
15		0.102(-20)	0.245(-17)	0.461(-13)
20			0.709(-21)	0.455(-15)
		r = 10		
2	0.647(-6)	0.727(-6)	0.100(-5)	0.153(-5)
4	0.106(-8)	0.148(-8)	0.366(-8)	0.125(-7)
6	0.187(-11)	0.393(-11)	0.229(-10)	0.249(-9)
10	0.269(-17)	0.466(-16)	0.268(-14)	0.547(-12)
15		0.944(-22)	0.127(-18)	0.133(-14)
20				0.790(-17)

TABLE 9.27

COEFFICIENTS FOR $K_{3/4}(z)$

$$K_{\frac{3}{4}}(z) = (\pi/2z)^{\frac{1}{2}}e^{-z}\left[\varphi_n(z)/f_n(z)+R_n(z)\right]$$

$$f_n(z) = b_o^{-1}\sum_{k=0}^{n} b_k z^k$$

n	$b_o, b_1, \ldots, b_n$
0	1
1	27, 64
2	819, 5824, 4096
3	1 53153, 21 78176, 38 29760, 13 10720
4	13 78377, 326 72640, 1034 03520, 825 75360, 167 77216
5	727 47675, 25865 84000, 1 27339 52000, 1 74325 76000, 79691 77600, 10737 41824
6	4 85226 99225, 241 53521 39200, 1698 70919 68000, 3488 25845 76000, 2657 72072 96000, 787 80437 62688, 75 59142 44096

$$\varphi_n(z) = b_o^{-1}\sum_{k=0}^{n} c_k z^k$$

n	$c_o, c_1, \ldots, c_n$
0	1
1	37, 64
2	1459, 6464, 4096
3	3 22943, 26 90176, 40 34560, 13 10720
4	32 95497, 444 34560, 1152 00000, 851 96800, 167 77216
5	1922 71205, 38270 78080, 1 49995 72480, 1 86069 81120, 81369 49760, 10737 41824
6	13 94082 10585, 384 75614 17280, 2110 88331 98080, 3856 32016 79360, 2775 83233 02400, 799 61553 63328, 75 59142 44096

TABLE 9.28

COEFFICIENTS FOR $K_1(z)$

$$K_1(z) = (\pi/2z)^{\frac{1}{2}} e^{-z}\left[\varphi_n(z)/f_n(z)+R_n(z)\right]$$

$$f_n(z) = b_o^{-1} \sum_{k=0}^{n} b_k z^k$$

n	$b_o, b_1, \ldots, b_n$
0	1
1	5, 16
2	35, 336, 256
3	315, 6048, 11520, 4096
4	3465, 1 10880, 3 80160, 3 15392, 65536
5	45045, 21 62160, 115 31520, 164 00384, 76 67712, 10 48576
6	6 75675, 454 05360, 3459 45600, 7380 17280, 5750 78400, 1730 15040, 167 77216

$$\varphi_n(z) = b_o^{-1} \sum_{k=0}^{n} c_k z^k$$

n	$c_o, c_1, \ldots, c_n$
0	1
1	11, 16
2	131, 432, 256
3	1653, 9888, 13056, 4096
4	23385, 2 23200, 4 90752, 3 39968, 65536
5	3 71595, 51 99600, 168 90624, 191 52896, 80 60928, 10 48576
6	65 87595, 1273 25520, 5706 31680, 9351 16800, 6379 92960, 1793 06496, 167 77216

TABLE 9.28 *(Continued)*

$|R_n(z)|$, $z = re^{i\theta}$

n/θ	0	π/4	π/2	3π/4
		r = 1		
2	0.758(-3)	0.170(-2)	0.584(-2)	0.312(-1)
4	0.362(-5)	0.249(-4)	0.310(-3)	0.653(-2)
6	0.255(-7)	0.725(-6)	0.258(-4)	0.178(-2)
10	0.270(-10)	0.165(-8)	0.348(-6)	0.190(-3)
15	0.908(-15)	0.228(-11)	0.330(-8)	0.170(-4)
20	0.550(-17)	0.627(-14)	0.510(-10)	0.210(-5)
		r = 2		
2	0.224(-3)	0.323(-3)	0.807(-3)	0.339(-2)
4	0.127(-6)	0.235(-5)	0.239(-4)	0.488(-3)
6	0.806(-9)	0.311(-7)	0.110(-5)	0.936(-4)
10	0.695(-13)	0.162(-10)	0.507(-8)	0.546(-5)
15	0.499(-17)	0.433(-14)	0.147(-10)	0.256(-6)
20	0.734(-21)	0.266(-17)	0.776(-13)	0.165(-7)
		r = 4		
2	0.334(-4)	0.417(-4)	0.774(-4)	0.198(-3)
4	0.914(-7)	0.199(-6)	0.108(-5)	0.124(-4)
6	0.347(-10)	0.114(-8)	0.237(-7)	0.134(-5)
10	0.923(-15)	0.107(-12)	0.284(-10)	0.323(-7)
15	0.193(-20)	0.383(-17)	0.183(-13)	0.623(-9)
20		0.391(-21)	0.248(-16)	0.188(-10)
		r = 6		
2	0.916(-5)	0.103(-4)	0.173(-4)	0.340(-4)
4	0.220(-7)	0.370(-7)	0.137(-6)	0.899(-6)
6	0.377(-10)	0.150(-9)	0.179(-8)	0.541(-7)
10	0.541(-16)	0.505(-14)	0.829(-12)	0.586(-9)
15		0.491(-19)	0.185(-15)	0.548(-11)
20			0.960(-19)	0.927(-13)
		r = 8		
2	0.345(-5)	0.396(-5)	0.578(-5)	0.967(-5)
4	0.658(-8)	0.989(-8)	0.287(-7)	0.127(-6)
6	0.122(-10)	0.315(-10)	0.248(-9)	0.422(-8)
10	0.975(-19)	0.562(-15)	0.537(-13)	0.204(-10)
15		0.212(-20)	0.512(-17)	0.965(-13)
20			0.120(-20)	0.949(-15)

TABLE 9.28 *(Concluded)*

$|R_n(z)|$, $z = re^{i\theta}$

n/θ	0	π/4	π/2	3π/4
		r = 10		
2	0.158(-5)	0.177(-5)	0.243(-5)	0.367(-5)
4	0.238(-8)	0.335(-8)	0.823(-8)	0.277(-7)
6	0.409(-11)	0.858(-11)	0.499(-10)	0.538(-9)
10	0.576(-17)	0.984(-16)	0.568(-14)	0.115(-11)
15		0.195(-21)	0.264(-18)	0.276(-14)
20				0.164(-16)
		r = 12		
2	0.821(-6)	0.908(-6)	0.119(-5)	0.167(-5)
4	0.996(-9)	0.133(-8)	0.290(-8)	0.801(-8)
6	0.151(-11)	0.280(-11)	0.130(-10)	0.982(-10)
10	0.292(-17)	0.226(-16)	0.844(-15)	0.957(-13)
15		0.259(-22)	0.209(-19)	0.116(-15)
20				0.419(-18)
		r = 16		
2	0.286(-6)	0.310(-6)	0.383(-6)	0.493(-6)
4	0.235(-9)	0.297(-9)	0.546(-9)	0.117(-8)
6	0.270(-12)	0.435(-12)	0.147(-11)	0.682(-11)
10	0.489(-18)	0.194(-17)	0.371(-16)	0.163(-14)
15			0.347(-21)	0.504(-18)
20				0.835(-21)
		r = 20		
2	0.124(-6)	0.133(-6)	0.158(-6)	0.193(-6)
4	0.735(-10)	0.890(-10)	0.147(-9)	0.268(-9)
6	0.647(-13)	0.958(-13)	0.262(-12)	0.893(-12)
10	0.877(-19)	0.256(-18)	0.304(-17)	0.672(-16)
15				0.570(-20)

9.10. Computation of Bessel Functions by Use of Recurrence Formulas

9.10.1. Introduction

Some general results on the application of recurrence formulas for the evaluation of functions are detailed in Chapter 12. For the case of Bessel functions, a considerable body of specific information is now known in analytical form. Such data due to Luke (1972c) are set down in the following pages. As previously remarked in 9.9.1, there is an intimate relation between certain rational approximations for $I_\nu(z)$ given in 9.9.2 and certain schemata for evaluation of $I_\nu(z)$ by use of the backward recurrence formula. This relationship is also presented. Rational approximations for $I_{\nu+1}(z)/I_\nu(z)$ are also a by product of the recurrence scheme evaluations for $I_\nu(z)$, and these data are the approximations given by 9.9.2(21).

In the main body of this section, it is convenient to deal with $I_\nu(z)$. We assume $-\pi < \arg z \leq \pi$ and $-\pi < \arg \nu \leq \pi$, ν not a negative integer. This is no burden for if ν is an integer n, then $I_{-n}(z) = I_n(z)$. Actually it is sufficient to have $0 \leq \arg z \leq \pi/2$ and $0 \leq \arg \nu \leq \pi$, $\nu \neq -n$, in view of known relations for the analytic continuation of $I_\nu(z)$. Even so, it is convenient to restate some of our results for the Bessel function $J_\nu(z)$.

9.10.2. Backward Recurrence Schemata for Generating $I_\nu(z)$

The technique for generating $I_\nu(z)$ by use of the recurrence formula for $I_\nu(z)$ employed in the backward direction, introduced by J.C.P. Miller (1952) is as follows. The difference equation

$$Q_{m,\nu}(z) = \{2(m+\nu+1)/z\}Q_{m+1,\nu}(z) + Q_{m+2,\nu}(z) \qquad (1)$$

is satisfied by

$$I_{m+\nu}(z) \text{ and } e^{i(m+\nu)\pi}K_{m+\nu}(z) . \qquad (2)$$

Here and throughout, m is a positive integer or zero.

Let N be a large positive integer with $m \leq N + 2$. Set

$$W_{N+2,\nu}(z) = 0, \; W_{N+1,\nu}(z) = 1 \qquad (3)$$

and evaluate $W_{m,\nu}(z)$ for $m = N, N-1,\ldots,1,0$ from (1) with

$Q_{m,\nu}(z)$ replaced by $W_{m,\nu}(z)$. Clearly $W_{m,\nu}(z)$ is a linear combination of the solutions (2) subject to the conditions (3), and we readily find that

$$W_{m,\nu}(z) = z\{I_{m+\nu}(z)K_{N+2+\nu}(z) + e^{-i\pi(N+1-m)}K_{m+\nu}(z)I_{N+2+\nu}(z)\} \tag{4}$$

in view of the Wronskian relation

$$I_\nu(z)K_{\nu+1}(z) + I_{\nu+1}(z)K_\nu(z) = 1/z \ . \tag{5}$$

We suppose that a normalization relation is known of the form

$$P(z) = \sum_{k=0}^{\infty} a_k I_{k+\nu}(z) \ . \tag{6}$$

Let

$$P_N(z) = \sum_{k=0}^{N+1} a_k W_{k,\nu}(z) \ , \tag{7}$$

and consider

$$i_{m+\nu}(z) = P(z)W_{m,\nu}(z)/P_N(z) \ , \quad m \le N + 1 \ . \tag{8}$$

Observe that both $W_{m,\nu}(z)$ and $i_{m+\nu}(z)$ are N dependent. We have omitted adding an N to the notation for the sake of simplicity. In view of (4) and (7), (8) can be written as

$$\begin{aligned} i_{m+\nu}(z) &= P(z)U_m(z)/\{\sum_{k=0}^{N+1} a_k U_k(z)\} \ , \\ U_k(z) &= I_{k+\nu}(z) - (-)^{N+k} I_{N+2+\nu}(z)K_{k+\nu}(z)/K_{N+2+\nu}(z) \ . \end{aligned} \tag{9}$$

From the latter and the known behavior of Bessel functions of large order which follows from 9.2(2,8,15), we find that

$$\lim_{N\to\infty} i_{m+\nu}(z) = I_{m+\nu}(z) \ , \ m = 0,1,\ldots, \tag{10}$$

provided that

$$\lim_{N\to\infty} \{ I_{N+2+\nu}(z)/K_{N+2+\nu}(z)\} \sum_{k=0}^{N+1} (-)^k a_k K_{k+\nu}(z) = 0 \ . \tag{11}$$

The preceding analysis shows clearly why the process converges, and also why it converges to $I_{m+\nu}(z)$ and not to $e^{i(m+\nu)\pi}K_{m+\nu}(z)$. The point is that in magnitude $I_{m+\nu}(z)$ is very small compared to $K_{m+\nu}(z)$ as $|m+\nu| \to \infty$ with z and ν fixed. This feature of the algorithm—that the solution of the difference equation to which the scheme converges, if it converges, must, in a certain sense, converge to the smallest solution—remains valid when the technique is applied to a homogeneous difference equation of arbitrary order. See Chapter 12.

A remarkable characteristic of the algorithm is that no tabular values of $I_{m+\nu}(z)$ are required provided we have a nontrivial normalization relationship like (6). Of course, in the trivial situation when $a_k = 0$ for $k > 0$ and $P(z) = I_\nu(z)$, then we are demanding that $I_\nu(z)$ be known. Closed form representations for $i_{m+\nu}(z)$, see (8), for two nontrivial normalization relations, called Cases I and II, are developed in 9.10.3. The situation $P(z) = I_\nu(z)$ is also treated there.

If (1) is used in the forward direction, then values of $I_\nu(z)$ and $I_{\nu+1}(z)$ are necessary. Further, small random errors which are inevitably introduced in the forward computation process grow rapidly as m increases. Such a phenomenon is called instability. On the other hand, evaluation of $e^{i(m+\nu)\pi}K_{m+\nu}(z)$ by use of (1) in the forward direction with known initial values of $K_\nu(z)$ and $K_{\nu+1}(z)$ is stable in the sense that even though roundoff errors introduced in the computation grow with m, the relative errors do not grow with m. In general, a difference equation can be effectively employed in the forward direction only to compute the 'largest' solution of the equation. For further discussion on forward computation with second order difference equations, see papers by Wimp (1972) and Gautschi (1972).

9.10.3. Closed Form Expressions

As previously noted, we distinguish three cases according to the normalization relation used. For each case, closed form expressions for the error are derived. The first two cases are equivalent to the rational approximations given in 9.9.2. To identify these data, we first present a closed form result for $W_{m,\nu}(z)$ and then proceed to the cases noted. We have

$$W_{N+1,\nu}(z) = 1, \quad W_{N,\nu}(z) = (N+\nu+1)y,$$

$$W_{N-1,\nu}(z) = 1 + (N+\nu)_2 y^2, \quad W_{N-2,\nu}(z) = 2(N+\nu)y + (N+\nu-1)_3 y^3,$$

$$W_{N-3,\nu}(z) = 1 + 3(N+\nu-1)_2 y^2 + (N+\nu-2)_4 y^4, \tag{1}$$

$$W_{N-4,\nu}(z) = 3(N+\nu-1)y + 4(N+\nu-2)_3 y^3 + (N+\nu-3)_5 y^5, \quad y = 2/z,$$

and in general

$$W_{m,\nu}(z) = \{(2/z)^{2p}\Gamma(2n+2-\delta+\nu)/\Gamma(m+\nu+1)\}$$
$$\times {}_2F_3^r(-p,\ \tfrac{1}{2}-p;\ -2n-1+\delta-\nu,\ m+\nu+1,\ -2p;\ z^2), \tag{2}$$

$$p = n+\tfrac{1}{2}(1-m-\delta),\ N=2n-\delta,\ \delta=0 \text{ or } \delta=1 \text{ and } r=[p] \tag{3}$$

is the largest integer $\leq p$. Clearly

$$W_{0,\nu}(z) = (2/z)^{N+1}(\nu+1)_{N+1}A_n(z), \tag{4}$$

$$W_{0,\nu}(z) = (2/z)^{N+1}(\nu+1)_{N+1}G_{N+1}(z) \tag{5}$$

where $A_n(z)$ and $G_n(z)$ are given by 9.9.2(7) and 9.9.2(15) respectively.

CASE I. Consider the normalization relation

$$P(z) = \frac{(z/2)^\nu}{\Gamma(\nu+1)} = \sum_{k=0}^{\infty} \frac{(-)^k(2k+\nu)\Gamma(k+\nu)}{\Gamma(\nu+1)k!} I_{2k+\nu}(z), \quad \nu \neq 0,$$
$$= 1 = I_0(z) + 2\sum_{k=1}^{\infty} (-)^k I_{2k}(z), \quad \nu = 0. \tag{6}$$

Then

$$P_N(z) = (2/z)^{N+1}(\nu+1)_{N+1}B_n(z), \tag{7}$$

where $B_n(z)$ is given by 9.9.2(6). It follows that

$$i_\nu(z) = P(z)A_n(z)/B_n(z), \tag{8}$$

$$i_{m+\nu}(z) = (z/2)^\nu W_{m,\nu}(z)/\Gamma(\nu+1)P_N(z). \tag{9}$$

That is, the result produced by use of the recurrence formula for $I_\nu(z)$ employed in the backward direction together with the normalization relation (6) and the rational approximation 9.9.2(5) are identical.

Next we turn to the error. Let

$$E_{m,\nu}(z) = I_{m+\nu}(z) - i_{m+\nu}(z). \tag{10}$$

If ν is not a positive integer or zero,

$${}_0F_1^n(-2n-1+\delta-\nu;\ z^2/4)E_{m,\nu}(z)$$

$$-\ Z_n(z)\ {}_1F_2(1;\ n+2,\ -n+\delta-\nu;\ z^2/4)$$

$$= (\pi/\sin\nu\pi)V_n(z)I_{-m-\nu}(z) \tag{11}$$

$$= 2(-)^m V_n(z)K_{m+\nu}(z) + (\pi/\sin\nu\pi)V_n(z)I_{m+\nu}(z),$$

$$Z_n(z) = \frac{(-)^n(z/2)^{2n+2}\Gamma(n+1-\delta+\nu)}{(n+1)!\,\Gamma(2n+2-\delta+\nu)}\,I_{m+\nu}(z),$$

$$V_n(z) = (-)^\delta(z/2)^{2n+2-\delta+\nu}I_{2n+2-\delta+\nu}(z)/\Gamma(2n+2-\delta+\nu).$$

Equation (11) can be rearranged so that with the aid of L'Hospital's theorem, we can get a representation of the error when ν becomes a positive integer or zero. We do not give this result. However, for arbitrary ν, we always have

$${}_0F_1^n(-2n-1+\delta-\nu;\ z^2/4)E_{m,\nu}(z)$$

$$= Z_n(z)\ {}_1F_2^s(1;\ n+2,\ -n+\delta-\nu;\ z^2/4) \tag{12}$$

$$+\ 2(-)^m V_n(z)K_{m+\nu}(z) + \frac{O\{(z/2)^{2n}\}}{\Gamma(2n+2-\delta+\nu)\Gamma(2n+3-\delta+\nu)},$$

where $s = n-\delta+\nu$ $(s=\infty)$ if ν is (is not) a positive integer or zero. Clearly the backward recurrence scheme is convergent. Further, for n sufficiently large, $n \gg m$, the relative error is essentially independent of m. For convenience in the applications we record the formula

$$E_{m,\nu}(z) = Z_n(z)\{1+O(n^{-1})\} + 2(-)^m V_n(z)K_{m+\nu}(z). \tag{13}$$

If ν, n and z are fixed so that $E_{m,\nu}(z)$ is conceived as a function of m only, then the latter satisfies the recurrence formula for $W_{m,\nu}(z)$.

CASE II. Consider the normalization relation

$$M(z) = \frac{(z/2)^\nu e^z}{\Gamma(\nu+1)} = \sum_{k=0}^{\infty}\frac{(2k+2\nu)\Gamma(k+2\nu)}{\Gamma(2\nu+1)k!}\,I_{k+\nu}(z),\quad \nu\neq 0,$$

$$= e^z = I_0(z) + 2\sum_{k=1}^{\infty} I_k(z), \quad \nu = 0. \tag{14}$$

Then

$$M_N(z) = (2/z)^q(\nu+1)_q H_q(z), \quad q = N+1, \tag{15}$$

where $H_n(z)$ is given by 9.9.2(14) and N is defined in (3). It follows that

$$i_\nu(z) = M(z)G_q(z)/H_q(z), \tag{16}$$

$$i_{m+\nu}(z) = (z/2)^\nu e^z W_{m,\nu}(z)/\Gamma(\nu+1)M_N(z). \tag{17}$$

That is, the result produced by use of the recurrence formula for $I_\nu(z)$ employed in the backward in the backward direction together with the normalization relation (14) and the rational approximation 9.9.2(13) are identical.

We now turn to the error. Let

$$F_{m,\nu}(z) = I_{m+\nu}(z) - i_{m+\nu}(z)$$

where $i_{m+\nu}(z)$ is given by (17). If neither ν nor $\nu + \frac{1}{2}$ is a positive integer or zero, then

$$\begin{aligned} e^{-z}{}_1F_1^q(-q-\tfrac{1}{2}-\nu;\ -2q-1-2\nu;\ 2z)F_{m,\nu}(z) \\ + L_n(z)\,{}_2F_2(\tfrac{1}{2}-\nu,\ 1;\ q+2,\ -q-2\nu;\ 2z) \\ = (\pi/\sin\nu\pi)V_n(z)I_{-m-\nu}(z), \end{aligned} \tag{19}$$

$$L_n(z) = \frac{(2z)^{q+1}(\nu+\frac{1}{2})_{q+1}}{(q+1)!\,(q+1+2\nu)_{q+1}} e^{-z} I_{m+\nu}(z), \quad q \text{ as in (15)}.$$

Notice that the right hand sides of (19) and (11) are the same.

If $\nu = r + \frac{1}{2}$, $r = -1$ or r is a positive integer or zero, then

$$\begin{aligned} e^{-z}{}_1F_1^q(-q-\tfrac{1}{2}-\nu;\ -2q-1-2\nu;\ 2z)F_{m,\nu}(z) \\ = -L_n(z)\,{}_2F_2^r(-r,\ 1;\ q+2,\ -q-2r-1;\ 2z) \\ + \tfrac{1}{2}\pi(-)^{r+1}N_n(z)\{\mu_r I_{m+r+\frac{1}{2}}(z) - I_{-m-r-\frac{1}{2}}(z)\}, \end{aligned} \tag{20}$$

$$N_n(z) = \frac{2(-)^{\delta}(z/2)^{q+1+\nu}}{\Gamma(q+1+\nu)} I_{q+1+\nu}(z),$$

$$\nu = r + \tfrac{1}{2}, \ \mu_r = 1 \text{ if } r = -1, \ \mu_r = 2 \text{ if } r \geq 0.$$

In particular,

$$F_{0,-\frac{1}{2}}(z) = (-)^{\delta}(\pi/2z)^{\frac{1}{2}}e^{-z}I_{q+\frac{1}{2}}(z)/K_{q+\frac{1}{2}}(z),$$

$$\simeq \frac{(-)^{\delta}(2\pi z)^{\frac{1}{2}}e^{-z}(z/2)^{2q+1}}{(q!)^2}\{1+O(n^{-1})\}, \tag{21}$$

and so $(2\pi z)^{\frac{1}{2}}e^{-z}i_{-\frac{1}{2}}(z)$ is the main diagonal Padé approximation to $1+e^{-2z}$. Again (19) can be rearranged to get a representation for the error when ν is a positive or zero. This result is omitted. However, for arbitrary ν, $\nu \neq -\frac{1}{2}$, we always have

$$e^{-z}{}_1F_1^q(-q-\tfrac{1}{2}-\nu;\ -2q-1-2\nu;\ 2z)F_{m,\nu}(z)$$

$$= -L_n(z)\,{}_2F_2^t(\tfrac{1}{2}-\nu,\ 1;\ q+2,\ -q-2\nu;\ 2z)$$

$$+ (-)^m N_n(z)K_{m+\nu}(z) + \frac{O\{(z/2)^{2n}\}}{\Gamma(q+1+\nu)\Gamma(q+2+\nu)}, \tag{22}$$

where $t = \nu - \frac{1}{2}(t = 2n+1-\delta+2\nu)$ if ν is half a positive integer (is a positive integer or zero) and where $t = \infty$ for all other ν. Clearly, the backward recurrence scheme is convergent. Further, for n sufficiently large, $n \gg m$, the relative error is independent of m. For convenience in the applications, we record the formula

$$F_{m,\nu}(z) = -L_n(z)\{1+O(n^{-1})\} + (-)^m N_n(z)K_{m+\nu}(z),\ \nu \neq -\tfrac{1}{2}. \tag{23}$$

With n, ν and z fixed and m variable, $F_{m,\nu}(z)$ and $W_{m,\nu}(z)$ satisfy the same recurrence formula.

We now present some comments on the relative merits of the Case I and Case II procedures. In (13) and (23), for z, m and ν fixed and n sufficiently large, the term involving $K_{m+\nu}(z)$ is of lower order than the term involving $I_{m+\nu}(z)$. Neglecting the former term in each equation, we have

$$\frac{E_{m,\nu}(z)}{F_{m,\nu}(z)} = \frac{(-)^{n+1}(\frac{1}{2})_{\nu}(z/2)^{\delta}e^{z}}{n^{\nu+\delta}}\{1+O(n^{-1})\},\quad \nu \neq -\tfrac{1}{2}. \tag{24}$$

This shows that there is little difference in the accuracy of the two schemes. Computation wise, if the backward recursion scheme is used, Case I requires less operations since the associated normalization relation uses the sequence $\{W_{k,\nu}(z)\}$, $k = 0,2,4,\ldots$, while the Case II normalization relation uses the same sequence but for $k = 0,1,2,\ldots$. Also to get $I_\nu(z)$ by the Case II scheme, e^z must be evaluated. On the other hand, if $|z|$ is large, $R(z) > 0$, one often wants $e^{-z}I_\nu(z)$ and this is furnished by the Case II technique. It appears that for the same n, the Case II procedure might be more accurate than the Case I scheme even for moderate values of $|z|$, $R(z) > 0$, in view of the presence of e^z in the numerator of (24). Also, Case II is favored when $R(\nu+\delta) < 0$. Improved information cannot be derived from (24) as the estimate is for fixed z, m and ν. For error analyses it is suggested that one use (13) or (23) as appropriate. Further discussion is deferred to a later part of the section where numerical examples are presented.

If z is pure imaginary and ν is real, then $z^{-\nu}I_\nu(z)$ is real and definitely the Case I procedure is better than the Case II scheme since the former requires real arithmetic while the latter demands complex arithmetic.

If only $I_\nu(z)$ or only $e^{-z}I_\nu(z)$ is required, use of the rational approximation scheme or the equivalent backward recursion scheme demands about the same number of operations. In the absence of a priori estimates of the error, the rational approximation scheme employed in the following fashion is preferred. It is sufficient to consider the Case I situation. Compute $A_n(z)$ and $B_n(z)$ from 9.9.2(8) and 9.9.2(9) respectively for $n = 0,1,2$ and compute subsequent values of these polynomials by use of 9.9.2(10). Comparison of $A_n(z)/B_n(z)$ with $A_{n+1}(z)/B_{n+1}(z)$ affords an estimate of the error. If one requires $I_{k+\nu}(z)$ or $e^{-z}I_{k+\nu}(z)$ for a given ν and $k = 0,1,2,\ldots,r$, then obviously the backward recursion scheme is highly advantageous.

CASE III.

Here we assume that $I_\nu(z)$ is known. Thus in the notation of 9.10.2(6), $a_0 = 1$ and $a_k = 0$ for $k > 0$. Put

$$i_{m+\nu}(z) = I_\nu(z)W_{m,\nu}(z)/W_{0,\nu}(z) , \tag{25}$$

and represent the error by

$$G_{m,\nu}(z) = I_{m+\nu}(z) - i_{m+\nu}(z) . \tag{26}$$

Then

$$G_{m,\nu}(z) = (-)^m (2/z)^{m+\nu}\Gamma(m+\nu)V_n(z)\Omega_m(z)/A_n(z),$$

$$\Omega_m(z) = {}_2F_3^s(\tfrac{1}{2}-\tfrac{1}{2}m,\ 1-\tfrac{1}{2}m;\ 1-\nu-m,\ \nu+1,\ 1-m;\ z^2),$$

$$s = [\tfrac{1}{2}m-\tfrac{1}{2}],\ 0 < m \leq N+2, \tag{27}$$

where $V_n(z)$ is given in (11) and $A_n(z)$ is the hypergeometric polynomial in 9.10.3(2) with $m = 0$. We also have

$$G_{m,\nu}(z) = \frac{(-)^{m+\delta}\Gamma(m+\nu)(z/2)^{2q+2+2\nu-m}\Omega_m(z)}{\Gamma(q+1+\nu)\Gamma(q+2+\nu)\Gamma(\nu+1)I_\nu(z)}$$

$$\times\left[1+\{z^2/2(q+1+\nu)\}+O(z^4/n^2)\right],\ m>0, \tag{28}$$

with q as in (15). Clearly, the backward recurrence is convergent. Note that with n, ν and z fixed, $(-)^m G_{m,\nu}(z)$ satisfies the same recurrence formula as does $W_{m,\nu}(z)$. Also $G_{m,\nu}(z)$ is 0 when $m = 0$.

Next, it is of interest to compare Cases I and II with Case III when z, m and ν are fixed and n is sufficiently large. Then

$$\frac{G_{m,\nu}(z)}{E_{m,\nu}(z)} = \frac{(z/2)^{2n}}{\Gamma(2n+3-\delta+\nu)}\,O(n^{\delta+1-\nu}), \tag{29}$$

and so Case III is superior to Case I. Similarly, Case III is superior to Case II. Now suppose that m is sufficiently large so that in (13), the term involving $K_{m+\nu}(z)$ dominates the term involving $I_{m+\nu}(z)$ which is certainly the case if $m = 2n+1-\delta-d$, $d \ll n$. Then

$$\frac{G_{m,\nu}(z)}{E_{m,\nu}(z)} = (2/z)^{\delta}\{1+O(n^{-1})\}\{1+O(m^{-1})\}, \tag{30}$$

and under these conditions there is little to choose between the two cases. Overall, it appears that Case III gives better accuracy than Case I. However, for Case III, one must know $I_\nu(z)$, while for Case I no such knowledge is required. For all positive z and all ν, $0 \leq \nu \leq 1$, coefficients are available to facilitate the rapid evaluation of $J_\nu(z)$ and $I_\nu(z)$, see Luke (1971-1972). All of this can often make the Case III approach rather attractive. See the numerical examples.

An analytical formulation of the round off error has been given by Luke (1972c) who shows that if ω is the round off error in a particular value of $W_{m,\nu}(z)$, r fixed, then the effect of ω on the evaluation of $W_{m,\nu}(z)$, $m > r$, approaches zero as $N \to \infty$. Unfortunately, it seems difficult to deduce a pragmatic assessment of the round off error from the analytical formulations. Heuristic evidence is abundant to indicate that if 2 or 3 extra decimals are carried beyond that required for the truncation error, then the round off error is insignificant. These statements hold for all three cases.

9.10.4. Expressions for $J_\nu(z)$

As previously remarked, the analyses for $I_\nu(z)$ hold everywhere in the cut complex z-plane, $-\pi < \arg z \le \pi$, and everywhere in the cut complex ν-plane, $-\pi < \arg \nu \le \pi$, ν not a negative integer, although it is sufficient to have $0 \le \arg z \le \pi/2$ and $0 \le \arg \nu \le \pi$, ν not a negative integer. Nonetheless, it is convenient for the applications to restate the key results for $J_\nu(z)$. These are readily deduced from our findings for $I_\nu(z)$ by application of the connecting relations 9.2(2,5,10,11). Then in the $J_\nu(z)$ analyses, $-\pi/2 < \arg z \le 3\pi/2$. We omit discussion of Case II since it requires complex arithmetic to generate $J_\nu(z)$ which is real when z and ν are real.

It is convenient to use the following notation. Unless indicated otherwise, if Q is used to signify a function in the developments for $I_\nu(z)$, then Q^* is used to signify the corresponding function in the developments for $J_\nu(z)$. For example, the difference equation

$$Q^*_{m,\nu}(z) = \{2(m+\nu+1)/z\}Q^*_{m+1,\nu}(z) - Q^*_{m+2,\nu}(z) , \quad (1)$$

is satisfied by

$$J_{m+\nu}(z) \text{ and } Y_{m+\nu}(z) . \quad (2)$$

With
$$W^*_{N+2,\nu}(z) = 0 , \; W^*_{N+1,\nu}(z) = 1 , \quad (3)$$

we use (1) with $Q^*_{m,\nu}(z)$ replaced by $W^*_{m,\nu}(z)$ to compute the latter for $m = N, N-1, \ldots, 1, 0$. Then

$$W^*_{m,\nu}(z) = \{(2/z)^{2p}\Gamma(2n+2-\delta+\nu)/\Gamma(m+\nu+1)\} \times {}_2F^r_3(-p, \tfrac{1}{2}-p; -2n-1+\delta-\nu, m+\nu+1, -2p; -z^2) , \quad (4)$$

where r and p are given in 9.10.3(3). Clearly,

$$W^*_{0,\nu}(z) = (2/z)^{N+1}(\nu+1)_{N+1}A^*_n(z)\,, \tag{5}$$

where $A^*_n(z)$ is the hypergeometric polynomial in (4) with $m=0$. We consider the normalization relation

$$\begin{aligned} P^*(z) &= \frac{(z/2)^\nu}{\Gamma(\nu+1)} = \sum_{k=0}^{\infty} \frac{(2k+\nu)\Gamma(k+\nu)}{\Gamma(\nu+1)\,k!} J_{2k+\nu}(z)\,, \quad \nu \neq 0\,, \\ &= 1 = J_0(z) + 2\sum_{k=1}^{\infty} J_{2k}(z)\,, \quad \nu = 0\,. \end{aligned} \tag{6}$$

Then

$$\begin{aligned} P^*_N(z) &= (2/z)^{N+1}(\nu+1)_{N+1}B^*_n(z)\,, \\ B^*_n(z) &= {}_0F^n_1(-2n-1+\delta-\nu;\,-z^2/4)\,. \end{aligned} \tag{7}$$

The approximations to $J_\nu(z)$ and $J_{m+\nu}(z)$ are given by

$$j_\nu(z) = P^*(z)A^*_n(z)/B^*_n(z)\,, \tag{8}$$

and

$$j_{m+\nu}(z) = (z/2)^\nu W^*_{m,\nu}(z)/\Gamma(\nu+1)P^*_N(z)\,, \tag{9}$$

respectively. For the error, let

$$E^*_{m,\nu}(z) = J_{m+\nu}(z) - j_{m+\nu}(z)\,. \tag{10}$$

Then if ν is not a positive integer or zero,

$$\begin{aligned} &{}_0F^n_1(-2n-1+\delta-\nu;\,-z^2/4)E^*_{m,\nu}(z) \\ &\qquad + Z^*_n(z)\,{}_1F_2(1;\,n+2,\,-n+\delta-\nu;\,-z^2/4) \\ &\qquad = (-)^m(\pi/\sin\nu\pi)V^*_n(z)J_{-m-\nu}(z) \\ &\qquad = -\pi V^*_n(z)Y_{m+\nu}(z) + (\pi/2\tan\nu\pi)V^*_n(z)J_{m+\nu}(z)\,, \end{aligned} \tag{11}$$

$$Z^*_n(z) = \frac{(z/2)^{2n+2}\Gamma(n+1-\delta+\nu)}{(n+1)!\,\Gamma(2n+2-\delta+\nu)} J_{m+\nu}(z)\,,$$

$$V^*_n(z) = (z/2)^{2n+2-\delta+\nu}J_{2n+2-\delta+\nu}(z)/\Gamma(2n+2-\delta+\nu)\,.$$

Equation (11) can be rearranged so that with the aid of L'Hospital's theorem, we can get a representation of the error when ν becomes a positive integer or zero. This result is omitted. However, for arbitrary ν, we always have

$$\begin{aligned}{}_0F_1(-2n-1+\delta-\nu;-z^2/4)E^*_{m,\nu}(z)\\ =-Z_n(z)\,{}_1F_2^s(1;\ n+2,-n+\delta-\nu;-z^2/4) \qquad (12)\\ -\pi V_n^*(z)Y_{m+\nu}(z)+\frac{O\{(z/2)^{2n}\}}{\Gamma(2n+2-\delta+\nu)\Gamma(2n+3-\delta+\nu)}\end{aligned}$$

where $s=n-\delta+\nu(s=\infty)$ if ν is (is not) a positive integer or zero. Clearly the backward recurrence scheme is convergent. Further, for n sufficiently large, $n \gg m$, the relative error is essentially independent of m. For convenience in the applications we record the formula

$$E^*_{m,\nu}(z)=-Z_n^*(z)\{1+O(n^{-1})\}-\pi V_n^*(z)Y_{m+\nu}(z)\,. \qquad (13)$$

Let ν, n and z be fixed so that $E^*_{m,\nu}(z)$ is treated as a function of m only. Then $E^*_{m,\nu}(z)$ satisfies the recurrence formula for $W^*_{m,\nu}(z)$.

We now turn to the Case III formulation. We have

$$G^*_{m,\nu}(z)=J_{m+\nu}(z)-\mathcal{J}_{m+\nu}(z)\,, \qquad (14)$$

$$\mathcal{J}_{m+\nu}(z)=J_\nu(z)W^*_{m,\nu}(z)/W^*_{0,\nu}(z)\,, \qquad (15)$$

where $W^*_{m,\nu}(z)$ is given in (4). Also

$$\begin{aligned}G^*_{m,\nu}(z)&=(2/z)^{m+\nu}\Gamma(m+\nu)V_n^*(z)\Omega_m^*(z)/A_n^*(z)\,,\\ \Omega_m^*(z)&={}_2F_3^s(\tfrac12-\tfrac12m,\ 1-\tfrac12m;\ 1-\nu-m,\ \nu+1,\ 1-m;-z^2)\,,\\ s&=[\tfrac12m-\tfrac12]\,,\quad 0<m\le N+2\,, \qquad (16)\end{aligned}$$

with $A_n^*(z)$ as in (5), and

$$\begin{aligned}G^*_{m,\nu}(z)=&\frac{\Gamma(m+\nu)(z/2)^{2q+2+2\nu-m}\Omega_m^*(z)}{\Gamma(q+1+\nu)\Gamma(q+2+\nu)\Gamma(\nu+1)J_\nu(z)}\\ &\times|1-\{z^2/2(q+1+\nu)\}+O(z^4/n^2)|\,,\ m>0, \qquad (17)\end{aligned}$$

with q as in 9.10.3(15). Clearly, the backward recurrence scheme is convergent. Note that with n, ν and z fixed, $G^*_{m,\nu}(z)$ satisfies the recurrence formula for $W^*_{m,\nu}(z)$. Also $G^*_{m,\nu}(z)$ is 0 when $m=0$.

9.10.5. Numerical Examples

Let

$$N = 5,\ n = 3,\ \delta = 1,\ z = 2/3,\ \nu = 1/3.$$

Values of $W_{m,\nu}(z)$, $P_N(z)$ and $M_N(z)$ are given in the table below.

TABLE A.

m	$W_{m,\nu}(z)$
6	1
5	19
4	305
3	3984
2	40145
1	2 84999
0	11 80141

$P_N(z) = 880\ 75120/81$

$M_N(z) = 1\ 38952\ 97360/6561$

Since $(z/2)^{1/3}/\Gamma(4/3) = 0.77645\ 82114$, $e^{-2/3} = 0.51341\ 71190$, the Case I and Case II approximations are 0.84272 08930 and 0.84272 10326, respectively. To 10 decimals, $I_{1/3}(2/3)$ $= 0.84272\ 08819$. Thus the errors in the Case I and Case II approximations are $-0.111\cdot 10^{-7}$ and $-0.151\cdot 10^{-6}$, respectively. Using 9.10.3(13,23) each with $O(n^{-1})$ and the term involving $K_{m+\nu}(z)$ neglected, the approximate Case I and Case II errors are $-0.110\cdot 10^{-7}$ and $-0.149\cdot 10^{-6}$, respectively.

For a second example, let

$$N = 5,\ n = 3,\ \delta = 1,\ z = 2,\ \nu = 0.$$

Again we illustrate the Case I and Case II schemes. We have the following data

TABLE B.

m	$W_{m,\nu}(z)$
6	1
5	6
4	31
3	130
2	421
1	972
0	1393

$P_N(z) = 611$

$M_N(z) = 4515$

$e^2 = 7.38905\ 6099$

m	$i_m(z)$ CASE I	$i_m(z)$ CASE II	$I_m(z)$
0	2.27986 9067	2.27972 4285	2.27958 5302
1	1.59083 4697	1.59073 3672	1.59063 6855
2	0.68903 4370	0.68899 0613	0.68894 8448
3	0.21276 5957	0.21275 2446	0.21273 9959
4	$0.50736\ 4975\cdot10^{-1}$	$0.50733\ 2755\cdot10^{-1}$	$0.50728\ 5700\cdot10^{-1}$
5	$0.98199\ 6727\cdot10^{-2}$	$0.98193\ 4365\cdot10^{-2}$	$0.98256\ 7932\cdot10^{-2}$
6	$0.16366\ 6121\cdot10^{-2}$	$0.16365\ 5728\cdot10^{-2}$	$0.16001\ 7336\cdot10^{-2}$

m	Error CASE I	Error CASE II	Relative Error CASE I	Relative Error CASE II
0	$-0.284\cdot10^{-3}$	$-0.139\cdot10^{-3}$	$-0.124\cdot10^{-3}$	$-0.610\cdot10^{-4}$
1	$-0.198\cdot10^{-3}$	$-0.968\cdot10^{-4}$	$-0.124\cdot10^{-3}$	$-0.609\cdot10^{-4}$
2	$-0.859\cdot10^{-4}$	$-0.422\cdot10^{-4}$	$-0.125\cdot10^{-3}$	$-0.612\cdot10^{-4}$
3	$-0.260\cdot10^{-4}$	$-0.125\cdot10^{-4}$	$-0.122\cdot10^{-3}$	$-0.587\cdot10^{-4}$
4	$-0.793\cdot10^{-5}$	$-0.471\cdot10^{-5}$	$-0.156\cdot10^{-3}$	$-0.929\cdot10^{-4}$
5	$0.571\cdot10^{-5}$	$0.634\cdot10^{-5}$	$0.581\cdot10^{-3}$	$0.645\cdot10^{-4}$
6	$-0.365\cdot10^{-4}$	$-0.364\cdot10^{-4}$	$-0.228\cdot10^{-1}$	$-0.228\cdot10^{-1}$

Use 9.10.3(13,23), each with $O(n^{-1})$ and the term involving $K_{m+\nu}(z)$ neglected. Then the approximate relative errors for Cases I and II, respectively, are $-0.116\ 10^{-3}$ and $-0.537\ 10^{-4}$, respectively.

In the table below, we record the approximate errors found by use of 9.10.3(3) with $O(n^{-1})$ omitted for $m = 6,5$ and by use of the recursion formula for the lower values of m. In this connection, see the remark following 9.10.3(23). We call this Case I-A. We also present the analogous Case II-A data based on 9.10.3(23) and the remark following this equation. In each instance known tabular values of $K_m(2)$ and $I_7(2)$ were used. In practice, we suggest using the approximations

$$I_{m+\nu}(z) = \{(z/2)^{m+\nu}/\Gamma(m+\nu+1)\}\{1+O(m^{-1})\}, \quad (1)$$

$$K_{m+\nu}(z) = \tfrac{1}{2}(z/2)^{-m-\nu}\Gamma(m+\nu)\{1+O(m^{-1})\}, \quad (2)$$

with $O(m^{-1})$ neglected. The appropriate form for $J_{m+\nu}(z)$ is the same as the right hand side of (1) while the appropriate form for $Y_{m+\nu}(z)$ is $(-2/\pi)$ times the right hand side of (2). Both (1) and (2) hold for z fixed and m large, and so are are not uniform in z. In some applications, use of the lead term of the uniform asymptotic expansion of these functions, see Olver (1954,1973) might be preferred. We also suggest that computation of the gamma functions be simplified as follows. With $|R(\alpha)| < 1$ and r a positive integer, we have

$$\Gamma(r+\alpha+1) = r!\Gamma(r+\alpha+1)/\Gamma(r+1) = r!r^{\alpha}\{1+O(r^{-1})\}$$

and for r sufficiently large, we neglect the order term. The approximation is of course superfluous if $\alpha = 0$. If $\alpha = \pm \frac{1}{2}$, the approximation may still be used though known tables of the gamma function for half an odd integer may be preferred. See also Chapter I.

TABLE C.

m	Approximate Error CASE I-A	Approximate Error CASE II-A
0	$-0.264\cdot10^{-3}$	$-0.114\cdot10^{-3}$
1	$-0.184\cdot10^{-3}$	$-0.765\cdot10^{-4}$
2	$-0.797\cdot10^{-4}$	$-0.373\cdot10^{-4}$
3	$-0.242\cdot10^{-4}$	$-0.110\cdot10^{-4}$
4	$-0.723\cdot10^{-5}$	$-0.410\cdot10^{-5}$
5	$-0.475\cdot10^{-5}$	$-0.536\cdot10^{-5}$
6	$-0.310\cdot10^{-4}$	$-0.309\cdot10^{-4}$

For a final example, we illustrate Case III using the data of our second example. We get the following numbers.

TABLE D.

m	$i_m(z)$	$G_m(z)$	Approximate Error
0	2.27958 5302	0	0
1	1.59063 6693	$0.162\cdot10^{-6}$	$0.155\cdot10^{-6}$
2	0.68894 8609	$-0.162\cdot10^{-6}$	$-0.155\cdot10^{-6}$
3	0.21273 9475	$0.484\cdot10^{-6}$	$0.467\cdot10^{-6}$
4	$0.50730\ 1826\cdot10^{-1}$	$-0.161\cdot10^{-5}$	$-0.155\cdot10^{-5}$
5	$0.98187\ 4502\cdot10^{-2}$	$0.693\cdot10^{-5}$	$0.669\cdot10^{-5}$
6	$0.16364\ 5750\cdot10^{-2}$	$-0.363\cdot10^{-4}$	$-0.350\cdot10^{-4}$

Here $G_m(z)$ is the true error while the approximate error means that $G_m(z)$ is approximated for $m = 1$ by use of 9.10.3(28) with the order neglected and subsequent values of the error are found by use of the recurrence formula as explained in the remark after 9.10.3(28). Use of the recurrence formula in this fashion is stable as the error is an increasing function of m.

A measure of the accuracy of the three cases treated can be had by use of normalization relations. Thus if the Case III procedure is employed, then 9.10.3(6,14) with $I_{k+\nu}(z)$ replaced by $\dot{i}_{k+\nu}(z)$ are available as checks. Similarly, 9.10.3(6 and 14) are available as checks for Cases II and I respectively. Some other useful normalization relations can be deduced from 5.11. For more details, see Luke (1969).

Analyses of the error in the backward recursion process for solution of a general second and higher order linear difference equation have been given by a number of authors. Some authors have studied the case of Bessel functions directly. We make no attempt to survey the various contributions here. Some pertinent references are given in Chapter 12. Suffice it to say that none of the analyses have the precision and simplicity of those given here. We deliberately chose N and as a consequence n small ($N = 5$, $n = 3$) in our numerical examples to put our asymptotic estimates under a severe test. The efficiency and realism of our error formulas is manifest.

9.11. Evaluation of Bessel Functions by Application of Trapezoidal Type Integration Formulas

The following developments stem from application of two trapezoidal type quadrature rules to some integral transforms which define certain Bessel functions. For further details, see Luke (1969) and the references quoted there. For general transforms, the rules are as follows.

$$I = hT_m + G, \tag{1}$$

$$I = \int_0^a f(t)\,dt, \quad T_m = \sum_{k=0}^{m} f(kh) - \tfrac{1}{2}\{f(0) + f(a)\}, \quad a = mh, \tag{2}$$

$$G = -2\sum_{r=1}^{\infty} G_r, \quad G_r = \int_0^a f(t)\cos(2\pi rt/h)\,dt, \tag{3}$$

and

$$I = hL_m + H, \quad L_m = \sum_{k=0}^{m-1} f(kh + \tfrac{1}{2}h), \quad a = mh, \tag{4}$$

$$H = -2\sum_{r=1}^{\infty} (-)^r G_r. \tag{5}$$

We assume that all integrals are defined and that the infinite series are convergent. We also suppose that the results hold in the limit when $a \to \infty$ whence also $m \to \infty$. Clearly $f(\infty) = 0$.

EXAMPLE I. BESSEL FUNCTIONS OF THE FIRST KIND OF ORDER N

For n a positive integer or zero,

$$J_n(z) = (-)^{\frac{1}{2}n}(2/\pi)\int_0^{\frac{1}{2}\pi} \cos(z \cos t)\cos(nt)dt, \ n \text{ even}, \tag{6}$$

$$J_n(z) = (2/\pi)\int_0^{\frac{1}{2}\pi} \sin(z \sin t)\sin(nt)dt, \quad n \text{ odd}. \tag{7}$$

For any n,

$$G_r = \tfrac{1}{2}[(-)^n J_{4mr-n}(z) + J_{4mr+n}(z)], \qquad mh = \pi/2. \tag{8}$$

If z is fixed, $J_n(z)$ decreases rapidly as n increases, and the series G and H are rapidly convergent. Evaluation of the first term of these series is usually sufficient to appraise the accuracy of the trapezoidal rules. The error can be estimated using well-known tables or by application of the inequalitiy

$$|J_q(z)| \leqslant \frac{|(z/2)^q| e^y}{q!}, \qquad z = x + iy, \quad y \geqslant 0, \tag{9}$$

Notice that for large $|z|$, one usually computes $e^{-y}J_n(z)$ and an efficient approximate bound for the error follows by use of (9). In illustration, suppose $n = 0$ and $|z| = 2$. Let $m = 3$. Now $J_{12}(2) = 0.193\cdot10^{-8}$ and $I_{12}(2) = 0.225\cdot10^{-8}$. Thus both $|G|$ and $|H|$ do not exceed $0.45\cdot10^{-8}$. Also $2/(12)! = 0.418\cdot10^{-8}$. Hence, on the basis of (9), if $|z| \leq 2$, 3 terms of either rule assure at least 8D accuracy for the computation of $e^{-y}J_0(z)$. Again, suppose $n = 0$ and $|z| = 10$. Let $m = 7$. From (9), $e^{-10}|J_{28}(z)| < 0.119\cdot10^{-9}$ for all $|z| \leq 10$. It follows that 7 terms of either rule assure at least 9D accuracy for the evaluation of $e^{-y}J_0(z)$ for all $|z| \leq 10$.

Next, consider the representation

$$J_n(z) = \frac{2(z/2)^n}{\pi(1/2)_n}\int_0^{\pi/2} \cos(z\cos\theta)\sin^{2n}\theta \, d\theta. \tag{10}$$

where n is a positive integer or zero. Here

$$G_r = \left(\frac{z}{2}\right)^n n!\left[J_{4mr}(z) + \sum_{s=1}^{n}\frac{(-)^s}{(n-s)!\,(n+s)!}\{J_{4mr+2s}(z) + J_{4mr-2s}(z)\}\right], \quad mh = \pi/2, \tag{11}$$

and a bound for the remainder can be deduced from (9).

A third useful representation for the Bessel function of the first kind is

$$J_n(w) = \frac{1}{\pi}\left(\frac{x-y}{x+y}\right)^{n/2}\int_0^\pi e^{y\cos\theta}\cos(x\sin\theta - n\theta)\,d\theta,$$

$$w = (x^2 - y^2)^{1/2}, \quad R(x+y) > 0, \qquad w = u + iv, \quad v \geqslant 0. \tag{12}$$

where again n is a positive integer or zero. In this case

$$G_r = \frac{1}{2}\left[(-)^n\left(\frac{x-y}{x+y}\right)^{mr} J_{2mr-n}(w) + \left(\frac{x+y}{x-y}\right)^{mr} J_{2mr+n}(w)\right], \quad mh = \pi. \tag{13}$$

Example II. The Modified Bessel Function of the Second Kind

Consider

$$K_\nu(z) = \int_0^\infty e^{-z\cosh t}\cosh \nu t\, dt, \qquad R(z) > 0. \tag{14}$$

More generally, we study

$$\mathscr{K}_{\alpha,\nu}(z) = \int_0^\infty e^{-z\cosh t}(\operatorname{sech} t)^\alpha \cosh \nu t\, dt,$$

$$R(z) > 0 \quad \text{or} \quad R(z) = 0 \quad \text{and} \quad R(\alpha) > |R(\nu)|. \tag{15}$$

Observe that if α is a positive integer or zero r, then

$$\mathscr{K}_{0,\nu}(z) = K_\nu(z), \qquad \mathscr{K}_{1,\nu}(z) = \int_z^\infty K_\nu(t)\,dt,\ldots, \qquad \mathscr{K}_{r,\nu}(z) = \int_z^\infty \mathscr{K}_{r-1,\nu}(t)\,dt. \tag{16}$$

Also,

$$\mathscr{K}_{-r,\nu}(z) = (-)^r \frac{d^r K_\nu(z)}{dz^r}. \tag{17}$$

We have the further representation

$$\mathscr{K}_{\alpha,\nu}(z) = [\Gamma(\alpha)]^{-1}\int_z^\infty (t-z)^{\alpha-1} K_\nu(t)\,dt,$$

$$R(z) > 0, \quad R(\alpha) > 0 \quad \text{or} \quad z = 0, \quad R(\alpha) > |R(\nu)|. \tag{18}$$

If either of the rules (1)-(3) or (4),(5) are applied to (14) and if G_r is now called $G_r(z;\alpha,\nu)$, then

$$G_r(z;\alpha,\nu) = \tfrac{1}{2}[\mathscr{K}_{\alpha,iq_1}(z) + \mathscr{K}_{\alpha,iq_2}(z)],$$

$$q_1 = p + i\nu, \qquad q_2 = p - i\nu, \qquad p = 2\pi r/h, \tag{19}$$

$$\mathscr{K}_{\alpha,iq}(z) = \left\{\frac{(\pi/2q)^{1/2}(z/q)^\alpha e^u}{\sinh q\pi}\left[\sin\theta + \frac{z^2}{4q}\cos\theta + \frac{z^2}{32q^2}\{8(2\alpha+1) - z^2\}\sin\theta\right]\right.$$
$$\left. + v_\alpha\left[1 + \frac{(\alpha-1)(\alpha-2)z^2}{2\{q^2+(\alpha-2)^2\}}\right] - zv_{\alpha-1}\left[1 + \frac{(\alpha-2)(\alpha-3)z^2}{6\{q^2+(\alpha-3)^2\}}\right]\right\}$$
$$\times [1 + O(q^{-3})],$$
$$\alpha \text{ and } z \text{ fixed}, \qquad |q| \to \infty, \qquad |\arg(\alpha + iq)| < \pi, \tag{20}$$

$$u = \frac{q\pi}{2} - \frac{\alpha(\alpha+1)(2\alpha+1)}{12q^2}, \qquad v_\alpha = \frac{\pi q^{\alpha-1}}{\Gamma(\alpha)}\exp\left\{-\frac{q\pi}{2} + \frac{\alpha(\alpha-1)(\alpha-2)}{6q^2}\right\},$$
$$\theta = q\ln\left(\frac{2q}{ez}\right) - \frac{\pi}{4}(2\alpha - 1) - \frac{(6\alpha^2+6\alpha+1)}{12q^2}. \tag{21}$$

In practice, there is also an error due to truncation of T_∞ and L_∞. This is easy to estimate and we suppose that the error due to this source is less than G or H as appropriate.

The representation (15) is advantageous for evaluation of $K_\nu(z)$ by the trapezoidal rules since ν need not be a positive integer or zero. However, z is restricted to lie in the right half-plane, so that it is not possible to recover values of $J_\nu(x)$ and $Y_\nu(x)$, $x > 0$, by use of 9.2(11). Further, the error analysis is valid for $|z|/h$ small. In fact, it can be shown that for a given h, the error increases as $|z|$ increases. We now turn to a suitable representation for $K_n(z)$ valid for $|\arg z| < \pi$, for which the error in the trapezoidal rules decreases as $|z|$ increases. Consider

$$K_n(z) = \frac{2e^{-z}}{(\frac{1}{2})_n(2z)^n}\int_0^\infty e^{-t^2}t^{2n}(2z+t^2)^{n-\frac{1}{2}}\,dt,$$
$$n \text{ a positive integer or zero}, \qquad |\arg z| < \pi. \tag{22}$$

Let $E_n(z,h)$ and $F_n(z,h)$ correspond to G of (1) and H of (3) respectively when these rules are used to approximate (22). Then for $R(z) \geq \pi^2/2h^2$,

$$E_n(z,h) \sim -\frac{2(-)^n\pi^{1/2}e^{-z}e^{-\pi^2/h^2}(\pi^2/2h^2z)^n(2z - \pi^2/h^2)^{n-\frac{1}{2}}}{(\frac{1}{2})_n(1 - e^{-2\pi^2/h^2})}, \tag{23}$$

$$F_n(z,h) \sim \frac{2(-)^n\pi^{1/2}e^{-z}e^{-\pi^2/h^2}(\pi^2/2h^2z)^n(2z - \pi^2/h^2)^{n-\frac{1}{2}}}{(\frac{1}{2})_n(1 + e^{-2\pi^2/h^2})}. \tag{24}$$

These estimates for the remainder are quite realistic. In illustration, for $z = 8$, $h = 2$, and $n = 0$, the right-hand side of (24) gives $0.272 \cdot 10^{-4}$, whereas the true error is $0.267 \cdot 10^{-4}$. Again for $z = 8$, $h = 2$, and $n = 1$, the right-hand side of (24) yields $-0.112 \cdot 10^{-3}$, while the true error is $-0.100 \cdot 10^{-3}$.

For references on the above, see Hunter (1964,1968), Krumhaar (1965) and Luke (1969).

To use (21), n must be a positive integer or zero and $R(z) \geq \pi^2/2h^2$. To avoid these restrictions Mechel (1966) studied evaluation of the Mellin-Barnes integral representation for $K_\nu(z)$ which follows from 8.3(13) and 5.3.1(1,2) by use of the trapezoidal rule. See this source for further details.

9.12. Inequalities

Since Bessel functions are a special case of confluent hypergeometric functions, inequalities for the former follow from 5.14(13-16), 7.9(1-3) and 5.14(17, 19-21) in view of the connecting relations 9.2(4) and 8.3(13). In particular

$$\{1 - \frac{x}{2(\nu+1)}\}\{1 + \frac{(2\nu+1)x}{2(\nu+1)}\}^{-1} < e^{-x}M_\nu(x)I_\nu(x)$$

$$< \frac{1-2\nu}{2\nu+3} + \frac{2(2\nu+1)}{2\nu+3}\{1+\frac{(2\nu+3)x}{2(2\nu+1)}\}^{-1}, \quad x>0, \quad -\tfrac{1}{2} \leq \nu \leq \tfrac{1}{2},$$

$$M_\nu(x) = (2/x)^\nu \Gamma(\nu+1), \tag{1}$$

$$\frac{2-x}{2+x} < e^{-x}M_\nu(x)I_\nu(x) < \frac{1}{2\nu+3} + \frac{2(\nu+1)}{2\nu+3}\{1+\frac{(2\nu+3)x}{2(\nu+1)}\}^{-1}, \quad x>0, \ \nu \geq -\tfrac{1}{2}, \tag{2}$$

$$e^{-x} < e^{-x}M_\nu(x)I_\nu(x) < \tfrac{1}{2}(1+e^{-2x}), \quad x>0, \quad \nu > -\tfrac{1}{2}. \tag{3}$$

The left hand sides of these inequalities are very weak unless x is near zero. If $\nu = 0$, (1) and (2) coincide. If $\nu = -\tfrac{1}{2}$, (3) becomes an equality. The inequalities become equalities if $x = 0$. For $K_\nu(x)$, we have

$$1 - \frac{\mu}{2x+\mu} < (2x/\pi)^{\frac{1}{2}}e^{x}K_\nu(x) < 1 - \frac{\mu}{2x+1+\frac{1}{2}\mu}, \quad x>0, \quad 0 \leq \nu < \tfrac{1}{2}, \tag{4}$$

$$1 - \frac{\mu}{2x+1+\frac{1}{2}\mu} < (2x/\pi)^{\frac{1}{2}}e^{x}K_\nu(x) < 1 - \frac{\mu(2\nu+1)}{4x(5-2\nu)} - \frac{3\mu(3-2\nu)}{4(5-2\nu)\{x+(2\nu+1)(5-2\nu)/24\}}, \quad \mu = \tfrac{1}{4}-\nu^2, \ x>0, \ \tfrac{1}{2} < \nu < 3/2. \tag{5}$$

The inequalities (4) and (5) have the correct asymptotic behavior as $x \to \infty$. They also become equalities if $\nu = \tfrac{1}{2}$ and (5) becomes an equality if $\nu = 3/2$. For improvement of these results and extension of them in the ν direction, see Luke (1972a). These inequalities are very sharp. In illustration, put $\nu = 0$ in (4). We then get the following inequality and table.

$$L(x) = \frac{8x}{8x+1} < F(x) = \left(\frac{2x}{\pi}\right)^{\frac{1}{2}} e^x K_0(x) < R(x) = \frac{16x+7}{16x+9}, \quad x > 0. \tag{7}$$

x	$L(x)$	$F(x)$	$R(x)$
0.01	0.07407	0.38049	0.78166
0.10	0.44444	0.67679	0.81132
0.50	0.80000	0.85989	0.88235
1.0	0.88889	0.91315	0.92000
2.0	0.94118	0.94961	0.95122
4.0	0.96970	0.97230	0.97260
10.0	0.98765	0.98814	0.98817

(8)

Notice that for $x \geq \frac{1}{2}$, the arithmetic mean of $L(x)$ and $R(x)$ approximates $F(x)$ to within about 2.2%. This is quite remarkable as $K_0(x)$ has a logarithmic singularity at $x = 0$.

The inequalities

$$|M_\nu(z) J_\nu(z)| \leq e^y, \quad z = x + iy, \quad y > 0, \quad \nu \geq -\tfrac{1}{2}, \tag{8}$$

$$|M_{\nu-1}(z) J'_\nu(z)| \leq \tfrac{1}{2} e^y \{1 + |\tfrac{1}{2} z^2 / \nu(\nu+1)|\}, \quad \nu > -\tfrac{1}{2}, \tag{9}$$

$M_\nu(z)$ as in (1), have been given by Watson (1945, p. 49). The statements

$$\partial I_\nu(x)/\partial\nu < 0, \quad x > 0, \quad \nu \geq 0; \quad \partial K_\nu(x)/\partial\nu > 0, \quad x > 0, \quad \nu \geq 0, \tag{10}$$

have been given by Cochran (1967) and Reudnik (1968). Lorch (1967) proved that

$$0 < I_{\nu+\varepsilon}(x)/I_\nu(x) < 1, \quad x > 0; \quad \varepsilon > 0, \quad \nu > -\tfrac{1}{2}\varepsilon; \quad \sin\tfrac{1}{2}\pi\varepsilon > 0, \quad \nu = -\tfrac{1}{2}\varepsilon,$$

$$0 < K_\nu(x)/K_{\nu+\varepsilon}(x) < 1, \quad x > 0, \quad \varepsilon > 0, \quad \nu \geq 0, \tag{11}$$

which includes results given by Soni (1965), S. L. Gupta (1966) and A. L. Jones (1967). For $\varepsilon = 1$, Nasell gave the improved inequality

$$\frac{1 + (2\nu+3)/2x}{1 + \{2(\nu+1)/x\} + \{(\nu+1)(2\nu+3)/x^2\}} < \frac{I_{\nu+1}(x)}{I_\nu(x)} < (1 + \nu/x)^{-1}, \quad x > 0, \quad \nu \geq -1, \tag{12}$$

which is asymptotically correct for ν fixed and $x \to \infty$. Two sided inequalities for $I_{\nu+1}(x)/I_\nu(x)$ have already been given by 9.9.2(30). These are quite sharp and are asymptotically correct in the directions $x \to 0$, $\nu \to \infty$.

Let

$$G_{\nu,k}(x) = 2\sum_{j=1}^{k} \frac{(-)^{j+1}k!\,(2\nu+1)_{j-1}(\nu+j)}{(k-j)\,(2\nu+k+1)_j} I_{\nu+j}(x)\,, \quad (13)$$

$$H_{\nu,k}(x) = G_{\nu,k}(x) + e^{-x}/M_\nu(x)\,, \quad (14)$$

$M_\nu(x)$ as in (1). Then Nasell (1974) proved that

$$G_{\nu,k}(x) < H_{\nu,k}(x),$$

$$0 < H_{\nu,k}(x) < H_{\nu,k+1}(x) < I_\nu(x),\ k \geq 0,$$

$$H_{\nu,k}(x) \sim I_\nu(x),\ x \to 0,\ k \geq 0, \quad (15)$$

$$I_\nu(x) - H_{\nu,k}(x) \sim \{(2\nu+1)_k/(2x)^k\}I_\nu(x),\ x \to \infty,\ k \geq 0,$$

$$\lim_{k\to\infty} H_{\nu,k}(x) = I_\nu(x)\,.$$

The basic equation equation used in the proof of these items is the expansion formula 7.8(7) with $a=\nu+\frac{1}{2}$, $c=2a+k$ and $z=2x$. We compare the inequality $H_{\nu,k}(x)<I_\nu(x)$, (call this Case I), with the right hand side inequality of (11) for $\varepsilon=1$, (call this Case II), and $\varepsilon>0$,(call this Case III), and the right hand side of (12), (call this Case IV). Notice that $G_{\nu,1}(x)=I_{\nu+1}(x)$. So Case I is sharper than Case II for all $k\geq 1$. Case I is sharper that Case III for x sufficiently small. Again Case I is sharper than Case III for $k\geq 2$ and x sufficiently large. Finally, Case I is sharper than Case IV for $k\geq 2$ and $\nu>-\frac{1}{2}$. It can be shown that

$$I_\nu(x) - G_{\nu,2}(x) = \{4(\nu+1)/(2\nu+3)\}\left[I_\nu(x) - \{1+(\nu+\tfrac{1}{2})/x\}I_{\nu+1}(x)\right]. \quad (16)$$

If we replace ν by $\nu+1$ and eliminate $I_{\nu+2}(x)$ with the aid of the recurrence formula for $I_\nu(x)$, then the left hand inequality of (12) follows. Stronger versions of this result obtain by use of the inequalities $H_{\nu,k}(x)<I_\nu(x)$ for $k\geq 2$ or $G_{\nu,k}(x) < I_\nu(x)$ for $k\geq 3$.

It is convenient to put

$$r_\nu(x) = I_{\nu+1}(x)/I_\nu(x)\,,\quad x \geq 0,\ \nu \geq 0 \quad (17)$$

and to define some parameters by the following table.

n	a_n	b_n	c_n	d_n
1	$\nu+1$	$\nu+1$	ν	$\nu+2$
2	$\nu+\frac{1}{2}$	$\nu+3/2$	$\nu+\frac{1}{2}$	$\nu+\frac{1}{2}$

(18)

Then Amos (1974) proved that

$$L_n(x,\nu) = \frac{x}{a_n + (x^2+b_n^2)^{\frac{1}{2}}} < r_\nu(x) < \frac{x}{c_n + (x^2+d_n^2)^{\frac{1}{2}}} = R_n(x,\nu), \quad x \geq 0, \ \nu \geq 0. \qquad (19)$$

Note that $L_1(x,\nu) \leq L_2(x,\nu)$ for $x \geq 0$ and $R_2(x,\nu) \leq R_1(x,\nu)$ provided that $4(2\nu+3)(\nu+1) < x^2$. Amos also showed that further improved bounds can be found by use of an iterative process. The inequalities (19) are quite sharp and have the correct asymptotic behavior in the directions $x\to 0$, $x\to\infty$ or $\nu\to\infty$. Now

$$I_\nu(x) = (x/z)^\nu I_\nu(z) \exp\{\int_z^x r_\nu(t)dt\}. \qquad (20)$$

Combining (19) and (20), Amos derived

$$L(x,z,\nu) \leq I_\nu(x) \leq R(x,z,\nu), \qquad (21)$$

$$L(x,z,\nu) = (x/z)^\nu I_\nu(z)\{zL_2(x,\nu)/xL_2(z,\nu)\}^p \exp\{(x^2-z^2)/f\},$$

$$R(x,z,\nu) = (x/z)^\nu I_\nu(z)\{zR_2(x,\nu)/xR_2(z,\nu)\}^p \exp\{(x^2-z^2)/g\}, \qquad (22)$$

$$p=\nu+\tfrac{1}{2}, \quad q=p+1, \quad f=(x^2+q^2)^{\frac{1}{2}}+(z^2+q^2)^{\frac{1}{2}}, \quad g=(x^2+p^2)^{\frac{1}{2}}+(z^2+p^2)^{\frac{1}{2}},$$

with the proviso that $z \leq x$ and with the inequality signs in (21) reversed when $x < z$. The most interesting cases are $z\to 0$ and $z\to\infty$. For these cases, we have

$$L(x,0,\nu) = N_\nu(x)\{sL_2(x,\nu)\}^p \exp\{xR_2(x,\nu+1)\},$$

$$R(x,0,\nu) = N_\nu(x)\{pR_2(x,\nu)\}^p \exp\{xR_2(x,\nu)\}, \qquad (23)$$

$$s = \nu+1, \quad N_\nu(x) = (2/x)^{\frac{1}{2}}/\Gamma(\nu+1),$$

$$L(x,\infty,\nu) = (2\pi x)^{-\frac{1}{2}}\{R_2(x,\nu)\}^p \exp\{x^2+p^2\}^{\frac{1}{2}},$$

$$R(x,\infty,\nu) = (2\pi x)^{-\frac{1}{2}}\{L_2(x,\nu)\}^p \exp\{x^2+(p+1)^2\}^{\frac{1}{2}}, \qquad (24)$$

where $L_2(x,\nu)$ and $R_2(x,\nu)$ are defined in (18), (19). The results (21-24) are quite sharp. They also have the correct asymptotic behavior as $x\to 0$, $x\to\infty$ or $\nu\to\infty$.

Further inequalities for $r_\nu(x)$ follow from (20) with $z=0$ and 9.9.2(30). Such inequalities are also quite sharp, but for moderately large yet bounded values of x and all $\nu > -1$. They have the correct asymptotic behavior as $x\to 0$, $\nu\to\infty$. Also, the inequalities become sharper as n increases in view of 9.9.2(27). For $n=0$ and 1, respectively, we find

$$0 < M_\nu(x)I_\nu(x) < \exp\{x^2/4(\nu+1)\}, \tag{25}$$

$$(1+x^2/\alpha^2)^\beta < M_\nu(x)I_\nu(x) < (1+x^2/\gamma)^\delta \exp\{\delta x^2/\gamma\}, \tag{26}$$

$$\alpha = \nu+2,\ \beta = (\nu+2)^2/4(\nu+1),\ \gamma = 2(\nu+2)(\nu+3),\ \delta = \gamma/8(\nu+1).$$

The above results are by no means complete. For further details, see the sources cited. For further results on inequalities relating to Bessel functions, see P.A.P. Moran (1957), Lorch and Szego (1966), Prohorov (1968), Petrovskaja (1970), B.S. Lee and Shah (1970), Elbert (1971), Steinig (1971), Grünbaum (1973) and Askey (1973).

9.13. Bibliographic and Numerical Data

9.13.1. References

For general references on the entire spectrum of Bessel functions, see Whittaker and Watson (1927), Watson (1945), Erdélyi et al (1953, 1954), Petiau (1955), Luke (1962, 1969), Abramowitz and Stegun (1964) and Tranter (1969). Since Bessel functions are a special case of confluent hypergeometric functions, the references given in 5.1 and 7.11.1 are also pertinent.

The zeros of Bessel functions as a function of the variable have a well established literature. Döring (1967) discusses this aspect of the subject and develops one all purpose algorithm for the successive computation of the coefficients in the McMahon expansions for real and complex zeros of cylinder functions and their derivatives. Hethcote (1970a) develops a procedure to get inequalities for zeros of solutions of second order differential equations. He gives inequalities for the n-th negative zero of $Ai(z)$ and the n-th positive zero of $J_\nu(x)$, $\nu \geq 0$. In recent years, a number of papers have been devoted to the zeros of Bessel functions as a function of the order. See Magnus and Kotin (1960), Franz (1961), D.S. Cohen (1964), Streifer (1965), Cochran (1964, 1965, 1966a, 1966b), Martinek, Thielman and Huebschman (1966), Hall (1967), Martinek (1968), Hethcote (1970b), Ferreira and Sesma (1970) and Cochran and Hoffspiegel (1970).

For evaluation of integrals of products of three Bessel functions, see P.J. Roberts (1970) and Jackson and Maximon (1972). A table of integrals of Bessel functions which can be expressed in terms of the complete elliptic integrals has been given by Okui (1974). For nonnumerical tables of summable series and integrals involving Bessel functions, see Wheelon (1968) and referencesin 5.1.

9.13.2. Description of and References to Tables

Bessel Functions of Integral and General Order

Fettis and Caslin (1967): $I_n(x)$, $e^{-x}I_n(x)$, $n = 0,1$, $x = 0(0.001)10$, 15S.

Fettis and Caslin (1969b): Same as above entry and also $x = 1(1)10$, $n = x(1)x + 25$, 16S.

Berger and McAllister (1970): $I_n(x)$, $K_n(x)$, $n = 0,1$, $x = 1(1)40$, 61S-98S.

Bark, Bol'shev, Kuznetsov and Cherenkov (1964): $e^{-x}I_0(x)$, $x = 0(0.001)3(0.01)15(0.1)24.9$, 7D; $\exp(-x^2)I_0(x^2)$, $x^{-1} = 0(0.001)0.2$, 7D.

McClain, Schoenig and Palladino (1962): $J_n(x)$, $Y_n(x)$, $I_n(x)$, $K_n(x)$, $n = 0,1$, $x = 0(0.1)85$, 4S.

Dempsey and Benson (1960): $K_n(\pi a q^{\frac{1}{2}})$, $n = 0(1)10$, $q = 1(1)250$ if $a = \frac{1}{2}$, $q = 1(1)300$ if $a = 1/3$, 10D

Olver (1962): This tome gives tables to facilitate the evaluation of $J_\nu(\nu x)$, $Y_\nu(\nu x)$, $I_\nu(\nu x)$ and $K_\nu(\nu x)$ and their first derivatives to 10S (except near zeros) when $\nu \geq 10$.

Wrench (1970): Converging factors $C_n(n)$ and its reduced derivatives for the evaluation of $K_\nu(x)$, $n = 10(1)40$, 30D. For similar data relating to $I_\nu(x)$, see Wrench (1971).

Young and Kirk (1964): Let $\rho_n = ber_n^2(x) + bei_n^2(x)$, $\theta_n = \arctan\{bei_n(x)/ber_n(x)\}$, $\sigma_n = ker_n^2(x) + kei_n^2(x)$, $\phi_n = \arctan\{kei_n(x)/ker_n(x)\}$. Table I: $ber_n(x)$, $bei_n(x)$ $ker_n(x)$, $kei_n(x)$, $n = 0,1$, $x = 0(0.1)10$, 15D. Table 2: Same functions as in Table 1 and ρ_n, θ_n, σ_n, ϕ_n, $n = 0,1,2$, $x = 0(0.01)2.5$, 7S or 8S. Table 3: Same functions as in Table 2, $n = 0(1)10$, $x = 0(0.1)10$, 6S or 7S.

Aggarwal and Sagherian (1969): Let $h_n(z) = zH_{n-1}^{(1)}(z)/H_n^{(1)}(z)$. $R\{h_n(z)\}$, $I\{h_n(z)\}$, $n = 0(1)10$, $z = 0(0.2)15$, 5D.

Bessel Functions of Fractional Order

Singh, Lumley and Betchov (1963): $h_n(z) = v^{1/3}H_{1/3}^{(n)}(v)$, $n = 1,2$, $v = (2/3)z^{3/2}$, $h(z)$, $\int_0^s h(iu)\,du$, $\int_0^s\int_0^t h(iu)\,du\,dt$, $z = is$ $s = -10(0.1)10$ where h stands for h_n, 4 to 8S.

Dougherty and Johnson (1964): $\pi^{\frac{1}{2}}Ai(t)$, $\pi^{\frac{1}{2}}Bi(t)$, their derivatives and related data, $t = -6(0.1)6$, mostly 8S.

Nosova and Tumarkin (1961): Airy functions and their derivatives for pure imaginary argument is, $s = 0(0.1)6$, 6D. Also some tables to facilitate the evaluation of $\int_0^t Bi(-x)dx$ and $\int_0^t Ai(-x)dx$ where t is pure imaginary. See also Osipova and Tumarkin (1963).

Jakovleva (1969): Airy functions and their derivatives, $t = -9(0.001)9$, 7D.

Bessel Functions of Pure Imaginary Order

S. Luke and Weissman (1964): $K_{iq}(v)$, $v = e^{-x}$, varying x and q. The range of the tables may be generally described as $q = 0.2(0.2)50$ where the tables are cut in the x direction for each set of q's where the oscillating amplitude appears to be a constant. The accuracy is at least 5S for $q < 40$ and 4S for higher q.

Zhurina and Karmazina (1967): $K_{iq}(x)$, $q = 0.01(0.01)10$, $x = 0.1(0.1)10.2$, 7S.

Kiyono and Murashima (1973): $K_{iq}(x)$, $q = 0.01(0.01)0.05, 0.1(0.1)2(0.5)6$, $x = 0.01(0.01)0.1(0.1)1(0.5)5$, 8S

Derivatives with Respect to the Order

K. Lee and Radosevich (1960): $\partial J_\nu(z)/\partial\nu$, $\nu = \alpha(1)4+\alpha$, $\alpha = 1/4, 1/3, 2/3, 3/4$, $z = 1(0.5)$, 4D.

Erber and Gordon (1963): $\partial I_\nu(z)/\partial\nu$, $\nu = \pm 1/3$, $z = 0.01(0.01)1(0.05)5$, 4D.

Infinite Series Involving Bessel Functions

Bark and Kuznetsov (1962): $\sum_{m=0}^{\infty} (y/x)^{n+2m} I_{n+2m}(x) = x(y/x)^n \int_0^1 I_{n-1}(xt)\{\cosh \tfrac{1}{2}y(1-t^2)\}t^n dt$, $n = 1,2$, $y = 0(0.01)1(0.1)20$, $x = 0(0.01)1(0.1)y$, 7S.

Hebermehl, Minkwitz and Schulz (1965): $\sum_{m=0}^{\infty} (-)^m (w/z)^{n+2m} \times J_{n+2m}(z) = z(w/z)^n \int_0^1 J_{n-1}(zt)\{\cos \tfrac{1}{2}w(1-t^2)\}t^n dt$, $w = 0(0.2,\pi)39.8$, $z = 0(0.2)4$; $w = 0(0.2,\pi)19.8$, $z = 4.2(0.2)12$, 7D.

Ng (1966): Let $Y_n(w,z)$ be the function treated by Bark and Kuznetsov above with $x = z$ and $y = w$, and put $Z_n(w,z) = Y_n(z,w)$. Y_1, Y_2, Z_0, Z_1, $w = 0.1(0.1)1$, $z = 0.1(0.1)1$; Y_1, Y_2, $w = 1(1)z$, $z = 2(1)20$; Z_0, Z_1, $w = 2(1)20$, $z = 1(1)w$, 6S.

Tibery and Wrench (1964): Let $T_{1,\nu} = S_{1,\nu}(ix) - \nu e^{i\nu\pi/2}K_\nu(x)$, $R = \{8(\mu^2+1)/\pi^2\mu^2\}\sum_{m=0}^{\infty}\{T_{1,\nu}(\nu\mu)/(2m+1)^2\}$, where $\nu = p(m+\frac{1}{2})$. R, $p = 2(1)10$, $\mu/\mu_0 = 0.2(0.1)0.8(0.05)0.95$, $\mu_0 = 0.25(0.25)6(1)12$, 5S.

Zeros and Extrema of Bessel Functions

Gerber (1964): First one hundred zeros of $J_0(x)$, 19S.

Morgenthaler and Reisman (1963): Zeros of $J_n'(x)$, $21 \leq n \leq 51$, $0 \leq x < 100$, 8D generally.

Roman (1969): Critical points less than 100 and corresponding extrema of the first four derivatives of $J_0(x)$, 10S.

Switzer (1965): First eight zeros of (a): $\tan S = C$, $C = 0.001(0.001)0.1(0.01)1(0.1)10(1)100(10)400$, 5D; (b): $S \cot S = C$, $C = -0.999(0.001)-0.1(0.01)1(0.1)10(1)100(10)400$, 5D; (c): $CJ_0(S) = SJ_1(S)$, C as in (a), 5D.

Parnes (1972): Complex zeros $z_{n,p}$ of $K_n(z)$, $n = 2(2)10$, $p = 1(1)n$; $n = 3(2)9$, $p = 1(1)n-1$, 9D.

Döring (1966b): All complex zeros of $Y_n(z)$ and $H_n^{(1)}(z)$, $n = 0,1$, $|z| < 158$, $|\arg z| \leq \pi$, and some complex zeros of these functions for $n = 2(1)5$, 15 and $|\arg z| \leq \pi$ are given to 10D.

G. F. Miller and Haines (1959): α_s and $Ai'(-\alpha_s)$ such that $Ai(-\alpha_s) = 0$, β_s and $Ai(-\beta_s)$ such that $Ai'(-\beta_s) = 0$, $s = 1(1)56$, 15D.

Sherry (1959): Same functions as in G. F. Miller and Haines above, $s = 1(1)50$, 25S. Also first eighteen coefficients in the asymptotic series expansion of $\arctan\{Ai(x)/Bi(x)\}$ and $\arctan\{Ai'(x)/Bi'(x)\}$, 25S.

Cochran and Hoffspiegel (1970): This gives tabular data to facilitate evaluation of ν-zeros of $H_\nu^{(1)}(z)$ and its derivative by use of the uniform asymptotic expansions.

Zeros of Functions Involving Products of Bessel Functions

Description of the data is facilitated by use of the notation

$$A_n(x) = J_n(x)Y_n(kx) - J_n(kx)Y_n(x)\ ,$$

$$B_n(x) = J_n'(x)Y_n'(kx) - J_n'(kx)Y_n'(x)\ .$$

Bridge and Angrist (1962): First eleven zeros of $B_n(x)$, $n = 1(1)12$, $k = 1.1, 1.2, 1.5(0.5)5$, 5D.

Laslett and Lewish (1962): This is an investigation of the zeros of $A_n(x)$ and $B_n(x)$ with tabular data in which n is large and $(k-1)/(k+1)$ is small.

Bauer (1964): First ten zeros of $B_n(x)$, $n = 0(1)25$, $k = 0(0.1)0.9$, 5D.

Weil, Murty and Rao (1967): First ten positive zeros of $A_n(x)$, $n = 0(1)10$, $k = 0(0.05)0.95$, 7S.

Weil, Murty and Rao (1968): First ten positive zeros of $B_n(x)$, $n = 0(1)10$, $k = 0(0.05)0.95$, 5D.

Fettis and Caslin (1966): Let $C(x) = J_0(x)Y_1(kx) - Y_0(x)J_1(kx)$. First five zeros of $A_n(x)$, x_m, $m = 0(1)4$ and normalized zeros $y_m = (1-k)x_m/m\pi$, $n = 0,1$, $k = 0.01(0.01)\ 0.99$, 10D. First five zeros x_m and normalized zeros $y_m = |1-k|/mr\pi$, $r = [m-\frac{1}{2}]$, of $C(x)$, $k = 0.01(0.01)0.99$, $1.01(0.01)20$, 10D. Also first five zeros and normalized equivalents for $C(x)$, $k^{-1} = 0.001(0.001)0.05$, 10D.

Fettis and Caslin (1968): See the previous entry for notation. The $m = 5(1)10$ zeros and normalized equivalents of $A_n(x)$, $n = 0,1$ and $C(x)$ for $k = 0.001(0.001)0.3$, 10D. Also zeros for $C(x)$, $k^{-1} = 0.001(0.001)0.3$, 10D.

Integrals Involving a Single Bessel Function

In addition to the following material, see Singh *et al*, p. 404; Nosova and Tumarkin, Osipova and Tumarkin, and Bark and Kuznetzov, p. 405; Hebermehl *et al*, p. 405; and Ng, p. 406.

Drane and McIlvena (1963): $\int_0^y J_0(t)dt$, $y = 0.04(0.04)5$, 6D; $\int_0^y I_0(t)dt$, $y = 0.04(0.04)10$, 6D; $\int_x^1 I_0(yw)dt/\int_0^1 I_0(yt)dt$, $w = (t^2 - x^2)^{\frac{1}{2}}$, $y = 0.04(0.04)10$, $x = 0(0.01)0.99$, 6D;

$\int_x^1 J_0(yw)dt / \int_0^1 J_0(yt)dt$ $y = 0.04(0.04)5, x = 0(0.01)$ 0.99, 6D.

Frisch-Fay (1965): $\int_0^z J_0(t)dt$, $\pm \nu = 1/6, 1/4, 1/3, 1/2, 2/3, 3/4, 5/6$, $z = 0(0.1)6.3$, 7D and $z = 6.3(0.1)10$, 5D.

Harvey (1965): $\int_0^x e^{-bt} I_0(t)dt$, $b = 0(0.1)1$, $x = 0(0.1)10$, 6S.

Chatelet, Lansraux and Million (1959): Let $L_{i,j}^{(n)}(x)$ be the be the polynomials which enter in the Lagrangian interpolating polynomial to $f(x)$. That is, $L_{i,j}^{(n)}(x) = \prod_{j=1}^{n} (x - x_j) / \prod_{\substack{j=1 \\ j \neq i}}^{n} (x_i - x_j)$. $\int_0^1 L_i^{(n)}(x) J_0(xy) d(x^2)$, $n = 11$, $i = 1(1)11$, $y = 0(1)12$, 10D.

DiDonato and Jarnagin (1962). Let $P(R,D) = \exp(-D^2/2) \times \int_0^R \exp(-r^2/2) I_0(rD) r dr$. Table of R for $P = 0.01, 0.05 (0.05)0.95, 0.97, 0.995$ and $1 - 10^{-m}$, $m = 3(1)6$, $D = 0.1, 0.5(0.05)2(1)6(2)10(10)30, 50, 80, 120$, 7S.

Weingarten and DiDonato (1961): Let $P(K,c) = c^{-1} \int_0^E F(u)du$, $F(u) = e^{-Bu} I_0(Au)$, $E = \frac{1}{2}K^2$, $A = (1 - c^2)/2c^2$, $B = A + 1$. Table of K for $P = 0.05(0.05)0.95(0.01)0.99$, $c = 0.05 (0.05)1$, 5D.

Bark, Bol'shev, Kuznetsov and Cherenkov (1964): $\int_u^\infty F(t)dt$, $F(t) = t\exp\{-\frac{1}{2}(v^2 + t^2)\} I_0(vt)$, $u = 0(0.02)y$, various y but max $y = 7.84$, $v = 0(0.02)3$, 6D.

Dishon and Weiss (1966): $A(y,w) = (2w)^{-1} \int_2^\infty u e^{-f} I_0(uy/2w) du$, $\partial A(y,w)/\partial y$, are tabulated for various y and w, 5D. Here $f = \{(u^2 + y^2)/4w\}$.

George (1962): $\int_0^\infty e^{-zu} u^n \sin u \, J_\nu(ur) du$, $\int_0^\infty e^{-zu} u^{n-1} (1 - \cos u) \times J_\nu(ur) du$, $\nu = 0$, $n = 0, 1$ or $\nu = 1$, $n = -1, 0, 1$, $r, z = 0 (0.1)2(1)10$, 5D.

Deutsch (1962): Explicit evaluation of $\int_0^\infty e^{-zt} t^{m-2} (\sin at - at \cos at) J_n(rt) dt$ for most values of $m = -2(1)2$, $n = 0 (1)3$.

Integrals Involving Products of Two or More Bessel Functions

M. Kumar and Dhawan (1970): $\int_0^\infty e^{-pt}t^\lambda J_\mu(at)J_\nu(bt)dt$ is tabulated to 6D for $p = 2$ and an assortment of μ, ν, a and b values.

Peavy (1967): Exact numerical coefficients to facilitate evaluation of $\int^z tf(t)C_0(at)D_0(bt)dt$ where $C_0(at)$ and $D_0(bt)$ are cylinder functions and $f(t)$ is a polynomial.

Newman and Frank (1963): $z^{-2m}\int_0^{\frac{1}{2}\pi} J_n^2(z\sin t)\sin^{1-2m}t\,dt$

$$= \frac{z^{2n-2m}\Gamma(\tfrac{1}{2})\Gamma(n+\tfrac{1}{2}-m)}{2^{2n+1}(n!)^2(n-m)!}\ {}_2F_3\left(\begin{matrix} n+\tfrac{1}{2},\ n+\tfrac{1}{2}-m \\ 2n+1,\ n+1,\ n+m+1 \end{matrix}\middle| -z^2\right),$$

all pairs (m,n) such that m, $n = 0(1)3$ and $m,n = \tfrac{1}{2}(1)3\tfrac{1}{2}$ with $m \le n$, $z = 0(0.2)10$, 5S.

Fettis (1963): $\int_0^\infty tJ_n(xt)\{J_0(t)\}^5dt$ $n = 0,1$, $x = 0(0.1)1$, 6D.

Bishop (1970): $\int_0^\infty \{x^\mu K_\nu(x)\}^4 x^n dx$, $n = 0(1)4$, $\nu = 0.1(0.1)5$, 8S.

Kölbig (1965): $\int_0^1 J_1(j_{n,m}t)J_0(xt)dt$, $j_{n,m}$ is the m-th positive zero of $J_n(x)$, $n = 0,1,2$, $m = 1(1)5$, $x = 0(0.1)20$, 6D.

Taggart and Schott (1970): Assorted tables of $\int_0^s r^k j_t(ar)j_n(br)dr$ and integrals involving derivatives of the spherical Bessel functions $j_\mu(z)$, $as = y_{n,p}$, the p-th zero of $j_n(ar)$, $k = 0,1,2$, $s = 1$, n and t are integers ≤ 5, $p = 1,2,3$, 5S

McQueary and Mack (1967): Let $J_0(K_n) = 0$, $J_{m,n} = J_m(K_n r)$, $m = 0,1$, $I(F) = \int_0^1 rF(r)dr$, where $F(r)$ represents a product of four Bessel functions of type $J_{m,n}$ for various $n = 1, 2,\ldots,10$. Tables of $I(F)$, 5D. Similar tables are given for the case $J_1(K_n) = 0$.

Kilpatrick, Katsura and Inoue (1966): An assortment of tables for $\int_0^\infty t^\alpha J_\mu(at)J_\nu(bt)J_\gamma(ct)f(t)dt$, where $f(t)$ is constant or proportional to $J_\delta(t)$. Accuracy ranges from 6S to 14S.

Katsura and Nishihara (1969): Coefficients p^u in the polynomial expression for $\int_0^\infty J_{\frac{1}{2}}(pt)J_{3/2+m}(bt)J_{3/2+n}(ct)t^{-3/2}dt$, $u = -\tfrac{1}{2}(1)m+n+7/2$, $m,n = 0(1)6$, $m+n$ even, $b,c = 1$ and 2, 16S.

Other Integrals Involving Bessel Functions

Troesch and Weidman (1972): $\int_0^\infty \{J_m(t)/J_m'(t)\}dt$, $m = 1$, $r = 0(0.1)1.8, 1.84$; $m = 2$, $r = 0(0.1)3, 3.04$; $m = 3$, $r = 0(0.2)4.2$; $m = 4$, $r = 0(0.2)5.2$, 5.28; 3D.

Cooke and Tranter (1959): Let $G(n,\nu) = f_n \int_0^\infty \{t^n K_\nu(t)/I_\nu(t)\}dt = \{f_n/(n+1)\}\int_0^\infty \{t^n/I_\nu^2(t)\}dt$. Put $f_n = 1$. $G(n,\nu)$, $\nu = \pm\frac{1}{2}$, $n = 1(2)9$, 5S; $\nu = 0$, $n = 0(2)10$, 5S; $\nu = 1$, $n = 2(2)8$, 5S.

J.A. Roberts (1965): $G(n,\nu)$ as above, $f_n = 1$, $\nu = 1$, $n = 1(1)100$, 7S. Coefficients in the asymptotic expansion of $K_1(t)/I_1(t)$ for large t and in the asymptotic expansion of $G(n,1)$ for large n are also given. See this reference for comments on the accuracy of the data given by Smythe and Brenner and Sonshine noted below.

Smythe (1960-1964): $G(2n,\nu)$ as above, $f_{2n} = (2n+1)/(n!)^2$, $\nu = 0,1$, $n = 0(1)84$, 6-7D or 8S.

Brenner and Sonshine (1964): $G(2n,\nu)$ as above, $f_{2n} = 2(-)^n/(2n)!\pi$, $\nu = 1$, $n = 2(2)24$, 8S.

Ling and Lin (1972): $G(n,\nu)$ as above, $f_n = 2^{n+1}/\pi(n!)$, $n \geq 2\nu$, $\nu = 0,1,2$, $n = 0(1)50$, 12D. Coefficients in an asymptotic expansion of $G(n,\nu)$ for large n are also tabulated.

Kölbig (1964): Let $A(x,t) = t^{-1}e^{-xt}/\{K_0^2(t) + \pi^2 I_0^2(t)\}$, $B(t) = a + bI_0(t)$, where $a = 0$ and $b = 1$ or $a = 1$ and $b = 0$, $C(t) = I_0(at)K_0(t) - I_0(t)K_0(at)$. $\int_0^\infty A(x,t)B(t)dt$, $\int_0^\infty A(x,t)C(t)dt$, various x and a, 4D.

G. F. Miller (1966b): $(2/\pi)\int_0^\infty [u^2\{A_n^2(u) + A_{n-1}^2(u)\}]^{-1}A_n^2(u)du$, $A_n(u) = J_{n+\frac{1}{2}}(u)$, $n = 1(1)10(5)40$, 5D.

9.13.3. Description of and References to Other Approximations and Expansions

Clenshaw (1962): $J_n(x)$, $Y_n(x)$, $I_n(x)$, $K_n(x)$, $CT(x)$, $n = 0,1$, $a = 8$, 20D. Also same functions, $CT(y)$, $a = 8$, 20D except for $I_n(x)$, $K_n(x)$, 14D.

Clenshaw and Picken (1966): $J_\nu(x)$, $Y_\nu(x)$, $I_\nu(x)$, $K_\nu(x)$, $CT(x)$, $CT(y)$, $\nu = 0, 1/4, 1/3, 1/2, 2/3, 3/4, 1$, $a = 8$, mostly 20D.

The authors also give coefficients for $x^{-\nu}\{\varepsilon J_\nu(x) + (1-\varepsilon)I_\nu(x)\} = \sum a_{rs}T_s(-\nu)\{\varepsilon T_r(-t) + (1-\varepsilon)T_r(t)\}$, $\varepsilon = 0,1$, $-1 \leq \nu \leq 1$, $t = x^2/64$, $x \leq 8$, 6D.

Luke (1971-1972): See the discussion on pp. 320, 321.

Arthurs and McCarroll (1961): $F_r(x) = (\pi/2x)^{1/2}(x/2)^{-r} J_{r+\frac{1}{2}}(x)$, $CT(x)$, $r = 0(1)15$, $a = 10$, 12D.

Burgoyne (1962): $I_0(x)$, $x^{-1}I_1(x)$, $\tau CP(x)$, $a = 4$, 9D.
$I_n(x) = (2\pi x)^{-1/2} e^x F_n(x)$, $F_n(x)$, $n = 0, 1$, $\tau CP(y)$, $a = 4$, 9D.

Wimp (1962): $x^{-\nu}J_\nu(x)$, $x^{-\nu}I_\nu(x)$, $CT(x)$, $\nu = 0, \pm\frac{1}{3}, \pm\frac{2}{3}, 1$, $a = 5$, 15D. $Y_n(x)$, $K_n(x)$, $CT(x)$, $n = 0, 1$, $a = 5$, 15D. The exponent for $B_1^{-(1/4)}$ should be $(+01)$ instead of $(+00)$.

Luke and Wimp (1963): $K_0(x)$, $K_1(x)$, $CT^*(y)$, $a = 2$, 20D.

Bhagwandin (1962): $x^{-\nu}J_\nu(x)$, $x^{-\nu}I_\nu(x)$, $\nu = \pm\frac{1}{3}, \pm\frac{2}{3}$, $BCP(x)$, $a = 4$, 10D. $K_\nu(x)$, $\nu = 1/3, 2/3$, $BCP(y)$, $a = 4$, 10D. Also a best Chebyshev rational approximation for $H_\nu^{(1)}(x)$ where $\nu = 1/3, 2/3$ valid for $x \geq 4$. Accuracy is about 10D.

Gloden (1964): $J_0(x)$, $x^{-1}J_1(x)$, $Y_n(x)$, $n = 0, 1$, $BCP(x)$, $a = 8$, 12D. $I_0(x)$, $x^{-1}I_1(x)$, $K_n(x)$, $n = 0, 1$, $BCP(x)$, $a = 8$, 12D. $J_n(x)$, $e^{-x}I_n(x)$, $Y_n(x)$, $K_n(x)$, $n = 0, 1$, $BCP(x)$, $8 \leqslant x \leqslant 13$, 12D. Further coefficients of a like nature are given so that the full infinite range $0 \leqslant x \leqslant \infty$ is covered.

Burgoyne (1963): ber, bei, ker, kei, and their first derivatives. $\tau CP(x)$, $a = 10$, 9D; $\tau CP(y)$, $a = 10$, 9D.

Gargantini and Pomentale (1964): Rational Chebyshev approximations for $\mathscr{K}_{\alpha,\nu}(z)$, see 9.11(16), over several segments to cover the entire positive real axis. $\nu = 0$, $\alpha = 1,2,3$, with accuracy from about 3D to at least 5D.

Hart et al (1969): $J_n(x)$, $Y_n(x)$, $n = 0,1$, $BCP(x)$, various approximations to cover the range $0 < x < \infty$, accuracy up to about 25D.

Russon and Blair (1969): Best rational Chebyshev approximations to $I_n(x)$ and $K_n(x)$, $n = 0,1$, for the ranges $0 \leq x \leq 1$ and $x \geq 1$. There are numerous approximations and accuracy goes from about 10^{-3} to 10^{-23}.

Blair (1974) and Blair and Edwards (1974): Best rational Chebyshev approximations for $I_0(x)$ and $I_1(x)$. Errors range down to about 10^{-23}.

Németh (1971): Chebyshev expansions of the Airy functions, their zeros, derivatives, first and second integrals. First 14 to 39 coefficients are given to 15D.

Luke (1973): See the discussion in 9.9.1 on p. 361.

Wimp and Luke (1969): $e^{-x}I_n(x)$, $CT^*(x)$, $n = 0,1$, $a = 8$, 16D; $x^{-1}e^{(b-1)x}\int_0^x e^{-bt}I_0(t)dt$, $CT^*(x)$, $b = 0(0.1)1$, $a = 8$, 16D.

Németh (1972): Let $j_{\nu,n}$ be the n-th positive zero of $J_\nu(x)$. This report gives the coefficients $c_{k,n}$ in the formulas

$$j_{\nu,n} = \sum_{k=0}^{\infty} c_{k,n}T_n^*(u), \qquad j_{\nu,n} = \nu\sum_{k=0}^{\infty} d_{k,n}T_n^*(w),$$

$$u = \nu/2 < 1, \qquad w = (2/\nu)^{3/2}, \quad \nu \geq 2,$$

$n = 0,1$, 15D. Similar coefficients are given for the zeros of $J'_\nu(x)$, $d\{x^{-\frac{1}{2}}J_\nu(x)\}/dx$, $d\{x^{\frac{1}{2}}J_\nu(x)\}/dx$.

Werner (1958): Various best Chebyshev polynomial and rational approximations for $J_n(x)$, $Y_n(x)$, $I_n(x)$, $K_n(x)$, $n = 0,1$ and the integrals of $I_0(x)$ and $K_0(x)$ with accuracy up to about 8D. For the most part approximations are given so that all the positive real axis is covered.

Gargantini (1966): Best Chebyshev rational approximations for $I_n(x)$, $K_n(x)$, $n = 0,1$, $\int_0^x I_0(t)dt$ and $\int_x^\infty K_0(t)dt$. The ranges covered are the closed intervals (0,70), (0,8), (0,30) and (0,70) for the first, second, third and fourth items, respectively. Accuracy is about 8S.

Golden, McGuire and Nuttal (1973): Discussion of numerical results on calculation of Bessel functions with Pade approximants. No tables of coefficients are given.

Gargantini and Henrici (1967). Continued fraction representations for $K_0(z)$.

CHAPTER X LOMMEL FUNCTIONS, STRUVE FUNCTIONS, AND ASSOCIATED BESSEL FUNCTIONS

10.1. Definitions, Connecting Relations and Power Series

$$s_{\mu,\nu}(z) = \frac{z^{\mu+1}}{(\mu-\nu+1)(\mu+\nu+1)} \times {}_1F_2\left(\begin{matrix}1\\ \frac{1}{2}(\mu-\nu+3),\ \frac{1}{2}(\mu+\nu+3)\end{matrix}\middle| \frac{-z^2}{4}\right). \tag{1}$$

$$S_{\mu,\nu}(z) = s_{\mu,\nu}(z) + \{2^{\mu-1}\Gamma[\tfrac{1}{2}(\mu-\nu+1)]\,\Gamma[\tfrac{1}{2}(\mu+\nu+1)]\} \times \{\sin[(\mu-\nu)\,\pi/2]\,J_\nu(z) - \cos[(\mu-\nu)\,\pi/2]\,Y_\nu(z)\}. \tag{2}$$

$$\mathbf{H}_\nu(z) = \frac{(z/2)^{\nu+1}}{\Gamma(3/2)\,\Gamma(3/2+\nu)}\,{}_1F_2\left(\begin{matrix}1\\ 3/2,\ 3/2+\nu\end{matrix}\middle| \frac{-z^2}{4}\right) = [\pi 2^{\nu-1}(\tfrac{1}{2})_\nu]^{-1}\,s_{\nu,\nu}(z). \tag{3}$$

$$\mathbf{H}_\nu(z) - Y_\nu(z) = [\pi 2^{\nu-1}(\tfrac{1}{2})_\nu]^{-1}\,S_{\nu,\nu}(z). \tag{4}$$

$$\mathbf{L}_\nu(z) = \frac{(z/2)^{\nu+1}}{\Gamma(3/2)\,\Gamma(3/2+\nu)}\,{}_1F_2\left(\begin{matrix}1\\ 3/2,\ 3/2+\nu\end{matrix}\middle| \frac{z^2}{4}\right) = \exp[-\tfrac{1}{2}(\nu+1)\,i\pi]\,\mathbf{H}_\nu(ze^{i\pi/2}). \tag{5}$$

Equations (1) and (2) are Lommel functions. $\mathbf{H}_\nu(z)$ is the Struve function and $\mathbf{L}_\nu(z)$ is the modified Struve function.

$$\mathbf{J}_\nu(z) = \frac{\sin\nu\pi}{\pi}\,s_{0,\nu}(z) - \frac{\nu\sin\nu\pi}{\pi}\,s_{-1,\nu}(z). \tag{6}$$

$$\mathbf{E}_\nu(z) = -\frac{(1+\cos\nu\pi)}{\pi}\,s_{0,\nu}(z) - \frac{\nu(1-\cos\nu\pi)}{\pi}\,s_{-1,\nu}(z). \tag{7}$$

$$\pi[\mathbf{E}_n(z) + \mathbf{H}_n(z)] = \sum_{k=0}^{<\frac{1}{2}n} \frac{\Gamma(k+\frac{1}{2})(\frac{1}{2}z)^{n-2k-1}}{\Gamma(n+\frac{1}{2}-k)}. \tag{8}$$

$$\pi[\mathbf{E}_{-n}(z) + \mathbf{H}_{-n}(z)] = (-)^{n+1} \sum_{k=0}^{<\frac{1}{2}n} \frac{\Gamma(n-k-\frac{1}{2})(\frac{1}{2}z)^{-n+2k+1}}{\Gamma(k+\frac{3}{2})}. \tag{9}$$

$\mathbf{J}_\nu(z)$ and $\mathbf{E}_\nu(z)$ are known as Anger–Weber functions.

The functions $S_{\mu,\nu}(z)$ and those related to it are often called associated Bessel functions since any linear combination of $S_{\mu,\nu}(z)$, $J_\nu(z)$ and $Y_\nu(z)$ satisfy the differential equation

$$\Delta w(z) = f(z), \quad \Delta = z^2D^2 + zD + (z^2 - \nu^2), \quad D = d/dz, \tag{10}$$

with $f(z) = z^{\mu+1}$. In general, the particular solution of (10) with arbitrary $f(z)$ is called an associated Bessel function. Two other associated Bessel functions are

$$h_{\mu,\nu}(z) = \frac{e^{-z}z^{\mu+1}}{(\mu-\nu+1)(\mu+\nu+1)}\,{}_2F_2\left(\begin{matrix}1, \mu+\frac{3}{2}\\ \mu-\nu+2, \mu+\nu+2\end{matrix}\middle|\, 2z\right), \tag{11}$$

$$H_{\mu,\nu}(z) = h_{\mu,\nu}(z) - \frac{\Gamma(\mu-\nu+1)\,\Gamma(\mu+\nu+1)}{2^\mu(\frac{3}{2})_\mu} \times \left[I_\nu(z) + \frac{K_\nu(z)\sin(\nu-\mu)\pi}{\pi\cos\mu\pi}\right], \tag{12}$$

which satisfy the differential equation

$$[z^2D^2 + zD - (z^2+\nu^2)]\,H_{\mu,\nu}(z) = e^{-z}z^{\mu+1}. \tag{13}$$

The importance of associated Bessel functions rests in part on the fact that they can be used to represent integrals of Bessel functions. Thus

$$\int^z t^{-1}f(t)C_\nu(t)\,dt = zC_\nu(z)w'(z) - zw(z)C'_\nu(z), \tag{14}$$

$$C_\nu(z) = AJ_\nu(z) + BY_\nu(z), \tag{15}$$

with A and B independent of z. $C_\nu(z)$ is called a cylinder function. The subject can be viewed in a different setting. There are numerous integral representations for Bessel functions of the form

$$I(z,a,b) = \int_a^b K(z,t)g(t)\,dt \tag{16}$$

where a and b are fixed and free of z. Usually $(a,b) = (0,1)$, $(0,\infty)$ or $(1,\infty)$. Integrals of this kind are referred to as complete integrals. In numerous applications, integrals of the type $I(z,a,y)$ arise. These are known as incomplete integrals or incomplete cylinder functions. They satisfy an equation like (10) and so can also be termed associated Bessel functions. References on these and related topics are detailed in 10.4.1.

10.2. Asymptotic Expansions

$$S_{\mu,\nu}(z) \sim z^{\mu-1}\,{}_3F_0(1, (1-\mu+\nu)/2, (1-\mu-\nu)/2; -4/z^2),$$
$$|z| \to \infty, \qquad |\arg z| \leq \pi - \delta, \qquad \delta > 0. \quad (1)$$

$$\mathbf{H}_\nu(z) - Y_\nu(z) \sim \frac{(z/2)^{\nu-1}}{\pi(1/2)_\nu}\,{}_3F_0(1, 1/2, 1/2-\nu; -4/z^2),$$
$$|z| \to \infty, \qquad |\arg z| \leq \pi - \delta, \qquad \delta > 0. \quad (2)$$

$$I_{-\nu}(z) - \mathbf{L}_\nu(z) \sim \frac{(z/2)^{\nu-1}}{\pi(1/2)_\nu}\,{}_3F_0(1, 1/2, 1/2-\nu; 4/z^2),$$
$$|z| \to \infty, \qquad |\arg z| \leq \pi/2 - \delta, \qquad \delta > 0. \quad (3)$$

$$H_{\mu,\nu}(z) \sim -\frac{e^{-z}z^\mu}{2\mu+1}\,{}_3F_1\left(\begin{matrix}1, -\mu+\nu, -\mu-\nu\\ \frac{1}{2}-\mu\end{matrix}\middle|\frac{-1}{2z}\right),$$
$$|z| \to \infty, \qquad |\arg z| \leq 3\pi/2 - \delta, \qquad \delta > 0. \quad (4)$$

10.3. Expansions in Series of Chebyshev Polynomials and Bessel Functions

$$s_{\mu,\nu}(ax) = \{(ax)^{\mu+1-\varepsilon}a^\varepsilon/2^{\varepsilon-1}\}\sum_{n=0}^{\infty} A_n T_{2n+\varepsilon}(x), \quad \varepsilon = 0,1, \quad 0 < x \leq 1, \quad (1)$$

$$A_n = \left[(-)^n(a/4)^{2n}/(\mu-\nu+1)(\mu+\nu+1)\{\tfrac{1}{2}(\mu-\nu+3)\}_n\{\tfrac{1}{2}(\mu+\nu+3)\}_n\right]$$
$$\times\,{}_2F_3\left(\begin{matrix}\tfrac{1}{2}+\varepsilon+n,\ 1+n\\ 1+\varepsilon+2n,\ \tfrac{1}{2}(\mu-\nu+3)+n,\ \tfrac{1}{2}(\mu+\nu+3)+n\end{matrix}\middle| -a^2/4\right). \quad (2)$$

A recursion formula for A_n follows from 5.13.4(19-21) upon noting that we can go from the ${}_4F_3$ there to the above ${}_2F_3$ by confluence. If $\mu = -1$ or $\mu = 0$, A_n can be expressed as a product of Bessel functions in view of 9.4(1). For expansion of the Lommel functions $(ax)^{1-\mu}S_{\mu,\nu}(ax)$ in series of Chebyshev polynomials $T_{2n}(1/x)$, see 5.10.1(19).

Expansions of $s_{\mu,\nu}(z)$ in series of Bessel Functions follow from 5.11.

TABLE 10.1

CHEBYSHEV COEFFICIENTS FOR $\mathbf{H}_\nu(x)$ AND $\mathbf{H}_\nu(x) - Y_\nu(x)$, $\nu = 0, 1$

$$\mathbf{H}_0(x) = \sum_{n=0}^{\infty} a_n T_{2n+1}(x/8) \qquad \mathbf{H}_0(x) = (x/8) \sum_{n=0}^{\infty} b_n T_{2n}(x/8)$$

$-8 \le x \le 8$

n	a_n	n	b_n
0	0.18231 19926 92574 06885	0	1.00215 84560 99119 80619
1	-0.06866 17653 15081 64320	1	-1.63969 29268 13091 47468
2	0.38875 91215 80854 73371	2	1.50236 93961 82928 18828
3	-0.26764 89396 55143 68269	3	-0.72485 11530 21218 72087
4	0.07944 13767 40525 67667	4	0.18955 32737 10931 35549
5	-0.01364 74528 78064 02930	5	-0.03067 05202 29880 00215
6	0.00155 29821 65312 96133	6	0.00337 56144 73751 94355
7	-0.00012 66377 63099 94869	7	-0.00026 96501 43126 02089
8	0.00000 77960 86420 52048	8	0.00001 63746 16926 12350
9	-0.00000 03761 14076 60050	9	-0.00000 07824 44085 08254
10	0.00000 00146 26332 71602	10	0.00000 00302 15931 88153
11	-0.00000 00004 68736 53930	11	-0.00000 00009 63266 44950
12	0.00000 00000 12602 41570	12	0.00000 00000 25793 37089
13	-0.00000 00000 00288 47808	13	-0.00000 00000 00588 53949
14	0.00000 00000 00005 69231	14	0.00000 00000 00011 58332
15	-0.00000 00000 00000 09785	15	-0.00000 00000 00000 19870
16	0.00000 00000 00000 00148	16	0.00000 00000 00000 00300
17	-0.00000 00000 00000 00002	17	-0.00000 00000 00000 00004

$$\mathbf{H}_1(x) = \sum_{n=0}^{\infty} c_n T_{2n}(x/8)$$

$-8 \le x \le 8$

n	c_n
0	0.55788 91446 48160 50428
1	-0.11188 32572 65698 16037
2	-0.16337 95812 52009 39277
3	0.32256 93207 24059 02436
4	-0.14581 63236 72442 42034
5	0.03292 67739 93740 35217
6	-0.00460 37214 20935 72841
7	0.00044 34706 16331 39592
8	-0.00003 14209 95293 41169
9	0.00000 17123 71993 80035
10	-0.00000 00741 69870 05204
11	0.00000 00026 18376 70705
12	-0.00000 00000 76858 39395
13	0.00000 00000 01906 70416
14	-0.00000 00000 00040 52291
15	0.00000 00000 00000 74633
16	-0.00000 00000 00000 01203
17	0.00000 00000 00000 00017

TABLE 10.1 *(Concluded)*

$\mathbf{H}_0(x)-Y_0(x) = (2/\pi x)\sum_{n=0}^{\infty} d_n T_{2n}(8/x)$ $\qquad$ $\mathbf{H}_1(x)-Y_1(x) = (2/\pi)\sum_{n=0}^{\infty} e_n T_{2n}(8/x)$

$x \geq 8$

n	d_n	n	e_n
0	0.99283 72757 64239 43189	0	1.00757 64729 38656 41255
1	-0.00696 89128 11386 24757	1	0.00750 31605 12482 57125
2	0.00018 20510 37870 37123	2	-0.00007 04393 32645 19049
3	-0.00001 06325 82528 44161	3	0.00000 26620 53933 82266
4	0.00000 09819 82942 86525	4	-0.00000 01884 11577 53405
5	-0.00000 01225 06454 44977	5	0.00000 00194 90149 58394
6	0.00000 00189 40833 11800	6	-0.00000 00026 12619 89905
7	-0.00000 00034 43582 25604	7	0.00000 00004 23626 90104
8	0.00000 00007 11191 01711	8	-0.00000 00000 79551 55531
9	-0.00000 00001 62887 44137	9	0.00000 00000 16799 73006
10	0.00000 00000 40656 80728	10	-0.00000 00000 03907 19821
11	-0.00000 00000 10915 04796	11	0.00000 00000 00985 43090
12	0.00000 00000 03120 05243	12	-0.00000 00000 00266 35794
13	-0.00000 00000 00942 02070	13	0.00000 00000 00076 45035
14	0.00000 00000 00298 47947	14	-0.00000 00000 00023 12961
15	-0.00000 00000 00098 72416	15	0.00000 00000 00007 33212
16	0.00000 00000 00033 93712	16	-0.00000 00000 00002 42334
17	-0.00000 00000 00012 07980	17	0.00000 00000 00000 83162
18	0.00000 00000 00004 43821	18	-0.00000 00000 00000 29528
19	-0.00000 00000 00001 67859	19	0.00000 00000 00000 10816
20	0.00000 00000 00000 65200	20	-0.00000 00000 00000 04076
21	-0.00000 00000 00000 25956	21	0.00000 00000 00000 01577
22	0.00000 00000 00000 10571	22	-0.00000 00000 00000 00625
23	-0.00000 00000 00000 04397	23	0.00000 00000 00000 00253
24	0.00000 00000 00000 01866	24	-0.00000 00000 00000 00105
25	-0.00000 00000 00000 00806	25	0.00000 00000 00000 00044
26	0.00000 00000 00000 00355	26	-0.00000 00000 00000 00019
27	-0.00000 00000 00000 00159	27	0.00000 00000 00000 00008
28	0.00000 00000 00000 00072	28	-0.00000 00000 00000 00004
29	-0.00000 00000 00000 00033	29	0.00000 00000 00000 00002
30	0.00000 00000 00000 00015	30	-0.00000 00000 00000 00001
31	-0.00000 00000 00000 00007		
32	0.00000 00000 00000 00004		
33	-0.00000 00000 00000 00002		
34	0.00000 00000 00000 00001		

Chebyshev coefficients for $Y_0(x)$ and $Y_1(x)$ are given in Tables 9.1 and 9.2 respectively.

TABLE 10.2

CHEBYSHEV COEFFICIENTS FOR $\int_0^x t^{-m}\mathbf{H}_0(t)\,dt$ AND $\int_0^x t^{-m}[\mathbf{H}_0(t) - Y_0(t)]\,dt$, $m = 0, 1$

$$\int_0^x \mathbf{H}_0(t)dt = \sum_{n=0}^{\infty} a_n T_{2n}(x/8) \qquad \int_0^x t^{-1}\mathbf{H}_0(t)dt = \sum_{n=0}^{\infty} b_n T_{2n+1}(x/8)$$

$-8 \le x \le 8$

n	a_n	n	b_n
0	1.61333 66756 84918 40385	0	1.82200 49195 05665 54353
1	0.50194 75160 15311 42410	1	-0.52367 70538 32669 94383
2	-0.45742 08868 95936 37691	2	0.22272 20549 20414 69091
3	0.43760 53741 57332 27760	3	-0.06531 46019 09439 29117
4	-0.17354 51581 97834 67968	4	0.01223 46552 18933 96431
5	0.03723 55318 47435 88239	5	-0.00154 75515 77437 81571
6	-0.00506 68116 81125 66354	6	0.00014 02024 85264 53709
7	0.00047 98914 08117 97429	7	-0.00000 95341 58668 40481
8	-0.00003 36084 62380 11729	8	0.00000 05046 19441 50606
9	0.00000 18160 44554 91577	9	-0.00000 00213 85789 92011
10	-0.00000 00781 48081 86330	10	0.00000 00007 42361 86502
11	0.00000 00027 44558 04642	11	-0.00000 00000 21501 30044
12	-0.00000 00000 80223 15917	12	0.00000 00000 00527 63821
13	0.00000 00000 01983 21443	13	-0.00000 00000 00011 11339
14	-0.00000 00000 00042 02434	14	0.00000 00000 00000 20314
15	0.00000 00000 00000 77202	15	-0.00000 00000 00000 00325
16	-0.00000 00000 00000 01242	16	0.00000 00000 00000 00005
17	0.00000 00000 00000 00018		

TABLE 10.2 *(Concluded)*

$$\int_0^x \left[\mathbf{H}_0(t)-Y_0(t)\right]dt = (2/\pi)\left[\gamma+\ln 2x\right]+(2/\pi)\sum_{n=0}^{\infty} c_n T_{2n}(8/x)$$

$$\int_x^{\infty} t^{-1}\left[\mathbf{H}_0(t)-Y_0(t)\right]dt = (2/\pi)\sum_{n=0}^{\infty} d_n T_{2n+1}(8/x)$$

$x \geq 8$

n	c_n	n	d_n
0	0.00373 19877 09364 06118	0	0.12454 02165 21241 56946
1	0.00367 82678 30067 44432	1	-0.00014 89784 13526 17956
2	-0.00005 13929 52719 56746	2	0.00000 24085 45254 98516
3	0.00000 21480 31746 57617	3	-0.00000 01037 01477 42238
4	-0.00000 01592 03126 01877	4	0.00000 00076 70065 25913
5	0.00000 00169 13561 08351	5	-0.00000 00008 03677 77027
6	-0.00000 00023 07360 27228	6	0.00000 00001 07617 38161
7	0.00000 00003 78850 56381	7	-0.00000 00000 17311 55530
8	-0.00000 00000 71820 91376	8	0.00000 00000 03213 52374
9	0.00000 00000 15280 89787	9	-0.00000 00000 00669 55345
10	-0.00000 00000 03575 59565	10	0.00000 00000 00153 48766
11	0.00000 00000 00906 36626	11	-0.00000 00000 00038 13886
12	-0.00000 00000 00246 03989	12	0.00000 00000 00010 15518
13	0.00000 00000 00070 88016	13	-0.00000 00000 00002 87153
14	-0.00000 00000 00021 51372	14	0.00000 00000 00000 85604
15	0.00000 00000 00006 83935	15	-0.00000 00000 00000 26746
16	-0.00000 00000 00002 26622	16	0.00000 00000 00000 08715
17	0.00000 00000 00000 77948	17	-0.00000 00000 00000 02950
18	-0.00000 00000 00000 27734	18	0.00000 00000 00000 01033
19	0.00000 00000 00000 10178	19	-0.00000 00000 00000 00373
20	-0.00000 00000 00000 03842	20	0.00000 00000 00000 00139
21	0.00000 00000 00000 01489	21	-0.00000 00000 00000 00053
22	-0.00000 00000 00000 00591	22	0.00000 00000 00000 00021
23	0.00000 00000 00000 00240	23	-0.00000 00000 00000 00008
24	-0.00000 00000 00000 00099	24	0.00000 00000 00000 00003
25	0.00000 00000 00000 00042	25	-0.00000 00000 00000 00001
26	-0.00000 00000 00000 00018	26	0.00000 00000 00000 00001
27	0.00000 00000 00000 00008		
28	-0.00000 00000 00000 00004		
29	0.00000 00000 00000 00002		
30	-0.00000 00000 00000 00001		

Chebyshev coefficients for the integrals of $Y_0(x)$ and $x^{-1}Y_1(x)$ are given in Table 9.3.

TABLE 10.3

CHEBYSHEV COEFFICIENTS FOR $\mathbf{L}_0(x)$, $\mathbf{L}_1(x)$, AND $\int_0^x t^{-m}\mathbf{L}_0(t)\,dt$, $m = 0, 1$

$$L_o(x) = \sum_{n=0}^{\infty} a_n T_{2n+1}(x/8) \qquad L_o(x) = (x/8) \sum_{n=0}^{\infty} b_n T_{2n}(x/8)$$

$-8 \le x \le 8$

n	a_n	n	b_n
0	237.05797 11456 54399 96542	0	140.17728 76879 03660 39157
1	133.58386 78793 36169 62517	1	193.76136 69155 01479 14769
2	45.27002 69600 85132 33772	2	73.40636 88431 70860 10264
3	9.90658 74987 31401 88030	3	17.13368 50769 99404 57281
4	1.48894 26172 89602 74209	4	2.67948 99204 63399 18779
5	0.16158 21914 20658 32857	5	0.29839 53141 15806 29639
6	0.01317 83421 19061 75538	6	0.02476 90687 25510 36075
7	0.00083 42134 01106 21985	7	0.00158 76155 12613 15002
8	0.00004 20764 71776 83639	8	0.00008 08112 89599 28967
9	0.00000 17280 19116 13112	9	0.00000 33416 53954 38311
10	0.00000 00588 38384 62453	10	0.00000 01143 84277 87914
11	0.00000 00016 86630 52513	11	0.00000 00032 92491 36992
12	0.00000 00000 41238 93168	12	0.00000 00000 80769 68034
13	0.00000 00000 00869 82019	13	0.00000 00000 01708 18303
14	0.00000 00000 00015 98312	14	0.00000 00000 00031 45736
15	0.00000 00000 00000 25808	15	0.00000 00000 00000 50888
16	0.00000 00000 00000 00369	16	0.00000 00000 00000 00729
17	0.00000 00000 00000 00005	17	0.00000 00000 00000 00009

$$L_1(x) = \sum_{n=0}^{\infty} c_n T_{2n}(x/8)$$

$-8 \le x \le 8$

n	c_n
0	117.97191 65767 37839 66655
1	177.95257 99117 97242 02790
2	77.76467 90022 95114 80902
3	21.17714 53021 88699 38687
4	3.84061 71794 08746 09635
5	0.49049 62905 07139 92664
6	0.04614 52641 00329 52307
7	0.00331 56522 13378 81808
8	0.00018 73519 59230 49366
9	0.00000 85269 54178 93899
10	0.00000 03188 63377 31614
11	0.00000 00099 61858 03738
12	0.00000 00002 63732 51790
13	0.00000 00000 05989 19489
14	0.00000 00000 00117 90860
15	0.00000 00000 00002 03100
16	0.00000 00000 00000 03086
17	0.00000 00000 00000 00042
18	0.00000 00000 00000 00001

TABLE 10.3 *(Concluded)*

$$\int_0^x \mathbf{L}_0(t)dt = \sum_{n=0}^{\infty} d_n T_{2n}(x/8) \qquad \int_0^x t^{-1}\mathbf{L}_0(t)dt = \sum_{n=0}^{\infty} e_n T_{2n+1}(x/8)$$

$-8 \le x \le 8$

n	d_n	n	e_n
0	138.48598 32031 10893 88703	0	43.29660 42301 52920 81773
1	206.94820 65326 36460 68050	1	20.05916 63453 88436 50751
2	88.31384 09192 51037 28744	2	5.62726 83766 17145 55298
3	23.57562 63075 69153 63828	3	1.03244 25111 81143 24179
4	4.20882 24407 20899 56910	4	0.13228 30336 85977 38286
5	0.53094 41703 47577 76541	5	0.01243 75566 08649 81526
6	0.04946 79497 67198 85773	6	0.00089 15943 54342 20041
7	0.00352 68939 19415 86730	7	0.00005 02268 07433 79534
8	0.00019 80342 32332 34586	8	0.00000 22785 18695 43843
9	0.00000 89663 22813 49006	9	0.00000 00849 28149 38168
10	0.00000 03338 36146 30132	10	0.00000 00026 45042 53593
11	0.00000 00103 91228 01807	11	0.00000 00000 69820 03673
12	0.00000 00002 74231 93224	12	0.00000 00000 01581 22995
13	0.00000 00000 06210 63254	13	0.00000 00000 00031 05048
14	0.00000 00000 00121 97672	14	0.00000 00000 00000 53359
15	0.00000 00000 00002 09667	15	0.00000 00000 00000 00809
16	0.00000 00000 00000 03180	16	0.00000 00000 00000 00011
17	0.00000 00000 00000 00043		
18	0.00000 00000 00000 00001		

10.4. Rational Approximations for $\mathbf{H}_\nu(z) - Y_\nu(z)$ and the Errors in These Approximations

We use 8.3(17) with $\omega = 1 - \nu$ and Theorem 6 of 5.12 with

$$p = 3,\ q = 0,\ \alpha_1 = \alpha_2 = \tfrac{1}{2},\ \alpha_3 = \tfrac{1}{2} - \nu,\ a = 0,$$

$$\alpha = \beta = 0,\ \text{so}\ \lambda = 1,\ \text{and}\ z\ \text{replaced by}\ z^2/4. \tag{1}$$

Then

$$\pi(z/2)^{1-\nu}(\tfrac{1}{2})_\nu\{\mathbf{H}_\nu(z) - Y_\nu(z)\} = \{\phi_n(z)/f_n(z)\} + R_n(z)\ , \tag{2}$$

$$\phi_n(z) = \sum_{k=0}^{n} \frac{(-n)_k(n+1)_k(\frac{1}{2})_k(\frac{1}{2}-\nu)_k}{(k!)^2(3/2)_k(3/2-\nu)_k}$$

$$\times {}_3F_4\left(\begin{matrix}-n+k,\ n+k+1,\ 1\\ k+1,\ k+1,\ k+3/2,\ k-\nu+3/2\end{matrix}\middle| -z^2/4\right) \tag{3}$$

and $f_n(z)$ is the ${}_3F_4$ in (3) with $k = 0$, whence the ${}_3F_4$ becomes a ${}_2F_3$.

That both $\phi_n(z)$ and $f_n(z)$ satisfy the same recurrence formula follows from the theory in 5.13.2 and 5.13.3. In particular, the recurrence formula for $f_n(z)$ can be deduced from 5.13.2(11-15) or 5.13.3(1-4). It can also be deduced from 5.13.2(29-31) by use of the confluence principle. There, put $\lambda = 1$, $b_1 = 1$, $b_2 = 3/2$, $b_3 = 3/2 - \nu$, replace z by $z^2/4a_1a_2$ and let both a_1 and $a_2 \to \infty$. The details are straight forward and we omit the final result. The explicit statement can be found in Luke (1969).

Coefficients for the polynomials $\phi_n(z)$ and $f_n(z)$ for $n = 0(1)6$, and $\nu = 0$ and $\nu = \pm 1$ are recorded in Tables 10.4 and 10.5 respectively. Further polynomials can be generated from the recurrence formula noted above. The corresponding values of $|R_n(z)|$ for $\nu = 0, \pm 1$, are presented in Table 10.6.

TABLE 10.4

COEFFICIENTS FOR $\mathbf{H}_0(z) - Y_0(z)$

$$\mathbf{H}_0(z) - Y_0(z) = (2/\pi z)\left[\varphi_n(z)/f_n(z) + R_n(z)\right]$$

$$f_n(z) = a_0^{-1} \sum_{k=0}^{n} a_k z^{2k}$$

n	$a_0, a_1, \ldots, a_n$
0	1
1	9, 2
2	75, 50, 2
3	2205, 2940, 294, 4
4	25515, 56700, 10206, 324, 2
5	17 15175, 57 17250, 16 00830, 87120, 1210, 4
6	790 53975, 3689 18550, 1475 67420, 120 46320, 2 78850, 2028, 4

$$\varphi_n(z) = a_0^{-1} \sum_{k=0}^{n} b_k z^{2k}$$

n	$b_0, b_1, \ldots, b_n$
0	1
1	7, 2
2	43, 48, 2
3	1011, 2682, 290, 4
4	9819, 48960, 9900, 322, 2
5	5 71545, 46 72350, 15 23700, 85946, 1206, 4
6	233 10855, 2858 13360, 1376 18550, 117 84822, 2 76858, 2024, 4

TABLE 10.5

COEFFICIENTS FOR $\mathbf{H}_{-1}(z) - Y_{-1}(z)$

$$\mathbf{H}_{-1}(z)-Y_{-1}(z) = -(2/\pi z^2)\left[\varphi_n(z)/f_n(z)+R_n(z)\right]$$

$$\mathbf{H}_{1}(z)-Y_{1}(z) = 2/\pi+(2/\pi z^2)\left[\varphi_n(z)/f_n(z)+R_n(z)\right]$$

$$f_n(z) = a_o^{-1}\sum_{k=0}^{n} a_k z^{2k}$$

n	$a_o, a_1, \ldots, a_n$
0	1
1	15, 2
2	175, 70, 2
3	6615, 5292, 378, 4
4	93555, 1 24740, 16038, 396, 2
5	74 32425, 148 64850, 29 72970, 1 25840, 1430, 4
6	3952 69875, 11067 55650, 3162 15900, 200 77200, 3 80250, 2340.

$$\varphi_n(z) = a_o^{-1}\sum_{k=0}^{n} b_k z^{2k}$$

n	$b_o, b_1, \ldots, b_n$
0	1
1	9, 2
2	55, 64, 2
3	1449, 4338, 366, 4
4	15795, 91296, 14940, 390, 2
5	10 22175, 97 53390, 26 53500, 1 21730, 1418, 4
6	458 84475, 6555 81600, 2698 06950, 190 35450, 3 73410, 2328, 4

TABLE 10.6

ERROR COEFFICIENTS

$|R_n(z)|$, $\nu=0$

n/z	1	2	4	6	8
2	0.223(-1)	0.115(-1)	0.906(-3)	0.597(-4)	0.626(-5)
4	0.106(-1)	0.537(-3)	0.566(-4)	0.678(-6)	0.763(-6)
6	0.169(-3)	0.230(-3)	0.612(-5)	0.744(-7)	0.267(-7)
10	0.453(-3)	0.168(-4)	0.110(-6)	0.286(-8)	0.108(-10)
15	0.538(-4)	0.101(-5)	0.759(-9)	0.595(-11)	0.385(-12)
20	0.100(-4)	0.482(-7)	0.120(-10)	0.232(-12)	0.215(-14)

n/z	10	12	16	20
2	0.699(-5)	0.400(-5)	0.117(-5)	0.384(-6)
4	0.201(-6)	0.489(-7)	0.323(-8)	0.239(-9)
6	0.440(-8)	0.495(-9)	0.934(-11)	0.228(-11)
10	0.845(-11)	0.495(-12)	0.109(-13)	0.235(-15)
15	0.139(-13)	0.315(-15)	0.304(-17)	0.382(-19)
20	0.563(-17)	0.192(-17)	0.295(-20)	0.179(-22)

$|R_n(z)|$, $\nu=1$

n/z	1	2	4	6	8
2	0.490(-1)	0.319(-1)	0.373(-2)	0.327(-3)	0.212(-4)
4	0.173(-1)	0.148(-2)	0.236(-3)	0.419(-5)	0.502(-5)
6	0.690(-4)	0.583(-3)	0.262(-4)	0.402(-6)	0.189(-6)
10	0.705(-3)	0.411(-4)	0.475(-6)	0.171(-7)	0.728(-10)
15	0.629(-4)	0.236(-5)	0.345(-8)	0.366(-10)	0.298(-11)
20	0.162(-4)	0.918(-7)	0.630(-10)	0.140(-11)	0.169(-13)

n/z	10	12	16	20
2	0.362(-4)	0.227(-4)	0.719(-5)	0.247(-5)
4	0.148(-5)	0.393(-6)	0.293(-7)	0.239(-8)
6	0.357(-7)	0.448(-8)	0.949(-10)	0.260(-10)
10	0.752(-10)	0.497(-11)	0.134(-12)	0.329(-14)
15	0.129(-12)	0.336(-14)	0.405(-16)	0.601(-18)
20	0.529(-16)	0.214(-16)	0.413(-19)	0.301(-21)

10.5. Bibliographic and Numerical Data

10.5.1. References

For material on associated Bessel functions, see Watson (1945), Erdelyi et al (1953), Luke (1962,1969), Abramowitz and Stegun (1964), Babister (1967) and Agrest and Maksimov (1965). The volume by Babister also analyzes functions which satisfy $\Delta w(z) = f(z)$ where Δ is the differential operator corresponding to other members of the hypergeometric family. The Agrest and Maksomov tome notes numerous physical applications. Some additional references are Vodicka (1959), Berljand, Kiricenko and Kogan (1965), Ditkin (1966), Agrest and Rikenglaz (1967), Erukhimovich and Pimenov (1969), Sikorsskii (1970), Agrest (1970), Sikorsky (1970) and Agrest (1971). In view of the remarks concerning integrals of Bessel functions made in 10.1, the references in 9.13.2 are also pertinent.

Two papers by Steinig (1970,1972) give results on the sign of Lommel and Struve functions.

10.5.2. Description of and References to Tables

Bierlein (1962): Let $Ai(A,x) = \pi^{-1}\int_0^A \cos\,(ux + u^3/3)du$. For each A, $\pm A = 0(0.25)5(0.5)6$, ∞, a total of from 28 to 33 interlacing zeros and turning points are tabulated to 4D. Also included are corresponding data for $\lim Ai(A,x)/A$ as $A\to 0$. This appears incorrectly in the column heading as Ai/x. The associated extremal values of these integrals are given to 5 or 6D.

Bernard and Ishimaru (1962): $\boldsymbol{J}_\nu(x)$, $\boldsymbol{E}_\nu(x)$, $\nu = -10(0.1)10$, $x = 0)0.1)10$, 5D.

Barrett (1964): $\boldsymbol{L}_0(x)$, $\boldsymbol{L}_1(x)$, $x = 0.02(0.005)4(0.05)10(0.1)19.2$, 5 and 6S; $x = 6(0.25)59.5(0.5)100$, 2S.

Agrest and Maksimov (1965): Let

$$\tfrac{1}{2}J_n(\alpha,\rho) = \int_0^\alpha c_n(\theta,\rho)\cos\,\phi d\theta,\quad \tfrac{1}{2}H_n(\alpha,\rho) = \int_0^\alpha c_n(\theta,\rho)\sin\,\phi d\theta,$$

$$F_n^{\pm}(\alpha,\rho) = \int_0^\alpha c_n(\theta,\rho)e^{\pm\phi}d\theta,\quad c_n(\theta,\rho) = (\tfrac{1}{2}\rho\sin^2\theta)^n/\Gamma(\tfrac{1}{2})\Gamma(n+\tfrac{1}{2}),$$

$\phi = \rho\cos\,\theta$. These integrals are tabulated for $n = 0,1$, $\alpha = 0.2(0.2)1.4,\tfrac{1}{2}\pi$, $\rho = 0.2(0.2)10$, 5D except that $F_n^{\pm}(\alpha,\rho)$ is given to 6S.

Ditkin (1966): Same functions as in Agrest and Maksimov above, $n = 0, 1$, $\rho = 0.1(0.1)5.0$, $\alpha = 0.01(0.01)1.57, \frac{1}{2}\pi$, 6D, except that $F_n^{\pm}(\alpha, \rho)$ is given to 6S.

Agrest, Bekauri, Orlov, Rikenglaz, Haihjan and Cacibaja (1966): Same functions etc. as in Ditkin above except that $\rho = 0.1(0.1)10$.

Nosova and Tumarkin (1961), Osipova and Tumarkin (1963): They give tables relating to the solution of $y'' + ty = 1$. These are the functions noted under the same authors on p. 405.

CHAPTER XI ORTHOGONAL POLYNOMIALS

11.1. Introduction

Much emphasis in this work is on expansion of functions in series of the classical orthogonal polynomials known as Jacobi polynomials, and in particular in series of Chebyshev polynomials of the first kind. In our work, where no confusion can arise, we speak of expansions in Chebyshev polynomials of the first kind simply as expansions in Chebyshev polynomials. Numerous tables of coefficients of this type for a wide variety of commonly used transcendental functions are provided in this volume. Here, we present formulae to facilitate use and evaluation of these expansions, and to build up similar type expansions of related functions from the basic expansions. The essential point is that these expansions can be evaluated and manipulated in just as simple a manner as expansions in powers of x without first converting a series of Chebyshev polynomials to an ordinary polynomial.

The standard book on orthogonal polynomials is by Szegö (1967). Other references are Kacmarz and Steinhaus(1951), Erdélyi et al (1953, Vol. 2, Ch. 10), Tricomi (1955), Sansone (1959), Geronimus (1958), Abramowitz and Stegun (1964) and Freud (1969).

In contradistinction to other chapters and with the exception of some listings in 6.12.2, we do not give recent references to tables of zeros and associated data of orthogonal polynomials as this kind of information is closely akin to the subject of numerical integration. A quick compilation of many references on the subject can be obtained by use of the index by Luke, Wimp and Fair (1972). See also a recent volume by Davis and Rabinowitz (1974).

11.2. Orthogonal Properties

In virtually all of our work, we are concerned with the classical orthogonal polynomials. However, to introduce the concept of orthogonality and its applications, it is convenient to consider real functions defined over real intervals, although the ideas are readily extended to complex functions defined over paths in the complex plane.

Consider an interval (a, b) and a weight function $w(x)$ which is nonnegative over this interval. Let $\{\theta_n(x)\}$ be a sequence of functions, such that $w\theta_n{}^2$ is integrable in (a, b). We define the scalar product.

$$(\theta_n, \theta_m) = \int_a^b w(x)\, \theta_n(x)\, \theta_m(x)\, dx. \tag{1}$$

If $a(x)$ is a nondecreasing function, we may generalize (1) by the Stieltjes integral

$$(\theta_n, \theta_m) = \int_a^b \theta_n(x)\, \theta_m(x)\, da(x). \tag{2}$$

So if $a(x)$ is absolutely continuous, (2) becomes (1) with $w(x) = a'(x)$. If $a(x)$ is a jump function which is constant except for jumps w_i at $x = x_i$, then (2) reduces to

$$(\theta_n, \theta_m) = \sum_{i=0}^{n} w_i \theta_n(x_i)\, \theta_m(x_i). \tag{3}$$

This is the appropriate definition for the scalar product for functions of a discrete variable.

The sequence of functions $\{\theta_n(x)\}$ is said to be orthogonal [with respect to a weight function $w(x)$] over the interval (a, b) if

$$(\theta_n, \theta_m) = h_n \delta_{mn} \tag{4}$$

where

$$h_n = (\theta_n, \theta_n) = \int_a^b w(x)[\theta_n(x)]^2\, dx, \tag{5}$$

and δ_{mn} is the Kronecker delta function. That is,

$$\begin{aligned} \delta_{mn} &= 0 \quad \text{if} \quad m \neq n \\ &= 1 \quad \text{if} \quad m = n. \end{aligned} \tag{6}$$

If $h_n = 1$ for all n, the system is said to be orthonormal. Clearly any orthogonal system can be made orthonormal if we replace $\theta_n(x)$ by $\theta_n(x)/h_n^{1/2}$.

A natural question concerns the possibility of representing an arbitrary function $f \equiv f(x)$ as a sum of orthogonal functions, thus

$$f = \sum_{k=0}^{\infty} a_k \theta_k . \tag{7}$$

Assuming this is so, then on a formal basis we have

$$(f, \theta_k) = \sum_{n=0}^{\infty} a_n(\theta_k, \theta_n) = a_k h_k, \qquad a_k = h_k^{-1}(f, \theta_k). \tag{8}$$

The a_k's are often called the Fourier coefficients associated with f, and the series on the right of (7) is called a generalized Fourier series. The fact that we can calculate a_k does not guarantee that the series on the right of (7) converges, or if the series converges that its sum is f. Throughout the discussion we suppose that $\int_a^b f^2(x)\, w(x)\, dx$ exists and is finite in the Lebesgue sense. The class of functions for which this is true is called L_w^2. We further suppose that $\{\theta_n(x)\}$ belongs to this class. Let

$$f_n = \sum_{k=0}^{n} b_k \theta_k . \tag{9}$$

We call f_n an approximation to f of order n. A measure of the accuracy of this approximation is afforded by the integral

$$I_n(b_h) \equiv I_n(b_0, \ldots, b_n) = \int_a^b \left[f(x) - \sum_{k=0}^{n} b_k \theta_k(x) \right]^2 w(x)\, dx. \tag{10}$$

A best choice for the b_h's is that which makes $I_n(b_h)$ a minimum, if such exists, in which event we speak of a best mean square approximation to $f(x)$. We have the following theorem.

Theorem 1. *Of all the* nth *order approximations to* $f(x)$, *the best in the mean square sense occurs when* $b_k = a_k$.

A measure of the accuracy of the approximation is given by

$$I_n(a_h) = \int_a^b f^2(x)\, w(x)\, dx - \sum_{k=0}^{n} a_k^2 h_k . \tag{11}$$

Since $I_n(a_h) \geqslant 0$, $\sum_{k=0}^{n} a_k^2 h_k$ converges as $n \to \infty$, and we have Bessel's inequality

$$\sum_{k=0}^{\infty} a_k^2 h_k \leqslant \int_a^b f^2(x)\, w(x)\, dx. \tag{12}$$

When there is equality (the formula then goes by the name of Parseval) for every function in L_w^2, then $\{\theta_n(x)\}$ is said to be closed in L_w^2. In this case

$$\lim_{n\to\infty} \int_a^b \left[f(x) - \sum_{k=0}^{n} a_k \theta_k(x) \right]^2 w(x)\, dx = 0, \tag{13}$$

and the partial sums of the generalized Fourier series are said to converge in the mean to $f(x)$.

In the case of functions of a discrete variable [see the discussion around (3)] suppose that

$$f_n(x) = \sum_{k=0}^{n} c_k \theta_k(x), \tag{14}$$

$$\sum_{i=0}^{n} \theta_k(x_i)\, \theta_m(x_i)\, W_i = H_m \delta_{km}\,, \tag{15}$$

where $W_i = W(x_i)$ is positive and δ_{km} is the Kronecker delta function. Now multiply both sides of (14) by $\theta_m(x_i)\, W_i$ and sum on i from 0 to n. Apply (15). Then

$$c_k = H_k^{-1} \sum_{i=0}^{n} \theta_k(x_i) f_n(x_i)\, W_i\,. \tag{16}$$

Thus, if $f(x)$ is known at the $(n + 1)$ distinct points x_i, $i = 0, 1, \ldots, n$, $f(x_i) = f_n(x_i)$, then $f_n(x)$ as given by (14) is a curve fit to $f(x)$.

Suppose we have another curve fit to $f(x)$ in the form

$$f_n^*(x) = \sum_{k=0}^{n} d_k \theta_k(x), \qquad f_n^*(x_i) = f_n(x_i) = f(x_i). \tag{17}$$

Then a measure of the accuracy of the curve-fitting process may be described by

$$I(d_k) = \sum_{i=0}^{n} \left\{ f_n(x_i) - \sum_{k=0}^{n} d_k \theta_k(x_i) \right\}^2 W_i\,. \tag{18}$$

The result analogous to Theorem 1 is as follows.

Theorem 2. *The best approximation to $f_n(x)$ in the sense that $I_n(d_k)$ is least happens when $c_k = d_k$, and in this instance*

$$I_n(c_k) = \sum_{i=0}^{n} W_i f_n^2(x_i) - \sum_{k=0}^{n} h_k c_k^2. \tag{19}$$

Such an approximation is said to be best in the sense of least squares.

Next we consider orthogonal polynomials for which we use the symbol $q_n(x)$. They possess some important properties (given by Theorems 3–5 below) which make them very suitable for use in approximation theory.

Theorem 3. *The zeros of $q_n(x)$ are simple and lie in the interior of $[a, b]$.*

Theorem 4. *Let*

$$f(x) = \sum_{k=0}^{\infty} a_k q_k(x), \qquad f_n(x) = \sum_{k=0}^{n} a_k q_k(x). \tag{20}$$

Then $[f(x) - f_n(x)]$ vanishes at least $(n + 1)$ times in $[a, b]$.

Theorem 5. *Any three consecutive orthogonal polynomials satisfy a recurrence formula of the form*

$$q_{n+1}(x) = (A_n x + B_n)\, q_n(x) - C_n q_{n-1}(x), \qquad n = 1, 2, \ldots . \tag{21}$$

Furthermore, with

$$q_n(x) = \sum_{k=0}^{n} a_{k,n} x^k,$$

$$A_n = \frac{a_{n+1,n+1}}{a_{n,n}}, \qquad B_n = A_n(r_{n+1} - r_n), \qquad r_n = \frac{a_{n-1,n}}{a_{n,n}},$$
$$C_n = \frac{A_n h_n}{A_{n-1} h_{n-1}} = \frac{a_{n+1,n+1} a_{n-1,n-1} h_n}{(a_{n,n})^2 h_{n-1}}, \qquad h_n = (q_n, q_n). \tag{22}$$

Theorem 6. *If a sequence of polynomials are orthogonal with respect to integration, they are also orthogonal with respect to summation. To be precise, if*

$$\int_a^b w(x) q_j(x) q_k(x)\, dx = h_k \delta_{jk},$$

then

$$A_n h_n \sum_{\alpha=0}^{n} \frac{q_j(x_\alpha) q_k(x_\alpha)}{q_n'(x_\alpha) q_n(x_\alpha)} = h_k \delta_{jk}, \tag{23}$$

$$j \le n, \; k \le n, \; q_{n+1}(x_\alpha) = 0 .$$

For details on Theorem 6, see Luke (1975).

The Christoffel-Darboux formulas are

$$\sum_{k=0}^{n} h_k^{-1} q_k(x)\, q_k(y) = (A_n h_n)^{-1} \frac{q_{n+1}(x)\, q_n(y) - q_n(x)\, q_{n+1}(y)}{x - y},$$

$$\sum_{k=0}^{n} h_k^{-1} q_k^{\,2}(x) = (A_n h_n)^{-1} [q_n(x)\, q'_{n+1}(x) - q'_n(x)\, q_{n+1}(x)]. \tag{24}$$

As previously remarked, we are primarily interested in the classical orthogonal polynomials. These polynomials are special cases of Gaussian or confluent hypergeometric functions and virtually all the results given in the later sections are special cases of formulae given in Chapters 6 and 7. From (25-34) below, we see that all the classical polynomials stem from the Jacobi polynomial $P_n^{(\alpha,\beta)}(x)$ of degree n. If, in the hypergeometric representation of this polynomial, the parameter n is allowed to be an arbitrary complex number, then the hypergeometric function is called a Jacobi function of the first kind. The Jacobi function of the second kind is essentially the second solution of the differential equation satisfied by the function of the first kind. Similarly, we speak of Legendre functions.

In Tables A and B, pp. 434-436, we list the classical orthogonal polynomials and supply the basic quantities entering equations (21,22). These formulae are equations (25-44).

The class of Bessel polynomials

$$P_n(\nu,z) = {}_2F_0(-n,\ n+\nu;\ z) \tag{45}$$

is an example of polynomials orthogonal over complex paths. Note that $P_n(\nu,z)$ is a confluent form of the ${}_2F_1$ which enters into the definition of $R_n^{(\alpha,\beta)}(x)$, see 11.2(25,26). The orthogonality relation is

$$(2\pi i)^{-1} \int_C \frac{e^z P_m(\nu,1/z) P_n(\nu,1/z)\, dz}{z^{\nu+1}} = \frac{(-)^n n!\, \delta_{mn}}{(2n+\nu)\Gamma(n+\nu)}, \tag{46}$$

where C is the path $c - i\infty$ to $c + i\infty$, $c > 0$. Also

$$A_n = -\frac{(2n+\nu)(2n+\nu+1)}{n+\nu},\quad B_n = \frac{(\nu-1)(2n+\nu)}{(n+\nu)(2n+\nu-1)},\quad C_n = -\frac{n(2n+\nu+1)}{(n+\nu)(2n+\nu-1)}. \tag{47}$$

TABLE A

Name	a	b	$w(x)$	Eq.
Jacobi:				
$P_n^{(\alpha,\beta)}(x) = \frac{(\alpha+1)_n}{n!}\,{}_2F_1\left(\begin{matrix}-n,\, n+\alpha+\beta+1\\ \alpha+1\end{matrix}\,\middle\vert\,\frac{1-x}{2}\right)$	-1	1	$(1-x)^\alpha(1+x)^\beta$	(25)
Jacobi (shifted):				
$R_n^{(\alpha,\beta)}(x) = P_n^{(\alpha,\beta)}(2x-1)$	0	1	$(1-x)^\alpha x^\beta$	(26)
Gegenbauer or ultraspherical:				
$C_n^{(\alpha+1/2)}(x) = \frac{(2\alpha+1)_n}{(\alpha+1)_n} P_n^{(\alpha,\alpha)}(x)$	-1	1	$(1-x^2)^\alpha$	(27)
Legendre:				
$P_n(x) = P_n^{(0,0)}(x)$	-1	1	1	(28)
Chebyshev (first kind):				
$T_n(x) = \frac{n!}{(1/2)_n} P_n^{(-1/2,-1/2)}(x)$	-1	1	$(1-x^2)^{-1/2}$	(29)
Chebyshev (first kind, shifted):				
$T_n^*(x) = T_n(2x-1)$	0	1	$[x(1-x)]^{-1/2}$	(30)
Chebyshev (second kind):				
$U_n(x) = \frac{(n+1)!}{(3/2)_n} P_n^{(1/2,1/2)}(x)$	-1	1	$(1-x^2)^{1/2}$	(31)
Chebyshev (second kind, shifted):				
$U_n^*(x) = U_n(2x-1)$	0	1	$[x(1-x)]^{1/2}$	(32)
Laguerre:				
$L_n^{(\alpha)}(x) = \lim_{\beta\to\infty} P_n^{(\alpha,\beta)}\left(1-\frac{2x}{\beta}\right)$	0	∞	$e^{-x}x^\alpha$	
$= \lim_{\beta\to\infty} R_n^{(\alpha,\beta)}\left(1-\frac{x}{\beta}\right)$				
$= (-)^n \lim_{\beta\to\infty} R_n^{(\beta,\alpha)}(x/\beta)$,				
$L_n^{(\alpha)}(x) = \frac{(\alpha+1)_n}{n!}\,{}_1F_1(-n;\alpha+1;x)$				(33)
The polynomial introduced by Laguerre is the case $\alpha = 0$.				
Hermite:				
$H_{2m+\epsilon}(x) = (-)^m 2^{2m+\epsilon} m!\, x^\epsilon L_m^{(\epsilon-1/2)}(x^2)$,	$-\infty$	∞	e^{-x^2}	
$\epsilon = 0$ or $\epsilon = 1$				(34)

TABLE B

$q_n(x)$	A_n	B_n	C_n	h_n	
$P_n^{(\alpha,\beta)}(x)$	$\dfrac{(2n+\lambda)(2n+\lambda+1)}{2(n+1)(n+\lambda)}$	$\dfrac{(\alpha^2-\beta^2)(2n+\lambda)}{2(n+1)(n+\lambda)(2n+\lambda-1)}$	$\dfrac{2(n+\alpha)(n+\beta)A_n}{(2n+\lambda-1)(2n+\lambda)}$	$\dfrac{2^\lambda\Gamma(n+\alpha+1)\Gamma(n+\beta+1)}{(2n+\lambda)n!\Gamma(n+\lambda)}$	(35)
$R_n^{(\alpha,\beta)}(x)$	$\dfrac{(2n+\lambda)(2n+\lambda+1)}{(n+1)(n+\lambda)}$	B_n-A_n for $P_n^{(\alpha,\beta)}(x)$	$\dfrac{(n+\alpha)(n+\beta)A_n}{(2n+\lambda-1)(2n+\lambda)}$	$\dfrac{\Gamma(n+\alpha+1)\Gamma(n+\beta+1)}{(2n+\lambda)n!\Gamma(n+\lambda)}$	(36)
$C_n^{(\alpha+\frac{1}{2})}(x)$	$\dfrac{2n+2\alpha+1}{n+1}$	0	$\dfrac{n+2\alpha}{n+1}$	$\dfrac{2^\lambda\Gamma(n+\lambda)}{(2n+\lambda)n!}\left(\dfrac{\Gamma(\frac{1}{2}\lambda+\frac{1}{2})}{\Gamma(\lambda)}\right)^2$	(37)

In equations (35) and (36), $\lambda = \alpha + \beta + 1$.

In equation (37), $\lambda = 2\alpha + 1$, $\alpha \neq -\frac{1}{2}$.

TABLE B is continued on the next page.

TABLE B *(Concluded)*

$q_n(x)$	A_n	B_n	C_n	h_n	
$P_n(x)$	$\frac{2n+1}{n+1}$	0	$\frac{n}{n+1}$	$\frac{2}{2n+1}$	(38)
$T_n(x)$	2	0	1	π/ε_n	(39)
$T_n^*(x)$	4	-2	1	π/ε_n	(40)
In equations (39) and (40), $\varepsilon_0 = 1$, $\varepsilon_n = 2$ if $n > 0$.					
$U_n(x)$	2	0	1	$\pi/2$	(41)
$U_n^*(x)$	4	-2	1	$\pi/8$	(42)
$L_n^{(\alpha)}(x)$	$-\frac{1}{n+1}$	$\frac{2n+\alpha+1}{n+1}$	$\frac{n+\alpha}{n+1}$	$\frac{\Gamma(n+\alpha+1)}{n!}$	(43)
$H_n(x)$	2	0	$2n$	$\pi^{\frac{1}{2}}2^n n!$	(44)

11.3. Jacobi Polynomials

11.3.1. Expansion Formulae

The Jacobi polynomial has been defined by 11.2(25). We suppose that $\alpha > -1$, $\beta > -1$ so that $w(x)$ is nonnegative and integrable in $[-1,1]$. However, many of the formal results are valid without this restriction. The term $\alpha + \beta + 1$ occurs very frequently and for simplification we put

$$\lambda = \alpha + \beta + 1. \tag{1}$$

In hypergeometric form, we have

$$P_n^{(\alpha,\beta)}(x) = \frac{(\alpha+1)_n}{n!}\, {}_2F_1\left(\begin{matrix}-n, n+\lambda\\ \alpha+1\end{matrix}\,\middle|\, \frac{1-x}{2}\right) \tag{2}$$

$$= \frac{(-)^n(\beta+1)_n}{n!}\, {}_2F_1\left(\begin{matrix}-n, n+\lambda\\ \beta+1\end{matrix}\,\middle|\, \frac{1+x}{2}\right). \tag{3}$$

$$P_n^{(\alpha,\beta)}(-x) = (-)^n P_n^{(\beta,\alpha)}(x),$$
$$P_n^{(\alpha,\beta)}(1) = \frac{(\alpha+1)_n}{n!}, \qquad P_n^{(\alpha,\beta)}(-1) = \frac{(-)^n(\beta+1)_n}{n!}. \tag{4}$$

Also

$$P_n^{(\alpha,-1/2)}(2x^2-1) = \frac{(2n)!\,(\alpha+1)_n}{n!\,(\alpha+1)_{2n}} P_{2n}^{(\alpha,\alpha)}(x), \tag{5}$$

$$P_n^{(\alpha,1/2)}(2x^2-1) = \frac{(2n+1)!\,(\alpha+1)_n}{n!\,(\alpha+1)_{2n+1}} x^{-1}P_{2n+1}^{(\alpha,\alpha)}(x). \tag{6}$$

Thus $P_n^{(\alpha,\alpha)}(x)$ is an even (odd) polynomial if n is even (odd).

Rodrigues' formula is

$$2^n n!\, P_n^{(\alpha,\beta)}(x) = (-)^n(1-x)^{-\alpha}(1+x)^{-\beta}\,(d^n/dx^n)\{(1-x)^{\alpha+n}(1+x)^{\beta+n}\}, \tag{7}$$

and from this we have

$$P_n^{(\alpha,\beta)}(x) = 2^{-n}\sum_{k=0}^{n}\binom{n+\alpha}{k}\binom{n+\beta}{n-k}(x-1)^{n-k}(x+1)^k. \tag{8}$$

Also,

$$P_n^{(\alpha,\beta)}(x) = \sum_{r=0}^{n} k_{r,n}^{(\alpha,\beta)} x^r,$$

$$k_{r,n}^{(\alpha,\beta)} = \frac{(-)^r(\alpha+1)_n(-n)_r(n+\lambda)_r}{r!\,(\alpha+1)_r\,2^r n!}\, {}_2F_1\left(\begin{matrix} r-n,\, n+\lambda+r \\ \alpha+1+r \end{matrix}\,\middle|\,\frac{1}{2}\right)$$
$$= \frac{(-)^n(\beta+1)_n(-n)_r(n+\lambda)_r}{r!\,(\beta+1)_r\,2^r n!}\, {}_2F_1\left(\begin{matrix} r-n,\, n+\lambda+r \\ \beta+1+r \end{matrix}\,\middle|\,\frac{1}{2}\right). \tag{9}$$

In particular,

$$k_{n,n}^{(\alpha,\beta)} = \frac{(n+\lambda)_n}{2^n n!}, \qquad k_{n-1,n}^{(\alpha,\beta)} = \frac{(\alpha-\beta)\,\Gamma(2n+\lambda-1)}{2^n(n-1)!\,\Gamma(n+\lambda)}. \tag{10}$$

In general a more simple expression for (9) is not known unless $\alpha = \beta$ or $\alpha = \beta \pm 1$. We have

$$k_{r,n}^{(\alpha,\alpha)} = \frac{(-)^n(\alpha+1)_n(-n)_r(n+2\alpha+1)_r\,\Gamma(\alpha+1+r)\,\Gamma(\frac{1}{2})}{2^r r!\,(\alpha+1)_r\,\Gamma[(r-n+1)/2]\,\Gamma[(r+n)/2+\alpha+1]\,n!} \tag{11}$$

and (11) vanishes whenever $n-r$ is an odd positive integer. Also

$$k_{r,n}^{(\beta+\varepsilon,\beta)} = \{(-)^r\Gamma(n+\beta+\tfrac{1}{2}\varepsilon+\tfrac{1}{2})(-n)_r\Gamma(\tfrac{1}{2})/2^r r!n!\}$$
$$\times\left[\{\Gamma(s)\Gamma(t+\tfrac{1}{2})\}^{-1} - \varepsilon\{\Gamma(s+\tfrac{1}{2})\Gamma(t)\}^{-1}\right],$$
$$s = \tfrac{1}{2}(r+n+2\beta+\varepsilon+1),\ t = \tfrac{1}{2}(r-n),\ \varepsilon = \pm 1. \tag{12}$$

For convenience, we record some results for the shifted Jacobi polynomials.

$$R_n^{(\alpha,\beta)}(x) = P_n^{(\alpha,\beta)}(2x-1), \qquad P_n^{(\alpha,\beta)}(x) = R_n^{(\alpha,\beta)}[(1+x)/2], \tag{13}$$

$$R_n^{(\alpha,\beta)}(x) = [(\alpha+1)_n/n!]\,{}_2F_1\left(\begin{matrix}-n, n+\lambda\\ \alpha+1\end{matrix}\,\middle|\,1-x\right)$$
$$= (-)^n[(\beta+1)_n/n!]\,{}_2F_1\left(\begin{matrix}-n, n+\lambda\\ \beta+1\end{matrix}\,\middle|\,x\right). \tag{14}$$

$$R_n^{(\alpha,\beta)}(1-x) = (-)^n R_n^{(\beta,\alpha)}(x). \tag{15}$$

$$R_n^{(\alpha,\beta)}(1) = (\alpha+1)_n/n!, \qquad R_n^{(\alpha,\beta)}(0) = (-)^n(\beta+1)_n/n!. \tag{16}$$

$$n!\,R_n^{(\alpha,\beta)}(x) = (-)^n(1-x)^{-\alpha}x^{-\beta}(d^n/dx^n)\{(1-x)^{\alpha+n}x^{\beta+n}\}. \tag{17}$$

Next, we turn to an asymptotic form of the Jacobi polynomial for large order.

$${}_2F_1(-n,\ n+\lambda;\ \rho;\ z) = \{\Gamma(\rho)/\Gamma(\tfrac{1}{2})\}(N\sin\tfrac{1}{2}\theta)^{-\frac{1}{2}\omega}(\cos\tfrac{1}{2}\theta)^{-\frac{1}{2}\eta-\frac{1}{2}}$$
$$\times\exp\{(N^{-2}/64)\left[(\eta^2-1)\sec^2\tfrac{1}{2}\theta+\omega(\omega-2)\csc^2\tfrac{1}{2}\theta\right]$$
$$+\ (N^{-2}/48)\left[(\omega^2-1)(\omega-3\lambda)-3\lambda^2\right]+O(N^{-4})\}$$
$$\times\cos\{N\theta-\tfrac{1}{4}\pi\omega+(N^{-1}/8)\left[(\eta^2-1)\cot\ \theta+2(\lambda-1)(\omega-\lambda)\cot\tfrac{1}{2}\theta+\lambda^2\theta\right]$$
$$+\ N^{-3}\left[a\ \cot^3\tfrac{1}{2}\theta+b\ \cot\tfrac{1}{2}\theta+c\ \cot^2\theta+d\ \cot\theta-(\lambda^4\theta/128)\right]$$
$$+\ O(N^{-5})\},\ \ \omega=2\rho-1,\ \eta=2\lambda-2\rho,\ N^2=n(n+\lambda),$$
$$\cos\ \theta=1-2z\ \text{or}\ z=\sin^2\tfrac{1}{2}\theta,$$
$$|\arg\ z|\le\pi-\varepsilon,\ |\arg\ (1-z)|\le\pi-\varepsilon,\ \varepsilon>0, \tag{18}$$

where

$$a=\{(\lambda-1)/768\}\{16\rho^3-24\rho^2(\lambda+1)+4\rho(4\lambda^2+4\lambda-9)-2(\lambda+1)(2\lambda^2-11)\},$$
$$b=\{(\lambda-1)/128\}\{8\rho^3-12\rho^2(\lambda+1)+2\rho(4\lambda-1)+(\lambda+1)(2\lambda^2+3)\},$$

$$c = (\eta^2-1)(\eta^2-25)/384 ,$$

$$d = (9c/4)+(512)^{-1}\{16\rho^4-64\rho^3-8\rho^2(12\lambda^2-16\lambda-23)$$

$$+ 16\rho(8\lambda^3-4\lambda^2-24\lambda+1)-(12\lambda^2-47)(4\lambda^2-1)\} . \qquad (19)$$

If $\rho = \frac{1}{2}$ and $\lambda = 0$, then the above is exact without the order terms.

In (18), n need not be an integer. The corresponding result for the extended Jacobi function ${}_{p+2}F_{q+1}(-n, n+\lambda, \alpha_p; \rho_{q+1}; z)$ including the case $z=1$ when $p=q$ is given by Luke (1969), where n is arbitrary if $p \leq q$, and is a positive integer, otherwise. See 5.12(10-14,20,21) for some related results. See also Fields (1972b,1973). Luke also gives asymptotic formulas for the extended Laguerre function which is the extended Jacobi function with the numerator parameter $n+\lambda$ suppressed. For a uniform asymptotic expansion of the extended Jacobi function when $0 \leq z \leq 1$, see Fields (1968). Uniform asymptotic expansions of the Jacobi polynomial and the assiciated function $Q_n^{(\alpha,\beta)}(z)$ have been given by Elliott (1971). An asymptotic result for $Q_n^{(\alpha,\beta)}(z)$ for large n is given by 11.3.6.3(6).

11.3.2. Difference-Differential Formulae

$$(1-x^2)\frac{d^2P_n^{(\alpha,\beta)}(x)}{dx^2} + [\beta-\alpha-(\lambda+1)x]\frac{dP_n^{(\alpha,\beta)}(x)}{dx} + n(n+\lambda)P_n^{(\alpha,\beta)}(x) = 0. \qquad (1)$$

$$(2n+\lambda-1)(1-x^2)\frac{dP_n^{(\alpha,\beta)}(x)}{dx} = n[(\alpha-\beta)-(2n+\lambda-1)x]P_n^{(\alpha,\beta)}(x)$$

$$+ 2(n+\alpha)(n+\beta)P_{n-1}^{(\alpha,\beta)}(x). \qquad (2)$$

$$2(n+1)(n+\lambda)(2n+\lambda-1)P_{n+1}^{(\alpha,\beta)}(x)$$

$$= (2n+\lambda)[(2n+\lambda-1)(2n+\lambda+1)x+\alpha^2-\beta^2]P_n^{(\alpha,\beta)}(x)$$

$$- 2(n+\alpha)(n+\beta)(2n+\lambda+1)P_{n-1}^{(\alpha,\beta)}(x). \qquad (3)$$

$$2^m\frac{d^mP_n^{(\alpha,\beta)}(x)}{dx^m} = (n+\lambda)_m P_{n-m}^{(\alpha+m,\beta+m)}(x), \qquad m = 1, 2,\ldots, n. \qquad (4)$$

11.3.3. Integrals

In addition to the following, see 11.2(35-46).

$$\int_{-1}^{x} P_n^{(\alpha,\beta)}(t)\,dt$$
$$= 2\left\{\frac{(n+\lambda)}{(2n+\lambda+1)(2n+\lambda)} P_{n+1}^{(\alpha,\beta)}(x) + \frac{(\alpha-\beta)}{(2n+\lambda+1)(2n+\lambda-1)} P_n^{(\alpha,\beta)}(x)\right.$$
$$\left. - \frac{(n+\beta)(n+\alpha)}{(2n+\lambda)(2n+\lambda-1)(n+\lambda-1)} P_{n-1}^{(\alpha,\beta)}(x)\right\} + \frac{2(-)^n \Gamma(n+\beta+1)}{(n+\lambda-1)(n+1)!\,\Gamma(\beta)}. \tag{1}$$

$$2n\int_0^x (1-t)^\alpha (1+t)^\beta P_n^{(\alpha,\beta)}(t)\,dt$$
$$= P_{n-1}^{(\alpha+1,\beta+1)}(0) - (1-x)^{\alpha+1}(1+x)^{\beta+1} P_{n-1}^{(\alpha+1,\beta+1)}(x). \tag{2}$$

$$2n\int_0^x (1-t^2)^\alpha P_n^{(\alpha,\alpha)}(t)\,dt$$
$$= \frac{(-)^{n-1}(\alpha+2)_{n-1}\,\Gamma(\alpha+2)\,\Gamma(\frac{1}{2})}{(n-1)!\,\Gamma[(2-n)/2]\,\Gamma[(n+3+2\alpha)/2]} - (1-x^2)^\alpha P_{n-1}^{(\alpha+1,\alpha+1)}(x). \tag{3}$$

$$\int_{-1}^{1} (1-x)^\rho (1+x)^\sigma P_n^{(\alpha,\beta)}(x)\,dx$$
$$= (-)^n \int_{-1}^{1} (1-x)^\sigma (1+x)^\rho P_n^{(\alpha,\beta)}(x)\,dx$$
$$= \frac{(-)^n (\beta+1)_n\, 2^{\rho+\sigma+1}\Gamma(\rho+1)\,\Gamma(\sigma+1)}{n!\,\Gamma(\rho+\sigma+2)}\, {}_3F_2\left(\begin{matrix}-n,\, n+\lambda,\, \sigma+1\\ \beta+1,\, \rho+\sigma+2\end{matrix}\,\middle|\, 1\right),$$
$$R(\rho) > -1, \qquad R(\sigma) > -1. \tag{4}$$

$$\int_{-1}^{1} (1-x)^\alpha (1+x)^\sigma P_n^{(\alpha,\beta)}(x)\,dx$$
$$= \frac{(-)^n\, 2^{\alpha+\sigma+1}\Gamma(\sigma+1)\,\Gamma(n+\alpha+1)(\beta-\sigma)_n}{n!\,\Gamma(\alpha+\sigma+n+2)},$$
$$R(\alpha) > -1, \qquad R(\sigma) > -1. \tag{5}$$

11.3.4. Expansion of x^ρ in Series of Jacobi Polynomials

If m is a positive integer,

$$x^m = \sum_{n=0}^{m} a_n P_n^{(\alpha,\beta)}(x),$$
$$a_n = \frac{2^n m!\,\Gamma(n+\lambda)}{(m-n)!\,\Gamma(2n+\lambda)}\, {}_2F_1\left(\begin{matrix}n-m,\, \alpha+n+1\\ 2n+\lambda+1\end{matrix}\,\middle|\, 2\right), \tag{1}$$

$$x^{2m+\epsilon} = \sum_{n=0}^{m} b_n P_{2n+\epsilon}^{(\alpha,\alpha)}(x), \qquad \epsilon = 0 \quad \text{or} \quad \epsilon = 1,$$

$$b_n = \frac{(2m+\epsilon)!\,(2n+\alpha+\frac{1}{2}+\epsilon)\,\Gamma(2n+2\alpha+1+\epsilon)\,\Gamma(\frac{1}{2})}{2^{2m+2\alpha+\epsilon}(m-n)!\,\Gamma(2n+\alpha+\epsilon+1)\,\Gamma(m+n+\alpha+\frac{3}{2}+\epsilon)}. \tag{2}$$

Also,

$$(1-x)^m = \sum_{n=0}^{m} c_n P_n^{(\alpha,\beta)}(x),$$

$$c_n = \frac{(-)^n 2^m m!\,\Gamma(\alpha+m+1)(2n+\lambda)\,\Gamma(n+\lambda)}{(m-n)!\,\Gamma(\alpha+n+1)\,\Gamma(m+n+\lambda+1)}. \tag{3}$$

For an arbitrary power,

$$x^\rho = \sum_{k=0}^{\infty} c_{k,\rho} P_k^{(\alpha,\beta)}(x),$$

$$\alpha > -1, \qquad \beta > -1, \qquad R(\rho) \geqslant 0, \qquad 0 < x < 1,$$

where

$$c_{k,\rho} = \frac{(-)^k(-\rho)_k 2^\rho (2k+\lambda)\,\Gamma(\beta+\rho+1)\,\Gamma(k+\lambda)}{\Gamma(k+\beta+1)\,\Gamma(k+\rho+\lambda+1)}\,{}_2F_1\left(\begin{matrix} k-\rho,\ -k-\rho-\lambda \\ -\rho-\beta \end{matrix}\,\middle|\,\frac{1}{2}\right)$$

$$+\frac{(-)^{k+1}\pi(2k+\lambda)\,\Gamma(k+\lambda)\,e^{i\pi(\rho+\beta)}}{2^{\beta+1}\Gamma(k+\alpha+1)\,\Gamma(\rho+\beta+2)\,\Gamma(-\rho)\cos\pi(\rho+\beta)}$$

$$\times\,{}_2F_1\left(\begin{matrix} -k-\alpha,\ k+\beta+1 \\ \rho+\beta+2 \end{matrix}\,\middle|\,\frac{1}{2}\right). \tag{4}$$

If $\alpha = \beta$,

$$c_{k,\rho} = \frac{(-\rho)_k(2k+2\alpha+1)\,\Gamma(k+2\alpha+1)\,\Gamma[(1+\rho-k)/2]}{2^{k+2\alpha+2}\Gamma(k+\alpha+1)\,\Gamma[(k+2\alpha+\rho+3)/2]}$$

$$\times\,[(-)^k + e^{i\pi\rho}]. \tag{5}$$

$$x^\mu = \Gamma(\mu+\beta+1)\sum_{n=0}^{\infty}\frac{(-)^n(2n+\lambda)\,\Gamma(n+\lambda)(-\mu)_n}{\Gamma(n+\beta+1)\,\Gamma(n+\lambda+\mu+1)}\,R_n^{(\alpha,\beta)}(x), \tag{6}$$

valid under the conditions (7), (8) or (9) which follow.

$$\alpha > -1, \quad \beta > -1, \quad -R(\mu) < \min(\beta+1, \beta/2+3/4),$$
$$\epsilon \leqslant x \leqslant 1-\epsilon, \quad 0 < \epsilon < 1. \tag{7}$$

$$\alpha > -1, \quad \beta > -1, \quad R(\mu) > 0, \quad 0 \leqslant x \leqslant 1-\epsilon, \quad 0 < \epsilon < 1. \tag{8}$$

$$\alpha > -1, \qquad \beta > -1, \qquad -R(\mu) < \min(0, \tfrac{1}{2}(\beta - \alpha)), \qquad 0 \leqslant x \leqslant 1. \quad (9)$$

Expansions for numerous special functions in series of Jacobi polynomials are given in 5.10.

11.3.5. Convergence Theorems for the Expansion of Arbitrary Functions in Series of Jacobi Polynomials

Suppose formally that

$$f(x) = \sum_{n=0}^{\infty} c_n P_n^{(\alpha,\beta)}(x). \tag{1}$$

$$c_n = (h_n)^{-1} \int_{-1}^{1} f(x)(1-x)^{\alpha}(1+x)^{\beta} P_n^{(\alpha,\beta)}(x)\, dx. \tag{2}$$

The c_n's are often called the Fourier coefficients associated with $f(x)$. Alternative integral representations are given in 11.3.6.1.

We state some theorems which give conditions to insure that the series in (1) converges and that its sum is $f(x)$.

Theorem 1. *Let $f(x)$ be Lebesgue-measurable in $-1 \leqslant x \leqslant 1$, and let the integrals*

$$\int_{-1}^{1} (1-x)^{\alpha}(1+x)^{\beta}\, |f(x)|\, dx, \qquad \int_{-1}^{1} (1-x)^{\alpha/2-1/4}(1+x)^{\beta/2-1/4}\, |f(x)|\, dx$$

exist. Let $s_n(x)$ denote the nth partial sum of the expansion of $f(x)$ in series of Jacobi polynomials, and $s_n^*(x)$ the nth partial sum of the Fourier cosine series of

$$(1-\cos\theta)^{\alpha/2+1/4}(1+\cos\theta)^{\beta/2+1/4} f(\cos\theta).$$

Then for $-1 < x < 1$,

$$\lim_{n\to\infty} \{s_n(x) - (1-x)^{-\alpha/2-1/4}(1+x)^{-\beta/2-1/4} s_n^*(x)\} = 0,$$

uniformly in $-1+\epsilon \leqslant x \leqslant 1-\epsilon$, $0 < \epsilon < 1$.

This is called an equiconvergence theorem. Clearly, application of the theorem requires knowledge of the convergence of the Fourier series for $f(x)$. In this connection, it is sufficient to quote the following result.

Theorem 2. *Let $f(x)$ be periodic and have period 2π. If $f'(x)$ is continuous in $-\pi \leqslant x \leqslant \pi$ except for a finite number of bounded jumps, then the Fourier series for $f(x)$ converges pointwise to $\frac{1}{2}[f(x-0)+f(x+0)]$. Furthermore, the convergence is uniform in any closed interval which does not include a point of discontinuity of $f(x)$.*

Rau (1950) has proved the following theorem.

Theorem 3. *If $f(x)$ is continuous in the closed interval $-1 \leqslant x \leqslant 1$ and has a piecewise continuous derivative there, then with $\alpha > -1$, $\beta > -1$, the Jacobi series (1) associated with $f(x)$ converges uniformly to $f(x)$ in $-1+\epsilon \leqslant x \leqslant 1-\epsilon$, $0 < \epsilon < 1$.*

In the case of analytic functions [see Szegö (1959, p. 243)] we have

Theorem 4. *If $f(x)$ is analytic in the closed interval $-1 \leqslant x \leqslant 1$, then the Jacobi series (1) associated with $f(x)$ is convergent in the interior of the largest ellipse with foci at ± 1 in which $f(x)$ is analytic.*

11.3.6. Evaluation and Estimation of the Coefficients in the Expansion of a Given Function F(x) in Series of Jacobi Polynomials

11.3.6.1. The Coefficients as an Integral Transform

Formally at least, we have

$$f(x) = \sum_{n=0}^{\infty} c_n P_n^{(\alpha,\beta)}(x), \tag{1}$$

$$c_n = (h_n)^{-1} \int_{-1}^{1} f(x)(1-x)^{\alpha}(1+x)^{\beta} P_n^{(\alpha,\beta)}(x)\, dx. \tag{2}$$

When $\alpha=\beta$, $\alpha=-\frac{1}{2}$, approximations for c_n, which follow from (2) by application of two trapezoidal-type rules of integration, are given by 11.7(3,9). Alternative forms for c_n are

$$c_n = \frac{(-)^n (h_n)^{-1}}{2^n n!} \int_{-1}^{1} f(x) \frac{d^n}{dx^n}\{(1-x)^{\alpha+n}(1+x)^{\beta+n}\}\, dx, \tag{3}$$

$$c_n = \frac{(-)^n (h_n)^{-1}}{2^n n!} \int_{-1}^{1} f^{(n)}(x)(1-x)^{\alpha+n}(1+x)^{\beta+n}\, dx. \tag{4}$$

The latter is called an Euler or beta transform. Thus if $f(x)$ is a member of the hypergeometric family, so also is c_n as readily follows from 5.2.2(1), 5.2.3(2) and 5.3.2(17), 5.6(22). Other types of transforms are useful to identify c_n.

For a general illustration, suppose

$$f(x) = \int_c^d h(x, t)\, g(t)\, dt, \tag{5}$$

$$h(x, t) = \sum_{n=0}^{\infty} b_n(t)\, P_n^{(\alpha,\beta)}(x). \tag{6}$$

Combine (5) and (6), interchange the order of integration and summation (which we assume is valid), and compare with (1) to get

$$c_n = \int_c^d b_n(t)\, g(t)\, dt. \tag{7}$$

Observe that both $f(x)$ and c_n are from the same family of transforms. Thus if $f(x)$ is defined by (5) and the kernel of the transform has a known expansion in series of Jacobi polynomials, then c_n is defined by a like transform with kernel $b_n(t)$.

For a more concrete but still general example, suppose $f(x)$ is the Laplace transform of a known function $g(t)$. Thus,

$$f(x) = \int^{\infty} e^{-xt} g(t)\, dt. \tag{8}$$

Since

$$\begin{aligned} e^{-xt} &= \sum_{n=0}^{\infty} \frac{(-)^n \Omega_n I_{n+\alpha+1/2}(t)}{t^{\alpha+1/2}}\, P_n^{(\alpha,\alpha)}(x), && -1 \leqslant x \leqslant 1, \\ \Omega_n &= \frac{(2\pi)^{1/2}(n+\alpha+\tfrac{1}{2})\,\Gamma(n+2\alpha+1)}{2^{\alpha}\Gamma(n+\alpha+1)}, && \alpha \neq -\tfrac{1}{2}, \\ \Omega_n &= n!\,\epsilon_n/(\tfrac{1}{2})_n, \quad \epsilon_0 = 1, \quad \epsilon_n = 2 \quad \text{if} \quad n > 0, && \text{when} \quad \alpha = -\tfrac{1}{2}. \end{aligned} \tag{9}$$

Thus,

$$c_n = (-)^n \Omega_n \int_0^{\infty} t^{-\alpha-1/2} I_{n+\alpha+1/2}(t)\, g(t)\, dt, \tag{10}$$

which is a Hankel transform, see 5.6(8). We write

$$\mathscr{H}\{F(t), y, \nu\} = \int_0^{\infty} F(t)\, J_\nu(yt)(yt)^{1/2}\, dt. \tag{11}$$

Then

$$c_n = \Omega_n \exp[i\pi(n-\alpha-1)/2]\, \mathscr{H}\{g(t)/t^{\alpha+1}, e^{i\pi/2}, n+\alpha+\tfrac{1}{2}\}. \tag{12}$$

The case $\alpha = -\tfrac{1}{2}$ is important for the applications. Thus with

$$f(x) = \sum_{n=0}^{\infty} a_n T_n(x), \tag{13}$$

and $f(x)$ also given by (8), then

$$a_n = \epsilon_n \exp[i\pi(n - \tfrac{1}{2})/2] \, \mathscr{H}\{g(t)/t^{1/2}, e^{i\pi/2}, n\}. \tag{14}$$

For expansions in series of the "shifted" polynomials, suppose

$$f(x) = \sum_{n=0}^{\infty} a_n R_n^{(\alpha,\alpha)}(x), \tag{15}$$

$$f(x) = \int_0^{\infty} e^{-xt} g(t) \, dt. \tag{16}$$

Since

$$e^{-xt} = \sum_{n=0}^{\infty} \frac{(-)^n \Omega_n e^{-t/2} I_{n+\alpha+1/2}(t/2)}{(t/2)^{\alpha+1/2}} R_n^{(\alpha,\alpha)}(x), \tag{17}$$

we find

$$\begin{aligned} a_n &= 2(-)^n \Omega_n \int_0^{\infty} \frac{e^{-t} I_{n+\alpha+1/2}(t)\, g(2t)}{t^{\alpha+1/2}} \, dt, \\ a_n &= 2\Omega_n \exp[i\pi(n - \alpha - 1)/2] \, \mathscr{H}\{[e^{-t} g(2t)]/t^{\alpha+1}, e^{i\pi/2}, n + \alpha + \tfrac{1}{2}\}. \end{aligned} \tag{18}$$

In particular, with (16) and

$$f(x) = \sum_{n=0}^{\infty} c_n T_n^*(x), \tag{19}$$

$$c_n = 2\epsilon_n \exp[i\pi(n - \tfrac{1}{2})/2] \, \mathscr{H}\{[e^{-t} g(2t)]/t^{1/2}, e^{i\pi/2}, n\}. \tag{20}$$

Next suppose that $f(x)$ is defined as a Fourier transform. Let

$$f(x) = \int_0^{\infty} e^{ixt} g(t) \, dt = f_1(x) + i f_2(x), \tag{21}$$

$$f_1(x) = \int_0^{\infty} (\cos xt)\, g(t) \, dt, \qquad f_2(x) = \int_0^{\infty} (\sin xt)\, g(t) \, dt, \tag{22}$$

$$f_1(x) = \sum_{n=0}^{\infty} C_n P_n^{(\alpha,\alpha)}(x), \qquad f_2(x) = \sum_{n=0}^{\infty} S_n P_n^{(\alpha,\alpha)}(x). \tag{23}$$

Then

$$C_{2n+1} = 0, \qquad C_{2n} = (-)^n \Omega_{2n} \mathscr{H}\{g(t)/t^{\alpha+1}, 1, 2n + \alpha + \tfrac{1}{2}\}, \tag{24}$$

$$S_{2n} = 0, \qquad S_{2n+1} = (-)^n \Omega_{2n+1} \mathscr{H}\{g(t)/t^{\alpha+1}, 1, 2n + \alpha + \tfrac{3}{2}\}. \tag{25}$$

In particular, using (21), (22), $\alpha = -\frac{1}{2}$, and

$$f_1(x) = \sum_{n=0}^{\infty} c_n T_{2n}(x), \qquad f_2(x) = \sum_{n=0}^{\infty} s_n T_{2n+1}(x), \tag{26}$$

we have

$$c_n = (-)^n \epsilon_n \mathscr{H}\{g(t)/t^{1/2}, 1, 2n\}, \qquad s_n = 2(-)^n \mathscr{H}\{g(t)/t^{1/2}, 1, 2n+1\}. \tag{27}$$

Similar type results follow for the inverse Laplace transforms and also other transforms. See Elliott and Szekeres (1965), Luke (1969) and Tuan and Elliott (1972).

11.3.6.2. Evaluation of the Coefficients when $F(x)$ is Defined by a Taylor Series

Suppose now that $f(x)$ is analytic at $x = 0$ and

$$f(\omega x) = \sum_{k=0}^{\infty} \xi_k \omega^k x^k, \qquad \xi_k = f^{(k)}(0)/k!\,. \tag{1}$$

Let $g_n(x)$ be a polynomial in x of degree n. Then there exist constants $\sigma_{k,n}$ such that

$$x^k = \sum_{n=0}^{k} \sigma_{k,n} g_n(x). \tag{2}$$

Combining (1) and (2), we have

$$f(\omega x) = \sum_{n=0}^{\infty} C_n g_n(x), \tag{3}$$

$$C_n = \sum_{k=n}^{\infty} \sigma_{k,n} \xi_k \omega^k. \tag{4}$$

Note that if $\sigma_{k,n} = O(1)$ uniformly in n as $k \to \infty$, then most certainly (4) converges if $|\omega|$ is less than the radius of convergence of $f(x)$. Equation (3) is called the basic series for $f(\omega x)$ corresponding to the set of functions $\{g_n(x)\}$. For a general discussion of basic series, see the work of Boas and Buck (1958). Here we consider the basic series where $g_n(x)$ is either the extended Jacobi polynomial

$$G_n(x, \lambda) = {}_{p+2}F_q\left(\begin{matrix}-n, n+\lambda, \alpha_p \\ \rho_q\end{matrix}\middle| x\right) \tag{5}$$

or its confluent form, the extended Laguerre polynomial

$$G_n(x) = \lim_{\lambda\to\infty} G_n(x/\lambda, \lambda) = {}_{p+1}F_q\left(\begin{matrix}-n, \alpha_p \\ \rho_q\end{matrix}\middle| x\right). \tag{6}$$

In the sequel, we use the notation

$$x^k = \sum_{n=0}^{k} \sigma_{k,n}(\lambda)\, G_n(x, \lambda),$$

$$f(\omega x) = \sum_{n=0}^{\infty} C_n(\omega, \lambda)\, G_n(x, \lambda),$$

$$C_n(\omega, \lambda) = \sum_{k=n}^{\infty} \sigma_{k,n}(\lambda)\, \xi_k \omega^k, \qquad \xi_k = f^{(k)}(0)/k!\,, \tag{7}$$

and if $G_n(x, \lambda)$ is replaced by $G_n(x)$, we simply replace $\sigma_{k,n}(\lambda)$ and $C_n(\omega, \lambda)$ by $\sigma_{k,n}$ and $C_n(\omega)$, respectively.

Let none of the quantities $\lambda + 1$, α_j, and ρ_j be a negative integer or zero. Then Fields and Wimp (1963) have proved that

$$\sigma_{k,n}(\lambda) = \frac{(-k)_n(2n+\lambda)(\rho_q)_k}{n!\,(n+\lambda)(n+\lambda+1)_k(\alpha_p)_k}, \tag{8}$$

$$C_n(\omega, \lambda) = \frac{(-)^n\omega^n(\rho_q)_n}{(n+\lambda)_n(\alpha_p)_n} \sum_{k=0}^{\infty} \frac{(n+1)_k(n+\rho_q)_k\xi_{k+n}\omega^k}{(2n+\lambda+1)_k(n+\alpha_p)_k k!}, \tag{9}$$

and for the confluent case,

$$\sigma_{k,n} = \frac{(-k)_n(\rho_q)_k}{n!\,(\alpha_p)_k}, \tag{10}$$

$$C_n(\omega) = \frac{(-)^n\omega^n(\rho_q)_n}{(\alpha_p)_n} \sum_{k=0}^{\infty} \frac{(n+1)_k(n+\rho_q)_k\xi_{k+n}\omega^k}{(n+\alpha_p)_k k!}. \tag{11}$$

Note that in view of (8),

$$x^k = \frac{(\rho_q)_k}{(\alpha_p)_k} \sum_{n=0}^{k} \frac{(-k)_n(2n+\lambda)}{(n+\lambda+1)_k(n+\lambda)\,n!}\, {}_{p+2}F_q\left(\begin{matrix}-n, n+\lambda, \alpha_p \\ \rho_q\end{matrix}\middle| x\right). \tag{12}$$

For the expansion of $f(z)$ in series of the shifted Jacobi polynomials, we have

$$f(\omega x) = \sum_{n=0}^{\infty} S_n(\omega)\, R_n^{(\alpha,\beta)}(x), \tag{13}$$

$$S_n(\omega) = \frac{n!\,\omega^n}{(n+\lambda)_n} \sum_{k=0}^{\infty} \frac{(n+\beta+1)_k (n+1)_k \xi_{k+n}\omega^k}{(2n+\lambda+1)_k k!}. \tag{14}$$

When $\alpha = \beta = -\frac{1}{2}$, that is, $\lambda = 0$, we get the important expansion in series of shifted Chebyshev polynomials of the first kind. Thus,

$$f(\omega x) = \sum_{n=0}^{\infty} A_n(\omega)\, T_n^*(x), \qquad A_0(\omega) = \sum_{k=0}^{\infty} [(\tfrac{1}{2})_k \xi_k \omega^k]/k!\,,$$

$$A_n(\omega) = 2\left(\frac{\omega}{4}\right)^n \sum_{k=0}^{\infty} \frac{(n+\frac{1}{2})_k (n+1)_k \xi_{k+n}\omega^k}{(2n+1)_k k!}, \qquad n \geqslant 1. \tag{15}$$

In a similar fashion,

$$x^{\epsilon} f(\omega x^2) = \sum_{n=0}^{\infty} E_n(\omega)\, P_{2n+\epsilon}^{(\alpha,\alpha)}(x), \qquad \epsilon = 0 \quad \text{or} \quad \epsilon = 1,$$

$$E_n(\omega) = \frac{2^{\epsilon}(2n+\epsilon)!\,(4\omega)^n}{(2n+2\alpha+1+\epsilon)_{2n+\epsilon}} \sum_{k=0}^{\infty} \frac{(n+\frac{1}{2}+\epsilon)_k (n+1)_k \xi_{k+n}\omega^k}{(2n+\alpha+\frac{3}{2}+\epsilon)_k k!}, \tag{16}$$

where ξ_k has the same meaning as in (1). For the case $\alpha = -\frac{1}{2}$, we have

$$x^{\epsilon} f(\omega x^2) = \sum_{n=0}^{\infty} F_n(\omega)\, T_{2n+\epsilon}(x), \qquad \epsilon = 0 \quad \text{or} \quad \epsilon = 1,$$

$$F_n(\omega) = 2^{1-\epsilon}\left(\frac{\omega}{4}\right)^n \sum_{k=0}^{\infty} \frac{(n+\frac{1}{2}+\epsilon)_k (n+1)_k \xi_{k+n}\omega^k}{(2n+1+\epsilon)_k k!}. \tag{17}$$

11.3.6.3. Asymptotic Estimates of the Coefficients

We consider the representation 11.3.6.1(1,2). Let C be a completely closed contour such that $f(x)$ is analytic on and within C and let the line segment $-1 \leq x \leq 1$ lie within C. We suppose that $\alpha > -1$, $\beta > -1$. Put Cauchy's formula

$$f(x) = (2\pi i)^{-1} \int_C (z-x)^{-1} f(z)\,dz \tag{1}$$

in 11.3.6.1(2), interchange the order of integration, and get

$$c_n = (2\pi i h_n)^{-1}\int_C (z-1)^{\alpha}(z+1)^{\beta}f(z)Q_n^{(\alpha,\beta)}(z)dz\,. \tag{2}$$

Here

$$Q_n^{(\alpha,\beta)}(z) = \tfrac{1}{2}(z-1)^{-\alpha}(z+1)^{-\beta}\int_{-1}^{1}\frac{(1-x)^{\alpha}(1+x)^{\beta}P_n^{(\alpha,\beta)}(x)dx}{z-x} \tag{3}$$

is an associated Jacobi function and in hypergeometric form

$$Q_n^{(\alpha,\beta)}(z) = \frac{2^{n+\lambda-1}\Gamma(n+\alpha+1)\Gamma(n+\beta+1)}{(z-1)^{\alpha}(z+1)^{n+\beta+1}\Gamma(2n+\lambda+1)}\,{}_2F_1\left(\begin{matrix}n+1,\ n+\beta+1\\ 2n+\lambda+1\end{matrix}\,\middle|\,\frac{2}{1+z}\right). \tag{4}$$

In (3) and (4), the complex z-plane is cut along the segment $-1\leq z\leq 1$. Let

$$e^{-\phi} = z\mp(z^2-1)^{\frac{1}{2}} \tag{5}$$

where the sign is chosen so that $|e^{-\phi}|<1$. This is possible for all z not on the cut $-1\leq z\leq 1$. It can be shown that, see Luke (1969, v. 1, p. 237),

$$Q_n^{(\alpha,\beta)}(z) = \frac{(\pi/n)^{\frac{1}{2}}e^{-(n+1)\phi}(1-e^{-\phi})^{\alpha-\frac{1}{2}}(1+e^{-\phi})^{\beta-\frac{1}{2}}}{(z-1)^{\alpha}(z+1)^{\beta}}\{1+O(n^{-1})\}\,, \tag{6}$$

whence

$$c_n = \frac{(n/\pi)^{\frac{1}{2}}}{2^{\lambda}i}\int_C\frac{f(z)(1-e^{-\phi})^{\alpha+\frac{1}{2}}(1+e^{-\phi})^{\beta+\frac{1}{2}}\{1+O(n^{-1})\}dz}{e^{n\phi}\sinh\phi}\,. \tag{7}$$

Notice that the character of c_n for n large is virtually independent of α and β. If

$$z = u+iv = \cosh\phi,\quad \phi = \gamma+i\delta\,, \tag{8}$$

then

$$(u/\cosh\alpha)^2+(v/\cosh\beta)^2 = 1,\quad |e^{\phi}| = e^{\gamma} = \rho\,. \tag{9}$$

Thus the latter equation generates an ellipse which we call E_ρ with foci at $z=\pm 1$ and semiaxes $\cosh\gamma$ and $\sinh\gamma$, respectively. Observe that if $z=x$, $-1\leq x\leq 1$, then $\rho=1$ and $\gamma=0$. Introduce the transformation

$$e^{\phi} = \sigma\,. \tag{10}$$

This maps the exterior of the ellipse E_ρ onto the exterior of the circle C_ρ of radius ρ with center at the origin in the σ-plane. In (7) let C be this circle. Then

$$c_n = \frac{(n/\pi)^{\frac{1}{2}}}{2^\lambda i}\int_{C_\rho}\frac{f(\frac{1}{2}\{\sigma+\sigma^{-1}\})(1-\sigma^{-1})^{\alpha+\frac{1}{2}}(1+\sigma^{-1})^{\beta+\frac{1}{2}}\{1+O(n^{-1})\}d\sigma}{\sigma^{n+1}} \tag{11}$$

and

$$|c_n| \leq (n\pi)^{\frac{1}{2}}M(\rho)(1+\mu/n)/2^{\lambda-1}\rho^n,$$

$$M(\rho) = \max_{z \text{ on } C_\rho} |f(z)(1-e^{-\phi})^{\alpha+\frac{1}{2}}(1+e^{-\phi})^{\beta+\frac{1}{2}}|. \tag{12}$$

Here μ depends on α, β and ρ but is independent of n. This bound is not very pragmatic. If $\rho > 1$ and n increases, the bound decreases for ρ fixed because of ρ^n. On the other hand if ρ increases, the bound increases for n fixed because of $M(\rho)$. For a given n there is a ρ for which the bound is a minimum. Obviously this is not easily realized in practice. For practical situations, an asymptotic estimate of c_n is much more informative. To achieve such we deform the path of integration in (7) so that it never passes through the branch points $z = \pm 1$, and then use the method of steepest descent. Let

$$e^{\omega(z)} = \frac{f(z)(1-e^{-\phi})^{\alpha+\frac{1}{2}}(1+e^{-\phi})^{\beta+\frac{1}{2}}}{e^{n\phi}\sinh\phi} \tag{13}$$

so that

$$\omega'(z) = \frac{f'(z)}{f(z)} - \frac{(n+\coth\phi)}{\sinh\phi} + \frac{(\alpha-\beta+\lambda e^{-\phi})}{2\sinh^2\phi},$$

$$\omega''(z) = \frac{f''(z)}{f(z)} - \frac{\{f'(z)\}^2}{\{f(z)\}^2} + \frac{\frac{1}{2}n\sinh 2\phi+\cosh^2\phi+1}{\sinh^4\phi}$$

$$- \frac{\lambda e^{-\phi}(\frac{1}{2}\sinh\phi+\cosh\phi)+(\alpha-\beta)\cosh\phi}{\sinh^4\phi}. \tag{14}$$

Let $\xi = \cosh\theta$ be a number such that $\omega'(\xi) = 0$. Upon application of the method of steepest descent, we have for sufficiently large n

$$c_n \sim -\frac{i(2n)^{\frac{1}{2}}\eta f(\xi)e^{-n\theta}(1-e^{-\theta})^{\alpha+\frac{1}{2}}(1+e^{-\theta})^{\beta+\frac{1}{2}}}{2^{\lambda}(\sinh\theta)|\omega''(\xi)|^{\frac{1}{2}}},$$

$$\eta = \exp\{\tfrac{1}{2}i\pi - \tfrac{1}{2}i\arg\omega''(\xi)\}, \tag{15}$$

or a sum of such relations, one for each point ξ which lies on the deformed contour. Note that if c_n is designated as $c_n^{(\alpha,\beta)}(P)$ and $c_n^{(\alpha,\beta)}(S)$ designates a like coefficient with P replaced by S where $S_n^{(\alpha,\beta)}(x) = U_n c_n^{(\alpha,\beta)}(x)$, then

$$c_n^{(\alpha,\beta)}(P) = U_n c_n^{(\alpha,\beta)}(S). \tag{16}$$

For an illustration, suppose $f(x) = e^{\sigma x}$ and $\alpha = \beta = 0$. Then from 5.10.2(8), see also 11.3.6.1(9), we have

$$c_n = (\pi/2\sigma)^{\frac{1}{2}}(2n+1)I_{n+\frac{1}{2}}(\sigma), \tag{17}$$

and with the aid of 9.2(2) and 1.3(13),

$$c_n \sim (n\pi)^{\frac{1}{2}}(\sigma/2)^n/n! \tag{18}$$

From (14), if $\omega'(\xi) = 0$, then for n sufficiently large

$$\cosh\theta \sim (1+n^2/\sigma^2)^{\frac{1}{2}},$$

$$c_n \sim \frac{(\sinh\theta)\exp\{\sigma\cosh\theta-(n+\frac{1}{2})\theta\}}{(\cosh\theta)^{\frac{1}{2}}}. \tag{19}$$

In particular, suppose $\sigma = 2$ and $n = 4$. Then the exact value deduced from (17) and standard tables is 0.1823. The approximate values given by (18) and (19) are 0.1477 and 0.1768 respectively. Again, if $n = 8$, the exact value is $0.1402\cdot 10^{-3}$ while the values given by (18) and (19) are $0.1243\cdot 10^{-3}$ and $0.1390\cdot 10^{-3}$ respectively.

Some applications of (15) when $f(x)$ is expanded in series of $T_n(x)$ have been given by Elliott and Szekeres (1965). For estimates of the coefficients in the Chebyshev case when $f(x)$ has branch points or poles, see Elliott (1964,1965), and in the situation when $f(x)$ has a logarithmic singularity, see Chawla (1967). For an excellent discussion on estimating general contour integrals like (2), see Donaldson and Elliott (1972). Asymptotic estimates of the c_n's for the case of Jacobi, Laguerre and Hermite polynomials is the subject of a paper by Elliott and Tuan (1974).

In the case of hypergeometric functions, excellent estimates of the coefficients c_n have been given by Luke (1969). Clearly, from 5.10.2(5), it is sufficient to examine the behavior of

$$F_n(z) = {}_{p+1}F_{q+1}(\beta+1+n,\ a_p+n;\ \lambda+1+2n,\ b_q+n;\ z),$$

$$p \leq q,\ |z| < \infty;\ p = q+1,\ z \neq 1,\ \arg(1-z) < \pi. \tag{20}$$

The results follow.

$$F_n(z) = 1 + \sum_{k=0}^{r-1} t_k + O(n^{(p-q)r}),\ p \leq q-1,$$

$$t_k = \frac{(a_p+n)_{k+1}(\beta+1+n)_{k+1}z^{k+1}}{(b_q+n)_{k+1}(\lambda+1+2n)_{k+1}(k+1)!}. \tag{21}$$

$$F_n(z) = e^{\frac{1}{2}z}\left[1 + \frac{z^2 - 4z(\lambda-1-2\beta-2\theta)}{8(\lambda+2n)} + O(n^{-2})\right],$$

$$p = q,\ \theta = \sum_{j=1}^{p} (a_j - b_j). \tag{22}$$

$$F_n(z) = \frac{(\pi/n)^{\frac{1}{2}}\Gamma(2n+\lambda+1)e^{-(n+\beta+1)\nu}(1+e^{-\nu})^{\theta-\beta-3/2}\{1+O(n^{-1})\}}{\Gamma(n+\lambda+1-\theta)\Gamma(n+\theta)z^{n+\beta+1}(1-e^{-\nu})^{\theta+\beta-\lambda+\frac{1}{2}}},$$

$$e^{-\nu} = \{2 - z \mp 2(1-z)^{\frac{1}{2}}\}/z,\ \theta = \sum_{j=1}^{q+1} a_j - \sum_{j=1}^{q} b_j,$$

$$z \neq 0,\ z \neq 1,\ |\arg(1-z)| < \pi,\ p = q+1, \tag{23}$$

where in $e^{-\nu}$ the sign is chosen so that $|e^{-\nu}| < 1$ which is possible for all z in the cut complex plane.

If in $e^{-\phi}$ as defined in (5) z is replaced by $2z+1$, we get the function e^{-w} discussed in 2.4.4. Likewise, if in $e^{-\nu}$ above, z is replaced by $-1/z$, we get the function e^{-w} of 2.4.4.

Asymptotic results for the coefficients when a certain G-function is expanded in series of Jacobi polynomials have been given in 5.10.1(9).

11.4. The Chebyshev Polynomials $T_n(x)$ and $U_n(x)$

In this section we present a collection of results for the Chebyshev polynomials of the first kind. A few formulae for the polynomials of the second kind are also given. Many of the corresponding formulae for the shifted polynomials are presented in 11.5.

$$T_n(x) = \frac{n!}{(\frac{1}{2})_n} P_n^{(-1/2,-1/2)}(x), \qquad U_n(x) = \frac{(n+1)!}{(\frac{3}{2})_n} P_n^{(1/2,1/2)}(x), \tag{1}$$

$$T_n(x) = \cos n\theta, \qquad U_n(x) = \csc\theta \sin(n+1)\theta, \qquad x = \cos\theta. \tag{2}$$

$$T_{2n}(\sin\theta) = (-)^n \cos 2n\theta, \qquad T_{2n+1}(\sin\theta) = (-)^n \sin(2n+1)\theta, \tag{3}$$

$$U_{2n}(\sin\theta) = \frac{(-)^n \cos(2n+1)\theta}{\cos\theta}, \qquad U_{2n+1}(\sin\theta) = \frac{(-)^n \sin(2n+2)\theta}{\cos\theta}. \tag{4}$$

$$T_n(x) = {}_2F_1\left(\begin{matrix}-n, n\\ \frac{1}{2}\end{matrix}\,\middle|\,\frac{1-x}{2}\right) = (-)^n\, {}_2F_1\left(\begin{matrix}-n, n\\ \frac{1}{2}\end{matrix}\,\middle|\,\frac{1+x}{2}\right). \tag{5}$$

$$T_n(x) = \tfrac{1}{2}n \sum_{k=0}^{[n/2]} \frac{(-)^k (n-k-1)!}{k!\,(n-2k)!} (2x)^{n-2k}, \qquad n = 1, 2, \ldots . \tag{6}$$

$$T_{2n}(x) = (-)^n\, {}_2F_1\left(\begin{matrix}-n, n\\ \frac{1}{2}\end{matrix}\,\middle|\,x^2\right) = {}_2F_1\left(\begin{matrix}-n, n\\ \frac{1}{2}\end{matrix}\,\middle|\,1-x^2\right). \tag{7}$$

$$T_{2n}(x) = T_n(2x^2-1). \tag{8}$$

$$T_{2n+1}(x) = (-)^n (2n+1)\, x\, {}_2F_1\left(\begin{matrix}-n, n+1\\ \frac{3}{2}\end{matrix}\,\middle|\,x^2\right) = x\, {}_2F_1\left(\begin{matrix}-n, n+1\\ \frac{1}{2}\end{matrix}\,\middle|\,1-x^2\right). \tag{9}$$

$$T_n(1) = 1, \qquad T_n(-1) = (-)^n, \qquad T_{2n}(0) = (-)^n, \qquad T_{2n+1}(0) = 0. \tag{10}$$

$$2^n (\tfrac{1}{2})_n T_n(x) = (-)^n (1-x^2)^{1/2} \frac{d^n}{dx^n} (1-x^2)^{n-1/2}. \tag{11}$$

$$(1-x^2) \frac{d^2 T_n(x)}{dx^2} - \frac{x\, dT_n(x)}{dx} + n^2 T_n(x) = 0. \tag{12}$$

$$(1-x^2) \frac{dT_n(x)}{dx} = n[T_{n-1}(x) - x T_n(x)]. \tag{13}$$

If

$$y_n(x) = A T_n(x) + B U_n(x),$$

where A and B are constants independent of n and x, then

$$\begin{aligned} y_{n+1}(x) &= 2xy_n(x) - y_{n-1}(x), & n > 0, \\ T_1(x) &= xT_0(x) = x, & U_1(x) = 2xU_0(x) = 2x, \end{aligned} \tag{14}$$

$$\begin{aligned} y_{2n+2}(x) &= 2(2x^2 - 1)\, y_{2n}(x) - y_{2n-2}(x), \qquad n > 0, \\ T_2(x) &= (2x^2 - 1)\, T_0(x) = 2x^2 - 1, \\ U_2(x) &= (4x^2 - 1)\, U_0(x) = 4x^2 - 1. \end{aligned} \tag{15}$$

$$\begin{aligned} y_{2n+3}(x) &= 2(2x^2 - 1)\, y_{2n+1}(x) - y_{2n-1}(x), \qquad n > 0, \\ T_3(x) &= 2(2x^2 - 1)\, T_1(x) - T_1(x) = (4x^2 - 3)\, T_1(x) = 4x^3 - 3x, \\ U_3(x) &= 2(2x^2 - 1)\, U_1(x) = 8x^3 - 4x. \end{aligned} \tag{16}$$

$$2T_m(x)\, T_n(x) = T_{m+n}(x) + T_{|m-n|}(x). \tag{17}$$

$$2(x^2 - 1)\, U_{m-1}(x)\, U_{n-1}(x) = T_{m+n}(x) - T_{|m-n|}(x). \tag{18}$$

$$2T_m(x)\, U_{n-1}(x) = U_{n+m-1}(x) + U_{n-m-1}(x), \qquad n > m. \tag{19}$$

$$2T_n(x)\, U_{m-1}(x) = U_{n+m-1}(x) - U_{n-m-1}(x), \qquad n > m. \tag{20}$$

$$T_m(x)\, U_n(x) - U_m(x)\, T_n(x) = T_1(x)\, U_{n-m-1}(x), \qquad n > m. \tag{21}$$

$$x^m = 2^{1-m} \sum_{k=0}^{[m/2]} a_k \binom{m}{k} T_{m-2k}(x), \qquad m > 0.$$

$$a_k = \tfrac{1}{2} \text{ if } k = \tfrac{1}{2}m; \; a_k = 1, \text{ otherwise.} \tag{22}$$

$$x^m T_n(x) = 2^{-m} \sum_{k=0}^{m} \binom{m}{k} T_{|n+m-2k|}(x). \tag{23}$$

$$\frac{dT_{2n}(x)}{dx} = 4n \sum_{k=0}^{n-1} T_{2k+1}(x). \tag{24}$$

$$\frac{dT_{2n+1}(x)}{dx} = (2n + 1) + 2(2n + 1) \sum_{k=1}^{n} T_{2k}(x). \tag{25}$$

$$x\,\frac{dT_{2n}(x)}{dx} = 2n(T_0(x) + T_{2n}(x)) + 4n \sum_{k=1}^{n-1} T_{2k}(x). \tag{26}$$

$$x\,\frac{dT_{2n+1}(x)}{dx} = 2(2n + 1) \sum_{k=0}^{n-1} T_{2k+1}(x) + (2n + 1)\, T_{2n+1}(x). \tag{27}$$

$$x^2\,\frac{dT_{2n}(x)}{dx} = 4n \left[\sum_{k=0}^{n-2} T_{2k+1}(x) + \tfrac{3}{4} T_{2n-1}(x) + \tfrac{1}{4} T_{2n+1}(x) \right]. \tag{28}$$

$$x^2\frac{dT_{2n+1}(x)}{dx} = 2(2n+1)\left[\tfrac{1}{2}T_0(x) + \sum_{k=1}^{n-1} T_{2k}(x) + \tfrac{3}{4}T_{2n}(x) + \tfrac{1}{4}T_{2n+2}(x)\right],$$

$$n \geqslant 1; \qquad x^2\frac{dT_1(x)}{dx} = \tfrac{1}{2}[T_0(x) + T_2(x)]. \tag{29}$$

$$\frac{d^2T_{2n}(x)}{dx^2} = 4n^3T_0(x) + 8n\sum_{k=1}^{n-1}(n^2-k^2)\,T_{2k}(x). \tag{30}$$

$$\frac{d^2T_{2n+1}(x)}{dx^2} = 4(2n+1)\sum_{k=0}^{n-1}(n-k)(n+1+k)\,T_{2k+1}(x),$$

$$n \geqslant 1, \qquad \frac{d^2T_1(x)}{dx^2} = 0. \tag{31}$$

$$x\frac{d^2T_{2n}(x)}{dx^2} = 4n\sum_{k=0}^{n-1}\{2n^2 - k^2 - (k+1)^2\}\,T_{2k+1}(x). \tag{32}$$

$$x\frac{d^2T_{2n+1}(x)}{dx^2} = 2(2n+1)\left[n(n+1)\,T_0(x) + 2\sum_{k=1}^{n}(n^2+n-k^2)\,T_{2k}(x)\right]. \tag{33}$$

$$x^2\frac{d^2T_{2n}(x)}{dx^2} = 2n\left[(2n^2-1)T_0(x) + 2\sum_{k=1}^{n-1}(2n^2-2k^2-1)T_{2k}(x) + (2n-1)T_{2n}(x)\right]. \tag{34}$$

$$x^2\frac{d^2T_{2n+1}(x)}{dx^2} = 2(2n+1)\left[\sum_{k=0}^{n-1}(2n^2+2n-2k^2-2k-1)\,T_{2k+1}(x) + nT_{2n+1}(x)\right],$$

$$n \geqslant 1. \tag{35}$$

$$\begin{aligned}\int_{-1}^{1}(1-x^2)^{-1/2}T_n(x)\,T_m(x)\,dx &= 0 \qquad &&\text{if} \quad m \neq n,\\ &= \pi/2 &&\text{if} \quad m = n \neq 0,\\ &= \pi &&\text{if} \quad m = n = 0.\end{aligned} \tag{36}$$

$$\int_{-1}^{1}(1-x^2)^{1/2}U_n(x)\,U_m(x)\,dx = (\pi/2)\delta_{mn}. \tag{37}$$

$$\int_{-1}^{1}(x-y)^{-1}(1-x^2)^{-1/2}[T_n(x) - T_n(y)]\,dx = \pi U_{n-1}(y). \tag{38}$$

$$\text{P.V.}\int_{-1}^{1}(x-y)^{-1}(1-x^2)^{-1/2}T_n(x)\,dx = \pi U_{n-1}(y), \qquad -1 < y < 1. \tag{39}$$

$$\text{P.V.}\int_{-1}^{1}(x-y)^{-1}(1-x^2)^{1/2}U_{n-1}(x)\,dx = -\pi T_n(y), \qquad -1 < y < 1. \tag{40}$$

$$\int_{-1}^{1} (z-x)^{-1}(1-x^2)^{-1/2}T_n(x)\,dx = \frac{\pi e^{-n\varphi}}{\sinh\varphi}, \tag{41}$$

$$\int_{-1}^{1} (z-x)^{-1}(1-x^2)^{1/2}U_{n-1}(x)\,dx = \pi e^{-n\varphi}, \qquad n > 0,$$

$$z = \cosh\varphi, \qquad z \text{ as in 11.6.3(5).} \tag{42}$$

$$\int_0^x T_{2n}(t)\,dt = \frac{1}{2}\left[\frac{T_{2n+1}(x)}{2n+1} - \frac{T_{|2n-1|}(x)}{2n-1}\right]. \tag{43}$$

$$\int_0^x T_{2n+1}(t)\,dt = \frac{1}{2}\left[\frac{T_{2n+2}(x)}{2n+2} - \frac{T_{2n}(x)}{2n}\right] + \frac{(-)^n}{4}\left(\frac{1}{n} + \frac{1}{n+1}\right), \qquad n > 0,$$

$$\int_0^x T_1(t)\,dt = \tfrac{1}{4}[T_2(x) + T_0(x)]. \tag{44}$$

$$(n+1)\int_0^x U_n(t)\,dt = T_{n+1}(x) - T_{n+1}(0). \tag{45}$$

$$\int_0^x\int_0^t T_{2n}(u)\,du\,dt = \frac{T_{2n+2}(x)}{4(2n+1)(2n+2)} - \frac{T_{2n}(x)}{2(4n^2-1)} + \frac{T_{2n-2}(x)}{4(2n-1)(2n-2)}$$

$$+ \frac{(-)^n}{4(n^2-1)}, \qquad n > 1,$$

$$\int_0^x\int_0^t T_0(u)\,du\,dt = \tfrac{1}{4}[T_0(x) + T_2(x)],$$

$$\int_0^x\int_0^t T_2(u)\,du\,dt = -\frac{3T_0(x)}{16} - \frac{T_2(x)}{6} + \frac{T_4(x)}{48}. \tag{46}$$

$$\int_0^x\int_0^t T_{2n+1}(u)\,du\,dt = \frac{T_{2n+3}(x)}{4(2n+2)(2n+3)} - \frac{T_{2n+1}(x)}{2(2n+2)(2n)} + \frac{T_{2n-1}(x)}{4(2n)(2n-1)}$$

$$+ \frac{(-)^n(2n+1)}{4n(n+1)}T_1(x), \qquad n > 0,$$

$$\int_0^x\int_0^t T_1(u)\,du\,dt = \tfrac{1}{24}[3T_1(x) + T_3(x)]. \tag{47}$$

Numerical tables which relate integral powers of x and $T_n(x)$ are provided on pages 457 and 458. Corresponding tables for the shifted polynomials are given on pages 461-463.

TABLE 11.1

COEFFICIENTS FOR THE EXPANSION OF x^n IN SERIES OF CHEBYSHEV POLYNOMIALS $T_k(x)$

$$x^n = d_n^{-1} \sum_{k=0}^{\frac{1}{2}(n-\epsilon)} a_k T_{2k+\epsilon}(x)$$

$\epsilon = 0$ or 1 according as n is even or odd, respectively; e.g.,

$$x^4 = 8^{-1}[3T_0(x) + 4T_2(x) + T_4(x)],$$

$$x^5 = 16^{-1}[10T_1(x) + 5T_3(x) + T_5(x)]$$

n	$d_n \cdot a_0, a_1, \ldots, a_{\frac{1}{2}(n-\epsilon)}$
0	1, 1
1	1, 1
2	2, 1, 1
3	4, 3, 1
4	8, 3, 4, 1
5	16, 10, 5, 1
6	32, 10, 15, 6, 1
7	64, 35, 21, 7, 1
8	128, 35, 56, 28, 8, 1
9	256, 126, 84, 36, 9, 1
10	512, 126, 210, 120, 45, 10, 1
11	1024, 462, 330, 165, 55, 11, 1
12	2048, 462, 792, 495, 220, 66, 12, 1

TABLE 11.2

Coefficients for the Expansion of the Chebyshev Polynomials $T_n(x)$ in Powers of x

$$T_n(x) = \sum_{k=0}^{m} a_k x^{n-2k}$$

$m = n/2$ if n is even, $m = (n-1)/2$ if n is odd; e.g.,

$T_4(x) = 8x^4 - 8x^2 + 1, \qquad T_5(x) = 16x^5 - 20x^3 + 5x$

n	$a_0, a_1, \ldots, a_m$
0	1
1	1
2	2, -1
3	4, -3
4	8, -8, 1
5	16, -20, 5
6	32, -48, 18, -1
7	64, -112, 56, -7
8	128, -256, 160, -32, 1
9	256, -576, 432, -120, 9
10	512, -1280, 1120, -400, 50, -1
11	1024, -2816, 2816, -1232, 220, -11
12	2048, -6144, 6912, -3584, 840, -72, 1
13	4096, -13312, 16640, -9984, 2912, -364, 13
14	8192, -28672, 39424, -26880, 9408, -1568, 98, -1
15	16384, -61440, 92160, -70400, 28800, -6048, 560, -15
16	32768, -1 31072, 2 12992, -1 80224, 84480, -21504, 2688, -128, 1
17	65536, -2 78528, 4 87424, -4 52608, 2 39360, -71808, 11424, -816, 17
18	1 31072, -5 89824, 11 05920, -11 18208, 6 58944, -2 28096, 44352, -4320, 162, -1
19	2 62144, -12 45184, 24 90368, -27 23840, 17 70496, -6 95552, 1 60512, -20064, 1140, -19
20	5 24288, -26 21440, 55 70560, -65 53600, 46 59200, -20 50048, 5 49120, -84480, 6600, -200, 1

11.5. The Chebyshev Polynomials $T_n^*(x)$ and $U_n^*(x)$

$$T_n^*(x) = T_n(2x-1), \qquad T_n(x) = T_n^*[(1+x)/2], \tag{1}$$

$$T_{2n}(x) = T_n^*(x^2), \tag{2}$$

$$U_n^*(x) = U_n(2x-1), \qquad U_n(x) = U_n^*[(1+x)/2], \tag{3}$$

$$U_{2n+1}(x) = 2xU_n^*(x^2). \tag{4}$$

$$T_n^*(x) = {}_2F_1\left(\begin{matrix}-n,\, n\\ \frac{1}{2}\end{matrix}\,\middle|\, 1-x\right) = (-)^n\, {}_2F_1\left(\begin{matrix}-n,\, n\\ \frac{1}{2}\end{matrix}\,\middle|\, x\right). \tag{5}$$

$$T_n^*(1-x) = (-)^n T_n^*(x). \tag{6}$$

$$T_n^*(1) = 1, \qquad T_n^*(0) = (-)^n. \tag{7}$$

$$T_{n+1}^*(x) = 2(2x-1)\, T_n^*(x) - T_{|n-1|}^*(x). \tag{8}$$

$$\begin{aligned}\int_0^1 \frac{T_n^*(x)\, T_m^*(x)}{[x(1-x)]^{1/2}}\, dx &= 0 \qquad \text{if} \quad m \neq n,\\ &= \tfrac{1}{2}\pi \qquad \text{if} \quad m = n \neq 0,\\ &= \pi \qquad \text{if} \quad m = n = 0.\end{aligned} \tag{9}$$

$$x^m = 2^{-2m}\left[\binom{2m}{m} T_0^*(x) + 2\sum_{k=1}^{m} \binom{2m}{m-k} T_k^*(x)\right]. \tag{10}$$

$$x^m T_n^*(x) = 2^{-2m} \sum_{k=0}^{2m} \binom{2m}{k} T_{|n+m-k|}^*(x). \tag{11}$$

$$\frac{dT_{2n}^*(x)}{dx} = 8n \sum_{k=0}^{n-1} T_{2k+1}^*(x). \tag{12}$$

$$\frac{dT_{2n+1}^*(x)}{dx} = 2(2n+1)\, T_0^*(x) + 4(2n+1) \sum_{k=1}^{n} T_{2k}^*(x). \tag{13}$$

$$x\,\frac{dT_n^*(x)}{dx} = 2n\left[\tfrac{1}{2}\{T_0^*(x) + T_n^*(x)\} + \sum_{k=1}^{n-1} T_k^*(x)\right]. \tag{14}$$

$$\begin{aligned}x^2\,\frac{dT_n^*(x)}{dx} &= 2n\left[\tfrac{1}{2}T_0^*(x) + \sum_{k=1}^{n-2} T_k^*(x) + \tfrac{7}{8}T_{n-1}^*(x) + \tfrac{1}{2}T_n^*(x)\right.\\ &\qquad \left. + \tfrac{1}{8}T_{n+1}^*(x)\right], \qquad n \geqslant 2;\\ x^2\,\frac{dT_1^*(x)}{dx} &= \tfrac{3}{4}T_0^*(x) + T_1^*(x) + \tfrac{1}{4}T_2^*(x), \qquad x^2\,\frac{dT_0^*(x)}{dx} = 0.\end{aligned} \tag{15}$$

$$\frac{d^2T_{2n}^*(x)}{dx^2} = 16n^3 T_0^*(x) + 32n \sum_{k=1}^{n-1} (n^2 - k^2)\, T_{2k}^*(x). \tag{16}$$

$$\frac{d^2T_{2n+1}^*(x)}{dx^2} = 16(2n+1)\sum_{k=0}^{n-1}(n-k)(n+1+k)\,T_{2k+1}^*(x). \tag{17}$$

$$x\frac{d^2T_{2n}^*(x)}{dx^2} = 8n\Big[n^2T_0^*(x) + 2\sum_{k=1}^{n-1}(n^2-k^2)\,T_{2k}^*(x) + \sum_{k=0}^{n-1}\{2n^2-k^2-(k+1)^2\}\,T_{2k+1}^*(x)\Big]. \tag{18}$$

$$x\frac{d^2T_{2n+1}^*(x)}{dx^2} = 8(2n+1)\Big[\frac{n(n+1)}{2}\,T_0^*(x) + \sum_{k=1}^{n}(n^2+n-k^2)\,T_{2k}^*(x) + \sum_{k=0}^{n-1}(n-k)(n+1+k)\,T_{2k+1}^*(x)\Big]. \tag{19}$$

$$x^2\frac{d^2T_{2n}^*}{dx^2} = 2n\Big[(4n^2-1)\,T_0^*(x) + 2\sum_{k=1}^{n-1}\{4(n^2-k^2)-1\}\,T_{2k}^*(x) + (2n-1)\,T_{2n}^*(x) + 4\sum_{k=0}^{n-1}\{2n^2-k^2-(k+1)^2\}\,T_{2k+1}^*(x)\Big]. \tag{20}$$

$$x^2\frac{d^2T_{2n+1}^*(x)}{dx^2} = (2n+1)\Big[4n(n+1)\,T_0^*(x) + 8\sum_{k=1}^{n}(n^2+n-k^2)\,T_{2k}^*(x) + 2\sum_{k=0}^{n-1}(4n^2+4n-4k^2-4k-1)\,T_{2k+1}^*(x) + 2nT_{2n+1}^*(x)\Big]. \tag{21}$$

$$\int_0^x T_n^*(t)\,dt = \frac{1}{4}\Big[\frac{T_{n+1}^*(x)}{n+1} - \frac{T_{|n-1|}^*(x)}{n-1}\Big] - \frac{(-)^n}{2(n^2-1)}, \qquad n = 0, 2, 3, \ldots,$$

$$\int_0^x T_1^*(t)\,dt = \tfrac{1}{8}[T_2^*(x) - T_0^*(x)], \qquad \int_0^x T_0^*(t)\,dt = \tfrac{1}{2}[T_1^*(x) + T_0^*(x)]. \tag{22}$$

$$\int_0^x\int_0^t T_n^*(u)\,du\,dt = \frac{T_{n+2}^*(x)}{16(n+2)(n+1)} - \frac{T_n^*(x)}{8(n^2-1)} + \frac{T_{n-2}^*(x)}{16(n-1)(n-2)} - \frac{(-)^nT_1^*(x)}{4(n^2-1)} - \frac{(-)^nT_0^*(x)}{4(n^2-4)}, \qquad n > 2,$$

$$\int_0^x\int_0^t T_0^*(u)\,du\,dt = \tfrac{1}{2}x^2 = \tfrac{1}{16}[3T_0^*(x) + 4T_1^*(x) + T_2^*(x)], \tag{23}$$

$$\int_0^x\int_0^t T_1^*(u)\,du\,dt = \tfrac{1}{96}[-8T_0^*(x) - 9T_1^*(x) + T_3^*(x)],$$

$$\int_0^x\int_0^t T_2^*(u)\,du\,dt = \tfrac{1}{192}[-9T_0^*(x) - 16T_1^*(x) - 8T_2^*(x) + T_4^*(x)].$$

TABLE 11.3

COEFFICIENTS FOR THE EXPANSION OF x^n IN SERIES OF CHEBYSHEV POLYNOMIALS $T_k^*(x)$

$$x^n = d_n^{-1} \sum_{k=0}^{n} a_k T_k^*(x)$$

e.g., $x^4 = 128^{-1}[35T_0^*(x) + 56T_1^*(x) + 28T_2^*(x) + 8T_3^*(x) + T_4^*(x)]$

n	$d_n, a_0, a_1, \ldots, a_n$
0	1, 1
1	2, 1, 1
2	8, 3, 4, 1
3	32, 10, 15, 6, 1
4	128, 35, 56, 28, 8, 1
5	512, 126, 210, 120, 45, 10, 1
6	2048, 462, 792, 495, 220, 66, 12, 1
7	8192, 1716, 3003, 2002, 1001, 364, 91, 14, 1
8	32768, 6435, 11440, 8008, 4368, 1820, 560, 120, 16, 1
9	1 31072, 24310, 43758, 31824, 18564, 8568, 3060, 816, 153, 18, 1
10	5 24288, 92378. 1 67960, 1 25970, 77520, 38760, 15504, 4845, 1140, 190, 20, 1
11	20 97152, 3 52716, 6 46646, 4 97420, 3 19770, 1 70544, 74613, 26334, 7315, 1540, 231, 22, 1
12	83 88608, 13 52078, 24 96144, 19 61256, 13 07504, 7 35471, 3 46104, 1 34596, 42504, 10626, 2024, 276, 24, 1

TABLE 11.4

COEFFICIENTS FOR THE EXPANSION OF THE CHEBYSHEV POLYNOMIALS $T_n^*(x)$ IN POWERS OF x

$$T_n^*(x) = \sum_{k=0}^{n} a_k x^{n-k}, \qquad \text{e.g.,} \quad T_4^*(x) = 128x^4 - 256x^3 + 160x^2 - 32x + 1$$

n	$a_0, a_1, \ldots, a_n$
0	1
1	2, -1
2	8, -8, 1
3	32, -48, 18, -1
4	128, -256, 160, -32, 1
5	512, -1280, 1120, -400, 50, -1
6	2048, -6144, 6912, -3584, 840, -72, 1
7	8192, -28672, 39424, -26880, 9408, -1568, 98, -1
8	32768, -1 31072, 2 12992, -1 80224, 84480, -21504, 2688, -128, 1
9	1 31072, -5 89824, 11 05920, -11 18208, 6 58944, -2 28096, 44352, -4320, 162, -1
10	5 24288, -26 21440, 55 70560, -65 53600, 46 59200, -20 50048, 5 49120, -84480, 6600, -200, 1
11	20 97152, -115 34336, 273 94048, -367 65696, 306 38080, -164 00384, 56 37632, -12 08064, 1 51008, -9680, 242, -1
12	83 88608, -503 31648, 1321 20576, -1992 29440, 1905 13152, -1203 24096, 506 92096, -140 57472, 24 71040, -2 56256, 13728, -288, 1
13	335 54432, -2181 03808, 6270 48448, -10496 24576, 11331 17440, -8255 56992, 4127 78496, -1412 13696, 323 61472, -47 59040, 4 16416, -18928, 338, -1

14	1342 17728, -9395 24096, 29360 12800, -54022 63552, 64995 98336, -53692 33408, 31117 14816, -12700 87620, 3611 81184, -697 01632, 87 12704, -6 52288, 25480, -392, 1
15	5368 70912, -40265 31840, 1 35895 44960, -2 72629 76000, 3 61758 72000, -3 34265 05728, 2 20522 08640, -1 04782 23360, 35721 21600, -8599 55200, 1418 92608, -152 75520, 9 90080, -33600, 450, -1
16	21474 83648, -1 71798 69184, 6 22770 25792, -13 52914 69824, 19 62934 27200, -20 06555 03360, 14 85622 47680, -8 06480 77312, 3 21332 18304, -93139 76320, 19262 99648, -2751 85664, 257 98656, -14 62272, 43520, -512, 1
17	85899 34592, -7 30144 44032, 28 29309 70624, -66 16933 99040, 104 21671 03488, -116 79458 91840, 95 93841 25440, -58 62902 98880, 26 77768 19200, -9 10441 18528, 2 27610 29632, -40933 86752, 5116 73344, -421 70880, 21 08544, -55488, 578, -1
18	3 43597 38368, -30 92376 45312, 127 56052 86912, -319 54556 68224, 542 97781 86240, -662 08263 04512, 597 71348 58240, -406 32739 43040, 209 51256 26880, -81 90820 35200, 24 09991 37280, -5 25816 29952, 83071 67232, -9168 44544, 669 77280, -29 76768, 69768, -648, 1
19	13 74389 53472, -130 56700 57984, 571 23065 03680, -1526 00188 02688, 2782 70931 10784, -3668 11681 91488, 3610 80249 38496, -2703 94195 96800, 1554 76662 68160, -688 02890 95680, 233 43838 00320, -60 12806 75840, 11 56308 99200, -1 61883 25888, 15899 24864, -1036 90752, 41 24064, -86640, 722, -1
20	54 97558 13888, -549 75581 38880, 2542 62063 92320, -7215 54505 72800, 14055 28047 61600, -19918 34033 19296, 21236 46579 50720, -17375 29019 59680, 11029 23694 08000, -5455 32149 76000, 2100 29877 65760, -625 48082 68800, 142 40858 11200, -24 34334 72000, 3 04291 84000, -26777 68192, 1569 00480, -56 17920, 1 06400, -800, 1

11.6. Coefficients for Expansion of Integrals of Functions in Series of Chebyshev Polynomials of the First Kind

11.6.1. Introduction

Suppose we know the coefficients b_n in the representation

$$f(x) = \sum_{n=0}^{\infty} b_n T_n^*(x/\lambda) . \tag{1}$$

Here λ is an arbitrary but fixed number and x is restricted so that $0 \leq x/\lambda \leq 1$. Then

$$\int_0^x t^u \{f(0) - f(t)\}dt, \int_0^x t^u f(t)dt, \int_0^x t^u (\ln t)f(t)dt, \tag{2}$$

have representations in terms of forms like (1). In 11.6.2, we formulate algorithms for evaluation of the coefficients in these series expansions in terms of the coefficients b_n. Similar formulae for expansions in series of the even and odd Chebyshev polynomials $T_{2n}(x)$ and $T_{2n+1}(x)$, respectively, are presented in 11.6.3 and 11.6.4, respectively.

In most applications, $f(x)$ is analytic in the neighborhood of the origin, and for purposes of devising checks, we assume the representation.

$$f(x) = \sum_{n=0}^{\infty} a_n x^n . \tag{3}$$

11.6.2. Series of Shifted Chebyshev Polynomials

CASE I. Let

$$\int_0^x t^u f(t)dt = x^{u+1} \sum_{n=0}^{\infty} e_n(u) T_n^*(x/\lambda) . \tag{1}$$

Then

$$(2e_n/\varepsilon_n) = \{(u-n)e_{n+1} + (2b_n/\varepsilon_n) - b_{n+1}\}/(n+u+1), \ n \geq 0 , \tag{2}$$

where

$$\varepsilon_0 = 1, \ \varepsilon_n = 2 \text{ for } n > 0 . \tag{3}$$

In (2) and elsewhere, when no confusion can result, we simply write e_n for $e_n(u)$. These coefficients are most readily found from (2) by use of the backward recurrence technique.

Let N be a large positive integer. Put $e_{N+1,N} = 0$, and calculate $e_{n,N}$ for $n=N,N-1,\ldots,1,0$ from (2) with e_n replaced by $e_{n,N}$. Then

$$\lim_{N\to\infty} e_{n,N} = e_n, \quad n = 0,1,\ldots \; . \tag{4}$$

We have the following check formulas.

$$\sum_{n=0}^{\infty} (-)^n e_n = f(0)/(u+1) \tag{5}$$

$$\sum_{n=1}^{\infty} (-)^n n^2 e_n = -a_1\lambda/2(u+2) , \tag{6}$$

$$\sum_{n=2}^{\infty} (-)^n n^2(n^2-1)e_n = 2a_2\lambda^2/3(u+3) , \tag{7}$$

$$(u+1)e_0 - b_0 + \tfrac{1}{4}\{b_1 - (u-2)e_1\} + \sum_{n=2}^{\infty} (-)^n(b_n - ue_n)/(n^2-1) = 0. \tag{8}$$

CASE II. Let $f(0) \neq 0$ and put

$$\int_0^x t^u\{f(t) - f(0)\}dt = x^{u+1}\sum_{n=0}^{\infty} g_n(u)T_n^*(x/\lambda) , \; R(u) > -1 . \tag{9}$$

Then

$$g_0(u) = e_0 - f(0), \; g_n(u) = e_n(u), \; n > 0 , \tag{10}$$

where the $e_n(u)$ are as (1). When $u=-1$, $e_n(-1)$ for $n>0$ only can be determined by the backward recurrence scheme described above. In this event $g_0(-1)$ can be found from either of the formulas

$$g_0(-1) = -b_1 + \sum_{n=2}^{\infty} (-)^n\{2\sum_{k=1}^{n-1} k^{-1} + n^{-1}\}b_n , \tag{11}$$

$$\sum_{n=0}^{\infty} (-)^n g_n(-1) = 0 . \tag{12}$$

The formula not used to evaluate $g_0(-1)$ can then be employed as a check.

CASE III. We consider Chebyshev series expansions for

$$\int_0^x t^u(\ln t)f(t)dt = (\ln x)\int_0^x t^u f(t)dt - \int_0^x t^{-1}\int_0^t v^u f(v)\,dv\,dt. \tag{13}$$

Again we assume $R(u) > -1$. Let

$$\int_0^x t^u (\ln t) f(t)\,dt = x^{u+1} \sum_{n=0}^{\infty} c_n(u) T_n^*(x) + x^{u+1} (\ln x) \sum_{n=0}^{\infty} e_n(u) T_n^*(x), \tag{14}$$

Then the coefficients e_n are as in (1) and

$$c_n(u) = \partial e_n(u)/\partial u . \tag{15}$$

Differentiating (2), we have

$$(2c_n/\varepsilon_n) = \{(u-n)c_{n+1} + e_{n+1} - (2e_n/\varepsilon_n)\}/(n+u+1). \tag{16}$$

The coefficients $c_n = c_n(u)$ are readily found by use of the backward recurrence scheme. Thus let N be a large positive integer. Put $c_{N+1,N} = 0$ and calculate $c_{n,N}$ for $n = N, N-1, \ldots, 1, 0$ from (16) with c_n replaced by $c_{n,N}$. Then

$$\lim_{N\to\infty} c_{n,N} = c_n, \quad n = 0, 1, \ldots . \tag{17}$$

The following formulae are useful for check purposes.

$$\sum_{n=0}^{\infty} (-)^n c_n = -a_0/(u+1)^2 , \tag{18}$$

$$\sum_{n=1}^{\infty} (-)^n n^2 c_n = a_1 \lambda/2(u+2)^2 , \tag{19}$$

$$\sum_{n=2}^{\infty} (-)^n n^2 (n^2 - 1) c_n = -3a_2 \lambda^2/2(u+3)^2 , \tag{20}$$

We now turn to an example illustrating the Case I formulae. Suppose we know the coefficients b_n in

$$e^x = \sum_{n=0}^{\infty} b_n T_n^*(x) , \tag{21}$$

and that we desire the coefficients e_n in

$$\int_0^x t^{-\frac{1}{2}} e^t\,dt = x^{\frac{1}{2}} \sum_{n=0}^{\infty} e_n T_n^*(x) . \tag{22}$$

Thus $\lambda = 1$, $u = -\frac{1}{2}$ and

$$(2e_n/\varepsilon_n) = -e_{n+1} + 2\{(2b_n/\varepsilon_n) - b_{n+1}\}/(2n+1) . \tag{23}$$

The b_n's that we use are the values of $C_n(a)$, $a=1$, from Table 3.3 rounded to 10D. Put $N=8$, $e_{9,8}=0$ and evaluate $e_{n,8}$ for $n=8,7,\ldots,0$ from the above recurrence formula with e_n replaced by $e_{n,8}$. We get the following table.

n	$e_{n,8}$	n	$e_{n,8}$
0	2.42617 33593	5	0.00000 45627
1	0.46042 05914	6	0.00000 01623
2	0.03636 80487	7	0.00000 00051
3	0.00222 65869	8	0.00000 00001
4	0.00011 01754		

In general, it can be shown that

$$(2e_{k,N}/\varepsilon_k) = \sum_{r=0}^{N-k} (-)^r (k-u)_r B_{k+r}/(k+u+1)_{r+1},$$

$$B_k = (2/\varepsilon_k)b_k - b_{k+1}, \quad k=0,1,\ldots,N, \tag{24}$$

and for $u=-\frac{1}{2}$,

$$(2e_{k,N}/\varepsilon_k) = \sum_{r=0}^{N-k} (-)^r B_{k+r}/(k+r+\tfrac{1}{2}). \tag{25}$$

It follows that the error in e_k, call it Δe_k, is given by

$$\Delta(2e_k/\varepsilon_k) = \sum_{r=N+1-k}^{\infty} (-)^r B_{k+r}/(k+r+\tfrac{1}{2}). \tag{26}$$

Thus in the present example, round-off error aside, the results are accurate to 10D. Allowing for round-off, we conclude that the entries are accurate to at least 9D. For checks on the calculations, appropriate use of the sums in (5) and (6) give 1.99999 99999 and 0.33333 33405, respectively, while the true values are 2.0 and 1/3, respectively. Again using (8), we get 0 to 10D.

We can use this example to illustrate the ideas in the discussion surrounding 11.3.6.3(20-23). In 5.10.2(10), put $p=q=1$, $a_1=1/2$, $b_1=3/2$. Then

$$e_n = \frac{2\varepsilon_n (1/2)_n}{2^{2n}(3/2)_n n!}\, {}_2F_2\left(\begin{matrix} n+1/2,\ n+1/2 \\ 2n+1,\ n+3/2 \end{matrix}\,\middle|\, 1\right) \tag{27}$$

and from 11.3.6.3(22),

$$e_n = \{e^{\frac{1}{2}}/2^{2n-2}n!(2n+1)\}\{1-(7/16n)+O(n^{-2})\}. \tag{28}$$

For $n=4,6$ and 7, with the order term neglected, we get the excellent estimates $0.106\cdot 10^{-3}$, $0.159\cdot 10^{-6}$ and $0.499\cdot 10^{-8}$, respectively.

11.6.3. Series of Chebyshev Polynomials of Even Order

All the results in 11.6.2 apply here since $T_n^*(x^2)=T_{2n}(x)$. Thus in 11.6.2(1-4), replace x by x^2, λ by λ^2 and t by t^2. So if

$$f(x^2) = \sum_{n=0}^{\infty} b_n T_{2n}(x/\lambda), \tag{1}$$

then

$$\int_0^x t^{2u+1}f(t^2)dt = \tfrac{1}{2}x^{2u+2}\sum_{n=0}^{\infty} e_n(u)T_{2n}(x/\lambda), \quad R(u)>-1, \tag{2}$$

and the $e_n(u)$'s are found from 11.6.2(2) using the prescription surrounding 11.6.2(4). Similarly, under the same conditions as for (2), we have

$$\int_0^x t^{2u+1}\{f(t^2)-f(0)\}dt = \tfrac{1}{2}x^{2u+2}\sum_{n=0}^{\infty} g_n(u)T_{2n}(x/\lambda), \tag{3}$$

$$\int_0^x t^{2u+1}(\ln t)f(t^2)dt = \tfrac{1}{4}x^{2u+2}\sum_{n=0}^{\infty} c_n(u)T_{2n}(x/\lambda) + \tfrac{1}{2}x^{2u+2}(\ln x) \times \sum_{n=0}^{\infty} e_n(u)T_{2n}(x/\lambda). \tag{4}$$

11.6.4. Series of Chebyshev Polynomials of Odd Order

Suppose we are given

$$h(x) = xf(x^2), \quad h(x) = \sum_{n=0}^{\infty} q_n T_{2n+1}(x/\lambda), \tag{1}$$

where $f(x^2)$ is given by 11.6.3(1). Then

$$q_n = \tfrac{1}{2}\lambda\{(2/\varepsilon_n)b_n + b_{n+1}\} \tag{2}$$

and

and

$$\lambda b_n/\varepsilon_n = \sum_{k=0}^{\infty} (-)^k q_{n+k}, \tag{3}$$

where ε_n is defined as in 11.6.2(3). Thus all the results previously derived are applicable. We have

$$\int_0^x t^{2u}h(t)dt = \tfrac{1}{2}x^{2u+2}\sum_{n=0}^{\infty} e_n(u)T_{2n}(x/\lambda), \quad R(u) > -\tfrac{1}{2}, \tag{4}$$

or under the same condition on u,

$$\int_0^x t^{2u}h(t)dt = \tfrac{1}{4}\lambda x^{2u+1}\sum_{n=0}^{\infty} \{(2/\varepsilon_n)e_n(u) + e_{n+1}(u)\}T_{2n+1}(x/\lambda), \tag{5}$$

$$\int_0^x t^{2u}h(t)dt = (\lambda^2/8)\{2e_0(u) + e_1(u)\}T_0(x/\lambda)$$

$$+ (\lambda^2/8)\sum_{n=0}^{\infty} \{(2/\varepsilon_n)e_n(u) + 2e_{n+1}(u) + e_{n+2}(u)\}T_{2n+2}(x/\lambda). \tag{6}$$

11.7. Orthogonality Properties of Chebyshev Polynomials with Respect to Summation

Here we cite results named in title of this section and formulas which follow therefrom for approximating a function by a polynomial. For a generalization of these results, see Theorem 6, p. 432, Luke (1975) and Luke, Ting and Kemp (1975).

Lemma 1. *Let*

$$T_n(x_\alpha) = \cos n\theta_\alpha = 0,$$

$$x_\alpha = \cos\theta_\alpha, \qquad \theta_\alpha = (\pi/2n)(2\alpha+1), \qquad \alpha = 0, 1, \ldots, n-1,$$

$$U_{j,k} = \sum_{\alpha=0}^{n-1} T_j(x_\alpha)\,T_k(x_\alpha) = \sum_{\alpha=0}^{n-1} \cos j\theta_\alpha \cos k\theta_\alpha.$$

Then

$$\begin{aligned} U_{j,k} &= 0 && \text{if } j, k < n, \quad j \neq k, \\ &= n/2 && \text{if } j = k, \quad 0 < j < n, \\ &= n && \text{if } j = k = 0. \end{aligned} \tag{1}$$

We also have some further properties of $U_{j,k}$. *For convenience put* $m = j + k$, $p = |j - k|$, *and let* s *and* t *be positive integers. Then*

$$\begin{aligned}
U_{j,0} &= (-)^s n && \textit{if } j = 2sn, \\
&= 0 && \textit{if } j \neq 2sn. \\
U_{j,j} &= n && \textit{if } j = 2sn, \\
&= n/2 && \textit{if } j \neq sn, \\
&= 0 && \textit{if } j = (2s+1)n. \\
U_{j,k} &= 0 && \textit{if } j \neq k,\ m \neq 2sn,\ p \neq 2tn, \\
&= n && \textit{if } j \neq k,\ m = 2sn,\ p = 2tn, \textit{ and } s \textit{ and } t \textit{ are both even}, \\
&= -n && \textit{if } j \neq k,\ m = 2sn,\ p = 2tn, \textit{ and } s \textit{ and } t \textit{ are both odd}, \\
&= 0 && \textit{if } j \neq k,\ m = 2sn,\ p = 2tn, \textit{ and } s \textit{ and } t \textit{ are of opposite parity}, \\
&= (-)^s n/2 && \textit{if } j \neq k,\ m = 2sn,\ p \neq 2tn, \\
&= (-)^t n/2 && \textit{if } j \neq k,\ m \neq 2sn,\ p = 2tn.
\end{aligned} \tag{2}$$

Theorem 1. *If $f_n(x)$ is an approximation to $f(x)$,*

$$f_n(\cos\theta) = \tfrac{1}{2}d_0 + \sum_{k=1}^{n-1} d_k \cos k\theta \qquad \textit{or} \qquad f_n(x) = \tfrac{1}{2}d_0 + \sum_{k=1}^{n-1} d_k T_k(x),$$

then

$$d_k = (2/n)\sum_{\alpha=0}^{n-1} f(\cos\theta_\alpha)\cos k\theta_\alpha = (2/n)\sum_{\alpha=0}^{n-1} f(x_\alpha)\, T_k(x_\alpha). \tag{3}$$

Suppose $f'(x)$ is continuous in the closed interval (-1,1) except for a finite number of bounded jumps. Then $f(x)$ can be expanded in a convergent series as

$$f(x) = \tfrac{1}{2}c_0 + \sum_{k=1}^{\infty} c_k T_k(x),$$

$$c_k = (2/\pi)\int_{-1}^{1} \frac{f(x)\, T_k(x)}{(1-x^2)^{1/2}}\, dx = (2/\pi)\int_0^\pi f(\cos\theta)\cos k\theta\, d\theta. \tag{4}$$

Lemma 2.

$$d_0 = c_0 + 2\sum_{r=1}^{\infty} (-)^r c_{2rn}$$

$$d_k = c_k + \sum_{r=1}^{\infty} (-)^r c_{2rn-k} + \sum_{r=1}^{\infty} (-)^r c_{2rn+k}\,, \qquad k = 1, 2, \ldots, n-1. \tag{5}$$

Theorem 2. *If*

$$\epsilon_n(x) = f(x) - f_n(x),$$

where $f(x)$ and $f_n(x)$ are given by (4) *and* (3), *respectively, then*

$$\begin{aligned}\epsilon_n(\cos\theta) = {} & \cos n\theta \left\{c_n + 2\sum_{r=1}^{2n-1} c_{n+r}\cos r\theta + c_{3n}\cos 2n\theta\right\} \\ & - \sin 2n\theta \left\{c_{3n}\sin n\theta + 2\sum_{r=1}^{2n-1} c_{3n+r}\sin(n+r)\theta + c_{5n}\sin 3n\theta\right\} \\ & + \cos 3n\theta \left\{c_{5n}\cos 2n\theta + 2\sum_{r=1}^{2n-1} c_{5n+r}\cos(2n+r)\theta + c_{7n}\cos 4n\theta\right\} - \cdots,\end{aligned} \tag{6}$$

$$\epsilon_n(\cos\theta_\alpha) = 0,$$

$$\epsilon_n(\cos\theta) \sim c_n \cos n\theta[1 + (2c_{n+1}/c_n)\cos\theta].$$

Obviously in (3) we restrict x so that $-1 \leqslant x \leqslant 1$. Observe that each x_α lies in the interior of this interval. In numerous applications, it is useful to have a curve fit which uses the end point $x = \pm 1$ and the points which lie midway between the points θ_α. Here again the Chebyshev polynomials possess an orthogonality property which leads to an expansion formula based on the abscissas just named. We state lemmas and theorems analogous to those above.

Lemma 3. *Let*

$$x_\alpha = \cos\varphi_\alpha, \qquad \varphi_\alpha = \alpha\pi/n, \qquad \alpha = 0, 1, \ldots, n,$$

$$\begin{aligned} V_{j,k} &= \tfrac{1}{2}\{T_j(x_0)\,T_k(x_0) + T_j(x_n)\,T_k(x_n)\} + \sum_{j=1}^{n-1} T_j(x_\alpha)\,T_k(x_\alpha) \\ &= \tfrac{1}{2}\{\cos j\varphi_0 \cos k\varphi_0 + \cos j\varphi_n \cos k\varphi_n\} + \sum_{j=1}^{n-1} \cos j\varphi_\alpha \cos k\varphi_\alpha .\end{aligned}$$

Then

$$\begin{aligned} V_{j,k} &= 0 && \textit{if} \quad j, k < n, \quad j \neq k, \\ &= n/2 && \textit{if} \quad j = k, \quad 0 < j < n, \\ &= n && \textit{if} \quad j = k = 0 \ \textit{or}\ j = k = n. \end{aligned} \tag{7}$$

Some further values of $V_{j,k}$ follow. It is convenient to put $m = j + k$, $p = |j - k|$, and let s and t be positive integers. Then

$$
\begin{aligned}
V_{j,0} &= n && \text{if } j = 2sn, \\
&= 0 && \text{if } j \neq 2sn. \\
V_{j,j} &= n && \text{if } j \neq 0, \quad j \neq n, \quad j = sn, \\
&= n/2 && \text{if } j \neq 0, \quad j \neq n, \quad j \neq sn. \\
V_{j,k} &= 0 && \text{if } j \neq k, \quad m \neq 2sn, \quad p \neq 2tn, \\
&= n && \text{if } j \neq k, \quad m = 2sn, \quad p = 2tn, \\
&= n/2 && \text{if } j \neq k, \quad m = 2sn, \quad p \neq 2tn, \\
&= n/2 && \text{if } j \neq k, \quad m \neq 2sn, \quad p = 2tn.
\end{aligned}
\tag{8}
$$

Theorem 3. *If $f_n(x)$ is an approximation to $f(x)$,*

$$f_n(\cos\varphi) = \tfrac{1}{2}e_0 + \sum_{k=1}^{n-1} e_k \cos k\varphi + \tfrac{1}{2}e_n \cos n\varphi$$

or

$$f_n(x) = \tfrac{1}{2}e_0 + \sum_{k=1}^{n-1} e_k T_k(x) + \tfrac{1}{2}e_n T_n(x),$$

then

$$
\begin{aligned}
e_k &= \frac{2}{n}\Big[\frac{f(1) + (-)^k f(-1)}{2} + \sum_{\alpha=1}^{n-1} f(\cos\varphi_\alpha)\cos k\varphi_\alpha\Big], \\
&\textit{or} \\
e_k &= \frac{2}{n}\Big[\frac{f(1) + (-)^k f(-1)}{2} + \sum_{\alpha=1}^{n-1} f(x_\alpha)\, T_k(x_\alpha)\Big].
\end{aligned}
\tag{9}
$$

Lemma 4.

$$
\begin{aligned}
e_0 &= c_0 + 2\sum_{r=1}^{\infty} c_{2rn}, \qquad e_n = 2c_n + 2\sum_{r=1}^{\infty} c_{(2r+1)n}, \\
e_k &= c_k + \sum_{r=1}^{\infty} c_{2rn-k} + \sum_{r=1}^{\infty} c_{2rn+k}, \qquad k = 1, 2, \ldots, n-1.
\end{aligned}
\tag{10}
$$

Theorem 4. *If*

$$\delta_n(x) = f(x) - f_n(x),$$

where $f(x)$ and $f_n(x)$ are given by (4) *and* (9), *respectively, then*

$$\delta_n(\cos\varphi) = -2\sin n\varphi \sum_{r=1}^{\infty} c_{n+r}\sin r\varphi,$$

$$\delta_n(\cos\varphi_\alpha) = 0, \tag{11}$$

$$\delta_n(\cos\varphi) \sim -2\sin n\varphi \sin\varphi\, c_{n+1}\left[1 + \frac{2c_{n+2}}{c_{n+1}}\cos\varphi\right].$$

If c_k in (4) is evaluated approximately by the trapezoidal rule, then e_k is an approximation to c_k and (10) may be interpreted as the error in this process. Similarly, if c_k is evaluated by what we call the modified trapezoidal rule, that is, if c_k is approximated by d_k [see (3)], then (5) may be interpreted as the error in this numerical integration scheme. Note that $\frac{1}{2}(d_k + e_k)$ is an improved approximation for c_k, $k = 0, 1,\ldots, n$.

In the above, we dealt with expansions in series of $T_n(x)$. We now give without proof analogous statements for expansions in series of Chebyshev polynomials of the second kind $U_n(x)$.

Lemma 5. *Let*

$$U_n(x_\alpha) = \frac{\sin(n+1)\,\theta_\alpha}{\sin\theta_\alpha},$$

$$x_\alpha = \cos\theta_\alpha, \qquad \theta_\alpha = \frac{\pi}{2n}(2\alpha+1), \qquad \alpha = 0, 1,\ldots, n-1,$$

$$W_{j,k} = \sum_{\alpha=0}^{n-1}(1 - x_\alpha^2)\, U_{j-1}(x_\alpha)\, U_{k-1}(x_\alpha) = \sum_{\alpha=0}^{n-1}\sin j\theta_\alpha \sin k\theta_\alpha .$$

Then

$$\begin{aligned} W_{j,k} &= 0 && \textit{if } j, k < n, \quad j \neq k,\\ &= n/2 && \textit{if } j = k, \quad 0 < j < n,\\ &= 0 && \textit{if } j = 0 \quad \textit{for all } k \textit{ or if } k = 0 \textit{ for all } j. \end{aligned} \tag{12}$$

Theorem 5. *If $f_n(x)$ is an approximation to $f(x)$,*

$$f_n(\cos\theta) = \sum_{k=1}^{n-1} q_k \sin k\theta \qquad \text{or} \qquad f_n(x) = (1-x^2)^{1/2}\sum_{k=1}^{n-1} q_k U_{k-1}(x),$$

then

$$q_k = \frac{2}{n}\sum_{\alpha=0}^{n-1} f(\cos\theta_\alpha)\sin k\theta_\alpha = \frac{2}{n}\sum_{\alpha=0}^{n-1}(1-x_\alpha^2)^{1/2} f(x_\alpha)\, U_{k-1}(x_\alpha). \tag{13}$$

Again we suppose that $f'(x)$ is defined as in the discussion surrounding (4). *Then*

$$f(x) = (1 - x^2)^{1/2} \sum_{k=1}^{\infty} p_k U_{k-1}(x) = \sum_{k=1}^{\infty} p_k \sin k\theta,$$

$$p_k = \frac{2}{\pi} \int_{-1}^{1} f(x)\, U_{k-1}(x)\, dx = \frac{2}{\pi} \int_{0}^{\pi} f(\cos\theta) \sin k\theta\, d\theta. \tag{14}$$

Lemma 6.

$$q_j = p_j - \sum_{r=1}^{\infty} (-)^r p_{2nr-j} + \sum_{r=1}^{\infty} (-)^r p_{2nr+j}\,, \qquad j = 1, 2, \ldots, n-1. \tag{15}$$

Theorem 6. *If*

$$\epsilon_n(x) = f(x) - f_n(x)$$

where $f(x)$ *and* $f_n(x)$ *are given by* (14) *and* (13), *respectively, then*

$$\begin{aligned} \epsilon_n(\cos\theta) &= p_n \sin n\theta + 2\cos n\theta \sum_{r=1}^{\infty} p_{n+r} \sin r\theta, \\ \epsilon_n(\cos\theta_\alpha) &= (-)^\alpha p_n\,, \\ \epsilon_n(\cos\theta) &\sim p_n \sin n\theta \left(1 + \frac{2p_{n+1}}{p_n} \cot n\theta \sin\theta\right). \end{aligned} \tag{16}$$

Lemma 7. *Let*

$$x_\alpha = \cos\varphi_\alpha\,, \qquad \varphi_\alpha = \alpha\pi/n\,, \qquad \alpha = 0, 1, \ldots, n,$$

$$X_{j,k} = \sum_{\alpha=1}^{n-1} (1 - x_\alpha{}^2)\, U_{j-1}(x_\alpha)\, U_{k-1}(x_\alpha) = \sum_{\alpha=1}^{n-1} \sin j\varphi_\alpha \sin k\varphi_\alpha. \tag{17}$$

Then

$$\begin{aligned} X_{j,k} &= 0 && \textit{if} \quad j, k < n, \quad j \neq k, \\ &= n/2 && \textit{if} \quad j = k, \qquad 0 < j < n, \\ &= 0 && \textit{if} \quad j = 0 \qquad \textit{for all } k \textit{ or if } k = 0 \textit{ for all } j. \end{aligned} \tag{18}$$

Theorem 7. *If* $f_n(x)$ *is an approximation to* $f(x)$,

$$f_n(\cos\varphi) = \sum_{k=1}^{n-1} r_k \sin k\varphi \qquad \textit{or} \qquad f_n(x) = (1 - x^2)^{1/2} \sum_{k=1}^{n-1} r_k U_{k-1}(x),$$

then

$$r_k = \frac{2}{n} \sum_{\alpha=1}^{n-1} f(\cos\varphi_\alpha) \sin k\varphi_\alpha = \frac{2}{n} \sum_{\alpha=1}^{n-1} (1 - x_\alpha{}^2)^{1/2} f(x_\alpha)\, U_{k-1}(x_\alpha). \tag{19}$$

Lemma 8.

$$r_j = p_j - \sum_{s=1}^{\infty} p_{2sn-j} + \sum_{s=1}^{\infty} p_{2sn+j} . \tag{20}$$

Theorem 8. *If*

$$\delta_n(x) = f(x) - f_n(x),$$

where $f(x)$ *and* $f_n(x)$ *are given by* (14) *and* (19), *respectively, then*

$$\begin{aligned} \delta_n(\cos\varphi) &= p_n \sin n\varphi + 2 \sin n\varphi \sum_{r=1}^{\infty} p_{n+r} \cos r\varphi \\ \delta_n(\cos\varphi_\alpha) &= 0, \\ \delta_n(\cos\varphi) &\sim p_n \sin n\varphi \left(1 + \frac{2p_{n+1}}{p_n} \cos\varphi\right). \end{aligned} \tag{21}$$

Note that (15) and (20) may be interpreted as the error when p_k in (14) is approximated by the modified trapezoidal rule and the trapezoidal rule, respectively. See the comments following (11). Also $\frac{1}{2}(q_k + r_k)$ is an improved approximation for p_k, $k = 0, 1,..., n$. Estimates of the error in these approximations follow from the discussion in 11.3.6.3. For further comments, see Elliott (1965).

11.8. A Nesting Procedure for the Computation of Expansions in Series of Functions Where the Functions Satisfy a Linear Finite Difference Equation

Suppose we have to evaluate

$$f_N(x) = \sum_{n=0}^{N} a_n x^n, \tag{1}$$

an approximation to

$$f(x) = \sum_{n=0}^{\infty} a_n x^n. \tag{2}$$

Here our functions are the monomials x^n. Define the backward recurrence scheme

$$B_n = a_n + xB_{n+1}, \qquad n = N, N-1,..., 1, 0, \qquad B_{N+1} = 0. \tag{3}$$

We then get the well-known result

$$f_N(x) = B_0 . \tag{4}$$

It is of interest to examine the effect of a round-off error in a_n when evaluating $f_N(x)$ in this manner. We note that the general solution of

$$u_n = a_n + xu_{n+1} \tag{5}$$

is

$$u_n = \alpha x^{-n} - x^{-n} \sum_{k=0}^{n-1} a_k x^k \tag{6}$$

where α is a constant. Thus a rounding error δ_n in a_n or B_n produces an error $\epsilon_s(n)$ in B_s for $s \leqslant n$ given by

$$\epsilon_s(n) = \alpha x^{-s} - x^{-s} \sum_{k=0}^{s-1} \delta_k x^k. \tag{7}$$

Also

$$0 = \alpha x^{-n-1} - x^{-n-1} \sum_{k=0}^{n-1} \delta_k x^k. \tag{8}$$

We can solve the latter for α and so find

$$\epsilon_s(n) = x^{-s} \sum_{k=s}^{n} \delta_k x^k. \tag{9}$$

The error in $f(x)$ is (9) with $s = 0$, which is exactly that produced by an error δ_k in a_k when the series is summed in the usual fashion. Thus, if B_n is rounded to the same number of places as a_n, the maximum rounding error in $f_N(x)$ is only doubled. If one or two guard units are retained in a_n or B_n, this error may be made negligible when compared with the truncation error due to the approximation of $f(x)$ by $f_N(x)$. The above analysis shows that although the error in B_n may become quite large when N is large, the error in $f_N(x)$ is the same as that obtained if we form the required powers of x and sum according to (1). The computation of $f_N(x)$ by the backward recurrence scheme requires only N additions and N multiplications. An alternative method for the computation of $f_N(x)$ is to use (1) where the needed powers of x are obtained by recurrence from $x^n = x \cdot x^{n-1}$. This approach requires $(2N - 1)$ multiplications and N additions and so is not as economical as the backward recursion scheme.

To generalize the preceding results, let

$$f_N(x) = \sum_{k=0}^{N} a_k p_k(x),$$

$$p_{n+1}(x) + \alpha_n p_n(x) + \beta_n p_{n-1}(x) = 0, \qquad n \geqslant 1, \tag{10}$$

where α_n and β_n may depend on both n and x. Here $p_k(x)$ need not be a polynomial. Consider the backward recursion system

$$\begin{aligned} B_n &= -\alpha_n B_{n+1} - \beta_{n+1} B_{n+2} + a_n, \qquad n = N, N-1, \ldots, 1, 0, \\ B_{N+1} &= B_{N+2} = 0. \end{aligned} \tag{11}$$

Then

$$f_N(x) = B_0 p_0(x) + B_1\{p_1(x) + \alpha_0 p_0(x)\}. \tag{12}$$

This is easily proved by solving for a_n from (11), substituting in $f_N(x)$ and using the recurrence formula in (10). The extension of this principle to the case where the sequence $\{p_k(x)\}$ satisfies a higher order difference equation is direct. Suppose

$$p_{n+1}(x) + \alpha_{n,0} p_n(x) + \alpha_{n,1} p_{n-1}(x) + \cdots + \alpha_{n,r} p_{n-r}(x) = 0, \qquad r < N. \tag{13}$$

Consider

$$\begin{aligned} B_n &= -\alpha_{n,0} B_{n+1} - \alpha_{n+1,1} B_{n+2} - \alpha_{n+2,2} B_{n+3} - \cdots - \alpha_{n+r,r} B_{n+r+1} + a_n, \\ B_{N+1} &= B_{N+2} = \cdots = 0. \end{aligned} \tag{14}$$

Then

$$f_N(x) = \sum_{s=0}^{r} B_s\{p_s + \alpha_{s-1,0} p_{s-1} + \cdots + \alpha_{s-1,s-1} p_0\}. \tag{15}$$

We now show that a similar nesting procedure is available for evaluation of $f_N'(x)$. It is sufficient to treat the system (10), (11). We have

$$\begin{aligned} f_N'(x) &= (B_0 + B_1\alpha_0)\, p_0'(x) + B_1 p_1'(x) + B_1 \alpha_0' p_0(x) - g_N(x), \\ g_N(x) &= \sum_{k=0}^{N-1} b_k p_k(x), \qquad b_k = B_{k+1}\alpha_k' + B_{k+2}\beta_{k+1}', \end{aligned} \tag{16}$$

and $g_N(x)$ is readily evaluated using (10)–(12). Thus, let

$$\begin{aligned} C_n &= -\alpha_n C_{n+1} - \beta_{n+1} C_{n+2} + b_n, \qquad n = N-1, N-2, \ldots, 1, 0, \\ C_N &= C_{N+1} = 0. \end{aligned} \tag{17}$$

Then

$$\begin{aligned} f_N'(x) = (B_0 + B_1\alpha_0)\, p_0'(x) + B_1 p_1'(x) + B_1 \alpha_0' p_0(x) \\ - C_0 p_0(x) - C_1\{p_1(x) + \alpha_0 p_0(x)\}. \end{aligned} \tag{18}$$

The algorithm (10)–(12) for the evaluation of $f_N(x)$ when $p_n(x)$ is the Chebyshev polynomial $T_n(x)$ is due to Clenshaw (1955). Its importance lies in the fact that series of Chebyshev polynomials can be evaluated in a manner very much like that a polynomial. Thus the conversion of a series of Chebyshev polynomials to an ordinary polynomial is not necessary for its evaluation. The extension of the Clenshaw procedure to the calculation of $f_N'(x)$ when $p_n(x)$ is a polynomial and has the recurrence formula (10) is due to Smith (1965). We should like to emphasize that the treatment given here is quite general since $p_n(x)$ need not be a polynomial. The applicability of the algorithm as regards stability in preserving significance of figures and growth of round-off errors is discussed later.

For the case of expansions in series of Chebyshev polynomials of the first kind, the algorithms (10)–(12) and (16)–(18) can be summarized as follows:

$$f_N(x) = \sum_{k=0}^{N} a_k p_k(x), \qquad f_N'(x) = \sum_{k=1}^{N} a_k p_k'(x),$$

$$p_{n+1}(x) + \alpha_n p_n(x) + p_{n-1}(x) = 0. \tag{19}$$

$$\begin{aligned} B_n &= -\alpha_n B_{n+1} - B_{n+2} + a_n, \qquad n = N, N-1, \ldots, 0, \\ & B_{N+1} = B_{N+2} = 0, \\ C_n &= -\alpha_n C_{n+1} - C_{n+2} + b_n, \qquad n = N-1, N-2, \ldots, 0, \\ & C_N = C_{N+1} = 0. \end{aligned} \tag{20}$$

$p_n(x)$	α_n	b_n	$f_N(x)$	$f_N'(x)$
$T_n(x)$	$-2x$	$-2B_{n+1}$	$B_0 - xB_1$	$-B_1 - C_0 + xC_1$
$T_{2n}(x)$	$-2(2x^2 - 1)$	$-8xB_{n+1}$	$B_0 - (2x^2 - 1)B_1$	$-4xB_1 - C_0 + (2x^2 - 1)C_1$
$T_{2n+1}(x)$	$-2(2x^2 - 1)$	$-8xB_{n+1}$	$x(B_0 - B_1)$	$B_0 - B_1 - x(C_0 - C_1)$
$T_n^*(x)$	$-2(2x - 1)$	$-4B_{n+1}$	$B_0 - (2x - 1)B_1$	$-2B_1 - C_0 + (2x - 1)C_1$

(21)

To understand the applicability of the algorithm, let q_n be a second solution of (10) which is linearly independent of p_n. Then the Wronskian

$$W_n = p_{n-1}q_n - p_n q_{n-1} \tag{22}$$

cannot vanish, and the quantities B_n can be evaluated as

$$B_n = W_n^{-1} \sum_{k=n}^{N} (p_{n-1}q_k - p_{k-1}q_n)a_k, \quad n = 0(1)N. \tag{23}$$

Let

$$Q_N = \sum_{k=0}^{N} a_k q_k . \tag{24}$$

Then

$$Q_N = q_0 B_0 + (q_1 + \alpha_0 q_0)B_1 . \tag{25}$$

It follows that with $f_N \equiv f_N(x)$,

$$\begin{aligned} B_0 &= W_1^{-1}\{(q_1 + \alpha_0 q_0)f_N - (p_1 + \alpha_0 p_0)Q_N\}, \\ B_1 &= -W_1^{-1}\{q_0 f_N - p_0 Q_N\} . \end{aligned} \tag{26}$$

This result is quite instructive since it tells when the algorithm should not be used. For suppose $|q_n| \gg |p_n|$ for most values of $n = 0(1)N$. Then in general $|Q_N| \gg |f_N|$. The sum f_N will then be obtained as the difference of two large quantities and a considerable loss of significance will happen. In illustration, the procedure should not be used to evaluate the sum of Bessel functions

$$f_{12} = J_0(x) + 2\sum_{n=1}^{6} J_{2n}(x), \quad x = 1,$$

for which the true value is 1 to 10D. Suppose the B_n's are evaluated to 10S using only single precision operations. Then B_n for $n = 4(1)12$ can be computed exactly because for such n, B_n is an integer with 10 or less digits. For the remaining B_n, rounded values must be used. In particular, we find that $B_0 = -0.73772\ 45906 \cdot 10^{11}$ and $B_1 = 0.12828\ 18767 \cdot 10^{12}$. Now use (12) with values of $J_0(1)$ and $J_1(1)$ correctly rounded to 10D and get $f_{12} = -30$ instead of 1. In this example, the second solution is $Y_n(x)$. Now $J_n(x) \sim (x/2)^n/n!$ and $Y_n(x) \sim -(2/x)^n \times (n-1)!/\pi$, and the reason for the above discrepancy is clear.

When $p_n(x)$ satisfies a three-term recurrence formula as in (10), we study the growth of round-off errors in the evaluation of f_N by (12) after the manner of Elliott (1968). The possible sources of error in f_N are

(1) the round-off error in a_n,
(2) errors in the computed values of α_n and β_n,

(3) errors in the computed values of B_n from (11), and
(4) error in the computed value of f_N from (12).

Let f_N^* be the computed value of f_N. Similarly, let a_n^* be the computed value of a_n, etc. From (11), we have

$$B_n^* = (a_n^* - \alpha_n^* B_{n+1}^* - \beta_{n+1}^* B_{n+2}^*) + r_n , \tag{27}$$

where r_n is the round-off error which occurs in evaluating the ( ) portion in (27). We rewrite (27) as

$$B_n^* = a_n^* - \alpha_n B_{n+1}^* - \beta_{n+1} B_{n+2}^* + \varepsilon_n + r_n , \tag{28}$$

where, upon neglecting second order effects,

$$\varepsilon_n = (\alpha_n - \alpha_n^*) B_{n+1} + (\beta_{n+1} - \beta_{n+1}^*) B_{n+2} . \tag{29}$$

Let ϕ_n and ψ_n denote the errors in a_n and B_n, respectively. So

$$\begin{gathered} a_n^* = a_n + \phi_n, \; B_n^* = B_n + \psi_n, \; n = 0(1)N, \\ \phi_n = \psi_n = 0 \text{ for } n = N+1, \, N+2. \end{gathered} \tag{30}$$

Then from (11) and (28), we find that

$$\psi_n + \alpha_n \psi_{n+1} + \beta_{n+1} \psi_{n+2} = \phi_n + \varepsilon_n + r_n . \tag{31}$$

Thus ψ_n and B_n satisfy difference equations which differ only in their nonhomogeneous parts. So comparing (31) and (11), we have from (12) that

$$p_0 \psi_0 + (p_1 + \alpha_0 p_0) \psi_1 = \sum_{k=0}^{N} (\phi_k + \varepsilon_k + r_k) p_k . \tag{32}$$

From (12),

$$f_N = \{p_0^* B_0^* + (p_1^* + \alpha_0^* p_0^*) B_1^*\} + s , \tag{33}$$

where s is the round-off error which occurs in the evaluation of the { } portion of (33). We can rewrite (33) as

$$f_N^* = p_0 B_0^* + (p_1 + \alpha_0 p_0) B_1^* + \xi + s , \tag{34}$$

$$\xi = (p_0^* - p_0) B_0^* + (p_1^* - p_1) B_1^* + \{\alpha_0^* (p_0^* - p_0) + (\alpha_0^* - \alpha_0) p_0\} B_1^* . \tag{35}$$

Now subtract (12) from (34), use (32) and so obtain the desired result

$$f_N^* - f_N = \sum_{k=0}^{N} (\phi_k + \varepsilon_k + r_k)p_k + \xi + s . \qquad (36)$$

We now apply this analysis to the case of summing a series of Chebyshev polynomials of the first kind. Thus $p_k(x) = T_k(x)$, $q_k(x) = U_k(x)$. It is easy to show that

$$(p_{n-1}q_k - p_{k-1}q_{n-1})/W_n = U_{n-k}(x), \; k \geq n, \qquad (37)$$

whence from (23),

$$B_n = \sum_{k=n}^{N} a_k U_{k-n}(x) . \qquad (38)$$

Suppose that the numerics are done in fixed point arithmetic to t decimal places. Further, suppose that $|r_n|$, $|s| \leq \frac{1}{2}\cdot 10^{-t}$. Let x^* be the approximate value of x used in the calculations. With $x^* - x = \gamma$, $\alpha_n - \alpha_n^* = 2\gamma$ for all n. Now $\beta_n = 1$ and so $\beta_n = \beta_n^*$ for all n. Then from (29), $\varepsilon_n = 2\gamma B_{n+1}$ approximately. Since $p_0 = 1$ and $p_1 = x$, $p_0^* = p_0$ and $p_1^* - p_1 = \gamma$. Then from (35), $\xi = -\gamma B_1$ approximately, since we neglect second order effects. Thus (36) becomes

$$f_N^* - f_N = \sum_{k=0}^{N} (\phi_k + r_k)T_k(x) + 2\gamma \sum_{k=0}^{N} B_{k+1}T_k(x) + \gamma B_1 + s. \qquad (39)$$

The latter can be simplified for in view of (38),

$$\sum_{k=0}^{N} B_{k+1}T_k(x) = \tfrac{1}{2}\sum_{k=1}^{N} (k+1)a_k U_{k-1}(x) . \qquad (40)$$

Now for $-1 \leq x \leq 1$, $|T_k(x)| \leq 1$ and $|U_k(x)| \leq k$. Suppose further that $|\phi_k|$ and $|\gamma|$ are less than $\frac{1}{2}\cdot 10^{-t}$. Then

$$|f_N^* - f_N| \leq \tfrac{1}{2}\cdot 10^{-t}\{(2N+3) + \sum_{k=1}^{N} k^2 |a_k|\} \qquad (41)$$

This generalizes a result of Clenshaw (1955) whose simplified error analysis leaves out the summation term in (41). This is certainly justified if N is sufficiently large for all the Chebyshev series expansions given in this volume, since in these situations a_k is an exponential decay. A slowly convergent series would be one where $a_k = O(k^{-2})$. Suppose we have a constant A such that $|a_k| \leq A/k^2$ for all $k \geq 1$. Then (41) states that

$$|f_N^* - f_N| \leq \tfrac{1}{2}\cdot 10^{-t}\{(A+2)N+3\}. \qquad (42)$$

For some further work on evaluation of sums like (10), see G.J. Cooper (1967), Hunter (1970) and Newberry (1973).

CHAPTER XII COMPUTATION BY USE OF RECURRENCE FORMULAS

12.1. Introduction

In this chapter we present a very short pragmatic but general discussion on the use of recurrence formulas for computation. The emphasis is on technique and the material is sufficient to enable one to reproduce the coefficients in expansions of Chebyshev polynomials given in this volume and to derive other similar coefficients as well. We have taken up the subject for particular cases in 4.5, 9.10, 11.6 and 13.6, and study of these sections provides an excellent introduction to the material in this chapter. For some references, see Luke (1969), Wimp (1969, 1970,1972) and Gautschi (1972) and the references given in these sources.

12.2. Homogeneous Difference Equations

Consider the homogeneous difference equation of order $s \geq 2$,

$$\sum_{v=0}^{s} C_v(n)y(n+v) = 0, \quad C_0(n) = 1, \quad C_s(n) \neq 0, \tag{1}$$

where n is a positive integer or zero. Let $\{y_r(n)\}$, $1 \leq r \leq s$, be a fundamental set of solutions of (1) and suppose that $y_1(n)$ is the particular solution wanted. If

$$\lim_{n\to\infty} \{y_1(n)/y_r(n)\} = 0, \quad r = 2,3,\ldots,s, \tag{2}$$

then $y_1(n)$ is called a minimal or antidominant solution of (1).

To generate the minimal solution, the first algorithm we consider proceeds as follows. Let m be a positive integer or zero. Put

$$A_{m+u-1}(m) = 0, \quad u = 2(1)s, \quad A_m(m) = 1, \tag{3}$$

and evaluate $A_r(m)$ for $r = m-1(-1)0$ from

$$\sum_{v=0}^{\infty} C_v(n)A_{n+v}(m) = 0. \tag{4}$$

Suppose we have the convergent normalization relation

$$\sum_{k=0}^{\infty} L_k y_1(k) = 1 . \qquad (5)$$

Define

$$B(m) = \sum_{k=0}^{m} L_k A_k(m) , \qquad (6)$$

$$E_n(m) = A_n(m)/B(m) . \qquad (7)$$

If

$$\lim_{m\to\infty} E_n(m) = y_1(n) , \qquad (8)$$

then computation of $y_1(n)$ by backward recursion based on (1) and (5) is said to converge.

We now state a result giving conditions for convergence of the above algorithm to the minimal solution. Let $(-)^{r-1}T_r(n)$ be the determinant of the $(s-1)\times s$ matrix (a_{ij}), $a_{ij} = y_j(n+i)$, with the $j = r$ column omitted. Let $T_1(m) \neq 0$ for m sufficiently large,

$$R_r \equiv R_r(m) = \{T_r(m)/T_1(m)\} ,$$

$$S_r \equiv S_r(m) = \sum_{k=0}^{m} L_k y_r(k), \quad 1 \leq r \leq s, \qquad (9)$$

$$\lim_{m\to\infty} R_r = \lim_{m\to\infty} R_r S_r = 0, \quad 2 \leq r \leq s.$$

Then computation of $y_1(n)$ by backward recursion based on (1) and (5) converges.

If $y_1(0)$ is known, then in (5), $L_k = 0$, $k > 0$ and $L_0 y_1(0) = 1$. In this event, if $\lim_{m\to\infty} R_r = 0$, $2 \leq r \leq s$, then

$$\lim_{m\to\infty} \{A_n(m)/A_0(m)\} = \{y_1(n)/y_1(0)\}, \quad n \geq 0 . \qquad (10)$$

The hypotheses of the results (9) and (10) guarantee that the described backward recurrence scheme converges. Nonetheless, the rate of convergence might be slow. The rate of convergence is a difficult concept to formulate. However, it is closely allied to the rate at which the ratio $w_r(m) = \{y_r(m)/y_1(m)\} \to 0$ as $m \to \infty$, $r = 2,3,\ldots,s$. The rate of convergence increases as the rate at which $w_r(m)$ decreases to 0 as $m \to \infty$. It can well be that $w_r(m)$ for $r = 2,3,\ldots,u$ decreases very slowly to zero while the same ratio for $r = u+1, u+2,\ldots,s$ decreases rapidly to zero as $m \to \infty$. Another possibility is that the hypotheses involving u of the R_r's in (9) are not valid. If $(u-1)$ additional normalization relations like (5) are known, then with weakened conditions on the R_r's, we can state another algorithm for the evaluation of $y_1(n)$. Further, this same algorithm is applicable to the situation where u of the w_r's decrease slowly as described above. The result is as follows.

Let m_j, $j = 1,2,\ldots,u$, be a set of positive integers subject to the restriction given in the statement surrounding (10) below such that $m = m_1 < m_2 < \ldots < m_u$. For each m_j evaluate $A_r(m_j)$ after the manner of $A_r(m)$ in (3)-(4) with m replaced by m_j. Let the constants $L_{k,j}$ be given for $0 \le k < \infty$, $1 \le j \le u$ such that

$$\sum_{k=0}^{\infty} L_{k,j} y_1(k) = 1, \; j = 1,2,\ldots,u \tag{11}$$

where $L_{k,1} = L_k$ for all $k \ge 0$. For each m_j evaluate $B(m_j)$ and $E_n(m_j)$ as in (6) and (7) with m replaced by m_j. Define

$$S_{r,j} \equiv S_{r,j}(m) = \sum_{k=0}^{m} L_{k,j} y_r(k) ,$$

$$2 \le r \le s, \; 1 \le j \le u, \; 2 \le u \le s . \tag{12}$$

Let R_r be bounded and bounded away from zero as $m \to \infty$ for $2 \le r \le u$, while $R_r \to 0$ as $m \to \infty$ for $u+1 \le r \le s$. Let the determinant $|R_r(m_j)|$, $r,j = 1,2,\ldots,u$ be bounded away from zero as $m_1 \to \infty$. Also let

$$\lim_{m\to\infty} S_{r,j} R_r = 0, \; u+1 \le r \le s, \; 1 \le j \le u,$$

$$\lim_{m\to\infty} S_{r,j} = B_{r,j} < \infty, \; 2 \le r \le u. \tag{13}$$

Further, with $B_{1,j} = 1$, $1 \le j \le u$, let the determinants $|B_{r,j}|$, $r,j = 1,2,\ldots,u$ not be zero. Then for $m = m_1$ sufficiently large, we can determine m_r for $2 \le r \le u$ which satisfy the above inequalities for the m_r's, so that the system of equations

$$\sum_{v=1}^{u} g_v \sum_{k=0}^{m_v} L_{k,j} E_k(m_j) = 1, \; j = 2,3,\ldots,u,$$

$$\sum_{v=1}^{u} g_v = 1 \tag{14}$$

has a unique solution $\{g_j\}$ which depends of course on $\{m_j\}$. Then

$$\lim_{m\to\infty} \sum_{v=1}^{u} g_v E_n(m_v) = y_1(n), \; n \geq 0. \tag{15}$$

This last result requires u normalization relations. If instead, we know u values of $y_1(n)$, then an alternative procedure is readily devised, but we omit details. See, however, the paragraph following (18) below. The analog of the statements (11)-(15) for $s = 2$ is as follows.

Let the relation (14) hold for $u = 2$ and let $S_{r,j}$ be as in (13) with $r = 2$ and $j = 1,2$. Let

$$\lim_{m\to\infty} S_{2,j} = B_j, \; j = 1,2, \; B_1 \neq B_2. \tag{16}$$

Let the quantities $q(n) = y_1(n)/y_2(n)$ and $q(m_1) - q(m_2)$ be bounded and bounded away from zero as $n \to \infty$ and $m_1 \to \infty$, respectively, where m_1 and m_2 are as below. Also let $B(m)$ be nonzero for m sufficiently large. For $m = m_1$ sufficiently large, we can determine $m_2 > m_1$, so that the equation

$$\mu M_1 + (1-\mu) M_2 = 1, \; M_j = \sum_{k=0}^{m_j} L_{k,2} E_k(m_j), \; j = 1,2, \tag{17}$$

has a unique solution μ which depends of course on m_1 and m_2. Then for $n \geq 0$,

$$\lim_{m\to\infty} E_n = y_1(n), \; E_n = \mu E_n(m_1) + (1-\mu) E_n(m_2). \tag{18}$$

In numerous practical applications, $y_1(0)$ is known a priori in which event $L_{k,2} = 0$ for $k > 0$ while $L_{0,2} = 1/y_1(0)$. It is on this basis, where applicable, that many of the tables given in this volume were derived. In this connection, see the discussion surrounding 9.7(24). See also 14.4.

We return to (17,18) and show how the actual evaluation of μ can be obviated. This is advantageous since considerable loss of significant figures occurs when M_1 is near M_2. Notice that if $M_1 = M_2$, any value of μ satisfies (17). The alternative and preferred form is

$$E_n = E_n(m_2) + (1 - M_2)(M_1 - M_2)^{-1}\{E_n(m_1) - E_n(m_2)\}, \quad (19)$$

or this same expression with the subscripts 1 and 2 interchanged.

12.3. Inhomogeneous Difference Equations

Consider for $s \geq 1$,

$$\sum_{v=0}^{s} C_v(n)g(n+v) = f(n),$$

$$C_0(n) = 1, \; C_s(n) \neq 0, \; f(n) \not\equiv 0, \; n \geq 0. \quad (1)$$

Suppose that $g(n)$ is the particular solution of (1) desired. The algorithm for evaluation of this solution proceeds as follows. With m a positive integer or zero, let $A_r(m)$ be as in 12.2(3,4), put

$$B_{m+r-1}(m) = 0, \; r = 1(1)s, \quad (2)$$

and calculate $B_n(m)$ for $0 \leq n \leq m-1$ by recursion from

$$\sum_{v=0}^{s} C_v(n)B_{n+v}(m) = f(n). \quad (3)$$

Suppose we have the convergent normalization relation

$$\sum_{k=0}^{\infty} P_k g(k) = 1. \quad (4)$$

Define

$$W_n(m) = \{U(m)/V(m)\}A_n(m) + B_n(m), \; n = 0(1)m,$$

$$U(m) = 1 - \sum_{k=0}^{m-1} P_k B_k(m) \quad V(m) = \sum_{k=0}^{m} P_k A_k(m), \quad (5)$$

If

$$\lim_{m\to\infty} W_n(m) = g(n), \quad (6)$$

then computation of $g(n)$ by backward recursion based on (1) and (4) is said to converge.

The following statements relate to the convergence of the scheme (1)-(5). Let $y_r(n)$ be as in the discussion around 12.2(1) such that for m sufficiently large

$$\sum_{k=0}^{m} P_k A_k(m) \neq 0. \tag{7}$$

Put

$$U_r(m) = \sum_{k=0}^{m} P_k y_r(k), \quad 1 \leq r \leq s,$$

$$U_{s+1}(m) = \sum_{k=m+1}^{\infty} P_k g(k). \tag{8}$$

Consider the $(s-1)\times(s+1)$ matrix (b_{ij}), where $b_{ij} = y_j(m+i)$ for $i = 1,2,\ldots,s-1$, $j = 1,2,\ldots,s$ and where $y_{s+1}(m+i) = g(m+i)$. Let $\Delta_{r,s}(m)$, $r \neq s$, be the determinant formed from this matrix after the r-th and s-th columns are deleted. Further, let

$$\lim_{m\to\infty} \frac{U_t(m)\Delta_{r,t}(m)}{U_1(m)\Delta_{s+1,1}(m)} = 0, \quad 1 \leq t \leq s+1, \quad 1 \leq r \leq s, \quad r \neq t,$$

$$\lim_{m\to\infty} \frac{U_t(m)\Delta_{s+1,t}(m)}{U_1(m)\Delta_{s+1,1}(m)} = 0, \quad 2 \leq t \leq s. \tag{9}$$

Then the computation of $g(n)$ by backward recursion as described by (1)-(5) converges.

If instead of having a normalization relation like (4), we know $g(0)$, then the above theorem can be simplified. For in this event, we can put $P_k = 0$ for $k > 0$ and $P_0 g(0) = 1$. Thus in (9), $U_r(m) = y_r(0)/g(0)$ for $1 \leq r \leq s$ and $U_{s+1}(m) = 0$.

Now any inhomogeneous difference equation of a given order can be converted into a homogeneous equation of order one higher. For consider (1) and the same equation with n replaced by $n+1$. Multiply the former by $f(n+1)$, multiply the latter by $f(n)$ and subtract. Then

$$\sum_{v=0}^{s+1} D_v(n) g(n+v) = 0,$$

$$D_v(n) = C_v(n) - \{f(n)/f(n+1)\} C_{v-1}(n+1), \quad 0 \leq v \leq s+1 \tag{10}$$

with the understanding that $C_v(n) = 0$ for $v < 0$ or $v > s$. Let $\{Y_r(n)\}$, $1 \leq r \leq s+1$ be a basis of solutions of (10). Then we can put

$$Y_1(n) = g(n),\ Y_{r+1}(n) = y_r(n),\ 1 \leq r \leq s, \quad (11)$$

where $\{y_r(n)\}$ is a basis of solutions of 12.2(1). Thus all previous results for homogeneous difference equations can be applied for the $g(n)$ solution of (10). The procedure can often be simplified as follows. Put

$$G_{m+r-1}(m) = 0,\ r = 1(1)s,\ G_{m-1}(m) = f(m-1), \quad (12)$$

and calculate $G_n(m)$ for $0 \leq n \leq m-2$ by recursion from (10) with $g(n+v)$ replaced by $G_{n+v}(m)$. Clearly $G_n(m) = B_n(m)$ where the $B_n(m)$'s are generated as in (2,3). A result like the statement (9) can be made, but we omit details.

We hasten to remark that there may be a gain in flexibility by computing with the nonhomogeneous equation (1) rather than the equivalent homogeneous equation (10) since the solution so evaluated need not be the minimal solution of (10). For a thorough analytical study of a particular third order nonhomogeneous difference equation which manifests this behavior, see Wimp and Luke (1969).

CHAPTER XIII SOME ASPECTS OF RATIONAL AND POLYNOMIAL APPROXIMATIONS

13.1. Introduction

As remarked in the preface, there are a number of topics which should be developed in depth and become part of a series of handbooks on the special functions and their approximations. However, space limitations dictate that such topics as computation with recurrence relations (see Chapter 12), rational and Padé approximations and other topics be treated briefly. This is done here to assure proper use and understanding of concepts and methods used herein. In 13.2, we consider various types of Chebyshev approximations and use of the word 'best' in various senses. The Padé table is the subject of 13.3. Approximations of functions defined by a differential equation and by a series, respectively, are treated in 13.4 and 13.5, respectively. Solution of differential equations and other functional equations in series of Chebyshev polynomials of the first kind are studied in 13.6.

13.2. Approximations in Series of Chebyshev Polynomials of the First Kind

In this volume much attention has been devoted to the problem of evaluating the coefficients in the expansion of a given functions in infinite series of Jacobi polynomials, and more particularly in infinite series of Chebyshev polynomials of the first kind. For convenience, we refer to such expansions as Chebyshev expansions and to the corresponding coefficients as Chebyshev coefficients. See 5.10 and the tables of Chebyshev coefficients given for numerous special functions throughout this tome. Our interest is primarily in analytic formulas, and so discussion and development of curve fit type approximations such as 'best' Chebyshev approximations of various types is omitted. However, attention of the reader is called to the availability of such approximations in the literature for particular special functions. In this connection, it is convenient to briefly outline the concept of 'best' Chebyshev approximation. Let $f(x)$ and $s(x)$ be real and continuous in the closed interval (a,b). Consider $Q_{mn}(x) = s(x)q_n(x)/p_m(x)$ where $q_n(x)$ and $p_m(x)$ are polynomials in x of degree n and m respectively. Given $f(x)$, $s(x)$, m, n, a and b, the problem posed by Chebyshev is that of finding the coefficients in the the polynomial so that the maximum deviation of $Q_{nm}(x)$ from $f(x)$ is least in the closed interval (a,b). Such an approximation is said to be best in the sense so described. If $m = 0$ ($m \neq 0$) we speak of a best polynomial (rational) approximation in the sense of Chebyshev. Often one simply says a best Chebyshev polynomial approximation, etc. In the above, when $s(x) = 1$, it is the error which is minimized. We can also choose to minimize the relative error. Thus, one can pose the problem of finding the coefficients in the polynomial so that the maximum deviation of $\{f(x) - Q_{nm}(x)\}/f(x)$ from zero is least in the closed interval (a,b). Here, again, usually $s(x) = 1$.

Despite use of the word 'best', best Chebyshev approximations are not very flexible. In only a few cases are they known in closed form. They are most often obtained by an iterative process. Further, an extensive table of values of the function being approximated must be available. Given a best Chebyshev polynomial approximation of order n, there is no known connection between it and those whose order are contiguous to n. Thus for every n, a and b, the curve fitting process must be redone. For many special functions, the coefficients in its representation in infinite series of Chebyshev polynomials are readily expressed in closed form and are easily determined by use of recurrence formulas, see Chapter 12. Even if the coefficients are not known in closed form, but the function satisfies a differential equation, then the coefficients again are easily determined by use of a recursion formula in the backward direction. For an example, see 13.6. Such infinite expansions when truncated are best in the mean square sense. In other cases, when tabular values are available, approximations in a finite series of Chebyshev polynomials are easily obtained using interpolation formulas based on orthogonal properties of the Chebyshev polynomials with respect to summation. These developments are outlined in 11.7. These approximations are best in the least squares sense. In 13.4, we discuss development of approximations by use of the τ-method which is best in another sense. Thus the word 'best' must always be qualified. For analytic functions, which is almost always the case in practice, the difference in accuracy of the best Chebyshev polynomial approximation over the other approximations noted is a trifle, and the additional labor required to achieve a best approximation in the Chebyshev sense is not worthwhile. In the sequel, we present some inequalities which confirm this position.

Suppose

$$f(x) = \sum_{n=0}^{\infty} a_n T_n(x) . \tag{1}$$

Let S_n be the magnitude of the maximum error when $f(x)$ is approximated by the first $(n+1)$ terms of the right hand side of (1), and let E_n be the magnitude of the corresponding error for the best Chebyshev polynomial approximation of degree n. Then it is known that

$$A_n \leq E_n \leq S_n \leq \sum_{k=n+1}^{\infty} |a_k|, \qquad (2)$$

where A_n is any one of the values

$$A_{n,1} = 2^{-\frac{1}{2}}\left(\sum_{k=n+1}^{\infty} a_k^2\right), \quad A_{n,2} = 2^{-\frac{1}{2}}|a_{n+1}|,$$

$$A_{n,3} = \left|\sum_{k=1}^{\infty} a_r\right|, \quad r = (2k-1)(n+1). \qquad (3)$$

If the error in $f(x)$ when (1) is truncated after $(n+1)$ terms is dominated by $a_{n+1}T_{n+1}(x)$ which is usually the case, then E_n and S_n cannot differ very much. A similar statement pertains when $f(x)$ is approximated by the best least squares approximation noted above.

Powell (1967) has shown that if $f(x)$ is continuous in the closed interval (-1,1), then

$$S_n/E_n \leq u(n), \quad u(n) \sim 1 + (4/\pi^2)\ln n, \qquad (4)$$

for n sufficiently large. $u(n) = 3.22$, 4.14 and 5.07 for $n = 10$, 100 and 1000, respectively.

Suppose that a polynomial $g(x)$ of degree $(n+r+1)$ is approximated by its Chebyshev series expansion truncated after $(n+1)$ terms whence the approximation is a polynomial of degree n. Let t_n be the error and let s_n be the maximum error. Further, let e_n be the maximum error in the best Chebyshev polynomial approximation to $g(x)$ of degree n. Suppose that n is large and all coefficients of t_n are either of the same sign or of strictly alternating sign. The latter supposition appears to be heuristically valid. Under these conditions, the ratio s_n/e_n has been studied by Clenshaw (1964), Lam and Elliott (1972) and Elliott and Lam (1973). It is shown that

$$s_n/e_n \leq \pi^{-1}\{\ln r + b + (3/4r) - (43/192r^2) + \ldots\},$$

$$b = \gamma + 4\ln 2 = 3.34980. \qquad (5)$$

In illustration, $s_n/e_n \leq 1.82$, 2.56 and 3.29 for $r = 10$, 100 and 1000, respectively. Comparison of (4), see Powell for further values of $u(n)$, and (5) shows that the former gives a better upper bound for $r < n$. But for $1 \leq r < n$, (5) is better.

Now suppose that V_n is the magnitude of the error when $f(x)$ is approximated by the Lagrangian type interpolation formula of degree n with abscissae at the zeros of the Chebyshev polynomial $T_{n+1}(x)$, see 11.7(3) with n replaced by $n+1$. Then Powell (1967) proved that

$$V_n/E_n \leq v(n), \; v(n) \sim 1 + (2/\pi)\ln n, \tag{6}$$

for n sufficiently large. $v(n) = 3.49$, 4.90 and 6.36 for $n = 10$, 100 and 1000, respectively.

As previously remarked, we are mostly concerned with analytical formulas (as opposed to curve fits) for the approximation of functions. For a survey on approximation of functions which deals mostly with those of the best Chebyshev type, see Cody (1970). A thorough survey on computational methods for the special functions has been prepared by Gautschi (1975).

13.3. The Padé Table

Consider the at least formal power series

$$F(z) = \sum_{r=0}^{\infty} c_r z^r, \; c_0 \neq 0, \tag{1}$$

and the doubly infinite array of rational functions

$$F_{mn}(z) = A_{mn}(z)/B_{mn}(z), \tag{2}$$

$$A_{mn}(z) = \sum_{r=0}^{\infty} a_{rn} z^r, \; B_{mn}(z) = \sum_{r=0}^{\infty} b_{rn} z^r, \tag{3}$$

determined in such a way that the power series representation of $F_{mn}(z)$ agrees with that of $F(z)$ for as high a power of z as possible. This uniquely determined rational function is called a Padé approximant to $F(z)$ and is said to occupy the (m,n) position in the Padé table. The Padé table can be conceived as a doubly infinite matrix (F_{mn}), $m,n = 0,1,\ldots$. Here m and n designate columns and rows respectively. Evidently the elements in the first row ($n = 0$) are the partial sums of the series for $F(z)$ while the elements in the first column ($m = 0$) are the partial sums of the series for $1/F(z)$. The positions (n,n) fill the main diagonal, while the positions $(n-r,n)$ and $(n+r,n)$ fill the r-th subdiagonal and r-th superdiagonal, respectively.

Let us write

$$B_{mn}(z)F(z) - A_{mn}(z) = z^{m+n+1}G(z),\ G(0) \neq 0. \tag{4}$$

We have the system of linear equations

$$\sum_{r=0}^{k} b_{rm}c_{k-r} = a_{km},\ k = 0(1)m, \tag{5}$$

$$\sum_{r=0}^{n} b_{rm}c_{k-r} = 0,\ k = n+1(1)m, \tag{6}$$

where it is understood that $c_r = 0$ for $r < 0$. The polynomials $A_{mn}(z)$ and $B_{mn}(z)$ can be taken as relatively prime. Also, without loss of generality, we can take $b_{0n} = 1$ whence $a_{0m} = c_0$. In this event, the approximation F_{mn} is said to be reduced. Let

$$c_{mn} = |c_{m+i-j}| \tag{7}$$

be the determinant of order n whose (i,j) element is c_{m+i-j}. It is convenient to extend the definition of c_{mn} so that m or n or both m and n can be 0. Thus we put

$$c_{m0} = 1,\ c_{m1} = c_m,\ c_{0n} = c_0^n. \tag{8}$$

Under certain circumstances it can happen, for example, that $F_{01} = F_{02} = F_{11} = F_{22}$ due to the vanishing of certain coefficients in the polynomials. The Padé approximant is called normal if it occurs only once in the Padé matrix. Also the power series $F(z)$ is called normal if all its Padé approximants are normal.

The following statements are equivalent.

(1) The Padé approximant F_{mn} is normal.
(2) The degrees of degrees of $A_{mn}(z)$ and $B_{mn}(z)$ are m and n, respectively, and (4) is true.
(3) The determinants c_{mn}, $c_{m\ n+1}$, $c_{m+1,n}$, and $c_{m+1,n+1}$ do not vanish. The power series $F(z)$ is normal if and only if $c_{mn} \neq 0$ for all $m,n \geq 0$. In particular, $c_m \neq 0$ for all m.

It follows that the Padé approximants for a normal power series $F(z)$ can always be found by solving the system (5,6), and can be expressed as a ratio of determinants. The numerator and denominator polynomials of normal Padé approximants satisfy the same linear three term recursion formulas. Indeed there exist constants α_m and β_m such that

$$A_{m,n+r} = \beta_{2n}A_{m-1,n+r} + \alpha_{2n}A_{n-1,m-1+r}, \quad m \geq 1, \tag{9}$$

$$A_{m,n+r+1} = \beta_{2n+1}A_{m,n+r} + \alpha_{2n+1}A_{m-1,n+r}, \quad m \geq 1, \tag{10}$$

and these expressions are valid with A_{mn} replaced by B_{mn}. In particular, the main diagonal elements of a normal Padé table satisfy a pure three term recursion relation. The same is true for the first subdiagonal entries. For illustrations, see 4.3 and 6.10 and their special cases treated in this tome. In these examples, the coefficients in the recursion formulas are explicitly known and are simple in form. Unfortunately, such a situation is atypical. The entries in a normal Padé table also satisfy certain nonlinear recursion relations. These are quite elegant if one desires a number. But if one wants closed form or analytic expressions for the polynomials, the nonlinear recursion formulas are clumsy and awkward to apply. In this connection, the polynomials in the rational approximations for the ${}_pF_q$ discussed in 5.12 obey simple recursion formulas, and this coupled with the known convergence properties of the rational approximations make them ideal for evaluation of the ${}_pF_q$. See the remarks following Theorem 2 of 5.12.

In June, 1972, an international conference was held at the University of Colorado, Boulder, Colorado on Continued Fractions, Padé Approximants and Related Topics. The proceedings of this conference were published in the Rocky Mountain Journal of Mathematics, Vol. 4, Number 2, Spring 1974, and the reader should examine this source for a wealth of material and references. Some additional references are Baker (1975), Baker and Gammel (1970), Fike (1968), Graves-Morris (1973), Hanscomb (1966), Khovanskii (1963), Luke (1969), Perron (1950, 1954, 1957) and Wall (1948).

For extension of scalar results on Padé approximations and continued fractions to operators, see Fair (1971, 1972a, 1972b), Fair and Luke (1970) and Wragg and Davies (1973).

13.4. Approximation of Functions Defined by a Differential Equation- The τ-Method

The τ-method, first introduced by Lanczos (1938, 1952, 1956), and exploited by Luke (1969) to achieve rational and polynomial approximations mostly for hypergeometric functions, is an excellent example of how to get such approximations for functions defined by a differential equation. We also consider two variants of the τ- method. Approximations of this kind are often called "economized approximations" and the technique for obtaining them is often termed an "economization process." We are mainly concerned with a single differential equation. However, the basic philosophy is applicable to systems of differential equations, partial differential equations, integral equations, etc. For references on these and related topics, see Bjorck (1961), Boersma (1960), Boggs and Smith (1971), Burgoyne (1962, 1963), Clenshaw (1957, 1960, 1962), Clenshaw and Norton (1963), Clenshaw and Picken (1966), El-gendi (1969), Elliott (1959/60a), 1963), Fair (1964), Fike (1968), Fox and Parker (1968), Kizner (1966), Mason (1967), Norton (1964), Oliver (1969), Sag (1970), Scraton (1972), Shimasaki and Kiyono (1973), Snyder (1966), Verbeek (1970), Wragg (1966) and Wright (1964).

Let $\mathscr{L}(D, y, z)$, $D = d/dz$, be a function of the derivatives of y (including y) with respect to z where the power to which each derivative is raised is an integer or zero, and let the coefficients of these derivatives be polynomials in z. Let r be the order of $\mathscr{L}(D, y, z)$ and consider the solution of

$$\mathscr{L}(D, y, z) = g(z), \tag{1}$$

where $g(z)$ is a polynomial in z, subject to the conditions that at

$$z = z_0, \qquad D^k y = c_k, \quad k = 0, 1, \ldots, r-1. \tag{2}$$

Without loss of generality, we can take z_0 as the origin. Observe that (1) may be linear or nonlinear. Suppose that the unique solution to (1) and (2) has the form

$$y(z) = \sum_{k=0}^{\infty} a_k z^k, \qquad |z| < c, \tag{3}$$

where $a_k = c_k/k!$ for $k = 0, 1, \ldots, r-1$ in view of (2). For $k \geqslant r$, the a_k's are found by putting (3) in (1) and equating like powers of z. This leads to a recursion relation between the a_k's and solution of this system yields the desired coefficients. If (3) is truncated after n terms, then

$$y_n^*(z) = \sum_{k=0}^{n} a_k z^k \tag{4}$$

is a polynomial approximation to (1). Here, and in the sequel we assume that $n \geqslant r$. Suppose we let

$$y_n(z) = \sum_{k=0}^{n} b_k z^k, \qquad b_k = a_k \quad \text{for} \quad k = 0, 1, \ldots, s-1, \quad s \leqslant r. \tag{5}$$

Clearly (5) does not satisfy (1) unless the solution to (1) is indeed a polynomial, in which case the problem is trivial.

Since (5) cannot satisfy the system (1), we attempt to construct an equation closely related to (1) which is satisfied by (5), with the hope that (5) is a good or even better approximation to (1) than is $y_n^*(x)$ in some qualified sense. Our philosophy is to append terms to the right-hand side of (1) so that this modified equation is satisfied by (5). Let us assume that any factor common to the terms in (1) has been removed and write

$$\mathscr{L}(D, y_n, z) = g(z) + \sum_{\mu=0}^{p} \tau_{m-\mu} h_{m-\mu}(z/\gamma) \tag{6}$$

where the τ_i's are constants, and the functions $h_\mu(z)$ are preassigned polynomials in z of degree μ. Also γ is a parameter whose role will emerge in our later discussion. The τ_i's enter the recursion relation involving the b_j's and both the τ_i's and b_j's are found to be the solution of this recursion system. The values of m and p depend on the nature of $\mathscr{L}(D, y, z)$, r, and s, and an attempt to make general statements in this regard is hazardous. The point is that p and m are selected so that the recursion system yields a well-defined set of b_j's and τ_i's. This much is certain: If $\mathscr{L}(D, y, z) \equiv \mathscr{L}(D)y$ is linear, $b_0 = a_0$, and the recursion relation which defines the a_j's is composed of $(s + 2)$ terms, then $p = s$.

To gain further insight, suppose that $\mathscr{L}(D)$ is a linear operator as described. Let $\epsilon_n(z)$ be the error incurred when $y(z)$ is approximated by $y_n(z)$. Then from (1) and (6),

$$\mathscr{L}(D)\, \epsilon_n(z) = H(z), \qquad H(z) = -\sum_{\mu=0}^{p} \tau_{m-\mu} h_{m-\mu}(z/\gamma), \tag{7}$$

$$\epsilon_n(z) = y(z) - y_n(z). \tag{8}$$

We now turn to the question of choosing the polynomials $h_\mu(z)$. The interpretation of (7) is that $H(z)$ is the result of operating on the error with $\mathscr{L}(D)$. This suggests that the polynomials $h_\mu(z)$ be chosen so that $H(z)$ is as small as possible over some range, say $0 \leqslant z/\gamma \leqslant 1$. Now of all polynomials of degree n defined over the range $0 \leqslant z \leqslant 1$ with leading coefficient 2^{2n-1}, the Chebyshev polynomial of the first kind $T_n^*(z)$ is best in the sense that its maximum deviations from zero are least. Thus, if $h_n(z) = T_n^*(z)$, the value of $\mathscr{L}(D)\, \epsilon_n(z)$ is best in the sense so described.

Rather than use the Chebyshev polynomials, it is more informative to use the Jacobi polynomials which include the Chebyshev polynomials and the Legendre polynomials as special cases.

The range parameter γ adds flexibility to the method. When $y_n(z)$ is at hand, γ can be taken equal to z and rational approximations to $y(z)$ result.

The representation (7) is advantageous for once a basis of solutions of the homogeneous equation $\mathscr{L}(D)y(z) = 0$ are known, then by the method of variation of parameters, we can construct the solution $\varepsilon_n(z)$ and so analyze the error. For some applications of this approach, see Luke (1969). See also 5.12.

There are two variants of the τ-method. For convenience in the discussion suppose that (6) is linear and that $h_n(z) = T_n^*(z)$. For a first variant, assume that

$$y_n(z) = \sum_{k=0}^{n} c_k T_k^*(z/\gamma). \tag{9}$$

Under the assumption that the coefficients of $y_n(z)$ and its derivatives in (6) are rational, we put (9) in (6), use data in 11.5 as appropriate, equate like coefficients of $T_k^*(z/\gamma)$, and thus obtain a system of equations for the determination of the c_k's and the τ_u's. The scheme is closely akin to the basic approach already delineated. If we append no τ_u terms in (6) and assume $y(z)$ has the form (9) with $n = \infty$, then we are directly seeking a solution of the differential equation in series of Chebyshev polynomials, a topic considered in 13.6.

The second variant is a generalization of the τ-scheme and is termed by Lanczos (1956) as the "method of selected points." Here we need not assume that $g(z)$ and the coefficients of $y(z)$ and its derivatives in (1) are rational. The technique is suggested by the method of collocation used to solve ordinary differential equations. Thus, assume

$$y(z) = y_n(z) + \varepsilon_n(z), \tag{10}$$

where $y_n(z)$ is given by (9). Put (10) in (1). Suppose we require $D^k y_n(z) = D^k y(z)$ at $z = 0$ for $k = 0,1,\ldots,s-1$, $s \le r$. This leaves $n+1-s$ conditions to be fulfilled to obtain a system of equations for the determination of the c_k's. To this end, require the collection of terms involving $\varepsilon_n(z)$ and its derivatives in (1) to vanish at the zeros of $T_{n+1-s}^*(z/\gamma)$. The two variants of the τ-method give identical results, but the second variant is more universal as neither the coefficients of y and its derivatives, nor $g(z)$ need be rational.

In our discussion of the τ-method and its variants, the boundary conditions are assumed to be at the same point. The extension of the ideas to the case of multipoint boundary conditions involving $y(z)$ and its derivatives is direct and we dispense with further details.

13.5. Approximations of Functions Defined by a Series

Suppose

$$F(z) = \sum_{r=0}^{n-1} b_r z^r + R_n(z) , \tag{1}$$

where b_r is independent of n and z, and $R_n(z)$ is the error in the above polynomial approximation to $F(z)$. Expressions of the type (1) can be obtained in a variety of ways -- viz. solutions of differential equations and from integral transforms such as those of Laplace and Mellin-Barnes integrals. In (1) replace n by $k+1-a$, $a = 0$ or $a = 1$, multiply both sides by $A_{n,k}\gamma^{-k}$ and sum from $k = 0$ to $k = n$. Here γ is a free parameter, $A_{n,k}$ is independent of z and γ, and $A_{n,k} = 0$ if $k > n$. Then

$$F(z)D_n(\gamma) = C_n(z,\gamma) + S_n(z,\gamma) , \tag{2}$$

$$D_n(\gamma) = \sum_{k=0}^{n} A_{n\ k}\gamma^{-k}, \tag{3}$$

$$C_n(z,\gamma) = \sum_{k=a}^{n} \gamma^{-k} \sum_{r=0}^{n-k} A_{n,r+k} b_r \delta^r, \quad \delta = z/\gamma , \tag{4}$$

$$S_n(z,\gamma) = \sum_{k=0}^{n} A_{n,k}\gamma^{-k} R_{k+1-a}(z) . \tag{5}$$

Thus $C_n(z,\gamma)/D_n(\gamma)$ is an approximation to $F(z)$ with remainder $S_n(z,\gamma)/D_n(\gamma)$. So long as $\gamma \neq z$, the approximation is a polynomial in z of degree n. But since γ is free, we can put $\gamma = z$ and so achieve an approximation in the form of a ratio of two polynomials in z, where the numerator and denominator polynomials are of degree $n-a$ and n, respectively. Thus the approximation process has great flexibility. Indeed, the inportance of the above summability scheme is that there exists a large class of functions for which the process is convergent not only for the case when the series in (1) when extended to infinity is convergent, but also in cases where this infinite series is divergent but asymptotic to $F(z)$ in some sector of the complex z-plane.

An important aspect of the above developments is the possibility of recurrence relations for the polynomials in the approximation. Suppose there exists constants K_m, L_m and M_n such that

$$\sum_{m=0}^{t} (K_m + L_m/\gamma) D_{n-m}(\gamma) = M_n ,$$

$$K_0 = 1, \; L_0 = 0, \; n \geq t . \tag{6}$$

Then

$$\sum_{m=0}^{t} (K_m + L_m/\gamma) C_{n-m}(z,\gamma) = Q_n(z,\gamma) ,$$

$$Q_n(z,\gamma) = \gamma^{-a} \sum_{r=0}^{n-a} b_r \delta^r \sum_{m=0}^{t} K_m A_{n-m,r+a} . \tag{7}$$

It follows that

$$\sum_{m=0}^{t} (K_m + L_m/\gamma) S_{n-m}(z,\gamma) = F(z) M_n - Q_n(z,\gamma) . \tag{8}$$

Thus $S_n(z,\gamma)$ can be characterized and studied once we establish a basis of solutions of the difference equation (6) with $M_n = 0$. For application of this scheme to ${}_pF_q(z)$ and a particular class of G-functions, see the first remark following 5.12(15) and Theorem 6 of 5.12. See also 5.13.3.

In the case of the ${}_pF_q$ proper application of the τ-method in 13.4 and the procedure of this section leads to identical approximations.

13.6. Solution of Differential Equations in Series of Chebyshev Polynomials of the First Kind

The ideas are best illustrated by an example which we discuss from three different points of view. We consider the differential equation

$$\{xD^2 + (2\nu + 1)D + ax^2\} y(x) = 0, \; D = d/dx ,$$

$$y \equiv y(x) = \Gamma(\nu + 1)(ax/2)^{-\nu} J_\nu(ax) , \tag{1}$$

where $J_\nu(ax)$ is the Bessel function of the first kind. As y is an even function in x and $y(0) = 1$, we take

$$y(x) = \sum_{k=0}^{\infty} b_k T_{2k}(x), \; \sum_{k=0}^{\infty} (-)^k b_k = 1 . \tag{2}$$

Method I

We follow in essence the scheme introuduced by Clenshaw (1957). Let

$$dy/dx = \sum_{k=0}^{\infty} b_k' T_{2k+1}(x), \qquad d^2y/dx^2 = \sum_{k=0}^{\infty} b_k'' T_{2k}(x). \tag{3}$$

From 11.4(43,44), we have

$$b_k' = \frac{[(2b_k''/\epsilon_k) - b_{k+1}'']}{2(2k+1)}, \tag{4}$$

$$b_k = \frac{b_{k-1}' - b_k'}{4k}, \qquad k > 0, \tag{5}$$

where here and throughout this section,

$$\epsilon_k = 1 \quad \text{if} \quad k = 0, \qquad \epsilon_k = 2 \quad \text{if} \quad k > 0. \tag{6}$$

Put (2,3) in (1), use 11.4(23,32) and equate the coefficients of $T_{2k+1}(x)$ to zero. We obtain

$$\left(\frac{2b_k''}{\epsilon_k} + b_{k+1}''\right) + 2(2\nu + 1)\, b_k' + a^2 \left(\frac{2b_k}{\epsilon_k} + b_{k+1}\right) = 0. \tag{7}$$

Replace k by $k + 1$ in (7), subtract, and use (4) to get

$$\frac{2b_k}{\epsilon_k} = b_{k+2} - \frac{4}{a^2}(k + 1 - \nu)\, b_{k+1}' - \frac{4}{a^2}(k + 1 + \nu)\, b_k' . \tag{8}$$

Elimination of the primed components in (8) with the aid of (5) leads to the four-term recurrence formula

$$\frac{2b_k}{\epsilon_k} = -\left[\frac{16(k+1)(k+\nu+1)}{a^2} - \frac{(k+1)}{k+2}\right] b_{k+1} - \left[\frac{16(k+1)(k+2-\nu)}{a^2} - 1\right] b_{k+2} - \frac{(k+1)}{k+2}\, b_{k+3} . \tag{9}$$

a result previously given by 9.7(1,2). All the solutions of the differential equation (9) are known, and it can be shown that the b_k's can be evaluated by use of the backward recurrence scheme for $R(\nu) > -1$. Let $a_{n,n} = 1$ and $a_{k+1,n} = 0$ if $k \geq n$. Evaluate $a_{k,n}$ by use of (9) with b_k replaced by $a_{k,n}$ for $k = n-1, n-2, \ldots, 0$. Let

$$b_{k,n} = a_{k,n}/u_n , \qquad u_n = \sum_{k=0}^{n} (-)^k a_{k,n} . \tag{10}$$

Then

$$b_k = \lim_{n\to\infty} b_{k,n} , \qquad R(\nu) > -1. \tag{11}$$

In illustration, take $\nu = 0$, $a = 4$, and $n = 5$. We get the following table (12).

k	$a_{k,5}$	$b_{k,5}$	k	$a_{k,5}$	$b_{k,5}$
0	− 525.25	0.05013	3	348.333	− 0.03325
1	6970.0	− 0.66524	4	− 24.167	0.00231
2	− 2608.667	0.24898	5	1.0	− 0.00010

$$u_5 = -10477.417 . \qquad (12)$$

In particular, for $x = 4$, we get the approximation $J_0(4) = -0.39717$ whereas the true value is -0.39715.

In the reference cited Clenshaw does not get pure recurrence relations such as (9), but instead proposes to simultaneously find the b_k and b'_k terms by use of (5) and (8) in the backward direction. In this connection let $a_{k,n}$ and $a'_{k,n}$ satisfy the latter equations with $a_{n,n} = 1$ and $a_{k+1,n} = a'_{k,n} = 0$ if $k \geqslant n$. Evaluate $a_{k,n}$ and $a'_{k,n}$ for $k = n - 1$, $n - 2, \ldots, 0$ by use of the recurrence formulas. Let

$$b_{k,n} = a_{k,n}/u_n , \qquad b'_{k,n} = a'_{k,n}\, u_n , \qquad u_n = \sum_{k=0}^{n} (-)^k a_{k,n} . \qquad (13)$$

Then

$$b_k = \lim_{n\to\infty} b_{k,n} , \qquad b'_k = \lim_{n\to\infty} b'_{k,n} . \qquad (14)$$

In illustration, suppose $\nu = 0$ and $a = 4$. Let $n = 5$. Thus $a_{5,5} = 1$ and $a_{k+1,5} = a'_{k,5} = 0$ if $k \geqslant 5$. From (5), $a'_{4,5} = 20$ and from (8), $a_{4,5} = -25$. Continuing in this fashion, we get the following table (15).

k	$a_{k,5}$	$a'_{k,5}$	$b_{k,5}$	$b'_{k,5}$
0	− 1089/2	11220	0.05014	− 1.03310
1	7225	− 17680	− 0.66525	1.62792
2	− 2704	3952	0.24898	− 0.36389
3	361	− 380	− 0.03324	0.03499
4	− 25	20	0.00230	− 0.00184
5	1	0	− 0.00009	0

$$u_5 = -21721/2 . \qquad (15)$$

Thus approximately,

$$J_0(4x) = 0.05014T_0(x) - 0.66525T_2(x) + 0.24898T_4(x) - 0.03324T_6(x) + 0.00230T_8(x) - 0.00009T_{10}(x), \qquad -1 \leqslant x \leqslant 1, \tag{16}$$

$$J_1(4x) = 0.25828T_1(x) - 0.40698T_3(x) + 0.09097T_5(x) - 0.00875T_7(x) + 0.00046T_9(x), \qquad -1 \leqslant x \leqslant 1. \tag{17}$$

In particular, we find approximately $J_0(4) = -0.39716$ and $J_1(4) = -0.06602$, whereas the true $5d$ values are -0.39715 and -0.06604, respectively.

Method II

Put (2) in (1) and use 11.4(23,24,32). Then with an obvious change of notation, we get

$$8\sum_{n=1}^{\infty} nb_n\left\{\sum_{k=0}^{n-1}[2n^2 - k^2 - (k+1)^2]\,T_{2k+1}(x)\right\} + 8(2\nu+1)\sum_{n=1}^{\infty} nb_n \sum_{k=0}^{n-1} T_{2k+1}(x) + a^2 b_0 T_1(x) + a^2\sum_{n=1}^{\infty} b_n\{T_{2n-1}(x) + T_{2n+1}(x)\} = 0. \tag{18}$$

Equating to zero the coefficient of $T_{2m+1}(x)$, we get

$$8\sum_{n=m+1}^{\infty} nb_n[2n^2 - m^2 - (m+1)^2] + 8(2\nu+1)\sum_{n=m+1}^{\infty} nb_n + a^2[(2b_m/\epsilon_m) + b_{m+1}] = 0. \tag{19}$$

Call the left-hand side of (19) u_m. We get a recurrence formula for the b_k's by computing $\Delta^2 u_m = u_{m+1} - 2u_m + u_{m-1}$ and setting this quantity to zero. We find the five-term recurrence formula

$$\frac{2b_m}{\epsilon_m} - \left[\frac{16(m+1)(m+\nu+1)}{a^2} - 2\right] b_{m+1} + \frac{32(m+2)(\nu+1)}{a^2}\, b_{m+2} + \left[\frac{16(m+3)(m+3-\nu)}{a^2} - 2\right] b_{m+3} + b_{m+4}\,. \tag{20}$$

If in (9) we replace k by $k+1$, multiply the resulting equation by $-(k+3)\,(k+2)$, and add it to (9), (20) follows. Employing the notation and idea around (10) for (20), we find that (11) is true. In illustration, we take the same data as for the development (12) and arrive at the following table (21). The values of $b_{k,5}$ in (12) and (21) are in close agreement. Using (21), we have $J_0(4) = -0.39747$ which is correct to three places.

k	$a_{k,5}$	$b_{k,5}$	k	$a_{k,5}$	$b_{k,5}$
0	− 499	0.04999	3	332	− 0.03326
1	6641	− 0.66537	4	− 23	0.00230
2	− 2485	0.24897	5	1	− 0.00010

$$u_5 = -9981 . \tag{21}$$

Method III

We first integrate (1) twice to get the integral equation

$$xy(x) + (2\nu - 1)\int_0^x y(t)\,dt + a^2\int_0^x\int_0^t uy(u)\,du\,dt = 2\nu x. \tag{22}$$

Now put (2) in (22) and use 11.6.2(3,6,13). Equating to zero the coefficients of $T_1(x)$, $T_3(x)$, and $T_{2k+1}(x)$, we get the equations

$$\tfrac{1}{2}(2b_0 + b_1) + \tfrac{1}{2}(2\nu - 1)(2b_0 - b_1) + \frac{a^2}{4}\left[\frac{b_0}{2} - \frac{b_1}{4} - \frac{b_2}{12} + \sum_{k=3}^{\infty}\frac{(-)^{k+1}b_k}{k^2-1}\right] = 2\nu, \tag{23}$$

$$\tfrac{1}{2}(b_1 + b_2) + \tfrac{1}{6}(2\nu - 1)(b_1 - b_2) + \frac{a^2}{96}(4b_0 - b_1 - 2b_2 + b_3) = 0, \tag{24}$$

and (9), respectively. Thus our first and third approaches lead to the recurrence formula of lowest order. Of the three procedures described, the one stemming from (22) seems easiest to apply in practice. The integral equation approach is advantageous since it leads to relations like (23) and (24) which are useful for check purposes.

The ideas of this section and those of the τ-method treated in 13.4 are closely related. The notions can also be applied to the solution of integral equations, intego-differential equations and certain partial differential equations. In addition to the references in the introduction of 13.4, see Basu (1972), Clenshaw (1957,1966), Clenshaw and Elliott (1960), E.A. Cohen (1971), Dastidar and Majumdar (1973), Dew and Scraton (1972,1973), Elliott (1959/60b,1961,1965), Knibb and Scraton (1971), Sakai and Ishikawa (1971) and Scraton (1965,1969).

CHAPTER XIV MISCELLANEOUS TOPICS

14.1. Introduction

In this chapter we present some topics which frequently arise in connection with the special functions, but which are rather peripheral to the subject matter of the previous chapters. In 14.2, some key properties of Bernoulli polynomials and numbers are presented. The D and δ operators which are frequently used in differential equations are considered in 14.3. The techniques used to compute and check the tables of Chebyshev and Padé coefficients given in this volume are discussed in 14.4. Certain mathematical constants such as π, e, γ, etc., which often arise in computations are presented in 14.5. Finally, some late references and comments are given in 14.6.

14.2. Bernoulli Polynomials and Numbers

The generalized Bernoulli polynomials $B_k^{(a)}(x)$ can be defined by the generating formula

$$\frac{t^a e^{xt}}{(e^t-1)^a} = \sum_{k=0}^{\infty} \frac{t^k}{k!} B_k^{(a)}(x), \qquad |t| < 2\pi. \tag{1}$$

If $a = 1$, we have the Bernoulli polynomials $B_k^{(1)}(x) \equiv B_k(x)$, and if, further, $x = 0$, we have the Bernoulli numbers $B_k(0) \equiv B_k$. Clearly

$$B_k^{(0)}(x) = x^k \qquad \text{and} \qquad B_0^{(a)}(x) = 1. \tag{2}$$

The first six generalized Bernoulli polynomials are as follows:

$$B_0^{(a)}(x) = 1, \qquad B_1^{(a)}(x) = x - \frac{a}{2}, \qquad B_2^{(a)}(x) = x^2 - ax + \frac{a(3a-1)}{12},$$

$$B_3^{(a)}(x) = x^3 - \frac{3a}{2}x^2 + \frac{a(3a-1)}{4}x - \frac{a^2(a-1)}{8},$$

$$B_4^{(a)}(x) = x^4 - 2ax^3 + \frac{a(3a-1)}{2}x^2 - \frac{a^2(a-1)}{2}x$$

$$+ \frac{a}{240}(15a^3 - 30a^2 + 5a + 2),$$

$$
\begin{aligned}
B_5^{(a)}(x) &= x^5 - \frac{5a}{2}x^4 + \frac{5a(3a-1)}{6}x^3 - \frac{5a^2(a-1)}{4}x^2 \\
&\quad + \frac{a(15a^3 - 30a^2 + 5a + 2)}{48}x - \frac{a^2(a-1)(3a^2 - 7a - 2)}{96},
\end{aligned} \tag{3}
$$

$$
\begin{aligned}
B_6^{(a)}(x) &= x^6 - 3ax^5 + \frac{5a(3a-1)}{4}x^4 \\
&\quad - \frac{5a^2(a-1)}{2}x^3 + \frac{a(15a^3 - 30a^2 + 5a + 2)}{16}x^2 \\
&\quad - \frac{a^2(a-1)(3a^2 - 7a - 2)}{16}x \\
&\quad + \frac{a(63a^5 - 315a^4 + 315a^3 + 91a^2 - 42a - 16)}{4032}.
\end{aligned}
$$

Recall that the Bernoulli polynomials are $B_k^{(1)}(x) \equiv B_k(x)$. Thus,

$$
\begin{aligned}
&B_0(x) = 1, \qquad B_1(x) = x - \tfrac{1}{2}, \qquad B_2(x) = x^2 - x + \tfrac{1}{6}, \\
&B_3(x) = x^3 - 3x^2/2 + x/2, \qquad B_4(x) = x^4 - 2x^3 + x^2 - 1/30, \\
&B_5(x) = x^5 - 5x^4/2 + 5x^3/3 - x/6, \\
&B_6(x) = x^6 - 3x^5 + 5x^4/2 - x^2/2 + 1/42.
\end{aligned} \tag{4}
$$

The Bernoulli numbers are $B_k \equiv B_k(0)$. Note that $B_k = 0$ when k is odd, $k > 1$. Some additional values of the Bernoulli numbers are

$$
\begin{gathered}
B_8 = -1/30, \qquad B_{10} = 5/66, \qquad B_{12} = -691/2730, \\
B_{14} = 7/6, \qquad B_{16} = -3617/510.
\end{gathered} \tag{5}
$$

For further tables of Bernoulli polynomials and numbers, see Fletcher *et al.* (1962, pp. 65–117). For a detailed study of Bernoulli polynomials and numbers, see Nörlund (1954, 1961). See also Gould (1972).

A short enumeration of $B_{2k}^{(2\rho)}(\rho)$, which are polynomials in ρ of degree k, follows:

$$
\begin{aligned}
&B_0^{(2\rho)}(\rho) = 1, \qquad B_2^{(2\rho)}(\rho) = -\frac{\rho}{6}, \qquad B_4^{(2\rho)}(\rho) = \frac{\rho(5\rho+1)}{60}, \\
&B_6^{(2\rho)}(\rho) = -\frac{\rho}{504}(35\rho^2 + 21\rho + 4), \\
&B_8^{(2\rho)}(\rho) = \frac{\rho}{2160}(175\rho^3 + 210\rho^2 + 101\rho + 18), \\
&B_{10}^{(2\rho)}(\rho) = -\frac{\rho}{3168}[385\rho^4 + 770\rho^3 + 671\rho^2 + 286\rho + 48], \\
&B_{12}^{(2\rho)}(\rho) = \frac{\rho}{7\,86240}[1\,75175\rho^5 + 5\,25525\rho^4 + 7\,15715\rho^3 + 5\,31531\rho^2 \\
&\qquad + 2\,07974\rho + 33168].
\end{aligned} \tag{16}
$$

To generate further polynomials, use

$$B_{k+1}^{(a)}(x) = B_1^{(a)}(x)B_k^{(a)}(x) - a \sum_{r=0}^{[(k-1)/2]} \frac{\binom{k}{2r+1} B_{2r+2}}{(2r+2)} B_{k-2r-1}^{(a)}(x), \tag{22}$$

$$B_{2k}^{(2\rho)}(\rho) = -2\rho \sum_{r=0}^{k-1} \frac{\binom{2k-1}{2r+1} B_{2r+2}}{2r+2} B_{2k-2r-2}^{(2\rho)}(\rho). \tag{17}$$

14.3. D and δ Operators

These convenient symbols are defined by

$$\delta = zD, \qquad D = d/dz. \tag{1}$$

The following elementary properties are easily proved.

$$\delta_z = \nu^{-1}\delta_x, \qquad \delta_z = z\, d/dz, \quad z = ax^\nu, \tag{2}$$

$$(\delta + a)z^\nu = z^\nu(\nu + a), \qquad \prod_{i=1}^{p} (\delta + a_i)z^\nu = z^\nu \prod_{i=1}^{p} (\nu + a_i), \tag{3}$$

$$\prod_{h=1}^{n} (\delta + \nu + h - 1)f(z) = z^{1-\nu}D^n\{z^{\nu+n-1}f(z)\}. \tag{4}$$

Thus from (3)

$$\delta(\delta - 1) \cdots (\delta - n + 1)z^\nu = z^\nu\nu(\nu - 1) \cdots (\nu - n + 1) = z^nD^nz^\nu$$

which implies that

$$z^nD^n = \delta(\delta - 1) \cdots (\delta - n + 1). \tag{5}$$

We have the representations

$$z^nD^n = \sum_{r=0}^{n} \binom{n-1}{r-1} B_{n-r}^{(n)} \delta^r = \sum_{r=0}^{n} \binom{n}{r} B_r^{(n)} \delta^{n-r}, \tag{6}$$

$$\delta^n = \sum_{k=0}^{n} \binom{n}{k} B_{n-k}^{(-k)} z^kD^k = \sum_{k=0}^{n} \binom{n}{k} B_k^{(k-n)} z^{n-k}D^{n-k}. \tag{7}$$

Thus,

$$\delta = zD, \qquad \delta^2 = z^2D^2 + zD, \qquad \delta^3 = z^3D^3 + 3z^2D^2 + zD,$$

$$\delta^4 = z^4D^4 + 6z^3D^3 + 7z^2D^2 + zD,$$

$$\delta^5 = z^5D^5 + 10z^4D^4 + 25z^3D^3 + 15z^2D^2 + zD,$$

$$\delta^6 = z^6D^6 + 15z^5D^5 + 65z^4D^4 + 90z^3D^3 + 31z^2D^2 + zD. \tag{8}$$

Higher powers of δ can be computed recursively since

$$\delta^n = \sum_{k=0}^{n-1} b_k^{(n)} z^{n-k} D^{n-k}, \qquad b_k^{(n)} = \binom{n}{k} B_k^{(k-n)}, \tag{9}$$

$$b_k^{(n+1)} = b_k^{(n)} + (n+1-k) b_{k-1}^{(n)}, \qquad b_0^{(n)} = b_{n-1}^{(n)} = 1, \quad b_k^{(n)} = 0 \quad \text{for} \quad k > n. \tag{10}$$

It can be readily shown by induction that

$$\prod_{i=1}^{p} (\delta + a_i) = \sum_{m=0}^{p} c_{p,m} z^m D^m, \qquad c_{p,m} = \sum_{t=0}^{p-m} \binom{p-t}{p-t-m} B_{p-t-m}^{(-m)} S_t(a_p),$$

$$c_{p,p} = 1, \qquad c_{p,p-1} = \frac{p(p-1)}{2} + S_1(a_p),$$

$$c_{p,p-2} = \frac{p(p-1)(p-2)(3p-5)}{24} + \frac{(p-1)(p-2)}{2} S_1(a_p) + S_2(a_p),$$

$$\begin{aligned} c_{p,p-3} = {} & \frac{p(p-1)(p-2)^2(p-3)^2}{48} \\ & + \frac{(p-1)(p-2)(p-3)(3p-8)}{24} S_1(a_p) \\ & + \frac{(p-2)(p-3)}{2} S_2(a_p) + S_3(a_p), \end{aligned} \tag{11}$$

$$\vdots$$

$$c_{p,0} = S_p(a_p),$$

where $S_0(a_p) = 1$, and for $m > 0$, $S_m(a_p)$ are the symmetric polynomials

$$S_m(a_p) = \sum a_{t_m} a_{t_{m-1}} \cdots a_{t_1}, \qquad t_m > t_{m-1} > \cdots > t_1,$$
$$t_j \in \{1, 2, \ldots, p\}, \qquad j = 1, 2, \ldots, m. \tag{12}$$

These polynomials may also be implicitly defined by

$$\prod_{i=1}^{p} (x + a_i) = \sum_{m=0}^{p} S_m(a_p)\, x^{p-m}. \tag{13}$$

We also have

$$c_{p,m} = \frac{(-)^m}{m!} \left(\prod_{i=1}^{p} a_i \right) {}_{p+1}F_p \left(\begin{matrix} -m, 1 + a_p \\ a_p \end{matrix} \middle| 1 \right). \tag{14}$$

With the aid of (8)-(14), 5.7.1(1) can be expressed as

$$\sum_{m=0}^{s} u_m z^m D^m U(z) = 0, \quad s = \max(p, q+1), \tag{15}$$

where the u_m's are readily deduced. An explicit expression for u_m for the more general function

$$z^{\alpha}(\mu + z)^{\beta} {}_pF_{q-1}\left(a_p;\ 1 + b_{q-1};\ \lambda(\nu + z)^{\gamma}\right) \tag{16}$$

has been given by Lavoie and Mongeau (1968).

14.4. Computation and Check of the Tables

Most of the tables of coefficients presented in this volume were taken from my previous work, see Luke (1969). The manner of computing and checking the tables was discussed in this source. Nonetheless, it is beneficial to take up the subject again in terms of the present context. We first describe the procedures for computing and checking the tables of Chebyshev coefficients. For those functions of hypergeometric type, the coefficients are also of hypergeometric type. In these cases the coefficients were almost always produced by use of the known closed form recurrence relations in the backward direction (see the material in 5.10.1, 5.10.2, 5.13.4 and Chapter 12). Where possible, particular coefficients were checked by evaluation of their hypergeometric series representations. For many functions, coefficients were developed by using the results given in 11.6. In illustration, from the coefficients for e^x valid for $0 \leq x \leq a$, coefficients for the functions

$$\int_0^x t^{-1}(1 - e^{\pm t})dt, \int_0^x t^{-\frac{1}{2}} e^{\pm t} dt, \tag{1}$$

follow from formulae given in 11.6.2. In this connection, see the numerical example, 11.6.2(21-28), and Tables 4.1 and 4.5.

In a number of cases, to secure convergence, the backward recursion scheme had to be modified as in the discussion surrounding 12.2(16-19). In this fashion, we obtained the coefficients for $(2\pi x)^{\frac{1}{2}} e^{-x} I_{\nu}(x)$ (see Tables 9.5, 9.6, 9.17, 9.18, 9.20, 9.21).

Evaluation of the Chebyshev coefficients for $\int_z^{\infty} t^{\mu} K_{\nu}(t) dt$, $z = x$ and $z = xe^{-i\pi/2}$, $x > 0$ (see Tables 9.3, 9.7, 9.8, 9.16, 9.22) was done in the following way. We have

$$E(z) = (2/\pi)^{\frac{1}{2}} e^z z^{\frac{1}{2}-\mu} \int_z^{\infty e^{i\delta}} t^{\mu} K_{\nu}(t) dt; \; z \neq 0, \; |\arg z| < \pi,$$

$$|\arg \delta| < \pi/2 \text{ or } |\arg \delta| = \pi/2 \text{ and } R(\mu) < \tfrac{1}{2}, \quad (2)$$

$$dE(z)/dz - \{1 + (\tfrac{1}{2} - \mu)/z\} E(z) = -(2z/\pi)^{\frac{1}{2}} e^z K_{\nu}(z). \quad (3)$$

Now we have a Chebyshev series expansion for the right hand side of (3). Assume a like series for $E(z)$. Put this in (3), use the pertinent equations in 11.5, equate the coefficients of the Chebyshev polynomials of like order, and so get a recursion formula for the coefficients in the Chebyshev series for $E(z)$. These coefficients can then be found by use of the backward recurrence scheme. The procedure is much akin to the ideas of 11.6.2(21-28) and 13.6. For complete details, see Luke (1969). Basically, this same approach was used to get the coefficients in the expansions for the integrals involving $I_0(t)$, (see Tables 9.7, 9.8) in descending series of Chebyshev polynomials. In the latter two cases, the backward recursion technique was modified in accordance with the discussion surrounding 12.2(16-19). The coefficients in the similar type expansions for P.V.$\int_{-\infty}^{x} t^{-1} e^t dt$ and $\int_0^x e^{t^2} dt$ (see Tables 4.1 and 4.5, respectively) were also found in this fashion.

For nonhypergeometric functions, evaluation of the Chebyshev coefficients was often accomplished from a knowledge of the Taylor series expansions for these functions with the aid of the basic series or rearrangement analysis developed in 11.3.6.2. In particular, the coefficients for $\{\Gamma(x+1)\}^{-1}$ in Table 1.1 were produced in this way. The coefficients for $\Gamma(x+1)$ in Table 1.1 were not found in this manner because of the weak convergence of the Taylor series for $\Gamma(x+1)$.

In Luke (1969), the basic series technique was used to get the Chebyshev coefficients for $x\zeta(1+x)$, $-\ln \xi(x)$ and a third function related to $\zeta(x)$. The necessary Taylor series coefficients are due to Gram (1895, 1903). The Taylor series coefficients for $x\zeta(1+x)$ are given by Gram (1903) to 16D. In my original computations, 20D were carried and unfortunately, the data was not rounded to 16D for presentation in the final table. The manner of correcting this table is explained on p. 23.

The basic series method was also used to get the Chebyshev coefficients for those portions of the ascending series expansions for $Y_n(x)$ and $K_n(x)$ not attached to $J_n(x)$ and $I_n(x)$, respectively, $n = 0,1$. For $n = 0$, these coefficients were also generated with the help of a recursion formula used in the backward direction. Derivation of this formula can be briefly described as follows. Start with

$$J_\nu(x) = x^\nu \sum_{n=0}^{\infty} B_n(\nu) T_{2n}(x/a). \tag{4}$$

Differentiate (4) with respect to ν and then put $\nu = 0$. Thus

$$(\pi/2)Y_0(x) = (\ln x)J_0(x) + \sum_{n=0}^{\infty} A_n T_{2n}(x/a), \quad A_n = \{\partial B_n(\nu)/\partial\nu\}_{\nu=0}. \tag{5}$$

Differentiate the known recursion formula for $B_n(\nu)$ as given by 9.7(1) with respect to ν and then set $\nu = 0$. This gives a an inhomogeneous recursion formula for A_n which depends on the coefficients $B_n(0)$. The backward recurrence scheme is outlined in 12.3.

For certain functions, special techniques were devised. This was the situation for $\ln \Gamma(x+3)$ and its derivatives, Tables 1.3, 1.4. For details on this and the computation of coefficients for $\Gamma(x+1)$, Table 1.2, $\tan \pi x/4$ and $\cot \pi x/2$, Table 3.7, see Luke (1969).

For each set of coefficients numerous checks were made by evaluating the Chebyshev series for specific arguments and comparing with known tabular values or by computing tabular values anew from power series and asymptotic series. Frequently, coefficients or sums of coefficients could be related to transcendentals known to many decimal places. This afforded an excellent check since our basic computations are generally much more accurate than those reported here. In illustration, $\cos \pi/4 = \sin \pi/4 = 2^{-\frac{1}{2}}$. Further, since for the dilogarithm function $L(x)$ of Table 3.12, we have $L(1) = \pi^2/12$, so also is the sum of the coefficients in this table (except possibly for roundoff). Again, from Table 3.12, the value of $T(1)$ is known as Catalan's constant, see Table 14.1. In many instances, the tables themselves could be used for checks. For example, it is easy to show that the coefficient f_0 in Table 4.4 is $4\pi\{\boldsymbol{H}_0(8) - Y_0(8)\}$. Thus at once we have a check on the pertinent coefficients in Tables 4.4, 9.1, 10.1, 10.4. For another example, in Table 3.10, $a_0 = 2^{\frac{1}{2}}/\boldsymbol{K}(k)$, $k^2 = \frac{1}{2}$, and $a_1 = a_0 - 16/9\pi a_0$. Also for $k^2 = \frac{1}{2}$, $\boldsymbol{K}(k) = (\pi^{3/2}/2)/\{\Gamma(3/4)^2\}$. Thus at once we obtain a simultaneous check on Table 3.10, the various tables for the gamma function, and the table of coefficients for $\boldsymbol{K}(k)$. The latter are given in Luke (1969), but not in the present volume. Again from Table 9.5, we have data for the evaluation of $K_0(x)$ for $0 < x \leq 8$ and for $x \geq 5$. So calculations in the region of overlap provided a check on both sets of coefficients as well as the corresponding material in Table 9.23. This idea applies to many other tables also. In those cases where the same coefficients have been given by other workers, comparison checks were made. This was most interesting, since in almost every instance our methods of computation differed from those of the other workers.

In general, the computations were designed to produce an accuracy of about 25D. The coefficients were then rounded to 20D (in some cases to 15D) for presentation. The error in each coefficient should not exceed a half-unit of the last decimal given. The coefficients for $\psi^{(6)}(x+3)$ in Table 1.4 are the only exception. For an explanation, see Luke (1969).

Next we turn to the rational approximations. Where the polynomials are known in closed form, recurrence relations are also known. Thus in each instance, the coefficients in the polynomials were derived from the defining representation and checked by use of the recurrence formula. The Padé approximations given in Table 4.2 were determined by solving systems of linear equations as discussed in 13.3. Here 25D were carried and the results rounded to 20D. For each rational approximation, numerous numerical checks were performed including comparisons with tabular data deduced from corresponding Chebyshev expansions.

The tables in Luke (1969) are to the best of my knowledge error free except for the coefficients for $\zeta(1+x)$ already noted.

The coefficients in this volume coupled with those in Luke (1969) are the most complete ever assembled. References to similar tables will be found in 1.6.3, 3.5.2, 4.10.3, and 9.13.3.

14.5. Mathematical Constants

In this section we first give some recent bibliographic data on mathematical constants. There then follows a table of numerous mathematical constants to 20D which frequently arise in connection with the evaluation of special functions.

Shanks and Wrench (1962): π, 100,000D. See also Wrench (1960).

Shanks and Wrench (1969): e, 100,000D.

Sweeney (1963): γ, 3566D: ln 2, 3683D.

Robinson and Potter (1972): Tables I and II give mathematical constants to 20D, arranged according to decimal size. They can be used, for example, for improvement of the accuracy of a constant on hand or for possible identification of a number obtained by empirical or other means. Table III gives some properties of the integers from 1 to 1000, including the binary and ternary forms and representations as sums of two and three squares, two and three cubes, and differences of squares and cubes. Table IV consists of 260 sets of coefficients of the first 10 terms of the Maclaurin series of selected algebraic and transcendental functions and combinations thereof.

TABLE 14.1

MATHEMATICAL CONSTANTS

π	3.14159 26535 89793 23846	$1/\pi$	0.31830 98861 83790 67154
$\pi/2$	1.57079 63267 94896 61923	$2/\pi$	0.63661 97723 67581 34308
$\pi/4$	0.78539 81633 97448 30962	$4/\pi$	1.27323 95447 35162 68615
$\pi^{\frac{1}{2}}$	1.77245 38509 05516 02730	$(\pi)^{-\frac{1}{2}}$	0.56418 95835 47756 28695
$(\pi/2)^{\frac{1}{2}}$	1.25331 41373 15500 25121	$(\pi/2)^{-\frac{1}{2}}$	0.79788 45608 02865 35588
$(2\pi)^{\frac{1}{2}}$	2.50662 82746 31000 50242	$(2\pi)^{-\frac{1}{2}}$	0.39894 22804 01432 67794
$2^{\frac{1}{2}}$	1.41421 35623 73095 04880	$2^{-\frac{1}{2}}$	0.70710 67811 86547 52440
$3^{\frac{1}{2}}$	1.73205 08075 68877 29353	$3^{-\frac{1}{2}}$	0.57735 02691 89625 76451
e	2.71828 18284 59045 23536	e^{-1}	0.36787 94411 71442 32160
$\ln 2$	0.69314 71805 59945 30942	$\ln 3$	1.09861 22886 68109 69140
$\ln 10$	2.30258 50929 94045 68402	$\log_{10} e$	0.43429 44819 03251 82765
$\log_2 e$	1.44269 50408 88963 40736	$\ln(1+2^{\frac{1}{2}})$	0.88137 35870 19543 02523

γ	0.57721 56649 01532 86061	$\ln \gamma$	-0.54953 93129 81644 82234
$\Gamma(1/4)$	3.62560 99082 21908 31193	$(\Gamma(1/4))^{-1}$	0.27581 56628 30209 31436
$\Gamma(3/4)$	1.22541 67024 65177 64513	$(\Gamma(3/4))^{-1}$	0.81604 89390 98262 98108
$\Gamma(1/3)$	2.67893 85347 07747 63366	$(\Gamma(1/3))^{-1}$	0.37328 21739 07395 22833
$\Gamma(2/3)$	1.35411 79394 26400 41695	$(\Gamma(2/3))^{-1}$	0.73848 81116 21648 31294

Let $S_k = \sum_{r=1}^{\infty} r^{-k}$, C = Catalan's constant $= \sum_{k=0}^{\infty} (-)^k (2k+1)^{-2}$

$S_2 = \pi^2/6$	1.64493 40668 48226 43647	S_3	1.20205 69031 59594 28540
$S_4 = \pi^4/90$	1.08232 32337 11138 19152	S_5	1.03692 77551 43369 92633
$S_6 = \pi^6/945$	1.01734 30619 84449 13971	S_7	1.00834 92773 81922 82684

C = 0.91596 55941 77219 01505

14.6. Late Bibliography

The principal bibliography for this volume begins on the next page. Here we present some additional references which arrived too late to incorporate in the text where it should occur.

Arsenin, V.Ya. (1974). "Methods of Mathematical Physics and Special Functions" (in Russian). Izdat. Nauka, Moscow.

Davies, B. (1973). "Complex zeros of linear combinations of spherical Bessel functions and their derivatives." *SIAM J. Math. Anal.* 4, 128-133.

Németh, G. (1974). Expansion of generalized hypergeometric functions in Chebyshev polynomials. Rept. Central Inst. for Physics, Budapest.

Riekstins, E.Ya. (1974). "Asymptotic Expansion of Integrals" (in Russian), Vol. I. Zinatne Press, Riga, Latvia

Wimp, J. (1974). "On the computation of Tricomi's ψ function." Computing (Arch. Elektron. Rechnen) 13, 195-203.

For the most part, the titles of these references are indicative of the contents. Some comments on the works of Németh and Wimp should prove helpful. Németh considers $f(z)$, a form like ${}_pF_q(z)$ with the k in $(\alpha_i)_k$ and $(\rho_i)_k$ replaced by $\beta_i k$ and $\gamma_i k$ respectively. Let $\mu = \Sigma_{i=1}^{p} \beta_i - 1 - \Sigma_{i=1}^{q} \gamma_i \leq 0$. Then the series converges. If $\mu > 0$, the series diverges, in which case it is the asymptotic expansion of a function, call it $F(z)$, in an appropriate region of the complex plane. Asymptotic estimates for the large index coefficients in the expansion of $f(z)$ and $F(z)$ in series of Chebyshev polynomials of the first kind are derived. Several examples with numerical coefficients are presented.

Wimp considers $U(a; c; z)$, see 7.2. The function

$$U_n = \{\Gamma(v)\Gamma(w)\}^{-1} G_{2,3}^{3,1}\left(\lambda \middle| \begin{matrix} 1-n, & n+u \\ u, & v, & w \end{matrix}\right),$$

a special case of the G-function in 5.10.1(5), satisfies a four term recurrence relation in n, see 5.13.4(1). U_n is a minimal solution and so can be generated by application of the backward recursion process described in 12.2. Thus we can obobtain $U_0 = \lambda^v U(v; c; \lambda)$, $c = v + 1 - w$. In the above u is arbitrary. Let $u \to \infty$ in $u^n U_n/n!$ The limit function satisfies a three term recurrence formula and application of the backward recursion process can be used to generate $\{\lambda^v (v)_n (w)_n / n!\} \times U(n+v; c; \lambda)$. Again we can get U_0. When u is free, it is usually taken as 1. Here the rate of convergence of the backward recurrence schema is much better than in the case $u \to \infty$. By specializing parameters, we can generate values of $K_\nu(z)$. This is quite intersting since the usual recurrence formula does not provide a stable algorithm when used in the backward direction to compute this function.

BIBLIOGRAPHY

Abramowitz, M., and Stegun, I. (eds.)(1964). "Handbook of Mathematical Functions with Formulas, Graphs and Mathematical Tables." Appl. Math. Ser. 55, U.S. Govt. Printing Office, Washington, D.C. There are a number of reprints. For errata, see Luke, Wimp and Fair (1972) and the journal Mathematics of Computation.

Aggarwal, H.R., and Sagherian, V. (1969). *Math. Comp.* 23, 456-457.

Agrest, M.M. (1970). *Z. Vycisl. Mat. i Mat. Fiz.* 10, 313-325.

Agrest, M.M. (1971). *Z. Vycisl. Mat. i Mat. Fiz.* 11, 1127-1138.

Agrest, M.M., Bekauri, I.N., Orlov, L.A., Rikenglaz, M.M., Haihjan, N.V., and Cacibaja, C.S. (1966). "Tables of Incomplete Cylindrical Functions" (in Russian). Vycisl. Centr Akad. Nauk SSSR, Moscow.

Agrest, M.M., and Maksimov, M.Z. (1965). "Theory of incomplete Cylindrical Functions and Their Applications" (in Russian). Atomizdat, Moscow. Also in English, Springer, New York, (1971).

Agrest, M.M., and Rikenglaz, M.M. (1967). *Z. Vycisl. Mat. i Mat. Fiz.* 7, 1370-1374.

Akademija Nauk SSSR. (1960). "Tables of the Normal Probability Integral, Normalized Density and Their Derivatives" (in Russian). Mat. Institut in V.A. Steklova. Izdat., Moscow.

Albasiny, E.L., Bell, R.J., and Cooper, J.R.A. (1963). A table for the evaluation of Slater coefficients and integrals of triple products of spherical harmonics. Rept. 49. National Physical Laboratory Mathematics Division, Teddington, Middlesex. See *Math. Comp.* 19(1965), 157-158.

Alexander, M.J., and Vok, C.A. (1963). Tables of the cumulative distribution of sample multiple coherence. Res. Rept. RR63-37. Rocketdyne Division of North American Aviation, Canoga Park, California. See *Math. Comp.* 21(1967), 503.

Amos, D.E. (1963). *Biometrika* 50, 449-457.

Amos, D.E. (1973). *Math. Comp.* 27, 413-427.

Amos, D.E. (1974). *Math. Comp.* 28, 239-252.

Ancker, C.J., and Gafarian, A.V. (1962). The function $J(x,y) = \int_0^x t^{-1} g(y,t)\,dt$ - some properties and a table. System

Development Corp., Santa Monica, California. See also *Math. Comp.* 17(1963), 206.
Anolik M.V., and Mil'ner, C.H. (1970). *Metody Vycisl.* 6, 137-140, 141-145.
Arsenin, V.Ya. (1968). "Basic Equations and Special Functions of Mathematical Physics" (transl. from Russian). Iliffe, London.
Arthurs, A.M., and McCarroll, R. (1961). *Math. Comp.* 15, 159-162.
Artin, E. (1964). "The Gamma Function." Holt, Rinehart and Winston, New York.
Ascari, A. (1968). A table of the repeated integrals of the error function. Rept. S/6339. Societa Richerche Impianti Nucleari (SORIN), Nuclear Research Center, Saluggia, Vercelli, Italy. See *Math. Comp.* 22(1968), 898-899.
Askey, R. (1973). *J. Math. Anal. Appl.* 41, 122-124.
Ayant, Y., and Borg, M. (1971). "Fonctions Spéciales." Dunod, Paris
Babister, A.W. (1967). "Transcendental Functions Satisfying Nonhomogeneous Linear Differential Equations." Macmillan, New York
Bailey, W.N. (1935). "Generalized Hypergeometric Series." Cambridge Univ. Press, London and New York.
Baker, G.A.. Jr. (1975). "Essentials of Padé Approximations." Academic Press, New York
Baker, G.A., Jr., and Gammel, J.L. (eds.) (1970). "The Padé Approximant in Theoretical Physics." Academic Press, New York.
Barakat, R. (1961). *Math. Comp.* 15, 7-11.
Bark, L.S., Bol'shev, L.N., Kuznetsov, P.I., and Cherenkov, A.P. (1964). "Tables of the Rayleigh-Rice Distribution" (in Russian). Vycisl. Centr Akak. Nauk SSSR, Moscow.
Bark, L.S., and Kuznetsov, P.I. (1962). "Tables of Cylinder Functions of Two Imaginary Variables" (in Russian). Vycisl. Centr Akad. Nauk SSSR, Moscow. In English, Pergamon Press, Macmillan, New York, (1965).
Bark, L.S., Zhurina, M.I., and Karmazina, L.N. (1962). "Tables des Fonctions de Legendre Associées" (transl. from French to Russian). Vycisl. Centr Akad. Nauk SSSR, Moscow.
Barrett, R.F. (1964). *Math. Comp.* 18, 332.
Basu, N.K. (1972). *Yokohama Math. J.* 20, 57-68.
Bauer, H.F. (1964). *Math. Comp.* 18, 128-135.
Bellman, R., Kashef, B.G., and Vasudevan, R. (1972). *Math. Comp.* 26, 233-236.

Ben Daniel, D.J., and Carr, W.E. (1960). Tables of solutions of Legendre's equation for indices of nonintegral order. Rept. UCRL-5859, Univ. of California Lawrence Radiation Laboratory, Livermore. Available from Office of Technical Services, Washington, D.C. See *Math. Comp.* 16(1962), 117-119.

Berg, L. (1968). "Asymptotische Darstellungen und Entwicklungen." Deutscher Verlag der Wissenschaften, Berlin.

Berger, B.S., and McAllister, H. (1970). *Math. Comp.* 24, 488.

Berlyand, O.S., Gavrilova, R.I., and Prudnikov, A.P. (1961). "Tables of Integral Error Functions and Hermite Polynomials." (in Russian). Byelorussian Akad. Sci., Minsk. Also in English, Pergamon Press, Macmillan, New York, (1962).

Berlyand, O.S., Kiricenko, L.K., and Kogan, R.M. (1965). *Dokl. Akad. Nauk SSSR.* 160, 306-307.

Bernard, G.D., and Ishimaru, A. (1962). Tables of the Anger and Lommel-Weber functions. Tech. Rept. 53, AFCRL 796, Univ. of Washington Press, Seattle. See also *Math. Comp.* 17(1963), 315-317.

Bhagwandin, K. (1962). L'approximation uniforme des fonctions d'Airy-Stokes et fonctions de Bessel d'indices fractionnaires. *2nd Congr. Assoc. Franc. Calcul et Traitement Information, Paris*, 1961, pp. 137-145. Gauthier-Villars, Paris.

Biedenharn, L.C., and van Dam, H. (1965). "Quantum Theory of Angular Momentum." Academic Press, New York

Bierlein, J.A. (1962). *J. Chem. Phys.* 36, 2793-2802.

Birkhoff, G.D., and Trjitzinsky, W.J. (1932). *Acta Math.* 60, 1-89.

Bishop, D.M. (1970). *Math. Comp.* 24, 479.

Björck, A. (1961). *Nordisk Mat. Tidskr.* 9, 65-70.

Blair, J.M. (1974). *Math. Comp.* 28, 581-583.

Blair, J.M., and Edwards, C.A. (1974). Stable rational minimax approximations to the modified Bessel functions $I_0(x)$ and $I_1(x)$. Rept. AECL-4928, Atomic Energy of Canada, Chalk River, Ontario.

Boas, R.P., Jr., and Buck, R.C. (1958). "Polynomial Expansion of Analytic Functions." Springer, Berlin.

Boas, R.P., Jr., and Wrench, J.W., Jr. (1971). *American Math. Monthly* 78, 864-870.

Boersma, J. (1960). *Math. Comp.* 14, 380.

Boggs, R.A.C., and Smith, F.J. (1971). *Comput. J.* 14, 270-271.

Booth, A.D. (1955). *MTAC* 9, 21-23.

Boyd, A.V. (1959). *Rep. Statist. Appl. Res. Un. Japan Sci. Engrs.* 6, 44-46.

Braaksma, B.L.J. (1963). *Compositio Math.* 15, 239-341.

Breig, W.F., and Crosbie, A.L. (1974). *Math. Comp.* 28, 575-579.

Brenner, H., and Sonshine, R.M. (1964). *Quart. J. Mech. Appl. Math.* 17, 55-63.
Bridge, J.F., and Angrist, S.W. (1962). *Math. Comp.* 16, 198-204 and 18(1964), 348.
Briones, F. (1964). *Chiffres* 7, 177-184.
Britney, R.R., and Winkler, R.L. (1970). *Math. Comp.* 24, 995.
Bruijn, de, N.G. (1958). "Asymptotic Methods in Analysis." North-Holland Publishing Co., Amsterdam.
Buchholz, H. (1953). "Die konfluente hypergeometrische Funktion." Springer, Berlin. Also in English as "The Confluent Hypergeometric Function with Special Emphasis on its Applications." Springer, New York, (1969).
Bulirsch, R. (1967). *Numer. Math.* 9, 380-385.
Bulirsch, R. (1969). *Numer. Math.* 13, 305-315.
Burgoyne, F.D. (1962). *Math. Comp.* 16, 497-498.
Burgoyne, F.D. (1963). *Math. Comp.* 17, 295-298.
Burunova, N.M. (1959). "Handbook of Mathematical Functions, Supplement No. 1" (in Russian). Izdat. Akad. Nauk SSSR, Moscow. In English, Pergamon Press, Oxford, (1960).
Byrd, P.F., and Friedman, M.D. (1954). "Handbook of Elliptic Integrals for Engineers and Physicists." Springer, Berlin. Also a second revised edition, (1971). For errata, see Luke, Wimp and Fair (1972) and the journal Mathematics of Computation.
Campbell, R. (1966). "Les Intégrales Eulériennes et Leurs Applications." Dunod, Paris.
Carlitz, L. (1963). *Boll. Un. Mat. Ital.* (3)18, 90-93.
Carlitz, L. (1965). *Boll. Un. Mat. Ital.* (3)19, 436-440.
Carlson, B.C. (1965). *Proc. Amer. Math. Soc.* 16, 759-766.
Carlson, B.C. (1966). *Proc. Amer. Math. Soc.* 17, 32-39.
Carlson, B.C. (1970a). *SIAM J. Math. Anal.* 1, 232-242.
Carlson, B.C. (1970b). *Proc. Amer. Math. Soc.* 25, 698-703.
Carlson, B.C. (1972). *Math. Comp.* 26, 543-550.
Carlson, B.C., and Tobey, M.D. (1968). *Proc. Amer. Math. Soc.* 19, 255-262.
Carpenter, L. (1962). An addition to the Yale tables for the development of the disturbing function. National Aeronautics and Space Administration, TN D-1290, Washington, D.C.
Cayley, A. (1961). "An Elementary Treatise on Elliptic Functions." Dover, New York.
Centre National d'Études des Telecommunications. (1966). Tables numériques des fonctions associées de Legendre. Fonction associée de premiére espèce $P_n^m(\cos\theta)$. Deuxiéme Fascicule, Troisiéme Fascicule, Éditions de la Revue d'Optique, Paris.
Champion, P.M., Danielson, L.R., and Miksell, S.G. (1969). *Ganita* 20, 47-48.

Chatelet, F., Lansraux, G., and Million, B. (1959). *Chiffres* 2, 233-237.

Chawla, M.M. (1967). *Comput. J.* 9, 413.

Chi, B.E. (1962). "Numerical Table of Clebsch-Gordan Coefficients." Rensselaer Polytechnic Institute, New York.

Chiarella, C., and Reichel, A. (1968). *Math. Comp.* 22, 137-143.

Chipman, D.M. (1972). *Math. Comp.* 26, 241-250.

Clenshaw, C.W. (1955). *MTAC* 9, 118-120.

Clenshaw, C.W. (1957). *Proc. Cambridge Philos. Soc.* 53, 134-149.

Clenshaw, C.W. (1960). The numerical solution of ordinary differential equations in Chebyshev series. *Symposium on the Numerical Treatment of Ordinary Differential Equations, Integral and Integro-Differential Equations, Rome, 1960,* pp. 222-227. Birkhäuser, Basel.

Clenshaw, C.W. (1962). Chebyshev series for mathematical functions. *Nat. Phys. Lab. Math. Tables,* Vol. V. H.M. Stationery Office, London.

Clenshaw, C.W. (1964). *SIAM J. Numer. Anal.* 1, 26-37.

Clenshaw, C.W. (1966). The solution of van der Pol's equation in Chebyshev series. *Numerical Solutions of Nonlinear Differential Equations (Proc. Adv. Sympos.) Madison, Wisconsin, 1966,* pp. 55-63. John Wiley, New York.

Clenshaw, C.W., and Elliott, D. (1960). *Quart. J. Mech. Appl. Math.* 13, 300-313.

Clenshaw, C.W., Miller, G.F., and Woodger, M. (1963). *Numer. Math.* 4, 403-419.

Clenshaw, C.W., and Norton, H.J. (1963). *Comput. J.* 6, 88-92.

Clenshaw, C.W., and Picken, S.M. (1966). Chebyshev series for Bessel functions of fractional order. *Nat. Phys. Lab. Math. Tables,* Vol. VIII. H.M. Stationery Office, London.

Cochran, J.A. (1964). *J. Soc. Indust. Appl. Math.* 12, 580-587.

Cochran, J.A. (1965). *Numer. Math.* 7, 238-250.

Cochran, J.A. (1966a). *Proc. Cambridge Philos. Soc.* 62, 215-226.

Cochran, J.A. (1966b). *Quart. J. Mech. Appl. Math.* 19, 511-522.

Cochran, J.A. (1967). *J. Math. and Phys.* 46, 220-222.

Cochran, J.A., and Hoffspiegel, J.N. (1970). *Math. Comp.* 24, 413-422.

Cody, W.J. (1968). *Math. Comp.* 22, 450-453.

Cody, W.J. (1969). *Math. Comp.* 23, 631-637.

Cody, W.J. (1970). *SIAM Rev.* 12, 400-423.

Cody, W.J., and Hillstrom, K.E. (1967). *Math. Comp.* 21, 198-203.
Cody, W.J., and Hillstrom, K.E. (1970). *Math. Comp.* 24, 671-677.
Cody, W.J., Hillstrom, K.E., and Thacher, H.C., Jr. (1971). *Math. Comp.* 25, 537-547.
Cody, W.J., Meinardus, G., and Varga, R.S. (1969). *J. Approx. Theory* 2, 50-65.
Cody, W.J., Paciorek, K.A., and Thacher, H.C., Jr. (1970). *Math. Comp.* 24, 171-178.
Cody, W.J., Strecok, A.J., and Thacher, H.C., Jr. (1973). *Math. Comp.* 27, 123-127.
Cody, W.J., and Thacher, H.C., Jr. (1968). *Math. Comp.* 22, 641-649.
Cody, W.J., and Thacher, H.C., Jr. (1969). *Math. Comp.* 23, 289-303.
Cohen, D.S. (1964). *J. Math. and Phys.* 43, 133-139. For errata, see the same journal, 44(1965), 410.
Cohen, E.A., Jr. (1971). *SIAM J. Numer. Anal.* 8, 754-756.
Colombo, S., and Lavoine, J. (1972). "Transformations de Laplace et de Mellin." Gauthier-Villars, Paris.
Consul, P.C. (1963). *Sankhya B* 25, 197-214.
Cook, J.L., and Elliott, D. (1960). *Australian J. Appl. Sci.* 11, 16-32.
Cooke, J.C., and Tranter, C.J. (1959). *Quart. J. Mech. Appl. Math.* 12, 379-386.
Cooper, G.J. (1967). *Comput. J.* 10, 94-100.
Copson, E.T. (1965). "Asymptotic Expansions." Cambridge Univ. Press, London and New York.
Corput, van der, J.G. (1955). *Nederl. Akad. Wetensch. Proc. Ser. A* 58, 139-150.
Corrington, M.S. (1961). *Math. Comp.* 15, 1-6, 225.
Curtis, A.R. (1964). "Coulomb Wave Functions," Royal Society Mathematical Tables, Vol. 11, Cambridge Univ. Press, London and New York.
Dastidar, D.G., and Majumdar, S.K. (1973). *Indian J. Pure Appl. Math.* 4, 155-160.
Davis, P.J. (1959). *Amer. Math. Monthly* 66, 844-869.
Davis, P.J., and Rabinowitz, P. (1974). "Methods of Numerical Integration." Academic Press, New York.
Dempsey, E., and Benson, G.C. (1960). *Canadian J. Phys.* 38, 399-424.
Deutsch, E. (1962). *Proc. Edinburgh Math. Soc. Ser. 2*, 13, 285-290.
Dew, P.M., and Scraton, R.E. (1972). *J. Inst. Math. Appl.* 9, 299-309.
Dew, P.M., and Scraton, R.E. (1973). *J. Inst. Math. Appl.* 11, 231-240.

Diaz, J., and Osler, T. (1974). *Math. Comp.* 27, 185-202.
DiDonato, A.R., and Jarnagin, M.P. (1962). *Math. Comp.* 16, 347-355.
Didry, J.R., and Guy, J. (1962). *Chiffres* 5, 1-3.
Dimsdale, B., and Inselberg, A. (1972). *J. Math. Anal. Appl.* 37, 640-649.
Dingle, R.B. (1973). "Asymptotic Expansions: Their Derivation and Interpretation." Academic Press, New York.
Dishon, M., and Weiss, G.H. (1966). *J. Res. Nat. Bur. Standards, Math. and Math. Phys.* 70B, 95-118.
Ditkin, V.A. (ed.) (1965). "Tables of the Logarithmic Derivative of the Gamma Function and its Derivatives in the Complex Plane" (in Russian). Vycsl. Centr Akad. Nauk SSSR, Moscow.
Ditkin, V.A. (ed.) (1966). "Tables of Incomplete Cylindrical Functions" (in Russian). Vycsl. Centr Akad. Nauk SSSR, Moscow.
Donaldson, J.D., and Elliott, D. (1972). *SIAM J. Numer. Anal.* 9, 573-602.
Döring, B. (1966a). *Numer. Math.* 8, 123-136.
Döring, B. (1966b). *Math. Comp.* 20, 215-222.
Döring, B. (1967). *Z. Angew. Math. Phys.* 18. 461-473.
Dougherty, H.F., and Johnson, M.E. (1964). A tabulation of Airy functions. National Bureau of Standards Tech. Note 228. U.S. Govt. Printing Office, Washington, D.C. See *Math. Comp.* 19(1965), 691-692.
Drane, C.J., Jr. and McIlvenna, J.F., Jr. (1963). Response function for Taylor antenna distributions. Rept. AFCRL-63-368, Air Force Cambridge Research Laboratories, Cambridge, Mass.
Dyson, F.J. (1960). *Phys. Fluids* 3, 155-157.
Elbert, A. (1971). *Studia Sci. Math. Hungar.* 6, 277-285.
El-gendi, S.E. (1969). *Comput. J.* 12, 282-287.
Elliott, D. (1959/60a). *J. Australian Math. Soc.* 1, 344-356.
Elliott, D. (1959/60b). *J. Australian Math. Soc.* 1, 428-438.
Elliott, D. (1961). *Proc. Cambridge Philos. Soc.* 57, 823-832.
Elliott, D. (1963). *Comput. J.* 6, 102-111.
Elliott, D. (1964). *Math. Comp.* 18, 274-284.
Elliott, D. (1965). *Math. Comp.* 19, 234-248.
Elliott, D. (1968). *J. Australian Math. Soc.* 8, 213-221.
Elliott, D. (1971). *Math. Comp.* 25, 309-316.
Elliott, D., and Lam, B. (1973). *SIAM J. Numer. Anal.* 10, 1091-1102.
Elliott, D., and Szekeres, G. (1965). *Math. Comp.* 19, 25-32.
Elliott, D., and Tuan, P.D. (1974). *SIAM J. Math. Anal.* 5, 1-10.

Erber, T. (1960). *Arch. Rational Mech. Anal.* 4, 341-351.
Erber, T., and Gordon, A. (1963). *Math. Comp.* 17, 162-169.
Erdélyi, A. (1956). "Asymptotic Expansions." Dover, New York.
Erdélyi, A., Magnus, W., Oberhettinger, F., and Tricomi, F.G. (1953). "Higher Transcendental Functions," Vols. I,II,III, McGraw-Hill, New York. For errata, see Luke, Wimp and Fair (1972).
Erdélyi, A., Magnus, W., Oberhettinger, F., and Tricomi, F.G. (1954). "Tables of Integral Transforms," Vols. I,II, McGraw-Hill, New York. For errata, see Luke, Wimp and Fair (1972).
Erukhimovich, Yu.A., and Pimenov, Yu.V. (1969). *Z. Vycisl. Mat. i Mat. Fiz.* 9, 691-197.
Evgrafov, M. (1961). "Asymptotic Estimates and Entire Functions" (transl. from Russian). Gordon and Breach, New York.
Faddeeva, V.N., and Terent'ev, N.M. (1954). "Tables of the function $w(z) = e^{-z^2}(1 + 2i\pi^{-\frac{1}{2}}\int_0^z e^{t^2}dt)$ for Complex Argument" (in Russian). Gosudarstv. Izdat. Tehn.-Teor. Lit., Moscow. Also in English, Pergamon Press, New York, (1961).
Fair, W. (1971). *SIAM J. Math. Anal.* 2, 226-232.
Fair, W. (1972a). *J. Math. Anal. Appl.* 39, 318-323.
Fair, W. (1972b). *J. Approx. Theory* 5, 74-76.
Fair, W., and Luke, Y.L. (1970). *Numer. Math.* 14, 379-382.
Ferreira, E.M., and Sesma, J. (1970). *Numer. Math.* 16, 278-284.
Fettis, H.E. (1963). On a conjecture of Karl Pearson, *Dr. Paul R. Rider Anniversary Volume*, Wright-Patterson Air Force Base, Ohio.
Fettis, H.E. (1970). *Math. Comp.* 24, 667-670.
Fettis, H.E., and Caslin, J.C. (1966). An extended table of zeros of cross products of Bessel functions. Rept. ARL66-0023, Aerospace Research Labs., Wright-Patterson Air Force Base, Ohio.
Fettis, H.E., and Caslin, J.C. (1967). Tables of the modified Bessel functions $I_0(x)$, $I_1(x)$, $e^{-x}I_0(x)$ and $e^{-x}I_1(x)$. Aerospace Research Labs., Wright-Patterson Air Force Base, Ohio. See also *Math. Comp.* 21(1967), 736-737.
Fettis, H.E., and Caslin, J.C. (1968). More zeros of Bessel function cross products. Rept. ARL68-0209, Aerospace Research Labs., Wright-Patterson Air Force Base, Ohio.
Fettis, H.E., and Caslin, J.C. (1969a). Tables of toroidal harmonics, Part I(Orders 0-5), Part II(Orders 5-10). Repts. ARL69-0025, 69-0209, Aerospace Research Labs., Wright-Patterson Air Force Base, Ohio.

Fettis, H.E., and Caslin, J.C. (1969b). Table of modified Bessel functions. Rept. ARL69-0032, Aerospace Research Labs., Wright-Patterson Air Force Base, Ohio.

Fettis, H.E., and Caslin, J.C. (1970). Tables of toroidal harmonics, Part III-functions of the first kind (Orders 0-10). Rept. ARL70-0127, Aerospace Research Labs., Wright-Patterson Air Force Base, Ohio.

Fettis, H.E., Caslin, J.C., and Cramer, K.R. (1972). An improved tabulation of the plasma dispersion function and its first derivative, Part I-argument with positive imaginary part, Part II- argument with negative imaginary part (Rept. ARL. 72-0056); zeros and saddle points (Rept. ARL. 72-0057). Aerospace Research Labs., Wright-Patterson Air Force Base, Ohio.

Fettis, H.E., Caslin, J.C., and Cramer, K.R. (1973a). *Math. Comp.* 27, 401-408.

Fettis, H.E., Caslin, J.C., and Cramer, K.R. (1973b). *Math. Comp.* 27, 409-412.

Fields, J.L. (1966). *Proc. Edinburgh Math. Soc.* (2)15, 43-45.

Fields, J.L. (1967). *Math. Comp.* 21, 189-197.

Fields, J.L. (1968). Uniform expansions for generalized Jacobi functions, in Haimo, D.T. (ed.), "Orthogonal Expansions and Their Continuous Analogs." Southern Illinois Univ. Press, Carbondale, pp. 37-42.

Fields, J.L. (1972a). *Math. Comp.* 26, 757-766.

Fields, J.L. (1972b). *J. Approx. Theory* 6, 161-175.

Fields, J.L. (1973). *SIAM J. Math. Anal.* 4, 482-507.

Fields, J.L., Luke, Y.L., and Wimp, J. (1968). *J. Approx. Theory* 1, 137-166.

Fields, J.L., and Wimp, J. (1963). *Proc. Cambridge Philos. Soc.* 59, 599-605.

Fields, J.L., and Wimp, J. (1970). Rational approximations to Tricomi's ψ-function, in Gilbert, R.P., and Newton, R.G. (eds.), "Analytic Methods in Mathematical Physics." Gordon and Breach, New York, pp. 427-434.

Fike, C.T. (1968). "Computer Evaluation of Mathematical Functions." Prentice-Hall, Englewood Cliffs, New Jersey.

Finn, G.D., and Mugglestone, D. (1965). *Monthly Not. Roy. Astr. Soc.* 129, 221-235.

Fleckner, O.L. (1968). *Math. Comp.* 22, 635-640.

Fletcher, A., Miller, J.C.P., Rosenhead, L., and Comrie, L.J. (1962). "An Index of Mathematical Tables," (2nd. ed.). Addison-Wesley, Reading, Massachusetts.

Flett, T. (1972/73). *Proc. Edinburgh Math. Soc.* (2)18, 31-34.

Ford, W.B. (1960). "Studies on Divergent Series and Summability and Asymptotic Developments of Functions Defined by Maclaurin Series." Chelsea, New York.

Fox, L., and Parker, I.B. (1968). "Chebyshev Polynomials in Numerical Analysis." Oxford Univ. Press, London.

Franz, W. (1960). *Z. Angew. Math. Mech.* 40, 385-396.

Freud, G. (1969). "Orthogonale Polynome." Birkhäuser, Basel. Also in English, "Orthogoanl Polynomials." Pergamon Press, New York, (1971).

Frevel, L.K. (1973). Evaluation of the generalized error function. Chemistry Dept., The Johns Hopkins Univ., Baltimore, Maryland. See Also *Math. Comp.* 27(1973), 440-441.

Fried, B.D., and Conte, S.D. (1961). "The Plasma Dispersion Function. The Hilbert Transform of the Gaussian." Academic Press, New York.

Frisch-Fay, R. (1965). Tables of integrals of fractional order Bessel functions. UNICIV Rept. R-9, Univ. of New South Wales, Kensington, N.S.W., Australia. See also *Math. Comp.* 20(1966), 338.

Fröman, N., and Fröman, P.O. (1965). "JWKB Approximation." North-Holland Publishing Co., Amsterdam.

Galant, D.C., and Byrd, P.F. (1968). *Math. Comp.* 22, 885-886.

Gargantini, I. (1966). *Comm. ACM.* 9, 859-863.

Gargantini, I., and Henrici, P. (1967). *Math. Comp.* 21, 18-29.

Gargantini, I., and Pomentale, T. (1964). *Comm. ACM* 7, 727-730.

Gautschi, W. (1959a). *J. Math. and Phys.* 38, 77-81.

Gautschi, W. (1959b). *J. Res. Nat. Bur. Standards* 62,123-125.

Gautschi, W. (1967). *SIAM Rev.* 9, 24-82.

Gautschi, W. (1970). *SIAM J. Numer. Anal.* 7, 187-198.

Gautschi, W. (1972). *Computing (Arch. Elektron. Rechnen)* 9, 107-126. This is in German. Also available in English. Numerical aspects of recurrence relations. Rept. ARL73-0005, Aerospace Research Labs., Wright-Patterson Air Force Base, Ohio, (1973).

Gautschi, W. (1975). Computational methods in special functions-A survey. Advanced Seminar on Special Functions, Math. Research Center, Univ. of Wisconsin, Madison, March 31-April 2, 1975.

Geller, M., and Ng, E.W. (1969). *J. Res. Nat. Bur. Standards Section B* 73, 191-200.

Geller, M., and Ng, E.W. (1971). *J. Res. Nat. Bur. Standards Section B* 75, 149-164.

George, D.L. (1962). *Proc. Edinburgh Math. Soc.* (2), 13, 87-113.

Gerber, H. (1964). *Math. Comp.* 18, 319-322.

Geronimus, Ya L. (1958). "Polynomials Orthogonal on a Circumference and on an Interval. Estimates, Asymptotic Formulas, Orthogonal Series" (in Russian). Gosudarstv. Izdat. Fiz.-

Mat. Lit., Moscow. Also in English, (a) Pergamon Press. New York, (1960), (b) Consultants Bureau, New York, (1961).
Gloden, R.F. (1964). Approximation des fonctions de Bessel. Optimisation des programmes correspondants, Communauté Européene de l'Energie Atomique-EURATOM, Centre Commun de Recherche Nucléaire, Etablissement d'Ispra-Italie. Centre de Traitement de l'Information-CETIS, Brussels.
Golden, J.E., McGuire, J.H., and Nuttal, J. (1973). *J. Math. Anal. Appl.* 43, 754-767.
Gould, H.W. (1972). *Amer. Math. Monthly* 79, 44-50.
Gradstehyn, I.S., and Ryzhik, I.M. (1965). "Table of Integrals, Series and Products" 4th ed. (transl. from Russian). Academic Press, New York.
Gram, J.P. (1895). *Kongl. Danske Videnskab. Selskabs Forhandl.*, Oversigt pp. 303-308.
Gram, J.P. (1903). *Acta Math.* 27, 289-304.
Graves-Morris, P.R. (ed.) (1973). "Padé Approximants and Their Applications." Academic Press, New York.
Gray, H.L., Thompson, R.W., and McWilliams, G.V. (1969). *Math. Comp.* 23, 85-89.
Greenwood, J.A., and Hartley, H.O. (1962). "Guide to Tables in Mathematical Statistics." Princeton Univ. Press, Princeton, New Jersey.
Gröbner, W., and Hofreiter, N. (1961). "Integraltafel, Teil I: Unbestimmte Integrale. Teil II: Bestimmte Integrale." Springer, Berlin.
Grünbaum, F.A. (1973). *J. Math. Anal. Appl.* 41, 115-122.
Gupta, B.K. (1963). *Ann. Soc. Sci. Bruxelles* Ser. 1, 77, 30-59.
Gupta, S.L. (1966). *Univ. Roorkee Res. J.* 9, 17-19.
Gutschick, V.P., and Ludwig, O.G. (1969). *Math. Comp.* 23, 210 and 893.
Halbgewachs, R.D., and Shah, S.M. (1967). *Proc. Indian Acad. Sci.* Sect. A 65, 227-232.
Hall, L.S. (1967). *Proc. Cambridge Philos. Soc.* 63, 141-146.
Hangelbroek, R.J. (1967). *J. Engr. Math.* 1, 37-50.
Hanscomb, D.C. (ed.) (1966). "Methods of Numerical Approximation." Pergamon Press, New York.
Hansen, E. (1975). "A Table of Series and Products." Prentice-Hall, Englewood Cliffs, New Jersey.
Hart, R.G. (1966). *Math. Comp.* 20, 600-602.
Hart, J.F., Cheney, E.W., Lawson, C.L., Maehly, H.J., Mesztenyi, C.K., Rice, J.R., Thacher, H.C. Jr., and Witzgall, C. (1969). "Computer Approximations." John Wiley, New York. For errata, see *Math. Comp.* 23(1969), 470-471; 28(1974), 885.
Harter, H.L. (1964). "New Tables of the Incomplete Gamma-Function, Ratio and of Percentage Points of the Chi-Square

and Beta Distributions." U.S. Govt. Printing Office, Washington, D.C.
Harvey, A. (1965). *Math. Comp.* 19, 155-156.
Hayakawa, M. (1962). *Math. Japan* 7, 137-139.
Heading, J. (1962). "An Introduction to Phase-Integral Methods." John Wiley, New York.
Heatley, A.H. (1964). Tables of the confluent hypergeometric function and the Toronto function. Univ. of Waterloo, Ontario. See *Math. Comp.* 18(1964), 687-688 and 19(1965), 343.
Heatley, A.H. (1965). *Math. Comp.* 19(1965), 118-123.
Hebermehl, G., Minkwitz, G., and Schulz, G. (1965). "Tabellen der Lommelschen sowie abgeleiteter Funktionen." Akademie-Verlag, Berlin.
Hethcote, H.W. (1970a). *Proc. Amer. Math. Soc.* 25, 72-74.
Hethcote, H.W. (1070b). *J. Mathematical Phys.* 11, 2501-2504.
Hetnarski, R.B. (1964). *Zastos. Mat.* 7, 399-405.
Hobson, E.W. (1955). "The Theory of Spherical and Ellipsoidal Harmonics." Chelsea, New York.
Hochstadt, H. (1961). "Special Functions of Mathematical Physics." Holt, Rinehart and Winston, New York.
Hochstadt, H. (1971). "The Functions of Mathematical Physics." John Wiley, New York.
Hornecker, G. (1958). *Chiffres* 1, 157-169.
Hummer, D.G. (1964). *Math. Comp.* 18, 317-319.
Hummer, D.G. (1965). *Mem. Roy. Astr. Soc.* 70, 1-32.
Hunter, D.B. (1964). *Math. Comp.* 18, 123-128.
Hunter, D.B. (1968). *Math. Comp.* 22, 440-444.
Hunter, D.B. (1970). *Comput. J.* 13, 378-381.
Hunter, D.B., and Regan, T. (1972). *Math. Comp.* 26, 539-541.
Ihm, P. (1961). *Sankhya* Ser. A 23, 197-204.
Inui, T. (1967). "Special Functions" (in Japanese). 2nd ed. Iwanami Collection No. 252, Iwanami Shoten, Tokyo.
Ishidzu, T. (1960). "Tables of the Racah Coefficients." Pan-Pacific Press, Tokyo.
Jackson, A.D., and Maximon, L.C. (1972). *SIAM J. Math. Anal.* 3, 446-460.
Jakovleva, G.D. (1969). "Tables of Airy Functions and Their Derivatives" (in Russian). Izdat. Nauka, Moscow.
Jankovic, Z. (1961). *Rad. Jugoslav Akad. Znan. Umjet. Odjel Mat. Fiz. Tehn. Nauke* 319, 59-119.
Jeffreys, H. (1962). "Asymptotic Approximations." Oxford Univ. Press, London and New York.
Johnston, J.R. (1964). Tables of values and zeros of the confluent hypergeometric function. Rept. 31901, Aircraft Division, Douglas Aircraft Co., Long Beach, California. See *Math. Comp.* 19(1965), 343.

Jones, A.L. (1968). *J. Math. and Phys.* 47, 220-221.
Joshi, C.M., and McDonald, J.B. (1972). *J. Math. Anal. Appl.* 40, 278-285.
Kacmarz, S. and Steinhaus, H. (1951). "Theorie der Orthogonalreihen." Chelsea, New York.
Karpov, K.A. (1954). "Tables of the Function $w(z) = \exp(-z^2)\int_0^z \exp(x^2)dx$ in the Complex Domain" (in Russian). Izdat. Akad. Nauk SSSR, Moscow. Also in English, Pergamon Press, Macmillan, New York, (1965).
Karpov, K.A. (1958). "Tables of the Function $F(z)=\int_0^z e^{x^2}dx$" (in Russian). Izdat. Akad. Nauk SSSR, Moscow. Also in English, Pergamon Press, Macmillan, New York, (1964).
Karpov, K.A., and Cistova, E.A. (1964). "Tables of the Weber Functions," Vol. II, (in Russian). Vycisl. Centr Akad. Nauk SSSR, Moscow.
Karpov, K.A., and Cistova, E.A. (1968). "Tables of the Weber Functions," Vol. III, (in Russian). Vycisl. Centr Akad. Nauk SSSR, Moscow.
Katsura, S., Inoue, Y., Yamashita, S., and Kilpatrick, J.E. (1965). *Tech. Rep. Tohoku Univ.* 30, No. 2, 93-163, 625-626.
Katsura, S., and Nishihara, K. (1969). Tables of integral products of Bessel functions, II. Department of Applied Physics, Tohoku Univ., Sendai, Japan. See also *Math. Comp.* 23(1969), 457.
Keckic, J.D., and Vasic, P.M. (1971). *Publ. Inst. Math. (Beograd)* N.S. 11(25), 107-114.
Keckic, J.D., and Stankovic, M.S. (1972). *Publ. Inst. Math. (Beograd)* N.S. 13(27), 51-54.
Kelisky, R.P., and Rivlin, T.J. (1968). *Math. Comp.* 22, 128-136.
Khamis, S.H. (1965). "Tables of the Incomplete Gamma Function Ratio." Justus von Liebig Verlag, Darmstadt, Germany.
Khovanskii, A.N. (1963). "The Application of Continued Fractions and Their Generalizations to Problems in Approximation Theory" (transl. from Russian). Noordhoff, Groningen.
Kilpatrick, J.E., Katsura, S., and Inoue, Y. (1966). Tables of integrals of products of Bessel functions. Rice Univ., Houston, Texas and Tohoku Univ., Sendai, Japan. See also *Math. Comp.* 21(1967), 267, 407-412.
Kim, S.K. (1972). *Math. Comp.* 26, 963.
Kireeva, I.E., and Karpov, K.A. (1959). "Tables of Weber Functions," Vol. 1 (in Russian). Vycisl. Centr Akad. Nauk SSSR, Moscow. Also in English, Pergamon Press, New York, (1961).
Kiyono, T., and Murashima, S. (1973). *Mem. Fac. Engr. Kyoto*

Univ. 35, 102-107.
Kizner, W. (1966). *Comput. J.* 8, 372-382.
Knibb, D., and Scraton, R.E. (1971). *Comput. J.* 14, 428-432.
Knottnerus, U.J. (1960). "Approximation Formulae for Generalized Hypergeometric Functions for Large Values of the Parameters with Applications to Expansion Theorems for the Function $G_{p,q}^{m,n}(z)$." J.B. Wolters, Groningen.
Knuth, D.E., and Buckholtz, F.J. (1967). *Math. Comp.* 21, 663-688.
Kölbig, K.S. (1964). A definite integral with modified Bessel functions. CERN, European Organization for Nuclear Research, Geneva.
Kölbig, K.S. (1965). A short table of a definite integral involving a product of Bessel functions. CERN, European Organization for Nuclear Research, Geneva.
Kölbig, K.S. (1970). *Math. Comp.* 24, 679-696.
Kölbig, K.S. (1971). *Nordisk Tidskr. Informations-Behandling* (BIT) 11, 21-28.
Kölbig, K.S. (1972). *Math. Comp.* 26, 751-756.
Kölbig, K.S., Mignaco, J.A., and Remiddi, E. (1970). *Nordisk Tidskr. Informations-Behandling* (BIT) 10, 38-73.
Korobochkin, B.I., and Filippov, Y.A. (1965). "Tables of Modified Whittaker Functions" (in Russian). Vycisl. Centr Akad. Nauk SSSR, Moscow.
Kratzer, A., and Franz, W. (1960). "Tranzendente Funktionen." Akad. Verlagsges, Leipzig.
Kreyszig, E. (1957a). *Canad. J. Math.* 9, 118-131.
Kreyszig, E. (1957b). *Canad. J. Math.* 9, 500-510.
Krishnan, T. (1965). *Ann. Inst. Statist. Math.* 17, 211-223.
Krishnan, T. (1966). *Math. Comp.* 20, 337-338.
Krugliak, Y.A., and Whitman, D.R. (1965). "Tables of Quantum Chemistry Integrals" (in Russian). Vycisl. Centr Akad. Nauk SSSR, Moscow.
Krumhaar, H. (1965). *Z. Angew. Math. Mech.* 45, 245-255.
Krylov, V.I., and Skoblja, N.S. (1968). "Handbook on the Numerical Inversion of the Laplace Transform" (in Russian). Izdat. Nauka i Tehn., Minsk. Also in English, Israel Program for Scientific Translation, Jerusalem, (1969).
Kublanovskaja, B.N., and Smirnova, T.N. (1959). *Trudy Mat. Inst. Steklov* 53, 186-192. This is in Russian. Also in English. Zeros of Hankel functions and some related funcions. Rept. TN-60-1128, Air Force Cambridge Research Lab., Cambridge, Massachusetts.
Kumar, M., and Dhawan, G.K. (1970). Numerical values of certain integrals involving a product of two Bessel functions. Maulana Azad College of Technology, Bhopal, India. See also *Math. Comp.* 24(1970), 490.

Kumar, S.S. (1962). *Smithsonian Contributuons to Astrophysics* 5, 151-185.

Kuznecov, D.S. (1965). "Special Functions" (in Russian). 2nd revised and augmented edition. Izdat. Vyss. Skola, Moscow.

Lam, B., and Elliott, D. (1972). *SIAM J. Numer. Anal.* 9, 44-52.

Lanczos, C. (1938). *J. Math. and Phys.* 17, 123-199.

Lanczos, C. (1952). "Tables of Chebyshev Polynomials $S_n(x)$ and $C_n(x)$, Introduction." Applied Math. Series 9, U.S. Govt. Printing Office, Washington, D.C.

Lanczos, C. (1956). "Applied Analysis." Prentice-Hall, Englewood Cliffs, New Jersey.

Lanczos, C. (1964). *J. Soc. Indust. Appl. Math. Ser. B Numer. Anal.* 1, 86-96.

Laslett, L.J., and Lewish, W. (1962). *Math. Comp.* 16, 226-232.

Lauwerier, H.A. (1966). "Asymptotic Expansions." Mathematisch Centrum, Amsterdam.

Lavoie, J.L., and Mongeau, G. (1968). *Duke Math. J.* 35, 747-752.

Lebedev, A.V., and Fedorova, R.M. (1956). "Handbook of Mathematical Functions" (in Russian). Also in English, Pergamon Press, New York, (1960).

Lebedev, N.N. (1963). "Special Functions and Their Applications." 2nd, revised and augmented edition (in Russian). Gosudarstv. Izdat. Fiz.-Mat. Lit., Moscow. Also in English, Prentice-Hall, Englewood Cliffs, New Jersey, (1965).

Lee, B.S., and Shah, S.M. (1970). *J. Math. Anal. Appl.* 30, 144-156.

Lee, K., and Radosevich, L.G. (1960). *J. Math. and Phys.* 39, 293-299.

Lewin, L. (1958). "Dilogarithms and Associated Functions." Macdonald, London.

Ling, C.B., and Lin, J. (1971). *Math. Comp.* 25, 402.

Ling, C.B., and Lin, J. (1972). *Math. Comp.* 26, 529-538.

Lombet-Goffart, J. (1962). *Acad. Roy. Belg. Bull. Cl. Sci.* (5) 48, 1062-1080 and 1261-1280.

Lorch, L. (1967). *Arch. Math. (Brno)* 3, 1-9.

Lorch, L,, and Szego, P. (1966). *Proc. Amer. Math. Soc.* 17, 330-332.

Low, R.D. (1966). *Math. Comp.* 20, 421-424.

Luke, S.K., and Weissman, S. (1964). Bessel functions of imaginary order and imaginary argument. Rept. DA-ARO(D)-31-124-G466 No. 1, Univ. of Maryland Institute for Molecular physics, College Park. See also *Math. Comp.* 19(1965), 343-344.

Luke, Y.L. (1962). "Integrals of Bessel Functions." McGraw-Hill, New York. For errata, see *Math. Comp.* 22(1968), 907-908.
Luke, Y.L. (1969). "The Special Functions and Their Approximations," Vols. I,II. Academic Press, New York.
Luke, Y.L. (1970a). *Math. Comp.* 24, 191-198.
Luke, Y.L. (1970b). *SIAM J. Math. Anal.* 1, 266-281.
Luke, Y.L. (1971). *Applicable Anal.* 1, 65-73.
Luke, Y.L. (1971-1972). *Math. Comp.* 25, 323-330, 789-795, and *Math. Comp.* 26, 237-240.
Luke, Y.L. (1972a). *J. Approx. Theory* 5, 41-65.
Luke, Y.L. (1972b). *Math. Balkanica* 2, 117-123.
Luke, Y.L. (1972c). On generating Bessel functions by use of the backward recurrence formula. Rept. ARL 72-0030, Aerospace Research Labs., Wright-Patterson Air Force Base, Ohio.
Luke, Y.L. (1975). *J. Australian Math. Soc.* 19, 196-210.
Luke, Y.L., Fair, W., Coombs, G., and Moran, R. (1965). *Math. Comp.* 19, 501-502.
Luke, Y.L., Ting, B.Y., and Kemp, M.J. (1975). *Math. Comp.* 29, in press.
Luke, Y.L., and Wimp, J. (1963). *Math. Comp.* 17, 395-404.
Luke, Y.L., Wimp, J., and Fair, W. (1972). "Cumulative Index to Mathematics of Computation,—Vols. 1-23, 1943-1969." Amer. Math. Soc., Providence, Rhode Island.
Luk'ianov, A.V., Teplov, I.B., and Akimova, M.K. (1961). "Tables of Coulomb Wave Functions" (in Russian), Moscow. Also in English as "Tables of Coulomb Wave Functions (Whittaker Functions)." Pergamon Press, Macmillan, New York, (1965).
Lutz, H.F., and Karvelis, M.D. (1963). *Nuclear Phys.* 43, 31-44.
Lyusternik, L.A., Chervonenkis, O.A., and Yanpol'skii, A.R. (1963). "Handbook for Computing Elementary Functions" (in Russian). Gosudarstv. Izdat. Fiz.-Mat. Lit., Moscow. Also in English, Pergamon Press, New York, (1965).
Mackiernan, D.D. (1970). Table of values of integrals for the longitudinal and lateral von Karman turbulence spectra. NASA TMX-64529, G.C. Marshall Space Flight Center, Huntsville, Alabama.
MacRobert, T.M. (1938). *Proc. Roy. Soc. Edinburgh* 58, 1-13.
MacRobert, T.M. (1961). *Proc. Glasgow Math. Assoc.* 5, 30-34.
MacRobert, T.M. (1962). *Pacific J. Math.* 12, 999-1002.
Magnus, W., and Kotin, L. (1960). *Numer. Math.* 2, 228-244.
Magnus, W., and Oberhettinger, F. (1948). "Formeln und Sätze für die speziellen Funktionen der mathematischen Physik." 2nd, enlarged and improved edition. Springer, Berlin. Also an English transl. of 1943 edition, "Formulas and Theorems

for the Special Functions of Mathematical Physics, Chelsea, New York, (1949). For discussions of these volumes and errata, see *MTAC* 3(1948), 103-105; 3(1949), 368-369 and 522-523; *Math. Comp.* 21(1967), 523-524.
Magnus, W., Oberhettinger, F., and Soni, P.R. (1966). "Formulas and Theorems for the Special Functions of Mathematical Physics." Springer, New York.
Mangulis, V. (1965). "Handbook of Series for Scientists and Engineers." Academic Press, New York. For errata, see *Math. Comp.* 21(1967), 118-119 and 750-751.
Martinek, J. (1968). *Acta. Mech.* 6, 203-207.
Martinek, J., Thielman, H.P., and Huebschman, E.C. (1966). *J. Math. Mech.* 16, 447-452.
Martz, C.W. (1964). Tables of the complex Fresnel integral. Rept. NASA Sp-3010, Scientific and Technical Information Division, National Aeronautics and Space Administration, Washington, D.C. Copies obtainable from the National Technical Information Service, Operations Division, Springfield, Virginia. See the review of this item in *Math. Comp.* 27 (1973), 214,215.
Mason, J.C. (1967). *SIAM J. Appl. Math.* 15, 172-181.
Matta, F., and Reichel, A. (1971). *Math. Comp.* 25, 339-344.
McLain, J.W., Schoenig, F.C., and Palladino, N.J. (1962). Table of Bessel functions to argument 85. Engr. Res. Bull. B-85, Pennsylvania State Univ., University Park. See also *Math. Comp.* 18(1964), 161-162 and 175-176.
McQueary, C.E., and Mack, L.R. (1967). *Math. Comp.* 21, 413-417.
Mechel, F. (1966). *Math. Comp.* 20, 407-412.
Meijer, C.S. (1936). *Nieuw. Arch. Wisk.* 18, 10-39.
Meijer, C.S. (1941a). *Nederl. Akad. Wetensch. Proc. Ser. A* 44, 81-92, 186-194, 298-307, 442-451 and 590-598.
Meijer, C.S. (1941b). *Nederl. Akad. Wetensch. Proc. Ser. A* 44, 1062-1070.
Meijer, C.S. (1946). *Nederl. Akad. Wetensch. Proc. Ser. A* 49, 227-237, 344-356, 457-469, 632-641, 765-772, 936-943, 1063-1072 and 1165-1175.
Meijer, C.S. (1952-1956). *Nederl. Akad. Wetensch. Proc. Ser. A* 55, 369-379 and 483-487; 56, 43-49, 187-193 and 349-357; 57, 77-82, 83-91 and 273-279; 58, 243-251 and 309-314; 59, 70-82.
Miller, G.F. (1966a). *SIAM J. Numer. Anal.* 3, 390-409.
Miller, G.F. (1966b). *Proc. Cambridge Philos. Soc.* 62, 453-457.
Miller, G.F., and Haines, P.H. (1959). Roots and turning values of $Ai(-\alpha)$ and $Ai'(-\beta)$, in Logan, N.A., General research in diffraction theory, Vol. I. Rept. 288087, Lockheed Missiles and Space Division, Sunnyvale, California.

See also *Math. Comp.* 15(1961), 200.
Miller, J., Gerhouser, J.M., and Matsen, F.A. (1959). "Quantum Chemistry Integrals and Tables." Univ. of Texas Press, Austin.
Miller, J.C.P. (1952). "Bessel Functions, Part II, Functions of Positive Integer Order, Mathematical Tables, Vol. 10." Cambridge Univ. Press, Cambridge.
Miller, J.C.P. (1955). "Tables of Weber Parabolic Cylinder Functions." H.M. Stationery Office, London.
Miller, J.C.P., Fletcher, A., Rosenhead, L., and Comrie, L.J. (1962). "An Index of Mathematical Tables," 2nd. edition, Vols. I and II. Addison-Wesley, Reading, Massachusetts.
Miller, K.L., Molmud, P., and Meecham, W.C. (1963). Tables of the functions $G(x+iy) = \int_0^\infty \{t-(x+iy)\}^{-1} e^{-t^2} dt$ and $F(x+iy) = \int_0^\infty t^4 \{t-(x+iy)\}^{-1} e^{-t^2} dt$. Rept. 6121-6249-RU-000, Space Technology Laboratories, Redondo Beach, California. See also *Math. Comp.* 26(1972), 299.
Miller, W., Jr. (1972). *J. Mathematical Phys.* 13, 648-654.
Milne-Thomson, L.M. (1950). "Jacobian Elliptic Function Tables." Dover, New York.
Minton, B.M. (1970). *J. Mathematical Phys.* 11, 3061-3062.
Mitra, S.C. (1943). *J. Indian Math. Soc. N.S.* 7, 102-109.
Mitrinovic, D.S. (1970). "Analytic Inequalities" (in cooperation with P.M. Vasic). Springer, New York.
Mitrinovic, D.S. (1972). "Uvod u Specijalne Funkcije." Izdavacko Preduzece Gradevinska Knjiga, Beograd.
Mitrinovic, D.S., and Djokovic, D.Z. (1964). "Specijalne Funkcije." Grakjevinska Knjiga, Belgrade.
Mitrinovic, D.S., Dokovic, S., and Dragomir, Z. (1964). "Specijalne Funkcije." Izdavacko Preduzece Gradevinska Knjiga, Belgrade.
Mitrinovic, D.S., Tosic, D.D., and Janic, R.R. (1972). "Specijalne Funkcije-Zbornik Zadataka i Problema." Naucna Knjiga, Beograd.
Moody, W.T. (1967). *Math. Comp.* 21, 112.
Moran, P.A.P. (1957). *Quart. J. Math. Oxford* (2)8, 287-290.
Morgenthaler, G.W., and Reismann, H. (1963). *J. Res. Nat. Bur. Standards* 67B, 181-183.
Morris, A.H., Jr. (1973). Tables of coefficients of the Maclaurin expansions of $1/\Gamma(z+1)$ and $1/\Gamma(z+2)$. Naval Weapons Lab., Dahlgren, Virginia. See *Math. Comp.* 27(1973), 674.
Mosier, C., and Shillady, D.D. (1972). A fast accurate approximation for $F_0(z)$ occuring in Gaussian lobe basis electronic calculations. Chemistry Dept., Virginia Common-

wealth Univ., Richmond. See also *Math. Comp.* 26(1972), 1022.
Murnaghan, F.D. (1965). Evaluation of the probability integral to high precision. DTMB Rept. 1861, David Taylor Model Basin, Washington, D.C.
Murnaghan, F.D., and Wrench, J.W., Jr. (1963). The converging factor for the exponential integral. DTMB Rept. 1535, David Taylor Model Basin, Washington, D.C.
Murray, J.D. (1974). "Asymptotic Analysis." Oxford Univ. Press, London and New York.
Murty, T.S. (1971). Tables of the conical functions $K_p(x)$. Marine Sciences Branch, Dept. of Energy, Mines and Resources, Ottawa, Ontario, Canada. See also *Math. Comp.* 25(1971), 402.
Murty, T.S., and Taylor, J.D. (1969). Zeros and bend points of the Legendre functions of the first kind for fractional orders. Oceanographic Research, Marine Sciences Branch, Dept. of Energy, Mines and Resources, Ottawa, Ontario, Canada. See also *Math. Comp.* 23(1969), 887-888.
Murzewshi, J., and Sowa, A. (1972). *Zastos. Mat.* 13, 261-273.
Narasimha, R. (1966). On the incomplete gamma function with one negative argument. Rept. AE 123A, Dept. of Aeronautical Engr., Indian Inst. of Sci., Bangalore. See also *Math. Comp.* 22(1966), 624.
Nasell, I. (1974). *Math. Comp.* 28, 253-256.
Nayfeh, A.H. (1973). "Perturbation Methods." John Wiley, New York.
Németh, G. (1963). *Magyar Tud. Akad. Mat. Kutató Int. Közl.* 8, 641-643.
Németh, G. (1964a). Polynomial approximations for the evaluation of $\int_x^\infty u^{-1}e^{-u}du$. Rept. Central Inst. for Physics, Budapest.
Németh, G. (1964b). Polynomial approximations for the evaluation of $\int_0^x e^{-u^2}du$. Rept. Central Inst. for Physics, Budapest.
Németh, G. (1965a). *Numer. Math.* 7, 310-312.
Németh, G. (1965b). Polynomial approximations to the function $\psi(a, c; x)$, Chebyshev expansions of integral sine and cosine functions. Rept. Central Inst. for Physics, Budapest.
Németh, G. (1965c). Chebyshev expansions of Gauss' hypergeometric function. Rept. Central Inst. for Physics, Budapest.
Németh, G. (1966). Chebyshev expansion of the function $e^{-x^2/2}\int_0^x e^{u^2/2}du$. Rept. Central Inst. for Physics, Budapest.
Németh, G. (1967). *Matem. Lapok* 18, 329-333.
Németh, G. (1971). *Magyar Tud. Akad. Mat. Fiz. Oszt. Közl.* 20, 13-33.

Németh, G. (1972). Tables of the expansions of the first ten zeros of Bessel's functions. Communication of the Unified Institute for Nuclear Studies, Rept. 5-6336. Dubna, Russia.

Neville, E.H. (1951). "Jacobian Elliptic Functions." Oxford Univ. Press, London and New York.

Newberry, A.C.R. (1973). *Math. Comp.* 27, 639-644.

Newman, J.N., and Frank, W. (1963). *Math. Comp.* 17, 64-70.

Ng, E.W. (1966). *Math. Comp.* 20, 624-625.

Ng, E.W., and Geller, M. (1969). *J. Res. Nat. Bur. Standards -B. Math. Sci.* 73B, 1-20. For additions and corrections, see the same journal, 75B(1971), 149-164.

Ng, E.W., and Geller, M. (1970). *J. Res. Nat. Bur. Standards -B. Math. Sci.* 74B, 85-98.

Nielsen, N. (1965). "Handbuch der Theorie der Gammafunktion," Band I; "Theorie des Integrallogarithmus und Verwandter Tranzendenten," Band II. Chelsea, New York.

Nikiforov, A.F., Uvarov, V.B., and Levitan, Yu. L. (1965). "Tables of Racah Coefficients" (transl. from Russian). Pergamon Press, New York.

Nörlund, N.E. (1954). "Vorlesüngen über Differenzenrechnung." Chelsea, New York.

Nörlund, N.E. (1955). *Acta Math.* 94, 289-349.

Nörlund, N.E. (1961). *Rend. Circ. Mat. Palermo* 10(2), 27-44.

Nörlund, N.E. (1963). *Mat.-Fys. Skr. Danske Vid. Selsk.* 2, No. 5.

Norton, H.J. (1964). *Comput. J.* 7, 76-85.

Nosova, L.N., and Tumarkin, S.A. (1961). "Tables of Generalized Airy Functions for the Asymptotic Solution of the Differential Equation $\varepsilon(py')' + (q+\varepsilon r)y = f$" (in Russian). Vycisl. Centr Akad. Nauk SSSR, Moscow. Also in English, Pergamon Press, New York, (1961).

Oberhettinger, F. (1957). "Tabellen zur Fourier Transformation." Springer, Berlin.

Oberhettinger, F. (1972). "Tables of Bessel Transforms." Springer, New York.

Oberhettinger, F. (1973a). "Fourier Expansions." Academic Press, New York.

Oberhettinger, F. (1973b). "Fourier Transforms of Distributions and Their Inverses." Academic Press, New York.

Oberhettinger, F. (1975). "Tables of Mellin Transforms." Springer, New York.

Oberhettinger, F., and Badii, L. (1973). "Tables of Laplace Transforms." Springer, New York.

Oberhettinger, F., and Higgins, T.P. (1961). Tables of Lebedev, Mehler and generalized Mehler transforms. Math. Note No. 246, Boeing Scientific Research Laboratories, Seattle,

Washington. See also *Math. Comp.* 17(1963), 95.
Oberhettinger, F., and Magnus, W. (1949). "Anwendung der Elliptischen Funktionen in Physik und Technik." Springer, Berlin.
Okui, S. (1974). *J. Res. Nat. Bur. Standards-B. Math. Sci.* 78B, 113-135.
Oldham, K.B. (1968). *Math. Comp.* 22, 454.
Oliver, J. (1969). *Comput. J.* 12, 57-62.
Olver, F.W.J. (1954). *Philos. Trans. Roy. Soc. London* 247, 328-368.
Olver, F.W.J. (ed.) (1960). "Royal Society Mathematical Tables," Vol. VII, Bessel Functions, Pt. 3, Zeros and Associated Values. Cambridge Univ. Press, London and New York.
Olver, F.W.J. (1962). "Tables for Bessel Functions of Moderate or Large Orders," National Physical Laboratory, Mathematical Tables, Vol. VI. H.M. Stationery Office, London.
Olver, F.W.J. (1974). "Asymptotics and Special Functions." Academic Press, New York.
Osborn, D., and Madey, R. (1968). *Math. Comp.* 22, 159-162.
Osipova, L.N., and Tumarkin, S.A. (1963). "Tables for the Computation of Toroidal Shells" (in Russian). Vycisl. Centr Akad. Nauk SSSR, Moscow. Also in English, Noordhoff, Groningen, (1965).
Osler, T.J. (1973). *Math. Comp.* 29(1975), July.
Pagurova, V.I. (1959). "Tables of the Exponential Integral function $E_v(x) = \int_1^\infty e^{-xu}u^{-v}du$" (in Russian). Vycisl. Centr Akad. Nauk SSSR, Moscow. Also in English, Pergamon Press, New York, (1961).
Pagurova, V.I. (1963). "Tables of the Incomplete Gamma Function $I(x,m) = \{\Gamma(m)\}^{-1}\int_0^x e^{-t}t^{m-1}dt$" (in Russian). Vycisl. Centr Akad. Nauk SSSR, Moscow.
Parnes, R. (1972). *Math. Comp.* 26, 949-953.
Pearson, K. (1968). "Tables of the Incomplete Beta-Function," 2nd edition (with a new introduction by E.S. Pearson and N.L. Johnson). Cambridge Univ. Press, London and New York.
Peavy, B.A. (1967). *J. Res. Nat. Bur. Standards-B. Math. and Math. Phys.* 71B, 131-141.
Perlin, I.E., and Garret, J.R. (1960). *Math. Comp.* 14, 270-274.
Perron, O. (1950). "Die Lehre von den Kettenbrüchen." Chelsea, New York.
Perron, O. (1954,1957). "Die Lehre von den Kettenbrüchen," Vol. I (1954), Vol. II (1957). Teubner, Leipzig.
Petiau, G. (1955). "La Theorie des Fonctions de Bessel." Centre National de la Recherche Scientifique, Paris.

Petrovskaja, M.S. (1970). *Bjull. Inst. Teoret. Astronom.* 12, No. 5(138), 401-421.
Pierre, D.A. (1964). *J. Soc. Indust. Appl. Math.* 12, 93-104.
Piessens, R. (1971). *J. Engr. Math.* 5, 1-9.
Piessens, R., and Branders, M. (1972). *Math. Comp.* 26, 1022.
Pimbley, W.T., and Nelson, C.W. (1964). Table of values of $2\sqrt{x}F(\sqrt{x})$. IBM Engineering Publications Dept. No. PTP773, Endicott, New York. See also *Math. Comp.* 18(1964), 678.
Powell, M.J.D. (1967). *Comput. J.* 9, 404-407.
Prohorov, A.V. (1968). *Teor. Verojatnost. i Primenen* 13, 525-531. This is in Russian. For an English transl. see *Theor. Probability Appl.* 13(1968), 496-501.
Rainville, E.D. (1960). "Special Functions." Macmillan, New York.
Rau, H. (1950). *Arch. Math.* 2, 251-257.
Ray, W.D., and Pitman, A.E.N.T. (1963). *Ann. Math. Statist.* 34, 892-902.
Reichel, A. (1967). *Math. Comp.* 21, 647-651.
Reichel, A. (1969). *Math. Comp.* 23, 645-649.
Reudink, D.O. (1968). *J. Res. Nat. Bur. Standards-B. Math. Sci.* 72B, 279-280.
Rice, S.O. (1944). *Philos. Mag.* (7)35, 686-693.
Richards, J.M., and Mullineux, N. (1963). *ICC Bull.* 2, 143-157.
Roberts, G.E., and Kaufman, H. (1966). "Table of Laplace Transforms." Saunders, Philadelphia.
Roberts, J.A. (1965). *Math. Comp.* 19, 651-654.
Roberts, P.J. (1970). *J. London Math. Soc.* (2)2, 736-740.
Robin, L. (1957,1958,1959). "Fonctions Sphérique de Legendre et Fonctions Sphéroidales," Vols. I(1957), II(1958), III (1959). Gauthier-Villars, Paris.
Robinson, H.P. (1973). Tables of the derivative of the Psi function to 58 decimals. Univ. of California, Lawrence Berkeley Lab., Berkeley. See also *Math. Comp.* 28(1974), 872.
Robinson, H.P., and Potter, E. (1971). Mathematical constants. Rept. UCRL-20418, Lawrence Radiation Lab., Univ. of California, Berkeley. Available from National Technical Information Service, Operations Division, Springfield, Virginia. See also *Math. Comp.* 26(1972), 300-301 and 305-307.
Roman, I. (1969). *Math. Comp.* 23, 887.
Rotenberg, M., Bivins, R., Metropolis, N., and Wooten, J.K., Jr. (1960). "The 3-j and 6-j Symbols." Technology Press, Cambridge, Massachusetts.
Runckel, H.J. (1971). *Math. Ann.* 191, 53-58.
Russel, W., and Lal, M. (1969). *Math. Comp.* 23, 211-212.
Russon, A.E., and Blair, J.M. (1969). Rational function minimax approximation for the Bessel functions $K_0(x)$ and $K_1(x)$. Rept. AECL-3461, Atomic Energy of Canada, Chalk River, On-

tario. See also *Math. Comp.* 24(1970), 992.

Ryshik, I.M., and Gradstein, I.S. Summen-Produkt- und Integraltafeln. "Tables of Series, Products and Integrals" (transl. from Russian into German; transl. from German into English (1957). Deutscher Verlag, Berlin. Also, "Tables of Integrals, Series, and Products," 4th ed. (prepared by Yu.V. Geronimus and M.Yu. Tseytlin) (transl. from Russian into English) (1965). Academic Press, New York.

Sag, T.W. (1970). *Math. Comp.* 24, 341-356.

Sakai, M., and Ishikawa, S. (1971). *Mem. Fac. Sci. Kyushu Univ. Ser. A* 25, 244-252.

Salzer, H.E. (1961). *J. Math. and Phys.* 40, 72-86.

Sansone, G. (1959). "Orthogonal Functions." John Wiley, New York.

Schäfke, F.W. (1963). "Einführung in die Theorie der speziellen Funktionen der mathematischen Physik." Springer, Berlin.

Schucany, W.R., and Gray, H.L. (1968). *Math. Comp.* 22, 201-202.

Schütte, K. (1966). "Index of Mathematical Tables from All Branches of Sciences/Index mathematischer Tafelwerke und Tabellen aus allen Gebieten der Naturwissen-schaften." 2nd ed. R. Oldenbourg, Munich and Vienna.

Scraton, R.E. (1965). *Comput. J.* 8, 57-61.

Scraton, R.E. (1969). *Math. Comp.* 23, 837-844.

Scraton, R.E. (1972). *Proc. Cambridge Philos. Soc.* 73, 157-166.

Shafer, R.E. (1973). *Publ. Elektro. Fak. Univ. of Beograd* No. 437, 125-126.

Shanks, D., and Wrench, J.W., Jr. (1962). *Math. Comp.* 16, 76-99.

Shanks, D., and Wrench, J.W., Jr. (1969). *Math. Comp.* 23, 679-680.

Shenton, L.R., and Bowman, K.O. (1971). *SIAM J. Appl. Math.* 20, 547-554.

Sherry, M.E. (1959). The zeros and maxima of the Airy function and its first derivative to 25 significant figures. Electronics Research Directorate, Air Force Cambridge Research Center, Bedford, Massachusetts. See also *Math. Comp.* 16(1962), 389.

Shimasaki, M., and Kiyono, T. (1973). *Numer. Math.* 21, 373-380.

Shimpuku, T. (1960). *Progr. Theoret. Phys.*, Kyoto, Japan, Supplement No. 13, pp. 1-135.

Sikorsky, Yu. I. (1970). *Kiev Univ. Visnyh. Ser. Mat. Mekh.* 12, 121-124.

Sikorsskii, J.G. (1970). *Visnik Kiev Univ.* No. 12, 121-124.

Simauti, T. (1964). *Comment. Math. Univ. St. Paul* 12, 23-25; 18(1970), 153.

Singh, K., Lumley, J.F., and Betchov, R. (1963). Mofidied Hankel functions and their integrals to argument 10. Engr. Res. Bull. B-87, Pennsylvania State Univ., University Park. See also *Math. Comp.* 18(1964), 522.

Sirovich, L. (1971). "Techniques of Asymptotic Analysis." Springer, New York.

Skoblja, N. (1964a). "Tables for the Numerical Inversion of Laplace Transforms" (in Russian). Izdat. Akad. Nauk SSSR, Moscow.

Skoblja, N. (1964b). "Tables for the Numerical Inversion of the Laplace transform" (in Russian). *Izdat. Nauka i Tehnika, Minsk.* Also available as Rept. AF CRL-TN 60-1128, Electron. Res. Directorate, Air Force Cambridge Res. Lab., Cambridge, Massachusetts.

Skoblja, N.S. (1965). *Dokl. Akad. Nauk BSSR* 9, 288-291.

Slater, L.J. (1960). "Confluent Hypergeometric Functions." Cambridge Univ. Press, London and New York. For errata see *Math. Comp.* 17(1963), 486-487.

Slater, L.J. (1966). "Generalized Hypergeometric Functions." Cambridge Univ. Press, London and New York.

Slavic, D.V. (1971). *Publ. Elektro. Fak. Ser. Mat. Fiz. Univ. of Beograd* No. 357-380, 69-74.

Smirnov, N.V. (ed.) (1960). "Tables of the Normal Probability Integral, the Normal Density, and its Normalized Derivatives" (in Russian). Izdat. Akad. Nauk SSSR, Moscow. Also in English, Macmillan, New York, (1965).

Smirnov, N.V. (ed.) (1961). "Tables for the Distribution and Densith Functions of t-Distribution." (transl. from Russian). Pergamon Press, New York.

Smirnov, N.V., and Bol'shev, L.N. (1962). "Tables for Evaluating a Function of a Two Dimensional Normal Distribution" (in Russian). Izdat. Akad. Nauk SSSR, Moscow.

Smith, F.J. (1965). *Math. Comp.* 19, 33-36.

Smythe, W.R. (1960). *J. Appl. Phys.* 31, 553-556.

Smythe, W.R. (1961). *Phys. Fluids* 4, 756.

Smythe, W.R. (1963). *J. Mathematical Phys.* 4, 833-837.

Smythe, W.R. (1964). *Phys. Fluids* 7, 633-638.

Sneddon, I.N. (1961). "Special Functions of Mathematical Physics and Chemistry." Oliver and Boyd, London. Also in German, "Spezielle Funktionen der mathematischen Physik und Chemie." Bibliographisches Institut, Mannheim, (1963).

Sneddon, I.N. (1972). "The Use of Integral Transforms." McGraw-Hill, New York.

Snow, C. (1952). "Hypergeometric and Legendre Functions with Applications to Integral Equations of Potential Theory." Appl. Math. Ser. 19, U.S. Govt. Printing Office, Washington, D.C.

Snyder, M.A. (1966). "Chebyshev Methods in Numerical Approximation." Prentice-Hall, Englewood Cliffs, New Jersey.

Soni, R.P. (1965). *J. Math. and Phys.* 44, 406-407.

Spain, B., and Smith, M.G. (1970). "Functions of Mathematical Physics." Van Nostrand - Reinhold, New York.

Spellucci, P. (1971). *Numer. Math.* 18, 127-143.

Spicer, H.C. (1963). "Tables of the Inverse Probability Integral $P = (2/\pi^{\frac{1}{2}})\int_0^{\beta} e^{-u^2} du$." U.S. Geological Survey, Washington, D.C. See also *Math. Comp.* 17(1963), 320-321.

Spiegel, M.R. (1962). *Amer. Math. Monthly* 69, 894-896.

Spielberg, K. (1961). *Math. Comp.* 15, 409-417.

Spielberg, K. (1962). *Math. Comp.* 16, 205-217.

Spira, R. (1971). *Math. Comp.* 25, 317-322.

Stegun, I., and Zucker, R. (1970). *J. Res. Nat. Bur. Standards-B Math. Sci.* 74B, 211-224.

Steidley, K.D. (1963). Table of ${}_2F_1(a, b; c; z)$ for $a = 0(\frac{1}{2})7$, $b = 0(\frac{1}{2})7$ and $c = \frac{1}{2}(\frac{1}{2})2\frac{1}{2}$ with comments on closed forms of ${}_2F_1(a, b; c; z)$." National Aeronautics and Space Administration TN D-1735, Washington, D.C.

Steinig, J. (1970). *SIAM J. Math. Anal.* 1, 365-375.

Steinig, J. (1971). *Math. Zeit.* 122, 363-365.

Steinig, J. (1972). *Trans. Amer. Math. Soc.* 163, 123-129.

Strand, O.N. (1965). *Math. Comp.* 19, 127-129.

Strecok, A.J. (1968). *Math. Comp.* 22, 144-158.

Strecok, A.J., and Gregory, J.A. (1972). *Math. Comp.* 26, 955-961.

Streifer, W. (1965). *J. Math. and Phys.* 44, 403-405.

Strömgren, L. (1962). *Kungl. Tekn. Högsk. Handl. Stockholm* No. 193.

Sweeney, D.W. (1963). *Math. Comp.* 17, 170-178.

Switzer, K.A. (1965). Table of roots of certain transcendental equations arising in eigenfunction expansions. Circular 23, College of Engr., Washington State Univ. Pullman. See also *Math. Comp.* 20(1966), 335-336.

Syrett, H.E., and Wilson, M.W. (1966). Computation of Fresnel integrals to 28 figures: approximations to 8 and 20 figures. Computer Sci. Dept., Univ. of Western Ontario, London, Ontario, Canada. See also *Math. Comp.* 20(1966), 181.

Szegö, G. (1967). "Orthogonal Polynomials." (Colloq. Publ. Vol. 23, 3rd edition). Amer. Math. Soc., Providence, Rhode Island.

Taggart, D.A., and Schott, F.W. (1970). Mathematical tables of integrals involving spherical Bessel functions. Elect. Sci. and Engr. Dept., School of Engr. and Appl. Sci., Univ. of California, Los Angeles. See also *Math. Comp.* 24, 1970, 993.

Thompson, G.T. (1965). *Math. Comp.* 19, 661-663.

Tibery, C.L., and Wrench, J.W., Jr. (1964). Tables of the Goldstein factor. Rept. 1534, Appl. Math. Lab., David Taylor Model Basin, Washington, D.C.

Tooper, R.F., and Mark, J. (1968). *Math. Comp.* 22, 448-449.

Tranter, C.J. (1969). "Bessel Functions with Some Physical Applications." Hart Publishing Co., New York.

Tricomi, F.G. (1948). "Elliptische Funktionen." Akad. Verlagsges., Leipzig.

Tricomi, F.G. (1952). "Lezioni sulle Funzioni Ipergeometriche Confluenti." Gheroni, Torino.

Tricomi, F.G. (1954). "Funzioni Ipergeometriche Confluenti." Ed. Cremonese, Rome.

Tricomi, F.G. (1955). "Vorlesungen über Orthogonalreihen." Springer, Berlin.

Tricomi, F.G. (1960). "Fonctions Hypergeometriques Confluentes." Mémorial des Sciences Mathématiques, Fasc. CXL. Gauthier-Villars, Paris.

Troesch, B.A., and Weidman, P.D. (1972). *SIAM J. Appl. Math.* 23, 477-489.

Tsai, L. (1970). *Tamkang J. Math.* 1, 29-40.

Tuan, P.D., and Elliott, D. (1972). *Math. Comp.* 26, 213-232.

Verbeeck, P. (1970). *Acad. Roy. Belg. Bull. Cl. Sci.* (5)56, 1064-1072.

Verma, A. (1967). *Math. Comp.* 21, 232-236.

Vionnet, M. (1959). *Chiffres* 2, 77-96.

Vionnet, M. (1960). *Chiffres* 3, 65-78.

Vodicka, V. (1959). *Z. Angew. Math. Phys.* 10, 603-608.

Wall, H.S. (1948). "The Analytic Theory of Continued Fractions." Van Nostrand, New York.

Walters, L.C., and Wait, J.R. (1964). Computation of a modified Fresnel integral arising in the theory of diffraction by a variable screen. Nat. Bur. Standards Tech. Rept. Note 224, U.S. Govt. Printing Office, Washington, D.C.

Wang, C., and Kuo, T. (1965). "General Treatise on Special Functions" (in Chinese). Science Press, Peking.

Wasow, W.R. (1965). "Asymptotic Expansions for Ordinary Differential Equations." John Wiley, New York.

Waterman, P.C. (1963). *J. Math. and Phys.* 42, 323-328.

Watson, G.N. (1945). "A Treatise on the Theory of Bessel Functions." Cambridge Univ. Press, London and New York. For errata, see Luke, Wimp and Fair (1972).

Weil, J., Murty, T.S., and Rao, D.B. (1967). *Math. Comp.* 21 722-727.
Weil, J., Murty, T.S., and Rao, D.B. (1968). *Math. Comp.* 22 230.
Weingarten, H. and DiDonato, A.R. (1961). *Math. Comp.* 15, 169-173, 436.
Werner, H. (1958). *Nucleonik* 1, 60-63.
Werner, H., and Collinge, R. (1961). *Math. Comp.* 15, 195.
Wheelon, A.D. (1968). "Tables of Summable Series and Integrals Involving Bessel Functions." Holden Day, San Francisco.
Whittaker, E.T., and Watson, G.N. (1927). "A Course of Modern Analysis." Cambridge Univ. Press, London and New York.
Wilcox, P.H. (1968). *Math. Comp.* 22, 205-208.
Wimp, J. (1961). *Math. Comp.* 15, 174-178.
Wimp, J. (1962). *Math. Comp.* 16, 446-448; 19(1965), 175.
Wimp, J. (1967). *Math. Comp.* 21, 639-646.
Wimp, J. (1968). *Math. Comp.* 22, 363-373.
Wimp, J. (1969). On recursive computation. Rept. ARL69-0186, Aerospace Research Labs., Wright-Patterson Air Force Base, Ohio.
Wimp, J. (1970). Recent developments in recursive computation. "SIAM Studies in Applied Mathematics," Vol. VI, pp. 110-123, Philadelphia. Also published under the same title, Rept. ARL69-0104, Aerospace Research Labs., Wright-Patterson Air Force Base, Ohio, (1969).
Wimp, J. (1972). *Applicable Anal.* 1, 325-329.
Wimp, J. (1975). *Math. Comp.* 29, in press.
Wimp, J., and Luke, Y.L. (1969). *Rend. Circ. Mat. Palermo* 18, 251-275.
Wood, V.E. (1967). *Math. Comp.* 21, 494-496.
Wood, V.E., Kenan, R.P., and Glasser, M.L. (1966). *Math. Comp.* 20, 610-611.
Wragg, A. (1966). *Comput. J.* 9, 106-109.
Wragg, A., and Davies, C. (1973). *J. Inst. Math. Appl.* 11, 369-375.
Wrench, J.W., Jr. (1960). *Math. Teacher* 53, 644-650.
Wrench, J.W., Jr. (1968). *Math. Comp.* 22, 617-626 and 27(1973), 681-682.
Wrench, J.W., Jr. (1970). The converging factor for the modified Bessel function of the second kind. Rept. 3268, Naval Ship Res. and Development Center, Washington, D.C.
Wrench, J.W., Jr. (1971). Converging factors for Dawson's integral and the modified Bessel function of the first kind. Rept. 3517, Naval Ship Res. and Development Center, Washington, D.C.

Wrench, J.W., Jr., and Alley, V. (1972). The converging factors for the sine and cosine integrals. Rept. 3980, Naval Ship Res. and Development Center, Bethesda, Maryland.

Wrench, J.W., Jr., and Alley, V. (1973). The converging factors for the Fresnel integrals. Rept. 4102, Naval Ship. Res. and Development Center, Bethesda, Maryland.

Wright, K. (1964). *Comput. J.* 6, 358-365.

Young, A., and Kirk, A. (1964). "Bessel Functions, Part IV, Kelvin Functions." Royal Society Mathematical Tables, Vol. 10, Cambridge Univ. Press, London and New York.

Zhurina, M.I., and Karmazina, L.N. (1960,1962). "Tables of the Legendre Functions $P_{-\frac{1}{2}+i\tau}(x)$," Vols. I,II (in Russian). Izdat. Akad. Nauk SSSR, Moscow. Also in English, Pergamon Press, New York, (1964,1965).

Zhurina, M.I., and Karmazina, L.N. (1962). "Tables and Formulas for the Spherical Functions $P^{m}_{-\frac{1}{2}+i\tau}(x)$" (in Russian). Izdat. Akad. Nauk SSSR, Moscow. Also in English, Pergamon Press, New York, 1966.

Zhurina, M.I., and Karmazina, L.N. (1963). "Tables of the Legendre Functions $P^{1}_{-\frac{1}{2}+i\tau}(x)$" (in Russian). Vycisl. Centr Akad. Nauk SSSR, Moscow.

Zhurina, M.I., and Karmazina, L.N. (1967). "Tables of the Modified Bessel Function of Imaginary Argument" (in Russian). Vycisl. Centr Akad. Nauk SSSR, Moscow.

Zhurina, M.I., and Osipova, L.N. (1964). "Tables of the Confluent Hypergeometric Function" (in Russian). Vycisl. Centr Akad. Nauk SSSR, Moscow.

NOTATION INDEX

The system of notation employed to locate specific material in the text lists, in the following order, chapter, section, and subsection (if any).

EXAMPLE. 11.3.4 means Chapter XI, third section, fourth subsection. A number in parentheses refers to an equation.

EXAMPLE. 11.3.4(5) refers to the fifth equation in 11.3.4. An equation number in parantheses standing by itself refers to the equation of the particular section or subsection in which the reference occurs.

EXAMPLE. A reference to (5) in 11.6.2 means the fifth equation in 11.6.2.

There is much ad hoc notation which is explained in the text near where it occurs. Data of this kind are excluded in the index.

In the listings below, the numbers refer to the page on which the notation is defined.

Tables of numerical coefficients for expansion of functions in series of Chebyshev polynomials of the first kind and coefficients for the polynomials in rational approximants of functions, etc., are numbered sequentially in the chapter in which they occur.

EXAMPLE. Tables 1.1, 1.2, 1.3, etc., are the first, second, and third tables, respectively, in Chapter I. These numbers have no bearing on the chapter and section numbers as noted above. The reader will find that the extensive subject index provided in this volume enables one to rapidly locate desired data. There are a few tables of an ad hoc nature which are presented to illustrate the ideas and procedures described. These are notated Table A,,Table B, etc. The designators do not refer to any particular chapter.

Greek letters

Miscellaneous Notations

$z = x + iy$, $i = (-)^{\frac{1}{2}}$, is a complex number, x and y real

$\bar{z} = x - iy$

$R(z) = x$ = real part of z

$I(z) = y$ = imaginary part of z

arg z = argument of z, $\tan(\arg z) = y/x$

ln z = principal value of the natural logarithm of z

$z^\alpha = e^{\alpha \ln z}$ with ln z defined as above

$\binom{m}{n}$ = binomial coefficient $= m!/\{n!(m - n)!\}$. Also used in the generalized sense where for example $n!$ is replaced by $\Gamma(n + 1)$, 12

m as in $A^{(m)}(z)$ usually means $d^m A(z)/dz^m$

$[x]$ = largest integer contained in x, $x > 0$

$(-)^n = (-1)^n$, n an integer or zero

P.V.$\int$ means Cauchy principal of the integral

$\sim$ means asymptotic equality

* as in ${}_pF_q\left(\begin{matrix}\alpha_p \\ 1 + \rho_q - \rho_h^*\end{matrix}\middle| z\right)$, means to omit the parameter $1 + \rho_j - \rho_h$ when $j = h$. Also $(\alpha_p - \alpha_h)^*$ stands for $\prod_{\substack{j=1 \\ j \neq h}}^{p} (\alpha_j - \alpha_h)$, etc., 171

SUBJECT INDEX

A

Approximation

In the following, we cite references to pages where approximations to specified functions by use of polynomials, ration-functions, etc., are considered. We also include page references to pertinent numerical coefficients. Further, citations giving other sources for numerical coefficients are delineated.

For polynomial and other series approximations of functions which can be deduced by truncation of infinite series expansions of Jacobi polynomials, Chebyshev polynomials, etc., see Expansions. For references to continued fraction representations and related data, see entries in this section under RFP.

It is convenient to employ the following abbreviations.

BCP	best polynomial in the Chebyshev sense
BCRF	best or nearly best rational function in the Chebyshev sense
P	polynomial
R	recurrence formula
RF	rational function
RFP	rational function which is also a Padé approximation
TR	trapezoidal or modified trapezoidal rule integration formula

Approximation (Continued)

By a rational function approximation, we mean an approximation in the form of the ratio of two polynomials except possibly for a multiplicative factor. When using the above descriptors, it is to be understood that the entire portions of the designated functions are approximated in the described sense. RFP is sometimes used to denote that an approximation for the particular function or calculation under consideration is based on a Padé approximation for a related function. The Debye functions are a case in point. The situation is made clear in the text where the descriptions are more detailed than in this index. Some illustrations of the above descriptors follow.

Thus polynomial approximations for $J_\nu(z)$ and $I_\nu(z)$ which are best in the Chebyshev sense are noted on pp. 411,412. Evaluation of these functions by use of a recurrence formula is discussed on pp. 380-395 while rational function approxima tions are considered on pp. 361-365, 380-395. Finally, Padé approximations for these functions are mentioned on p. 412. Best Chebyshev polynomial and rational function approximations for $K_\nu(z)$ are noted on pp. 411, 412, and computation of this function by use of a recurrence formula is taken up on p. 381. Also a rational function approximation for $K_\nu(z)$ which is also a Padé approximation is noted on p. 412. Best Chebyshev rational function approximation for repeated integrals of $K_\nu(z)$ is mentioned on p. 411, and trapezoidal rule approximations for these functions are treated on pp. 397-399.

B

C

D

E

Expansion

In the following, we cite references to pages where specified functions are represented by infinite series of Jacobi polynomials, Chebyshev polynomials, Bessel functions, etc. We also include page references to pertinent numerical coefficients. Further, citations giving other sources for numerical coefficients are delineated. For polynomial and rational approximations to many of the same functions listed here, *see*

Approximation.

It is convenient to employ the following abbreviations.

- BF Bessel function, $J_\nu(z)$ or $I_\nu(z)$
- CP Chebyshev polynomials of the first kind
- GP Gegenbauer or ultrashperical polynomials
- JP Jacobi polynomials including extended and generalized Jacobi polynomials
- LP Laguerre polynomials including extended and generalized Laguerre polynomials

When using these descriptors, it is to be understood that the entire portions of the designated functions are expanded in the manner described. In the text, the notations $CT(x)$, etc. are used in connection with expansions of functions in series of Chebyshev polynomials of the first kind. These notations are explained in the Notation Index.

Some examples of the above descriptors follow.

Expansion
  Bessel functions
    $J_\nu(z)$, $I_\nu(z)$, CP, 317, 319, 500-504
    $J_n(z)$, $I_n(z)$, BF, 317

means that expansions of $J_\nu(z)$ and $I_\nu(z)$ in series of Chebyshev polynomials of the first kind are treated on pp. 317, 319, 500-504, while expansions of $J_n(z)$ and $I_n(z)$ in series of Bessel functions are found on p. 317.

F

G

H

I

A 5
B 6
C 7
D 8
E 9
F 0
G 1
H 2
I 3
J 4

[redacted]

MATHEMATICAL FUNCTIONS and their APPROXIMATIONS

Academic Press
Rapid Manuscript Reproduction